Editorial Director: Vernon R. Anthony
Acquisitions Editor: Lindsey Prudhomme Gill
Editor, Digital Projects: Nichole Caldwell
Editorial Assistant: Amanda Cerreto
Director of Marketing: David Gesell
Marketing Manager: Stacey Martinez
Senior Marketing Coordinator: Alicia Wozniak
Senior Marketing Assistant: Les Roberts
Senior Managing Editor: JoEllen Gohr
Senior Project Manager: Rex Davidson
Senior Operations Supervisor: Pat Tonneman
Creative Director: Andrea Nix
Art Director: Jayne Conte
Cover Image: Fotolia
Full-Service Project Management: Peggy Kellar/iEnergizer Aptara®, Inc.
Composition: Aptara®, Inc.
Printer/Binder: R. R. Donnelley & Sons
Cover Printer: Lehigh/Phoenix Color Hagerstown
Text Font: Times Roman

Credits and acknowledgments for materials borrowed from other sources and reproduced, with permission, in this textbook appear on the credits page.

Cataloging-in-Publication data is available from the Publisher on request.

10 9 8 7 6 5 4 3 2 1

ISBN 10: 0-13-298863-1
ISBN 13: 978-0-13-298863-6

ELECTRONIC COMMUNICATIONS

A Systems Approach

JEFFREY S. BEASLEY
New Mexico State University

JONATHAN D. HYMER
Mt. San Antonio College

GARY M. MILLER

PEARSON

Boston   Columbus   Indianapolis   New York   San Francisco   Upper Saddle River
Amsterdam   Cape Town   Dubai   London   Madrid   Milan   Munich   Paris   Montreal   Toronto
Delhi   Mexico City   São Paulo   Sydney   Hong Kong   Seoul   Singapore   Taipei   Tokyo

To the experimenter, tinkerer, and lifelong
learner in all of us.

BRIEF CONTENTS

CONTENTS

8   DIGITAL MODULATION AND DEMODULATION   255

10 WIRELESS COMMUNICATIONS SYSTEMS   337

11 COMPUTER COMMUNICATION AND THE INTERNET   384

12 TRANSMISSION LINES   416

13   WAVE PROPAGATION   465

15   WAVEGUIDES AND RADAR   536

16   FIBER OPTICS   562

PREFACE

The electronic communications field, the largest business sector in the industry and the original application from which the discipline has evolved, is undergoing a fundamental transformation. Thanks to the computer revolution and advances in large-scale integration, functions traditionally performed by analog circuits built from discrete components are now largely carried out within integrated circuits executing digital signal-processing operations. Such wholesale changes in system implementation bring with them an increasing need for a new approach to study. In contrast with the traditional emphasis on individual circuits embodied by many texts, the primary objective of this book is to foster an in-depth understanding of communications systems by providing a comprehensive description of how functional blocks work together to perform their intended tasks.

Notwithstanding the shift to digitally implemented systems, however, the underlying concepts, constraints, and themes that have informed the communications art for the last century have remained the same. A comprehensive introduction to communications technology in all its forms must emphasize thematic elements that highlight relationships among seemingly isolated concepts. To this end, the early chapters in particular have a narrative structure that provides readers with an overall conceptual framework for the development of foundational concepts and major themes. For example, all communications systems can be examined in terms of certain overriding characteristics, such as bandwidth and power distribution, as well as in terms of constraints on system operation such as noise. When viewing systems from the twin perspectives of characteristics and constraints, readers can begin to forge relationships among concepts that initially may seem very disconnected. The inevitable conclusion is that even the most advanced, highly integrated systems are composed of subsystems employing well-established ideas that are often familiar from the analog context. For this reason, the early chapters are largely given over to a study of modulation techniques and analog circuits, even as the conversion from the analog to the digital realm continues at an accelerating pace. A solid understanding of analog fundamentals provides the platform for a conceptual understanding of how equivalent actions are carried out in the digital domain. With such a foundation, the conceptual leap from analog to digital is less daunting than it may first appear.

Features and Audience

This text is intended for a one- or two-semester course sequence in electronic communications, wireless communications, communications maintenance technology, or introductory telecommunications. The text is suitable for students in two-year programs at community colleges or technical institutes as well as for students in some four-year programs in electronics engineering technology or industrial technology. Math analysis has been kept to the level of algebra and trigonometry but is sufficient to enhance understanding of key, underlying concepts. For completeness, a discussion of Fourier series and complex-exponential representations, topics not often found in books intended for two-year programs, has been included in Chapters 1 and 8, respectively. I have tried throughout to strike a middle ground between calculus-intensive communications texts intended for four-year engineering programs and the math-avoidance path followed by some texts intended for two-year programs.

Several chapters, many illustrations, and most end-of-chapter problems have been adapted from *Modern Electronic Communication* by Jeff Beasley and Gary Miller. This venerable text, now in its 9th edition, has been a standard in its field for over 25 years. As such, it provides an outstanding foundation for the systems-level approach of this book. I am deeply grateful to the authors for entrusting me with the task of adapting their exemplary work for students entering the communications field in the second decade of the 21st century. Students today are entering a field in which entire communications systems are now on single integrated circuits, rather than consisting of multiple stages with many integrated circuits and discrete components. The trend over the past five years has been to the widespread adoption of digital signal processing (DSP) techniques and software-defined radio. Both topics are given more extensive coverage in this text than is the case in competing texts or in previous editions of *Modern Electronic Communication*. As recently as five years ago, many of these techniques were more akin to laboratory curiosities rather than mainstream consumer products. In short, this text takes a top-down view of the discipline rather than a "bottom-up" view implied by a text focusing on circuits built from discrete components.

Topics covered include modulation; communications circuits; transmitters and receivers; digital communications techniques, including digital modulation and demodulation; telephone and wired computer networks; wireless communications systems, both short-range and wide area; transmission lines, impedance matching, and Smith charts; wave propagation; antennas; waveguides and radar; and fiber-optic systems. Considerable attention has been given to providing a narrative structure throughout, and particularly in the fundamentals chapters, to allow readers to put the many facts and concepts encountered into a larger, coherent whole. By explicitly tying together concepts that may have been left unstated before, I hope to help students see the big picture and make sense of topics at a conceptual level, rather than asking them to rely on rote memorization of a large number of seemingly unrelated facts.

Other key features are as follows:

- Review of some basic electronics concepts. Scheduling constraints or differences in curriculum policies may cause students to take the communications courses some significant amount of time after completing the fundamental electronics (dc/ac) and circuits courses. Recognizing this reality, I have expanded coverage of some concepts initially introduced in electronics fundamentals and devices (circuits) courses, including the nature of a sine wave, reactance and resonance, and classes of amplification, to allow opportunities for instructor-led or self-paced review.

- Inclusion of topics and end-of-chapter questions specifically directed at preparation for the U.S. Federal Communications Commission's (FCC) General Radiotelephone Operator License (GROL) exam. The FCC GROL is still valued in industry; many employers use it as a resume filter to screen applicants, and in some industries (particularly avionics) possession of the GROL is mandatory.

- Enhanced coverage of digital communications and DSP. This text has the most up-to-date treatment of enabling technologies behind latest-generation wireless systems, including third- and fourth-generation (3G and 4G) wireless data networks. In addition, sections on the following topics have been significantly expanded and enhanced: digital modulation, DSP, finite-impulse response filters, spread-spectrum implementations, orthogonal frequency-division multiplexing, and multiple-input/multiple-output configurations.

- Discussions of topics not found in other communications texts. The following topics, which are either covered superficially or not at all in other texts, have been included or significantly expanded: SINAD (receiver sensitivity) testing, squelch system operation, DSP modulation/demodulation, spread-spectrum techniques, wireless networks including 802.11n, Bluetooth and ZigBee, enhanced coverage of digital cellular voice networks (GSM and CDMA), coverage of two-way and trunked radio systems, software-defined and cognitive radio, cavity filters/duplexers/combiners, impedance matching and network analysis (including S parameters), Maxwell's equations, and system link budgeting and path-loss calculations.

- Introduction to the concept of analytic frequency and the complex exponential. A section has been added describing some mathematical concepts behind many DSP-based implementations of digital modulation and demodulation that are now becoming mainstream implementations.

Every discipline has as a core part of its narrative the concepts, constraints, and challenges that define it as an intellectual endeavor and a field of inquiry. Electronic communications is no exception. I hope to have conveyed in the pages of this text some sense of the magnitude of ingenuity and scientific accomplishment that is embodied in the technologies described herein, for it is the simple, overriding need of people to be able to talk to each other that has not only brought us to where we are today but that also represents what is surely a transformative achievement of human civilization.

Supplements

- Laboratory Manual to accompany *Electronic Communications* (ISBN: 0-13-301066-X)
- TestGen (Computerized Test Bank): This electronic bank of test questions can be used to develop customized quizzes, tests, and/or exams.
- Online PowerPoint® Presentation
- Online Instructor's Resource Manual

To access supplementary materials online, instructors need to request an instructor access code. Go to **www.pearsonhighered.com/irc**, where you can register for an instructor access code. Within 48 hours after registering, you will receive a confirming e-mail, including an instructor access code. Once you have received your code, go to the site and log on for full instructions on downloading the materials you wish to use.

A Note to the Student

Over the years, as I have taught communications electronics, I have noticed that many students approach the subject with a sense of trepidation, perhaps because the subject matter may seem overly mathematical, esoteric, or just plain "hard." I have also noticed that many students treat the subject strictly as one they are learning in school rather than as one with which they engage outside the classroom as an avocation or hobby. The joy of electronics, however, is in its hands-on nature and in the opportunities it provides for tinkering and experimentation. The opportunity to work with my hands, to explore and to experiment, is what initially attracted me to electronics as a vocation and a field of study. To this end, I encourage you to explore on your own the many opportunities you will have outside of class time to engage with electronic communications systems. There are many well designed radio and communications kits from vendors such as Elenco, TenTec, and Elecraft that will not only allow you to explore the fundamental communications concepts described in this text on your own but that will also give you the opportunity to experience the thrill of seeing something you created with your own hands work the first time you turn it on. In addition, radio amateurs or "hams" experience the enjoyment of worldwide communication with others, often using radios and antennas of their own design. The national organization of amateur radio operators in the United States is the American Radio Relay League (ARRL), located in Newington, Connecticut, and reachable on the internet at www.arrl.org. ARRL publications rank among the best anywhere for providing a rich introduction to communications systems design and operation. I encourage students using this text to explore the field outside the classroom as well, for it is this personal engagement with the subject matter that will make the topic come alive in a way that no book or classroom lecture possibly can.

Acknowledgments

I would again like to acknowledge Jeff Beasley and Gary Miller, authors of *Modern Electronic Communication,* for entrusting me with stewardship of their text. I trained with the 3rd edition of this standard work, never thinking for a moment that I would some day become part of the "MEC family." I am humbled to carry on the tradition it established and can only hope to maintain its standards of excellence. I would also like to thank my colleague, Steve Harsany, for his in-depth review of the end-of-chapter problems and for contributing Appendix A on GROL preparation, as well as former student Scott Cook for his invaluable research assistance. Colleagues Ken Miller, Joe Denny, Sarah Daum, and

Jemma Blake-Judd, all of Mt. San Antonio College, also provided invaluable assistance by reviewing portions of the manuscript, for which I am also deeply grateful. Of course, any errors are mine alone. Finally, no acknowledgment section is complete unless I recognize my mentor, Mr. Clarence E. "Pete" Davis, who, over the years, really taught me what I know about radio and who helped me get to where I am today. Thank you.

JONATHAN D. HYMER
Mt. San Antonio College
Walnut, California

CHAPTER 1

FUNDAMENTAL COMMUNICATIONS CONCEPTS

CHAPTER OUTLINE

KEY TERMS

channel
modulation
carrier
intelligence
demodulation
detection
frequency-division
multiplexing
amplitude modulation
(AM)
frequency modulation
(FM)
phase modulation (PM)
transducer
transceiver
dBm
bandwidth
Hartley's law
information theory
fundamental
harmonic
time domain
frequency domain
spectrum analyzer

fast Fourier transform
(FFT)
digitized
noise
external noise
internal noise
atmospheric noise
space noise
solar noise
cosmic noise
ionosphere
Johnson noise
thermal noise
white noise
low-noise resistor
shot noise
excess noise
transit-time noise
signal-to-noise ratio (*S/N*
ratio)
noise figure (NF)
noise ratio (NR)
octave
Friiss's formula

1-1   INTRODUCTION

The harnessing of electrical energy to enable long-distance communication represents not only a milestone in technological progress but also a transformative achievement of human civilization. This foundational application of electronics technology continues to be as dynamic and exciting today as it ever was, for societies all over the world are in the midst of yet another communications revolution. The ongoing transition away from exclusively analog technologies toward digital systems promises to continue unabated. Advances in digital communications are key to satisfying the seemingly never-ending demand for ever-higher rates of information transfer in ever more portable devices. The explosion in demand for latest-generation smart phones and other wireless devices, as well as the wide-spread availability of high-definition television sets, are but two recent entries in a never-ending parade of advances in communications technology.

This book provides a comprehensive overview of wireless and wired, analog and digital electronic communications technologies at the systems level. As with any discipline, the study of communication systems is informed by underlying principles and recurring themes. These themes will become apparent in the first three chapters of this text and will manifest themselves fully in the chapters that follow. Further, all electronic communication systems, whether analog or digital, can be analyzed in terms of certain overriding characteristics such as power distribution and bandwidth, as well as in terms of the fundamental constraints that bound their operation, among them noise. Those commonalities and constraints are the focus of this first chapter.

Communications Systems and Modulation

The function of any communication system is to transfer information from one point to another. All systems fundamentally consist of three elements: a transmitter, a receiver, and a **channel**, or link for information transfer. The channel can be, and often is, wireless. Either the Earth's atmosphere or free space—a vacuum—can form the path between transmitter and receiver. Alternatively, channels can be formed from physical media such as copper wires, transmission lines, waveguides, or optical fibers. As we shall soon see, the characteristics of the channel largely determine the maximum information capacity of the system.

Radio and television stations, whose transmitters use "the airwaves" to broadcast programs to widely dispersed receivers, are familiar examples of communications systems. Other systems make use of both wired and wireless links at the same time. For example, the link between a cellular phone and the base station serving it is wireless, but that interface is only one part of a much larger infrastructure consisting of many base stations, switching centers, and monitoring facilities. Links between base stations and the carrier's mobile switching centers, where equipment that routes calls within and outside the carrier's network is located, may be wired or wireless. Interconnections between switching centers belonging to different wireless and wireline carriers that, taken together, form the global telephone system are most likely made with fiber optic cable; other possibilities include copper wires or satellite interconnections. Other familiar examples abound. For example, subscription-based satellite television and radio services convey information wirelessly through outer space and the atmosphere to Earth-based receivers. Broadband internet capability to the home is provided through networks consisting of either copper wires or coaxial cables, or, in some areas, fiber-optic links. Regardless of their simplicity or complexity, however, all systems can be considered as consisting of the fundamental elements of transmitter, link, and receiver.

Basic to the field of communications is the concept of modulation. **Modulation** is the process of impressing relatively low-frequency voltages that represent information, such as voices, images, or data, onto a high-frequency signal called a **carrier** for transmission. The carrier does just what its name implies: It "carries" the information from the transmitter through the channel to the receiver. The low-frequency information, often termed the **intelligence**, is placed onto the carrier in such a way that its meaning is preserved but that it occupies a band of frequencies much higher than it did before modulation took place. Once received, the intelligence must be recovered, that is, separated, from the high-frequency carrier, a process known as **demodulation** or **detection**.

At this point, you may be thinking: "Why bother to go through this modulation/demodulation process? Why not just transmit the information directly?" As an example, if we wanted to use radio waves to send voice messages to receivers in the surrounding area, could we not just use a microphone to convert the messages from acoustical vibrations to electrical signals and then apply these signals to an antenna for transmission? Though theoretically possible, direct transmission of signals at such low frequencies presents two practical problems: first, the inability to share, and second, the size of the antenna that would be needed. The frequency range of the human voice is from about 20 to 3000 Hz. If more than one party in a given geographic area attempted at the same time to transmit those frequencies directly as radio waves, interference between the transmissions would cause them all to be ineffective. Also, for reasons that will become clear in Chapter 14, the antennas required for efficient propagation of radio signals at such low frequencies would be hundreds if not thousands of miles long—far too impractical for use.

The solution is modulation, in which a high-frequency carrier is used to propagate the low-frequency intelligence through a transmission medium. Through a process known as **frequency-division multiplexing**, in which each transmitter in a given area is assigned exclusive use of a carrier frequency, communications channels (in this case, bands of radio frequencies) are allocated for simultaneous use by multiple transmitters. Modulation enables multiplexing, thereby allowing access to a single communication medium by many different users at the same time. Also, and equally important, the frequencies employed by the modulated signal are high enough to permit the use of antennas of reasonable length. Among the recurring themes in the study of communications is certainly the concept of modulation, and it is for this reason that Chapters 2 and 3 are largely given over to an in-depth analysis of this important topic.

CHARACTERISTICS OF THE CARRIER   Because the carrier is often a sine wave, a review of the characteristics of such waves is in order. Recall from fundamental AC theory that a sine wave looks like the waveform shown in Figure 1-1. A periodic waveform, such as that produced by an electronic oscillator or by a mechanical generator whose armature is rotating within a magnetic field, can be represented by a vector OB whose length represents the peak voltage produced within the conductor. In the case of the generator, this would be the voltage created as the conductor cuts across the magnetic lines of force produced by the field. If we assume that the vector OB started at the location represented by the line OA, we see that a continuously increasing angle θ is created as vector OB moves counterclockwise around the circle from OA; the speed or velocity at which the vector rotates is directly related to the frequency of rotation: the faster the speed of rotation, the higher the frequency of the resulting waveform.

Why is this waveform called a sine wave? Referring back to Figure 1-1, we place a point t along a horizontal line extended from point OA that is the same distance that point B moved along a circle from A. Extending a line horizontally from point B to a point directly above t and calling it point B_1, we see that the point B_1t is the same height as an imaginary vertical line drawn from the end of radius point B to the line defined by OA. Put another way, by drawing a perpendicular line from B to the line OA, we have formed a right triangle, the hypotenuse of which is the line OB and which is equivalent in length to B_1t. The line B_1t represents the magnitude of the instantaneous voltage produced by our

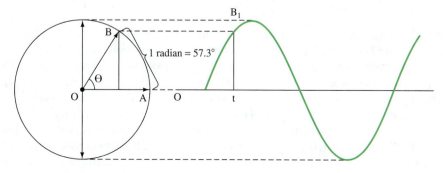

FIGURE 1-1   Sine wave represented as a rotating phasor.

rotating vector at the point B. Since, by definition, the sine of the angle θ is the ratio of the length of the side of a right triangle opposite to that of θ divided by the hypotenuse of that triangle, it follows that, if we define the length of OB to have an arbitrary length of 1 unit, the vertical line B_1t, and therefore the instantaneous voltage at point t, will have a value defined by the sine of θ. The familiar sine-wave representation of an alternating voltage is thus traced out by creating a vertical line created from each point where the vector OB intersects with the circle created by its counterclockwise rotation to the line OA and projecting that vertical line onto a time representation whose horizontal axis is represented by the length from A_1 to A_2. Each trip that the rotating vector makes around the circle represents one complete cycle of the alteration.

If in Figure 1-1 the vector OB is placed such that the distance along the curve from A to B equals the length OB, the size of the angle θ is said to equal 1 unit of angular measurement known as a radian. By definition, the radian is the angle that subtends an arc equal to the radius of the circle and is equal to approximately 57.3 degrees. Because the circumference of a circle equals 2π times the radius of the circle, it follows that there are 2π radians around the circle. (The quantity π is defined as the ratio of the circumference of a circle to its diameter and is a constant.) Therefore, if the vector OB has completed one cycle, it has swept through 2π radians of angle. The frequency of the alternating voltage is the number of cycles that occur in one second; alternatively, it is also the number of times per second that the vector OB goes around the circle. Therefore, the number of radians per second that the vector goes through the circle will be 2π times the frequency. This quantity, called the *angular frequency* or *angular velocity*, is customarily represented by ω, the Greek letter omega. Thus,

$$\omega = 2\pi f.$$

In general, the angular velocity has units of radians per second or degrees per second, and, in Figure 1-1, the distance from points A_1 to A_2 represents a time scale.

When applied to an electrical quantity such as voltage, the angular velocity specifies the time rate of change of the quantity under consideration. Put another way, ω indicates that the total variation of that quantity (such as voltage or current) will be 2π times the cyclic frequency, f, or number of cycles completed per second. Angular velocity and cyclic frequency are clearly related: the higher the angular velocity, the higher the frequency. The concept of angular velocity is useful because many electrical phenomena (in particular, capacitive and inductive reactance) involve rates of change and, therefore, their formulas each contain a π term. In addition, by expressing waveforms in terms of rates of change, we can invoke a number of trigonometric principles to describe the behavior of sine waves both singly and in combination. Since the essence of modulation is the analysis of combinations of sine waves, the ability to express the results of such combinations as trigonometric relationships will become very useful.

From the foregoing, it follows that any sine wave can be represented by the following expression:

$$v = V_P \sin(\omega t + \Phi), \tag{1-1}$$

where $v =$ instantaneous value

$V_P =$ peak value

$\omega =$ angular velocity $= 2\pi f$

$\Phi =$ phase angle.

The cyclic frequency is contained within the ω term, but the Φ term for phase angle may not yet be familiar. The *phase angle* represents the instantaneous number of electrical degrees by which a sine wave is advanced or delayed from some arbitrary starting time $t = 0$. As we shall see in Chapter 3, frequency and phase angle are interrelated: an instantaneous frequency change will create a change in the phase angle, and vice versa.

If the expression of Equation (1-1) represents a carrier, it follows that for modulation to take place, one or more characteristics of the carrier must be modified. **Amplitude modulation (AM)** occurs when the amplitude term, V_P, is varied. **Frequency modulation (FM)** occurs when the frequency term (contained within ω) is varied. Varying the phase angle, Φ, results in **phase modulation (PM)**. Because of the relationship between frequency and phase, these latter two forms of modulation are sometimes classified under the umbrella of "angle" modulation. One overriding fact should be kept in mind at all times: amplitude, frequency, and phase are the only characteristics of a sine-wave carrier that can be modified. The essence of modulation in any system, no matter how outwardly complex it appears, ultimately involves modifying one or more of those three parameters.

Though modulation is certainly not exclusive to wireless systems, the concept is perhaps most familiar in the context of AM and FM broadcast radio. In large part, electronics as a discipline within the field of physical science emerged from the discovery of radio waves, and many core ideas are adapted from those first developed for radio communications. Wireless systems will be the primary focus of many chapters of this text not only because of their historical importance but also because many circuits first developed for early radio systems are used in modified form in other areas of electronics even today.

The Electromagnetic Spectrum

What actually travels between transmitter and receiver in a wireless system? Recall from your study of electrical fundamentals that electricity and magnetism are intertwined. One gives rise to the other. Magnetic fields surround moving electric charges (i.e., currents); likewise, currents are generated in a circuit whenever relative motion occurs between magnetic fields and conductors. Electric and magnetic fields both result from voltage potentials and current flows. In ordinary conductors as well as in "free space," that is, in a vacuum, the electric and magnetic fields form at right angles to each other as well as at right angles to the direction of travel. This form of energy is therefore *electromagnetic* energy. For nonvarying—that is, direct—currents and voltages, the magnitudes of both the electric and magnetic fields are constant, and, therefore, do not reproduce in free space, whereas for alternating currents, the electric and magnetic fields take on the characteristics of the voltages and currents that generated them. A sinusoidal source, therefore, generates at its operating frequency both electric (voltage) and magnetic (current) fields that are sinusoidal in shape as well as at right angles to each other.

Electromagnetic energy is present as the result of electric charge moving within a conductor, but, in the case of alternating currents, the energy also exists outside the confines of the conductor and, indeed, propagates away from its source. With an appropriate **transducer**, a device that converts energy from one form to another, alternating currents flowing in a conductor are converted into waves that continue to exist beyond the physical confines of the conductor. (A wave is a mechanism for the transfer of energy that does not depend on matter.) As in a conductor, the electromagnetic wave in free space exists as both electric and magnetic fields. The voltage potentials defining the electric field and created by accelerating electric charges also create current flows, which in turn give rise to a moving magnetic field at right angles to the electric field. The moving magnetic field so created begets another electric field, and so on. The wave that is created from the moving electric and magnetic fields thereby propagates from its point of origin through space to its ultimate destination. In a wireless system, the transducers are antennas at the transmitter and receiver. Currents generated by the transmitter and applied to its antenna are converted to electromagnetic energy, whereas at the destination, the moving electromagnetic field impinging upon the conductors of the receiving antenna will generate currents within that antenna for subsequent application to the receiver input.

Electromagnetic energy exists at all frequencies from DC (0 Hz) to the frequencies represented by visible light and beyond. Indeed, light is an electromagnetic wave. The *electromagnetic spectrum*, therefore, is composed of the entire range of signals occupying all frequencies. Many familiar activities and services reside along the electromagnetic spectrum. *Audio frequencies*, those that can be heard by the human ear when converted to acoustical form, range from about 20 Hz up to approximately 20 kHz. Frequencies above 50 kHz or so are termed *radio frequencies*, for it is here that electromagnetic energy can be produced and radiated using antennas of reasonable length. The AM radio broadcast band occupies the frequency range from 540 kHz to 1.7 MHz; FM broadcasting is assigned the 20 MHz band of frequencies from 88 to 108 MHz. Cellular telephones use bands of frequencies at either 800 MHz or 1.8 to 2.1 GHz, depending on carrier and geographic region.

Household microwave ovens operate at 2.4 GHz, as do wireless networks of personal computers. Other household wireless devices, among them some cordless phones and newer-vintage local-area networks, operate at 5.8 GHz. These frequencies are well within the microwave region, characterized by very short wavelengths and necessitating specialized techniques that will be covered in subsequent chapters.

Among the ways communication systems can be categorized is by the frequency of the carrier. Table 1-1 shows the designations commonly applied to services within the radio-frequency portion of the electromagnetic spectrum. Note, however, that the electromagnetic spectrum extends beyond even the highest frequencies shown in the table. Above the extra-high-frequency range shown in the table reside the so-called "millimeter wave" bands, which are of particular interest to physicists and astronomers. Above that resides the optical spectrum, consisting of infrared, visible light, and ultraviolet waves. At the very highest frequencies are found X rays, gamma rays, and cosmic rays.

TABLE 1-1 • Radio-Frequency Spectrum

FREQUENCY	DESIGNATION	ABBREVIATION
30–300 Hz	Extremely low frequency	ELF
300–3000 Hz	Voice frequency	VF
3–30 kHz	Very low frequency	VLF
30–300 kHz	Low frequency	LF
300 kHz–3 MHz	Medium frequency	MF
3–30 MHz	High frequency	HF
30–300 MHz	Very high frequency	VHF
300 MHz–3 GHz	Ultra high frequency	UHF
3–30 GHz	Super high frequency	SHF
30–300 GHz	Extra high frequency	EHF

Communications Systems

Figure 1-2 represents a simple communication system in block diagram form. The modulated stage accepts two inputs—the carrier and the information (intelligence) signal—and combines them to produce the modulated signal. This signal is subsequently amplified—often by a factor of thousands, in the case of high-power wireless systems—before transmission. Transmission of the modulated signal can be wireless, or it can involve physical media such as copper wire, coaxial cable, or optical fibers. The receiving unit picks up the transmitted signal and reamplifies it to compensate for attenuation that occurred during transmission. The amplified signal is then applied to the demodulator (often referred to as a detector), where the information is extracted from the high-frequency carrier. The demodulated intelligence is then fed to the amplifier and raised to a level enabling it to drive a speaker or any other output transducer.

All communications systems, from the most basic to the most complex, are fundamentally limited by two factors: bandwidth and noise. For this reason, we will devote considerable space to the study of these important considerations, for these are the themes that inform and unify the development of the communications art. Also, one of the overriding themes of this text is that all communications systems, from the most basic to the most complex, make use of certain principles that have formed the building blocks of communications engineering for over a century, and, particularly in the first three chapters, much space will be given over to the study of these themes because they inform discussion of the system-level topics covered in following chapters. First, however, we must discuss decibel units because of their extreme utility in addressing issues that are common to all communications systems.

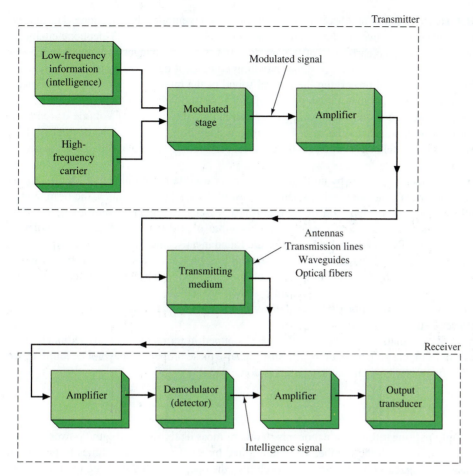

FIGURE 1-2   **Communication system block diagram.**

1-2 THE DECIBEL IN COMMUNICATIONS WORK

A defining characteristic of any communication system is the wide range of power levels it will encounter. For example, a broadcast station's transmitter might supply tens of thousands of watts to its antenna, but a receiver within the station's coverage area would encounter a power level in the picowatt range at its antenna input. (One picowatt is 10^{-12} watts.) A single **transceiver** (combination transmitter and receiver, such as a mobile two-way radio) will have power levels from femtowatts (10^{-15} W) within the receiver to a substantial fraction of a kilowatt (10^{3} W) or more at the transmitter output. Within a receiver, signal voltages are at the millivolt (10^{-3}) or microvolt (10^{-6}) level. Such wide differences in any quantity under consideration are conceptually difficult to envision with ordinary units of measure, yet expansive ranges of powers and voltages, along with the need to make computations involving very large and very small numbers at the same time, are routinely encountered in the analysis of communications systems. For these reasons, we employ units of measure that not only compress an extremely wide range of quantities to a more manageable span but that also make computations involving the multiplication and division of very large or very small quantities easier to manage. Such measurements can be made with relative ease when quantities of interest—power, voltage, or current—are represented as ratios in logarithmic form.

The term *decibel* (dB) may be familiar as a unit of sound intensity. In acoustics, decibels represent ratios related to sound pressure levels, where 0 dB is considered to be absolute silence, and the range from 140 to 160 dB represents the sound pressures encountered in the immediate vicinity of a jet engine. The term is derived from a unit called the

Bel, named in honor of Alexander Graham Bell, the inventor of the telephone. The historical relationship between telephones and sound levels is no accident. Telephone engineers and others realized early on that the ear does not perceive changes in volume in a linear fashion. Amplifiers and other signal-handling equipment deployed in telephone systems must be capable of preserving the natural sound of the human voice over long distances. Researchers learned that human perception of increased volume is more accurately modeled as an exponential relationship, where an apparent doubling of volume or loudness results from a ten-times increase, rather than a doubling, of power. The decibel ($\frac{1}{10}$ Bel) was originally defined to represent the smallest perceivable change in sound level in acoustic or electronic systems.

Decibel notation is by no means exclusively used to represent sound levels or other signals for eventual conversion to acoustic form. As used in electronics, the term simply allows for easy comparison of two, perhaps widely divergent, quantities of interest. Though derived from power ratios, decibels are also used to represent ratios of voltages and currents. Decibel-based calculations are found in noise analysis, audio systems, microwave system gain calculations, satellite system link-budget analysis, antenna power gains, light-budget calculations, and many other communications system measurements. Expressed as ratios, decibels represent system gains and losses; when referenced to absolute levels, decibel units can be used in conjunction with those levels to represent absolute powers, voltages, or currents.

As we shall see shortly, the decibel is defined in terms of, and derives much of its utility from, the properties of logarithms. Because these properties may not be familiar, we shall first describe their characteristics in some detail.

Logarithms

Simply put, logarithms are exponents. You are most likely familiar with "powers-of-10" notation, in which the number 10 is multiplied by itself one or more times to produce a result that increases by a factor of 10 each time the operation is carried out. (You may also have heard the term *order of magnitude*; when properly used in scientific contexts, this expression refers to a power-of-10 relationship.) A raised (superscript) number termed the *exponent* denotes the number of times 10 appears in such expressions. Thus, the expression $10 \times 10 = 100$ can be represented as $10^2 = 100$ because the number 10 has to be multiplied once by itself to produce the answer, 100. Likewise, $10 \times 10 \times 10 = 1000$ is represented in exponential form as $10^3 = 1000$, and so on. As mentioned, in the expression $10^2 = 100$, the raised 2 is called the *exponent*. In that same expression, the 10 is called the *base*, and the result, 100, is the *number*. The exponent (2) can also be called the *logarithm* of the number 100 to the base 10. Such an expression would be expressed in writing as $\log_{10} 100 = 2$ and would be read aloud as "the logarithm to the base 10 of the number 100 is 2." Put in general terms, the logarithm (log) of a number to a given base is the power to which the base must be raised to give the number. While any number could appear as a base in logarithms, those used in decibel expressions are always base-10, or *common,* logarithms. For common logarithms, the base may not be expressed explicitly. Thus, the above expression could be written simply as $\log 100 = 2$ and read aloud as "the log of 100 is 2."

Following the same line of reasoning as for an exponent of 2, the expression $10^3 = 1000$ could be expressed as "$\log 1000 = 3$." The common logarithm of any number between 100 and 1000—that is, the power to which the base, 10, has to be raised to give a result between 100 and 1000—will fall between 2 and 3. Stated another way, the logarithm of a number between 100 and 1000 will be 2 plus some decimal fraction. The whole-number part is called the *characteristic*, and the decimal fraction is the *mantissa;* both of these values historically have been available in published tables of logarithms but are now most easily determined with scientific calculators. The common logarithm is denoted with the "log" key on scientific calculators, and it is distinct from the *natural logarithm,* which has as its base a number denoted as e, equal to 2.71828 . . . . The natural logarithm, represented by the "ln" key on scientific calculators, is the exponent of the function e^x. These terms describe a number of natural phenomena, among them the charge and discharge rates of capacitors and the rates at which magnetic fields expand and collapse around inductors. The natural logarithm is *not* used in decibel calculations.

Conversion of very large and very small numbers into exponential form allows for two very useful properties of exponents, known as the *product rule* and the *quotient rule,* to come into play. The product rule states that when two numbers in exponential form are multiplied, the result is the *product* of the multipliers but the *sum* of the exponents. That is,

$$(A \times 10^n)(B \times 10^m) = (A)(B) \times 10^{n+m}$$

In the above expression, note that, to find the answer, one would add the exponents n and m. There is no need to multiply very large numbers if those numbers are converted to their exponential equivalents. Likewise, division of large numbers in exponential form follows the quotient rule:

$$\frac{A \times 10^n}{B \times 10^m} = \frac{A}{B} \times 10^{n-m}$$

The quotient rule allows for the division of large and small numbers to be reduced to the simple subtraction of their exponents. These properties, though still extremely useful, were indispensable when all calculations had to be carried out with nothing more than a pencil and paper.

Because logarithms are based on exponents, it follows that rules pertaining to exponents apply also to logarithms. Indeed they do, and these useful "log rules" can be summarized as follows:

$$\log ab = \log a + \log b, \quad \text{(Rule 1)}$$
$$\log a/b = \log a - \log b, \quad \text{(Rule 2)}$$

and

$$\log a^b = b \log a. \quad \text{(Rule 3)}$$

Such it is with logarithms that we are able to accomplish our original objective of stating and comparing either very large or very small numbers in a more convenient form. Another advantage is expressed in the above rules: Because quantities in decibel units are expressed as logarithms of power, voltage, or current ratios, these decibel results can simply be added and subtracted. Multiplication and division of very large and very small numbers is thus reduced to the addition or subtraction of their logarithmic, that is, decibel, equivalents.

The inverse of a logarithm is the *antilogarithm.* The antilogarithm (antilog) of a number to a given base is simply the base raised to that number. Therefore, for common logarithms, the antilog is the expression 10^x. For example (again, assuming common logarithms), antilog $2 = 10^2$, or 100. Both the log and antilog can be easily determined with a calculator. On some calculators, the antilog is indeed denoted as 10^x, while others may have an INV key that provides the antilog function when paired with the log key.

The Decibel as a Power Ratio

In electronics, the decibel is fundamentally defined as a power ratio:

$$dB = 10 \log \frac{P_2}{P_1}. \tag{1-2}$$

In words, Equation (1-2) states that one decibel is equal to 10 times the logarithm of the ratio of two power levels, P_1 and P_2.[*] By convention, the numerator of the expression, P_2, is the higher power level so that the result is expressed as a positive number. This convention made determining the logarithm easier when published tables were the norm, but it is no longer absolutely necessary now that calculators are commonplace. Note, however, that if

---

[*] Astute readers may note an apparent contradiction between the prefix "deci-," which means "$\frac{1}{10}$," and the definition in Equation (1-2), in which the Bel ratio, originally defined as $\log(P_2/P_1)$ is multiplied, rather than divided, by 10. It would appear at first that the Bel should be multiplied by 0.1 to create the decibel. The original problem was one of scale: Use of the Bel ratio without modification would cause the very large range of numbers from 0.00000001 to 100,000,000 to be represented by a range of only −8 to +8 Bels. The original ratio produced results that were far too compressed to represent small changes reasonably. Multiplying the log ratio by 10 gives a range of −80 to +80, with each whole-number change representing an increment $\frac{1}{10}$ as large as before (hence the term "deci-"), and allowing for much smaller changes to be represented with numbers of reasonable size.

the smaller value appears in the numerator, the result will be a negative value in decibels, thus representing a power loss (negative gain). It is important, in any event, to keep track of negative signs and to show them in the result when denoting a loss.

EXAMPLE 1-1

A certain amplifier has an input power of 0.5 *W*. What is the gain in decibels if the output power is (a) 1 *W*, (b) 2 *W*, and (c) 5 *W*?

SOLUTION

In each case, the power gain is determined by applying Equation (1-2):

(a) $P_{dB} = 10 \log \left(\dfrac{1 \text{ W}}{0.5 \text{ W}} \right) = 3.01 \approx 3 \text{ dB}$

(b) $P_{dB} = 10 \log \left(\dfrac{2 \text{ W}}{0.5 \text{ W}} \right) = 6.02 \approx 6 \text{ dB}$

(c) $P_{dB} = 10 \log \left(\dfrac{5 \text{ W}}{0.5 \text{ W}} \right) = 10 \text{ dB}$

The Decibel as a Voltage or Current Ratio

Though the decibel is properly thought of as a power ratio, voltages and currents can be expressed as decibel relationships *provided that input and output impedances are taken into account*. Since $P = V^2/R$, one can substitute as follows into the power relationship (because only the resistive [real] portion of impedance is significant, the variable R rather than Z will be used throughout):

$$dB = 10 \log \frac{P_2}{P_1} = 10 \log \frac{\dfrac{V_2^2}{R_2}}{\dfrac{V_1^2}{R_1}}.$$

Clearing the fractions in the numerator and denominator, we obtain

$$dB = 10 \log \frac{V_2^2 R_1}{V_1^2 R_2}.$$

For clarity, this expression can be rewritten as:

$$dB = 10 \log \left(\frac{V_2}{V_1} \right)^2 \frac{R_1}{R_2},$$

which, from log rule 1, is equivalent to

$$dB = 10 \log \left(\frac{V_2}{V_1} \right)^2 + 10 \log \frac{R_1}{R_2}.$$

From log rule 3, the squared term can be removed, giving

$$dB = 20 \log \frac{V_2}{V_1} + 10 \log \frac{R_1}{R_2}.$$

The preceding expression can be further simplified by recognizing that $10 \log \dfrac{R_1}{R_2}$ is equal to $20 \log \dfrac{\sqrt{R_1}}{\sqrt{R_2}}$ (a square root is equivalent to raising a base to the $\frac{1}{2}$ power). Thus, one now has

$$dB = 20 \log \frac{V_2}{V_1} + 20 \log \frac{\sqrt{R_1}}{\sqrt{R_2}},$$

which, by factoring, simplifies to a final result of

$$dB = 20 \log \frac{V_2\sqrt{R_1}}{V_1\sqrt{R_2}}.$$

(1-3)

EXAMPLE 1-2

A certain amplifier has an input impedance, Z_{in}, of 500 Ω and an output imped-ance, Z_{out}, of 4.5 kΩ. If a 100-mV input signal produces an output of 350 V, what is the gain of this amplifier in decibels?

SOLUTION

The decibel gain can be computed directly with Equation (1-3):

$$dB = 20 \log\left(\frac{350\sqrt{500}}{0.1\sqrt{4500}}\right) = 20 \log 1167 = 61 \text{ dB}$$

This result can be confirmed by determining the power at both the input and output of the amplifier, and then using the power form of the decibel equation, Equation (1-2):

$$P_{out} = \frac{(V_{out})^2}{Z_{out}} = \frac{350^2}{4500} = 27.22 \text{ W}$$

$$P_{in} = \frac{(V_{in})^2}{Z_{in}} = \frac{0.1^2}{500} = 20 \text{ }\mu\text{W}.$$

The gain in decibels could then be computed as $10 \log\left(\dfrac{27.22 \text{ W}}{20 \times 10^{-6} \text{ W}}\right) = 61 \text{ dB}$,

thus confirming the result obtained from Equation (1-3).

Example 1-2 demonstrates that input and output impedances must be taken into account because voltages and currents are affected by changes in impedance. This result occurs because, if the impedance at the output of a circuit is increased, the voltage increases (because $V = IR$), but the total *power* in the circuit remains the same; that is, the voltage gain does not, in and of itself, produce a power gain. If the impedance change had not been taken into account, the unmodified voltage form of the decibel formula would falsely indicate that a power increase had occurred. In Example 1-2, not correcting for impedance differences would result in a calculated voltage gain of 70.8 dB. Power is independent of impedance, however. Impedance changes cause changes in voltage or cur-rent, but power remains the same. Put another way, the resistance terms cancel when P_1 and P_2 in the power ratio form of the decibel equation are expressed in terms of their volt-ages and resistances.

Currents can be represented in decibel form by remembering that $P = I^2R$. Similarly,

$$dB = 10 \log \frac{I_2^2 R_2}{I_1^2 R_1},$$

which simplifies to

$$dB = 20 \log \frac{I_2}{I_1} + 10 \log \frac{R_2}{R_1},$$

or

$$dB = 20 \log \frac{I_2\sqrt{R_2}}{I_1\sqrt{R_1}}.$$

(1-4)

For the special case where $R_1 = R_2$ (that is, equal input and output resistances), the relationships for voltage and current simplify to

$$\text{dB} = 20 \log \frac{V_2}{V_1} \qquad \qquad \textbf{(1-5)}$$

and

$$\text{dB} = 20 \log \frac{I_2}{I_1}. \qquad \qquad \textbf{(1-6)}$$

Often, the assumption is made that input and output impedances are equal, for this is the condition that ensures maximum power transfer. In such cases, the forms of Equation (1-5) or Equation (1-6) may be used without further modification.

Reference Levels

By itself, the decibel has no absolute value; decibels represent ratios of powers, voltages, or currents. Therefore, one could not determine, say, the output power of an amplifier without knowing the input power, even if the decibel gain were specified. That is, a "23-dB amplifier" says nothing about its absolute output power level without knowledge of the power applied to the input. For convenience, however, one often assigns a 0-dB reference power or voltage level and represents all other quantities with respect to that reference. (Why 0 dB? Remember that 0 dB is $10 \log 10^0$. The value 10^0 equals 1, and the logarithm of 1 is 0.) Most commonly encountered for both audio- and radio-frequency systems is a power reference of 1 mW (one-thousandth of a watt, or 0.001 W). Decibels referenced to 1 mW are represented by the term **dBm** (where the lower-case m represents milliwatt). That is, 0 dBm = 1 mW. For audio applications, including wireline telephone networks, 0 dBm represents 1 mW across a 600-Ω system impedance, whereas for radio work the same 0 dBm represents 1 mW across 50 Ω. Though the term dBm is standard, the impedance may not be explicitly stated and sometimes must be inferred from the context (audio, video, or radio). Though 1 mW may seem like a small number to use as a reference, it turns out to be an extremely high power level for receivers, as we shall see in subsequent chapters.

Confusion between the relative nature of decibel notation versus the absolute power levels represented by decibels expressed with an implied or explicitly stated reference level such as dBm can arise. Particularly in casual usage, dB is sometimes used incorrectly in place of dBm. Some of this casual usage may have historical precedent: most analog volt-ohm meters (VOMs), such as the one shown in Figure 1-3, have a "dB"

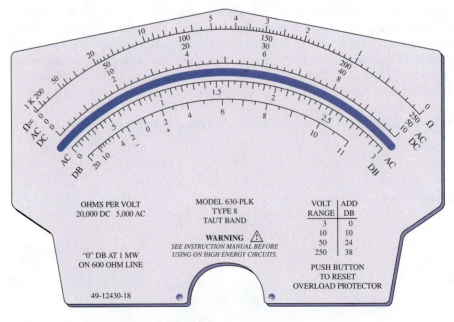

FIGURE 1-3 **Decibel (dB) scale on analog meter face.**

(a)   (b)

FIGURE 1-4   **(a) DMM with dBm scale. (b) Adjustable impedance feature.**

scale. These types of meters have been around for many years; generations of technicians have trained with them and have used them throughout their careers, and perhaps
users have become accustomed to thinking of decibel scales as absolute measures when,
in fact, they are not. Close examination of the meter scale in the figure reveals that what
is labeled as "0 dB" is referenced to a 600-Ω system impedance. By rearranging the
equation $P = V^2 \div R$ to solve for voltage, we see that 1 mW in a 600-Ω system produces
a root-mean-square (rms) AC voltage of approximately 0.775 V. The 0 dB value on the
meter face should be interpreted as 0 dBm because that point lines up with 0.775 V on
the AC volts scale. In contrast, a radio-frequency (rf) millivoltmeter would most likely
be referenced to a 50-Ω impedance, so its 0-dB point would be different from that for the
VOM, which is calibrated for audio-frequency work. Note that each of these dBm values
corresponds to a different voltage, so a meter with a dB scale calibrated to read "0 dB"
(actually 0 dBm) across 600 Ω would not read correctly if used in a 50-Ω system. A
device intended for video systems, such as cable television, is most likely calibrated for
75 Ω, so its 0-dB point would be different from those of either the audio or rf instruments. Some newer digital multimeters (DMMs), such as the one shown in Figure 1-4,
have a dBm measurement capability with adjustable impedances to ensure accurate
measurements in different system environments. Again, the essential point to remember
is that, in modern practice, only those decibel measurements with an explicit or implied
reference can be expressed in terms of absolute values; all other expressions represent
only gains or losses.

   While dBm is the most often encountered reference in radio communications work,
other references are in wide use. Some of these include dBμV and dBmV, where 0 dB represents 1 μV or 1 mV, respectively, and which are frequently encountered in video systems
and cable television; dBV (1-V reference); dBW (1-W reference); dBk (1 kW reference,
used in broadcasting); and dBc, representing levels referenced to the system carrier (often
used in interference studies to represent the ratio of undesired-to-desired signals).

   Yet another reference sometimes encountered is expressed as dBu. Here context is
important as well. In the field of audio engineering, the u means "unloaded," in which the
amplitude measurement is made with a high-impedance instrument so as not to reduce the
signal amplitude to half its original value (i.e., a 6-dB voltage reduction), which would
happen if the impedance of the measuring instrument were equal to that of the system. In
the field of radio engineering, the term dBu represents decibels referenced to a field-
strength intensity of 1 microvolt-per-meter (1 μV/m), which is a measurement of the voltage levels induced in an antenna by a received signal.

In all cases, levels may be above or below the reference. Values above the 0-dB point should be—but are not always—denoted with a leading plus sign to minimize confusion; values below 0 dB *must* be shown with a minus sign.

With a specified reference level and use of the antilogarithm, one can determine exact values. For example, a microwave transmitter specified at +27 dBm (27 dB above 0 dBm) would have a power output of 0.5 W, shown as follows:

$$\text{Pout (dB)} = 10 \log \frac{P_2}{0.001 \text{ W}}$$

$$27 \text{ dBm} = 10 \log \frac{P_2}{0.001 \text{ W}}$$

$$2.7 \text{ dBm} = \log \frac{P_2}{0.001 \text{ W}}.$$

To clear the log on the right-hand side of the equation in the above step, we take the antilogarithm (antilog) of the number on the left. Recall that the antilog of a common logarithm is the expression 10^x, so in the next step the antilog is the expression $10^{2.7}$, which equals 501.2.

$$\text{Antilog } 2.7 = \frac{P_2}{0.001 \text{ W}}$$

$$501.2 = \frac{P_2}{0.001 \text{ W}}$$

$$(501.2)(0.001 \text{ W}) = P_2$$

$$0.5 \text{ W} = P_2$$

Approximating with Decibels

Though Equations (1-2), (1-5), and (1-6) can always be used for exact calculations, the precision so implied is rarely needed in practice. The true utility of decibels lies in their ability to provide quick approximations. By memorizing only a very few relationships and with a little practice, the communications technician or engineer can quickly become proficient not only in estimating system gains and losses but also in converting levels shown in dBm to watts, and vice versa.

From inspection of the results obtained in Example 1-1, one sees that a 10-dB power increase is not a doubling of power, but rather an increase of 10 times. A doubling of power corresponds to a 3-dB increase. Similarly, a halving of power is a 3-dB decrease. A 1-dB increase in power is approximately a 25% increase. By committing some common values to memory, one can easily estimate power or voltage changes with decibels without resorting to pencil-and-paper calculations or calculators. In the microwave transmitter example just given, one can arrive at the same result by knowing that +30 dBm, 30 dB above 1 mW, is 10^3 above 1 mW, or 1 W. The transmitter output, at +27 dBm, is 3 dB below 1 W, which is a power reduction of half; thus, the power output is 0.5 W. Alternatively, the power output can be arrived at by recognizing that the output consists of multiple 10-dB, 3-dB, and 1-dB steps above 0 dBm: The first 10-dB step, from 0 to +10 dBm, increases power from 1 mW to 10 mW; the second 10-dB step is an additional ten-times power increase, to 100 mW, or +20 dBm. Three dB above that is 200 mW; the next 3 dB brings the power to 400 mW, or +26 dBm. The final 1-dB increase can be approximated as a 25% increase over 400 mW. Twenty-five percent of 400 is 100, so +27 dBm is 500 mW, or 0.5 W. With a bit of practice, the communications technician can become very proficient at determining power outputs based on decibel readings.

Table 1-2 shows several decibel relationships worth memorizing. Note in particular how the 10-, 20-, 30-, and so forth, dB power relationships correspond with powers of 10. That is, a 10-dB power increase relates to 10^1 (i.e., log 10, therefore 10 times), 20 dB to 10^2 (i.e., log 100, therefore one hundred times), and 30 dB to 10^3 (i.e., log 1000, therefore one thousand times). Decreases follow the same pattern: A 3-dB decrease is a halving of power; a 10-dB decrease is 1/10, or 10^{-1}; a 20-dB decrease is 1/100 or 10^{-2} (i.e., $1/10^2$), and so

TABLE 1-2 • Common Decibel Relationships

CHANGE (DB)	POWER	VOLTAGE
+1	~1.25×	
+3	2×	1.414×
+6	4×	2×
+10	10×	
+20	100×	10×
+30	1000×	
+40	10,000×	100×
−3	0.5×	0.707×
−6	0.25×	0.5×
−10	0.1×	
−20	0.01×	0.1×
−30	0.001×	
−40	0.0001×	0.01×

forth. Similar statements can be made for voltages or currents, but because power scales as the square of either of these quantities, a doubling of voltage (or current) equates with a 6-dB increase, and a ten-times increase in voltage or current is equal to a 20-dB increase.

EXAMPLE 1-3

You suddenly find yourself in a country where those found in possession of calculators with "log" keys are shot on sight. Your calculator doesn't work anyway because the battery is dead (it is a very poor country, and there are no replacement batteries to be had). Based on what you know about decibel relationships, explain to the ruler of this country what the output power of his only radio transmitter would be if you knew it had 46 dB of gain and a power input of 1 mW. Demonstrate how you could determine the output power without resorting to any "pencil-and-paper" calculations (that is, no calculator, slide rule, or published table of logarithms).

SOLUTION

A one-milliwatt level is equal to 0 dBm. Therefore, one can state that an amplifier with 46 dB of gain has an output of +46 dBm. This output can be broken down into increments of 10 and 3 dB. The first four 10-dB increments, from 0 to +40 dBm, represent four successive factor-of-10 power increases. Therefore, +40 dBm equals 10 W. The remaining 6 dB can be broken down into two 3-dB increments. The first increment, from +40 to +43 dBm, doubles the power from 10 to 20 W, and the second doubles it again, to 40 W.

Note in Example 1-3 that the individual decibel steps were added. This is the product rule for exponents at work. Also note that the absolute power level (40 W) could only be specified because the power input was given. Had only the gain been given, the only valid statement that could be made would be that the output was 40,000 times the input (that is, the decibel gain would be equal to 10 times the common logarithm of 40,000) without reference to an absolute power level.

Stage Gains and Losses

Decibel relationships are particularly useful for determining gains (or losses) through pieces of equipment with multiple stages. An example involving a superheterodyne radio receiver (which will be studied more thoroughly in Chapter 6) will illustrate how output power can be easily calculated if the input power (or, in this case, the input voltage and impedance) and stage gains are known.

EXAMPLE 1-4

An 8-microvolt signal is applied to the 50-Ω input of a receiver whose stages exhibit the following gains (all values in decibels):

RF amplifier:	8
Mixer:	3
First IF amplifier:	24
Second IF amplifier:	26
Third IF amplifier:	26
Detector:	−2
Audio amplifier:	34

Calculate the input power in watts and dBm, and calculate the power driven into the speaker (i.e., output of the audio amplifier.)

SOLUTION

Because the input signal levels are given as voltage and impedance, it is first necessary to convert to equivalent power:

$$P = \frac{V^2}{R} = \frac{(8\ \mu\text{V})^2}{50\ \Omega} = 1.28 \times 10^{-12}\ \text{W}.$$

With input power known, it is easy to convert to dBm (remember that the denominator of the decibel power formula is known—it is 1 mW):

$$\text{dBm} = 10 \log \frac{P}{1\ \text{mW}},$$

$$= 10 \log \left(\frac{1.28 \times 10^{-12}}{1 \times 10^{-3}} \right) = -89\ \text{dBm}.$$

With the input power expressed in dBm, determination of the output power is simply a matter of adding the individual stage gains or losses:

$$-89\ \text{dBm} + 8\ \text{dB} + 3\ \text{dB} + 24\ \text{dB} + 26\ \text{dB} + 26\ \text{dB} - 2\ \text{dB} + 34\ \text{dB} = +30\ \text{dBm}.$$

From the approximations given in Table 1-2, it should be apparent that +30 dBm is 30 dB above 1 mW, or a one-thousand-times increase, so the drive power to the speaker is 0.001 W $\times$ 1000 = 1 W.

Example 1-4 also makes a subtle but important point. Note that the input *level* is referenced to a specific value (and is thus denoted by dBm) but that individual stage *gains* are not. When two values expressed with respect to a reference are compared, the resulting gain or loss is properly expressed in dB, *not* in dB with respect to that reference. In other words, the references cancel.

EXAMPLE 1-5

The power input to an amplifier is measured as −47 dBm, and the output is measured as −24 dBm. What is the gain? Is the gain properly expressed in dB or dBm?

SOLUTION

The gain is simply the output minus the input, or −24 dBm −(−47 dBm) = 23 dB. The difference in input and output simply represents the *ratio* between two power *ratios*, not the ratio with respect to a fixed level (1 mW in this case), so the gain is properly expressed in dB.

1-3 INFORMATION AND BANDWIDTH

As already mentioned, one of two fundamental limitations encountered by all communications systems involves **bandwidth**, defined as the difference between the highest and lowest frequencies occupied by signals of interest. Put another way, bandwidth defines the frequency range over which a circuit or system operates. As an example, voice frequencies found on telephone circuits occupy the range of frequencies approximately between 300 Hz and 3 kHz. The bandwidth of voice frequencies would, therefore, be 2.7 kHz. In Section 1-1 we established that electrical signals reside on the electromagnetic spectrum, so another way to look at bandwidth in the communications context is to define it as the span of frequencies occupied by signals within the communication channel; that is, bandwidth describes the portion of the electromagnetic spectrum occupied by the entire frequency range of signals conveyed from transmitter to receiver.

Also established in Section 1-1 was the idea that information transfer occurs as a result of modulation. Bandwidth and information transfer are interrelated: the greater the bandwidth, the greater the amount of information that can be transferred from source to destination. In 1928, R. Hartley of Bell Laboratories developed a formal relationship between the two quantities. Expressed as an equation, **Hartley's law** states that

$$\text{information} \propto \text{bandwidth} \times \text{time of transmission}$$

In words, Hartley's law states that information transmitted is directly proportional to the product of the bandwidth used and the time of transmission. Thus, if a given amount of information is to be sent within a given time interval, the channel must have a bandwidth sufficient to permit the information to be conveyed without introducing distortion. Hartley's law is perhaps one of the foundational relationships in the study of information transfer in communications systems. The science of **information theory** is concerned, in part, with providing for the most efficient use of a band of frequencies in electrical communications. Though beyond our intentions in this chapter to embark on a detailed examination of this highly theoretical field of study, one aspect becomes immediately apparent: Because modulation is concerned with the transfer of information, a logical consequence of Hartley's law is that modulated signals necessarily occupy wider bandwidths than do unmodulated ones. Much of the analyses of Chapters 2 and 3 are concerned with predicting the occupied bandwidths of modulated signals.

At this juncture you may wonder why bandwidth is so important. In many cases, bandwidth equals money. The description of frequency-division multiplexing in Section 1-1, which established that modulated information signals occupy bands of frequencies much higher than do unmodulated ones, also established that different users could share bands of frequencies provided that each user was somehow assigned exclusive use of a frequency range within that band. Implied in that description was the notion that some entity actively manages the allocation of frequencies among competing users to minimize the potential for interference between them. Such activities are, in fact, ongoing for wireless systems. Governments of the United States and other countries recognized in the early decades of the twentieth century that the radio-frequency spectrum is a scarce and valuable public resource. Only so many frequency bands are in existence. No new ones can be created, so for the radio-frequency spectrum to be useful to anyone, its use has to be regulated so that frequencies can be efficiently allocated among many competing users while minimizing the potential for destructive interference. In 1934, the U.S. Congress passed the Communications Act, a law establishing the present-day Federal Communications Commission (FCC) and giving it the authority to establish and enforce regulations regarding the use of radio transmitters and other generators of radio-frequency energy. FCC rules have the force of law and apply to all nonfederal uses of the radio spectrum. (Another federal agency, the National Telecommunications and Information Administration, regulates use of the radio spectrum by federal government users, including the military services.) Other countries have agencies that perform an equivalent function. For example, the communications branch of Industry and Science Canada and the Secretaria de Comunicaciones y Transportes (Transport and Communications Ministry) have regulatory jurisdiction over radio communications in Canada and Mexico, respectively. Because radio waves can travel great distances, telecommunications regulation is an international matter, and the International Telecommunication Union (ITU), a body under the auspices of the United Nations, coordinates spectrum affairs among nations.

Because spectrum allocations are a regulatory function, and because spectrum is scarce, several questions immediately arise, among them the following: How are allocation decisions made? How do regulatory bodies decide who gets what spectrum? In the United States, spectrum-allocation decisions have historically been made either as part of the political process or have been deliberative in nature, often involving adversarial hearings among competing interests that in many ways resembled court trials. The intended outcome of these hearings was somehow to discern the "best" use of spectrum and to identify the "best" custodians of scarce public resources. The process was, at best, cumbersome and inefficient. Today, spectrum-allocation decisions involving private-sector users are made on economic grounds under the theory that free-market pricing mechanisms are sufficient to determine the best use of scarce resources. Since 1994, the FCC has had the authority to conduct periodic auctions of chunks of spectrum, wherein private parties bid for licenses authorizing them to use slices of bandwidth, either nationally or within a defined geographic area. Sometimes the auctions involve reallocating spectrum away from one group of users and awarding it to others. The nationwide transition from analog to digital television broadcasting, completed in 2009, is illustrative. As television broadcast stations relinquished their licenses for analog channels, those frequencies, many of which were in the highly desirable VHF and UHF portions of the frequency spectrum, were reclaimed by the federal government and subsequently auctioned off to the highest bidders. Revenue to the tune of $20 billion accrued to the U.S. Treasury as a result. The auction mechanism has raised over $60 billion since its inception in 1994.

Whether the allocation mechanism is economic or political, the bottom line is the same: Bandwidth is valuable because frequencies are scarce. Particularly in the radio-frequency portions of the electromagnetic spectrum, conservation and efficient use of bandwidth are of paramount importance. Competition for radio frequencies is fierce; nearly all frequencies up to 30 GHz are completely spoken for, and any repurposing of radio spectrum often comes at the expense of incumbent users. Any entity licensed to occupy a band of radio frequencies must ultimately be concerned with making the most efficient use of those frequencies, either to maximize revenue or to provide opportunities for others. For private-sector users, such as wireless carriers, efficiency of spectrum use equals revenue. Ultimately, licensed bandwidth is the factor that limits the maximum number of calls, text messages, or downloads any one carrier can accommodate in a given geographic area. Carriers also implicitly acknowledge Hartley's law when they limit download speeds, for high data-transfer rates necessarily imply high bandwidths. Because the electromagnetic spectrum is one of our most precious natural resources, regulatory agencies and others devote much effort to ensuring the equitable distribution of spectrum among competing interests. The discipline of communications engineering is ultimately concerned with making the best use of finite spectrum, and many techniques described in this text evolved specifically to promote spectrum efficiency through the minimization of transmission bandwidth.

Understanding Frequency Spectra

For reasons that will be examined more fully in the following chapters, the process of modulation necessarily involves the production of frequencies in addition to those of the carrier and intelligence. These additional frequencies reside both above and below the carrier and represent the physical manifestation of Hartley's law; that is, the locations on the spectrum of all the frequencies produced as the result of modulation are what determine the bandwidth of the modulated signal. The amplitude and frequency characteristics of both the carrier, which is usually a sine wave, and the modulating signal, which is often not a sine wave, therefore play a large role in determining the overall bandwidth characteristics of the modulated signal. Thus, to develop a complete picture of occupied bandwidth, one must have a means of resolving (identifying) all the significant frequency components and amplitude characteristics of modulated signals. To do so, we make use of a mathematical technique known as Fourier analysis.

In 1822, the French mathematician Jean Baptiste Joseph Fourier developed a means to break down any periodic waveform into a series of sine and/or cosine waves at multiples of the fundamental frequency. (A cosine wave has the same appearance as a sine wave when plotted on a time axis, such as that shown in Figure 1-1, but the cosine wave is shifted in phase by 90 degrees. Thus, at some time $t = 0$, when a sine wave has an amplitude of 0 V, a cosine wave will have maximum amplitude.) Though Fourier applied his ideas to the field of heat transfer, his ideas apply well to the study of other phenomena that can be represented in terms of complex waveforms, among them the signals encountered in electronic communications systems. Modulated signals produce complex, periodic waveforms. Understanding the bandwidth implications of these signals necessarily involves breaking them down to their individual frequency components. A Fourier series, as applied to a time-varying signal, can be expressed as follows:

$$f(t) = A_0 + A_1 \sin \omega t + A_2 \sin 2\omega t + A_3 \sin 3\omega t + \cdots + A_n \sin n\omega t$$
$$+ B_1 \cos \omega t + B_2 \cos 2\omega t + B_3 \cos 3\omega t + \cdots + B_n \cos n\omega t \qquad \text{(1-7)}$$

where

$$f(t) = \text{any function of time, such as a voltage } v(t) \text{ or current } i(t)$$
$$A_n \text{ and } B_n = \text{real-number coefficients (either positive, negative, or zero)}$$
$$\omega = \text{radian frequency of the fundamental.}$$

Waveforms with harmonic content may require a large number of the terms shown in Equation (1-7) to describe them fully in terms of their amplitudes, frequencies, and phase relationships; however, the underlying implication that even the most complex signals can be identified in terms of the frequencies of individual sine and cosine waves and the phase relationships between them remains true. The essence of the Fourier transform is that any signal can be decomposed mathematically into its constituent frequencies and amplitudes and that these constituents bear a definite relationship to each other.

The Fourier series shown in Equation (1-7) consists of three parts: a dc term, a series of sine terms, and a series of cosine terms. The dc term, A_0, is the average value of the waveform over one full cycle. The dc term may be zero, in which case the area of the waveform above the horizontal axis equals the area below the axis, or it may have a positive or negative average value. If A_0 is positive, the waveform has an area above the horizontal axis that is greater than the area below the axis; the opposite situation occurs for negative values of A_0. The first term of each of the sine and cosine series represents the minimum frequency of the waveform being represented. This first frequency term, called the **fundamental**, is the frequency of the waveform being represented and must be present in any Fourier series representation. The sine and cosine terms following the fundamental are called **harmonic** terms because they are integer multiples of the frequency of the first term in their respective series. Thus, a frequency that is twice the fundamental is the second harmonic; three times, the third harmonic, and so on. The number of sine or cosine terms is not fixed. An ideal system, that is, one with infinite bandwidth, will have an infinite number of sine and/or cosine terms, each bearing a harmonic relationship to the fundamental. Any or all of the sine or cosine terms may be zero, and there are no restrictions on their amplitudes. The frequencies following the fundamental will bear a harmonic relationship to the fundamental term,

TABLE 1-3   Fourier Expressions for Selected Periodic Waveforms, $f = 1/T$, $2\pi f = \omega$

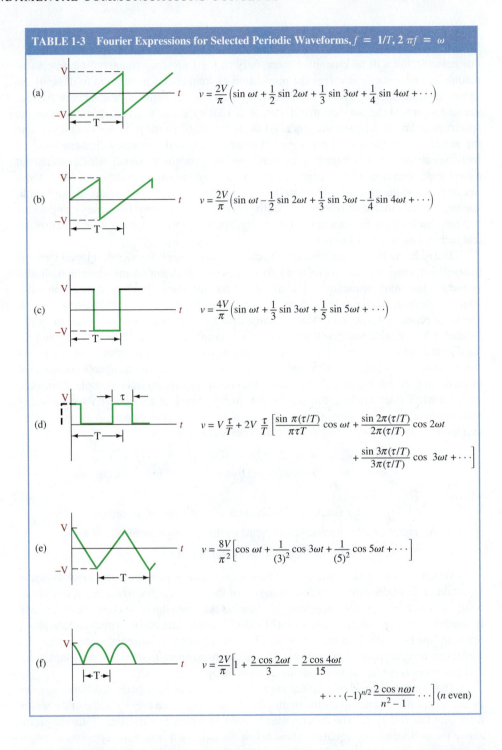

(a)
$$v = \frac{2V}{\pi}\left(\sin \omega t + \frac{1}{2}\sin 2\omega t + \frac{1}{3}\sin 3\omega t + \frac{1}{4}\sin 4\omega t + \cdots\right)$$

(b)
$$v = \frac{2V}{\pi}\left(\sin \omega t - \frac{1}{2}\sin 2\omega t + \frac{1}{3}\sin 3\omega t - \frac{1}{4}\sin 4\omega t + \cdots\right)$$

(c)
$$v = \frac{4V}{\pi}\left(\sin \omega t + \frac{1}{3}\sin 3\omega t + \frac{1}{5}\sin 5\omega t + \cdots\right)$$

(d)
$$v = V\frac{\tau}{T} + 2V\frac{\tau}{T}\left[\frac{\sin \pi(\tau/T)}{\pi\tau T}\cos \omega t + \frac{\sin 2\pi(\tau/T)}{2\pi(\tau/T)}\cos 2\omega t\right.$$
$$\left. + \frac{\sin 3\pi(\tau/T)}{3\pi(\tau/T)}\cos\ 3\omega t + \cdots\right]$$

(e)
$$v = \frac{8V}{\pi^2}\left[\cos \omega t + \frac{1}{(3)^2}\cos 3\omega t + \frac{1}{(5)^2}\cos 5\omega t + \cdots\right]$$

(f)
$$v = \frac{2V}{\pi}\left[1 + \frac{2\cos 2\omega t}{3} - \frac{2\cos 4\omega t}{15}\right.$$
$$\left. + \cdots (-1)^{n/2}\frac{2\cos n\omega t}{n^2 - 1}\cdots\right]\ (n\ \text{even})$$

however. The presence of harmonics has a strong influence on the bandwidth requirements of any transmitted signal, and for the signal to be transmitted without severe distortion, the communications channel must have sufficient bandwidth to pass all significant frequency components without excessive attenuation.

Table 1-3 shows the Fourier expressions for selected periodic waveforms. Of particular interest in communications studies is the Fourier series for a square wave, because impulse-like waveforms approximating square or rectangular waves often appear either as the result of modulation or as the modulating signals themselves. The Fourier series for a square wave, shown in Table 1-3, is made up of a summation of sinusoids multiplied by a constant $4V/\pi$. The sinusoids are represented by

$$\sin \omega t + \frac{1}{3}\sin 3\omega t + \frac{1}{5}\sin 5\omega t + \cdots$$

where the frequency sin ωt is recognized as the radian frequency of the fundamental (recall that $\omega = 2\pi f$), and the components $\frac{1}{3} \sin 3\omega t$ and $\frac{1}{5} \sin 5\omega t$ represent the third and fifth harmonics of the signal, respectively. This process continues until the system bandwidth is reached. The $\frac{1}{3}$ and $\frac{1}{5}$ values simply indicate that the amplitude of each harmonic decreases as the frequency increases.

One aspect of this analysis becomes immediately apparent: Square waves are made up of the fundamental and its *odd*-numbered harmonics, and the more harmonics that are present, the more ideal the square wave looks. (Other types of waveforms consist of the fundamental and either or both of the even and odd harmonics of both sine and cosine waves.) Figure 1-5 shows how the construction of a square wave can be seen to be made up of a number of sinusoids, each bearing a harmonic relationship to the fundamental. Figure 1-5(a) shows the fundamental frequency. Figure 1-5(b) shows the addition of the first and third harmonics, and Figure 1-5(c) shows the addition of the first, third, and fifth harmonics. Though somewhat distorted, the signal of Figure 1-5(c) begins to resemble a square wave. With the addition of more odd harmonics, the wave rapidly approaches the ideal: Figures 1-5(d) and 1-5(e) show signals with 13 and 51 harmonics, respectively. The figure and accompanying analysis also imply that a tradeoff between bandwidth and preservation of harmonic content must take place; that is, if a square wave is to be transmitted over a communication channel with its shape preserved, the channel bandwidth must be sufficient to pass not only the fundamental but also the harmonics. Limiting bandwidth necessarily implies limiting or filtering the harmonics, and the more severe the filtering, the more the square wave of Figure 1-5(e) begins to resemble the sine wave of Figure 1-5(a). We shall see that the transmission of square waves (or any waveform with sharp, square-wave-like edge transitions) through any bandwidth-limited medium without distortion is often not desirable or even possible, and filtering techniques must be employed to "smooth out" the edges of the square wave while still preserving enough harmonic content to preserve the information contained therein.

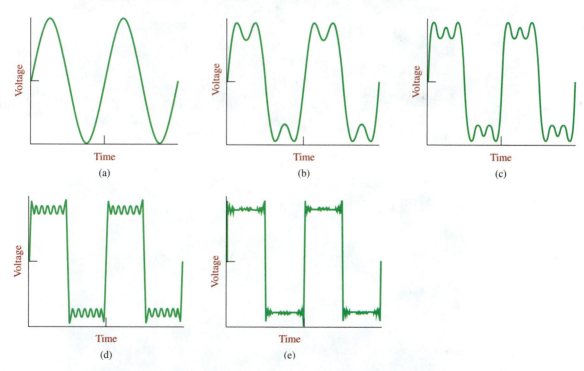

FIGURE 1-5 (a) Fundamental frequency (sin ωt); (b) the addition of the first and third harmonics (sin $\omega t + \frac{1}{3} \sin 3\omega t$); (c) addition of the first, third, and fifth harmonics (sin $\omega t + \frac{1}{3} \sin 3\omega t + \frac{1}{5} \sin 5\omega t$); (d) addition of the first 13 odd harmonics; (e) addition of the first 51 harmonics.

Time- and Frequency-Domain Representations

Fourier analysis reveals that waveforms can be viewed from either of two perspectives. In the **time domain**, the waveform amplitude, that is, the instantaneous magnitude of the quantity under consideration (usually voltage), is shown as a function of time. The familiar graphical display produced by an oscilloscope is a time-domain representation. From an oscilloscope display, the frequency of a sine wave is easily determined as the reciprocal of its period, and Fourier analysis confirms that only one frequency is present. However, for signals such as square waves that consist of multiple frequencies, time-domain representations produce an incomplete picture. Only the period associated with the fundamental frequency can be directly ascertained from an oscilloscope display, for example; all other frequency and amplitude components, though present, are effectively hidden from view. Consideration of these other components requires that the signal be viewed in the **frequency domain**, where amplitude is viewed as a function of frequency rather than of time. Fourier analysis essentially allows for a time-domain signal to be analyzed in the frequency domain, and vice versa.

As mentioned, the oscilloscope displays waveforms in the time domain. A **spectrum analyzer** represents signals in the frequency domain and is an indispensable tool in communications work. The spectrum analyzer allows for the complete characterization of the frequency and amplitude attributes of signals under study as well as for the identification and elimination of unwanted, interfering signals. The device is, in effect, an automatic frequency-selective voltmeter that provides information about both frequency and amplitude, usually expressed in dBm, on a visual display. Figure 1-6 shows a typical hardware spectrum analyzer and its associated display. The amplitude of each frequency component is displayed on the vertical axis in decibels with respect to a user-selected reference level in dBm, which is usually represented by the horizontal line at the top of the display. Each of the eight major divisions on the vertical axis corresponds to a reduction in amplitude that is user-selectable

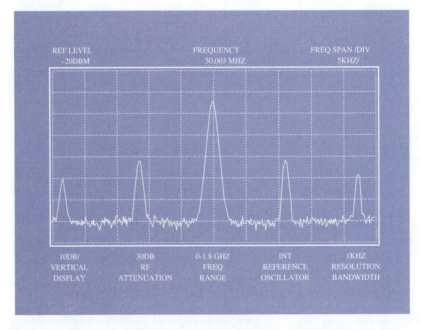

FIGURE 1-6   **Spectrum analyzer and display.**   (Copyright Tektronix. Reprinted with permission. All rights reserved.)

but that is usually set to be 10 dB per division. Each square shown in Figure 1-6 in the vertical direction thus corresponds to a reduction in signal amplitude of 10 dB from the top-line reference. By displaying amplitudes in terms of decibel units, the spectrum analyzer is able to compress signals spanning 80 dB—a full eight orders of magnitude—onto a reasonable-sized display. Magnitudes associated with both the vertical and horizontal axes are user-selectable. By selecting the frequency span per division as well as the center frequency, the user defines the frequency range to be displayed and can zoom in or out as necessary to examine and compare the amplitude characteristics of signals that are either closely spaced in frequency or that have a harmonic relationship to the signal under consideration.

The Fast Fourier Transform

Conversion from the time to the frequency domain can be approximated with the **fast Fourier transform (FFT)**, an algorithm that allows for spectrum analysis to be accomplished relatively easily in software. The FFT is particularly well suited to computer-based implementation because it allows for the necessarily large number of calculations associated with the full Fourier series to be reduced to a more manageable number through the elimination of redundancies. In addition, when an analog signal is **digitized**—that is, when a series of multiple-bit digital words, each representing the amplitude characteristics of the original signal at a particular instant in time, is created—the sheer amount of data produced as a result of the process mandates that subsequent signal processing operations be as computationally efficient as possible. Though also applicable in the analog domain, the economies of calculation afforded by the FFT find particular utility in the spectral analysis of digital signals, which, because of their impulse-like nature, display harmonic and other spectral characteristics similar to those of square waves.

Many math software packages have an FFT function, as do modern digital sampling oscilloscopes, such as the Tektronix TDS 340. An oscilloscope with "FFT math" software built into it is effectively functioning as a spectrum analyzer when it performs the FFT operation. With this feature we can confirm the predictions already described regarding the fundamental and harmonic frequencies of square waves, and we can also examine the distortion effects produced by a bandwidth-limited channel.

Figure 1-7 shows the FFT-equipped oscilloscope display in both the time and frequency domains for a 1-kHz square wave. The repetitive square wave with sharp edges is

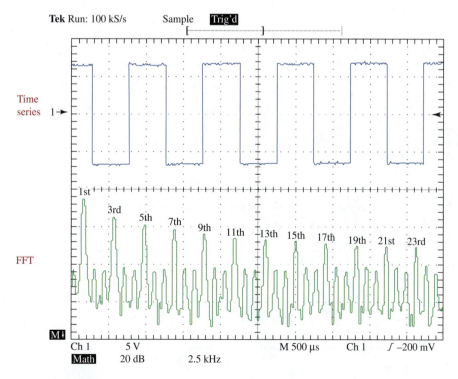

FIGURE 1-7   **A 1-kHz square wave and its FFT representation.**

shown at the top of Figure 1-7 with a time base setting of 500 μs/div. The upper trace is, therefore, a time-domain representation. The FFT, with a horizontal scale of 2.5 kHz per division, is shown at the bottom. See how the FFT reveals the frequency components. The fundamental is at 1 kHz, as expected. The second frequency component visible is the third harmonic at 3 kHz (which is the first odd harmonic of the fundamental), and the subsequent odd harmonics are displayed, each with progressively lower amplitude than that of the frequency component before it. This is indeed the result predicted by Equation (c) in Table 1-3 and shown in the frequency-domain representation of Figure 1-7. A square wave is constructed as a series of sine waves that are shown to be the fundamental and its odd harmonics, each of decreasing amplitude.

If we were to extend the display of Figure 1-7 to the right, we would see that the number of odd harmonics present goes on indefinitely. This result is a consequence of the nice, sharp edges of the square wave under consideration. The high-frequency components shown in the FFT indicate the contribution of the higher harmonics to the creation of a well-shaped square wave. What would happen if such a square wave were transmitted through a bandwidth-limited medium such as a voice-grade telephone channel, which has an upper frequency limit of about 3 kHz? Hartley's law predicts that distortion would occur, and visible proof of the distortion, with its effects on higher-order harmonics, is visible in the time- and frequency-domain representations of Figure 1-8. The bandwidth-limiting effects of the channel can be simulated with a low-pass filter (Figure 1-8(a)). When a 1-kHz square wave is applied to the filter, the result is severe attenuation, and the waveform appearing at the filter output bears little resemblance in the time domain to a square wave. Some features of the original signal (specifically, the sharpness of the square wave edges) are lost. The FFT of the waveform shows that the higher harmonic values are severely attenuated, meaning that there is a loss of detail (i.e., information) through this bandwidth-limited system. Such distorted data can still often be used in communications applications, but if the frequency information lost in the channel is needed for proper

FIGURE 1-8 **(a) A low-pass filter simulating a bandwidth-limited communications channel; (b) the resulting time series and FFT waveforms after passing through the low-pass filter.**

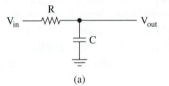

(a)

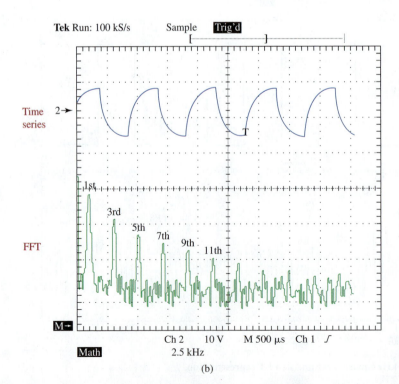

(b)

signal or data representation, the lost information may not be recoverable. For an analog system, the lost information would result in noise and distortion. For a digital system, the loss of information could result in increased bit errors, which in turn could prevent the digital information from being reconverted to analog form with sufficient integrity to preserve the original meaning. In cases where distortion is present—a reality with any bandwidth-limited channel—subsequent signal processing may be required to regenerate signals with the requisite characteristics for use in their intended application.

The signals of Figure 1-8 show the effects of filtering. The bandwidths of many signals are theoretically infinite, but transmission of such signals through any bandwidth-limited channel, particularly wireless channels, is impractical. Filtering causes distortion, both by removing higher-order harmonics and also by creating uneven frequency responses or nonlinear phase shifts, all of which affect the time-domain representations. Fourier analysis provides the means to go back and forth at will between the time and frequency domains, and its utility lies in the ability to show in very real terms the effects on signals passing through and being processed by the many stages that make up a complete communications system. In short, Fourier analysis is more than just a useful mathematical equation representing a wave. The sine or cosine waves represented by Equation (1-7) are, indeed, physically real and measurable, as the previous examples demonstrate.

1-4 NOISE

If one of two fundamental themes unifying the study of communications systems is the notion of bandwidth, surely the other is noise. Electrical **noise** may be defined as any undesired voltages or currents that ultimately end up appearing at the output of a receiver. Noise is broadband, consisting of a mixture of signals at all frequencies and at randomly occurring amplitudes. Noise is generated not only from within all types of electronic equipment but is also produced from external sources and introduced into the communications channel. To the listener of an AM or FM radio, noise manifests itself as static and may only represent a passing annoyance. To a television viewer, noise corrupts the received digital signal and causes pixilation and dropouts if the noise is severe enough. However noise is manifested, an understanding of its sources and effects is a crucial element in the study of communications fundamentals because the ability of a system to perform satisfactorily in the presence of noise is the ultimate arbiter of the success of its underlying design strategy, for noise is the one unavoidable factor that ultimately determines the success or failure of the communications endeavor.

Noise signals at their point of origin are generally very small, often at the microvolt level. Why, then, do they create so much trouble? The answer is that the level of the desired signal at the receiver input is comparable to the noise level. A communications receiver is a very sensitive instrument that will greatly amplify a very small input signal to a level that can drive a speaker. If the block diagram of Figure 1-2 is considered to be representative of a standard FM radio (receiver), the first amplifier block, which forms the "front end" of the radio, must amplify a signal obtained from the antenna that is often much less than $10\ \mu\text{V}$. Therefore, even a small dose of undesired signal (noise) is sufficient to ruin reception. Even though the transmitter may have an output power of many thousands of watts, the received signal will be severely attenuated because the transmitted power between transmit and receive antennas falls off in an inverse-square relationship to distance. If the distance between transmit and receive antennas is doubled, the power falls off by a factor of 2^2, or four, a tripling of distance causes the power to fall off by 3^2, or nine times, and so on. The inverse-square relationship between power and distance causes signals to be attenuated by a factor of 120 dB or more between transmitter and receiver. A desired signal of the same order of magnitude as an undesired signal would probably be unintelligible. This situation is made even worse because the receiver itself introduces additional noise.

Noise present in a received radio signal can be classified as being of either of two major types, depending on origin. **External noise** is induced in the transmitting medium by sources external to the communications system, whereas **internal noise** is introduced by the receiver itself, or by other equipment in the communication system. The important implications of both types of noise in the study of communications systems cannot be overemphasized.

External Noise

External noise can be categorized based on source: human-made (caused by electronic or mechanical devices operating in proximity to a communications system) or caused by natural phenomena occurring either within the Earth's atmosphere or in outer space. The frequency characteristics as well as the amplitudes of each source determine its overall importance as a contributor to the total noise burden faced by communications systems.

HUMAN-MADE NOISE   The most troublesome form of external noise is usually the human-made variety. It is often produced by spark-producing mechanisms such as engine ignition systems, fluorescent lights, and commutators in electric motors. The sparks create electromagnetic radiation that is transmitted through the atmosphere in the same fashion that a transmitting antenna radiates desirable electrical signals to a receiving antenna. If the human-made noise exists in the vicinity of the transmitted radio signal and is within its frequency range, these two signals will "add" together. This is obviously an undesirable phenomenon. Human-made noise occurs randomly at frequencies up to around 500 MHz.

Another common source of human-made noise is contained in the power lines that supply the energy for most electronic systems. In this context the ac ripple in the dc power supply output of a receiver can be classified as noise because it is an unwanted electrical signal, and steps must be taken to minimize its effects in receivers. Additionally, ac power lines contain surges of voltage caused by the switching on and off of highly inductive loads such as motors. (It is certainly ill-advised to operate sensitive electrical equipment in close proximity to an elevator!) Human-made noise is weakest in sparsely populated areas, which explains why extremely sensitive communications equipment, such as satellite tracking stations, are found in desert locations.

ATMOSPHERIC NOISE   **Atmospheric noise** is caused by naturally occurring disturbances in the Earth's atmosphere, with lightning discharges being the most prominent contributors. The frequency content is spread over the entire radio spectrum but with an intensity that is inversely related to frequency. Atmospheric noise is, therefore, most troublesome at lower frequencies and is manifested as the static heard on standard AM radio receivers. Storms nearest the receiver produce the highest-amplitude atmospheric noise, but the additive effect of noise produced by distant disturbances is also a factor. This cumulative noise effect is often most apparent when listening to a distant AM station at night. Atmospheric noise is not a significant factor for frequencies exceeding about 20 MHz.

SPACE NOISE   The third form of external noise, **space noise**, arrives from outer space and is pretty evenly divided in origin between the sun and all the other stars. That originating from the sun is termed **solar noise**. Solar noise is cyclical and reaches very annoying peaks about every eleven years.

All other stars generate space noise also; their collective contribution is termed **cosmic noise**. Being much farther from Earth than the sun, the contribution produced by individual stars to total cosmic noise is small, but the countless numbers of stars and the additive effects of their noise contributions can be significant. Space noise occurs at frequencies from about 8 MHz up to 1.5 GHz (1.5×10^9 Hz). Energy below 8 MHz is absorbed by the Earth's **ionosphere**, a region from about sixty to several hundred miles above the Earth's surface in which free ions and electrons exist in sufficient quantity to have an appreciable effect on wave travel. These effects will be studied further in Chapter 13.

Internal Noise

Components and circuits within the receiver itself introduce internal noise that adds to the external noise applied to the receiving antenna. The biggest effects occur in the first stage of amplification, where the desired signal is at its lowest level; noise signals injected at that point will be greatest in proportion to intelligence signals. Figure 1-9 illustrates the relative effects of noise introduced at the first and subsequent receiver stages. Note in particular that noise contributed by the second and following receiver stages is comparatively negligible when compared with the first stage because the signal level is much higher than it is at the first stage. In particular, the noise injected between amplifiers 1 and 2 has not appreciably increased the noise on the desired signal, even though it is of the same magnitude as the noise injected into amplifier 1. For this reason, the first receiver stage must be carefully designed to have low noise characteristics, with noise considerations in the design of the following stages becoming less important as the desired signal gets larger and larger.

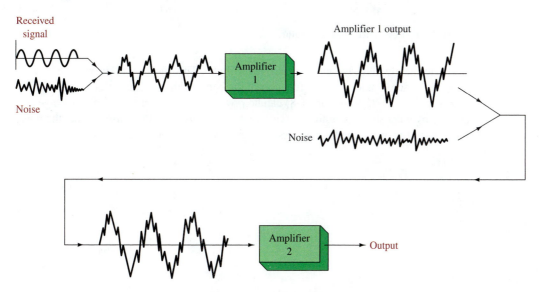

FIGURE 1-9   **Noise effect on a receiver's first and second amplifier stages.**

THERMAL NOISE   Electronic circuits produce two types of noise. The first is created as the result of thermal interactions between free electrons and vibrating ions in a conductor. These vibrations, which occur at all temperatures above absolute zero, cause electrons to arrive at the ends of a resistor at random rates, which in turn create random potential differences (voltages) across the resistor. Put another way, resistors, and the resistances present within all electronic devices, constantly produce noise voltages even in the absence of an external voltage source. J. B. Johnson of Bell Laboratories thoroughly studied this form of noise in 1928, so it is often termed **Johnson noise**. He found that the phenomenon was both temperature-dependent and directly related to bandwidth. Since it is dependent on temperature, Johnson noise is also referred to as **thermal noise**. Its frequency content is spread equally throughout the usable spectrum, which leads to a third designator: **white noise** (from optics, where white light contains all frequencies or colors). The terms *Johnson*, *thermal*, and *white noise* may be used interchangeably. Johnson demonstrated that the power of this generated noise is

$$P_n = kT\,\Delta f \tag{1-8}$$

where k = Boltzmann's constant (1.38×10^{-23} J/K)

   T = resistor temperature in Kelvin (K)

   Δf = system bandwidth.

Equation (1-8) demonstrates an extremely important point: White noise is directly proportional to both temperature and bandwidth. Receiver bandwidths should, therefore, be kept just wide enough to accept the required intelligence but without introducing additional, unwanted noise. Noise and bandwidth are related as follows: Noise is an ac voltage with random instantaneous amplitudes but with predictable rms values and with a frequency that is just as random as the voltage peaks. The more frequencies allowed into the measurement

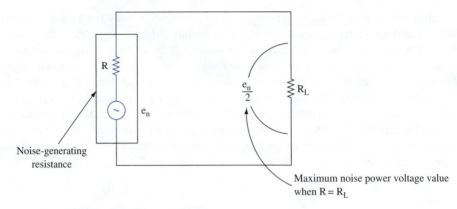

Noise-generating
resistance

Maximum noise power voltage value
when R = R$_L$

FIGURE 1-10   **Resistance noise generator.**

(i.e., greater bandwidth), the greater the noise voltage. This means that the rms noise voltage measured across a resistor is a function of the bandwidth of frequencies included.

Since $P = E^2/R$, Equation (1-8) can be rewritten to determine the noise voltage (e_n) generated by a resistor. Assuming maximum power transfer of the noise source, the noise voltage is split between the load and itself, as shown in Figure 1-10:

$$P_n = \frac{(e_n/2)^2}{R} = kT \, \Delta f$$

Therefore,

$$\frac{e_n^2}{4} = kT \, \Delta f R$$

$$e_n = \sqrt{4kT \Delta f R} \tag{1-9}$$

where e_n is the rms noise voltage and R is the resistance generating the noise. The instantaneous value of thermal noise is not predictable but has peak values generally less than ten times the rms value from Equation (1-9). The thermal noise associated with all nonresistor devices, including capacitors and inductors, is a direct result of their inherent resistance and, to a much lesser extent, their composition. The rms noise voltage predicted by Equation (1-9) applies to copper wire-wound resistors, with all other types exhibiting slightly greater noise voltages. Thus, dissimilar resistors of equal value exhibit different noise levels, which gives rise to the term **low-noise resistor**; you may have heard this term before but not understood it. Standard carbon resistors are the least expensive variety, but unfortunately they also tend to be the noisiest. Metal film resistors offer a good compromise between cost and noise performance and can be used in all but the most demanding low-noise designs. The ultimate noise performance (lowest noise generated, that is) is obtained with the most expensive and bulkiest variety: the wire-wound resistor. We use Equation (1-9) as a reasonable approximation for all calculations in spite of these variations.

EXAMPLE 1-6

Determine the noise voltage produced by a 1-MΩ resistor at room temperature (17°C) over a 1-MHz bandwidth.

SOLUTION

It is helpful to know that $4kT$ at room temperature (17°C) is 1.60×10^{-20} Joules.

$$e_n = \sqrt{4kT \, \Delta f R}$$
$$= [(1.6 \times 10^{-20})(1 \times 10^6)(1 \times 10^6)]^{\frac{1}{2}}$$
$$= (1.6 \times 10^{-8})^{\frac{1}{2}}$$
$$= 126 \, \mu V \text{ rms.} \tag{1-9}$$

The preceding example illustrates that a device with an input resistance of 1 MΩ and a 1-MHz bandwidth generates 126 μV of rms noise. A 50-Ω resistor under the same conditions would generate only about 0.9 μV of noise. This result explains why low impedances are desirable in low-noise circuits.

EXAMPLE 1-7

An amplifier operating over a 4-MHz bandwidth has a 100-Ω source resistance. It is operating at 27°C, has a voltage gain of 200, and has an input signal of 5 μV rms. Determine the rms output signals (desired and noise), assuming external noise can be disregarded.

SOLUTION

To convert °C to Kelvin, simply add 273°, so that K = 27°C + 273° = 300 K. Therefore,

$$e_n = \sqrt{4kT\,\Delta f R}$$
$$= \sqrt{4 \times 1.38 \times 10^{-23}\,\text{J/K} \times 300\,\text{K} \times 4\,\text{MHz} \times 100\,\Omega}$$
$$= 2.57\,\mu\text{V rms} \tag{1-9}$$

After multiplying the input signal e_s (5 μV) and noise signal by the voltage gain of 200, we find that the output signal consists of a 1-mV rms signal and 0.514-mV rms noise. This situation would most likely be unacceptable because the intelligence would probably be unintelligible!

TRANSISTOR NOISE   The analysis of Example 1-7 did not take into consideration the noise (other than thermal noise) that is introduced by transistors in the amplifier. The discrete-particle nature of current carriers in all semiconductors gives rise to **shot noise**, so called because it sounds like a shower of lead shot falling on a metallic surface when heard through a speaker. In a bipolar-junction transistor, shot noise results from currents flowing within the emitter-base and collector-base diodes. Even under dc conditions, current carriers do not flow in a steady and continuous fashion. Instead, the distance of travel is variable because the paths in which the charge carriers move are random. Shot noise and thermal noise are additive. There is no valid formula for calculating the shot noise contributed for a complete transistor; these values must be determined empirically. Users must usually refer to manufacturer data sheets for an indication of shot noise characteristics of specific devices. Shot noise generally increases proportionally with dc bias currents except in metal-oxide semiconductor field-effect transistors, where shot noise seems to be relatively independent of dc current levels.

FREQUENCY NOISE EFFECTS   Two little-understood forms of device noise occur at opposite frequency extremes. The low-frequency effect is called **excess noise** and occurs below about 1 kHz. Excess noise is likely caused by crystal surface defects in semiconductors and is inversely proportional to frequency and directly proportional to temperature and dc current levels. Excess noise, often referred to as *flicker noise, pink noise,* or 1/*f* noise, is present in both bipolar junction and field-effect transistors.

At high frequencies, device noise starts to increase rapidly in the vicinity of the device's high-frequency cutoff. When the transit time of carriers crossing a junction is comparable to the signal's period (i.e., at high frequencies), some carriers may diffuse back to the source or emitter. This effect is termed **transit-time noise**. These high- and low-frequency effects are relatively unimportant in the design of receivers since the critical stages (the front end) will usually be working well above 1 kHz and below the device's high-frequency cutoff area. The low-frequency effects are important, however, to the design of low-level, low-frequency amplifiers encountered in certain instrument and biomedical applications.

The overall noise-intensity versus frequency curves for semiconductor devices (and tubes) have a bathtub shape, as represented in Figure 1-11. At low frequencies excess

FIGURE 1-11   Device noise versus frequency.

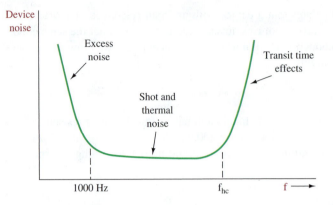

noise is dominant, while in the midrange, shot noise and thermal noise predominate, and above that, the high-frequency effects take over. Of course, tubes are now seldom used, and, fortunately, their semiconductor replacements offer better noise characteristics. However, semiconductors are not noise-free. Since semiconductors possess inherent resistances, they generate thermal noise in addition to shot noise, as indicated in Figure 1-11. The noise characteristics provided in manufacturers' data sheets take into account both the shot and thermal effects. At the device's high-frequency cutoff, f_{hc}, the high-frequency effects take over, and the noise increases rapidly.

1-5 NOISE DESIGNATION AND CALCULATION

Signal-to-Noise Ratio

We have thus far dealt with different types of noise without showing how to deal with noise in a practical way. The most fundamental relationship is the **signal-to-noise ratio** (**S/N ratio**), which is a relative measure of the desired signal power to the noise power. Often designated simply as S/N, the ratio is expressed mathematically at any point as

$$\frac{S}{N} = \frac{\text{signal power}}{\text{noise power}} = \frac{P_S}{P_N} \qquad (1\text{-}10)$$

The ratio is often expressed in decibel form as

$$\frac{S}{N} = 10 \log_{10} \frac{P_S}{P_N} \qquad (1\text{-}11)$$

For example, the output of the amplifier in Example 1-7 was 1 mV rms and the noise was 0.514 mV rms, and thus (remembering that $P = V^2/R$)

$$\frac{S}{N} = \frac{1^2/R}{0.514^2/R} = 3.79 \quad \text{or} \quad 10 \log_{10} 3.79 = 5.78 \text{ dB}$$

Noise Figure

The S/N ratio identifies the noise content at a specific point but is not useful in relating how much additional noise a particular device (either an individual component such as a transistor or an entire stage such as an amplifier) has injected into the signal path. The **noise figure (NF)** specifies exactly how noisy a device is. It is defined as follows:

$$NF = 10 \log_{10} \frac{S_i/N_i}{S_o/N_o} = 10 \log_{10} NR \qquad (1\text{-}12)$$

where S_i/N_i is the signal-to-noise power ratio at the device's input, and S_o/N_o is the signal-to-noise power ratio at its output. The term $(S_i/N_i)/(S_o/N_o)$ is called the **noise ratio (NR)**. If the device under consideration were ideal (that is, it injected no additional noise), then S_i/N_i and S_o/N_o would be equal, the NR would equal 1, and NF $= 10 \log 1 = 10 \times 0 = 0$ dB. Of course, this result cannot be obtained in practice.

EXAMPLE 1-8

A transistor amplifier has a measured S/N power of 10 at its input and 5 at its output.

(a) Calculate the NR.

(b) Calculate the NF.

(c) Using the results of part (a), verify that Equation (1-12) can be rewritten mathematically as

$$NF = 10 \log_{10} \frac{S_i}{N_i} - 10 \log_{10} \frac{S_o}{N_o}$$

SOLUTION

(a)
$$NR = \frac{S_i/N_i}{S_o/N_o} = \frac{10}{5} = 2$$

(b)
$$NF = 10 \log_{10} \frac{S_i/N_i}{S_o/N_o} = 10 \log_{10} NR \qquad \textbf{(1-12)}$$

$$= 10 \log_{10}\frac{10}{5} = 10 \log_{10} 2$$

$$= 3 \text{ dB}$$

(c)
$$10 \log \frac{S_i}{N_i} = 10 \log_{10} 10 = 10 \times 1 = 10 \text{ dB}$$

$$10 \log \frac{S_o}{N_o} = 10 \log_{10} 5 = 10 \times 0.7 = 7 \text{ dB}$$

Their difference (10 dB $-$ 7 dB) is equal to the result of 3 dB determined in part (b).

The result of Example 1-8 is a typical transistor NF. However, for low-noise requirements, devices with NFs down to less than 1 dB are available at a price premium. The graph in Figure 1-12 shows the manufacturer's NF versus frequency characteristics for the 2N4957 transistor. As you can see, the curve is flat in the mid-frequency range (NF $\simeq$ 2.2 dB) and has a slope of -3 dB/octave at low frequencies (excess noise) and 6 dB/octave in the high-frequency area (transit-time noise). An **octave** is a range of frequency in which the upper frequency is double the lower frequency.

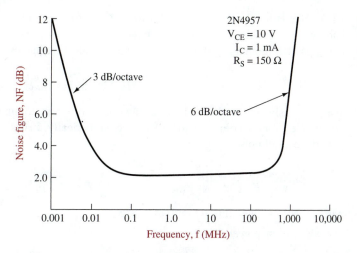

FIGURE 1-12 Noise figure versus frequency for a 2N4957 transistor. (Used with permission from SCILLC dba ON Semiconductor. All rights reserved.)

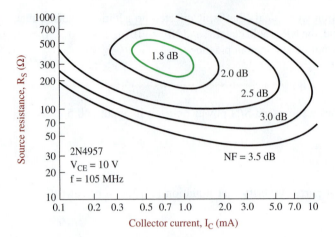

Manufacturers of low-noise devices usually supply a whole host of curves to exhibit their noise characteristics under as many varied conditions as possible. One of the more interesting curves provided for the 2N4957 transistor is shown in Figure 1-13. It provides a visualization of the contours of NF versus source resistance and dc collector current for a 2N4957 transistor at 105 MHz. It indicates that noise operation at 105 MHz will be optimum when a dc (bias) collector current of about 0.7 mA and source resistance of 350 Ω is utilized because the lowest NF of 1.8 dB occurs under these conditions.

The current state of the art for low-noise transistors offers some surprisingly low numbers. The leading edge for room temperature designs at 4 GHz is an NF of about 0.5 dB using gallium arsenide (GaAs) FETs. At 144 MHz, amplifiers with NFs down to 0.3 dB are being employed. The ultimate in low-noise-amplifier (LNA) design utilizes cryogenically cooled circuits (using liquid helium). Noise figures down to about 0.2 dB at microwave frequencies up to about 10 GHz are thereby made possible.

Reactance Noise Effects

In theory, reactive components do not introduce noise into systems of which they are a part because ideal capacitors and inductors contain no resistance. This ideal cannot be attained in practice, but, fortunately, resistances within capacitors and inductors produce negligible noise effects as compared with semiconductors and other resistances.

Reactive circuits do limit frequency response, however, which in turn has a significant effect on noise characteristics. The previous discussion assumed an ideal bandwidth that is rectangular in response. In practice, *RC*-, *LC*-, and *RLC*-generated passbands are not rectangular but slope off gradually, with the bandwidth defined as a function of half-power frequencies. The equivalent bandwidth (Δf_{eq}) to be used in noise calculations with reactive circuits is given by

$$\Delta f_{\text{eq}} = \frac{\pi}{2} \text{BW} \tag{1-13}$$

where BW is the 3-dB bandwidth for *RC*, *LC*, or *RLC* circuits. The fact that the "noise" bandwidth is greater than the "3-dB" bandwidth is not surprising. Significant noise is still being passed through a system beyond the 3-dB cutoff frequency.

Noise Created by Amplifiers in Cascade

We previously specified that the first stage of a system is dominant with regard to noise effect. We are now going to show that effect numerically. **Friiss's formula** is used to provide the overall noise effect of a multistage system.

$$\text{NR} = \text{NR}_1 + \frac{\text{NR}_2 - 1}{P_{G_1}} + \cdots + \frac{\text{NR}_n - 1}{P_{G_1} \times P_{G_2} \times \cdots \times P_{G(n-1)}} \tag{1-14}$$

where   NR = overall noise ratio of *n* stages

P_G = power gain ratio

EXAMPLE 1-9

A three-stage amplifier system has a 3-dB bandwidth of 200 kHz determined by an LC-tuned circuit at its input, and operates at 22°C. The first stage has a power gain of 14 dB and an NF of 3 dB. The second and third stages are identical, with power gains of 20 dB and NF = 8 dB. The output load is 300 Ω. The input noise is generated by a 10-kΩ resistor. Calculate

(a) the noise voltage and power at the input and the output of this system assuming ideal noiseless amplifiers.

(b) the overall noise figure for the system.

(c) the actual output noise voltage and power.

SOLUTION

(a) The effective noise bandwidth is

$$\Delta f_{eq} = \frac{\pi}{2} BW$$

$$= \frac{\pi}{2} \times 200 \text{ kHz} \qquad\qquad \text{(1-13)}$$

$$= 3.14 \times 10^5 \text{ Hz}.$$

Thus, at the input,

$$P_n = kT \, \Delta f$$

$$= 1.38 \times 10^{-23} \text{ J/K} \times (273 + 22) \text{ K} \times 3.14 \times 10^5 \text{ Hz} \qquad \text{(1-8)}$$

$$= 1.28 \times 10^{-15} \text{ W}$$

and

$$e_n = \sqrt{4kT \, \Delta fR}$$

$$= \sqrt{4 \times 1.28 \times 10^{-15} \times 10 \times 10^3} \qquad\qquad \text{(1-9)}$$

$$= 7.15 \, \mu\text{V}.$$

The total power gain is 14 dB + 20 dB + 20 dB = 54 dB.

$$54 \text{ dB} = 10 \log P_G$$

Therefore,

$$P_G = 2.51 \times 10^5.$$

Assuming perfect noiseless amplifiers,

$$P_{n \text{ out}} = P_{n \text{ in}} \times P_G$$

$$= 1.28 \times 10^{-15} \text{ W} \times 2.51 \times 10^5$$

$$= 3.22 \times 10^{-10} \text{ W}.$$

Remembering that the output is driven into a 300-Ω load and $P = V^2/R$, we have

$$3.22 \times 10^{-10} \text{ W} = \frac{(e_{n \text{ out}})^2}{300 \, \Omega}$$

$$e_n = 0.311 \text{ mV}.$$

Notice that the noise has gone from microvolts to millivolts without considering the noise injected by each amplifier stage.

(b) Recall that to use Friiss's formula, ratios and not decibels must be used. Thus,

$$P_{G_1} = 14 \text{ dB} = 25.1$$
$$P_{G_2} = P_{G_3} = 20 \text{ dB} = 100$$
$$\text{NF}_1 = 3 \text{ dB} \qquad \text{NR}_1 = 2$$
$$\text{NF}_2 = \text{NF}_3 = 8 \text{ dB} \qquad \text{NR}_2 = \text{NR}_3 = 6.31$$

$$\text{NR} = \text{NR}_1 + \frac{\text{NR}_2 - 1}{P_{G_1}} + \cdots + \frac{\text{NR}_n - 1}{P_{G_1}P_{G_2}\cdots P_{G(n-1)}} \qquad \text{(1-14)}$$

$$= 2 + \frac{6.31 - 1}{25.1} + \frac{6.31 - 1}{25.1 \times 100}$$

$$= 2 + 0.21 + 0.002 = 2.212$$

Thus, the overall noise ratio (2.212) converts into an overall noise figure of $10 \log_{10} 2.212 = 3.45$ dB:

$$\text{NF} = 3.45 \text{ dB}$$

(c)
$$\text{NR} = \frac{S_i/N_i}{S_o/N_o}$$

$$P_G = \frac{S_o}{S_i} = 2.51 \times 10^5$$

Therefore,

$$\text{NR} = \frac{N_o}{N_i \times 2.51 \times 10^5}$$

$$2.212 = \frac{N_o}{1.28 \times 10^{-15} \text{ W} \times 2.51 \times 10^5}$$

$$N_o = 7.11 \times 10^{-10} \text{ W}$$

To get the output noise voltage, since $P = V^2/R$,

$$7.11 \times 10^{-10} \text{ W} = \frac{e_n^2}{300 \text{ }\Omega}$$

$$e_n = 0.462 \text{ mV}$$

Notice that the actual noise voltage (0.462 mV) is about 50% greater than the noise voltage when we did not consider the noise effects of the amplifier stages (0.311 mV).

Equivalent Noise Temperature

Another way of representing noise is through the concept of equivalent noise temperature. The concept has nothing to do with actual operating temperature. Instead, it is a means of representing the noise produced at the output of a real-world device (such as an amplifier) or system (such as a receiver, transmission line, and antenna) in terms of the noise generated by a resistor placed at the input to a noiseless (that is, theoretically perfect) amplifier with the same gain as the device or system under consideration. Since noise is directly proportional to temperature, the temperature at which a single resistor produces an equivalent amount of noise to that measured at the output of a real-world receiver becomes a proxy for the total amount of noise produced by all sources. The concept of equivalent noise temperature conveniently allows for the handling of noise calculations involved with microwave receivers (1 GHz and above) and their associated antenna systems, especially space communication systems. It also allows for easy calculation of noise power at a receiver using the decibel relationships of Equation (1-2) since the equivalent noise temperature (T_{eq}) of microwave antennas and their coupling networks are then simply additive.

The T_{eq} of a receiver is related to its noise ratio, NR, by

$$T_{eq} = T_0(NR - 1) \qquad \text{(1-15)}$$

where $T_0 = 290$ K, a reference temperature in Kelvin. The use of noise temperature is convenient since microwave antenna and receiver manufacturers usually provide T_{eq} information for their equipment. Additionally, for low noise levels, noise temperature shows greater variation of noise changes than does NF, making the difference easier to comprehend. For example, an NF of 1 dB corresponds to a T_{eq} of 75 K, while 1.6 dB corresponds to 129 K. Verify these comparisons using Equation (1-15), remembering first to convert NF to NR. Keep in mind that noise temperature is not an actual temperature but is employed because of its convenience.

EXAMPLE 1-10

A satellite receiving system includes a dish antenna ($T_{eq} = 35$ K) connected via a coupling network ($T_{eq} = 40$ K) to a microwave receiver ($T_{eq} = 52$ K referred to its input). What is the noise power to the receiver's input over a 1-MHz frequency range? Determine the receiver's NF.

SOLUTION

$$P_n = kT \, \Delta f \qquad \text{(1-8)}$$
$$= 1.38 \times 10^{-23} \text{J/K} \times (35 + 40 + 52) \text{ K} \times 1 \text{ MHz}$$
$$= 1.75 \times 10^{-15} \text{W}$$

$$T_{eq} = T_0(NR - 1) \qquad \text{(1-15)}$$
$$52 \text{ K} = 290 \text{ K}(NR - 1)$$

$$NR = \frac{52}{290} + 1$$
$$= 1.18$$

Therefore, $NF = 10 \log_{10}(1.18) = 0.716$ dB.

Equivalent Noise Resistance

Manufacturers sometimes represent the noise generated by a device with a fictitious resistance termed the equivalent noise resistance (R_{eq}). It is the resistance that generates the same amount of noise predicted by $\sqrt{4kT \, \Delta fR}$ as the device does. The device (or complete amplifier) is then assumed to be noiseless in making subsequent noise calculations. The latest trends in noise analysis have shifted away from the use of equivalent noise resistance in favor of using the noise figure or noise temperatures.

1-6 TROUBLESHOOTING

Because of the increasing complexity of electronic communications equipment, you must have a good understanding of communication circuits and concepts. To be an effective troubleshooter, you must also be able to isolate faulty components quickly and repair the defective circuit. Recognizing the way a circuit may malfunction is a key factor in speedy repair procedures.

After completing this section you should be able to

- Explain general troubleshooting techniques,
- Recognize major types of circuit failures, and
- List the four troubleshooting techniques.

General Troubleshooting Techniques

Troubleshooting requires asking questions such as: *What could cause this to happen? Why is this voltage so low/high? Or, if this resistance were open/shorted, what effect would it have on the operation of the circuit I'm working on?* Each question calls for measurements to be made and tests to be performed. The defective component(s) is isolated when measurements give results far different than they would be in a properly operating unit, or when a test fails. The ability to ask the right questions makes a good troubleshooter. Obviously, the more you know about the circuit or system being worked on, the more quickly you will be able to correct the problem.

Always start troubleshooting by doing the easy things first:

- Be sure the unit is plugged in and turned on.
- Check fuses.
- Check if all connections are made.
- Ask yourself, *Am I forgetting something?*

Basic troubleshooting test equipment includes:

- a digital multimeter (DMM) capable of reading at the frequencies you intend to work at;
- a broadband oscilloscope, preferably dual trace;
- signal generators, both audio and RF (the RF generator should have internal modulation capabilities); and
- a collection of probes and clip leads.

Advanced test equipment would include a spectrum analyzer to observe frequency spectra and a logic analyzer for digital work. Time spent learning your test equipment, its capabilities and limitations, and how to use it will pay off with faster troubleshooting.

Always be aware of any possible effects that the test equipment you connect to a circuit may have on the operation of that circuit. Do not let the measuring equipment change what you are measuring. For example, a scope's test lead may have a capacitance of several hundred picofarads. Should that lead be connected across the output of an oscillator, the oscillator's frequency could be changed to the point that any measurements are worthless.

If the equipment you are troubleshooting employs dangerous voltages, do not work alone. Turn off all power switches before entering equipment. Keep all manuals that came with the equipment. Such manuals usually include troubleshooting procedures. Check them before trying any other approaches.

Maintain clear, up-to-date records of all changes made to equipment.

Replace a suspicious unit with a known good one—this is one of the best, most commonly used troubleshooting techniques.

Test points are often built into electronic equipment. They provide convenient connections to the circuitry for adjustment and/or testing. There are various types, from jacks or sockets to short, stubby wires sticking up from PC boards. Equipment manuals will diagram the location of each test point and describe (and sometimes illustrate) the condition and/or signal that should be found there. The better manuals indicate the proper test equipment to use.

Plot a game plan or strategy with which you will troubleshoot a problem (just as you might with a car problem).

Use all your senses when troubleshooting:

Look—discolored or charred components might indicate overheating.

Smell—some components, especially transformers, emit characteristic odors when overheated.

Feel—for hot components. Wiggle components to find broken connections.

Listen—for "frying" noises that indicate a component is about to fail.

Reasons Electronic Circuits Fail

Electronic circuits fail in many ways. Let's look at some major types of failures that you will encounter.

1. **Complete Failures** Complete failures cause the piece of equipment to go totally dead. Equipment with some circuits still operating has not completely failed. Normally this type of failure is the result of a major circuit path becoming open. Blown (open) fuses, open power resistors, defective power supply rails, and bad regulator transistors in the power supply can cause complete failures. Complete failures are often the easiest problems to repair.

2. **Intermittent Faults** Intermittent faults are characterized by sporadic circuit operation. The circuit works for a while and then quits working. It works one moment and does not work the next. Keeping the circuit in a failed condition can be quite difficult. Loose wires and components, poor soldering, and effects of temperature on sensitive components can all contribute to intermittent operation in a piece of communications equipment. Intermittent faults are usually the most difficult to repair since troubleshooting can be done only when the equipment is malfunctioning.

3. **Poor System Performance** Equipment that is functioning below specified operational standards is said to have poor system performance characteristics. For example, a transmitter is showing poor performance if the specifications call for 4 W of output power but it is putting out only 2 W. Degradation of equipment performance takes place over a period of time due to deteriorating components (components change in value), poor alignment, and weakening power components. Regular performance checks are necessary for critical communications systems. Commercial radio transmitters require performance checks to be done on a regular basis.

4. **Induced Failures** Induced failures often come from equipment abuse. Unauthorized modifications may have been performed on the equipment. An inexperienced technician without supervision may have attempted repairs and damaged the equipment. Exercising proper equipment care can eliminate induced failures. Repairs should be done or supervised by experienced technicians.

Troubleshooting Plan

Experienced technicians have developed a method for troubleshooting. They follow certain logical steps when looking for a defect in a piece of equipment. The following four troubleshooting techniques are popular and widely used to find defects in communications equipment.

1. **Symptoms as Clues to Faulty Stages** This technique relates a particular fault to a circuit function in the electronic equipment. For example, if a white horizontal line were displayed on the screen of a TV brought in for repair, the service technician would associate this symptom with the vertical output section. Troubleshooting would begin in that section of the TV. As you gain experience in troubleshooting you will start associating symptoms with specific circuit functions.

2. **Signal Tracing and Signal Injection** Signal injection is supplying a test signal at the input of a circuit and looking for the test signal at the circuit's output or listening for an audible tone at the speaker (Figure 1-14). This test signal is usually composed of an RF signal modulated with an audible frequency. If the signal is good at the circuit's output, then move to the next stage down the line and repeat the test. Signal tracing, as illustrated in Figure 1-15, is actually checking for the normal output signal from a stage. An oscilloscope is used to check for these signals. However, other test equipment is available that can be used to detect the presence of output signals. Signal tracing is monitoring the output of a stage for the presence of the expected signal.

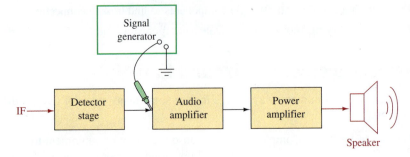

FIGURE 1-14 Signal injection.

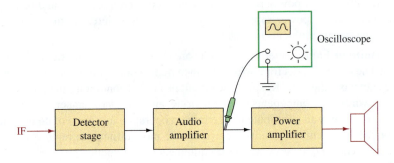

FIGURE 1-15 Signal tracing.

If the signal is there, then the next stage in line is checked. The malfunctioning stage precedes the point where the output signal is missing.

3. **Voltage and Resistance Measurements** Voltage and resistance measurements are made with respect to chassis ground. Using the DMM (digital multimeter), measurements at specific points in the circuit are compared to those found in the equipment's service manual. Service manuals furnish equipment voltage and resistance charts, or print the values right on the schematic diagram. Voltage and resistance checks are done to isolate defective components once the trouble has been pinpointed to a specific stage of the equipment. Remember that resistance measurements are done on circuits with the power turned off.

4. **Substitution** Another method often used to troubleshoot electronic circuits is to swap a known good component for the suspected bad component. A warning is in order here: The good component could get damaged in the substitution process. Do not get into the habit of indiscriminately substituting parts. The substitution method works best when you have narrowed the failure down to a specific component.

SUMMARY

A recurring theme of this text is that a relatively small number of fundamental concepts serves to unify and inform the study of electronic communications systems. Certainly foremost among the unifying themes are the concepts of modulation, bandwidth, and noise. This first chapter laid out these overriding ideas in broad form, and subsequent chapters will examine how the ideas are implemented in practical systems as well as the tradeoffs inherent therein. Among the most salient points are the following:

- Modulation, in which low-frequency information or intelligence signals modify one or more aspects of a high-frequency sine-wave carrier, is central to any communications system. The use of modulated carriers at different frequencies permits multiplexing, which, when carriers are modulated, allows for bands of frequencies to be shared among multiple users and, for wireless systems, for the use of reasonable-length antennas.

- The carrier is most often a sine wave, and it can be modulated by modifying one or more of three characteristics: amplitude, frequency, or phase. Thus, even the most advanced communications systems involve some variation of amplitude, frequency, or phase modulation.

- Most fundamentally, all communications systems employ three components: a transmitter, a receiver, and a communications channel. The channel may either consist of a physical medium (such as copper wire, coaxial cable, waveguide, or optical fiber), or it may consist of a wireless link between transmitting and receiving antennas, in which case the channel is either Earth's atmosphere or "free space," in other words, a vacuum.

- Propagating through free space is electromagnetic energy, consisting of alternating electric and magnetic fields that exist at right angles to each other as well as at right angles (transverse) to the direction of travel. Electromagnetic energy exists on a spectrum from 0 Hz (dc) up to and beyond the frequencies associated with visible light.

- Very large ranges of powers and voltages are routinely encountered in communications systems, and consideration of such a wide expanse requires a notation system capable of compressing the extremes into a manageable range. Such is the virtue of decibel notation, in which power, voltage, or current ratios are expressed in terms of common logarithms.

- Bandwidth, defined as the range of frequencies occupied by signals within the communication channel, is directly proportional to the information conveyed and the time of transmission. The more information that must be conveyed within a given time interval, the wider the bandwidth. Because modulated signals convey information, they occupy a wider bandwidth than unmodulated signals.

- Bandwidth equals money. Because the radio-frequency spectrum is a scare, natural resource, its use is regulated, and spectrum users must be concerned with efficient use of spectrum.

- Signals can be analyzed in both the time and frequency domains. Fourier analysis allows one to move from the time to the frequency domain, and vice versa. An oscilloscope is a time-domain instrument, and a spectrum analyzer is a frequency-domain device.

- The ability of a communications system to perform in the presence of noise is the ultimate arbiter of success. Because noise amplitudes are often of the same order of magnitude as those of received signals, identification and management of noise sources is extremely important to communication system design.

QUESTIONS AND PROBLEMS

SECTION 1-1

1. Define *modulation.*

*2. What is *carrier frequency*?

3. Describe the two reasons that modulation is used for communications transmissions.

4. List the three parameters of a high-frequency carrier that may be varied by a low-frequency intelligence signal.

5. What are the frequency ranges included in the following frequency subdivisions: MF (medium frequency), HF (high frequency), VHF (very high frequency), UHF (ultra high frequency), and SHF (super high frequency)?

SECTION 1-2

6. A signal level of $0.4\mu V$ is measured on the input to a satellite receiver. Express this voltage in terms of dBμV. Assume a 50-Ω system. (-7.95 dB$_\mu$V)

7. A microwave transmitter typically requires a +8-dBm audio level to drive the input fully. If a +10-dBm level is measured, what is the actual voltage level measured? Assume a 600-Ω system. (2.45 V)

8. If an impedance matched amplifier has a power gain (P_{out}/P_{in}) of 15, what is the value for the voltage gain (V_{out}/V_{in})? (3.87)

9. Convert the following powers to their dBm equivalents:
   (a) $p = 1$ W (30 dBm)
   (b) $p = 0.001$ W (0 dBm)
   (c) $p = 0.0001$ W (-10 dBm)
   (d) $p = 25\mu$W (-16 dBm)

---

*An asterisk preceding a number indicates a question that has been provided by the FCC as a study aid for licensing examinations.

10. The output power for an audio amplifier is specified to be 38 dBm. Convert this value to (a) watts and (b) dBW. (6.3 W, 8 dBW)

11. A 600-Ω microphone outputs a −70-dBm level. Calculate the equivalent output voltage for the −70-dBm level. (0.245 mV)

12. Convert 50-μV to a dBμV equivalent. (34 dBμV)

13. A 2.15-V rms signal is measured across a 600-Ω load. Convert this measured value to its dBm equivalent. (8.86 dBm (600))

14. A 2.15-V rms signal is measured across a 50-Ω load. Convert this measured value to its dBm(50) equivalent. (19.66 dBm(50))

SECTION 1-3

15. Define *information theory*.

16. What is Hartley's law? Explain its significance.

17. What is a *harmonic*?

18. What is the seventh harmonic of 360 kHz? (2520 kHz)

19. Why does transmission of a 2-kHz square wave require greater bandwidth than a 2-kHz sine wave?

20. Draw time- *and* frequency-domain sketches for a 2-kHz square wave. The time-domain sketch is a standard oscilloscope display, while the frequency domain is provided by a spectrum analyzer.

21. Explain the function of Fourier analysis.

22. A 2-kHz square wave is passed through a filter with a 0- to 10-kHz frequency response. Sketch the resulting signal, and explain why the distortion occurs.

23. A triangle wave of the type shown in Table 1-3(e) has a peak-to-peak amplitude of 2 V and $f = 1$ kHz. Write the expression $v(t)$, including the first five harmonics. Graphically add the harmonics to show the effects of passing the wave through a low-pass filter with cutoff frequency equal to 6 kHz.

24. The FFT shown in Figure 1-16 was obtained from a DSO.
    (a) What is the sample frequency?
    (b) What frequency is shown by the FFT?

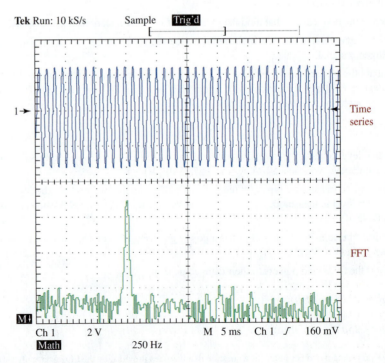

FIGURE 1-16   **FFT for Problem 1-24.**

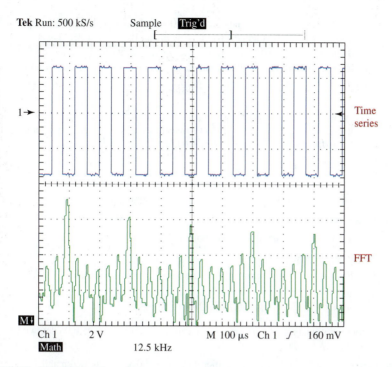

FIGURE 1-17   **FFT for Problem 1-25.**

25. Figure 1-17 was obtained from a DSO.
    (a) What are the frequencies of the third and fifth harmonics?
    (b) This FFT was created by inputting a 12.5-kHz square wave into a DSO. Explain where 12.5 kHz is located within the FFT spectrum.

SECTION 1-4

26. Define *electrical noise*, and explain why it is so troublesome to a communications receiver.

27. Explain the difference between external and internal noise.

28. List and briefly explain the various types of external noise.

29. Provide two other names for Johnson noise and calculate the noise voltage output of a 1-MΩ resistor at 27°C over a 1-MHz frequency range. (128.7 μV)

30. The noise produced by a resistor is to be amplified by a noiseless amplifier having a voltage gain of 75 and a bandwidth of 100 kHz. A sensitive meter at the output reads 240 μV rms. Assuming operation at 37°C, calculate the resistor's resistance. If the bandwidth were cut to 25 kHz, determine the expected output meter reading. (5.985 kΩ, 120μV)

31. Explain the term *low-noise resistor*.

32. Determine the noise current for the resistor in Problem 29. What happens to this noise current when the temperature increases? (129 pA)

33. The noise spectral density is given by $e_n^2/\Delta f = 4kTR$. Determine the bandwidth Δf of a system in which the noise voltage generated by a 20-kΩ resistor is 20μV rms at room temperature. (1.25 MHz)

SECTION 1-5

34. Calculate the *S/N* ratio for a receiver output of 4 V signal and 0.48 V noise both as a ratio and in decibel form. (69.44, 18.42 dB)

35. The receiver in Problem 34 has an *S/N* ratio of 110 at its input. Calculate the receiver's noise figure (NF) and noise ratio (NR). (1.998 dB, 1.584)

36. An amplifier with NF = 6 dB has S_i/N_i of 25 dB. Calculate the S_o/N_o in dB and as a ratio. (19 dB, 79.4)

37. A three-stage amplifier has an input stage with noise ratio (NR) = 5 and power gain (P_G) = 50. Stages 2 and 3 have NR = 10 and P_G = 1000. Calculate the NF for the overall system. (7.143 dB)

38. A two-stage amplifier has a 3-dB bandwidth of 150 kHz determined by an *LC* circuit at its input and operates at 27°C. The first stage has $P_G = 8$ dB and NF = 2.4 dB. The second stage has $P_G = 40$ dB and NF = 6.5 dB. The output is driving a load of 300 Ω. In testing this system, the noise of a 100-kΩ resistor is applied to its input. Calculate the input and output noise voltage and power and the system noise figure. (19.8μV, 0.206 mV, 9.75×10^{-16} W, 1.4×10^{-10} W, 3.6 dB)

39. A microwave antenna ($T_{eq} = 25$ K) is coupled through a network ($T_{eq} = 30$ K) to a microwave receiver with $T_{eq} = 60$ K referred to its output. Calculate the noise power at its input for a 2-MHz bandwidth. Determine the receiver's NF. (3.17×10^{-15} W, 0.817 dB)

SECTION 1-6

40. List and briefly describe the four basic troubleshooting techniques.

41. Describe the disadvantages of using substitution at the early stages of the troubleshooting plan.

42. Explain why resistance measurements are done with power off.

43. Describe the major types of circuit failures.

44. Describe when it is more appropriate to use the signal injection method.

QUESTIONS FOR CRITICAL THINKING

45. You cannot guarantee perfect performance in a communications system. What two basic limitations explain this?

46. You are working on a single-stage amplifier that has a 200-kHz bandwidth and a voltage gain of 100 at room temperature. The external noise is negligible. A 1-mV signal is applied to the amplifier's input. If the amplifier has a 5-dB NF and the input noise is generated by a 2-kΩ resistor, what output noise voltage would you predict? (458μV)

47. How does equivalent noise resistance relate to equivalent noise temperature? Explain similarities and/or differences.

CHAPTER 2

AMPLITUDE MODULATION

CHAPTER OUTLINE

KEY TERMS

modulation
mixing
linear device
nonlinear device
overmodulation
modulation index
modulation factor
percentage of
modulation
sideband splatter
intermodulation
distortion
intermod

broadband
baseband
aliasing
foldover distortion
upper sideband
lower sideband
peak envelope power
(PEP)
pilot carrier
twin-sideband
suppressed carrier
independent sideband
(ISB) transmission

2-1 OVERVIEW OF AMPLITUDE MODULATION

Chapter 1 established the importance of modulation in communications systems. **Modulation** involves combining an information signal, often one whose instantaneous peak voltage is constantly changing and whose range of occupied frequencies is relatively low, with a carrier, which is a relatively high-frequency sine wave of constant amplitude. In the case of amplitude modulation (AM), the carrier is also of constant frequency. Under certain conditions combining any two (or more) signals of different frequencies will cause them to interact in a way that creates additional frequencies, each bearing a mathematical relationship to those being combined. This interaction is called **mixing**, and modulation of the type to be studied in this chapter is essentially a special case of mixing. Some of the frequencies created as the result of mixing, together with the carrier, form the modulated signal. The modulated signal preserves the low-frequency characteristics of the information as it occupies the higher frequency range of the carrier. Modulated signals at high frequencies can be propagated wirelessly with antennas of reasonable size, and the available frequency spectrum can be shared by assigning a specific carrier frequency to each user. Voltage and frequency characteristics of the modulated waveform allow for subsequent recovery of the original information at the receiver, a process known as detection or demodulation and that will be studied extensively in Chapter 6.

Also identified in Chapter 1 were the three characteristics of a sine wave carrier—amplitude, frequency, or phase—that can be modified to convey information. Recall also that the information signal is generally termed the intelligence. Other terms for intelligence are modulating signal, audio signal, or modulating wave. AM, which conveys information through changes in the instantaneous peak voltage (amplitude) of the modulated carrier, is the oldest and most straightforward of the approaches and is the focus of this chapter.

AM has the virtue of simplicity, both in generation at the transmitter and detection at the receiver, and simplicity is one reason that AM is still widely used a full century after practical systems were first demonstrated. AM takes several forms, each defined principally in terms of the portion of the modulated signal that ends up being transmitted. The most basic form, called double-sideband, full-carrier (DSBFC) AM, transmits the entire modulated signal at full power and is used in broadcasting and aviation (aircraft-to-tower) radio. A modified form, single-sideband (SSB), is extensively used by the military services and also finds application in marine and citizens band (CB) radio as well as use by hobbyists known as amateur (ham) radio operators. Other forms of AM, in which the carrier is partially or completely removed but the sidebands are present, are used in such applications as stereo broadcasting, analog video, and analog color television. Perhaps most important to recognize for the study of modern communications systems is that many techniques used to encode and recover digital data derive directly from the circuits and methods developed for the generation and detection of analog AM and SSB signals. A solid understanding of fundamentals in the analog realm provides the platform for a conceptual understanding of how equivalent actions are carried out in the digital domain. With such a foundation, the conceptual leap from analog to digital is less daunting than it may first appear.

2-2 DOUBLE-SIDEBAND AM

Combining two sine waves in a linear fashion results in their simple algebraic addition, as shown in Figure 2-1. A circuit that would perform this function is shown in Figure 2-1(a), where two signals of two different frequencies are combined in a linear device such as a resistor. A **linear device** is one in which current is a linear function of voltage. Ohm's law confirms this relationship for a resistor: Current flow through the device increases in direct proportion to voltage. The result, shown in Figure 2-1(d), is *not* suitable for transmission as an AM waveform, however. If it were transmitted, the receiving antenna would detect

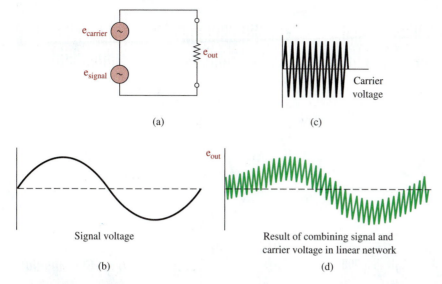

(a)

(c)

Carrier voltage

Signal voltage

(b)

e_{out}

Result of combining signal and carrier voltage in linear network

(d)

FIGURE 2-1   **Linear addition of two sine waves.**

just the carrier signal because the low-frequency intelligence component cannot be propagated efficiently as a radio wave.

Instead, for modulation to occur, a specific kind of interaction must take place between the input frequencies. Such an interaction occurs when the carrier and intelligence are combined within a **nonlinear device**. A nonlinear device or circuit is one where changes in current are *not* directly proportional to changes in applied voltage. An ordinary silicon diode is one example: It has the nonlinear response shown in Figure 2-2. Other components such as field-effect transistors also exhibit nonlinear characteristics. The net effect of the interaction so produced is for the carrier and intelligence signals to be *multiplied* rather than added. As a result, interference of a desirable sort takes place between the high-frequency carrier and the low-frequency intelligence. The modulated waveform carries useful information because of the instantaneous addition and subtraction of the constant-amplitude carrier with the varying-amplitude modulating signal.

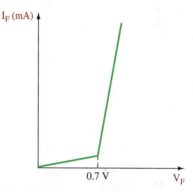

I_F (mA)

0.7 V

V_F

FIGURE 2-2   **Nonlinear characteristic curve of voltage versus current for a silicon diode.**

AM Waveforms

Figure 2-3 shows the modulated AM waveform in the time domain as the amplitude and frequency of a sine-wave intelligence signal are varied. The AM waveform shown in Figure 2-3(a) is a signal at the carrier frequency whose amplitude is changing at the rate of the intelligence. As the intelligence amplitude reaches a maximum positive value, the AM waveform has a maximum amplitude. The AM waveform is minimum when the intelligence amplitude is at a maximum negative value. In Figure 2-3(b), the intelligence frequency remains the same, but its amplitude has been increased. The resulting AM waveform reacts by reaching a larger maximum value and smaller minimum value. In Figure 2-3(c), the intelligence amplitude is reduced and its frequency has gone up. The resulting AM waveform, therefore, has reduced maximums and minimums, and the rate at which it swings between these extremes has increased to the intelligence frequency.

The shape created when the upper and lower contours of the individual cycles of the modulated carrier are joined together represents the frequency and amplitude of the intelligence, but with a 180° phase shift between the upper and lower contours. However, the modulated AM waveform does *not* include a low-frequency component—that is, it does not include any component at the intelligence frequency. Put another way, the low-frequency intelligence signals have been shifted upward to the frequency range of the carrier.

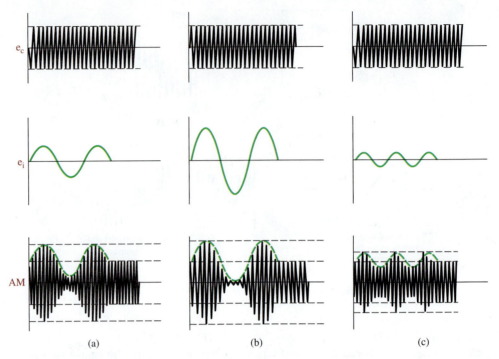

e_c

e_i

AM

(a)   (b)   (c)

FIGURE 2-3   **AM waveform under varying intelligence signal (e_i) conditions.**

Modulation Index

Figure 2-3 shows that the instantaneous amplitude of the modulated AM waveform varies as a result of changes in intelligence amplitude. Is there a limit to the maximum amplitude of the intelligence? There is, and close examination of Figure 2-3 shows why. If the modulating signal amplitude continues to increase, a situation known as **overmodulation** may result. Figure 2-4 shows overmodulation in the time domain. Overmodulation must never be allowed to occur and, in fact, is a violation of law in licensed systems. It is important, therefore, to have a means of quantifying the amount of modulation and to ensure that the situation depicted in Figure 2-4 does not occur. A figure of merit known as the **modulation index** or **modulation factor**, symbolized by m and defined as the ratio of intelligence amplitude to carrier amplitude, quantifies the extent to which the intelligence varies the carrier voltage. Expressed as a ratio

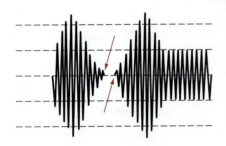

FIGURE 2-4   **Overmodulation.**

$$m = \frac{E_i}{E_c},\qquad(2\text{-}1)$$

where E_i is the intelligence voltage and E_c is the carrier voltage.

In AM systems the modulation index is most often expressed in percentage terms:

$$\%m = \frac{E_i}{E_c} \times 100\%,\qquad(2\text{-}2)$$

where %m is defined as the **percentage of modulation**.

Figure 2-5 illustrates the two most common methods for determining the percentage of modulation when a carrier is modulated with a sine wave. Notice that when the intelligence signal is zero, the carrier is unmodulated and has a peak amplitude labeled as E_c. When the intelligence reaches its first peak value (point w), the AM signal reaches a peak value labeled E_i (the increase from E_c.). Percentage modulation is then given by Equation (2-2).

The same result can be obtained by utilizing the maximum peak-to-peak value of the AM waveform (point w), which is shown as B, and the minimum peak-to-peak value (point x), which is A in the following equation:

$$\%m = \frac{B - A}{B + A} \times 100\%\qquad(2\text{-}3)$$

This method is usually more convenient in graphical (oscilloscope) solutions.

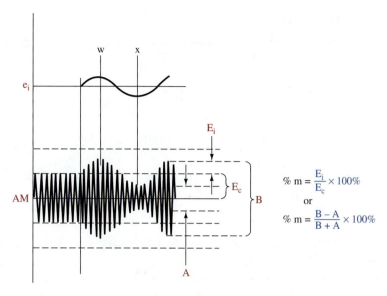

$$\% m = \frac{E_i}{E_c} \times 100\%$$

or

$$\% m = \frac{B - A}{B + A} \times 100\%$$

FIGURE 2-5 Percentage modulation determination.

Overmodulation

If the AM waveform's minimum value *A* falls to zero as a result of an increase in the intelligence amplitude, the percentage modulation becomes

$$\%m = \frac{B - A}{B + A} \times 100\% = \frac{B - O}{B + O} \times 100\% = 100\%$$

This is the maximum permissible degree of modulation. In this situation the carrier is being varied between zero and double its unmodulated value. Any further increase in the intelligence amplitude will cause overmodulation, in which the modulated carrier will go to more than double its unmodulated value but will fall to zero for an interval of time, as shown in Figure 2-4. This "gap" produces distortion termed **sideband splatter**, which results in the transmission of frequencies outside a station's normal allocated range. Fourier analysis (Chapter 1) of the sharp edges that are created as the modulated signal goes to zero (shown by the arrows in Figure 2-4) would reveal the presence of high-frequency components added to the original intelligence. Recall that square waves consist of the fundamental frequency and odd harmonics of that frequency. In general, any signal with sharp edges will have harmonics, which in this case are unwanted high-frequency components that will increase the bandwidth of the modulated signal beyond its authorized maximum. This condition is unacceptable because it causes severe interference to other stations and causes a loud splattering sound to be heard at the receiver. For this reason, AM transmitters have limiting circuits to prevent overmodulation.

EXAMPLE 2-1

Determine the %m for the following conditions for an unmodulated carrier of 80 V peak-to-peak (p-p).

	MAXIMUM P-P CARRIER (V)	MINIMUM P-P CARRIER (V)
(a)	100	60
(b)	125	35
(c)	160	0
(d)	180	0
(e)	135	35

SOLUTION

(a)
$$\%m = \frac{B - A}{B + A} \times 100\%$$

$$= \frac{100 - 60}{100 + 60} \times 100\% = 25\% \qquad (2\text{-}3)$$

(b)
$$\%m = \frac{125 - 35}{125 + 35} \times 100\% = 56.25\%$$

(c)
$$\%m = \frac{160 - 0}{160 + 0} \times 100\% = 100\%$$

(d) This is a case of overmodulation since the modulated carrier reaches a value more than twice its unmodulated value.

(e) The increase is greater than the decrease in the carrier's amplitude. This is a distorted AM wave.

Amplitude Modulation and Mixing in the Frequency Domain

When two signals at different frequencies are combined in a nonlinear device a phenomenon known as **mixing** takes place, where additional frequencies are created, among them new signals residing at the algebraic sum and difference of the applied signal frequencies. Mixing is effectively a translation process whereby signals over a range of frequencies can be shifted up or down to a different rage as needed. This need arises often in communications systems, so mixer circuits—sometimes called by different names but performing the same general functions—are encountered in both transmitters and receivers. Modulation is mixing of a desired sort, but keep in mind that mixing is not always desirable. It can occur whenever multiple signals and device or circuit nonlinearities are present. Undesired mixing is sometimes known as **intermodulation distortion** or **intermod**, and the additional frequencies created are called *intermod products*. Intermod is a serious concern of system designers and will be studied in more detail in subsequent chapters.

A transmitter *modulator* circuit combines carrier and intelligence signals that are widely separated in frequency. The signal at the modulator output can be described as having been distorted in a desirable way because, in the case of AM, the additional sum and difference frequencies created as the result of mixing will be shown shortly to contain the intelligence information. In accordance with Hartley's law, the modulated signal necessarily has a wider bandwidth than that of the carrier or intelligence alone because information is contained within the modulation signal. Modulation, therefore, produces a **broadband** signal, in contrast with the **baseband**, which in this case is the band of frequencies occupied by the intelligence before modulation. Recall from Chapter 1 that all communications systems can be analyzed, in part, in terms of their bandwidth requirements, and to do that in the case of AM, we must examine the modulated signal in the frequency domain.

Combining any two sine waves through a nonlinear device produces the following frequency components:

1. A dc level
2. Components at each of the two original frequencies
3. Components at the sum and difference frequencies of the two original frequencies
4. Harmonics of the two original frequencies

Figure 2-6 shows this process pictorially with two sine waves labeled f_c and f_i, to represent the carrier and intelligence. If all but the components surrounding the carrier are removed (perhaps with a bandpass filter), the remaining frequencies make up the modulated AM waveform. They are referred to as:

1. The *lower-side frequency*, which is the difference between the carrier and the intelligence ($f_c - f_i$)

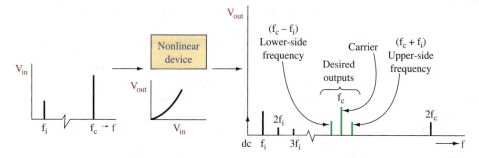

FIGURE 2-6 Nonlinear mixing.

2. The *carrier frequency* (f_c)

3. The *upper-side frequency*, which is the sum of the carrier and intelligence ($f_c + f_i$)

The first entry in the above list helps explain why the carrier and intelligence must be widely separated in frequency. If f_c and f_i are too close together, then the lower-side (difference) frequency created as the result of modulation would fall within the range of baseband frequencies occupied by the intelligence. This undesired form of mixing is called **aliasing** or **foldover distortion** and must be avoided because undesired frequencies in close proximity to desired ones, such as a difference frequency appearing within the baseband, are difficult if not impossible to identify and remove after they have been created. Aliasing is a serious concern generally, and when analog signals are converted to digital form, the conversion process itself can be considered to be a form of mixing. Steps must be taken to prevent aliasing, and these will be studied in subsequent chapters.

Fourier analysis demonstrates that, in the frequency domain, the modulated signal is composed of the spectral components just identified. However, a similar result can be obtained more simply by defining the instantaneous value of both the sine wave carrier, e_c, and intelligence signal, e_i, in forms derived from the general expression for a sine wave first introduced in Chapter 1. Expressed as an equation,

$$e_c = E_c \sin \omega_c t, \tag{2-4}$$

where e_c is the instantaneous value of the carrier; E_c is the maximum peak value of the carrier when unmodulated, $\omega = 2\pi f$ (f is the carrier frequency), and t is a particular point in time. Likewise, the intelligence (modulating) signal, if it is a pure sine wave, can be expressed as

$$e_i = E_i \sin \omega_i t, \tag{2-5}$$

where e_i is the instantaneous value and E_i the peak amplitude of the intelligence. The amplitude of a modulated AM waveform can be written as the carrier peak amplitude, E_c, plus the intelligence signal, e_i. Thus, the amplitude E is

$$E = E_c + e_i$$

but $e_i = E_i \sin \omega_i t$, so that

$$E = E_c + E_i \sin \omega_i t$$

Rearranging Equation (2-1) shows that $E_i = mE_c$, so that

$$E = E_c + mE_c \sin \omega_i t$$
$$= E_c(1 + m \sin \omega_i t)$$

The instantaneous value of the AM wave is the amplitude term E just developed times $\sin \omega_c t$. Thus,

$$e = E \sin \omega_c t$$
$$= E_c(1 + m \sin \omega_i t) \sin \omega_c t \tag{2-6}$$

Notice that the AM wave (e) is the result of the *product* of two sine waves. This result confirms that circuits intended to produce amplitude modulation must mathematically multiply the signals applied to them.

This product can next be expanded by invoking the trigonometric identity,

$$(\sin x)(\sin y) = 0.5 \cos(x - y) - 0.5 \cos(x + y) \qquad \text{(2-7)}$$

where x is the carrier frequency, ω_c, and y is the intelligence frequency, ω_i. The product of the carrier and intelligence sine waves will produce the sum and difference of the two frequencies. If we rewrite Equation (2-7) as $\sin x \sin y = \frac{1}{2}\left[\cos(x - y) - \cos(x + y)\right]$, then the product of the carrier and sine waves is

$$e = \overbrace{E_c \sin \omega_c t}^{(1)} + \overbrace{\frac{mE_c}{2}\cos(\omega_c - \omega_i)t}^{(2)} - \overbrace{\frac{mE_c}{2}\cos(\omega_c + \omega_i)t}^{(3)} \qquad \text{(2-8)}$$

The preceding equation proves that the AM wave contains the three terms previously listed: the carrier ①, the upper sideband at $f_c + f_i$ ③, and the lower sideband at $f_c - f_i$ ②. If, for example, a 1-MHz carrier were modulated by a 5-kHz intelligence signal, the AM waveform would include the following components:

$$1 \text{ MHz} + 5 \text{ kHz} = 1{,}005{,}000 \text{ Hz} \quad \text{(upper-side frequency)}$$
$$1 \text{ MHz} = 1{,}000{,}000 \text{ Hz} \quad \text{(carrier frequency)}$$
$$1 \text{ MHz} - 5 \text{ kHz} = 995{,}000 \text{ Hz} \quad \text{(lower-side frequency)}$$

A sketch in both the time and frequency domains of an amplitude-modulated signal is shown in Figure 2-7. Note in particular that the frequency-domain sketch shows the side frequencies separated from the carrier by an amount equal to the modulating signal frequency (in this example, 5 kHz). The bandwidth of this modulated signal is equal to the upper-side frequency minus the lower-side frequency, or 10 kHz in this example, demonstrating that the bandwidth required for AM transmission is twice the highest intelligence frequency. Equation (2-8) also proves that the carrier amplitude does not vary but that that the instantaneous amplitude of the side frequencies does vary by an amount determined by the modulation index. Specifically, the instantaneous amplitudes of both the upper- and lower-side frequencies is $mE_c/2$. Because information is represented by

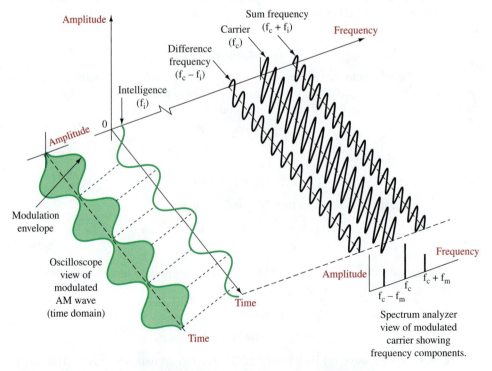

FIGURE 2-7   Amplitude modulation in the time and frequency domains.

changes in amplitude, the above analysis shows conclusively that the information is carried within the side frequencies, not in the carrier. Finally, Equation (2-8) shows that two side frequencies are always created as the result of mixing, one above and one below the carrier, and that the side frequencies are redundant, that is, the same information is carried within each.

Modulation of a carrier with a pure sine-wave intelligence signal has thus far been shown. However, in most systems the intelligence is a rather complex waveform that contains many frequency components. For example, the human voice contains components from roughly 200 Hz to 3 kHz and has a very erratic shape. If voice frequencies from 200 Hz to 3 kHz were used to modulate the carrier, two whole *bands* of side frequencies would be generated, one above and one below the carrier. The band of frequencies above the carrier is termed the **upper sideband**, while that below the carrier is termed the **lower sideband**. This situation is illustrated in Figure 2-8 for a 1-MHz carrier modulated by a band of frequencies ranging from 200 Hz up to 3 kHz.

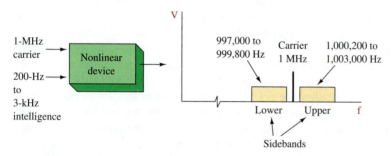

FIGURE 2-8   **Modulation by a band of intelligence frequencies.**

The upper sideband is from 1,000,200 to 1,003,000 Hz, and the lower sideband ranges from 997,000 to 999,800 Hz. The bandwidth of the modulated signal is still twice that of the highest intelligence signal, or 6 kHz in this example.

E X A M P L E   2 - 2

A 1.4-MHz carrier is modulated by a music signal that has frequency components from 20 Hz to 10 kHz. Determine the range of frequencies generated for the upper and lower sidebands.

SOLUTION

The upper sideband is equal to the sum of carrier and intelligence frequencies. Therefore, the upper sideband (usb) will include the frequencies from

$$1,400,000 \text{ Hz} + 20 \text{ Hz} = 1,400,020 \text{ Hz}$$

to

$$1,400,000 \text{ Hz} + 10,000 \text{ Hz} = 1,410,000 \text{ Hz}$$

The lower sideband (lsb) will include the frequencies from

$$1,400,000 \text{ Hz} - 10,000 \text{ Hz} = 1,390,000 \text{ Hz}$$

to

$$1,400,000 \text{ Hz} - 20 \text{ Hz} = 1,399,980 \text{ Hz}$$

This result is shown in Figure 2-9 with a frequency spectrum of the AM modulator's output.

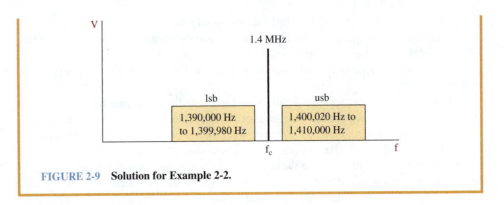

FIGURE 2-9 **Solution for Example 2-2.**

Amplitude Modulation in the Time Domain

The frequency-domain analysis just concluded demonstrates that the modulated AM signal produces side frequencies or sidebands, one on each side of the carrier. Figure 2-10 shows the modulated AM signal in the time domain. The intelligence envelope shown in the figure results from connecting a line from each RF peak value to the next one for both the top and bottom halves of the AM waveform. The lines so defined make up the *modulation envelope* and appear as a replica of the original intelligence signal. The envelopes are drawn in for clarity; they are not actual components of the waveform and would not be seen as such on an oscilloscope display. In addition, the top and bottom envelopes are *not* the upper- and lower-side frequencies—remember that the oscilloscope display is a time-domain representation, and observation of the side frequencies would require a frequency-domain display, such as that provided by a spectrum analyzer. The envelopes result from the nonlinear combination of a carrier with two lower-amplitude signals spaced in frequency by equal amounts above and below the carrier frequency. The increase and decrease in the AM waveform's amplitude is caused by the frequency difference in the side frequencies, which allows them alternately to add to and subtract from the carrier amplitude, depending on their instantaneous phase relationships.

FIGURE 2-10 **Carrier and side-frequency components result in AM waveform.**

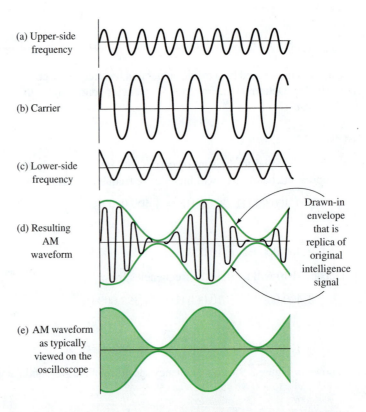

The AM waveform in Figure 2-10(d) does not show the relative frequencies to scale. If, for example, the ratio of f_c to the envelope frequency (which is also f_i) is 1 MHz to 5 kHz, then the fluctuating RF would show 200 cycles of carrier for each cycle of envelope variation, or a ratio of 200:1. To do that in a sketch is not possible, and an oscilloscope display of this example, and most practical AM waveforms, results in a well-defined envelope but with so many RF variations that they appear as a blur, as shown in Figure 2-10(e).

Phasor Representation of AM

It is often helpful to use a phasor representation to help understand generation of an AM signal. For simplicity, let us consider a carrier modulated by a single sine wave with a 100% modulation index ($m = 1$). As already established, upper- and lower-side frequencies will be created. Conventionally, however, these side frequencies are referred to as sidebands, even though only one frequency is present, and that convention will be followed here. The AM signal will therefore be composed of the carrier, the upper sideband (usb) at one-half the carrier amplitude with frequency equal to the carrier frequency plus the modulating signal frequency, and the lower sideband (lsb) at one-half the carrier amplitude at the carrier frequency minus the modulation frequency. With the aid of Figure 2-11 we will now show how these three sine waves combine to form the AM signal.

1. The carrier phasor represents the peak value of its sine wave. The upper and lower sidebands are one-half the carrier amplitude at 100% modulation.

2. A phasor rotating at a constant rate will generate a sine wave. One full revolution of the phasor corresponds to the full 360° of one sine-wave cycle. The rate of phasor rotation is called angular velocity (ω) and is related to sine-wave frequency ($\omega = 2\pi f$).

3. The sideband phasors' angular velocity is greater and less than the carrier's by the modulating signal's angular velocity. This means they are just slightly different from the carrier's because the modulating signal is such a low frequency compared to the carrier. You can think of the usb as always slightly gaining on the carrier and the lsb as slightly losing angular velocity with respect to the carrier.

4. If we let the carrier phasor be the reference (stationary with respect to the sidebands) the representation shown in Figure 2-11 can be studied. Think of the usb phasor as

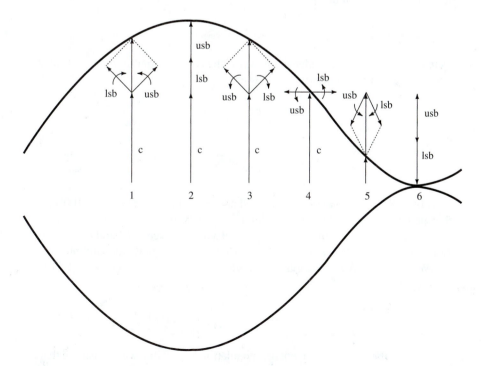

FIGURE 2-11   AM representation using vector addition of phasors.

rotating counterclockwise and the lsb phasor rotating clockwise with respect to the "stationary" carrier phasor.

5. The instantaneous amplitude of the AM waveform in Figure 2-11 is the vector sum of the phasors we have been discussing. At the peak value of the AM signal (point 2) the carrier and sidebands are all in phase, giving a sum of carrier + usb + lsb or twice the carrier amplitude, since each sideband is one-half the carrier amplitude.

6. At point 1 the vector sum of the usb and lsb is added to the carrier and results in an instantaneous value that is also equal to the value at point 3. Notice, however, that the positions of the usb and lsb phasors are interchanged at points 1 and 3.

7. At point 4 the vector sum of the three phasors equals the carrier since the sidebands cancel each other. At point 6 the sidebands combine to equal the opposite (negative) of the carrier, resulting in the zero amplitude AM signal that theoretically occurs with exactly 100% modulation.

The phasor addition concept helps in understanding how a carrier and sidebands combine to form the AM waveform. It is also helpful in analyzing other communication concepts.

Power Distribution in Carrier and Sidebands

In the case where a carrier is modulated by a pure sine wave, it can be shown that at 100% modulation, the upper- and lower-side frequencies are one-half the amplitude of the carrier. In general, as just developed,

$$E_{SF} = \frac{mE_c}{2} \qquad \text{(2-9)}$$

where E_{SF} = side-frequency amplitude

   m = modulation index

   E_c = carrier amplitude

In an AM transmission, the carrier amplitude and frequency always remain constant, while the sidebands are usually changing in amplitude and frequency. The carrier contains no information since it never changes. However, it does contain the most power since its amplitude is always at least double (when $m = 100\%$) the amplitude of each sideband. It is the sidebands that contain the information.

EXAMPLE 2-3

Determine the maximum sideband power if the carrier output is 1 kW and calculate the total maximum transmitted power.

SOLUTION

Since

$$E_{SF} = \frac{mE_c}{2} \qquad \text{(2-9)}$$

it is obvious that the maximum sideband power occurs when $m = 1$ or 100%. At that percentage modulation, each side frequency is $\frac{1}{2}$ the carrier amplitude. Since power is proportional to the square of voltage, each sideband has $\frac{1}{4}$ of the carrier power or $\frac{1}{4} \times 1$ kW, or 250 W. Therefore, the total sideband power is 250 W $\times$ 2 = 500 W and the total transmitted power is 1 kW + 500 W, or 1.5 kW.

Importance of High-Percentage Modulation

It is important to use as high a percentage modulation as possible while ensuring that overmodulation does not occur. The sidebands contain the information and have maximum power at 100% modulation. For example, if 50% modulation were used in Example 2-3, the sideband amplitudes are $\frac{1}{4}$ the carrier amplitude, and since power is proportional to E^2, we have $(\frac{1}{4})^2$, or $\frac{1}{16}$ the carrier power. Thus, total sideband power is now $\frac{1}{16} \times 1$ kW $\times 2$, or 125 W. The actual transmitted intelligence is thus only $\frac{1}{4}$ of the 500 W sideband power transmitted at full 100% modulation. These results are summarized in Table 2-1. Even though the total transmitted power has only fallen from 1.5 kW to 1.125 kW, the effective transmission has only $\frac{1}{4}$ the strength at 50% modulation as compared to 100%. Because of these considerations, most AM transmitters attempt to maintain between 90 and 95% modulation as a compromise between efficiency and the chance of drifting into overmodulation.

TABLE 2-1 • Effective Transmission at 50% versus 100% Modulation				
MODULATION INDEX, m	CARRIER POWER (kW)	POWER IN ONE SIDEBAND (W)	TOTAL SIDEBAND POWER (W)	TOTAL TRANSMITTED POWER, P_t (kW)
1.0	1	250	500	1.5
0.5	1	62.5	125	1.125

A valuable relationship for many AM calculations is

$$P_t = P_c\left(1 + \frac{m^2}{2}\right) \tag{2-10}$$

where P_t = total transmitted power (sidebands and carrier)

P_c = carrier power

m = modulation index

Equation (2-10) can be manipulated to utilize current instead of power. This is a useful relationship since current is often the most easily measured parameter of a transmitter's output to the antenna.

$$I_t = I_c\sqrt{1 + \frac{m^2}{2}} \tag{2-11}$$

where I_t = total transmitted current

I_c = carrier current

m = modulation index

Equation (2-11) can also be used with E substituted for I ($E_t = E_c\sqrt{1 + m^2/2}$).

EXAMPLE 2-4

A 500-W carrier is to be modulated to a 90% level. Determine the total transmitted power.

SOLUTION

$$P_t = P_c\left(1 + \frac{m^2}{2}\right) \tag{2-10}$$

$$P_t = 500 \text{ W}\left(1 + \frac{0.9^2}{2}\right) = 702.5 \text{ W}$$

EXAMPLE 2-5

An AM broadcast station operates at its maximum allowed total output of 50 kW and at 95% modulation. How much of its transmitted power is intelligence (sidebands)?

SOLUTION

$$P_t = P_c\left(1 + \frac{m^2}{2}\right)$$ (2-10)

$$50\,\text{kW} = P_c\left(1 + \frac{0.95^2}{2}\right)$$

$$P_c = \frac{50\,\text{kW}}{1 + (0.95^2/2)} = 34.5\,\text{kW}$$

Therefore, the total intelligence signal is

$$P_i = P_t - P_c = 50\,\text{kW} - 34.5\,\text{kW} = 15.5\,\text{kW}$$

EXAMPLE 2-6

The antenna current of an AM transmitter is 12 A when unmodulated but increases to 13 A when modulated. Calculate %m.

SOLUTION

$$I_t = I_c\sqrt{1 + \frac{m^2}{2}}$$ (2-10)

$$13\,\text{A} = 12\,\text{A}\sqrt{1 + \frac{m^2}{2}}$$

$$1 + \frac{m^2}{2} = \left(\frac{13}{12}\right)^2$$

$$m^2 = 2\left[\left(\frac{13}{12}\right)^2 - 1\right] = 0.34$$

$$m = 0.59$$

$$\%m = 0.59 \times 100\% = 59\%$$

EXAMPLE 2-7

An intelligence signal is amplified by a 70% efficient amplifier before being combined with a 10-kW carrier to generate the AM signal. If you want to operate at 100% modulation, what is the dc input power to the final intelligence amplifier?

SOLUTION

You may recall that the efficiency of an amplifier is the ratio of ac output power to dc input power. To modulate a 10-kW carrier fully requires 5 kW of intelligence. Therefore, to provide 5 kW of sideband (intelligence) power through a 70% efficient amplifier requires a dc input of

$$\frac{5\,\text{kW}}{0.70} = 7.14\,\text{kW}$$

If a carrier is modulated by more than a single sine wave, the effective modulation index is given by

$$m_\text{eff} = \sqrt{m_1^2 + m_2^2 + m_3^2 + \cdots}$$ (2-12)

The total effective modulation index must not exceed 1 or distortion (as with a single sine wave) will result. The term m_eff can be used in all previously developed equations using m.

EXAMPLE 2-8

A transmitter with a 10-kW carrier transmits 11.2 kW when modulated with a single sine wave. Calculate the modulation index. If the carrier is simultaneously modulated with another sine wave at 50% modulation, calculate the total transmitted power.

SOLUTION

$$P_t = P_c\left(1 + \frac{m^2}{2}\right) \qquad (2\text{-}10)$$

$$11.2\,\text{kW} = 10\,\text{kW}\left(1 + \frac{m^2}{2}\right)$$

$$m = 0.49$$

$$m_{\text{eff}} = \sqrt{m_1^2 + m_2^2}$$

$$= \sqrt{0.49^2 + 0.5^2}$$

$$= 0.7$$

$$P_t = P_c\left(1 + \frac{m^2}{2}\right)$$

$$= 10\,\text{kW}\left(1 + \frac{0.7^2}{2}\right)$$

$$= 12.45\,\text{kW}$$

Summary of Amplitude Modulation

Much is revealed when the amplitude-modulated carrier is analyzed in both the time and frequency domains. Among the most important are the following:

1. The total power of the modulated AM signal is higher than that of the carrier alone. Maximum power is achieved at 100% modulation. In every case, most of the power is contained within the carrier.

2. No information resides in the carrier because its amplitude is constant.

3. All information is contained within the varying-amplitude sidebands.

4. The modulated AM waveform is the result of multiplication of the carrier and intelligence frequencies and is a special instance of mixing.

5. Mixing (hence modulation) is achieved when signals are applied to a nonlinear device.

6. The sidebands are redundant, mirror images of each other, and the same information is contained in each.

7. Bandwidth of the modulated waveform is always twice the highest modulating frequency.

8. Modulation index must not exceed 1 (or 100% modulation), but maintaining as close to 100% modulation as possible is desirable because power in the sidebands (thus information-carrying capacity of the modulated signal) is greatest with high modulation percentages.

This list of characteristics will prove useful when we compare and contrast the salient characteristics of amplitude modulation with those of angle (i.e., frequency and phase) modulation in Chapter 3. In addition, these characteristics highlight disadvantages of AM, some of which are addressed by the modified forms to be considered next.

2-3 SUPPRESSED CARRIER AND SINGLE-SIDEBAND AM

The modulated signal produced as the result of mixing, composed of the full-power carrier and two redundant sidebands, is the most basic form of AM. Called, logically enough, double-sideband, full-carrier (DSBFC) AM, it is the form used in AM radio broadcasting and in applications such as aircraft-to-tower communication where simplicity of receiver tuning is desired. Creation of the DSBFC signal is also relatively straightforward, and practical modulator circuits will be studied in Chapters 4 and 5. Simplicity of generation and reception is the great virtue of DSBFC AM. The technique has its drawbacks, however, and a review of the summary just presented reveals what they are: AM in its native form represents an inefficient use of both power and bandwidth.

Recall that, at 100% modulation, the power of *each* sideband is 25% of the carrier power and that the total output power is the sum of the carrier and sideband powers. Since there are two sidebands, it follows that the total power at 100% modulation is 150% of the carrier power, or, to look at the situation another way, two-thirds of the total modulated power resides in the carrier. However, recall also that the carrier *never varies in amplitude*. The changes in instantaneous modulated-signal amplitude seen in the time domain, and confirmed in the frequency domain with the result shown in Equation (2-8), result from the instantaneous phasor addition and subtraction of the constant-amplitude carrier with the varying-amplitude modulating signals. The varying amplitude modulating signals create the varying amplitude sidebands. Therefore, *all* information contained at the output of the AM modulator is contained within the *sidebands*, and this is the primary reason that maintaining a high average percentage of modulation is desirable in DSBFC AM systems.

If the carrier, in and of itself, does not contain information, the question naturally arises: Could one or both of the information-containing sidebands be sent without the carrier? The answer, perhaps counterintuitively, is yes, for the carrier really is not needed to "carry" information or anything else through the channel. More than anything, the carrier's job in the modulator is one of frequency translation. The mixing carried out in the modulator to produce sum and difference frequencies (i.e., the sidebands) causes the baseband frequencies of the intelligence to be converted to the radio-frequency range of the sidebands, and once there the intelligence is suitable for transmission through a communication channel without necessarily being associated with the carrier. The carrier has done its job once the translation has been accomplished. Considerable savings in power—indeed, up to two-thirds of the total power applied to the antenna—can be achieved by simply eliminating the carrier from the transmitted signal. Circuits that produce sidebands without the carrier are called balanced modulators, and their outputs are called DSBSC (double-sideband, suppressed-carrier) signals. Creation of a DSBSC signal is the first step in the development of a single-sideband (SSB) signal.

Also recall from the previous section that the DSBFC signal occupies twice the bandwidth of the highest modulating frequency. That is, a 5-kHz modulating signal will produce a modulated signal with a 10-kHz bandwidth. However, the sidebands are redundant: The same information resides in each. Elimination of one sideband reduces the bandwidth by half with no loss of information. Transmission of a modulated signal with both the carrier and one sideband removed produces an SSB signal. This specialized form of AM, though more difficult to generate and demodulate than DSBFC, is widely used in the crowded high-frequency (1.8 MHz to 30 MHz) portion of the radio spectrum. High-frequency spectrum is highly desirable because of its propagation characteristics. Communication over many thousands of miles is possible, so the HF bands are home to shortwave broadcasters, military agencies, radio amateurs, and others requiring long-distance communications capabilities without the need for infrastructure support such as satellites or repeater stations.

One other aspect of DSBFC AM systems that was not addressed in the previous discussion is susceptibility to noise. Recall from Chapter 1 that, in addition to bandwidth requirements, communications systems can be analyzed in terms of their behavior in the presence of noise. Noise often manifests itself in the form of undesired variations in the amplitude of a modulated signal; in other words, noise is often an AM signal. In fact, the search for a communications technique that was immune to noise was one of the primary reasons for the development of FM, which will be studied more extensively in Chapter 3. Another characteristic of noise presents itself in Equation (1-8). Noise is directly proportional to bandwidth. By reducing the bandwidth by half, we reduce the noise present at the receiver by half as well.

Power Measurement

The total power output of a conventional, DSBFC AM transmitter is equal to the carrier power plus the sideband power. Conventional AM transmitters are rated in terms of carrier power output. Consider a low-power AM system with 4 W of carrier power operating at 100 percent modulation. Each sideband would have 1 W of power, so the total transmitted power, again at 100 percent modulation, would be 6 W (4 W + 1 W + 1 W), but the AM transmitter is rated at 4 W—just the carrier power. Conversion of this system to SSB would require only one sideband at 1 W to be transmitted. This, of course, assumes a sine-wave intelligence signal. SSB systems are most often used for voice communications, which certainly do not generate a sinusoidal waveform.

What would be the output power of an SSB transmitter without modulation? A moment's reflection would show that since the output of an SSB transmitter consists only of one sideband, and because the sideband amplitude (hence, power) varies in response to the modulation, the output power in the absence of modulation would be 0 W. In the presence of a varying-amplitude modulating signal, the output power would also be varying, which gives rise to a difficulty in defining the output power of an SSB transmitter. Unlike conventional double-sideband AM transmitters, which are rated in terms of constant carrier power (for the carrier, which does not vary in power, is always present, even in the absence of modulation), SSB transmitters, and linear power amplifiers in general, are usually rated in terms of **peak envelope power (PEP)**. To calculate PEP, multiply the maximum (peak) envelope voltage by 0.707, square the result, and divide by the load resistance. For instance, an SSB signal with a maximum level (over time) of 150 V p-p driven into a 50-Ω antenna results in a PEP rating of $(150/2 \times 0.707)^2 \div 50\ \Omega = 56.2$ W. This is the same power rating that would be given to the 150-V p-p sine wave, but there is a difference. The 150-V p-p level in the SSB voice transmission may occur only occasionally, while for the sine wave the peaks occur with every cycle. These calculations are valid no matter what type of waveform the transmitter is providing. This could range from a series of short spikes with low average power (perhaps 5 W out of the PEP of 56.2 W) to a sine wave that would yield 56.2 W of average power. With a normal voice signal an SSB transmitter develops an average power of only one-fourth to one-third its PEP rating. Most transmitters cannot deliver an average power output equal to their peak envelope power capability because their power supplies and/or output-stage components are designed for a lower average power (voice operation) and cannot continuously operate at higher power levels. This consideration is important in a maintenance context because SSB transmitters are often not rated for "continuous duty" operation at maximum PEP, which is what would happen if a sinusoidal test tone were applied continuously. Instead, practical SSB transmitters are often rated in terms of a duty cycle for test applications, where a sinusoidal test tone can be applied as the intelligence input for a maximum time interval (perhaps 30 seconds) followed by removal for a longer time interval (say, 3 minutes) before it is reapplied. This interval is necessary to prevent overheating of components in the transmitter by giving them time to cool.

Advantages of SSB

The most important advantage of SSB systems is a more effective utilization of the available frequency spectrum. The bandwidth required for the transmission of one conventional AM signal contains two equivalent SSB transmissions. This type of communications is especially adaptable, therefore, to the already overcrowded high-frequency spectrum.

A second advantage of this system is that it is less subject to the effects of selective fading. In the propagation of conventional AM transmissions, if the upper-sideband frequencies strike the ionosphere and are refracted back to earth at a different phase angle from that of the carrier and lower-sideband frequencies, distortion is introduced at the receiver. Under extremely bad conditions, complete signal cancellation may result. The two sidebands should be identical in phase with respect to the carrier so that when passed through a nonlinear device (i.e., a diode detector), the difference between the sidebands and carrier is identical. That difference is the intelligence and will be distorted in AM systems if the two sidebands have a phase difference.

Another major advantage of SSB is the power saved by not transmitting the carrier and one sideband. The resultant lower power requirements and weight reduction are especially important in mobile communication systems.

The power-savings and noise advantages of SSB systems can be quantified in decibel terms by comparing the power of one information-containing sideband to the total power produced by a full-carrier DSB AM transmission. The peak amplitude of full-carrier DSB AM at 100% modulation is twice that of the carrier alone. Because power scales as the square of voltage, peak power at 100% modulation is four times the power without modulation. For example, if an AM broadcast station has a carrier power of 50 kW (the maximum permitted in the United States), the peak power at 100% modulation would be 200 kW. Also at 100% modulation, the maximum power in each sideband would be 25% of the carrier power, or 12.5 kW in this example. An SSB transmitter, on the other hand, would need to produce a maximum power of 12.5 kW—just the power associated with one sideband. Comparing the two cases and converting to decibels,

$$10 \log\left(\frac{200}{12.5}\right) = 12 \text{ dB}$$

This result means that to have the same overall effectiveness, a full-carrier DSB AM system must transmit 10 to 12 dB more power than SSB. Also, as already established, the SSB system has a noise advantage over AM because the bandwidth required by SSB has been cut in half. Taking into account the selective fading improvement, noise reduction, and power savings, SSB offers about a 10- to 12-dB advantage over DSB AM. Some controversy exists on this issue because of the many variables that affect the savings, and the above analysis illustrates the situation strictly for peak powers and strictly at 100% modulation. At lower modulation percentages the decibel advantage would be less dramatic. Suffice it to say that a 10-W SSB transmission is at least equivalent to a 100-W AM transmission (10-dB difference).

Types of Sideband Transmission

With all the advantages presented by SSB, you may wonder why double-sideband AM is still so widely used. There are two main reasons. The first is that SSB systems are more complex than those associated with conventional AM, both in the generation of the sideband to be transmitted and in the subsequent recovery of the intelligence at the receiver. These issues will be covered more fully in Chapters 5 and 6. The second reason has to do with the function of the carrier itself. Although it is not needed to convey information through the channel, the carrier *is* needed for mixing to occur. Mixing occurs in both the transmitter and the receiver. In the transmitter, mixing produces the sum and difference frequencies that form the sidebands. In the receiver, for reasons that will be developed more fully in Chapter 6, the received signal must be mixed with a signal whose frequency is the same as that of the carrier in order to recover the intelligence. In short, although the carrier is not needed in the channel, it is needed at both the transmitting and receiving ends of the system. Therefore, if the carrier is not transmitted, it must be recreated at the receiver, and the recreated carrier must be of exactly the same frequency and phase as that used to create the sidebands for the intelligence to be recovered properly. Because of the need for carrier reinsertion, SSB receivers are more difficult to tune than conventional AM receivers, where the transmitted carrier is received along with the sidebands.

The relative difficulty of tuning is both the second primary drawback of SSB systems and the reason for a number of variations on the basic concept. In these variations, the carrier may be suppressed but not fully eliminated, thus making subsequent demodulation easier than it otherwise would be, or one sideband may not be fully eliminated, thereby making the transmitter circuitry easier to construct. The type of system employed depends primarily on the constraints (bandwidth limitations, ease of use) faced by designers and end users and will depend on the intended application. The major types include the following:

1. In the standard single sideband, or simply SSB, system the carrier and one of the sidebands are completely eliminated at the transmitter; only one sideband is transmitted. This configuration is quite popular with amateur radio operators. The chief advantages of this system are maximum transmitted signal range with minimum transmitter power, bandwidth efficiency, and elimination of carrier interference. The primary drawbacks are system complexity and relative difficulty of tuning.

2. Another system, called single-sideband suppressed carrier (SSBSC), eliminates one sideband and suppresses, but does not fully eliminate, the carrier. The suppressed carrier can then be used at the receiver for a reference and, in some cases, for demodulation of the intelligence-bearing sideband. The suppressed carrier is sometimes called a **pilot carrier**. This system retains fidelity of the received signal and minimizes carrier interference.

3. The type of system often used in military communications is referred to as **twin-sideband suppressed carrier**, or **independent sideband (ISB) transmission**. This system involves the transmission of two independent sidebands, each containing different intelligence, with the carrier suppressed to a desired level.

4. Vestigial sideband is used for analog television video transmissions. In it, a vestige (trace) of the unwanted sideband and the carrier are included with one full sideband.

5. A more specialized system is called amplitude-compandored single sideband (ACSSB). It is actually a type of SSBSC because a pilot carrier is usually included. In ACSSB the amplitude of the speech signal is compressed at the transmitter and expanded at the receiver. ACSSB is primarily used by radio amateurs but has been employed also in some air-to-ground telephone systems (including the now-defunct Airfone system).

SUMMARY

Amplitude modulation occurs as the result of mixing, in which additional frequency components are produced when two or more signals are multiplied rather than added. Multiplication occurs within a nonlinear device or circuit, whereas addition occurs when signals are summed within a linear device. Modulation is mixing of a desired sort involving signals widely separated in frequency. Signals applied to the modulator circuit of an AM transmitter are the low-frequency, or baseband, intelligence signals and the high-frequency, sine-wave carrier. The result of modulation is a broadband signal occupying a range of frequencies on either side of the carrier, among them side frequencies residing at the algebraic sum and difference of the carrier and intelligence frequencies. The occupied bandwidth of the modulated signal is, therefore, greater than that of the baseband and is directly related to the highest modulating signal frequency.

Useful information can be obtained by analyzing the modulated signal in both the time and frequency domains. In the time domain, the modulated signal changes in amplitude as the result of modulation. The positive and negative peaks of the modulated carrier form the modulation envelope, which exhibits the amplitude and frequency characteristics of the intelligence. The shape of the modulated signal results from the instantaneous phasor addition of the constant-amplitude carrier with the varying-amplitude side frequencies. Power from the side frequencies adds to the carrier power, creating a modulated signal of higher power than that of the carrier alone. In the frequency domain, the frequencies and amplitudes of the individual components of the modulated signal become apparent, and their most important characteristics are summarized as follows:

1. The total power of the modulated AM signal is higher than that of the carrier alone. Maximum power is achieved at 100% modulation. In every case, most of the power is contained within the carrier.

2. No information resides in the carrier because its amplitude is constant.

3. All information is contained within the varying-amplitude sidebands.

4. The modulated AM waveform is the result of multiplication of the carrier and intelligence frequencies and is a special instance of mixing.

5. Mixing (hence modulation) is achieved when signals are applied to a nonlinear device.

6. The sidebands are redundant, mirror images of each other, and the same information is contained in each.

7. Bandwidth of the modulated waveform is always twice the highest modulating frequency.

8. Modulation index must not exceed 1 (or 100% modulation), but maintaining as close to 100% modulation as possible is desirable because power in the sidebands (thus information-carrying capacity of the modulated signal) is greatest with high modulation percentages.

Other forms of AM involve suppressing or completely eliminating the carrier and one of the sidebands. The advantages of single-sideband systems are that they make more efficient use of transmitted power and reduce the bandwidth required for the transmitted signal, but their disadvantages are that they are more complex, and single-sideband receivers are harder to tune than their full-carrier counterparts. SSB systems are widely deployed in both the analog and digital realms, and many techniques originally developed to generate and detect SSB signals in the analog realm find wide application in digital systems. Such systems will be explored more extensively in Chapter 10.

QUESTIONS AND PROBLEMS

SECTION 2-2

1. A 1500-kHz carrier and 2-kHz intelligence signal are combined in a *nonlinear* device. List *all* the frequency components produced.

*2. If a 1500-kHz radio wave is modulated by a 2-kHz sine-wave tone, what frequencies are contained in the modulated wave (the actual AM signal)?

*3. If a carrier is amplitude-modulated, what causes the sideband frequencies?

*4. What determines the bandwidth of emission for an AM transmission?

5. Explain the difference between a sideband and a side frequency.

6. What does the phasor at point 6 in Figure 2-11 imply about the modulation signal?

7. Explain how the phasor representation can describe the formation of an AM signal.

8. Construct phasor diagrams for the AM signal in Figure 2-11 midway between points 1 and 2, 3 and 4, and 5 and 6.

*9. Draw a diagram of a carrier wave envelope when modulated 50% by a sinusoidal wave. Indicate on the diagram the dimensions from which the percentage of modulation is determined.

*10. What are some of the possible results of overmodulation?

*11. An unmodulated carrier is 300 V p-p. Calculate %m when its maximum p-p value reaches 400, 500, and 600 V. (33.3%, 66.7%, 100%)

12. If $A = 60$ V and $B = 200$ V as shown in Figure 2-5, determine %m. (53.85%)

13. Determine E_c and E_m from Problem 12. ($E_c = 65$ Vpk, $E_m = 35$ Vpk)

14. Given that the amplitude of an AM waveform can be expressed as the sum of the carrier peak amplitude and intelligence signal, derive the expression for an AM signal that shows the existence of carrier and side frequencies.

15. A 100-V carrier is modulated by a 1-kHz sine wave. Determine the side-frequency amplitudes when $m = 0.75$. (37.5 V)

16. A 1-MHz, 40-V peak carrier is modulated by a 5-kHz intelligence signal so that $m = 0.7$. This AM signal is fed to a 50-Ω antenna. Calculate the power of each spectral component fed to the antenna. ($P_c = 16$ W, $P_{usb} = P_{lsb} = 1.96$ W)

17. Calculate the carrier and sideband power if the total transmitted power is 500 W in Problem 15. (390 W, 110 W)

18. The ac rms antenna current of an AM transmitter is 6.2 A when unmodulated and rises to 6.7 A when modulated. Calculate %m. (57.9%)

*19. Why is a high percentage of modulation desirable?

*20. During 100% modulation, what percentage of the average output power is in the sidebands? (33.3%)

---

*An asterisk preceding a number indicates a question that has been provided by the FCC as a study aid for licensing examinations.

21. An AM transmitter has a 1-kW carrier and is modulated by three different sine waves having equal amplitudes. If $m_{eff} = 0.8$, calculate the individual values of m and the total transmitted power. (0.462, 1.32 kW)

22. A 50-V rms carrier is modulated by a square wave as shown in Table 1-3(c). If only the first four harmonics are considered and $V = 20$ V, calculate m_{eff}. (0.77)

SECTION 2-3

23. An AM transmission of 1000 W is fully modulated. Calculate the power transmitted if it is transmitted as an SSB signal. (167 W)

24. An SSB transmission drives 121 V peak into a 50-Ω antenna. Calculate the PEP. (146 W)

25. Explain the difference between rms and PEP designations.

26. Provide detail on the differences between ACSSB, SSB, SSBSC, and ISB transmissions.

27. List and explain the advantages of SSB over conventional AM transmissions. Are there any disadvantages?

28. A sideband technique called doubled-sideband suppressed carrier (DSBSC) is similar to a regular AM transmission, double-sideband full carrier (DSBFC). Using your knowledge of SSBSC, explain the advantage DSBSC has over regular AM.

QUESTIONS FOR CRITICAL THINKING

29. Would the *linear* combination of a low-frequency intelligence signal and a high-frequency carrier signal be effective as a radio transmission? Why or why not?

30. You are analyzing an AM waveform. What significance do the upper and lower envelopes have?

31. An AM transmitter at 27 MHz develops 10 W of carrier power into a 50-Ω load. It is modulated by a 2-kHz sine wave between 20% and 90% modulation. Determine:
    (a) Component frequencies in the AM signal.
    (b) Maximum and minimum waveform voltage of the AM signal at 20% and 90% modulation. (25.3 to 37.9 V peak, 3.14 to 60.1 V peak)
    (c) Sideband signal voltage and power at 20% and 90% modulation. (2.24 V, 0.1 W, 10.06 V, 2.025 W)
    (d) Load current at 20% and 90% modulation. (0.451A, 0.530A)

32. Compare the display of an oscilloscope to that of a spectrum analyzer.

33. If a carrier and one sideband were eliminated from an AM signal, would the transmission still be usable? Why or why not?

34. Explain the principles involved in a single-sideband, suppressed-carrier (SSBSC) emission. How does its bandwidth of emission and required power compare with that of full carrier and sidebands?

35. You have been asked to provide SSB using a DSB signal, $\cos \omega_i t$, $\cos \omega_c t$. Can this be done? Provide mathematical proof of your judgment.

36. If, in an emergency, you had to use an AM receiver to receive an SSB broadcast, what modifications to the receiver would you need to make?

CHAPTER 3

ANGLE MODULATION

KEY TERMS

deviation
deviation constant
frequency deviation
direct FM
indirect FM
modulation index
Bessel function
Carson's rule
wideband
guard bands
specialized mobile radio (SMR)
narrowband FM (NBFM)

deviation ratio (DR)
predistorter
frequency-correcting network
$1/f$ filter
integrator
limiter
capture effect
capture ratio
threshold
preemphasis
deemphasis

3-1 INTRODUCTION TO ANGLE MODULATION

As established previously, modulation involves varying, either singly or in combination, the amplitude, frequency, or phase of a sine-wave carrier in step with the instantaneous value of the intelligence. Describing the bandwidth and power-distribution characteristics of the amplitude-modulated carrier is relatively straightforward: Bandwidth is twice the highest intelligence frequency, and modulation increases the total transmitted power because power developed in the sidebands adds to that of the carrier. Another important characteristic of amplitude modulation (AM) is also its biggest disadvantage: AM receivers respond to unwanted amplitude variations in exactly the same way they respond to desired ones. In other words, the receiver is inherently susceptible to external electrical noise produced by everything from lightning strikes to malfunctioning mechanical equipment. This drawback became apparent in the 1920s as AM became a well-established medium for broadcasting news and entertainment to a widely dispersed population. The desire for a noise-resistant, higher-fidelity alternative to AM motivated researchers to develop a transmission scheme whereby a modulating signal would cause the instantaneous carrier *frequency*, rather than its amplitude, to depart from its reference value. First practically postulated as an alternative to AM in 1931—by which time commercial broadcasting was well established, having been in existence for over ten years—Edwin H. Armstrong demonstrated the first working frequency-modulation (FM) system in 1936, and in July 1939 he began the first regularly scheduled FM broadcast from Alpine, New Jersey.

Though the broadcasting of high-fidelity radio programming to the public was the first application of FM radio and is probably still the most obvious, analog FM is also used in two-way communication, such as for emergency services and business dispatch, as well as for the transmission of audio information in analog television. A very close relation, phase modulation (PM), in which the instantaneous phase angle of the carrier is caused to depart from its reference value by an amount proportional to the modulating signal amplitude, is extensively used in digital communications and can be used to generate FM indirectly. Frequency and phase modulation are interrelated: A shift in the phase angle of a carrier will create an instantaneous frequency change, and a change in carrier frequency will inevitably create a change in phase angle. FM and PM, therefore, are both forms of *angle* modulation where the *angle* so defined is the instantaneous number of electrical degrees by which the high-frequency carrier has been advanced or delayed with respect to some pre-defined reference.

As was done with AM in Chapter 2, analyzing FM and PM in terms of power distribution and bandwidth provides us with a useful framework for identifying the benefits as well as the costs of each modulation scheme. Comparison and contrast proves useful in this regard, for, as we shall soon see, FM and PM potentially have the great advantages of noise suppression and high fidelity, but these advantages are purchased with the currency of bandwidth. Since all communications systems are ultimately constrained by bandwidth availability and performance in the presence of noise, a solid conceptual understanding of how practical systems behave in these critical dimensions will prove invaluable in the analysis of system behavior in both the analog and digital realms.

Like AM, either form of angle modulation involves multiplying low-frequency intelligence signals with a high-frequency carrier. The combination produces pairs of sidebands above and below the carrier that, in turn, define the occupied bandwidth of the modulated signal. However, unlike AM, the angle-modulated carrier constantly shifts above and below a nominal (reference) frequency or phase in step with the modulating signal, which in itself may consist of a range of frequencies. The result is that any form of angle modulation, including that involving a pure (i.e., single-frequency) sine wave, produces a theoretically infinite number of side frequencies above and below the carrier, thereby implying an infinite bandwidth. However, the total power transmitted by FM or PM systems is *distributed* among the carrier and side frequencies rather than added as for AM, so the side-frequency amplitudes generally decrease as they move farther away from the carrier, and at some point the amplitudes of the side frequencies become insignificant.

For this reason practical FM and PM systems can be said to have a finite bandwidth. Defining the occupied bandwidth, though somewhat more involved than for AM, rests largely on identifying which side frequencies have enough power to be significant. This exercise forms the core of FM analysis in the frequency domain. Before embarking on a detailed analysis of the bandwidth and power-distribution characteristics of angle-modulated systems in general, however, let us look first at how a simple frequency-modulated transmitter can be built from a basic oscillator circuit. With such a circuit we can see how the two most important characteristics of the intelligence are preserved in the modulated carrier.

3-2 FREQUENCY MODULATION IN THE TIME DOMAIN

A Simple FM Generator

To gain an intuitive understanding of FM, consider the system illustrated in Figure 3-1. This is actually a very simple, yet highly instructive, FM transmitting system. It consists of an *LC* tank circuit, which, as part of an oscillator, generates a sine-wave output at its resonant frequency, which is determined by the inductor and capacitor in the tank. The capacitance section of the *LC* tank is not a standard capacitor but is a capacitor microphone. This popular type of microphone is often referred to as a condenser mic and is, in fact, a variable capacitor. When no sound waves reach its plates, the capacitor microphone presents a constant capacitance at its output. Recall from electrical fundamentals that capacitance is determined in part by the distance between the plates of a capacitor. When sound waves reach the microphone, the pressure variations created cause the capacitor plates to move in and out, which, in turn, causes the capacitance to increase and decrease from its center value. The *rate* of this capacitance change is equal to the frequency of the sound waves striking the mike, and the *amount* of capacitance change is proportional to the amplitude of the sound waves.

FIGURE 3-1 Capacitor microphone FM generator.

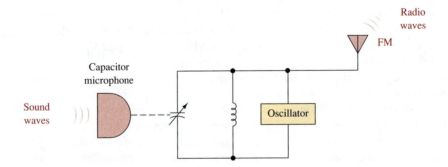

Because this capacitance value has a direct effect on the oscillator's frequency, the following two *important* conclusions can be made concerning the system's output frequency:

1. The frequency of impinging sound waves determines the *rate* of frequency change.
2. The amplitude of impinging sound waves determines the *amount* of frequency change, or **deviation** from the frequency produced by the oscillator in the absence of modulation.

The Concept of Deviation

The relationship for the FM signal generated by the capacitor microphone can be written as shown in Equation (3-1).

$$f_{out} = f_c + ke_i \tag{3-1}$$

where f_{out} = instantaneous output frequency

   f_c = output carrier frequency

   k = deviation constant [kHz/V]

   e_i = modulating (intelligence) input

Equation (3-1) shows that the output carrier frequency (f_c) depends on the amplitude and the frequency of the modulating signal (e_i) and also on the frequency deviation produced for a given input level. The unit k, called the **deviation constant** and defined in units of kHz/V, defines how much the carrier frequency will deviate (change) for a given input voltage level. The deviation constant is dependent on system design. For example, one system may produce a frequency shift of 10 kHz for every 1 V peak input voltage applied to it, while another may produce a frequency shift of 1 kHz per volt of peak applied voltage. The deviation constant is system specific, but it must be linear. For example, if $k = 1$ kHz/10 mV, then a modulating signal input level with a positive peak voltage of 20 mV will cause a 2-kHz (increasing) frequency shift. A modulating signal input level with a negative peak of −20 mV will cause a −2-kHz (decreasing) frequency shift. Assuming that the ±20-mV level is from a sinusoid (remember that a sinusoid contains both a positive and a negative peak, so the peak-to-peak voltage in this example is 40 mV), then the total deviation is 4 kHz, of which 2 kHz is above and 2 kHz below the carrier. The rate at which the carrier frequency deviates above and below the original carrier frequency depends on the frequency of the modulating signal. Using the same value for the deviation constant just defined, if a ±20-mV peak, 500-Hz input signal is applied to the microphone, then the carrier frequency will deviate ±2 kHz at a rate of 500 Hz. The symbol ± should be read as "plus or minus" and is conventionally used with values of deviation. Note in this example that although the total frequency deviation is 4 kHz, the symmetry about the carrier frequency implied by the equal amount of shift in either direction means that, in practice, the frequency deviation is expressed in terms of the amount of shift *either* above or below the carrier and is half the total, thus the expression of deviation in this example is ±2 kHz. The symmetry is important because a system that did not deviate equally on either side of the carrier would produce a distorted (nonlinear) output. For reasons that will soon become clear, knowing the deviation on either side of the carrier is essential for determining the occupied bandwidth of the modulated signal.

EXAMPLE 3-1

A sinusoid with positive and negative peaks of 25 mV at a frequency of 400 Hz is applied to a capacitor microphone FM generator. If the deviation constant for the capacitor microphone FM generator is 750 Hz/10 mV, determine

(a) The frequency deviation.

(b) The rate at which the carrier frequency is being deviated.

SOLUTION

(a)   positive frequency deviation $= 25 \text{ mV} \times \dfrac{750 \text{ Hz}}{10 \text{ mV}} = 1875 \text{ Hz or } 1.875 \text{ kHz}$

   negative frequency deviation $= -25 \text{ mV} \times \dfrac{750 \text{ Hz}}{10 \text{ mV}}$

   $= -1875 \text{ Hz}$ or -1.875 kHz

The total deviation is 3.75 kHz, but it is written as ±1.875 kHz for the given input signal level.

(b)   The input frequency (f_i) is 400 Hz; therefore, by Equation (3-1)

$$f_{out} = f_c + ke_i \qquad (3\text{-}1)$$

The carrier will deviate ±1.875 kHz at a rate of 400 Hz.

Time-Domain Representation

The Chapter 2 discussion of AM demonstrated that valuable information can be gleaned by analyzing the modulated carrier in both the time and frequency domains. The same is certainly true for FM; indeed, we will soon see that an in-depth picture of the modulated FM signal really requires the bulk of our discussion to involve frequency-domain analysis and use of the spectrum analyzer. However, because time-domain representations are perhaps still more intuitively familiar, let us first examine how the frequency-modulated carrier would appear in that context. Figure 3-2(a) shows a low-frequency sine wave (say 1 kHz, for this frequency is widely used in communications test applications), which will represent the intelligence. (Of course, an actual intelligence signal consisting of voice or music contains frequencies potentially spanning the entire audio range from 20 Hz to 20 kHz, but considering the intelligence to consist of a single-frequency sinusoid simplifies the analysis considerably.) Figure 3-2(b) represents the high-frequency carrier both before and after modulation is applied. Up until time T_1 both the frequency and amplitude of the waveform in Figure 3-2(b) are constant. The frequency until time T_1 corresponds to the carrier (f_c) or *rest* frequency in FM systems. To take a familiar example, f_c in broadcast FM radio is the station's position on the radio dial (for example, 101.1 MHz or 106.5 MHz), which represents the carrier frequency of the station's transmitter in the absence of modulation. The carrier amplitude will never change, but its frequency will only rest at f_c in the absence of modulation. At T_1 the sound wave in Figure 3-2(a) starts increasing sinusoidally and reaches a maximum positive value at T_2. During this interval, the oscillator also increases in frequency at a rate defined by the slope of the sine-wave modulating signal (i.e., sinusoidally). The carrier oscillator reaches its highest frequency when the sound wave has maximum amplitude at time T_2. In the time-domain representation of the carrier shown in Figure 3-2(b), the increase in carrier frequency is represented by the individual carrier cycles appearing compressed (drawn closer together—the period of each cycle is shorter, hence the frequency is increased). From time T_2 to T_4 the sound wave goes from its maximum positive peak to its maximum negative peak. As a result, the oscillator frequency goes from a maximum frequency shift *above* the rest value to a maximum frequency shift *below* the rest frequency. At time T_4 the modulated waveform appears with its individual cycles appearing spread out, signifying a longer period and hence a lower frequency. At time T_3 the sound wave is passing through zero, and therefore the oscillator output is instantaneously equal to the carrier frequency. Two very important points can be made as the result of this analysis. The first is that the amplitude of the carrier never changes, but its frequency will only equal f_c either in the absence of modulation or when the modulating signal is passing through zero. In other words, the carrier frequency is constantly changing in the presence of modulation. Second, modulation causes the carrier to shift (deviate) both above and below its rest or center frequency in a manner that preserves both the amplitude and frequency characteristics of the intelligence.

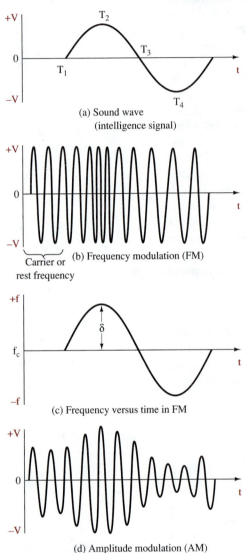

(a) Sound wave
(intelligence signal)

(b) Frequency modulation (FM)
Carrier or rest frequency

(c) Frequency versus time in FM

(d) Amplitude modulation (AM)

FIGURE 3-2 **FM representation in the time domain.**

The Two Major Concepts

As just established, the amount by which the oscillator frequency increases and decreases from f_c is called the **frequency deviation**, δ. This deviation is shown in Figure 3-2(c) as a function of time. Notice that Figure 3-2(c) is a graph of frequency versus time—not the usual voltage versus time. It is ideally shown as a sine-wave replica of the original intelligence signal. It shows that the oscillator output is indeed an FM waveform. Recall that FM is defined as a sine-wave carrier that changes in frequency by an *amount* proportional to the instantaneous value of the intelligence wave and at a *rate* equal to the intelligence frequency.

For comparison, Figure 3-2(d) shows the AM wave resulting from the intelligence signal shown in Figure 3-2(a). This should help you to see the difference between an AM and FM signal. In the case of AM, the carrier's amplitude is varied (by its sidebands) in step with the intelligence, while in FM, the carrier's frequency is varied in step with the intelligence.

If the sound-wave intelligence striking the microphone were doubled in frequency from 1 kHz to 2 kHz with constant amplitude, the rate at which the FM output swings above and below the center frequency (f_c) would change from 1 kHz to 2 kHz. Because the intelligence amplitude was not changed, however, the *amount* of frequency deviation (δ) above and below f_c will remain the same. On the other hand, if the 1-kHz intelligence frequency were kept the same but its amplitude were doubled, the *rate* of deviation above and below f_c would remain at 1 kHz, but the *amount* of frequency deviation would double.

DIRECT AND INDIRECT FM   The capacitor microphone FM generation system (and, indeed, the *LC* oscillator generally) is seldom used in practical radio-frequency transmission applications because, among other drawbacks, it is inherently unstable; its importance is derived from its relative ease of providing an understanding of FM basics. Such a system is a good example of what is known as direct FM, however. In a **direct FM** transmitter, the modulating signal is applied directly to the frequency-determining element of the carrier oscillator (in this case, the capacitor of the oscillator tank circuit), thus causing the oscillator frequency to vary directly as a result of the modulation. Effectively, then, the modulation function of a direct-FM transmitter is contained within the oscillator. Direct FM is widely employed in modern practice, albeit in somewhat more elaborate forms than our simple capacitor-microphone example, particularly in such applications as mobile two-way radio.

The drawback to direct FM systems is the challenge in maintaining the frequency stability of the transmitter, particularly over wide frequency deviations. Wide frequency deviations are encountered in broadcasting, where noise suppression and high fidelity are important design considerations. Recall from Chapter 1 that the operation of radio transmitters is tightly regulated. In many cases a license from the appropriate regulatory authority (in the United States, the Federal Communications Commission) is required before the transmitter can be put on the air, and in almost all cases the equipment itself must be subjected to a series of regulatory compliance tests and results submitted for approval by the regulatory body before the equipment can be legally sold or imported. Among the prescribed technical standards for radio transmitters are those for carrier frequency stability. For example, in the U.S. the FCC requires that FM transmitters in two-way radio applications (such as the walkie-talkies used by public safety personnel) maintain a frequency stability of 0.0005%, or five parts per million. This specification means that the carrier must be kept to within 5 Hz for every 1 MHz of its assigned frequency. Thus, a transmitter designed to operate at 100 MHz, for example, must be capable of being controlled to within 500 Hz. Maintaining such a high level of frequency stability is a design challenge whenever the oscillator will be directly affected by external circuitry, and attaining such stability is impractical with *LC*-only oscillators.

This design challenge was not lost on Edwin Armstrong, the primary inventor of FM. He realized that FM can also be produced indirectly by applying the intelligence to a modulator stage that is separate from and placed subsequent to the oscillator. **Indirect FM** causes the instantaneous phase angle of the carrier to be varied in response to the modulating signal and is effectively an example of phase modulation. As will be shown in Section 3-4, changing the phase angle of a sine wave changes its instantaneous frequency, so frequency and phase modulation (hence, direct and indirect FM) are essentially two sides of the same coin. Indirect FM systems are widely employed in "wideband" FM applications such as broadcasting, in contrast with "narrowband" FM such as two-way radio. Advantages of indirect FM systems are that oscillator stability can be maintained while also producing linear deviation over a wide frequency range. As will be shown in Section 3-5, this latter consideration is essential in explaining the fidelity advantage of FM versus AM in broadcasting applications. Phase modulation can also be produced directly and is widely used in digital communications systems.

As you continue through your study of FM, whenever you start getting bogged down on basic theory, it will often be helpful to review the capacitor mike FM generator. *Remember:*

1. The intelligence amplitude determines the *amount* of carrier frequency deviation.
2. The intelligence frequency (f_i) determines the *rate* of carrier frequency deviation.

EXAMPLE 3-2

An FM signal has a center frequency of 100 MHz but is swinging between 100.001 MHz and 99.999 MHz at a rate of 100 times per second. Determine:

(a) The intelligence frequency f_i.

(b) The intelligence amplitude.

(c) What happened to the intelligence amplitude if the frequency deviation changed to between 100.002 and 99.998 MHz.

SOLUTION

(a) Because the FM signal is changing frequency at a 100-Hz rate, $f_i = 100$ Hz.

(b) There is no way of determining the actual amplitude of the intelligence signal. Every FM system has a different proportionality constant between the intelligence amplitude and the amount of deviation it causes.

(c) The frequency deviation has now been doubled, which means that the intelligence amplitude is now double whatever it originally was.

3-3 FM IN THE FREQUENCY DOMAIN

As was the case for AM in Chapter 2, our analysis of FM is primarily concerned with determining the bandwidth requirements of the modulated carrier. Determining bandwidth rests in turn on determining where the power resides in the modulated signal. We saw how a change in carrier frequency as a result of modulation is represented in the time domain: As seen in Figure 3-2 or on an oscilloscope, the individual carrier cycles appear alternately compressed and expanded as the carrier increases and decreases in frequency in response to the modulation. Beyond that, however, little can be gleaned from the time-domain representation. Determining the frequency components and occupied bandwidth of the modulated signal necessitates a frequency-domain analysis.

The equation for the instantaneous voltage of a frequency-modulated signal is as follows:

$$e = A \sin(\omega_c t + m_f \sin \omega_i t) \tag{3-2}$$

where $e =$ instantaneous voltage

$A =$ peak value of original carrier wave

$\omega_c =$ carrier carrier angular velocity ($2\pi f_c$)

$\omega_i =$ modulating modulating (intelligence) signal angular velocity ($2\pi f_i$)

The term m_f in the above equation is defined as the **modulation index** for FM and is a measure of the extent to which the carrier is varied by the intelligence. It is defined as the ratio of maximum deviation to intelligence frequency. That is,

$$m_f = \text{FM modulation index} = \frac{\delta}{f_i} \tag{3-3}$$

where $\delta =$ maximum frequency shift caused by the intelligence signal (deviation) either above or below the carrier; therefore, deviation written as ± 3 kHz, for example, has $\delta = 3$ kHz (not 6 kHz) in Equation (3-3)

$f_i =$ of the intelligence (modulating) signal

As will be shown shortly, determining the modulation index is the first step in determining the occupied bandwidth of the modulated carrier. Equation (3-3) also points out two interesting facts about the FM modulation index that either partially or fully contrast with the AM counterpart introduced in Chapter 2. The first is that *two* aspects of a complex modulating signal such as voice or music cause the FM modulation index to be constantly changing, rather than just one, as would be the case for AM. For FM, both the modulating signal frequency and its amplitude affect the index. The frequencies affect the denominator of the ratio, while the changing amplitudes of f_i affect the instantaneous deviation, and hence the numerator. The AM modulation index, though also instantaneously varying in the presence of complex modulating signals, is really a measure of modulation depth; that is, the index shows the depth of the valley created when the modulating signal instantaneously subtracts from the carrier. In AM, only the instantaneous modulating signal *amplitude,* not its frequency, affects the modulation index. The second fact points out a full contrast between the AM and FM modulation indices. Whereas in AM, the modulation index can never exceed one without causing overmodulation (an illegal condition in licensed systems), in FM the index can exceed one and often does.

Bandwidth Determination: Bessel Function Solutions

Equation (3-2) shows that FM involves multiplying carrier and intelligence frequencies, as does AM. Mixing occurs as a result, so it is logical to predict that additional frequency components will be produced. Indeed that is the case, and the additional components appear as sum and difference frequencies on either side of the carrier, just as they did for AM. The equation is more complex than it looks, however, because it contains the sine of a sine. Solving for the frequency components of the frequency-modulated signal requires the use of a high-level mathematical tool known as the **Bessel function**. Though its derivation is beyond the scope of this text, the Bessel function solution to the FM equation can be shown as

$$
\begin{aligned}
f_c(t) = {}& J_0(m_f)\cos\omega_c t - J_1(m_f)[\cos(\omega_c - \omega_i)t - \cos(\omega_c + \omega_i)t] \\
& + J_2(m_f)[\cos(\omega_c - 2\omega_i)t + \cos(\omega_c + 2\omega_i)t] \\
& - J_3(m_f)[\cos(\omega_c - 3\omega_i)t + \cos(\omega_c + 3\omega_i)t] \\
& + \cdots
\end{aligned}
\tag{3-4}
$$

where

$$f_c(t) = \text{FM frequency components}$$
$$J_0(m_f)\cos\omega_c t = \text{carrier component}$$
$$J_1(m_f)[\cos(\omega_c - \omega_i)t - \cos(\omega_c + \omega_i)t] = \text{first set of side frequencies at}$$
$$\pm f_i \text{ above and below the carrier}$$
$$J_2(m_f)[\cos(\omega_c - 2\omega_i)t + \cos(\omega_c + 2\omega_i)]t = \text{second set of side frequencies}$$
$$\text{at } \pm 2f_i \text{ above and below}$$
$$\text{the carrier, etc.}$$

To solve for the amplitude of any side-frequency component, J_n, the following equation should be applied:

$$
J_n(m_f) = \left(\frac{m_f}{2}\right)^n \left[\frac{1}{n!} - \frac{(m_f/2)^2}{1!(n+1)!} + \frac{(m_f/2)^4}{2!(n+2)!} - \frac{(m_f/2)^6}{3!(n+3)!} + \cdots \right]
\tag{3-5}
$$

Equations (3-4) and (3-5) are presented mainly to make two points: The first is that solving for the amplitudes is a very tedious process and is strictly dependent on the modulation index, m_f. Note that every component within the brackets is multiplied by the coefficient $(m_f/2)^n$. Any change in the modulation index would change the solutions. The second conclusion flows from Equations (3-3) and (3-4). These equations show that if the angular velocity of the carrier wave (the ω_c term in Equation (3-3)) is caused to vary continuously as the result of the cyclical amplitude variations of another periodic wave (the ω_i term), the rate at which the frequency of the modulated wave, f_c, is repeated passes through an infinite number of values. Put another way, the above expressions predict that an infinite number of side-frequency components will be produced. This result occurs even if a single-frequency sine wave is used as the modulating signal. Side frequencies will appear both above and below the carrier and will be spaced at multiples of the intelligence frequency, f_i. In general, the side-frequency amplitudes will decrease the farther removed they appear from the carrier.

This result implies that FM and PM systems produce modulated signals occupying infinite bandwidths. In theory, they do, and in reality, the modulated FM signal will often occupy a wider bandwidth than its equivalent AM counterpart. With AM, because the carrier frequency stays constant, the multiplication process produced by mixing creates only two significant sidebands, one at the sum of and the other at the difference between the frequencies of the constant-amplitude carrier and the varying-amplitude modulating signal. The AM bandwidth, therefore, is always equal to twice the highest frequency of the modulating signal, regardless of its amplitude. Limiting the AM bandwidth, then, is purely a function of limiting the highest modulating signal frequency. With both FM and PM, though, the angular velocity of the carrier, hence its frequency and phase, are constantly changing as a result of modulation. Therefore, the carrier shifts through an infinite number of frequency or phase values, creating an infinite number of sum and difference products. Upon first examination, FM and PM systems would appear to be impractical forms of communication in any bandwidth-limited medium such as radio. However, Equation (3-5) demonstrates that the amplitudes of the individual frequency components (both the carrier and sidebands) are dependent on the modulation index; therefore, the power residing within those components is also dependent on m_f. Further, the total power is distributed among the carrier and sidebands, so it is reasonable to assume that sidebands far removed from the carrier have such low amplitudes (and, therefore, powers) as to be insignificant. This fact also proves true and ultimately explains how angle modulation is practical in a bandwidth-limited context. It also shows that the bandwidth of an FM signal is dependent on the modulation index, which in turn is dependent on *both* the modulating signal frequency and the permissible carrier deviation.

To determine the bandwidth of the modulated signal, then, we must determine how the power is distributed among the carrier and sidebands, and this determination in turn depends on identifying the number of significant sideband pairs. This is a Bessel function solution, and although Bessel functions are difficult to solve, there really is no need to do so. The Bessel coefficients of Equation (3-5), expressed as normalized voltages, are widely available in tabular form and are standard engineering references. Table 3-1 gives the solutions for several modulation indices, and Figure 3-3, which is a plot of the Bessel function data from Table 3-1, shows that the relative amplitudes form curves that are characteristic of the general form of first-order Bessel functions. Bessel tables show the number of significant side frequencies and the relative amplitudes in each set of side frequencies as well as in the carrier. The amplitudes are shown as normalized values of the form V_2/V_1, that is, as values referenced to the amplitude of the carrier with no modulation applied. The unmodulated-carrier reference voltage is the V_1 of the ratio, while V_2 represents the voltage of the spectral element (i.e., carrier or sideband) under consideration with modulation applied. Numbers shown with a minus sign represent signals with a 180° phase shift. If the unmodulated carrier had an amplitude of 1 V, then the actual voltages for the carrier and sidebands at any modulation index, m_f, would be read directly from the table. For any unmodulated carrier amplitude other than 1 V, the actual voltage for the frequency component of interest when modulation is applied would be determined by multiplying the decimal fraction shown in the table for that frequency component and modulation index by the voltage of the unmodulated carrier. In short, then, the Bessel table is a table of percentages expressed in decimal form.

Table 3-1 shows the amplitudes of the carrier (J_0) and of each side frequency (J_1 through J_{15}) for several modulation indices when the modulating signal, f_i, is a sine wave at a single, fixed frequency. If f_i is 1 kHz, m_f is determined by, and will be equal to, the deviation in kHz. For example, a deviation, δ, of ±3 kHz produces a m_f of 3.0 with a 1-kHz modulating signal. (This is one reason 1 kHz is frequently used as the modulating signal frequency in test applications.) Remember that there will only be one carrier but that the side frequencies *always appear in pairs*, one above and one below the carrier.

TABLE 3-1 • FM Side Frequencies from Bessel Functions

									n OR ORDER								
× (m_f)	(CARRIER) J_0	J_1	J_2	J_3	J_4	J_5	J_6	J_7	J_8	J_9	J_{10}	J_{11}	J_{12}	J_{13}	J_{14}	J_{15}	J_{16}
0.00	1.00	—	—	—	—	—	—	—	—	—	—	—	—	—	—	—	—
0.25	0.98	0.12	—	—	—	—	—	—	—	—	—	—	—	—	—	—	—
0.5	0.94	0.24	0.03	—	—	—	—	—	—	—	—	—	—	—	—	—	—
1.0	0.77	0.44	0.11	0.02	—	—	—	—	—	—	—	—	—	—	—	—	—
1.5	0.51	0.56	0.23	0.06	0.01	—	—	—	—	—	—	—	—	—	—	—	—
1.75	0.37	0.58	0.29	0.09	0.02	—	—	—	—	—	—	—	—	—	—	—	—
2.0	0.22	0.58	0.35	0.13	0.03	—	—	—	—	—	—	—	—	—	—	—	—
2.4	0.00	0.52	0.43	0.20	0.06	0.02	—	—	—	—	—	—	—	—	—	—	—
2.5	−0.05	0.50	0.45	0.22	0.07	0.02	—	—	—	—	—	—	—	—	—	—	—
3.0	−0.26	0.34	0.49	0.31	0.13	0.04	0.01	—	—	—	—	—	—	—	—	—	—
4.0	−0.40	−0.07	0.36	0.43	0.28	0.13	0.05	0.02	—	—	—	—	—	—	—	—	—
5.0	−0.18	−0.33	0.05	0.36	0.39	0.26	0.13	0.05	0.02	—	—	—	—	—	—	—	—
5.5	0.00	−0.34	−0.12	0.26	0.40	0.32	0.19	0.09	0.03	0.01	—	—	—	—	—	—	—
6.0	0.15	−0.28	−0.24	0.11	0.36	0.36	0.25	0.13	0.06	0.02	—	—	—	—	—	—	—
7.0	0.30	0.00	−0.30	−0.17	0.16	0.35	0.34	0.23	0.13	0.06	0.02	—	—	—	—	—	—
8.0	0.17	0.23	−0.11	−0.29	−0.10	0.19	0.34	0.32	0.22	0.13	0.06	0.03	—	—	—	—	—
8.65	0.00	0.27	0.06	−0.24	−0.23	0.03	0.27	0.34	0.28	0.18	0.10	0.05	0.02	0.01	—	—	—
9.0	−0.09	0.24	0.14	−0.18	−0.27	−0.06	0.20	0.33	0.30	0.21	0.12	0.06	0.03	0.01	—	—	—
10.0	−0.25	0.04	0.25	0.06	−0.22	−0.23	−0.01	0.22	0.31	0.29	0.20	0.12	0.06	0.03	0.01	—	—
12.0	0.05	−0.22	−0.08	0.20	0.18	−0.07	−0.24	−0.17	0.05	0.23	0.30	0.27	0.20	0.12	0.07	0.03	0.01
15.0	−0.01	0.21	0.04	−0.19	−0.12	0.13	0.21	0.03	−0.17	−0.22	−0.09	0.10	0.24	0.28	0.25	0.18	0.12

73

FIGURE 3-3 Normalized amplitudes of carrier and sidebands as a function of modulation index. The curves produced are representative of Bessel functions of the first kind.

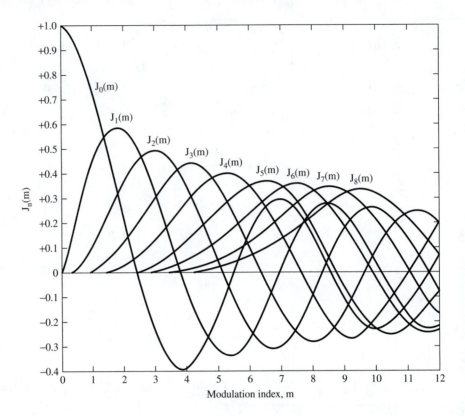

In a frequency-domain representation such as that of Figure 3-4, the first pair of side frequencies (J_1) is spaced on either side the carrier by an amount equal to the modulating signal frequency, f_i. Each higher-order side frequency (J_2 and above) is in turn spaced from its nearest neighbor by the same amount. The total *occupied bandwidth* of the signal, therefore, is twice the spacing of the side frequencies on either side of the carrier, or, put another way, the frequency difference between the highest-order significant side frequencies on either side of the carrier.

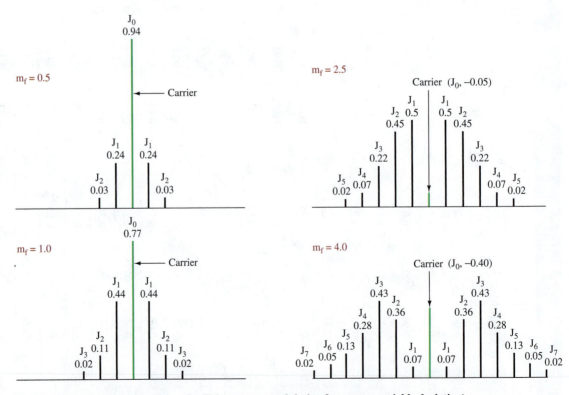

FIGURE 3-4 Frequency spectrum for FM (constant modulating frequency, variable deviation).

EXAMPLE 3-3

Determine the bandwidth required to transmit an FM signal with $f_i = 10$ kHz and a maximum deviation $\delta = 20$ kHz.

SOLUTION

$$m_f = \frac{\delta}{f_i} = \frac{20 \text{ kHz}}{10 \text{ kHz}} = 2 \qquad (3\text{-}3)$$

From Table 3-1 with $m_f = 2$, the following significant components are obtained:

$$J_0, J_1, J_2, J_3, J_4$$

This means that besides the carrier, J_1 will exist ± 10 kHz around the carrier, J_2 at ± 20 kHz, J_3 at ± 30 kHz, and J_4 at ± 40 kHz. Therefore, the total required bandwidth is 2×40 kHz $= 80$ kHz.

EXAMPLE 3-4

Repeat Example 3-3 with f_i changed to 5 kHz.

SOLUTION

$$m_f = \frac{\delta}{f_i} \qquad (3\text{-}3)$$

$$= \frac{20 \text{ kHz}}{5 \text{ kHz}}$$

$$= 4$$

In Table 3-1, $m_f = 4$ shows that the highest significant side-frequency component is J_7. Since J_7 will be at $\pm 7 \times 5$ kHz around the carrier, the required BW is 2×35 kHz $= 70$ kHz.

Examples 3-3 and 3-4 point out a potentially confusing aspect of FM analysis. Deviation and bandwidth are related *but different*. They are related because deviation, in part, determines modulation index, which in turn determines significant sideband pairs. The bandwidth, however, is computed by sideband pairs and *not* deviation frequency. Notice in Example 3-3 that the maximum deviation was ± 20 kHz, yet the bandwidth was 80 kHz. The deviation is *not* the bandwidth but *does* have an effect on the bandwidth.

As indicated previously, FM generates an infinite number of sidebands, but not all of them are significant. The question immediately arises: What is a *significant* sideband? In general, a sideband or side frequency is considered to be significant if its amplitude is 1% (0.01) or more of the unmodulated carrier amplitude. Table 3-1 is constructed to show amplitudes of 1% or greater. Though this definition is somewhat arbitrary, an amplitude reduction to 1% represents a 40-dB reduction in voltage, or a reduction to 1/10,000 the original power. If a numerical entry appears in the table, the side frequency associated with that entry is significant for determination of occupied bandwidth. Also, as already established, as the modulation index climbs, the number of significant sideband pairs (those for which an entry is shown in the table) increases as well. This statement confirms the results of Examples 3-3 and 3-4: Higher modulation indices produce wider-bandwidth signals.

Also as mentioned previously, the Bessel table is essentially a table of percentages. The carrier or sideband amplitude at any modulation index is determined by multiplying the appropriate decimal fraction shown in the table (that is, the number appearing at the intersection of the row containing the m_f and the column for the component under consideration) by the amplitude of the unmodulated carrier. For no modulation ($m_f = 0$), the carrier (J_0) is the only frequency present and is at its full amplitude value of 1.0, or 100%. However, as the carrier is modulated, energy is shifted into the sidebands. For $m_f = 0.25$, the carrier amplitude has dropped to 98% (0.98) of full amplitude, and the first side frequencies (J_1) at $\pm f_i$ around the carrier have an amplitude of 12% (0.12). (The table entries labeled J_1 through J_{15} have been adjusted to show the relative amplitudes for their respective side frequencies on *either* side of the carrier, not for the pair. In other words, there is no need to divide the table entries by two to determine individual side-frequency amplitudes.) In the presence of modulation, the carrier amplitude decreases and energy appears in the side frequencies. For higher values of m_f, more sideband pairs appear, and, in general, the carrier amplitude decreases as more energy is transferred to the sidebands. Figure 3-4 shows the spectral representation of the modulated FM signal for several modulation indices with the relative amplitudes shown for each.

FM Spectrum Analysis

Though the Bessel table shows carrier and side-frequency levels in terms of normalized amplitudes, expressing these levels in decibel terms, specifically as a reduction in decibels below the full-carrier amplitude, is perhaps more useful from a practical standpoint because it matches the measurement scale of the spectrum analyzer. Each table entry can be converted into decibel form as follows:

$$P_{dB} = 20 \log (V_2/V_1).$$

Remember that each Bessel table entry represents the ratio V_2/V_1 for its respective carrier (J_0) or sideband (J_1 and above) signal component, for a given modulation index. For example, the carrier power at a modulation index of 1.0 is

$$P_{dB} = 20 \log 0.77 = -2.27 \text{ dB}.$$

For a reference level of 0 dBm, the carrier would, therefore, show a power of -2.27 dBm. For any other reference level, the displayed power would appear 2.27 dB below that reference. If this procedure were carried out for each entry in Table 3-1, the result would be as shown in Figure 3-5.

FIGURE 3-5   FM carrier and sideband amplitudes expressed in decibels below the unmodulated carrier amplitude as a function of modulation index.

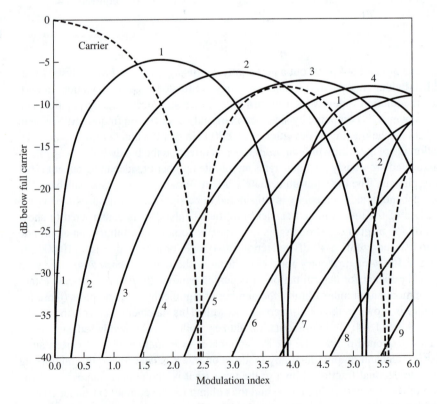

It is also possible to determine the modulation index from the spectrum-analyzer display. Example 3-5 shows the process.

EXAMPLE 3-5

From the display shown in Figure 3-6, determine

(a) The modulating signal frequency, f_i.

(b) The modulation index for the displayed waveform.

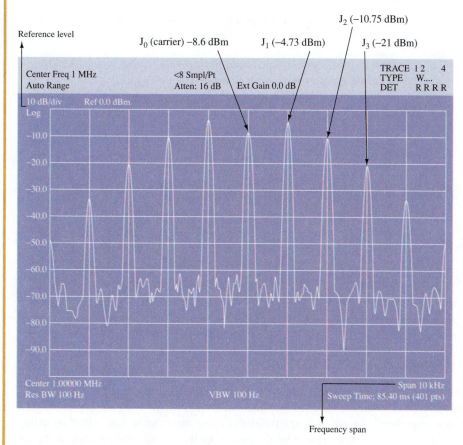

FIGURE 3-6 Spectrum analyzer display for Example 3-5.

SOLUTION

(a) The spectrum analyzer display indicates that the frequency span is set for 10 kHz. Therefore, each major division represents a 1-kHz frequency span. Since the amplitude of each side frequency falls on a major division line, the frequency separation between side frequencies is 1 kHz. Therefore, the modulating signal is a single-frequency sine wave of 1 kHz.

(b) The modulation index is found by converting the observed amplitudes in dBm to their voltage-ratio equivalents. The process is as follows:

1. Determine the reference level, which is the top horizontal line on the analyzer display. All observed amplitudes are, therefore, expressed in terms of a reduction in decibels (negative number) from the reference level. The reference level in the figure is shown to be 0 dBm.

2. Determine the vertical scale factor in decibels per division. This setting on the analyzer shows the number of decibels by which the amplitude is reduced for each major division in the vertical direction. The vertical scale factor is shown to be 10 dB/div.

3. Determine amplitude in dBm for each frequency component, J_0 through J_4. In Figure 3-6, the amplitude of the carrier is -8.6 dBm, that of the first side frequency, J_1, is -4.73 dBm, that of the second side frequency, J_2, is -10.75 dBm, and of the third, J_3, is -21 dBm.

4. Determine the number of decibels by which each frequency component on the display is reduced in decibels from the reference value determined in step 1. Since the reference level was determined to be 0 dBm, the decibel reduction for each is as follows:

$$J_0: \quad 0 \text{ dBm} - 8.6 \text{ dBm} = -8.6 \text{ dB}$$
$$J_1: \quad 0 \text{ dBm} - 4.73 \text{ dBm} = -4.73 \text{ dB}$$
$$J_2: \quad 0 \text{ dBm} - 10.75 \text{ dBm} = -10.75 \text{ dB}$$
$$J_3: \quad 0 \text{ dBm} - 21 \text{ dBm} = -21 \text{ dB}$$

5. Convert each decibel value found in step 4 to a voltage ratio. The form for decibels expressed as voltage ratios was given in Chapter 1 as

$$V_{\text{dB}} = 20 \log (V_2/V_1)$$

To convert the decibels to voltage ratios, we first divide each decibel value found in step 4 by 20 and then take the antilogarithm of the result. Recall also from Chapter 1 that the antilogarithm is the expression 10^x, so the expressions appear as follows:

$$J_0: \quad 10^{(-8.6/20)} = 0.372$$
$$J_1: \quad 10^{(-4.73/20)} = 0.582$$
$$J_2: \quad 10^{(-10.75/20)} = 0.290$$
$$J_3: \quad 10^{(-21/20)} = 0.089$$

Examination of Table 3-1 shows that the decimal fractions for J_0 through J_3 are associated with a modulation index of 1.75.

Power Distribution

The Bessel table can be used to determine the power distribution in the carrier and sidebands. Recall that power scales as the square of the voltage ($P = V^2/R$). Therefore, for any system impedance, the power in any frequency component can be determined by multiplying the transmitter output power by the square of the voltage ratio shown in the table. Recall also that there is one carrier but that the sidebands appear in pairs, so the powers calculated for the sidebands have to be doubled and added to the carrier to produce the total power. Example 3-6 shows the procedure.

EXAMPLE 3-6

Determine the relative total power of the carrier and side frequencies when $m_f = 0.25$ for a 10-kW FM transmitter.

SOLUTION

For $m_f = 0.25$, the carrier is equal to 0.98 times its unmodulated amplitude and the only significant sideband is J_1, with a relative amplitude of 0.12 (from Table 3-1). Therefore, because power is proportional to the voltage squared, the carrier power is

$$(0.98)^2 \times 10 \text{ kW} = 9.604 \text{ kW}$$

and the power of each sideband is

$$(0.12)^2 \times 10 \text{ kW} = 144 \text{ W}$$

The total power is

$$9604 \text{ W} + 144 \text{ W} + 144 \text{ W} = 9.892 \text{ kW}$$
$$\cong 10 \text{ kW}$$

The result of Example 3-6 is predictable. In FM, the transmitted waveform never varies in amplitude, just frequency. Therefore, the total transmitted power must remain constant regardless of the level of modulation. It is thus seen that whatever energy is contained in the side frequencies has been obtained from the carrier. No additional energy is added during the modulation process. The carrier in FM is not redundant as in AM because, in FM, the amplitude of the carrier is dependent on the intelligence signal.

Carson's Rule Approximation

Though the Bessel table provides the most accurate measure of bandwidth and is needed to determine power distribution in the carrier and sidebands, an approximation known as **Carson's rule** is often used to predict the bandwidth necessary for an FM signal:

$$BW \simeq 2(\delta_{max} + f_{i_{max}}) \tag{3-6}$$

This approximation includes about 98% of the total power—that is, about 2% of the power is in the sidebands outside its predicted BW. Reasonably good fidelity results when limiting the BW to that predicted by Equation (3-6). With reference to Example 3-3, Carson's rule predicts a BW of 2(20 kHz + 10 kHz) = 60 kHz versus the 80 kHz shown. In Example 3-4, BW = 2(20 kHz + 5 kHz) = 50 kHz versus 70 kHz. It should be remembered that even the 70-kHz prediction does not include all the sidebands created.

EXAMPLE 3-7

An FM signal, $2000 \sin(2\pi \times 10^8 t + 2\sin\pi \times 10^4 t)$, is applied to a 50-$\Omega$ antenna. Determine

(a) The carrier frequency.

(b) The transmitted power.

(c) m_f.

(d) f_i.

(e) BW (by two methods).

(f) Power in the largest and smallest sidebands predicted by Table 3-1.

SOLUTION

(a) By inspection of the FM equation, $f_c = (2\pi \times 10^8)/2\pi = 10^8 = 100$ MHz.

(b) The peak voltage is 2000 V. Thus,

$$P = \frac{(2000/\sqrt{2})^2}{50\ \Omega} = 40\ \text{kW}$$

(c) By inspection of the FM equation, we have

$$m_f = 2 \tag{3-2}$$

(d) The intelligence frequency, f_i, is derived from the $\sin \pi 10^4 t$ term [Equation (3-2)]. Thus,

$$f_i = \frac{\pi \times 10^4}{2\pi} = 5\ \text{kHz}.$$

(e)
$$m_f = \frac{\delta}{f_i} \tag{3-3}$$

$$2 = \frac{\delta}{5\ \text{kHz}}$$

$$\delta = 10\ \text{kHz}$$

From Table 3-1 with $m_f = 2$, significant sidebands exist to J_4 (4×5 kHz = 20 kHz). Thus, BW = 2×20 kHz = 40 kHz. Using Carson's rule yields

$$\text{BW} \simeq 2(\delta_{\max} + f_{i\max}) \qquad (3\text{-}6)$$
$$= 2(10 \text{ kHz} + 5 \text{ kHz}) = 30 \text{ kHz}.$$

(f) From Table 3-1, J_1 is the largest sideband at 0.58 times the unmodulated carrier amplitude.

$$p = \frac{(0.58 \times 2000/\sqrt{2})^2}{50 \ \Omega} = 13.5 \text{ kW}$$

or 2×13.5 kW = 27 kW for the two sidebands at ± 5 kHz from the carrier. The smallest sideband, J_4, is 0.03 times the carrier or $(0.03 \times 2000/\sqrt{2})^2/50 \ \Omega = 36$ W.

Zero-Carrier Amplitude

Figure 3-4 shows the FM frequency spectrum for various levels of modulation while keeping the modulation frequency constant. The relative amplitude of all components is obtained from Table 3-1. Notice from the table that between $m_f = 2$ and $m_f = 2.5$, the carrier goes from a plus to a minus value. The minus sign simply indicates a phase reversal, but when $m_f = 2.4$, the carrier component has zero amplitude and all the energy is contained in the side frequencies. This also occurs when $m_f = 5.5$, 8.65, and between 10 and 12, and 12 and 15.

The zero-carrier conditions are known as *carrier nulls* and suggest a convenient means of determining the deviation produced in an FM modulator. A carrier is modulated by a single sine wave at a known frequency. The modulating signal's amplitude is varied while observing the generated FM on a spectrum analyzer. At the point where the carrier amplitude goes to zero, the modulation index, m_f, is determined based on the number of sidebands displayed. If four or five sidebands appear on both sides of the nulled carrier, you can assume that $m_f = 2.4$. The deviation, δ, is then equal to $2.4 \times f_i$. The modulating signal could be increased in amplitude, and the next carrier null should be at $m_f = 5.5$. A check on modulator linearity is thereby possible because the frequency deviation should be directly proportional to the amplitude of the modulating signal.

Wideband and Narrowband FM

One of the original applications for FM, and perhaps still its best-known use, is in broadcasting. Broadcast FM allows for a true high-fidelity modulating signal up to 15 kHz and offers superior noise performance over AM. An ordinary, double-sideband AM broadcast transmission with a 15-kHz maximum intelligence frequency would occupy a bandwidth of 30 kHz. The equivalent FM transmission is allocated a 200-kHz bandwidth, thus making standard broadcast FM the premier example of **wideband** FM. Wideband FM transmissions are defined as those requiring more bandwidth than would be occupied by AM transmissions with the same maximum intelligence frequency. Clearly, the allocation of 200 kHz per FM broadcast station is very large indeed when one considers that the bandwidth occupied by a single FM station could accommodate many standard AM stations, and this allowed bandwidth helps explain why FM broadcasts occupy higher frequency bands (in the 100-MHz region) than AM transmissions (in the 1-MHz region).

Figure 3-7 shows the FCC allocation for broadcast FM stations. The maximum allowed deviation around the carrier is ± 75 kHz, and 25-kHz **guard bands** at the upper and lower ends are also provided. The carrier is required to maintain a ± 2-kHz stability. Recall that an infinite number of side frequencies is generated during frequency modulation but that the side-frequency amplitudes gradually decrease as they become farther removed from the carrier. In FM broadcasting, the significant side frequencies exist up to

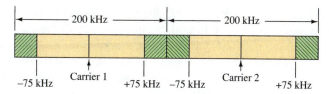

FIGURE 3-7   Commercial FM bandwidth allocations for two adjacent stations.

± 75 kHz around the carrier, and the guard bands (along with regulations governing how closely spaced stations can be within a given geographic area) ensure that adjacent channel interference will not be a problem.

Communications systems used by public-safety agencies such as police and fire departments, as well as military agencies, aircraft, taxicabs, weather service, and private industry networks also employ frequency modulation. Such systems fall under the general regulatory category of **specialized mobile radio (SMR)**, and their primary purpose is two-way voice communication rather than entertainment. You may have seen inexpensive walkie-talkies sold for personal or small-business use. Low-power models sold strictly for personal, nonbusiness applications and not requiring an operator license belong to the so-called Family Radio Service (FRS), while somewhat higher-power models sold primarily for business use and requiring a Federal Communications Commission (FCC) station license are regulated as part of the General Mobile Radio Service (GMRS). Yet another application for FM two-way radio is in the Amateur Radio Service, where radio hobbyists commonly known as *hams* communicate purely for personal enjoyment using equipment that is often home-made and sometimes self-designed. These are all considered to be **narrowband FM (NBFM)** systems, although the term does not completely comport with the formal definition of NBFM. Strictly speaking, NBFM systems are those whose occupied bandwidths are no greater than those of equivalent AM transmissions (that is to say, those containing the same intelligence frequencies). Because the bandwidth of an AM signal would be defined by the frequencies occupied by its single pair of sidebands, an NBFM signal adhering to the strict definition just given would also produce only one significant sideband pair. Examination of the Bessel table (Table 3-1) shows that this result definitely occurs with a modulation index of 0.25, and at $m_f = 0.5$, the second pair of sidebands is at a nearly negligible 3% of the unmodulated amplitude. Formally speaking, an NBFM system is defined as one with a modulation index of less than 1.57 ($\pi/2$). In the practical systems just described, the maximum permissible modulating signal frequency, f_i, is 3 kHz (the upper range of the frequency spectrum occupied by the human voice), and the maximum permissible frequency deviation is ± 5 kHz (or ± 3 kHz for the FRS). The modulation index of 1.67 that results for all but the FRS is not strictly NBFM, but because the bandwidth allocation of 10 to 30 kHz accorded to such systems is so much less than that allocated to wideband applications such as broadcasting, communications-oriented applications such as SMR, GMRS, and their amateur-radio equivalents are regarded as narrowband transmissions in practice, even though their characteristics do not accord strictly with the textbook definition of NBFM.

Percentage of Modulation and Deviation Ratio

In AM, the modulation index and the percentage of modulation were shown to be related measures of the same quantity: Both gave a numerical, easily determined measure of *how much* the intelligence amplitude instantaneously adds to or subtracts from the carrier amplitude. Higher percentages of modulation, hence, higher amplitudes, manifest themselves as louder sounds at the receiver output. In FM, the modulation index is partly a measure of loudness, insofar as that quantity is captured by the frequency deviation, δ, but the index is also affected by the maximum intelligence frequency. Therefore, if we were to define the FM *percentage of modulation* strictly in terms of the modulation index as was done for AM, we would find that the actual modulation index at 100% would vary inversely with the intelligence frequency. This result contrasts with that for AM, where full or 100% modulation means a modulation index of 1 regardless of intelligence frequency.

The percentage of modulation for FM, therefore, does not derive directly from the modulation index as it does for AM. Instead, the FM percentage of modulation, while still an expression of loudness manifested as deviation, *describes the maximum deviation permitted by law or regulation.* Thus, for broadcast FM, 100% modulation is defined to be ±75 kHz. For narrowband FM, 100% modulation would be (in most cases) achieved at ±5 kHz. As with AM, 100% modulation must not be exceeded, often under penalty of law, because the extra significant sideband pairs that would otherwise result would create excessive bandwidths and would cause prohibited interference to other licensed communications services. Broadcast stations have modulation monitors to ensure compliance, while transmitters in narrowband applications are equipped with instantaneous deviation controls to ensure that modulation limits are not exceeded under any circumstances.

Recall also that in FM the modulation index is constantly varying in the presence of a complex modulating signal such as voice or music. Another way to describe the phenomena captured by the modulation index, but in constant, unvarying terms, is to define them as the ratios of their maximum values. Such a definition is known as the **deviation ratio** (**DR**). Deviation ratio is the *maximum* frequency deviation divided by the *maximum* input frequency, as shown in Equation (3-7).

$$\text{DR} = \frac{\text{maximum possible frequency deviation}}{\text{maximum input frequency}} = \frac{f_{\text{dev(max)}}}{f_{i\text{(max)}}} \tag{3-7}$$

Deviation ratio is a commonly used term in both television and FM broadcasting. For example, broadcast FM radio permits a maximum carrier frequency deviation, $f_{\text{dev(max)}}$, of ±75 kHz and a maximum audio input frequency, $f_{i\text{(max)}}$ of 15 kHz. Therefore, for broadcast FM radio, the deviation ratio (DR) is

$$\text{DR(broadcast FM radio)} = \frac{75 \text{ kHz}}{15 \text{ kHz}} = 5,$$

and for analog broadcast television in the United States, the maximum frequency deviation of the aural carrier, $f_{\text{dev(max)}}$ is ±25 kHz with a maximum audio input frequency, $f_{i\text{(max)}}$, of 15 kHz. Therefore, for broadcast TV, the deviation ratio (DR) is

$$\text{DR(TV NTSC)} = \frac{25 \text{ kHz}}{15 \text{ kHz}} = 1.67.$$

The deviation ratio is a notion of convenience that allows for the comparison of various FM systems without the constant variation encountered with the modulation index. In addition, deviation ratio provides for the convenient characterization of FM systems as wideband or narrowband. Systems with a deviation ratio greater than or equal to 1 (DR $\geq$ 1) are considered to be **wideband** systems, whereas those with a ratio of less than 1 (DR < 1) are considered to be narrowband FM systems.

EXAMPLE 3-8

(a) Determine the permissible range in maximum modulation index for commercial FM that has 30-Hz to 15-kHz modulating frequencies.

(b) Repeat for a narrowband system that allows a maximum deviation of 1-kHz and 100-Hz to 2-kHz modulating frequencies.

(c) Determine the deviation ratio for the system in part (b).

SOLUTION

(a) The maximum deviation in broadcast FM is 75 kHz.

$$m_f = \frac{\delta}{f_i} \tag{3-3}$$

$$= \frac{75 \text{ kHz}}{30 \text{ Hz}} = 2500$$

For $f_i = 15$ kHz:

$$m_f = \frac{75 \text{ kHz}}{15 \text{ kHz}} = 5$$

(b)

$$m_f = \frac{\delta}{f_i} = \frac{1 \text{ kHz}}{100 \text{ Hz}} = 10$$

For $f_i = 2$ kHz:

$$m_f = \frac{1 \text{ kHz}}{2 \text{ kHz}} = 0.5$$

(c)

$$DR = \frac{f_{\text{dev(max)}}}{f_{i(\text{max})}} = \frac{1 \text{ kHz}}{2 \text{ kHz}} = 0.5 \qquad (3\text{-}7)$$

3-4 PHASE MODULATION

As already established, frequency and phase modulation are interrelated and fall under the umbrella of angle modulation because the modulating signal varies the angular velocity of the carrier. In fact, the two forms are so similar that the same receiver can be used to demodulate both. Phase modulation (PM) can be used to create FM indirectly when a stage subsequent to the oscillator is modulated. Conversely, direct PM can be thought of as indirect FM. The two forms of angle modulation, though related, are not identical, however, and this section is devoted to describing the one significant difference between them.

Phase modulation occurs when the modulating signal causes the instantaneous carrier phase, rather than its frequency, to shift from its reference (unmodulated) value. Put another way, the carrier phase angle in radians is either advanced or delayed from its reference value by an amount proportional to the modulating signal amplitude. The notion of a sine wave as a circularly rotating phasor is particularly useful here because phase shifts can be expressed in terms of rates of change of the angular velocity of the phasor. Recall that the *angular velocity* is the speed at which the phasor rotates around the circle and it defines the cyclic frequency of the wave because it expresses how fast the sine wave is created. One full revolution of 2π radians or 360° is one cycle. The essence of phase modulation is that, although the number of cycles per second (the frequency) does not change when modulation is applied, the individual cycles created by the rotating phasor appear alternately compressed or expanded in response to the modulating signal. In other words, the phasor speeds up or slows down in response to the modulating signal.

To visualize this concept in an everyday context, consider an ordinary circular wall clock with marks on its face to denote seconds and a sweep second hand. For this example, one full revolution of 60 seconds represents the period, T. The rotation of the second hand past each mark at a constant rate of one second per mark is analogous to the constant angular velocity of the unmodulated carrier. Now ask yourself: Could one revolution still take exactly 60 seconds even if the clock mechanism were somehow modified to make the second hand either speed up or slow down in response to some external influence? In other words, could the *rate* at which the hand passes by the individual marks either be faster or slower than one second per mark while still having one full revolution take 60 seconds? Sure it could, but if the second hand were caused to speed up early on, for example, it would have to be slowed down sometime later (before the revolution was completed) or else the period would be less than 60 seconds, implying an increase in frequency. If the period were kept at 60 seconds, however, the *frequency* could be derived from it in terms of the *number* of marks on the clock face. That term could be held constant even as the *instantaneous frequency,* represented by the increasing and decreasing speed of the hand passing over the individual marks, was caused to vary within the period. Therefore, the individual cycles can be "sped up" or "slowed down," even while keeping the frequency constant. This concept is essentially what phase modulation is all about.

You are most likely used to thinking of *phase shifts* in terms of the number of electrical degrees by which one wave has either been advanced or delayed with respect to another, reference wave. In this case, though, the comparison is made with respect to the unmodulated carrier rather than with respect to a wave in a different part of the circuit. The rotating phasor description just given is simply another way of looking at the same thing. Look at Figure 3-8.

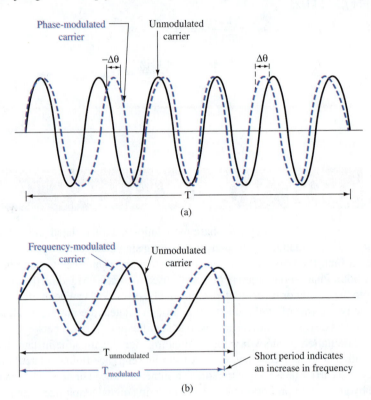

FIGURE 3-8   **Representation of phase modulation in the time domain. (a) Phase-modulated signal in the time domain. (b) Frequency-modulated carrier.**

The *reference wave* in this case is the unmodulated carrier. With modulation, the carrier can be seen to advance or delay by some number of electrical degrees from the reference. Figure 3-8(a) shows the phase-modulated signal in the time domain, while Figure 3-8(b) shows the frequency-modulated carrier. With the phase-modulated signal, over the given period, T, (which may be more than one cycle) the number of electrical degrees by which the phase changes with respect to the unmodulated value is proportional to the amplitude of the modulating signal. This is called the phase deviation, and is represented by the symbol $\Delta\theta$ in the figure. A positive-going modulating signal causes the phase shift to occur in one direction (leading, or shifting leftward, in the figure), whereas a negative-going modulating signal would cause the phase shift to occur in the other direction (lagging in the figure). Note, however, that after the number of cycles given by the period, which defines the frequency, both the modulated and unmodulated waves have returned to their original, unmodulated values. Compare the result with Figure 3-8(b), which shows the frequency-modulated signal in the time domain. Here, the modulating signal does change the period, thus changing the frequency as well as the phase. If these changes are continuous, the wave no longer is at a single frequency—that is, it would appear compressed or expanded in the time domain compared with the unmodulated waveform.

Figure 3-9 shows a carrier both frequency- and phase-modulated by a low-frequency sine wave. Note first the similarities between the two signals. They appear identical in the time domain save for a 90° phase shift. Specifically, the PM signal *leads* the FM signal by 90°. Close examination of the figure reveals that maximum frequency deviation of the FM carrier occurs at the positive and negative *peaks* of the modulating signal, whereas for the PM carrier the maximum phase shift occurs when the modulating signal polarity changes

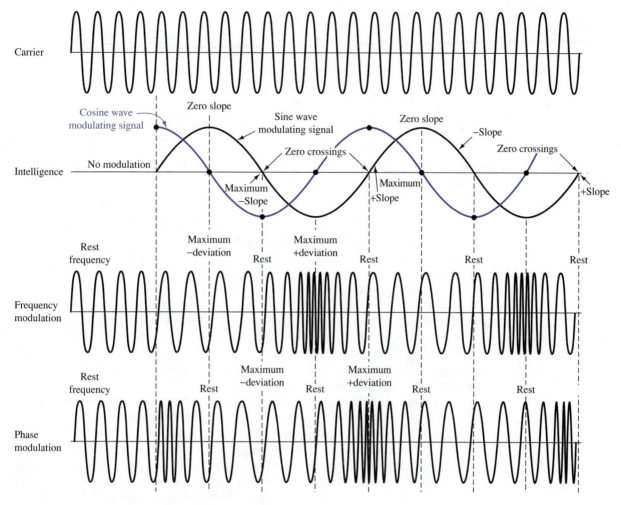

FIGURE 3-9 **Comparison of PM and FM in the time domain.**

from negative to positive or vice versa. At the zero-crossing points the slope of the sine-wave modulating signal is most nearly vertical, thus indicating that the maximum rate of change of its amplitude occurs at those points. Therefore, in PM the amount of carrier deviation (in radians) is proportional to the *rate of change* of intelligence amplitude, whereas in FM the frequency deviation (in radians per second) is directly proportional to the *magnitude* of the intelligence amplitude.

The 90° phase shift between FM and PM shown in Figure 3-9 may be a predictable result for readers familiar with the language of calculus. As just established, the amount of carrier deviation for the PM signal is directly proportional to the rate of change of the intelligence amplitude. In calculus, rates of change involve the derivative function, and the derivative of a sine wave is a cosine wave. Thus, if the intelligence is a sine wave, the PM carrier appears as if it were *frequency*-modulated by a *cosine* wave modulating signal. Recall from Chapter 1 that the cosine wave has the same appearance in the time domain as a sine wave but leads it by 90°, and that is why the phase modulated carrier in the figure appears shifted to the left compared with its frequency-modulated counterpart: PM looks like FM except for the derivative relationship among the modulating signals that manifests itself as a phase-shifted modulated carrier. This result provides a clue not only as to the only real difference between FM and PM but also reveals how to convert one to the other.

Other than the phase shift, are frequency and phase modulation identical? Not exactly, but they are so similar that a receiver would be hard-pressed to tell the difference. The practical difference between the two forms expresses itself most clearly in terms of how the carrier behaves in the presence of an increasing *frequency* intelligence signal, as distinct from one of increasing *amplitude*. Figure 3-10 summarizes the differences between the two forms of angle modulation graphically. As already described, with FM the intelligence amplitude determines the amount of deviation around the carrier, but the

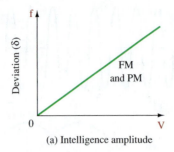
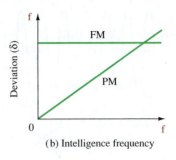

(a) Intelligence amplitude

(b) Intelligence frequency

FIGURE 3-10 Deviation effects on FM/PM by intelligence parameters: **(a)** intelligence amplitude; **(b)** intelligence frequency.

intelligence frequency determines only the rate at which the carrier swings above and below its rest frequency, not the amount of deviation. To repeat: The *frequency* of the modulating signal does *not* affect deviation in FM. In PM, however, it does. In fact, intelligence amplitude and frequency both affect deviation because both quantities have an effect on amplitude rates of change, the modulating signal characteristic to which the phase-modulated carrier responds. In PM, if the modulating signal (intelligence) frequency is increased, thus shortening its period, the sine wave swings from one peak to the other more quickly, thereby increasing the rate of change. In addition, an increased intelligence amplitude means that the modulating signal has a steeper slope for a larger voltage range than would otherwise be the case. Thus, the amplitude of the modulating signal also has an effect on the rate of change.

A PM signal can be made to look like FM by correcting for the deviation-increasing effect of higher intelligence frequencies. If a low-pass filter is placed before the modulating stage of a transmitter to attenuate higher frequency modulating signals, as shown in Figure 3-11, the effect is to offset exactly the greater frequency deviation (because of the greater rate of change) created at higher frequencies by reducing the amplitude, hence the phase shift and consequent instantaneous frequency deviation, by the same amount. Put another way, the low-pass filter, also called a **predistorter**, a **frequency-correcting network**, or **1/f filter**, is causing the modulation index of the phase-modulated signal to behave in exactly the same way as that of the frequency-modulated signal.

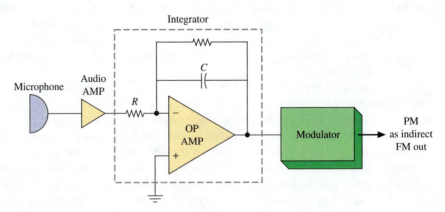

FIGURE 3-11 Addition of integrator to convert PM to FM.

If the phase modulated carrier of Figure 3-9 is produced by a cosine wave intelligence signal, and if the cosine wave is the derivative of the sine wave, then another way of explaining how to make the PM signal look like FM would be to ask what mathematical operation would be needed to convert the intelligence cosine wave back into an intelligence sine wave—thus converting PM back to FM. Mathematically speaking, the sine wave is the calculus *integral* of the cosine wave, and, not coincidentally, the low-pass filter shown in Figure 3-11 is also known as an **integrator**. Thus, addition of the integrator effectively converts PM to FM. The frequency-corrected signal appearing at the modulator output, though created directly as the result of phase modulation, is known as indirect FM because the oscillator frequency is not modulated directly.

Returning briefly to the clock analogy for angular velocity and borrowing again from the language of calculus, we can make one final point about the relationship between frequency and phase. Recall from the previous clock discussion that we defined the period to be one full revolution of the second hand and defined the frequency in terms of the number of second marks swept by the hand as it rotated. We could have just as easily defined the period differently, perhaps in terms of multiple revolutions (say, sixty) completed over a longer period, say, an hour. Then, each cycle could be defined as one full revolution of the second hand, and if the second hand rotated at a constant rate (a constant angular velocity) each cycle would take exactly one minute. The conclusion reached earlier still holds true, though: There is no inherent contradiction in stating that the second hand could speed up or slow down—thus making each cycle take either more or less than a minute, and even having this change in angular velocity last for more than one revolution—but still have the overall period be an hour as long as the second hand completes its sixtieth full revolution by the end of the hour. However, clearly the period *per cycle* would have changed because some cycles were more and some less than a minute. Since period and frequency are reciprocals, the frequency *per cycle* would have changed as well. Granted, a clock behaving in the manner just described probably would not be very useful as a clock, but the point is that the notions of period and frequency were matters of definition in that context. In electronics, a transmitter's carrier frequency is *defined* as the number of cycles it generates per second, and, just like our (perhaps useless) clock, the number of cycles per second can be kept unchanging even if the period associated with the individual cycles is somehow caused to vary.

The clock example also illustrates that if the second hand speeds up or slows down, the rate at which it passes by the second marks also increases or decreases. The second hand is passing by the marks at a changing rate *as* it speeds up or *as* it slows down. "Rate of change" is a term we have encountered before: It describes the calculus derivative. If the angular velocity of our second hand were continually caused to increase (as an example, for the converse is also true) and if it were not subsequently decreased before the end of our time interval, then clearly the frequency—measured in cycles per second, cycles per hour, or whatever other time period we care to define—would be changed as well. (In other words, frequency would increase in the clock example if the second hand completes its sixty revolutions in less than an hour.) Therefore, the rate of change of the angular velocity of the second hand has an effect on the frequency. The second hand is analogous to the rotating phasor, and since phase shifts manifest themselves as changing phasor angular velocities, the rate of change of angular velocity clearly affects the frequency. This relationship demonstrates that frequency is the calculus derivative of phase.

3-5 NOISE SUPPRESSION

The most important advantage of FM over AM is that FM has superior noise characteristics. You are probably aware that static noise is rarely heard on FM, although it is quite common in AM reception. You may be able to guess a reason for this improvement. The addition of noise to a received signal causes a change in its amplitude. Since the amplitude changes in AM contain the intelligence, any attempt to get rid of the noise adversely affects the received signal. However, in FM, the intelligence is *not* carried by amplitude changes but instead by frequency changes. The spikes of external noise picked up during transmission are clipped off by a **limiter** circuit and/or through the use of detector circuits that are insensitive to amplitude changes. Chapter 6 provides more detailed information on these FM receiver circuits.

Figure 3-12(a) shows the noise removal action of an FM limiter circuit, while in Figure 3-12(b) the noise spike feeds right through to the speaker in an AM system. The advantage for FM is clearly evident; in fact, you may think that the limiter removes all the effects of this noise spike. While it is possible to clip the noise spike off, it still causes an

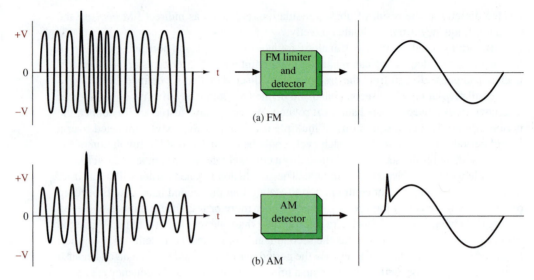

FIGURE 3-12 **FM, AM noise comparison.**

undesired phase shift, which creates an instantaneous frequency shift as well. This frequency shift *cannot* be removed. Therefore, it is not accurate to say that FM is completely insensitive to noise, just that noise in FM takes on a different characteristic than it would for AM.

The noise signal frequency will be close to the frequency of the desired FM signal because the receiver-tuned circuits in a receiver are frequency-selective. In other words, if you are tuned to an FM station at 96 MHz, the receiver's selectivity provides gain only for frequencies near 96 MHz. The noise that will affect this reception must, therefore, also be around 96 MHz because all other frequencies will be greatly attenuated. The effect of adding the desired *and* noise signals will give a resultant signal with a different phase angle than the desired FM signal alone. Therefore, the noise signal, even though it is clipped off in amplitude, will cause phase modulation (PM), which indirectly causes undesired FM. The amount of frequency deviation (FM) caused by PM is

$$\delta = \phi \times f_i \tag{3-8}$$

where δ = frequency deviation

ϕ = phase shift (radians)

f_i = frequency of intelligence signal.

FM Noise Analysis

The phase shift caused by the noise signal results in a frequency deviation that is predicted by Equation (3-8). Consider the situation illustrated in Figure 3-13. Here the noise signal is one-half the desired signal amplitude, which provides a voltage S/N ratio of 2:1. This is an intolerable situation in AM but, as the following analysis will show, is not so bad in FM.

Because the noise (N) and desired signal (S) are at different frequencies (but in the same range, as dictated by a receiver's tuned circuits), the noise is shown as a rotating vector using the S signal as a reference. The phase shift of the resultant (R) is maximum when R and N are at right angles to one another. At this worst-case condition

$$\phi = \sin^{-1}\frac{N}{S} = \sin^{-1}\frac{1}{2}$$

$$= 30°$$

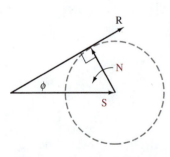

FIGURE 3-13 **Phase shift (ϕ) as a result of noise.**

or $30°/(57.3°$ per radian) $= 0.52$ rad, or about $\frac{1}{2}$ rad.

If the intelligence frequency, f_i, were known, then the deviation (δ) caused by this severe noise condition could now be calculated using Equation (3-8). Given that $\delta = \phi \times f_i$, the worst-case deviation occurs for the maximum intelligence frequency.

Assuming an f_i maximum of 15 kHz, the absolute worst case δ caused by this severe noise signal is

$$\delta = \phi \times f_i = 0.5 \times 15 \text{ kHz} = 7.5 \text{ kHz}.$$

In standard broadcast FM, the maximum modulating frequency is 15 kHz and the maximum allowed deviation is 75 kHz above and below the carrier. Thus, a 75-kHz deviation corresponds to maximum modulating signal amplitude and full volume at the receiver's output. The 7.5-kHz worst-case deviation output that results from the $S/N = 2$ condition is

$$\frac{7.5 \text{ kHz}}{75 \text{ kHz}} = \frac{1}{10}$$

and, therefore, the 2:1 S/N ratio results in an output S/N ratio of 10:1. This result assumes that the receiver's internal noise is negligible. Thus, FM is seen to exhibit a very strong capability to nullify the effects of noise! In AM, a 2:1 S/N ratio at the input essentially results in the same ratio at the output. Thus, FM is seen to have an inherent noise reduction capability not possible with AM.

EXAMPLE 3-9

Determine the worst-case output S/N for a broadcast FM program that has a maximum intelligence frequency of 5 kHz. The input S/N is 2.

SOLUTION

The input $S/N = 2$ means that the worst-case deviation is about $\frac{1}{2}$ rad (see the preceding paragraphs). Therefore,

$$\delta = \phi \times f_i \tag{3-8}$$
$$= 0.5 \times 5 \text{ kHz} = 2.5 \text{ kHz}.$$

Because full volume in broadcast FM corresponds to a 75-kHz deviation, this 2.5-kHz worst-case noise deviation means that the output S/N is

$$\frac{75 \text{ kHz}}{2.5 \text{ kHz}} = 30.$$

Example 3-9 shows that the inherent noise reduction capability of FM is improved when the maximum intelligence (modulating) frequency is reduced. A little thought shows that this capability can also be improved by increasing the maximum allowed frequency deviation from the standard 75-kHz value. An increase in allowed deviation means that increased bandwidths for each station would be necessary, however. In fact, many FM systems utilized as communication links operate with decreased bandwidths—narrowband FM systems. It is typical for them to operate with a $\pm$5-kHz maximum deviation. The inherent noise reduction of these systems is reduced by the lower allowed δ but is somewhat offset by the lower maximum modulating frequency of 3 kHz usually used for voice transmissions.

EXAMPLE 3-10

Determine the worst-case output S/N for a narrowband FM receiver with $\delta_{\text{max}} = 10$ kHz and a maximum intelligence frequency of 3 kHz. The S/N input is 3:1.

SOLUTION

The worst-case phase shift (ϕ) caused by noise occurs when $\phi = \sin^{-1}(N/S)$.

$$\phi = \sin^{-1}\frac{1}{3} = 19.5°, \text{ or } 0.34 \text{ rad}$$

and

$$\delta = \phi \times f_i \tag{3-8}$$
$$= 0.34 \times 3 \text{ kHz} \cong 1 \text{ kHz}$$

The S/N output will be

$$\frac{10 \text{ kHz}}{1 \text{ kHz}} = 10,$$

and thus the input S/N ratio of 3 is transformed to 10 or higher at the output.

Capture Effect

This inherent ability of FM to minimize the effect of undesired signals (noise in the preceding paragraphs) also applies to the reception of an undesired station operating at the same or nearly the same frequency as the desired station. This is known as the **capture effect**. You may have noticed that sometimes, when you are listening to an FM station in a moving vehicle, the original station you are listening to is suddenly replaced by a different one broadcasting on the same frequency. You may also have found that the receiver alternates abruptly back and forth between the two. This occurs because the two stations are presenting a variable signal as you drive. The capture effect causes the receiver to lock on the stronger signal by suppressing the weaker but can fluctuate back and forth when the two are nearly equal. When they are not nearly equal, however, the inherent FM noise suppression action is very effective in preventing the interference of an unwanted (but weaker) station. The weaker station is suppressed just as noise was in the preceding noise discussion. FM receivers typically have a **capture ratio** of 1dB—this means suppression of a 1-dB (or more) weaker station is accomplished. In AM, it is not uncommon to hear two separate broadcasts at the same time, but this is certainly a rare occurrence with FM.

The capture effect can also be illustrated by Figure 3-14. Notice that the S/N before and after demodulation for SSB and AM is linear. Assuming noiseless demodulation schemes, SSB (and DSB) has the same S/N at the detector's input and output. The degradation shown for AM comes about because so much of the signal power is being wasted in the redundant carrier. FM systems with m_f greater than 1 show an actual improvement in S/N, as illustrated in Examples 3-12 and 3-13. For example, consider $m_f = 5$ in Figure 3-14. When S/N before demodulation is 20, the S/N after demodulation is about 38—a significant improvement. An important tradeoff is also apparent. Although an increased maximum deviation is capable of an improved S/N ratio, the improvement comes at the expense of increased bandwidth.

Insight into the capture effect is provided by consideration of the inflection point (often termed **threshold**) shown in Figure 3-14. Notice that a rapid degradation in S/N after demodulation results when the noise approaches the same level as the desired signal. This threshold situation is noticeable when one is driving in a large city. The fluttering noise

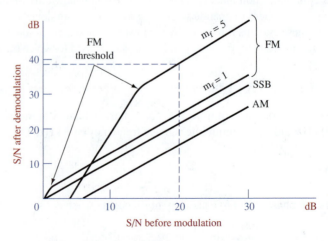

FIGURE 3-14 S/N **for basic modulation schemes.**

often heard is caused when the FM signal is reflecting off various structures. The signal strength fluctuates widely because of the additive or subtractive effects on the total received signal. The effect can cause the output to blank out totally and resume at a rapid rate as the *S/N* before demodulation moves back and forth through the threshold level.

Preemphasis

The ability of FM to suppress noise decreases with higher intelligence frequencies. This result is unfortunate because the higher intelligence frequencies tend to be of lower amplitude than the low frequencies. Thus, a high-pitched violin note that the human ear may perceive as being at the same "sound" level as the crash of a bass drum may have only half the electrical amplitude as the low-frequency signal representing the drum. In FM, half the amplitude means half the deviation and, subsequently, half the noise-reduction capability. To counteract this effect, almost all FM transmissions provide an artificial boost to the electrical amplitude of the higher frequencies. This process is termed *preemphasis*.

By definition, **preemphasis** involves increasing the relative strength of the high-frequency components of the audio signal before it is fed to the modulator. Thus, the relationship between the high-frequency intelligence components and the noise is altered. While the noise remains the same, the desired signal strength is increased.

A potential disadvantage, however, is that the natural balance between high- and low-frequency tones at the receiver would be altered. A **deemphasis** circuit in the receiver, however, corrects this defect by reducing the high-frequency audio by the same amount as the preemphasis circuit increased it, thus regaining the original tonal balance. In addition, the deemphasis network operates on both the high-frequency signal and the high-frequency noise; therefore, there is no change in the improved *S/N* ratio. The main reason for the preemphasis network, then, is to prevent the high-frequency components of the transmitted intelligence from being degraded by noise that would otherwise have more effect on the higher than on the lower intelligence frequencies.

The deemphasis network is normally inserted between the detector and the audio amplifier in the receiver. This ensures that the audio frequencies are returned to their original relative level before amplification. The preemphasis characteristic curve is flat up to 500 Hz, as shown in Figure 3-15. From 500 to 15,000 Hz, there is a sharp increase in gain up to approximately 17 dB. The gain at these frequencies is necessary to maintain the *S/N* ratio at high audio frequencies. The frequency characteristic of the deemphasis network is directly opposite to that of the preemphasis network. The high-frequency response decreases in proportion to its increase in the preemphasis network. The characteristic curve of the deemphasis circuit should be a mirror image of the preemphasis characteristic curve. Figure 3-15 shows the pre- and deemphasis curves as used by standard FM broadcasts in

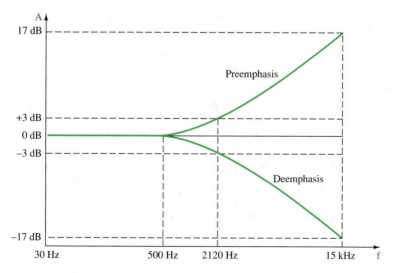

FIGURE 3-15   **Emphasis curves ($\tau = 75$ μs).**

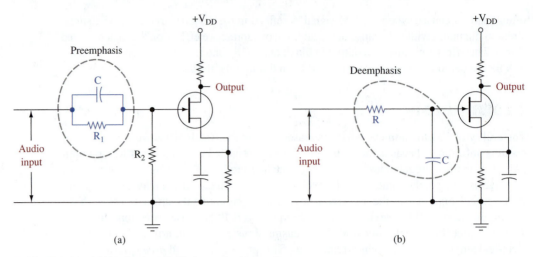

FIGURE 3-16 **(a) Emphasis and (b) de-emphasis circuits.**

the United States. As shown, the 3-dB points occur at 2120 Hz, as predicted by the RC time constant (τ) of 75 μs used to generate them.

$$f = \frac{1}{2\pi RC} = \frac{1}{2\pi \times 75\ \mu\text{s}} = 2120\ \text{Hz}$$

Figure 3-16(a) shows a typical preemphasis circuit. The impedance to the audio voltage is mainly that of the parallel circuit of C and R_1 because the effect of R_2 is small in comparison to that of either C or R_1. Since capacitive reactance is inversely proportional to frequency, audio frequency increases cause the reactance of C to decrease. This decrease of X_C provides an easier path for high frequencies as compared to R. Thus, with an increase of audio frequency, there is an increase in signal voltage. The result is a larger voltage drop across R_2 (the amplifier's input) at the higher frequencies and thus greater output.

Figure 3-16(b) depicts a typical deemphasis network. Note the physical position of R and C in relation to the base of the transistor. As the frequency of the audio signal increases, the reactance of capacitor C decreases. The voltage division between R and C now provides a smaller drop across C. The audio voltage applied to the base decreases; therefore, a reverse of the preemphasis circuit is accomplished. For the signal to be exactly the same as before preemphasis and deemphasis, the time constants of the two circuits must be equal to each other.

SUMMARY

Angle modulation results from the multiplication of an intelligence signal with a high-frequency carrier, as does amplitude modulation. However, in contrast with amplitude modulation, the frequency-modulated carrier exhibits the following characteristics:

- Total power of the modulated signal remains the same as that of the unmodulated carrier. Instead, power is transferred from the carrier to the side frequencies.

- Theoretically, an infinite number of side-frequency pairs is created, even with a single-frequency modulating signal. In the frequency domain, each side-frequency component would be spaced apart from its nearest neighbor by an amount equal to the frequency of the modulating signal. This result implies that FM signals have unlimited bandwidths.

- The amplitudes of the side frequencies become progressively lower the farther removed they are from the carrier; therefore, determination of occupied bandwidth requires determination of the power resident in the carrier and each sideband pair.

- Determination of significant sideband pairs is dependent on the modulation index, which is the ratio of deviation to modulating-signal frequency.

- The modulation index for FM systems is constantly varying; it can exceed 1.0 and often does.

- Percentage of modulation for FM systems is a function of deviation. One hundred percent modulation is defined as the maximum deviation permitted by law or regulation.

- Frequency and phase modulation are interrelated. Both are forms of angle modulation, and frequency can be shown to be the mathematical derivative of phase, and phase modulation can be used to generate a frequency-modulated signal indirectly.

Frequency-modulated signals have an inherent noise-reduction capability over their amplitude-modulated counterparts. FM receivers are insensitive to amplitude variations. In addition, the FM receiver is capable of improving the S/N ratio of the received signal. The degree of improvement is directly related to deviation. Wideband FM systems are capable of much noise suppression, but the degree by which S/N ratio at the receiver output is improved is directly related to deviation, and hence bandwidth. Thus, improved S/N ratio is achieved at the expense of increased bandwidth. Narrowband FM systems occupy less bandwidth but demonstrate no improvement in S/N ratio at the receiver output.

QUESTIONS AND PROBLEMS

SECTION 3-1

1. Define *angle modulation* and list its subcategories.

*2. What is the difference between frequency and phase modulation?

*3. What are the merits of an FM communications system compared to an AM system?

*4. Why is FM undesirable in the standard AM broadcast band?

5. What advantage does FM have over AM and SSB?

SECTION 3-2

6. Explain how a condenser microphone can be used very easily to generate FM.

7. Define *deviation constant*.

8. A 50-mV sinusoid, at a frequency of 1 kHz, is applied to a capacitor microphone FM generator. If the deviation constant for the capacitor microphone FM generator is 500 Hz/20 mV, determine:
   (a) The total frequency deviation. (±1.25 kHz)
   (b) The rate at which the carrier frequency is being deviated. (1 kHz)

9. Explain how the intelligence signal modulates the carrier.

10. In an FM transmitter, the output is changing between 90.001 and 89.999 MHz 1000 times a second. The intelligence signal amplitude is 3 V. Determine the carrier frequency and intelligence signal frequency. If the output deviation changes to between 90.0015 and 89.9985 MHz, calculate the intelligence signal amplitude. (90 MHz, 1 kHz, 4.5 V)

*11. What determines the rate of frequency swing for an FM broadcast transmitter?

12. Without knowledge of Section 3-3 and using Figure 3-1, write an equation that expresses the output frequency, f, of the FM generator. *Hint:* If there is no input into the microphone, then $f = f_c$, where f_c is the oscillator's output frequency.

SECTION 3-3

13. Define *modulation index* (m_f) as applied to an FM system.

*14. What characteristic(s) of an audio tone determines the percentage of modulation of an FM broadcast transmitter?

15. Explain what happens to the carrier in FM as m_f goes from 0 up to 15.

16. Calculate the bandwidth of an FM system (using Table 3-1) when the maximum deviation (δ) is 15 kHz and $f_i = 3$ kHz. Repeat for $f_i = 2.5$ and 5 kHz. (48 kHz, 45 kHz, 60 kHz)

17. Explain the purpose of the *guard bands* for broadcast FM. How wide is an FM broadcast channel?

*18. What frequency swing is defined as 100% modulation for an FM broadcast station?

*19. What is the meaning of the term *center frequency* in reference to FM broadcasting?

*20. What is the meaning of the term *frequency swing* in reference to FM broadcast stations?

*21. What is the frequency swing of an FM broadcast transmitter when modulated 60%? (±45 kHz)

*22. An FM broadcast transmitter is modulated 40% by a 5-kHz test tone. When the percentage of modulation is doubled, what is the frequency swing of the transmitter?

*23. An FM broadcast transmitter is modulated 50% by a 7-kHz test tone. When the frequency of the test tone is changed to 5 kHz and the percentage of modulation is unchanged, what is the transmitter frequency swing?

*24. If the output current of an FM broadcast transmitter is 8.5 A without modulation, what is the output current when the modulation is 90%?

25. An FM transmitter delivers, to a 75-Ω antenna, a signal of $v = 1000 \sin(10^9 t + 4 \sin 10^4 t)$. Calculate the carrier and intelligence frequencies, power, modulation index, deviation, and bandwidth. (159 MHz, 1.59 kHz, 6.67 kW, 4, 6.37 kHz, ~16 kHz)

26. Assuming that the 9.892-kW result of Example 3-6 is exactly correct, determine the total power in the J_2 sidebands and higher. (171 W)

27. Determine the deviation ratio for an FM system that has a maximum possible deviation of 5 kHz and the maximum input frequency is 3 kHz. Is this narrow- or wideband FM? (1.67, wideband)

SECTION 3-5

*28. What types of radio receivers do not respond to static interference?

*29. What is the purpose of a limiter stage in an FM broadcast receiver?

30. Explain why the limiter does *not* eliminate all noise effects in an FM system.

31. Calculate the amount of frequency deviation caused by a limited noise spike that still causes an undesired phase shift of 35° when f_i is 5 kHz. (3.05 kHz)

32. In a broadcast FM system, the input $S/N = 4$. Calculate the worst-case S/N at the output if the receiver's internal noise effect is negligible. (19.8:1)

33. Explain why narrowband FM systems have poorer noise performance than wideband systems.

34. Explain the *capture effect* in FM, and include the link between it and FM's inherent noise reduction capability.

*35. Why is narrowband FM rather than wideband FM used in radio communications systems?

*36. What is the purpose of preemphasis in an FM broadcast transmitter? Of deemphasis in an FM receiver? Draw a circuit diagram of a method of obtaining preemphasis.

*37. Discuss the following for frequency modulation systems:
   (a) The production of sidebands.
   (b) The relationship between the number of sidebands and the modulating frequency.
   (c) The relationship between the number of sidebands and the amplitude of the modulating voltage.
   (d) The relationship between percentage modulation and the number of sidebands.
   (e) The relationship between modulation index or deviation ratio and the number of sidebands.
   (f) The relationship between the spacing of the sidebands and the modulating frequency.
   (g) The relationship between the number of sidebands and the bandwidth of emissions.
   (h) The criteria for determining the bandwidth of emission.
   (i) Reasons for preemphasis.

QUESTIONS FOR CRITICAL THINKING

38. Analyze the effect of an intelligence signal's amplitude and frequency when it frequency-modulates a carrier.

39. Contrast the modulation indexes for PM versus FM. Given this difference, could you modify a modulating signal so that allowing it to phase-modulate a carrier would result in FM? Explain your answer.

40. Does the maximum deviation directly determine the bandwidth of an FM system? If not, explain how bandwidth and deviation are related.

41. An FM transmitter puts out 1 kW of power. When $m_f = 2$, analyze the distribution of power in the carrier and all significant sidebands. Use Bessel functions to verify that the sum of these powers is 1 kW.

42. Why is the FCC concerned if an FM broadcast station overmodulates (deviation exceeds ±75 kHz)?

CHAPTER 4

COMMUNICATIONS CIRCUITS

KEY TERMS

conduction angle
class A
class B
crossover distortion
class D
oscillator
flywheel effect
damped
continuous wave (CW)
Barkhausen criteria
quality
leakage
dissipation
resonance
tank circuit
pole
constant-k filters
m-derived filters
roll-off

stray capacitance
phasing capacitor
rejection notch
shape factor
peak-to-valley ratio
ripple amplitude
balanced modulator
double-sideband
suppressed carrier
ring modulator
lattice modulator
product detector
varactor
varactor diodes
varicap diodes
VVC diodes
direct digital synthesis (DDS)
phase noise

Though somewhat simplified, the block diagram of Figure 1-2 is representative of an actual transmitter and receiver. Subsequent chapters will examine the overall architecture of practical systems in more detail, but one statement about the circuits found there can be made at this point: With limited exceptions, communications circuits in the analog realm consist largely of amplifiers, oscillators, and frequency-selective elements, such as filters and tuned circuits. Some basic characteristics of these circuits will be covered in this chapter. This material may be a review for you, but its importance to subsequent communication circuit study merits inclusion at this time. Also found in communication systems are mixing circuits, such as the balanced modulator and product detector, as well as the phase-locked loop and the frequency synthesizer. These essential building blocks will also be covered in this chapter. In addition, because it is so central to the study of communications systems, the mixing principle will also be examined in more detail.

4-1   AMPLIFIERS

An amplifier uses one or more active devices (transistors or vacuum tubes) to increase either the voltage or current amplitude of an electrical signal applied to its input. Recall from your study of circuit fundamentals that a bipolar-junction transistor (BJT) is a current-amplification device. When biased for linear operation, the current in the collector circuit of the transistor is equal to the base current multiplied by an *amplification factor* or *current-gain factor,* defined as the ratio of collector current to base current and denoted as β (beta). Depending on the bias arrangement used, the value of beta may be an important design parameter. Generally, when the transistor is to be operated linearly, voltage-divider bias is employed because the steady-state operating point of the transistor is independent of beta. This consideration is important because β is extremely variable; the parameter varies from device to device (even devices with a common part number from the same manufacturer) and varies also as a function of operating temperature. Therefore, a well-designed amplifier does not rely on a constant β to define its quiescent, or steady-state operating condition.

Field-effect transistors (FETs) can be and are often used as amplifiers as well; however, because the FET is a voltage-controlled device, the relevant amplification factor for the FET (and corollary to BJT β) is known as *transconductance.* Denoted as g_m, transconductance is defined as the ratio of output current change to input voltage change and has units of Siemens. As we will see, FETs have a number of important advantages that make them ideal for use as amplifiers and mixers in communications applications.

Amplifiers operating at radio frequencies have design considerations that make them distinct from their audio-frequency counterparts. Some radio-frequency amplifiers are designed to operate over a narrow range of frequencies, while others may require a wide bandwidth. In addition, component layout and stray inductances and capacitances become a major concern. Finally, radio-frequency amplifiers may be required to generate very high output powers, in the hundreds if not thousands of watts, necessitating special design considerations specific to high-power operation. These issues will be studied in more detail in the following two chapters. All amplifiers, however, have considerations regarding linearity and efficiency. These considerations often involve tradeoffs, regardless of the frequency range involved. These operational considerations will be considered next.

Classes of Amplification

Amplifiers are classified in part by the amount of time during each input cycle that the active device within the circuit conducts current. This **conduction angle**, expressed in degrees, defines the portion of the input signal during which the active device is turned on. In **class A** operation, the most linear form of amplification, the active device conducts current over the full 360° of the input cycle. In fact, the active device is turned on whenever power is applied, even in the absence of an input signal. Figure 4-1(a) shows the schematic of a common-emitter (CE) amplifier with voltage-divider bias. The bias resistors, R_1 and R_2

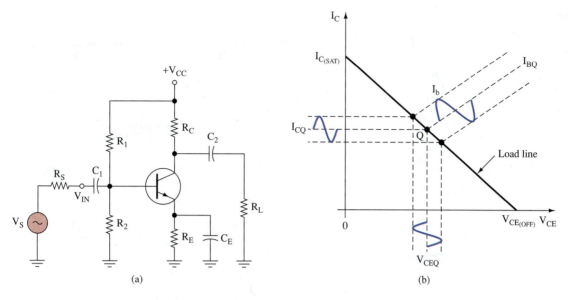

FIGURE 4-1   Common-emitter amplifier schematic (a) and load line (b).

in the figure, determine the "quiescent" or steady-state operating point of the device. The resistors forward-bias the base-emitter junction of the transistor and, by doing so, provide a path for current to flow from emitter to collector and through collector resistor R_C, whereupon a voltage is developed for subsequent application to a load. In this way, a small, varying current in the base circuit causes a much larger collector current to vary in a like manner, thus producing an output signal that is a faithful reproduction of the input but with greater amplitude. The load-line diagram of Figure 4-1(b) shows that the most linear operation—that is, a linear output over the largest positive and negative excursion of voltages applied to the input—is achieved by placing the quiescent point in the center of the line at the point marked Q in the figure. Applying a time-varying (that is, ac) signal on top of the dc bias essentially moves the Q point up and down the load line.

Class A operation is linear but inefficient. Close examination of Figure 4-1(b) reveals why. The V_{CE} term, represented along the horizontal axis, represents the voltage dropped across the transistor as collector-circuit current, I_C, flows through it. Changes in forward bias cause the Q point to move along the load line. At every point along the line, power dissipated by the transistor is given by the product of V_{CE} and I_C. If the transistor is biased such that the Q point rests at one end of the load line or the other, the result is either that a relatively high I_C is multiplied by a low V_{CE} or that a low I_C (which is zero, in the case of a fully turned-off transistor) is multiplied by a high V_{CE}. In either event, the power dissipated would be less than that expected when the Q point is in the center: The midpoint values of I_C and V_{CE} produce the greatest power dissipation. The inescapable conclusion is that the most linear operating region of the transistor is also the least efficient. The class A amplifier, though ideally suited for *small-signal* operation, where voltage gain is the primary goal, is nonetheless a poor choice for power amplification.

Efficiency can be improved at the expense of linearity by changing the bias conditions such that the Q point is moved to the end of the load line where $V_{CE} = V_{CC}$, as shown in Figure 4-2. The transistor is said to be *in cutoff* because the bias applied to its base-emitter junction is less than about 0.7 V, the barrier potential for any silicon diode. A transistor in cutoff has no current flow through it, so the voltage between collector and emitter is the open-circuit source voltage, V_{CC}. The transistor will conduct only when the instantaneous voltage applied to its base from the input signal is sufficient to produce forward bias. **Class B** operation is defined by a conduction angle of 180°, that is, the active device conducts for exactly half of each input cycle. Such operation is obviously nonlinear, but efficiency is improved over class A operation because the transistor is turned off (and thus dissipates no power) for half of each cycle.

Linear class B operation can be achieved by employing two active devices in a *push-pull* configuration such as that shown in Figure 4-3. Transistor Q_1 in the figure is an NPN

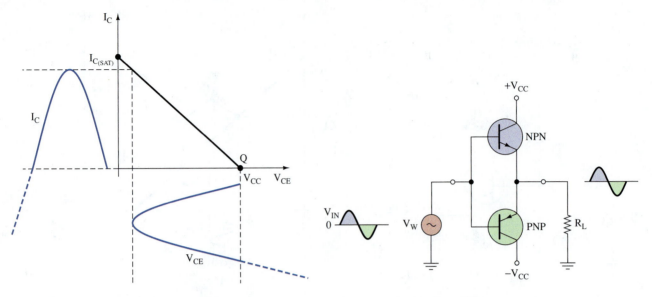

FIGURE 4-2 Load line for Class B amplifier.   **FIGURE 4-3** Push-pull amplifier.

device, while Q_2, said to be *complementary*, is a PNP transistor whose electrical character-istics match those of Q_1 but with opposite-polarity bias requirements. The transistors con-duct over alternating half-cycles of the input signal: During the first 180° of the sine wave input, NPN transistor Q_1 is forward-biased because its base is positive with respect to its emitter, and during the second half-cycle, PNP transistor Q_2 is biased on because its base is *negative* with respect to its emitter. The output of a true class-B push-pull amplifier resembles that of the input sine wave but with **crossover distortion**, shown in Figure 4-4, that occurs during the portion of the input waveform where its voltage is less than that of the base-emitter forward-breakdown potential in either the positive or negative directions, and where neither transistor is biased on. (Recall the discussion from Section 1-3 of Fourier series and square waves. The sharp edges created as the result of crossover distortion rep-resent harmonic content, and hence distortion.)

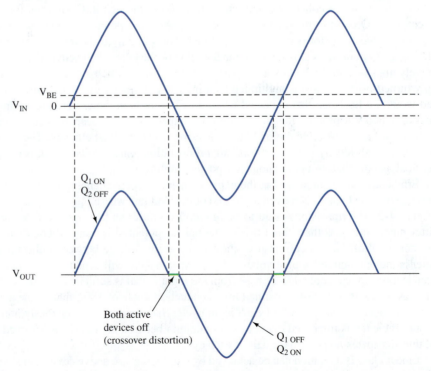

FIGURE 4-4 Crossover distortion.

The problem of crossover distortion can be alleviated by so-called class AB operation, where both transistors are biased into slight conduction. Put another way, the conduction angle of a class AB amplifier is slightly greater than 180°. The circuit in Figure 4-5 achieves the needed biasing through the use of diodes whose characteristics are closely matched to those of the transistors with which they are paired. The result is a *current mirror*, where the currents in the diodes and the currents in the transistors are the same; any tendency for current in the transistors to increase or decrease is met with a matching change in the diodes. The diodes create voltage drops of 0.7 V from anodes to cathodes, thereby producing sufficient bias voltages to turn on each transistor. Most audio power amplifiers are operated class AB because of their much greater efficiency over the class-A configuration. The maximum theoretical efficiency of a class-B amplifier is 79%, much better than the 25% theoretical maximum achievable with class A.

Further increases in efficiency can be attained by biasing the transistor "beyond cutoff," that is, by setting the Q point such that the conduction angle is significantly less than 180° (often, only 60° or so). In an amplifier operated class C, the transistor is biased on only for the small amount of time that both the base-emitter barrier potential and an additional negative bias are overcome by the amplitude of the input signal; a representative schematic and load-line diagram are shown in Figure 4-6. Class C amplifiers produce brief, high-energy pulses at the output of the active device. When the pulses are applied to a parallel-resonant *LC* tank circuit, sine waves are produced at the resonant frequency as a result of the flywheel effect. Even with its nonlinear output, the high efficiency and frequency selectivity of class C amplifiers make them very useful in communications circuits, particularly in high-power applications where high linearity is not required. Class C amplifiers have practical efficiencies in excess of 75%, and suitable applications for their use arise quite often, so class C operation will be studied more extensively in subsequent chapters.

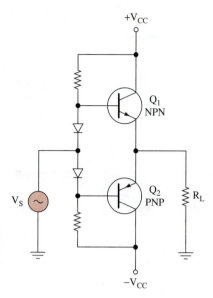

FIGURE 4-5 Push-pull amplifier with current mirror.

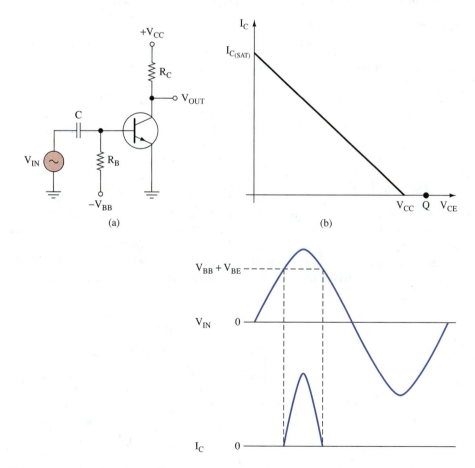

FIGURE 4-6 (a) Class C amplifier. (b) Load line and conduction curves.

Finally, near-100% efficiency can be obtained by not operating the active device in any part of its linear region but by alternately switching it between cutoff and saturation. As established, minimum power dissipation within the device occurs when its Q point is at either end of the load line, so it follows that an amplifier operating at the extremes of saturation or cutoff (or alternately between the extremes) will be more efficient than one whose active device spends any time in the linear region of operation. Such is the province of **class D**, or switching, amplifiers, which rely in part on the principle of pulse-width modulation, to be described more fully in Chapter 7.

4-2   OSCILLATORS

A second basic building block of any communication system is the **oscillator**. An oscillator generates a waveform by converting direct-current energy to an alternating current. The waveform can be of any type but occurs at some repetitive frequency. Among the many places they would be found in communications systems, oscillators are the first stages of transmitters: The transmitter oscillator creates the sine-wave carrier, which will subsequently be modulated by the intelligence.

Virtually all oscillators used in communications applications are of the feedback type because feedback oscillators are capable of producing highly pure sine waves (that is, waves free of harmonic distortion). Feedback oscillators are distinct from relaxation oscillators, which use *RC* timing circuits to generate nonsinusoidal waveforms. In contrast, a feedback oscillator is very similar to an amplifier but with the addition of a path by which energy at a specific frequency is fed from the amplifier output back to the input in phase with energy present at the input. This regenerative process sustains the oscillations. A number of different forms of sine-wave oscillators are available for use in electronic circuits. The choice of an oscillator type is based on the following criteria:

- Output frequency required.
- Frequency stability required.
- Range of frequency variability, if needed.
- Allowable waveform distortion.
- Power output required.

These performance considerations, combined with economic factors, will dictate the form of oscillator to be used in a given application.

LC Oscillator

Many oscillators used in communications applications employ some variation of the *LC tank,* or parallel-resonant circuit. The effect of charging the capacitor in Figure 4-7 to some voltage potential and then closing the switch results in the waveform shown in Figure 4-7(b). The switch closure starts a current flow as the capacitor begins to discharge through the inductor. The inductor, which resists a change in current flow, causes a gradual sinusoidal

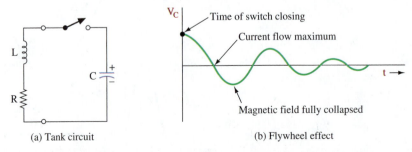

(a) Tank circuit

(b) Flywheel effect

FIGURE 4-7   **(a) Tank (parallel resonant) circuit and (b) illustration of flywheel effect.**

current buildup that reaches maximum when the capacitor is fully discharged. At this point the potential energy (i.e., voltage) is zero, but since current flow is maximum, the magnetic field energy around the inductor is maximum. The magnetic field, no longer maintained by capacitor voltage, then starts to collapse, and its counter electromotive force will keep current flowing in the same direction as before the collapse, thus charging the capacitor to the opposite polarity of its original charge. This repetitive exchange of energy is known as the **flywheel effect**. The circuit losses (mainly the dc winding resistance of the coil) cause the output to become gradually smaller as this process repeats itself after the complete collapse of the magnetic field. The resulting waveform, shown in Figure 4-7(b), is termed a **damped** sine wave. The energy of the magnetic field has been converted into the energy of the capacitor's electric field, and vice versa. The process repeats itself at the natural or resonant frequency, f_r, as predicted by Equation (4-1):

$$f_r = \frac{1}{2\pi\sqrt{LC}}$$ (4-1)

For an *LC* tank circuit to function as an oscillator, an amplifier restores the lost energy to provide a constant-amplitude sine-wave output. In radio work, the resulting *undamped* waveform is known as a **continuous wave (CW)**. The most straightforward method of restoring this lost energy is now examined, and the general conditions required for oscillation are introduced.

The *LC* oscillators are basically feedback amplifiers, with the feedback serving to increase or sustain the self-generated output. This is called positive feedback, and it occurs when the fed-back signal is in phase with (reinforces) the input signal. It would seem, then, that the regenerative effects of this positive feedback would cause the output to increase continually with each cycle of fed-back signal. However, in practice, component nonlinearity and power supply constraints limit the theoretically infinite gain.

The criteria for oscillation are formally stated by the **Barkhausen criteria** as follows:

1. The loop gain must equal 1.

2. The loop phase shift must be $n \times 360°$, where $n = 1, 2, 3, \ldots$.

An oscillating amplifier adjusts itself to meet both of these criteria. The initial surge of dc power or noise in the circuit creates a sinusoidal voltage in the tank circuit at its resonant frequency, and it is fed back to the input and amplified repeatedly until the amplifier works into the saturation and cutoff regions. At this time, the flywheel effect of the tank is effective in maintaining a sinusoidal output. This process shows us that too much gain would cause excessive distortion; therefore, the gain should be limited to a level that is just greater than or equal to 1.

Hartley Oscillator

Figure 4-8 shows the basic Hartley oscillator in simplified form. The inductors L_1 and L_2 are a single tapped inductor. Positive feedback is obtained by mutual inductance effects between L_1 and L_2 with L_1 in the transistor output circuit and L_2 across the base-emitter circuit. A portion of the amplifier signal in the collector circuit (L_1) is returned to the base circuit by means of inductive coupling from L_1 to L_2. As always in a common-emitter (CE) circuit, the collector and base voltages are 180° out of phase. Another 180° phase reversal between these two voltages occurs because they are taken from opposite ends of an inductor tap that is tied to the common transistor terminal—the emitter. Thus the in-phase feedback requirement is fulfilled and loop gain is, of course, provided by Q_1. The frequency of oscillation is approximately given by

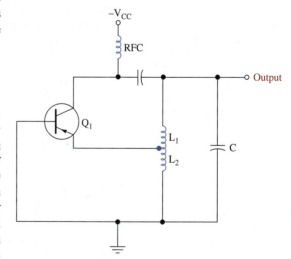

FIGURE 4-8 Simplified Hartley oscillator.

$$f \cong \frac{1}{2\pi\sqrt{(L_1 + L_2)C}}$$ (4-2)

and is influenced slightly by the transistor parameters and by the amount of coupling between L_1 and L_2.

Figure 4-9 shows a practical Hartley oscillator. A number of additional circuit elements are necessary to make a workable oscillator over the simplified one used for explanatory purposes in Figure 4-8. Naturally, the resistors R_A and R_B are for biasing purposes. The radio-frequency choke (RFC) is effectively an open circuit to the resonant frequency and thus allows a path for the bias (dc) current but does not allow the power supply to short out the ac signal. The coupling capacitor C_3 prevents dc current from flowing in the tank, and C_2 provides dc isolation between the base and the tank circuit. Both C_2 and C_3 can be considered as short circuits at the oscillator frequency.

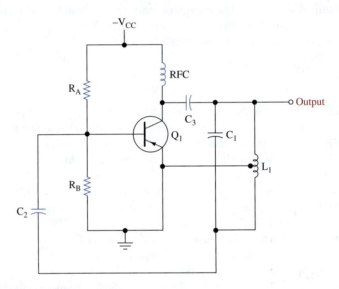

FIGURE 4-9   **Practical Hartley oscillator.**

Colpitts Oscillator

Figure 4-10 shows a Colpitts oscillator. It is similar to the Hartley oscillator except that the tank circuit elements have interchanged their roles. The capacitor is now split, so to speak, and the inductor is single-valued with no tap. The details of circuit operation are identical with the Hartley oscillator and therefore will not be explained further. The frequency of oscillation is given approximately by the resonant frequency of L_1 and C_1 in series with the C_2 tank circuit:

$$f \cong \frac{1}{2\pi\sqrt{[C_1C_2/(C_1 + C_2)]L_1}} \qquad (4\text{-}3)$$

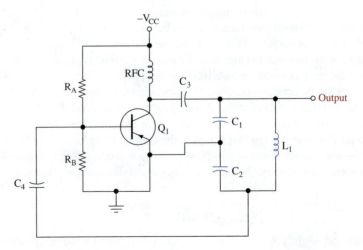

FIGURE 4-10   **Colpitts oscillator.**

The performance differences between these two oscillators are minor, and the choice between them is usually made on the basis of convenience or economics. They may both provide variable oscillator output frequencies by making one of the tank circuit elements variable.

Clapp Oscillator

A variation of the Colpitts oscillator is shown in Figure 4-11. The Clapp oscillator has a capacitor C_3 in series with the tank circuit inductor. If C_1 and C_2 are made large enough, they will "swamp" out the transistor's inherent junction capacitances, thereby negating transistor variations and junction capacitance changes with temperature. The frequency of oscillation is

$$f = \frac{1}{2\pi\sqrt{L_1 C_3}} \tag{4-4}$$

and an oscillator with better frequency stability than the Hartley or Colpitts versions results. The Clapp oscillator does not have as much frequency adjustment range, however.

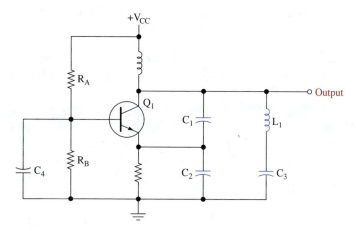

FIGURE 4-11   Clapp oscillator.

The *LC* oscillators presented in this section are the ones most commonly used. However, many different forms and variations exist and are used for special applications.

Crystal Oscillator

When greater frequency stability than that provided by *LC* oscillators is required, a crystal-controlled oscillator is often used. A crystal oscillator is one that uses a piezoelectric crystal as the inductive element of an *LC* circuit. The crystal, usually quartz, also has a resonant frequency of its own, but optimum performance is obtained when it is coupled with an external capacitance.

The electrical equivalent circuit of a crystal is shown in Figure 4-12. The crystal is effectively a series-resonant circuit (with resistive losses) in parallel with a capacitance C_p. The resonant frequencies of the series and parallel resonant circuits are within 1% of each other, hence the crystal impedance varies sharply within a narrow frequency range. This is equivalent to a very high Q circuit, and, in fact, crystals with a Q-factor of 20,000 are common; a Q of up to 10^6 is possible. This result compares with a maximum Q of about 1000 that can be obtained with high-quality *LC* resonant circuits. For this reason, and because of the good time- and temperature-stability characteristics of quartz, crystals are capable of maintaining a frequency to $\pm0.001\%$ over a fairly wide temperature range. The $\pm0.001\%$ term is equivalent to saying ±10 parts per million (ppm), and this is a preferred way of expressing such very small percentages. Note that $0.001\% = 0.00001 = 1/100,000 = 10/1,000,000 = 10$ ppm. Over very narrow temperature ranges or by maintaining the crystal in a small temperature-controlled oven, stabilities of ±0.01 ppm are possible.

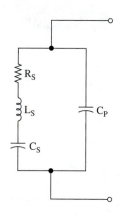

FIGURE 4-12   Electrical equivalent circuit of a crystal.

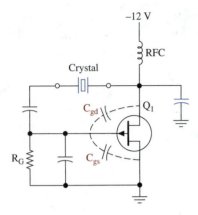

FIGURE 4-13 **Pierce oscillator.**

Crystals are fabricated by "cutting" the crude quartz in a very exacting fashion. The method of "cut" is a science in itself and determines the crystal's natural resonant frequency as well as its temperature characteristics. Crystals are available at frequencies of about 15 kHz and up, with higher frequencies providing the best frequency stability. However, at frequencies above 100 MHz, they become so small that handling is a problem.

Crystals may be used in place of the inductors in any of the *LC* oscillators discussed previously. A circuit especially adapted for crystal oscillators is the Pierce oscillator, shown in Figure 4-13. The use of an FET is desirable because its high impedance results in light loading of the crystal, provides for good stability, and does not lower the *Q*. This circuit is essentially a Colpitts oscillator with the crystal replacing the inductor and the inherent FET junction capacitances functioning as the split capacitor. Because these junction capacitances are generally low, this oscillator is effective only at high frequencies.

Crystal oscillators are available in various forms depending on the frequency stability required. The basic oscillator shown in Figure 4-13 (often referred to as CXO) may be adequate as a simple clock for a digital system. Increased performance can be attained by adding temperature compensation circuitry (TCXO). Further improvement is afforded by including microprocessor (digital) control in the crystal oscillator package (DTCXO). The ultimate performance is attained with oven control of the crystal's temperature and sometimes also includes the microprocessor control (OCXO). These obviously require significant power to maintain the oven at some constant elevated temperature. A comparison of the four types of commonly available crystal oscillators is provided in Table 4-1.

TABLE 4-1 • Typical Performance Comparison for Crystal Oscillators				
	BASIC CRYSTAL OSCILLATOR (CXO)	**TEMPERATURE COMPENSATED (TCXO)**	**DIGITAL TCXO (DTCXO)**	**OVEN-CONTROLLED CXO (OCXO)**
Frequency stability from 0 to 70°C	100 ppm	1 ppm	0.5 ppm	0.05 ppm
Frequency stability for 1 year at constant temperature	1 ppm	1 ppm	1 ppm	1 ppm

4-3 FREQUENCY-SELECTIVE CIRCUITS

A defining feature of any electronic communications system is its ability to respond selectively to applied stimuli. Because many different information signals reside at all points along the electromagnetic spectrum, any useful communications system must be capable of identifying and extracting desired signals while rejecting all others. Filters, which can couple (pass) or attenuate (reject) signals from one stage to the next, make use of the frequency-dependent behavior of one or more *reactive* components, either in the form of physical inductors and capacitors or in the form of circuit elements exhibiting those properties, perhaps as an undesired consequence of their construction or operation. The *reaction* implied by the word is the tendency of inductors to "react to," specifically to oppose, changes in current flowing through them and, likewise, the tendency of capacitors to oppose changes in voltage across their plates as they charge and discharge. Reactance is, therefore, a time-varying phenomenon; that is, reactance is an effect that manifests itself in ac circuits. Reactance and resistance together create an overall opposition to current flow in the circuit (impedance, Z), but unlike resistance, the magnitude of reactance is frequency dependent. Also used in communication systems are *tuned* circuits, which make use of the properties of inductors and capacitors to resonate at certain frequencies. Because of their importance in communications contexts, these types of circuits will be studied in some detail.

Reactance

Inductors and capacitors exhibit the property of reactance, which, as mentioned, manifests itself as an opposition to current flow in ac circuits and which is measured in ohms. Recall that an alternating current flowing in an inductor (which is most often a coil of wire that may or may not be wound on an iron core) creates a magnetic flux that is proportional to the current; in the presence of an alternating current, the *rate of change* of the flux will be proportional to the *rate of change* of the current. From Faraday's law of electromagnetic induction, we can state that the voltage induced in the coil (for this is one definition of inductance: the ability of a coil of wire to induce a voltage across itself), is proportional to the rate of change of the magnetic flux and, hence, the current. A natural outgrowth of this statement is the self-induced voltage equation, which states that the voltage across an inductor is related to the inductance and the rate of change of current. That is,

$$V_{\text{ind}} = L\frac{di_L}{dt}.$$

The "d" in the above equation should be interpreted to mean "change in"; you may have seen the concept of changing quantities represented with the Greek letter delta, Δ. From Section 1-1, we know that ω, when applied to a quantity like voltage, expresses the time rate of change of that quantity. We also know from Ohm's law that impedance, expressed in ohms, is the ratio of the voltage across the terminals of a coil to the current flowing in the coil. This point bears repeating: *Any* quantity expressed in ohms, reactance and impedance among them, is ultimately defined as the ratio of voltage to current. The voltage across the coil, and thus the impedance, are both proportional to ω, or $2\pi f$. Borrowing the calculus tool of differentiation, we can demonstrate that the peak voltage across the inductor is equal to

$$V_{\text{p}} = \omega L I_{\text{p}}.$$

Since impedance (Z) is opposition to current flow in ohms and is the ratio of voltage to current, we have

$$Z = \frac{Vp}{Ip} = \frac{\omega L Ip}{Ip} = \omega L.$$

Finally, making the simplifying assumption that the coil has no resistance, so its total impedance is only composed of reactance, we can state that the reactance is

$$X_L = \omega L = 2\pi f L.$$

The above expression states categorically that inductive reactance is *directly* proportional to both inductance and frequency.

Following a similar line of reasoning as for inductors, we can develop an expression for capacitive reactance. Capacitance is an expression of the rate at which a capacitor can store charge on its plates. Charge is stored as the result of current flowing into and out of the capacitor, and this current is proportional to the rate of change of voltage impressed upon the plates of the capacitor:

$$i_c = C\left(\frac{dV_c}{dt}\right).$$

The above equation tells us that the greater the rate of change of voltage across the capacitor, the greater the capacitive current will be. An increase in frequency corresponds to an increase in the rate of voltage change and, therefore, to a decrease in opposition to current flowing into and out of the capacitor. Again by applying the process of differentiation, we find that

$$I_p = \omega C V_p,$$

and by recognizing that capacitive reactance is an expression of opposition to current flow, we can state (again ignoring resistance effects) that

$$X_c = \frac{V_p}{I_p} = \frac{V_p}{\omega C V_p} = \frac{1}{\omega C} = \frac{1}{2\pi f C}.$$

We thus find that the capacitive reactance is *inversely* proportional to both frequency and capacitance.

Practical Inductors and Capacitors

Inductors (also referred to as *chokes* or *coils*) have an inductance rating in henries and a maximum current rating. Similarly, capacitors have a capacitance rating in farads and a maximum voltage rating. When selecting coils and capacitors for use at radio frequencies and above, an additional characteristic must be considered—the **quality** (Q) of the component. The Q is a ratio of energy stored in a component to energy lost.

Inductors store energy in the surrounding magnetic field and lose (dissipate) energy in their winding resistances. A capacitor stores energy in the electric field between its plates and primarily loses energy from **leakage** between the plates.

For an inductor,

$$Q = \frac{\text{reactance}}{\text{resistance}} = \frac{\omega L}{R} \tag{4-5}$$

where R is the series resistance distributed along the coil winding. The required Q for a coil varies with circuit application. Values up to about 500 are generally available.

For a capacitor,

$$Q = \frac{\text{susceptance}}{\text{conductance}} = \frac{\omega C}{G} \tag{4-6}$$

where G is the value of conductance through the dielectric between the capacitor plates. Good-quality capacitors used in radio circuits have typical Q factors of 1000.

At higher radio frequencies (VHF and above—see Table 1-1) the Q for inductors and capacitors is generally reduced by factors such as radiation, absorption, lead inductance, and package/mounting capacitance. Occasionally, an inverse term is used rather than Q. It is called the component **dissipation** (D) and is equal to 1/Q. Thus D = 1/Q, a term used more often in reference to a capacitor.

Resonance

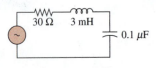

FIGURE 4-14 **Series *RLC* circuit.**

Resonance is the condition where inductive and capacitive reactances are equal ($X_L = X_C$). Consider the series *RLC* circuit shown in Figure 4-14. In this case, the total impedance, Z, is provided by the formula

$$Z = \sqrt{R^2 + (X_L - X_C)^2}.$$

An interesting effect occurs at the frequency where X_L is equal to X_C. That frequency is termed the resonant frequency, f_r. At f_r the circuit impedance is equal to the resistor value (which might only be the series winding resistance of the inductor). That result can be shown from the equation above because when $X_L = X_C$, $X_L - X_C$ equals zero, so that $Z = \sqrt{R^2 + 0^2} = \sqrt{R^2} = R$. The resonant frequency can be determined by finding the frequency where $X_L = X_C$.

$$X_L = X_C$$

$$2\pi f_r L = \frac{1}{2\pi f_r C}$$

$$f_r^2 = \frac{1}{4\pi^2 LC}$$

$$f_r = \frac{1}{2\pi\sqrt{LC}} \, (\text{Hz}) \qquad\qquad (4\text{-}1)$$

EXAMPLE 4-1

Determine the resonant frequency for the circuit shown in Figure 4-14. Calculate its impedance when $f = 12$ kHz.

SOLUTION

$$f_r = \frac{1}{2\pi\sqrt{LC}} \qquad\qquad (4\text{-}1)$$

$$= \frac{1}{2\pi\sqrt{3 \text{ mH} \times 0.1 \text{ } \mu\text{F}}}$$

$$= 9.19 \text{ kHz}$$

At 12 kHz,

$$X_L = 2\pi fL$$

$$= 2\pi \times 12 \text{ kHz} \times 3 \text{ mH}$$

$$= 226 \text{ } \Omega$$

$$X_C = \frac{1}{2\pi fC}$$

$$= \frac{1}{2\pi \times 12 \text{ kHz} \times 0.1 \text{ } \mu\text{F}}$$

$$= 133 \text{ } \Omega$$

$$Z = \sqrt{R^2 + (X_L - X_C)^2}$$

$$= \sqrt{30^2 + (226 - 133)^2}$$

$$= 97.7 \text{ } \Omega$$

This circuit contains more inductive than capacitive reactance at 12 kHz and is therefore said to look inductive.

The impedance of the series *RLC* circuit is minimum at its resonant frequency and equal to the value of *R*. A graph of its impedance, *Z*, versus frequency has the shape of the curve shown in Figure 4-15(a). At low frequencies the circuit's impedance is very high because X_C is high.

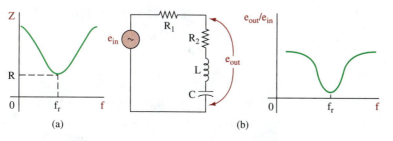

FIGURE 4-15   Series *RLC* circuit effects.

At high frequencies X_L is very high and thus Z is high. At resonance, when $f = f_r$, the circuit's $Z = R$ and is at its minimum value. This impedance characteristic can provide a filter effect, as shown in Figure 4-15(b). At f_r, $X_L = X_C$ and thus

$$e_{\text{out}} = e_{\text{in}} \times \frac{R_2}{R_1 + R_2}$$

by the voltage-divider effect. At all other frequencies, the impedance of the *LC* combination goes up (from 0 at resonance) and thus e_{out} goes up. The response for the circuit in Figure 4-15(b) is termed a *band-reject*, or *notch*, filter. A "band" of frequencies is being "rejected" and a "notch" is cut into the output at the resonant frequency, f_r.

Example 4-2 shows that the filter's output increases as the frequency is increased. Calculation of the circuit's output for frequencies below resonance would show a similar increase and is left as an exercise at the end of the chapter. The band-reject, or notch, filter is sometimes called a trap because it can "trap" or get rid of a specific range of frequencies near f_r. A trap is commonly used in a television receiver, where rejection of some specific frequencies is necessary for good picture quality.

EXAMPLE 4-2

Determine f_r for the circuit shown in Figure 4-15(b) when $R_1 = 20\ \Omega$, $R_2 = 1\ \Omega$, $L = 1$ mH, $C = 0.4\ \mu\text{F}$, and $e_{\text{in}} = 50$ mV. Calculate e_{out} at f_r and 12 kHz.

SOLUTION

The resonant frequency is

$$f_r = \frac{1}{2\pi\sqrt{LC}} \qquad (4\text{-}1)$$
$$= 7.96 \text{ kHz.}$$

At resonance,

$$e_{\text{out}} = e_{\text{in}} \times \frac{R_2}{R_1 + R_2}$$
$$= 50 \text{ mV} \times \frac{1\ \Omega}{1\ \Omega + 20\ \Omega}$$
$$= 2.38 \text{ mV.}$$

At $f = 12$ kHz,

$$X_L = 2\pi fL$$
$$= 2\pi \times 12 \text{ kHz} \times 1 \text{ mH}$$
$$= 75.4\ \Omega$$

and

$$X_C = \frac{1}{2\pi fC}$$
$$= \frac{1}{2\pi \times 12 \text{ kHz} \times 0.4\ \mu\text{F}}$$
$$= 33.2\ \Omega.$$

Thus,

$$Z_{\text{total}} = \sqrt{(R_1 + R_2)^2 + (X_L - X_C)^2}$$
$$= \sqrt{(20\ \Omega + 1\ \Omega)^2 + (75.4\ \Omega - 33.2\ \Omega)^2}$$
$$= 47.1\ \Omega$$

and

$$Z_{out} = \sqrt{R_2^2 + (X_L - X_C)^2} = 42.2 \ \Omega$$

$$e_{out} = 50 \ mV \times \frac{42.2 \ \Omega}{47.1 \ \Omega}$$

$$= 44.8 \ mV.$$

LC Bandpass Filter

If the filter's configuration is changed to that shown in Figure 4-16(a), it is called a band-pass filter and has a response as shown at Figure 4-16(b). The term f_{lc} is the low-frequency cutoff where the output voltage has fallen to 0.707 times its maximum value and f_{hc} is the high-frequency cutoff. The frequency range between f_{lc} and f_{hc} is called the filter's bandwidth, usually abbreviated BW. The BW is equal to $f_{hc} - f_{lc}$, and it can be shown mathematically that

$$BW = \frac{R}{2\pi L} \tag{4-7}$$

where BW = bandwidth (Hz)

R = total circuit resistance

L = circuit inductance.

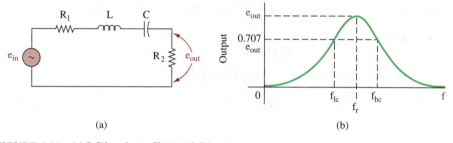

(a)   (b)

FIGURE 4-16   **(a) LC bandpass filter and (b) response.**

The filter's quality factor, Q, provides a measure of how selective (narrow) its passband is compared to its center frequency, f_r. Thus,

$$Q = \frac{f_r}{BW}. \tag{4-8}$$

As stated earlier, the quality factor, Q, can also be determined as

$$Q = \frac{\omega L}{R} \tag{4-9}$$

where ωL = inductive reactance at resonance

R = total circuit resistance.

As Q increases, the filter becomes more selective; that is, a smaller passband (narrower bandwidth) is allowed. A major limiting factor in the highest attainable Q is the resistance factor shown in Equation (4-9). To obtain a high Q, the circuit resistance must be low. Quite often, the limiting factor becomes the winding resistance of the inductor itself. The turns of wire (and associated resistance) used to make an inductor provide this limiting factor. To obtain the highest Q possible, larger wire (with less resistance) could be used, but then greater cost and physical size to obtain the same amount of inductance is required. Quality factors (Q) approaching 1000 are possible with very high-quality inductors.

EXAMPLE 4-3

A filter circuit of the form shown in Figure 4-16(a) has a response as shown in Figure 4-17.

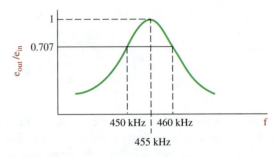

FIGURE 4-17   **Response curve for Example 4-3.**

Determine the

(a) bandwidth.

(b) Q.

(c) value of inductance if $C = 0.001 \ \mu F$.

(d) total circuit resistance.

SOLUTION

(a) From Figure 4-17, the BW is simply the frequency range between f_{hc} and f_{lc} or

$$460 \ \text{kHz} - 450 \ \text{kHz} = 10 \ \text{kHz}.$$

(b) The filter's peak output occurs at 455 kHz.

$$Q = \frac{f_r}{\text{BW}} \tag{4-8}$$

$$= \frac{455 \ \text{kHz}}{10 \ \text{kHz}}$$

$$= 45.5 \ \text{kHz}$$

(c) Equation (4-1) can be used to solve for L because f_r and C are known.

$$f_r = \frac{1}{2\pi\sqrt{LC}} \tag{4-1}$$

$$455 \ \text{kHz} = \frac{1}{2\pi\sqrt{L \times 0.001 \ \mu F}}$$

$$L = 0.12 \ \text{mH}$$

(d) Equation (4-7) can be used to solve for total circuit resistance because the BW and L are known.

$$\text{BW} = \frac{R}{2\pi L} \tag{4-7}$$

$$10 \ \text{kHz} = \frac{R}{2\pi \times 0.12 \ \text{mH}}$$

$$R = 10 \times 10^3 \ \text{Hz} \times 2\pi \times 0.12 \times 10^{-3} \ \text{H}$$

$$= 7.52 \ \Omega$$

The frequency-response characteristics of *LC* circuits are affected by the *L/C*. Different values of *L* and *C* can be used to exhibit resonance at a specific frequency. A high *L/C* ratio yields a more narrowband response, whereas lower *L/C* ratios provide a wider frequency response. This effect can be verified by examining the effect of changing *L* in Equation (4-7).

Parallel *LC* Circuits

A parallel *LC* circuit and its impedance versus frequency characteristic is shown in Figure 4-18.

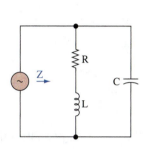
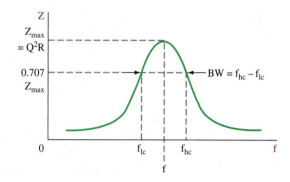

FIGURE 4-18   Parallel *LC* circuit and response.

The only resistance shown for this circuit is the inductor's winding resistance and is effectively in series with the inductor as shown. Notice that the impedance of the parallel *LC* circuit reaches a maximum value at the resonant frequency, f_r, and falls to a low value on either side of resonance. As shown in Figure 4-18, the maximum impedance is

$$Z_{max} = Q^2 \times R \qquad (4\text{-}10)$$

Equations (4-1), (4-5), (4-7), and (4-8) for series *LC* circuits also apply to parallel *LC* circuits when *Q* is greater than 10 ($Q > 10$), the usual condition.

The parallel *LC* circuit is sometimes called a **tank circuit** because energy is transferred from one reactive element to the other in much the same way that water sloshes around in a tank. Energy is stored in each reactive element (*L* and *C*), first in one and then released to the other. The transfer of energy between the two elements will occur at a natural rate equal to the resonant frequency and is sinusoidal in form.

E X A M P L E   4 - 4

A parallel *LC* tank circuit is made up of an inductor of 3 mH and a winding resistance of 2 Ω. The capacitor is 0.47 μF. Determine

(a) f_r.

(b) Q.

(c) Z_{max}.

(d) BW.

SOLUTION

(a)
$$f_r = \frac{1}{2\pi \sqrt{LC}} \qquad (4\text{-}1)$$

$$= \frac{1}{2\pi \sqrt{3 \text{ mH} \times 0.47 \, \mu\text{F}}}$$

$$= 4.24 \text{ kHz}$$

(b)
$$Q = \frac{X_L}{R} \tag{4-9}$$

where $X_L = 2\pi fL$

$$= 2\pi \times 4.24 \text{ kHz} \times 3 \text{ mH}$$
$$= 79.9 \text{ } \Omega$$
$$Q = \frac{79.9 \text{ } \Omega}{2 \text{ } \Omega}$$
$$= 39.9$$

(c)
$$Z_{max} = Q^2 \times R \tag{4-10}$$
$$= (39.9)^2 \times 2 \text{ } \Omega$$
$$= 3.19 \text{ k}\Omega$$

(d)
$$BW = \frac{R}{2\pi L} \tag{4-7}$$
$$= \frac{2 \text{ } \Omega}{2\pi \times 3 \text{ mH}}$$
$$= 106 \text{ Hz}$$

Types of *LC* Filters

An almost endless variety of filters is in use. At frequencies below 100 kHz, *RC* circuit configurations predominate because inductors suitable for use at low frequencies are bulky and expensive. Above 100 kHz, inductors become physically small enough that *LC* combinations become practical. Filters often employ multiple *RC* or *LC* sections (each of which is known as a **pole**) to achieve the desired effect.

The two basic types of *LC* filters are the constant-*k* and the *m*-derived filters. The **constant-*k* filters** have the capacitive and inductive reactances made equal to a constant value, *k*. The **m-derived filters** use a tuned circuit in the filter to provide nearly infinite attenuation at a specific frequency. The rate of attenuation is the steepness of the filter's response curve and is sometimes referred to as the **roll-off**. The severity of roll-off depends on the ratio of the filter's cutoff frequency to the frequency of near infinite attenuation— the *m* of the *m*-derived filter.

The configuration of an *LC* or *RC* filter also defines its response characteristics, that is, its roll-off rate and phase-shift characteristics over a specified range of frequencies. The four major configurations, named in honor of their developers, are as follows:

1. **Butterworth:** Flat amplitude response in passband and a roll-off rate of −20 dB/decade/pole with a phase shift that varies nonlinearly with frequency.

2. **Chebyshev:** Rapid roll-off rate (greater than −20 dB/decade/pole) but with greater passband ripple and more nonlinear phase shift than Butterworth.

3. **Cauer (often referred to as elliptical):** Sharper transition from passband to stopband than other configurations, but with ripple on both.

4. **Bessel (also called Thomson):** Linear phase characteristic (phase shift increases linearly with frequency) that produces no overshoot at output with pulse input. Pulse waveforms can be filtered without distorting pulse shape.

The study and design of *LC* filters is a large body of knowledge, the subject of many textbooks dedicated to filter design. Numerous software packages are also available to aid in their design and analysis.

High-Frequency Effects

At the very high frequencies encountered in communications, the small capacitances and inductances created by wire leads pose a problem. Even the capacitance of the wire windings of an inductor can cause problems. Consider the inductor shown in Figure 4-19. Notice the capacitance shown between the windings. This is termed **stray capacitance**. At low frequencies it has a negligible effect, but at high frequencies capacitance no longer appears as an open circuit and starts affecting circuit performance. The inductor is now functioning like a complex *RLC* circuit.

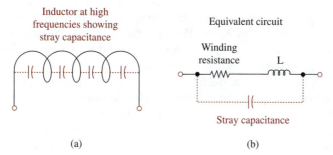

(a)   (b)

FIGURE 4-19   Inductor at high frequencies.

A simple wire exhibits a small amount of inductance; the longer the wire, the greater the inductance. At low frequencies this small inductance (usually a few nanohenries) looks like a short circuit and has no effect. Because inductive reactance increases as a function of frequency, however, the unwanted reactances created by these stray inductances can become significant at radio frequencies. Similarly, the stray capacitances created between wires do not look like open circuits at high frequencies. For these reasons it is important to minimize all lead lengths in RF circuits. The use of surface-mount components that have almost no leads except metallic end pieces to solder to the printed circuit board are very effective in minimizing problems associated with high-frequency effects.

The high-frequency effects just discussed for inductors and capacitors also cause problems with resistors. In fact, as shown in Figure 4-20, at high frequencies the equivalent circuit for a resistor is the same as that for an inductor.

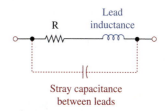

FIGURE 4-20   Resistor at high frequencies.

Crystal Filters

Crystal filters find many applications in communications because they can be designed to exhibit very high values of Q. As one example, the crystal filter is commonly used in single-sideband systems to attenuate the unwanted sideband. Because of its very high Q, the crystal filter passes a much narrower band of frequencies than the best *LC* filter. Crystals with a Q up to about 50,000 are available.

The equivalent circuit of the crystal and crystal holder is illustrated in Figure 4-21(a).

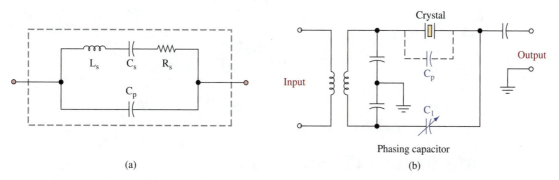

(a)   (b)

FIGURE 4-21   Crystal equivalent circuit (a) and filter (b).

The components L_s, C_s, and R_s represent the series resonant circuit of the crystal itself. C_p represents the parallel capacitance of the crystal holder. The crystal offers a very low-impedance path to the frequency to which it is resonant and a high-impedance path to other frequencies. However, the crystal holder capacitance, C_p, shunts the crystal and offers a path to other frequencies. For the crystal to operate as a bandpass filter, some means must be provided to counteract the shunting effect of the crystal holder. This is accomplished by placing an external variable capacitor in the circuit (C_1 in Figure 4-21(b)).

In Figure 4-21(b), a simple bandpass crystal filter is shown. The variable capacitor C_1, called the **phasing capacitor**, counteracts holder capacitance C_p. C_1 can be adjusted so that its capacitance equals the capacitance of C_p. Then both C_p and C_1 pass undesired frequencies equally well. Because of the circuit arrangement, the voltages across C_p and C_1 resulting from undesired frequencies are equal and 180° out of phase. Therefore, undesirable frequencies are canceled and do not appear in the output. This cancellation effect is called the **rejection notch**.

For circuit operation, assume that a lower sideband with a maximum frequency of 99.9 kHz and an upper sideband with a minimum frequency of 100.1 kHz are applied to the input of the crystal filter in Figure 4-21(b). Assume that the upper sideband is the unwanted sideband. By selecting a crystal that will provide a low-impedance path (series resonance) at about 99.9 kHz, the lower-sideband frequency will appear in the output. The upper sideband, as well as all other frequencies, will have been attenuated by the crystal filter. Improved performance is possible when two or more crystals are combined in a single filter circuit.

Ceramic Filters

Ceramic filters utilize the piezoelectric effect just as crystals do. However, they are normally constructed from lead zirconate-titanate. While ceramic filters do not offer Qs as high as a crystal, they do outperform LC filters in that regard. A Q of up to 2000 is practical with ceramic filters. They are lower in cost, more rugged, and smaller in size than crystal filters. They are used not only as sideband filters but also as replacements for the tuned IF transformers for superheterodyne receivers.

The circuit symbol for a ceramic filter is shown in Figure 4-22(a) and a typical attenuation response curve is shown in Figure 4-22(b). Note that the bandwidths at 60 dB and at 6 dB are shown. The ratio of these two bandwidths (8 kHz/6.8 kHz = 1.18) is defined as the **shape factor**. The shape factor (60-dB BW divided by a 6-dB BW) provides an indication of the filter's selectivity. The ideal value of 1 would indicate a vertical slope at both frequency extremes. The ideal filter would have a horizontal slope within the passband with zero attenuation. The practical case is shown in Figure 4-22(b), where a variation is illustrated. This variation is termed the **peak-to-valley ratio** or **ripple amplitude**. The shape factor and ripple amplitude characteristics also apply to the mechanical filters discussed next.

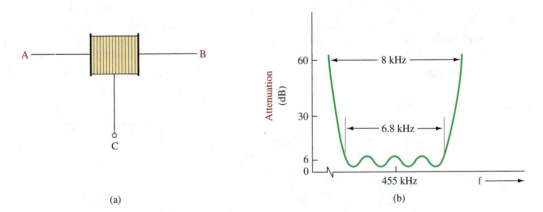

(a)   (b)

FIGURE 4-22   **Ceramic filter and response curve.**

Mechanical Filters

Mechanical filters have been used in single-sideband equipment since the 1950s. Some of the advantages of mechanical filters are their excellent rejection characteristics, extreme ruggedness, size small enough to be compatible with the miniaturization of equipment, and a Q in the order of 10,000, which is about 50 times that obtainable with LC filters.

The mechanical filter is a device that is mechanically resonant; it receives electrical energy, converts it to mechanical vibration, then converts this mechanical energy back into electrical energy as the output. Figure 4-23 shows a cutaway view of a typical unit. There are

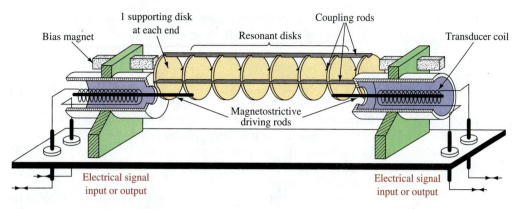

FIGURE 4-23 **Mechanical filter.**

four elements constituting a mechanical filter: (1) an input transducer that converts the electrical energy at the input into mechanical vibrations, (2) metal disks that are manufactured to be mechanically resonant at the desired frequency, (3) rods that couple the metal disks, and (4) an output transducer that converts the mechanical vibrations back into electrical energy.

Not all the disks are shown in the illustration. The shields around the transducer coils have been cut away to show the coil and magnetostrictive driving rods. As you can see by its symmetrical construction, either end of the filter may be used as the input.

Figure 4-24 is the electrical equivalent of the mechanical filter. The disks of the mechanical filter are represented by the series resonant circuits L_1C_1, while C_2 represents the coupling rods.

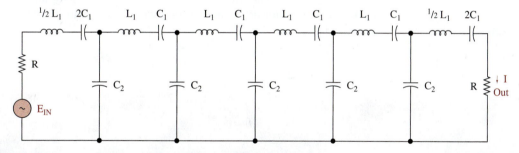

FIGURE 4-24 **Electrical analogy of a mechanical filter.**

The resistance R in both the input and output represents the matching mechanical loads. Phase shift of the input signal is introduced by the L and C components of the mechanical filter. For digital applications, a phase shift can affect the quality of the digital pulse. This can lead to an increase in data errors or bit errors. In analog systems, the voice transmission is not affected as much because the ear is very forgiving of distortion.

Let us assume that the mechanical filter of Figure 4-23 has disks tuned to pass the frequencies of the desired sideband. The input to the filter contains both sidebands, and the transducer driving rod applies both sidebands to the first disk. The vibration of the disk will be greater at a frequency to which it is tuned (resonant frequency), which is the desired sideband, than at the undesired sideband frequency. The mechanical vibration of the first disk is transferred to the second disk, but a smaller percentage of the unwanted sideband frequency is transferred. Each time the vibrations are transferred from one disk to the next, there is a smaller amount of the unwanted sideband. At the end of the filter there is practically none of the undesired sideband left. The desired sideband frequencies are taken off the transducer coil at the output end of the filter.

Varying the size of C_2 in the electrical equivalent circuit in Figure 4-24 varies the bandwidth of the filter. Similarly, by varying the mechanical coupling between the disks (Figure 4-23), that is, by making the coupling rods either larger or smaller, the bandwidth of the mechanical filter is varied. Because the bandwidth varies approximately as the total cross-sectional area of the coupling rods, the bandwidth of the mechanical filter can be increased by using either larger coupling rods or more coupling rods. Mechanical filters with bandwidths as narrow as 500 Hz and as wide as 35 kHz are practical in the range 100 to 500 kHz.

SAW Filters

A modern variant of the mechanical filter just described is the surface-acoustic-wave (SAW) filter. Such filters found widespread use in high-quality, analog color television sets before the conversion to digital television. They are also found in many other applications. One of these is radar systems (to be covered more extensively in Chapter 15). SAW filters have an advantage in radar applications because the filter characteristics can be matched to the pulse reflected by a target.

Recall that crystals rely on the effects in an entire solid piezoelectric material to develop a frequency sensitivity. SAW devices instead rely on the surface effects in a piezoelectric material such as quartz or lithium niobate. It is possible to cause mechanical vibrations (i.e., surface acoustic waves) that travel across the solid's surface at about 3000 m/s.

The process for setting up the surface wave is illustrated in Figure 4-25. A pattern of interdigitated metal electrodes is deposited by the same photolithography process used to produce integrated circuits, and great precision is therefore possible. Because the frequency characteristics of the SAW device are determined by the geometry of the electrodes, an accurate and repeatable response is provided. When an ac signal is applied, a surface wave is set up and travels toward the output electrodes. The surface wave is converted back to an electrical signal by these electrodes. The length of the input and output electrodes determines the strength of a transmitted signal. The spacing between the electrode *fingers* is approximately one wavelength for the center frequency of interest. The number of fingers and their configuration determines the bandwidth, shape of the response curve, and phase relationships.

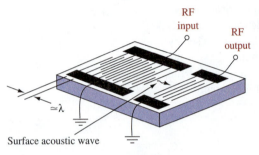

RF input

RF output

$\simeq \lambda$

Surface acoustic wave

FIGURE 4-25 **Surface-acoustic-wave (SAW) filter.**

4-4 MIXING AND MULTIPLICATION CIRCUITS

As discussed in Chapter 2, amplitude modulation (AM) is essentially a form of mixing. The modulator stage of an AM transmitter produces mixing of a desirable sort by multiplying carrier and intelligence frequencies, and mixing occurs when two or more signals are applied to a nonlinear device. Frequency and phase modulation were also seen to involve the multiplication of carrier and intelligence signals, as described in Chapter 3. As will be seen in Chapter 6, the recovery of intelligence in the receiver also involves mixing. In fact, the mixing action in a superheterodyne receiver will be shown to occur in several stages. Mixing operations in the transmitter cause baseband frequencies to be up-converted to the frequency range of the carrier in the form of sidebands, whereas mixing operations in the receiver reverse the process by down-converting the received, modulated signal back to the original frequency range of the intelligence. Many mixing circuits, though referred to by different names, are conceptually similar if not functionally identical, and because mixing is so prevalent in communications applications, we will examine the concept further here.

The mixing products produced in a nonlinear device can be represented as a *power series* of the form

$$v_o = Av_i + Bv_i^2 + Cv_i^3 + \cdots \tag{4-11}$$

where v_o = instantaneous output voltage

v_i = instantaneous input voltage

$A, B, C, \ldots$ = constants.

The above expression predicts that the application of a *single* input frequency into a non-linear device produces even and odd harmonics as well as the fundamental. The higher-order harmonics will have progressively lower amplitudes.

The application of two different input frequencies produces cross products. If the two input frequencies are f_1 and f_2, the mixing process produces $mf_1 \pm nf_2$, where the integer coefficients m and n represent harmonics. Often the most important products are the second-order products, $f_1 + f_2$ and $f_1 - f_2$, which represent the sum and difference frequencies that make up the sidebands of an amplitude-modulated signal. Higher-order products (for example, $2f_1 \pm f_2$ or $f_1 \pm 2f_2$, and so on) are also produced, but at lower amplitudes than the fundamental or second-order products. Cross products of any order, when undesired, produce a form of mixing known as intermodulation distortion.

To see why the cross products are created in a nonlinear device, let us consider the mixing action produced in a *square-law* device such as a field-effect transistor (FET). The transfer characteristic curve of a FET, shown in Figure 4-26, has a parabolic relationship because the equation for the curve has a squared term. It is, therefore, nonlinear. In general, a square-law device produces an output that is the square of the input. Though not perfect, an actual FET comes very close to the behavior of an ideal square-law device. Mixing of two signals in a FET produces an output that is a simplified form of the power series of Equation (4-11), consisting only of the fundamental and a squared term:

$$v_o = Av_i + Bv_i^2. \qquad (4\text{-}12)$$

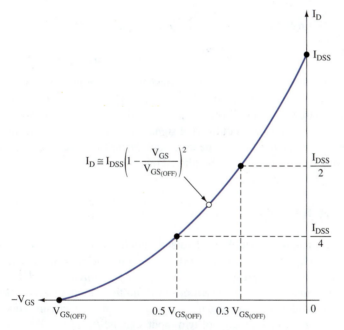

$$I_D \cong I_{DSS}\left(1 - \frac{V_{GS}}{V_{GS(OFF)}}\right)^2$$

FIGURE 4-26   Characteristic curve of N-Channel JFET.

If two sine waves at different frequencies are applied to such a device, the input will be their sum,

$$v_i = \sin \omega_1 t + \sin \omega_2 t,$$

and the output will follow the form of the power series given above, which is equal to the following:

$$
\begin{aligned}
v_o &= Av_i + Bv_i^2 \\
&= A(\sin \omega_1 t + \sin \omega_2 t) + B(\sin \omega_1 t + \sin \omega_2 t)^2 \\
&= A \sin \omega_1 t + A \sin \omega_2 t + B \sin^2 \omega_1 t + B \sin^2 \omega_2 t + 2B \sin \omega_1 t \sin \omega_2 t. \quad (4\text{-}13)
\end{aligned}
$$

The above result predicts the frequency components present at the output. The first two terms are the input signals themselves (multiplied by A, a gain factor). The third and fourth terms have within them the squares of the input signals and, therefore, reside at twice the input frequency. This result can be seen by invoking the following trigonometric identity:

$$\sin^2 A = \tfrac{1}{2} - \tfrac{1}{2}\cos 2A,$$

which causes the third term in Equation (4-13) to appear as follows:

$$\frac{B}{2} - \frac{B}{2}\cos 2\omega_1 t.$$

The output, therefore, has among its constituents a signal at twice the frequency of the input signal $\omega_1 t$. Likewise, the fourth term appears as

$$\frac{B}{2} - \frac{B}{2}\cos 2\omega_2 t,$$

which creates a similar outcome for the input signal $\omega_2 t$.

Close examination of the final term in Equation (4-13) shows that the expression is the product of two sines. Thus we can make use of the trigonometric identity first introduced as Equation 2-7,

$$\sin A \sin B = \tfrac{1}{2}\left[\cos\left(A - B\right) - \cos\left(A + B\right)\right],$$

to expand the final term in Equation (4-13) to have it appear as follows:

$$2B \sin \omega_1 t \sin \omega_2 t = \frac{2B}{2}\left[\cos\left(\omega_1 - \omega_2\right)t - \cos\left(\omega_1 + \omega_2\right)t\right]$$

$$= B\left[\cos\left(\omega_1 - \omega_2\right)t - \cos\left(\omega_1 + \omega_2\right)t\right]. \qquad \text{(4-14)}$$

The output of the square-law device thus consists of the input frequencies themselves, components at the sum and difference of the input frequencies, and the second harmonics. A mixing circuit used as an amplitude modulator will produce the sum and difference frequencies that form the sidebands of the AM signal. In other mixing applications, either the sum or difference frequencies would be used, and the unused side frequency as well as all the other mixing components would be filtered out.

Balanced Modulator

The purpose of a **balanced modulator** is to suppress (cancel) the carrier, leaving only the two sidebands. Such a signal is called a DSBSC (**double-sideband suppressed carrier**) signal. A balanced modulator is a special case of a multiplier circuit, which can be achieved through nonlinear mixing or as the result of linear multiplication, in which case the output signal is truly the product of sine-wave input signals. A multiplier circuit is one in which the output is proportional to the product of two input signals. An ideal balanced mixer is one that would produce only the sum and difference of the input frequencies without harmonics.

To see how a multiplier functions as a balanced mixer, let us examine its output. The output signal has the equation

$$v_o = A v_{i_1} v_{i_2},$$

where $v_o =$ instantaneous output voltage

$v_{i_1}, v_{i_2} =$ instantaneous voltages applied to input of multiplier

$A =$ a constant.

If the inputs are two sine waves of different frequencies, then

$$v_{i_1} = \sin \omega_1 t$$
$$v_{i_2} = \sin \omega_2 t.$$

The output will be their product:

$$v_o = A \sin \omega_1 t \sin \omega_2 t.$$

The result is the product of two sines, just as we have seen before. Using the trigonometric identity shown as Equation 2-7, we create an output of

$$v_o = \frac{A}{2}\left[\cos(\omega_1 - \omega_2)t - \cos(\omega_1 + \omega_2)t\right]. \qquad (4\text{-}15)$$

The only difference from the result predicted earlier is that the sum and difference frequencies (i.e., the sidebands) are produced, but the harmonics are not.

A very common balanced modulator configuration, known as a **ring modulator** or a **lattice modulator**, is shown in Figure 4-27. Consider the carrier with the instantaneous conventional current flow as indicated by the arrows. The current flow through both halves of L_5 is equal but opposite, and thus the carrier is canceled in the output. This is also true on the carrier's other half-cycle, only now diodes B and C conduct instead of A and D.

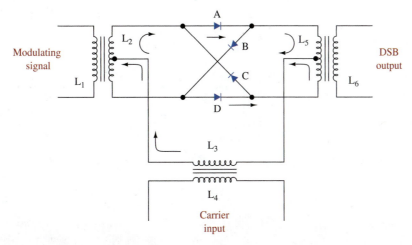

FIGURE 4-27   Balanced ring modulator.

Considering just the modulating signal, current flow occurs from winding L_2 through diodes C and D or A and B but not through L_5. Thus, there is no output of the modulating signal either. Now with both signals applied, but with the carrier amplitude much greater than the modulating signal, the conduction is determined by the polarity of the carrier. The modulating signal either aids or opposes this conduction. When the modulating signal is applied, current will flow from L_2 and diode D will conduct more than A, and the current balance in winding L_5 is upset. This causes outputs of the desired sidebands but continued suppression of the carrier. This modulator is capable of 60 dB carrier suppression when carefully matched diodes are utilized. It relies on the nonlinearity of the diodes to generate the sum and difference sideband signals.

LIC Balanced Modulator

A balanced modulator of the type previously explained requires extremely well matched components to provide good suppression of the carrier (40 or 50 dB suppression is usually adequate). This suggests the use of linear integrated circuits (LICs) because of the superior component-matching characteristics obtainable when devices are fabricated on the same silicon chip because this approach does not require the use of transformers or tuned circuits. The balanced modulator function is achieved with matched transistors in the differential amplifiers or *diff-amps*, with the modulating signal controlling the emitter current of the amplifiers. The carrier signal is applied to switch the diff-amps' bases, resulting in a mixing process with the mixing product signals out of phase at the collectors. This is an extremely versatile device since it can be used not only as a balanced modulator but also as an amplitude modulator, synchronous detector, FM detector, or frequency doubler.

Product Detector

In Chapter 6 we will demonstrate that recovering the intelligence in any signal in which the carrier has been suppressed at transmission (whether SSB or DSBSC) requires that the carrier be recreated and reinserted at the receiver. The balanced modulators used to create DSB can also be used to recover the intelligence in an SSB signal. When a balanced modulator is used in this fashion, it is usually called a **product detector**. This is the most common method of detecting an SSB signal.

Figure 4-28 shows another IC balanced modulator being used as a product detector. It is the Plessey Semiconductor SL640C. The capacitor connected to output pin 5 forms the low-pass filter to allow just the audio (low)-frequency component to appear in the output. The simplicity of this demodulator makes its desirability clear.

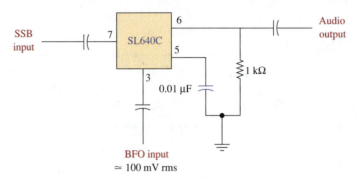

FIGURE 4-28   **SL640C SSB detector.**

4-5   THE PHASE-LOCKED LOOP AND FREQUENCY SYNTHESIS

The phase-locked loop (PLL) is an electronic feedback control system with many uses in communications, including frequency synthesis and FM demodulation. Originated in 1932, the PLL was an idea whose full realization was made possible by integrated-circuit technology. Prior to its availability in 1970 as a single IC, its complexity as a system composed of discrete circuits made the PLL economically unfeasible for most applications.

The PLL consists of three functional blocks and a feedback loop, as shown in the block diagram of Figure 4-29. The blocks are the phase detector (or phase comparator), loop filter, and voltage-controlled oscillator (VCO). The VCO produces an output frequency that varies in response to changes in a control voltage applied to components in the frequency-determining stage of the oscillator (usually a varactor diode, which behaves as a voltage-variable capacitor). The VCO output is applied through the feedback loop as one

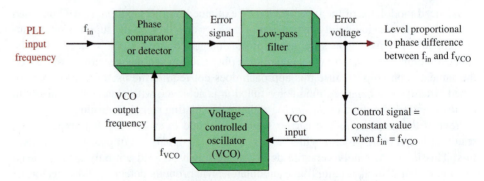

FIGURE 4-29   **Block diagram of a phase-locked loop.**

of two inputs to a phase detector. The second phase detector input is the reference frequency from an external source such as a stable crystal oscillator. The phase detector compares the two input frequencies (or phases, for phase changes create instantaneous frequency changes) and produces an output voltage, known as the *error voltage*, that varies in response to the difference in frequency or phase between the inputs. This error voltage, when filtered, is the VCO control voltage. If the VCO frequency or phase tries to drift either above or below the reference, the inputs to the phase detector become unequal, causing it to produce an error voltage of sufficient magnitude and polarity to counteract the effects of the drift and return the VCO output to the reference frequency.

Varactor Diodes

The VCO must be able to change frequency in response to a DC control voltage applied to it. A **varactor** diode achieves this by acting as a variable capacitor in an oscillator tank circuit. The varactor relies on the capacitance offered by a reverse-biased diode. Since this capacitance varies with the amount of reverse bias, a potentiometer can be used to provide the variable capacitance required for tuning. Diodes that have been specifically fabricated to enhance this variable capacitance versus reverse bias characteristic are referred to as **varactor diodes, varicap diodes,** or **VVC diodes**. Figure 4-30 shows the two generally used symbols for these diodes and a typical capacitance versus reverse bias characteristic.

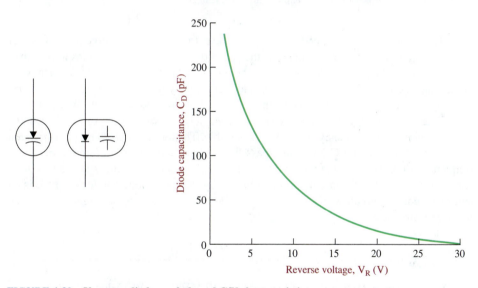

FIGURE 4-30   Varactor diode symbols and C/V characteristic.

The amount of capacitance exhibited by a reverse-biased silicon diode, C_d, can be approximated as

$$C_d = \frac{C_0}{(1 + 2|V_R|)^{\frac{1}{2}}} \qquad (4\text{-}16)$$

where C_0 = diode capacitance at zero bias

V_R = diode reverse bias voltage.

The use of Equation (4-16) allows the designer to determine accurately the amount of reverse bias needed to provide the necessary tuning range. The varactor diode can also be used to generate FM, as explained in Chapter 5.

PLL Capture and Lock

The PLL feedback loop ensures that the VCO frequency tracks the reference because the VCO output frequency changes in response to changes in the reference. The PLL is said to be *in lock* whenever the VCO frequency can be matched to the reference. The *lock range* defines the highest and lowest frequencies over which the PLL can track the reference signal once lock has occurred. If the reference frequency travels outside the lock range, or if the feedback loop is broken, the PLL will lose lock and the VCO will operate at a set frequency, called the *free-running* frequency, determined by an RC circuit in the VCO. The free-running frequency is usually at the center of the lock range. The *capture range,* which cannot exceed and which is generally less than the lock range, contains the frequencies over which the PLL circuit can initially acquire lock. Once locked, the PLL can remain locked over a wider frequency range than frequencies defined by the capture range. Figure 4-31 shows the relationship between free-running frequency, lock range, and capture range.

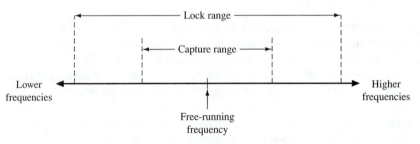

FIGURE 4-31   **Comparison of PLL capture and lock ranges.**

The error voltage produced by the phase detector tends to converge around, rather than go directly to, the value that causes the phase-detector input frequencies to be equal. In other words, the phase detector causes the VCO to "hunt" for the right frequency by producing an error voltage slightly above and then slightly below the target and repeating the process in ever smaller increments until the target is reached. In effect, these converging voltages represent high-frequency noise that must be removed if the PLL is to achieve lock at the intended frequency as quickly as possible. The loop, or *anti-hunt,* filter between the phase detector and VCO is a low-pass filter whose purpose is to remove the high-frequency components of the error voltage. The capture and lock ranges as well as other characteristics of the PLL are determined in part by the loop filter. The locked PLL is an automatic control circuit with an output signal identical to the reference signal except that it is free from noise.

EXAMPLE 4-5

A PLL is set up so that its VCO free-runs at 10 MHz. The VCO does not change frequency until the input is within 50 kHz of 10 MHz. After that condition, the VCO follows the input to ±200 kHz of 10 MHz before the VCO starts to free-run again. Determine the lock and capture ranges of the PLL.

SOLUTION

The capture occurred at 50 kHz from the free-running VCO frequency. Assume symmetrical operation, which implies a capture range of 50 kHz × 2 = 100 kHz. Once captured, the VCO follows the input to a 200-kHz deviation, implying a lock range of 200 kHz × 2 = 400 kHz.

Frequency Synthesis

The PLL is an essential subsystem within the *frequency synthesizer.* The PLL frequency synthesizer allows a range of frequencies to be generated from a stable, single-frequency reference (usually a crystal-controlled oscillator). Synthesizers are found in any modern device that must operate over a range of frequencies, including cellular phones, radio transmitters and receivers of all types, cordless telephones, wireless networks, communications equipment, and test equipment. Only one crystal in the reference oscillator is needed to produce many frequencies, which can either be "dialed in" manually with a keypad or with thumbwheel switches or controlled automatically with software or remote-control devices. An ordinary television set is an example of a device whose design has been revolutionized with low-cost frequency synthesizer ICs—no longer is a rotary dial and collection of resonant tank circuits necessary for channel selection. Instead, local oscillator frequencies (hence, received channels) can be changed inexpensively and automatically by remote controlled frequency synthesis. The PLL synthesizer saves labor and money because only one crystal in one oscillator circuit is required to generate many frequencies. The concept of frequency synthesis has been around since the 1930s, but the cost of the circuitry necessary was prohibitive for most designs until integrated circuit technology enabled semiconductor suppliers to offer the PLL in the form of a single, low-cost chip.

A basic frequency synthesizer is shown in Figure 4-32. Besides the PLL, the synthesizer includes a very stable crystal oscillator and the divide-by-N programmable divider. The output frequency of the VCO is a function of the applied control voltage. In basic terms, the PLL frequency synthesizer multiplies the frequency applied to the phase detector reference input by a given number, N. With reference to Figure 4-32, note that the divide-by-N counter divides the VCO output frequency, which will be an integer multiple of the reference frequency, by the appropriate integer so that the frequency fed back to the phase detector equals the reference. The loop adjusts to any changes in the reference frequency and produces a direct change in the VCO output frequency. Thus, reference frequency stability determines overall synthesizer frequency stability. In critical applications such as test equipment, the crystal oscillator may be housed in a temperature-stabilized oven, or highly accurate atomic standards may be employed as reference oscillators for the highest possible accuracy and stability.

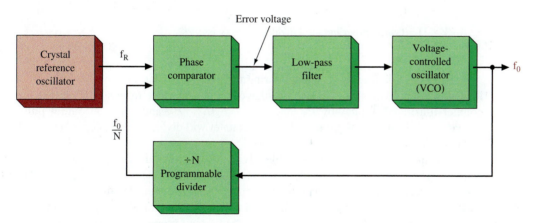

FIGURE 4-32   Basic frequency synthesizer.

The VCO frequency output (hence, synthesizer output) depends on the value of N in the divide-by-N counter. For example, if the reference frequency is 5 kHz, and the divide-by-N counter is set to divide by 2, the VCO output will be 10 kHz because 10 kHz ÷ 2 = 5 kHz. The 5-kHz frequency must be fed back to the phase detector to maintain a locked loop. Likewise, if the divide-by-N counter is set to divide by 4, the VCO output will be 20 kHz because 20 kHz ÷ 4 = 5 kHz. The synthesizer frequency range is ultimately determined by the range of frequencies over which the VCO can operate. In general, the VCO frequency is maintained at Nf_R, where f_R is the reference frequency of the master oscillator.

The design of frequency synthesizers using the principle just described, known as *indirect synthesis*, involves the design of various subsystems including the VCO, the phase comparator, any low-pass filters in the feedback path, and the programmable dividers. Its stability is directly governed by the stability of the reference input f_R, although it is also related to noise in the phase comparator, noise in any dc amplifier between the phase comparator and the VCO, and the characteristics of the low-pass filter usually placed between the phase comparator and the VCO.

The indirect synthesizer shown in Figure 4-32 is the most basic form of phase-locked synthesizer. The principal drawback of this configuration is that the frequencies produced must always be whole-number multiples of f_R. Consider the case where the programmable divider in Figure 4-31 can divide by any integer, N, from 1 to 10. If the reference frequency is 100 kHz and $N = 1$, then the output should be 100 kHz. If $N = 2$, f_0 must equal 200 kHz to provide a constant phase difference for the phase comparator. Similarly, for $N = 5$, $f_0 = 500$ kHz. The pattern, and the problem, should be apparent now. A synthesizer with outputs of 100 kHz, 200 kHz, 300 kHz, and so on, is not useful for most applications. Much smaller spacing between output frequencies is necessary, and means to attain that condition are discussed next.

Programmable Division

A typical *programmable divider* is shown in Figure 4-33. It consists of three stages with division ratios K_1, K_2, and K_3, which may be programmed by inputs P_1, P_2, and P_3, respectively.

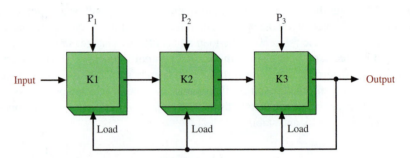

FIGURE 4-33   Typical programmable divider.

Each stage divides by K_n except during the first cycle after the program input P_n is loaded, when it divides by P (which may have any integer value from n to K). Hence the counter illustrated divides by $P_3 \times (K_1 K_2) + P_2 K_1 + P_1$, and when an output pulse occurs, the program inputs are reloaded. The counter will divide by any integer between 1 and $(K_1 K_2 K_3 - 1)$.

The most common programmable dividers are either decades or divide-by-16 counters. These are readily available in various logic families, including CMOS and TTL. CMOS devices are preferred when power consumption is a consideration. Using such a package, one can program a value of N from about 3 to 9999. The theoretical minimum count of 1 is not possible because of the effects of circuit propagation delays. The use of such counters permits the design of frequency synthesizers that can be programmed with decimal thumbwheel switches or numeric keypads with a minimum number of components. If a synthesizer is required with an output of nonconsecutive frequencies and steps, a custom programmable counter may be made using some custom devices such as programmable logic devices (PLDs) or programmable divider ICs.

The maximum input frequency of a programmable divider is limited by the speed of the logic used, and more particularly by the time taken to load the programmed count. The power consumption of high-frequency digital circuitry can be an issue in low-power applications (e.g., cellular phones). The output frequency of the simple synthesizer in Figure 4-32 is limited, of course, to the maximum frequency of the programmable divider.

There are many ways of overcoming this limitation on synthesizer frequency. The VCO output may be mixed with the output of a crystal oscillator and the resulting difference frequency fed to the programmable divider, or the VCO output may be multiplied from a low value in the operating range of the programmable divider to the required high output frequency. Alternatively, a fixed ratio divider capable of operating at a high frequency may be interposed between the VCO and the programmable divider. These methods are shown in Figures 4-34(a), (b), and (c), respectively.

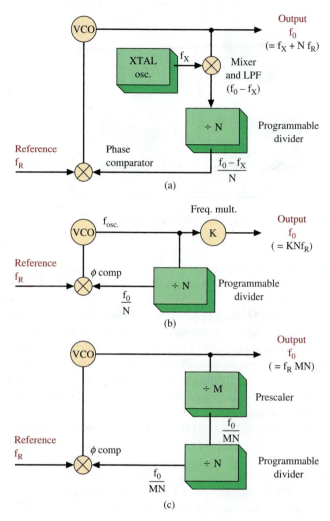

FIGURE 4-34 **Synthesizer alternatives.**

All the methods discussed above have their problems, although all have been used and will doubtless continue to be used in some applications. Method (a) is the most useful technique because it allows narrower channel spacing or high reference frequencies (hence faster lock times and less loop-generated jitter) than the other two, but it has the drawback that, because the crystal oscillator and the mixer are within the loop, any crystal oscillator noise or mixer noise appears in the synthesizer output. Nevertheless, this technique has much to recommend it.

The other two techniques are less useful. Frequency multiplication introduces noise, and both techniques must either use a very low reference or rather wide channel spacing. What is needed is a programmable divider that operates at the VCO frequency—one can then discard the techniques described above and synthesize directly at whatever frequency is required.

Two-Modulus Dividers

Considerations of speed and power make it impractical to design programmable counters of the type described above, even using relatively fast emitter-coupled logic, at frequencies much into the VHF band (30 to 300 MHz) or above. A different technique exists, however, using two-modulus dividers; that is, in one mode the synthesizer divides by N and in the other mode, by $N + 1$.

Figure 4-35 shows a divider using a two-modulus prescaler. The system is similar to the one shown in Figure 4-34(c), but in this case the prescaler divides by either N or $N + 1$, depending on the logic state of the control input. The output of the prescaler feeds two

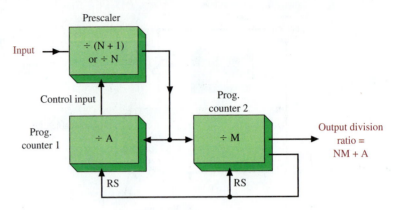

FIGURE 4-35 **Divider system with two-modulus prescaler.**

normal programmable counters. Counter 1 controls the two-modulus prescaler and has a division ratio A. Counter 2, which drives the output, has a division ratio M. In operation the $N/(N + 1)$ prescaler (Figure 4-35) divides by $N + 1$ until the count in programmable counter 1 reaches A and then divides by N until the count in programmable counter 2 reaches M, when both counters are reloaded, a pulse passes to output, and the cycle restarts. The division ratio of the entire system is $A(N + 1) + N(M - A)$, which equals $NM + A$. There is only one constraint on the system—because the two-modulus prescaler does not change modulus until counter 1 reaches A, the count in counter 2 (M) must never be less than A. This limits the minimum count the system may reach to $A(N + 1)$, where A is the maximum possible value of count in counter 1.

The use of this system entirely overcomes the problems of high-speed programmable division mentioned earlier. A number of $\div 10/11$ counters working at frequencies of up to 500 MHz and also $\div 5/6$, $\div 6/7$, and $\div 8/9$ counters working up to 500 MHz are now readily available. There is also a pair of circuits intended to allow $\div 10/11$ counters to be used in $\div 40/41$ and $\div 80/81$ counters in 25-kHz and 12.5-kHz channel VHF synthesizers. It is not necessary for two-modulus prescalers to divide by $N/(N + 1)$. The same principles apply to $\div N/(N + Q)$ counters (where Q is any integer), but $\div N/(N + 1)$ tends to be most useful.

Direct Digital Synthesis

Direct digital synthesis (DDS) systems became economically feasible in the late 1980s. They offer some advantages over the analog synthesizers discussed in the previous section but, until recently, generally tended to be somewhat more complex and expensive. The digital logic used can improve on the repeatability and drift problems of analog units that often require select-by-test components. These advantages also apply to digital filters that have replaced some standard analog filters in recent years. The disadvantages of DDS (and digital filters) are the relatively limited maximum output frequency and greater complexity/cost considerations.

A block diagram for a basic DDS system is provided in Figure 4-36. The numerically controlled oscillator (NCO) contains the phase accumulator and read-only memory (ROM) look-up table. The NCO provides the updated information to the digital-to-analog converter (DAC) to generate the RF output.

The phase accumulator generates a phase increment of the output waveform based on its input (Δ phase in Figure 4-36). The input (Δ phase) is a digital word that, in conjunction with the reference oscillator (f_{CLK}), determines the frequency of the output waveform. The output of the phase accumulator serves as a variable-frequency oscillator generating a digital ramp. The frequency of the signal is defined by the Δ phase as

$$f_{out} = \frac{(\Delta \text{ phase}) f_{CLK}}{2^N}$$

(4-17)

for an N-bit phase accumulator.

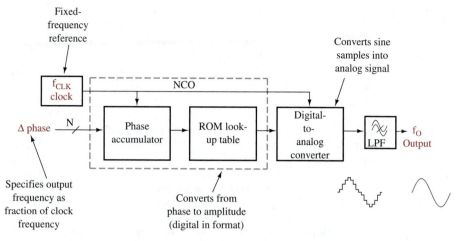

FIGURE 4-36 **Block diagram of a direct digital synthesis system.**

Translating phase information from the phase accumulator into amplitude data is accomplished by means of the look-up table stored in memory. Its digital output (amplitude data) is converted into an analog signal by the DAC. The low-pass filter provides a spectrally pure sine-wave output.

The final output frequency is typically limited to about 40% of f_{CLK}. The phase accumulator size is chosen based on the desired frequency resolution, which is equal to $f_{CLK} \div 2^N$ for an N-bit accumulator.

EXAMPLE 4-6

Calculate the maximum output frequency and frequency resolution for a DDS when operated at $f_{CLK\ MAX}$.

SOLUTION

The maximum output frequency is approximately 40% of $f_{CLK\ MAX}$:

$$= 0.40 \times 100\ \text{MHz}$$
$$= 40\ \text{MHz}$$

The frequency resolution is given by $f_{CLK} \div 2^N$:

$$= \frac{100\ \text{MHz}}{2^{32}} \cong 0.023\ \text{Hz}$$

The preceding example shows that DDS offers the possibility for extremely small frequency increments. This is one of the advantages offered by DDS over the analog synthesizers described earlier. Another DDS advantage is the ability to shift frequencies quickly. This characteristic is useful for *spread-spectrum* systems, which will be described in more detail in Chapter 8.

The disadvantages of DDS include the limit on maximum output frequency and higher **phase noise**. Spurious changes in phase of the synthesizer's output result in energy at frequencies other than the desired one. This phase noise is often specified for all types of oscillators and synthesizers. It is usually specified in $\text{dB}/\sqrt{\text{Hz}}$ at a particular offset from center frequency. A specification of $-90\ \text{dB}/\sqrt{\text{Hz}}$ at a 10-kHz offset means that noise energy at a 1-Hz bandwidth 10 kHz away from the center frequency should be 90 dB lower than the center frequency output. In a sensitive receiver, phase noise will mask out a weak signal that would otherwise be detected.

SUMMARY

Obsevation of the block diagram of a basic communications system reveals that communications circuits in the analog context are primarily built up from only a few basic building blocks: amplifiers, oscillators, and filters. One other receiver circuit, the detector, responsible for recovering the intelligence information from the modulated carrier, will be studied in Chapter 6.

Amplifier operation can be classified in part in terms of the conduction angle of the input signal, that is, the number of degrees over which the amplifier active device is conducting current. The most linear operation, class A, is also the least efficient, so there is an inherent tradeoff between linearity and efficiency. It is possible to improve efficiency while still maintaining linearity. This is the purview of *push-pull* amplification, in which more than one active device is used. Still greater strides in efficiency can be attained through class C amplification, where the active device conducts for only a small portion of the applied input. A sine-wave output is created through the flywheel effect created by a resonant tank circuit at the output of the active device; class C amplifiers thus have a bandwidth characteristic and are often referred to as *tuned* amplifiers.

Many oscillators used in communications applications are amplifiers with a regenerative feedback path to sustain oscillations at a desired frequency. Amplifiers and oscillators are closely related: The gain produced by the amplifier stage is exactly offset by the attenuation produced by the voltage-divider action in the frequency-determining tank circuit. There are a wide variety of LC oscillators, each defined primarily by the means by which the voltage divider is implemented. The highest frequency stability can be achieved by replacing the LC tank circuit with a quartz crystal, which functions as a series RLC circuit. At its resonant frequency, the crystal oscillator is capable of frequency-stable operation over a prolonged period.

All communications systems make extensive use of frequency-selective circuits in the form of filters and resonant circuits. Filters rely on the reactive properties of inductors and capacitors either to pass or reject a band of frequencies. Filters may be of the low-pass or high-pass type, or they may be bandpass or band-reject (notch) filters whose bandwidth is defined by the Q of the resonant circuit.

One important building block of any communications system is the mixer circuit, which can be an amplifier that is deliberately made to be nonlinear in operation. In fact, mixing can be seen to take place in any nonlinear device. Whether called a modulator (in the transmitter), a balanced modulator, a product detector, or a mixer, the circuit works on the principle of nonlinear multiplication of two signals. This operation produces cross products that, for example, produce the sidebands of an AM signal. In addition, harmonics are created, and these harmonics may produce additional, higher-order products that may be undesired. It is also possible to produce sum and difference frequencies as the result of linear multiplication, and a multiplier can function as a balanced mixer, where only the sum and difference frequencies are produced without the carrier.

Finally, any modern, "frequency-agile" communication system owes its existence to the phase-locked loop and the system derived from it, the frequency synthesizer. The phase-locked loop, when locked, is an automatic control circuit that produces an output signal identical to its reference (input) signal except that it is free from noise. When used as a frequency synthesizer, the phase-locked loop is capable of producing a wide range of frequencies from a single-frequency reference. Only one crystal in a reference oscillator is needed to produce many frequencies, which can be set manually or controlled automatically with software or remote-control devices. The newest incarnation of the frequency synthesizer concept is direct digital synthesis, which marries digital logic and signal-processing techniques to the production of stable signals well into the radio-frequency range.

QUESTIONS AND PROBLEMS

SECTION 4-2

1. Draw schematics for Hartley and Colpitts oscillators. Briefly explain their operation and differences.

2. Describe the reason that a Clapp oscillator has better frequency stability than the Hartley or Colpitts oscillators.

3. List the major advantages of crystal oscillators over the *LC* varieties. Draw a schematic for a Pierce oscillator.

4. The crystal oscillator time base for a digital wristwatch yields an accuracy of ± 15 s/month. Express this accuracy in parts per million (ppm). (± 5.787 ppm)

SECTION 4-3

5. Explain the makeup of a practical inductor and capacitor. Include the quality and dissipation in your discussion.

6. Define *resonance* and describe its use.

7. Calculate an inductor's Q at 100 MHz. It has an inductance of 6 mH and a series resistance of 1.2 k. Determine its dissipation. ($3.14 \times 10, 0.318 \times 10^{-3}$)

8. Calculate a capacitor's Q at 100 MHz given 0.001 μF and a leakage resistance of 0.7 MΩ. Calculate D for the same capacitor. ($4.39 \times 10^5, 2.27 \times 10^{-6}$)

9. The inductor and capacitor for Problems 7 and 8 are put in series. Calculate the impedance at 100 MHz. Calculate the frequency of resonance (f_r) and the impedance at that frequency. (3.77 MΩ, 65 kHz, 1200 Ω)

10. Calculate the output voltage for the circuit shown in Figure 4-16 at 6 kHz and 4 kHz. Graph these results together with those of Example 4-2 versus frequency. Use the circuit values given in Example 4-2.

11. Sketch the e_{out}/e_{in} versus frequency characteristic for an *LC* bandpass filter. Show f_{lc} and f_{hc} on the sketch and explain how they are defined. On this sketch, show the bandwidth (BW) of the filter and explain how it is defined.

12. Define the quality factor (Q) of an *LC* bandpass filter. Explain how it relates to the "selectivity" of the filter. Describe the major limiting value on the Q of a filter.

13. An FM radio receiver uses an *LC* bandpass filter with $f_r = 10.7$ MHz and requires a BW of 200 kHz. Calculate the Q for this filter. (53.5)

14. The circuit described in Problem 13 is shown in Figure 4-18. If $C = 0.1$ nF (0.1×10^{-9} F), calculate the required inductor value and the value of R. (2.21 μH, 2.78 Ω)

15. A parallel *LC* tank circuit has a Q of 60 and coil winding resistance of 5 Ω. Determine the circuit's impedance at resonance. (18 kΩ)

16. A parallel *LC* tank circuit has $L = 27$ mH, $C = 0.68$ μF, and a coil winding resistance of 4 Ω. Calculate f_r, Q, Z_{max}, the BW, f_{lc}, and f_{hc}. (1175 Hz, 49.8, 9.93 kΩ, 23.6 Hz, 1163 Hz, 1187 Hz)

17. Explain the significance of the k and m in constant-k and m-derived filters.

18. Describe the criteria used in choosing either an *RC* or *LC* filter.

19. Explain the importance of keeping lead lengths to a minimum in RF circuits.

20. Describe a pole.

21. Explain why Butterworth and Chebyshev filters are called constant-k filters.

*22. Draw the approximate equivalent circuit of a quartz crystal.

23. What are the undesired effects of the crystal holder capacitance in a crystal filter, and how are they overcome?

*24. What crystalline substance is widely used in crystal oscillators (and filters)?

25. Using your library or some other source, provide a schematic for a four-element crystal lattice filter and explain its operation.

*26. What are the principal advantages of crystal control over tuned circuit oscillators (or filters)?

27. Explain the operation of a ceramic filter. What is the significance of a filter's shape factor?

28. Define *shape factor*. Explain its use.

29. A bandpass filter has a 3-dB ripple amplitude. Explain this specification.

30. Explain the operation and use of mechanical filters.

31. Why are SAW filters not often used in SSB equipment?

SECTION 4-4

32. What are the typical inputs and outputs for a balanced modulator?

33. Briefly describe the operation of a balanced ring modulator.

34. Explain the advantages of using an IC for the four diodes in a balanced ring modulator as compared with four discrete diodes.

---

*An asterisk preceding a number indicates a question that has been provided by the FCC as a study aid for licensing examinations.

35. Referring to the specifications for the AD630 LIC balanced modulator in Figure 4-28, determine the channel separation at 10 kHz, explain how a gain of +1 and +2 are provided.

36. Explain how to generate an SSBSC signal from the balanced modulator.

SECTION 4-5

37. Draw a block diagram of a phase-locked loop (PLL) and briefly explain its operation.

38. List the three possible states of operation for a PLL and explain each one.

39. A PLL's VCO free-runs at 7 MHz. The VCO does not change frequency until the input is within 20 kHz of 7 MHz. After that condition, the VCO follows the input to ± 150 kHz of 7 MHz before the VCO starts to free-run again. Determine the PLL's lock and capture ranges. (300 kHz, 40 kHz)

40. Explain the operation of a basic frequency synthesizer as illustrated in Figure 4-32. Calculate f_0 if $f_R = 1$ MHz and $N = 61$. (61 MHz)

41. Discuss the relative merits of the synthesizers shown in Figures 4-34(a), (b), and (c) as compared to the one in Figure 4-32.

42. Describe the operation of the synthesizer divider in Figure 4-35. What basic problem does it overcome with respect to the varieties shown in Figures 4-32 and 4-34?

43. Calculate the output frequency of a synthesizer using the divider technique shown in Figure 4-35 when the reference frequency is 1 MHz, $A = 26$, $M = 28$, and $N = 4$. (138 MHz)

44. Briefly explain DDS operation based on the block diagram shown in Figure 4-36.

45. A DDS system has $f_{\text{CLK MAX}} = 60$ MHz and a 28-bit phase accumulator. Calculate its approximate maximum output frequency and frequency resolution when operated at $f_{\text{CLK MAX}}$. (24 MHz, 0.223 Hz)

CHAPTER 5

TRANSMITTERS

KEY TERMS

modulated amplifier
base modulation
neutralizing capacitor
parasitic oscillations
high-level modulation
low-level modulation
low excitation
downward modulation
spectrum analyzer
spurious frequencies
spurs
relative harmonic distortion
total harmonic distortion (THD)

dummy antenna
conversion frequency
varactor diode
Crosby systems
frequency multipliers
exciter
discriminator
Armstrong
pump chain
frequency multiplexing
multiplex operation
frequency-division multiplexing
matrix network

The communications transmitter is responsible for generating a carrier and then for modulating, filtering, and amplifying the modulated signal for delivery to an antenna. At the block-diagram level, transmitters for the various forms of modulation are more alike than not. For example, except for the simplest of low-power applications, the carrier-generating oscillator must be extremely accurate and stable in frequency to prevent destructive interference to systems on adjacent frequencies. The modulating signal may be analog or digital, but regardless of the modulation type or form of intelligence used, the mixing that occurs in the transmitter modulator produces an output composed of frequencies in addition to those applied. This result implies the need for filtering so that the modulated signal has the appropriate spectral characteristics for use within a bandwidth-limited medium. Also, power outputs may span levels from milliwatts for close-range applications like wireless networks to hundreds of kilowatts for applications like radar or international short-wave broadcasting. Therefore, transmitters will often have many amplifiers to boost the low-level oscillator output in stages up to the final output power level. Finally, the transmitter must deliver its power to an antenna, which in turn requires impedance-matching networks to ensure maximum power transfer. This chapter surveys the various transmitter configurations along with a number of modulator and other circuit implementations.

5-1   AM TRANSMITTER SYSTEMS

Figure 5-1 is a block diagram of a typical double-sideband, full-carrier (DSBFC) AM transmitter, such as that for a broadcast radio station. The figure shows that the modulating signal (the intelligence) can be applied in either of two locations, depending on whether the transmitter employs high-level or low-level modulation. Where the modulation is applied has implications for amplifier efficiency (hence, class of operation) as well as linearity requirements and will be studied in more detail shortly. Before turning our attention to modulators and modulation levels, let us look at the transmitter as a complete system.

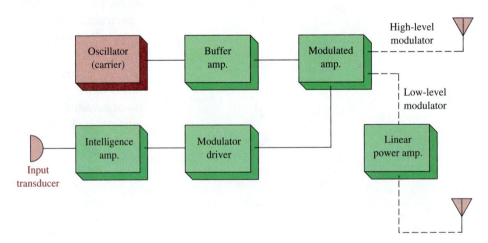

FIGURE 5-1   Simple AM transmitter block diagram.

The oscillator that generates the carrier signal will invariably be crystal-controlled to maintain the high accuracy required by the Federal Communications Commission (FCC). For example, AM broadcast stations must maintain a carrier accuracy within ± 20 Hz of the assigned frequency. Following the oscillator is the buffer amplifier, which provides a high-impedance load for the oscillator to minimize drift or undesired frequency shift as a result of circuit-loading effects. It also provides enough gain to drive the modulated amplifier sufficiently. Thus, the buffer amplifier could be a single class-A stage, or the buffering function could be combined with the number of gain stages needed to drive the following stage, the modulated amplifier.

The intelligence amplifier receives its signal from the input transducer (often a micro-phone), which then boosts the intelligence amplitude, and perhaps its power, through one or more stages of amplification. The number of stages depends on how much power needs to be developed, and this is, in turn, a function of whether *high-level* or *low-level* modulation is used. These terms describe the stage where the intelligence is mixed with the carrier. The choice involves a number of tradeoffs to be discussed shortly. However, for the present discussion it is sufficient to note that if the intelligence is to be boosted to a relatively high power level, as would be required for high-level modulation, then the amplification will take place over more than one stage. The term *driver* is sometimes applied to describe amplifiers intended to produce intermediate levels of power amplification, as distinct from the *final amplifier* stages, which produce the (often very high) power outputs destined for the antenna. The driver is generally a push-pull (class B or AB) amplifier intended to pro-duce watts or tens of watts of power, whereas the final amplifiers may be responsible for producing power outputs in the hundreds if not thousands of watts. The last stage of intel-ligence amplification occurs in the *modulator*, which, as we shall see shortly, is essentially a nonlinear amplifier stage. The nonlinearity is what creates the modulated AM signal con-sisting of carrier and sidebands. The modulator is sometimes termed the **modulated amplifier** and is the output stage for transmitters employing high-level modulation.

Modulator Circuits

As established in Chapter 2, amplitude modulation (AM) is generated as the result of mix-ing or combining carrier and intelligence frequencies within a nonlinear device or circuit. A transmitter mixer circuit designed to produce modulation directly is called a modulator. The Chapter 2 discussion of modulation also pointed out that diodes have nonlinear areas, and could act as modulators, but they are not often used for this purpose in practical trans-mitters because, being passive devices, they offer no gain. Transistors also offer nonlinear operation if properly biased and provide amplification, thus making them ideal for this application. Figure 5-2(a) shows an input/output relationship for a typical bipolar-junction transistor. Notice that nonlinear areas exist at both low and high values of current. Between these two extremes is the linear area that should be used for normal class A linear amplifi-cation. One of the nonlinear areas must be used to generate AM.

Figure 5-2(b) shows a very simple transistor modulator. It has no fixed bias circuit, in contrast with a conventional amplifier, thus the modulator depends on the positive peaks of

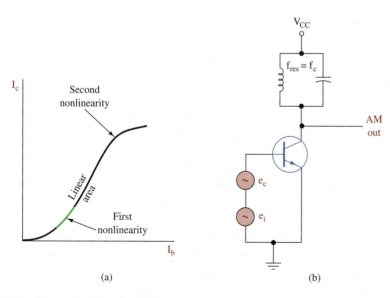

(a)

(b)

FIGURE 5-2 **Simple transistor modulator.**

e_c and e_i to bias the transistor into the first nonlinear area shown in Figure 5-2(a). Proper adjustment of the levels of e_c and e_i is necessary for good operation. Their levels must be low to stay in the first nonlinear area, and the intelligence power must be one-half the carrier power (or less) for 100% modulation (or less). In the collector a parallel LC tank circuit, resonant at the carrier frequency, acts as a bandpass filter. Because the tank impedance is highest at resonance, the voltages produced across the tank circuit are highest for the carrier and sideband frequencies, but the very low impedance of the tank at all other frequencies effectively shorts out unwanted harmonic components. Recall that the mixing of two frequencies through a nonlinear device generates more than just the desired AM components; the power-series representation of Equation (4-11) predicts that harmonics as well as side frequencies of the harmonics will also be produced. The relatively narrow bandwidth characteristics of the high-Q tank circuit selects only the desired AM components and rejects all others.

In practice, amplitude modulation can be obtained in several ways. The modulation type is generally described in terms of where the intelligence is injected. For example, in Figure 5-2(b) the intelligence is injected into the base; hence the term **base modulation**. *Collector* and *emitter modulation* are also used. In vacuum-tube modulators, which are mentioned mainly because some tube-based transmitters are bound to be in service even now, the most common form was *plate modulation*, but *grid, cathode,* and (for pentodes) *suppressor-grid* and *screen-grid* modulation schemes were also used.

Neutralization

Though certainly not used widely at low powers in new designs, vacuum tubes are still found in some areas of electronic communications, particularly in existing high-power transmitter installations. In fact, one of the last remaining applications where tubes still offer some advantages over solid-state devices is in high-power radio transmitters, where kilowatts of output power are required at high frequencies. In particular, vacuum tubes are still often encountered in the final power-amplification stages of high-frequency transmitters. As for other stages, vacuum-tube modulators with configurations similar to that shown in Figure 5-3 are still encountered.

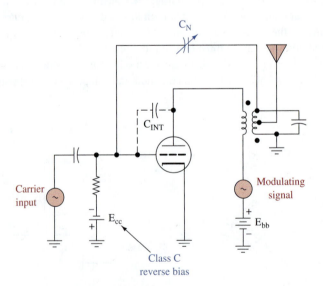

FIGURE 5-3   **Plate-modulated class-C amplifier.**

Note the variable capacitor, C_N, connected from the plate tank circuit back to the grid. It is termed the **neutralizing capacitor**. It provides a path for the return of a signal that is 180° out of phase with the signal returned from plate to grid via the internal interelectrode capacitance (C_{INT}) of the tube. C_N is adjusted to cancel the internally fed-back signal to reduce the tendency for self-oscillation. The transformer in the plate circuit is made to introduce a 180° phase shift by appropriate wiring.

Self-oscillation is a problem for all RF amplifiers (both linear and class C), both tube-based and solid state. The transistor amplifier shown in Figure 5-4 also has a neutralization capacitor (C_N). The self-oscillations can be at the tuned frequency or at a higher frequency. Self-oscillations at the tuned frequency prevent amplification from taking place. Higher-frequency self-oscillations are called **parasitic oscillations**. In any event these oscillations are undesirable because they introduce distortion, reduce desired amplification, and potentially radiate energy on unauthorized frequencies and causing destructive interference to other systems.

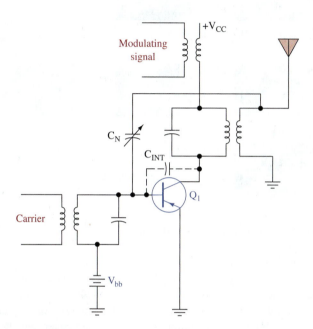

FIGURE 5-4   Collector modulator.

High- and Low-Level Modulation

Another common designator for AM transmitters involves the stage at which modulation occurs. Both the plate-modulated circuit of Figure 5-3 and the transistor circuit of Figure 5-4 are *high-level* modulators. In a **high-level modulation** scheme the intelligence is added at the last possible point before the transmitting antenna. In contrast, **low-level modulation** means that the intelligence is injected at a point before the final output stage, such as at the base or emitter of the final-amplifier transistor (or grid or cathode of a vacuum-tube final amplifier), or to a previous amplifier stage. Block diagrams for typical high- and low-level modulator systems are shown in Figure 5-5(a) and (b), respectively. Note that in the high-level modulation system [Figure 5-5(a)] the majority of power amplification takes place in the highly efficient class-C amplifier. The low-level modulation scheme requires that power amplification take place in the much less efficient linear final amplifier.

The choice of high- or low-level modulation is largely driven by the required power output, and, as expected, trade-offs are involved. For high-power applications such as standard radio broadcasting, where outputs are measured in terms of kilowatts instead of watts, high-level modulation is the most economical approach. Recall that class-C bias (device conduction for less than 180°) allows for the highest possible efficiency. It realistically provides 70 to 80% efficiency as compared to about 50 to 60% for the next most-efficient configuration, class-B amplification. However, class-C amplification cannot be used for reproduction of the complete AM signal, and hence large amounts of intelligence power must be injected at the final output to provide a high percentage of modulation. Recall from Chapter 2 that the total sideband power at 100% modulation is half the carrier power and that the total power output is additive. Thus, for a high-power transmitter employing high-level modulation, the modulator stages would be required to contribute as much as 50% of the carrier power.

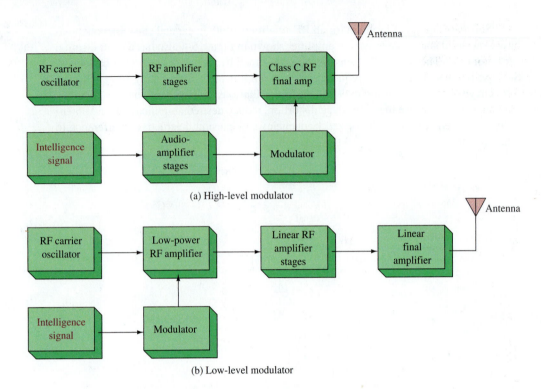

(a) High-level modulator

(b) Low-level modulator

FIGURE 5-5 **(a) High- and (b) low-level modulation.**

The modulation process is accomplished in a nonlinear device, but all circuitry that follows must be linear so that the AM signal is reproduced without distortion. The class-C amplifier is not linear but can reproduce (and amplify) the single-frequency carrier because of the flywheel effect exhibited by the output tank circuit. However, the class-C amplifier would distort a modulated AM signal because the tank circuit would distort rather than preserve the varying amplitudes of the sidebands.

In summary, then, high-level modulation requires larger intelligence power to produce modulation but allows extremely efficient amplification of the higher-powered carrier. Low-level schemes allow low-powered intelligence signals to be used, but all subsequent output stages must use less efficient linear (not class-C) configurations. Low-level systems usually offer the most economical approach for low-power transmitters.

Transistor High-Level Modulator

Figure 5-4 shows a transistorized class-C, high-level modulation scheme. Class-C operation provides an abrupt nonlinearity when the device switches on and off, which allows for the generation of the sum and difference frequencies. This is in contrast to the use of the gradual nonlinearities offered by a transistor at high and low levels of class A bias, as previously shown in Figure 5-2(a). Generally, the operating point is established to allow half the maximum ac output voltage to be supplied at the collector when the intelligence signal is zero. The V_{bb} supply provides a reverse bias for Q_1 so that it conducts on only the positive peak of the input carrier signal. This, by definition, is class C bias because Q_1 conducts for less than 180° per cycle. The tank circuit in Q_1's collector is tuned to resonate at f_c, and thus the full carrier sine wave is reconstructed there by the flywheel effect at the extremely high efficiency afforded by class-C operation.

The intelligence (modulating) signal for the collector modulator of Figure 5-4 is added directly in series with the collector supply voltage. The net effect of the intelligence signal is to vary the energy available to the tank circuit each time Q_1 conducts on the positive peaks of carrier input. This causes the output to reach a maximum value when the intelligence is at its peak positive value and a minimum value when the intelligence is at its peak negative value. Since the circuit is biased to provide one-half of the maximum possible carrier output when the intelligence is zero, theoretically an intelligence signal level exists where the carrier will swing between twice its static value and zero. This is a fully modulated (100% modulation) AM waveform. In practice, however, the collector modulator cannot achieve 100% modulation because the transistor's knee in its characteristic curve changes at the intelligence frequency rate. This limits the region over which the collector voltage can vary, and slight collector modulation of the preceding stage is necessary to allow the high modulation indexes that are usually desirable. This is sometimes not a necessary measure in the tube-type high-level modulators.

Figure 5-6(a) shows an intelligence signal for a collector modulator, and Figure 5-6(b) shows its effect on the collector supply voltage. In Figure 5-6(c), the resulting collector current variations that are in step with the available supply voltages are shown. Figure 5-6(d) shows the collector voltage produced by the flywheel effect of the tank circuit as a result of the varying current peaks that are flowing through the tank.

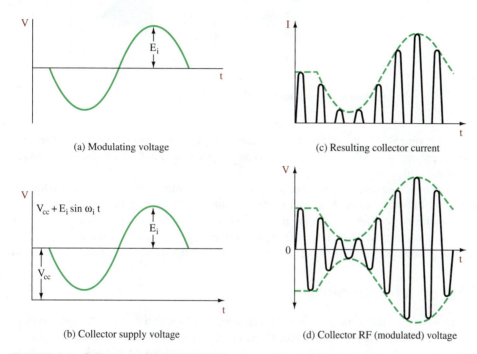

(a) Modulating voltage

(c) Resulting collector current

(b) Collector supply voltage

(d) Collector RF (modulated) voltage

FIGURE 5-6   Collector modulator waveforms.

5-2 AM TRANSMITTER MEASUREMENTS

Trapezoid Patterns

Several techniques are available to make operational checks on a transmitter's performance. A standard oscilloscope display of the transmitted AM signal will indicate any gross deficiencies. This technique is all the better if a dual-trace scope is used to allow the intelligence signal to be superimposed on the AM signal. An improvement to this method is known as the *trapezoidal pattern*. It is illustrated in Figure 5-7. The AM signal is connected to the vertical input and the intelligence signal is applied to the horizontal input with the scope's internal sweep disconnected. The intelligence signal usually must be applied through an adjustable *RC* phase-shifting network, as shown, to ensure that it is exactly in phase with the modulation envelope of the AM waveform. Figure 5-7(b) shows the resulting scope display with improper phase relationships, and Figure 5-7(c) shows the proper in-phase trapezoidal pattern for a typical AM signal. It easily allows percentage modulation calculations by applying the B and A dimensions to the following formula, first introduced in Chapter 2:

$$\%m = \frac{B - A}{B + A} \times 100\%. \qquad (2\text{-}3)$$

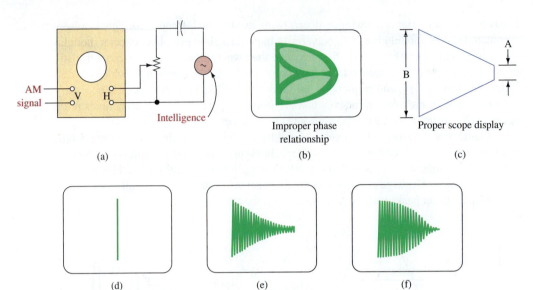

FIGURE 5-7   **Trapezoidal pattern connection scheme and displays.**

In Figure 5-7(d), the effect of 0% modulation (just the carrier) is indicated. The trapezoidal pattern is simply a vertical line because there is no intelligence signal to provide horizontal deflection.

Figures 5-7(e) and (f) show two more trapezoidal displays indicative of some common problems. In both cases the trapezoid's sides are not straight (linear). The concave curvature at (e) indicates poor linearity in the modulation stage, which is often caused by improper neutralization or by stray coupling in a previous stage. The convex curvature at (f) is usually caused by improper bias or low carrier signal power (often termed **low excitation**).

Meter Measurement

It is possible to make some meaningful transmitter checks with a dc ammeter in the collector (or plate) of the modulated stage. If the operation is correct, this current should not change as the intelligence signal is varied between zero and the point where full modulation is attained. This is true because the increase in current during the crest of the modulated wave should be exactly offset by the drop during the trough. A distorted AM signal will usually cause a change in dc current flow. In the case of overmodulation, the current will increase further during the crest but cannot decrease below zero at the trough, and a net increase in dc current will occur. It is also common for this current to decrease as modulation is applied. This malfunction is termed **downward modulation** and is usually the result of insufficient excitation. The current increase during the modulation envelope crest is minimized, but the decrease during the trough is nearly normal.

Spectrum Analyzers

Transmitter troubleshooting in the frequency domain is heavily reliant on use of spectrum analyzers, both for determining the spectral characteristics of the output as well as for identifying unwanted and potentially interfering frequency components. Recall from Chapter 1 that a **spectrum analyzer** visually displays the amplitudes of the components of a wave as a function of frequency rather than as a function of time. In Figure 5-8(a) the frequency-domain representation of a 1-MHz carrier amplitude-modulated by a 5-kHz intelligence signal is shown. Proper operation is indicated since only the carrier and upper- and lower-side frequencies are present. During malfunctions, and to a lesser extent even under normal conditions, transmitters will often generate **spurious frequencies** as shown in Figure 5-8(b), where components other than just the three desired are present. These spurious undesired components are usually called **spurs**, and their amplitude is tightly controlled by FCC regulation to minimize interference on adjacent channels. The coupling stage between the transmitter and its antenna is

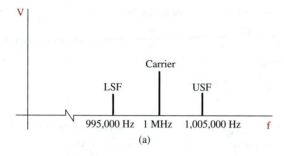

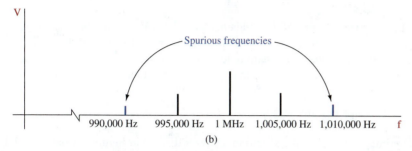

FIGURE 5-8   Spectrum analysis of AM waveforms.

designed to attenuate the spurs, but the transmitter's output stage must also be carefully designed to keep the spurs to a minimum level. The spectrum analyzer is obviously a very handy tool for use in evaluating transmitter performance.

Harmonic Distortion Measurements

Harmonic distortion measurements can be made easily by applying a spectrally pure signal source to the device under test (DUT). The quality of the measurement is dependent on the harmonic distortion of both the signal source and spectrum analyzer. The source provides a signal to the DUT and the spectrum analyzer is used to monitor the output. Figure 5-9 shows the results of a typical harmonic distortion measurement. The distortion can be

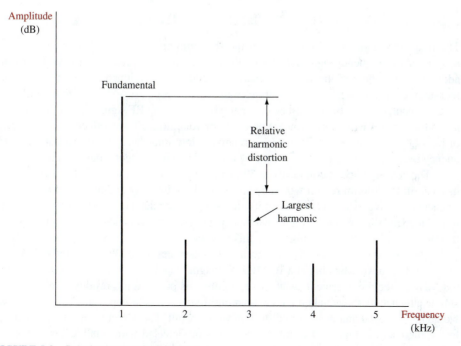

FIGURE 5-9   Relative harmonic distortion.

specified by expressing the fundamental with respect to the largest harmonic in dB. This is termed the **relative harmonic distortion**.

If the fundamental in Figure 5-9 is 1 V and the harmonic at 3 kHz (the largest distortion component) is 0.05 V, the relative harmonic distortion is

$$20 \log \frac{1 \text{ V}}{0.05 \text{ V}} = 26 \text{ dB}$$

A somewhat more descriptive distortion specification is **total harmonic distortion (THD)**. THD takes into account the power in all the significant harmonics:

$$\text{THD} = \sqrt{(V_2^2 + V_3^2 + \cdots)/V_1^2} \tag{5-1}$$

where V_1 is the rms voltage of the fundamental and V_2, V_3, … are the rms voltages of the harmonics. An infinite number of harmonics is theoretically created, but in practice the amplitude falls off for the higher harmonics. Virtually no error is introduced if the calculation does not include harmonics less than $\frac{1}{10}$ of the largest harmonic.

EXAMPLE 5-1

Determine the THD if the spectrum analyzer display in Figure 5-9 has $V_1 = 1$ V, $V_2 = 0.03$ V, $V_3 = 0.05$ V, $V_4 = 0.02$ V, and $V_5 = 0.04$ V.

SOLUTION

$$\begin{aligned}
\text{THD} &= \sqrt{(V_2^2 + V_3^2 + V_4^2 + V_5^2)/V_1^2} \tag{5-1}\\
&= \sqrt{(0.03^2 + 0.05^2 + 0.02^2 + 0.04^2)/1^2}\\
&= 0.07348\\
&= 7.35\%
\end{aligned}$$

THD calculations are somewhat tedious when a large number of significant harmonics exist. Some spectrum analyzers include an automatic THD function that does all the work and prints out the THD percentage.

Special RF Signal Measurement Precautions

The frequency-domain measurements of the spectrum analyzer provide a more thorough reading of RF frequency signals than does the time-domain oscilloscope. The high cost and additional setup time of the spectrum analyzer dictates the continued use of more standard measurement techniques—mainly voltmeter and oscilloscope usage. Whatever the means of measurement, certain effects must be understood when testing RF signals as compared to the audio frequencies with which you are probably more familiar. These effects are the loading of high-Q parallel-resonant circuits by a relatively low impedance instrument and the frequency response shift that can be caused by test lead and instrument input capacitance.

The consequence of connecting a 50-Ω signal generator into an RF-tuned circuit that has a Z_p in the kilohm region would be a drastically reduced Q and increased bandwidth. The same loading effect would result if a low-impedance detector were used to make RF impedance measurements. Using resistors, capacitors, or transformers in conjunction with the measurement instrument minimizes this loading.

The consequence of test lead or instrument capacitance is to shift the circuit's frequency response. If you were looking at a 10-MHz AM signal that had its resonant circuit shifted to 9.8 MHz by the measurement capacitance, an obvious problem has resulted. Besides some simple attenuation, the equal amplitude relationship between the upper and lower sidebands would be destroyed and waveform distortion would result. This effect is minimized by using low-valued series-connected coupling capacitors or canceled with small series-connected inductors. If more precise readings with inconsequential loading are necessary, specially designed resonant matching networks are required. They can be used as an add-on with the measuring instrument or built into the RF system at convenient test locations.

Measuring Transmitter Output Power

A dummy antenna is often a necessity when testing a transmitter. A **dummy antenna** is a resistive load used in place of the regular antenna. It is used to prevent undesired transmissions that may otherwise occur. The dummy antenna (also called a *dummy load*) also prevents damage to the output circuits that may occur under unloaded conditions. Figure 5-10 shows the circuit to be used when measuring and troubleshooting the output power of a transmitter. The dummy load acts like an antenna because it absorbs the energy output from the transmitter without allowing that energy to radiate and interfere with other stations. Its input impedance must match (be equal to) the transmitter's output impedance; this is usually 50 Ω.

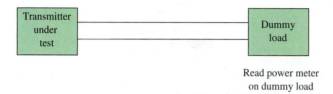

Read power meter
on dummy load

FIGURE 5-10   **Checking transmitter output power.**

Suppose we are checking a low-power commercial transmitter that is rated at 250-W output (the dummy load and wattmeter must be rated for this level). If the output power is greater than the station license allows, the drive control must be adjusted to bring the unit within specs. What if the output is below specs? Let us consider possible causes.

Remember the suggestion in Chapter 1: Do the easy tasks first. Perhaps the easiest thing to do in the case of low-power output is to check the drive control: Is it set correctly? Assuming it is, check the power supply voltage: Is it correct? Observe that voltage on a scope: Is it good, pure dc or has a rectifier shorted or opened, indicated by excessive ripple?

Once the easy tasks have been done, check the tuning of each amplifier stage between the carrier oscillator and the last or final amplifier driving the antenna. If the output power is still too low after peaking the tuning controls, use an oscilloscope to check the output voltage of each stage to see if they are up to specs. Are the signals good sinusoids? If a stage has a clean, undistorted input signal and a distorted output, there may be a defective component in the bias network. Or perhaps the tube/transistor needs checking and/or replacement.

5-3 SSB TRANSMITTERS

Chapter 2 introduced single-sideband as a form of amplitude modulation in which the carrier and one sideband are suppressed, and only the remaining sideband is transmitted. The balanced modulator, whose operation was described in Chapter 4, produces a suppressed-carrier, double-sideband output. The unwanted sideband can then be suppressed by either the filter method or the phasing method. The filter method is the traditional approach, and it is widely used in analog SSB applications such as amateur radio because it is well suited to the suppression of sidebands consisting of multiple frequencies. The phasing method has become popular in recent years. Though not as widely used for analog SSB as the filter method, the phasing method is ideal for digital modulation techniques, which will be shown in Chapter 8 to share many characteristics of SSB signals.

Filter Method

Figure 5-11 is a block diagram of a single-sideband transmitter using a balanced modulator to generate DSB and the filter method of eliminating one of the sidebands. For illustrative

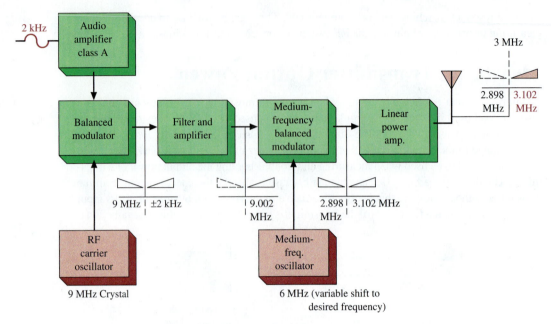

FIGURE 5-11 SSB transmitter block diagram.

purposes, a single-tone 2000-Hz intelligence signal is used, but the intelligence is normally "complex," in that it consists of a band of frequencies, such as those produced by the human voice.

A 9-MHz crystal frequency is used because of the excellent operating characteristics of monolithic filters at that frequency. The 2-kHz signal is amplified and mixed with a 9-MHz carrier (**conversion frequency**) in the balanced modulator. Remember, neither the carrier nor audio frequencies themselves appear in the output of the balanced modulator; the sum and difference frequencies (9 MHz ± 2 kHz) are its output. As illustrated in Figure 5-11, the two sidebands from the balanced modulator are applied to the filter. Only the desired sideband is passed. In this example, the upper sideband is shown as being passed and the lower sideband attenuated, although the lower sideband could also be used, and many SSB transmitters provide for switchable sideband filtering. The dashed lines in the figure show that the carrier and lower sideband have been removed.

The output of the first balanced modulator is filtered and mixed again with a new conversion frequency to adjust the output to the desired transmitter frequency. After mixing the two inputs to get two new sidebands, the balanced modulator removes the new 3-MHz carrier and applies the two new sidebands (3102 kHz and 2898 kHz) to a tunable linear power amplifier.

EXAMPLE 5-2

For the transmitter system shown in Figure 5-11, determine the filter Q required in the linear power amplifier.

SOLUTION

The second balanced modulator created another DSB signal from the SSB signal of the preceding high-Q filter. However, the frequency translation of the second balanced modulator means that a low-quality filter can be used once again to create SSB. The new DSB signal is at about 2.9 MHz and 3.1 MHz. The required filter Q is

$$\frac{3\text{ MHz}}{3.1\text{ MHz} - 2.9\text{ MHz}} = \frac{3\text{ MHz}}{0.2\text{ MHz}} = 15.$$

The input and output circuits of the linear power amplifier are tuned to reject one sideband and pass the other to the antenna for transmission. A standard *LC* filter is now adequate to remove one of the two new sidebands. The new sidebands are about 200 kHz apart ($\cong$3100 kHz − 2900 kHz), so the required *Q* is quite low. (See Example 5-2 for further illustration.) The high-frequency oscillator is variable so that the transmitter output frequency can be varied over a range of transmitting frequencies. Since both the carrier and one sideband have been eliminated, all the transmitted energy is in the single sideband.

Phase Method

The phase method of SSB generation relies on the fact that the upper and lower sidebands of an AM signal differ in the sign of their phase angles. This means that phase discrimination may be used to cancel one sideband of the DSB signal. The phase method offers the following advantages over the filter method:

1. There is greater ease in switching from one sideband to the other.
2. SSB can be generated directly at the desired transmitting frequency, which means that intermediate balanced modulators are not necessary.
3. Lower intelligence frequencies can be economically used because a high-*Q* filter is not necessary.

Despite these advantages, the filter method is widely encountered in many systems because it performs adequately and because the phase method has been difficult to implement until fairly recently. However, the phase method has truly come into its own because the essential characteristics of sine and cosine waves lend themselves to digital processing techniques, which are now a mainstay of modern communications system design. Digital signal processing will be introduced in Chapters 7 and 8.

Consider a modulating signal $f(t)$ to be a pure cosine wave. A resulting balanced modulator output (DSB) can then be written as

$$f_{DSB1}(t) = (\cos \omega_i t)(\cos \omega_c t) \tag{5-2}$$

where $\cos \omega_i t$ is the intelligence signal and $\cos \omega_c t$ the carrier. The term $\cos A \cos B$ is equal to $\frac{1}{2}[\cos(A + B) + \cos(A - B)]$ by trigonometric identity, and therefore Equation (5-2) can be rewritten as

$$f_{DSB1}(t) = \frac{1}{2}[\cos(\omega_c + \omega_i)t + \cos(\omega_c - \omega_i)t]. \tag{5-3}$$

If another signal,

$$f_{DSB2}(t) = \frac{1}{2}[\cos(\omega_c - \omega_i)t - \cos(\omega_c + \omega_i)t] \tag{5-4}$$

were added to Equation (5-3), the upper sideband would be canceled, leaving just the lower sideband,

$$f_{DSB1}(t) + f_{DSB2}(t) = \cos(\omega_c - \omega_i)t.$$

Since the signal in Equation (5-4) is equal to

$$\sin \omega_i t \sin \omega_c t$$

by trigonometric identity, it can be generated by shifting the phase of the carrier and intelligence signals by exactly 90° and then feeding them into a balanced modulator. Recall that sine and cosine waves are identical except for a 90° phase difference.

A block diagram for the system just described is shown in Figure 5-12. The upper balanced modulator receives the carrier and intelligence signals directly, while the lower balanced modulator receives both of them shifted in phase by 90°. Thus, combining the outputs of both balanced modulators in the adder results in an SSB output that is subsequently amplified and then driven into the transmitting antenna.

FIGURE 5-12   Phase-shift SSB generator.

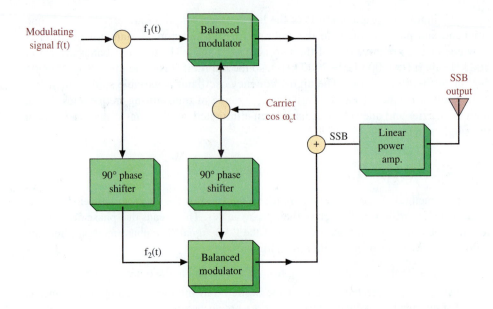

A major disadvantage of this system is the 90° phase-shifting network required for the intelligence signal. The *carrier* 90° phase shift is easily accomplished because it is a single frequency, but the audio signals residing in the sidebands cover a wide range of frequencies. Maintaining exactly 90° of phase shift for a complete range of frequencies is difficult but critical: even a 2° error (i.e., an 88° phase shift) for a given audio frequency results in about 30 dB of unwanted sideband suppression instead of the desired complete suppression obtained at 90°.

The block diagram of Figure 5-12 is in many respects identical in concept to the modulator used to transmit data in digital systems and will be encountered again in Chapter 8. In fact, it will be demonstrated that phase-shifted data display many characteristics of a DSB suppressed-carrier signal, and many considerations pertaining to suppressed-carrier signals in the analog domain apply in the digital realm as well.

5-4   FM TRANSMITTERS

Transmitters for FM share many of the functional blocks already described for AM transmitters. FM transmitters are composed primarily of oscillators, amplifiers, filters, and modulating stages. There are more variations in the basic block diagram, however, depending on factors such as whether the carrier oscillator is modulated directly (direct FM) or a subsequent modulator stage is employed (indirect FM), the amount of deviation required, and the frequency stability and degree of frequency "agility" (i.e., ability to be tuned over a frequency range) needed. The block diagram of Figure 5-13 shows the basic topology of an FM transmitter, and the following sections expand on the refinements found in practice.

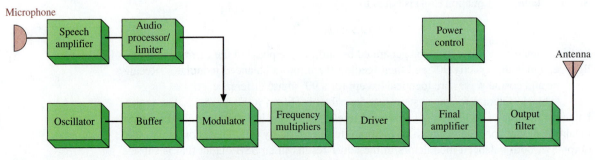

FIGURE 5-13   Block diagram of FM transmitter.

Direct FM Generation

The capacitance microphone system introduced in Chapter 3 to illustrate fundamental concepts can be used to generate FM directly. Recall that a direct FM system is one in which the frequency of the carrier oscillator is directly varied by the modulating signal. The Chapter 3 discussion illustrated that the capacitance of the microphone varies in step with the sound wave striking it. Figure 3-1 showed that if the amplitude of sound striking the microphone is increased, the oscillator frequency deviation is increased. If the frequency of sound waves is increased, the rate at which the oscillator deviates above and below the center frequency is increased. The capacitor-microphone oscillator is useful in explaining the principles of an FM signal but is not normally used in practical systems because of a problem previously mentioned: It does not exhibit sufficient frequency stability for use in licensed applications. It is also not able to produce enough deviation for practical use. However, it does illustrate that FM can be generated by directly modulating the frequency-determining state of the oscillator.

VARACTOR DIODE A varactor diode may be used to generate FM directly. All reverse-biased diodes exhibit a junction capacitance that varies inversely with the amount of reverse bias. A diode that is physically constructed to enhance this characteristic is termed a **varactor diode**. Figure 5-14 shows a schematic of a varactor diode modulator. With no intelligence signal (E_i) applied, the parallel combination of C_1, L_1, and D_1's capacitance forms the resonant carrier frequency. The diode D_1 is effectively in parallel with L_1 and C_1 because the $-V_{CC}$ supply appears as a short circuit to the ac signal. The coupling capacitor, C_C, isolates the dc levels and intelligence signal while looking like a short to the high-frequency carrier. When the intelligence signal, E_i, is applied to the varactor diode, its reverse bias is varied, which causes the diode's junction capacitance to vary in step with E_i. The oscillator frequency is subsequently varied as required for FM, and the FM signal is available at Q_1's collector. For simplicity, the dc bias and oscillator feedback circuitry is not shown in Figure 5-14.

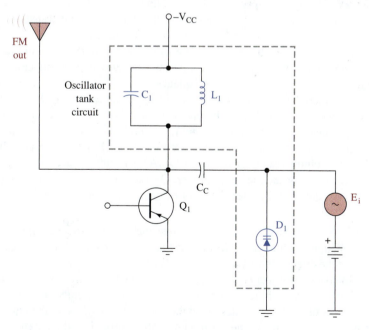

FIGURE 5-14 **Varactor diode modulator.**

MODULATOR CIRCUITS The previous discussions illustrate that much of the circuitry found in an FM transmitter is composed of the same functional blocks that were described in Chapter 4 and also found in SSB and AM transmitters. One circuit that has not been covered in detail, however, is the modulator stage for direct FM transmitter. Though largely obsolete for new designs in favor of phase-modulation approaches, one widely used configuration in the past has been the *reactance modulator*, and this design is encountered in many transmitters still in use. In a reactance modulator, a transistor is effectively made to function as a variable capacitance. The variable capacitance in turn affects the resonant frequency of the carrier oscillator. The reactance modulator is efficient and provides a large deviation.

Figure 5-15 illustrates a typical reactance modulator circuit. Refer to this figure throughout the following discussion. The circuit consists of the reactance circuit and the master oscillator. The reactance circuit operates on the master oscillator to cause its resonant frequency to shift up or shift down depending on the modulating signal being applied. The reactance circuit appears capacitive to the master oscillator. In this case, the reactance looks like a variable capacitor in the oscillator's tank circuit.

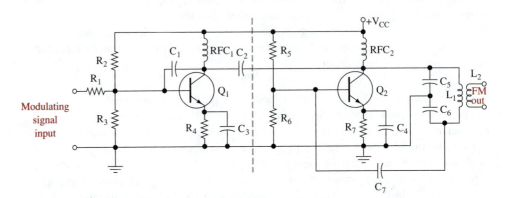

FIGURE 5-15 **Reactance modulator.**

Transistor Q_1 makes up the reactance modulator circuit. Resistors R_2 and R_3 establish a voltage divider network that biases Q_1. Resistor R_4 furnishes emitter feedback to thermally stabilize Q_1. Capacitor C_3 is a bypass component that prevents ac input signal degeneration. Capacitor C_1 interacts with transistor Q_1's inter-electrode capacitance to cause a varying capacitive reactance directly influenced by the input-modulating signal.

The master oscillator is a Colpitts oscillator built around transistor Q_2. Coil L_1, capacitor C_5, and capacitor C_6 make up the resonant tank circuit. Capacitor C_7 provides the required regenerative feedback to cause the circuit to oscillate. Q_1 and Q_2 are impedance coupled, and capacitor C_2 effectively couples the changes at Q_1's collector to the tank circuit of transistor Q_2 while blocking dc voltages.

When a modulating signal is applied to the base of transistor Q_1 via resistor R_1, the reactance of the transistor changes in relation to that signal. If the modulating voltage goes up, the capacitance of Q_1 goes down, and if the modulating voltage goes down, the reactance of Q_1 goes up. This change in reactance is felt on Q_1's collector and also at the tank circuit of the Colpitts oscillator transistor Q_2. As capacitive reactance at Q_1 goes up, the resonant frequency of the master oscillator, Q_2, decreases. Conversely, if Q_1's capacitive reactance goes down, the master oscillator resonant frequency increases.

CROSBY SYSTEMS   In addition to their inability to produce large amounts of frequency deviation, such as the ± 75-kHz deviation required for FM broadcasting, another weakness of the direct-FM methods discussed up to this point was in their lack of frequency stability. In no case was a crystal oscillator used as the basic reference or carrier frequency. The FCC controls the stability of the carrier frequency very tightly, and that stability is not attained by any of the methods described thus far. Because of the high Q of crystal oscillators, it is not possible to frequency-modulate them directly—their frequency cannot be made to deviate sufficiently to provide workable wideband FM systems. It is possible to modulate directly a crystal oscillator in some narrowband applications. If a crystal is modulated to a deviation of ± 50 Hz around a 5-MHz center frequency and both are multiplied by 100, a narrowband system with a 500-MHz carrier and ± 5-kHz of deviation results. One method of circumventing this dilemma for wideband systems is to provide some means of automatic frequency control (AFC) to correct any carrier drift by comparing the carrier frequency to that of a reference crystal oscillator.

FM systems utilizing direct generation with AFC are called **Crosby systems**. A Crosby direct FM transmitter for a standard broadcast station at 90 MHz is shown in Figure 5-16. Notice that the reactance modulator starts at an initial center frequency of 5 MHz and has a maximum deviation of ±4.167 kHz. This is a typical situation because reactance modulators cannot provide deviations exceeding about ±5 kHz and still offer a high degree of linearity (i.e., Δf directly proportional to the modulating voltage amplitude). Consequently, **frequency multipliers** are utilized to provide a × 18 multiplication up to a carrier frequency of 90 MHz (18 × 5 MHz) with a ±75-kHz (18 × 4.167 kHz) deviation. Notice that both the carrier and deviation are multiplied by the multiplier.

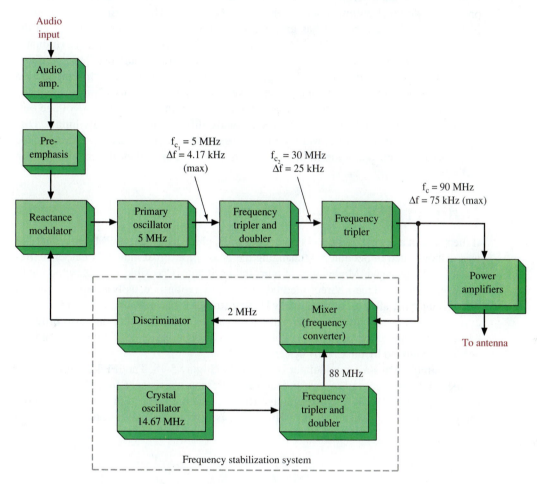

FIGURE 5-16   **Crosby direct FM transmitter.**

Frequency multiplication is normally obtained in steps of ×2 or ×3 (doublers or triplers). The principle involved is to feed a frequency rich in harmonic distortion from a class C amplifier into an *LC* tank circuit tuned to two or three times the input frequency. The harmonic is then the only significant output, as illustrated in Figure 5-17.

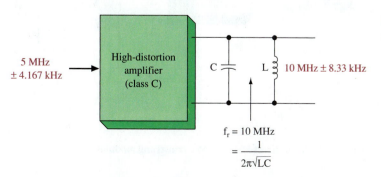

FIGURE 5-17   **Frequency multiplication (doubler).**

After the ×18 multiplication ($3 \times 2 \times 3$) shown in Figure 5-16, the FM *exciter* function is complete. The term **exciter** is often used to denote the circuitry that generates the modulated signal. The excited output goes to the power amplifiers for transmission *and* to the frequency stabilization system. The purpose of this system is to provide a control voltage to the reactance modulator whenever it drifts from its desired 5-MHz value. The control (AFC) voltage then varies the reactance of the primary 5-MHz oscillator slightly to bring it back on frequency.

The mixer in Figure 5-16 has the 90-MHz carrier and 88-MHz crystal oscillator signal as inputs. The mixer output accepts only the difference component of 2MHz, which is fed to the discriminator. A **discriminator** is the opposite of a VCO, because it provides a dc level output based on the frequency input. The discriminator output in Figure 5-16 will be zero if it has an input of exactly 2 MHz, which occurs when the transmitter is at precisely 90 MHz. Any carrier drift up or down causes the discriminator output to go positive or negative, resulting in the appropriate primary oscillator readjustment. Effectively, the discriminator is performing the same function in this application that a phase detector performs in a phase-locked loop. This similarity is no accident. As will be described in more detail in Chapter 6, the discriminator produces a voltage output that is proportional to the frequency or phase difference between its inputs. Discriminator circuits, therefore, were the first types of circuits to be used in phase-locked loops because their operation in that context is very similar to that of demodulating an FM signal.

Indirect FM Generation

If the phase of a crystal oscillator's output is varied, phase modulation (PM) will result. As was demonstrated in Section 3-4, changing the phase of a signal indirectly causes its frequency to be changed. Modulation of a crystal oscillator is possible via PM, which indirectly creates FM. This indirect method of FM generation is usually referred to as the **Armstrong** type, after its originator, E. H. Armstrong. It permits modulation of a stable crystal oscillator without the need for the cumbersome AFC circuitry and also provides carrier accuracies identical to the crystal accuracy, as opposed to the slight degradation of accuracy exhibited by the Crosby system.

A simple Armstrong modulator is depicted in Figure 5-18. The JFET is biased in the ohmic region by keeping V_{DS} low. In that way it presents a resistance from drain to source that is made variable by the gate voltage (the modulating signal). In the ohmic region, the drain-to-source resistance for a JFET transistor behaves like a voltage-controlled

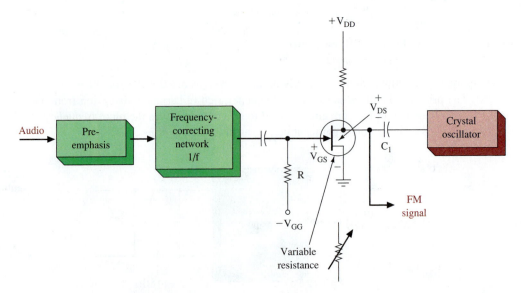

FIGURE 5-18 **Indirect FM via PM (Armstrong modulator).**

resistance (a variable resistor). The resistance value is controlled by the gate voltage (V_{GS}), where a change in the gate voltage will create a change in the drain-to-source resistance. Notice that the modulating signal is first given the standard preemphasis and then applied to a frequency-correcting network. This network is a low-pass RC circuit (an integrator) that makes the audio output amplitude inversely proportional to its frequency. This is necessary because in phase modulation, the frequency deviation created is not only proportional to modulating signal amplitude (as desired for FM) but also to the modulating signal frequency (undesired for FM). Thus, in PM if a 1-V, 1-kHz modulating signal caused a 100-Hz deviation, a 1-V, 2-kHz signal would cause a 200-Hz deviation instead of the same deviation of 100 Hz if that signal were applied to the $1/f$ network.

In summary, the Armstrong modulator of Figure 5-18 indirectly generates FM by changing the phase of a crystal oscillator's output. That phase change is accomplished by varying the phase angle of an RC network (C_1 and the JFET's resistance), in step with the frequency-corrected modulating signal.

The indirectly created FM is not capable of much frequency deviation. A typical deviation is 50 Hz out of 1 MHz (50 ppm). Thus, even with a ×90 frequency multiplication, a 90-MHz station would have a deviation of 90×50 Hz = 4.5 kHz. This may be adequate for narrowband communication FM but falls far short of the 75-kHz deviation required for broadcast FM. A complete Armstrong FM system providing a 75-kHz deviation is shown in Figure 5-19. It uses a balanced modulator and 90° phase shifter to phase-modulate a crystal oscillator. Sufficient deviation is obtained by a combination of multipliers and mixing. The ×81 multipliers ($3 \times 3 \times 3 \times 3$) raise the initial 400-kHz ± 14.47-Hz signal to 32.4 MHz ± 1172 Hz. The carrier *and* deviation are multiplied by 81. Applying this signal to the mixer, which also has a ± crystal oscillator signal input of 33.81 MHz, provides an output component (among others) of 33.81 MHz − (32.4 MHz ± 1172 Hz), or 1.41 MHz ± 1172 Hz. Notice that the mixer output changes the center frequency *without* changing the deviation. Following the mixer, the ×64 multipliers accept only the mixer difference output component of 1.41 MHz ± 1172 Hz and raise that to (64×1.41 MHz) ± (64×1172 Hz), or the desired 90.2 MHz ± 75 kHz.

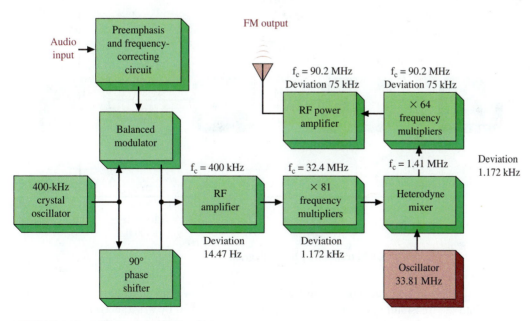

FIGURE 5-19   Wideband Armstrong FM.

The electronic circuitry used to increase the operating frequency of a transmitter up to a specified value is called the **pump chain**. A block diagram of the pump chain for the wideband Armstrong FM system is shown in Figure 5-20.

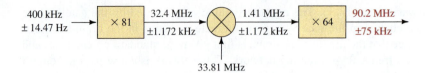

PHASE-LOCKED LOOP FM TRANSMITTER   The block diagram shown in Figure 5-21 illustrates a very practical FM transmitter. The amplified audio signal is used to frequency-modulate a crystal oscillator. The crystal frequency is pulled slightly by the variable capacitance exhibited by the varactor diode. The approximate ± 200-Hz deviation possible in this fashion is adequate for narrowband systems. The FM output from the crystal oscillator is then divided by 2 and applied as one of the inputs to the phase detector of a phase-locked-loop (PLL) system. As indicated in Figure 5-21, the other input to the phase detector is the same, and its output is therefore (in this case) the original audio signal. The input control signal to the VCO is therefore the same audio signal, and its output will be its free-running value of 125 MHz $\pm$ 5 kHz, which is set up to be exactly 50 times the 2.5-MHz value of the divided-by-2 crystal frequency of 5 MHz.

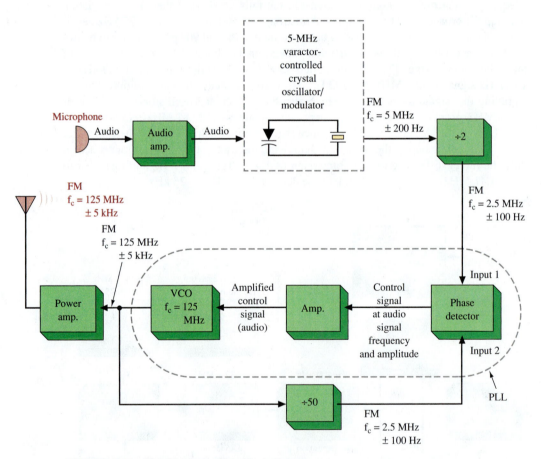

FIGURE 5-21   **PLL FM transmitter block diagram.**

The FM output signal from the VCO is given power amplification and then driven into the transmitting antenna. This output is also sampled by a $\div 50$ network, which provides the other FM signal input to the phase detector. The PLL system effectively provides the required $\times 50$ multiplication but, more important, provides the transmitter's required frequency stability. Any drift in the VCO center frequency causes an input to the phase detector (input 2 in Figure 5-21) that is slightly different from the exact 2.5-MHz crystal reference value. The phase-detector output therefore develops an error signal that corrects the VCO center frequency output back to exactly 125 MHz. This dynamic action of the phase detector/VCO and feedback path is the basis of a PLL.

The advent of stereo records and tapes and the associated high-fidelity playback equipment in the 1950s led to the development of stereo FM transmissions as authorized by the FCC in 1961. Stereo systems involve generating two separate signals, as from the left and right sides of a concert hall performance. When played back on left and right speakers, the listener gains greater spatial dimension or directivity.

A stereo radio broadcast requires that two separate 30-Hz to 15-kHz signals be used to modulate the carrier so that the receiver can extract the left and right channel information and amplify them separately into their respective speakers. In essence, then, the amount of information to be transmitted is doubled in a stereo broadcast. Hartley's law (Chapter 1) tells us that either the bandwidth or time of transmission must therefore be doubled, but this is not practical. The problem was solved by making more efficient use of the available bandwidth (200 kHz) by **frequency multiplexing** the two required modulating signals. **Multiplex operation** is the simultaneous transmission of two or more signals on one carrier.

Modulating Signal

The system approved by the FCC is *compatible* because a stereo broadcast received by a monaural FM receiver will provide an output equal to the sum of the left plus right channels (L + R), while a stereo receiver can provide separate left and right channel signals. The stereo transmitter has a modulating signal, as shown in Figure 5-22. Notice that the sum of the L + R modulating signal extends from 30 Hz to 15 kHz as does the full audio signal used to modulate the carrier in standard FM broadcasts. However, a signal corresponding to the left channel minus right channel (L − R) extends from 23 to 53 kHz. In addition, a 19-kHz pilot subcarrier is included in the composite stereo modulating signal.

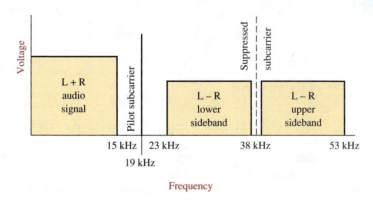

FIGURE 5-22 **Composite modulating signals.**

The reasons for the peculiar arrangement of the stereo modulating signal will become more apparent when the receiver portion of stereo FM is discussed in Chapter 6. For now, suffice it to say that two different signals (L + R and L − R) are used to modulate the carrier. The signal is an example of **frequency-division multiplexing** because two different signals are multiplexed together by having them coexist in two different frequency ranges.

Frequency-Domain Representation

The block diagram in Figure 5-23 shows how the composite modulating signal is generated and applied to the FM modulator for subsequent transmission. The left and right channels are picked up by their respective microphones and individually preemphasized. They are then applied to a **matrix network** that inverts the right channel, giving a −R signal, and then combines (adds) L and R to provide an (L + R) signal and also combines L and −R to provide the (L − R) signal. The two outputs *are still* 30-Hz to 15-kHz audio signals at this point. The (L − R) signal and a 38-kHz carrier signal are then applied to a balanced modulator that suppresses the carrier but provides a double-sideband (DSB) signal at its output. The upper and lower sidebands extend from 30 Hz to 15 kHz above and below the suppressed 38-kHz carrier and therefore range from 23 kHz (38 kHz − 15 kHz) up to 53 kHz (38 kHz + 15 kHz). Thus, the (L − R) signal has been translated from audio up to a higher frequency to keep it separate from the 30-Hz to 15-kHz (L + R) signal. The (L + R) signal is given a slight delay so that both signals are applied to the FM modulator in time phase. This step is necessary because of the slight delay encountered by the (L − R) signal in the balanced modulator. The 19-kHz master oscillator in Figure 5-23 is applied directly to the FM modulator and also doubled in frequency, to 38 kHz, for the balanced modulator carrier input.

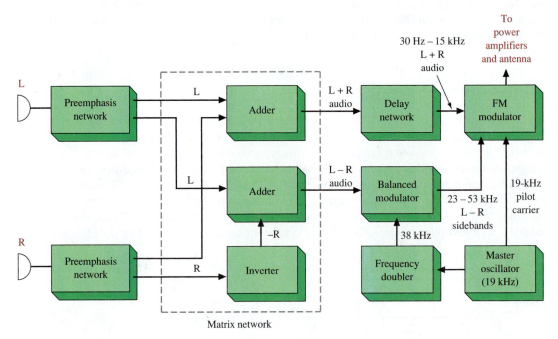

FIGURE 5-23 **Stereo FM transmitter.**

Stereo FM is more prone to noise than are monophonic (monaural) broadcasts. The (L − R) signal is weaker than the (L + R) signal, as shown in Figure 5-22. The (L − R) signal is also at a higher modulating frequency (23 to 53 kHz), and both of these effects cause poorer noise performance. The net result to the receiver is an *S/N* of about 20 dB less than the monaural signal. Because of this, some receivers have a mono/stereo switch so that a noisy (weak) stereo signal can be changed to monophonic for improved reception. A stereo signal received by a monophonic receiver is only about 1 dB worse (*S/N*) than an equivalent monaural broadcast because of the presence of the 19-kHz pilot carrier.

SUMMARY

Transmitters for various forms of modulation are more alike than not. This statement will also prove true for the various forms of digital modulation to be introduced in the following chapters. There we will see that most functions have been integrated into transceiver integrated circuits. At the block diagram level, it can be seen that all types of transmitters consist of but a few essential building blocks. These include oscillators, amplifiers, and various forms of filters and impedance-matching circuits. The oscillator, either directly or through various stages of frequency multiplication, determines the carrier frequency. The oscillator must be very stable, meaning that crystal reference oscillators are used in all cases.

Also common to all transmitters is the modulator. Although the circuit will look different depending on the type of modulation and the frequency range, the purpose is the same: to impress the intelligence onto the carrier. The modulator thus performs a mixing function. In

AM, modulation may be either high-level or low level. High-level modulation takes place at the last possible point before the carrier is applied to the antenna and allows for the use of efficient class-C final amplifier stages. The drawback in high-power applications is that the modulating stages must be able to produce up to half the carrier power at full modulation. Low-level modulation introduces the intelligence at a stage earlier in the block diagram. Though often preferred, particularly in low-power operation, the drawback is that all subsequent amplification must be linear, thus reducing efficiency.

Suppressed-carrier AM and SSB transmissions make use of the balanced modulator to generate sidebands without the carrier. The balanced modulator is an example of a linear multiplier. Two approaches are in use to suppress the unwanted sideband. The filter method is the traditional approach. Here, a very sharp filter is used to remove the sideband while leaving the desired sideband alone. These types of filters require very sharp *skirts*, or transitions from passband to stopband. The second method is the phase method. While difficult to achieve for voice-band transmissions because of the need to maintain a precise 90° phase shift at all frequencies, the phase method is gaining in popularity and will be shown to be the basis for virtually all digital transmitter implementations.

FM transmitters share many of the characteristics as their AM and SSB counterparts. They will often operate at higher frequency ranges, and wideband FM transmitters must be capable of maintaining frequency stability over wide frequency deviations. FM transmitters may either be of the direct or indirect type. In direct FM transmission, the frequency-determining stage (oscillator) is modulated directly. While possible, the difficulty is in obtaining sufficiently wide frequency deviations from frequency-stabilized oscillators, which essentially appear as very-high-Q tuned circuits. Direct modulation requires a number of frequency-multiplication stages to increase deviation as well as frequency. There are a number of direct modulation schemes. One is the reactance modulator. Though largely obsolete in favor of phase-modulation approaches, the circuit is presented to demonstrate the underlying concept. Its inclusion is also in recognition of the many installed, analog FM transmitters in service. A type of FM transmitter seen in FM broadcasting is the Crosby transmitter. This configuration allows wide and linear frequency deviations with the high frequency stability required for commercial service.

Other types of FM transmitters are of the indirect FM modulation variety. In such transmitters a stage subsequent to the frequency-determining oscillator stage is modulated. Since frequency and phase are related, this arrangement has the effect of changing frequency by changing the instantaneous carrier phase angle with the modulating signal. Such indirect FM systems must employ a low-pass filter configured as an integrator to make the modulator insensitive to modulating signal frequency. Other types of FM transmitter configurations, primarily those based on the phase-locked-loop, are also finding widespread application.

QUESTIONS AND PROBLEMS

SECTION 5-1

1. Describe two possible ways that a transistor can be used to generate an AM signal.

*2. What is *low-level modulation*?

*3. What is *high-level modulation*?

4. Explain the relative merits of high- and low-level modulation schemes.

*5. Why must some radio-frequency amplifiers be neutralized?

6. Describe the difference in effect of self-oscillations at a circuit's tuned frequency and parasitic oscillations.

7. Define *parasitic oscillation*.

8. How does self-oscillation occur?

9. Draw a schematic of a class-C transistor modulator and explain its operation.

*10. What is the principal advantage of a class-C amplifier?

*11. What is the function of a quartz crystal in a radio transmitter?

*12. Draw a block diagram of an AM transmitter.

*13. What is the purpose of a buffer amplifier stage in a transmitter?

*14. Draw a simple schematic diagram showing a method of coupling the radio-frequency output of the final power amplifier stage of a transmitter to an antenna.

15. Describe the functions of an antenna coupler.

*16. A ship radio-telephone transmitter operates on 2738 kHz. At a certain point distant from the transmitter the 2738-kHz signal has a measured field of 147 mV/m. The second harmonic field at the same point is measured as 405 μV/m. To the nearest whole unit in decibels, how much has the harmonic emission been attenuated below the 2738-kHz fundamental? (51.2 dB)

17. What is a *tune-up procedure*?

SECTION 5-2

*18. Draw a sample sketch of the trapezoidal pattern on a cathode-ray oscilloscope screen indicating low percentage modulation without distortion.

19. Explain the advantages of the trapezoidal display over a standard oscilloscope display of AM signals.

20. A spectrum analyzer display shows that a signal is made up of three components only: 960 kHz at 1 V, 962 kHz at $\frac{1}{2}$ V, 958 kHz at $\frac{1}{2}$ V. What is the signal and how was it generated?

21. Define *spur*.

22. Provide a sketch of the display of a spectrum analyzer for the AM signal described in Problem 31 from Chapter 2 at both 20% and 90% modulation. Label the amplitudes in dBm.

23. The spectrum analyzer display shown in Figure 1-6 from Chapter 1 is calibrated at 10 dB/vertical division and 5 kHz/horizontal division. The 50.0034-MHz carrier is shown riding on a −20-dBm noise floor. Calculate the carrier power, the frequency, and the power of the spurs. (2.51 W, 50.0149 MHz, 49.9919 MHz, 50.0264 MHz, 49.9804 MHz, 6.3 mW, 1 mW)

*24. What is the purpose of a *dummy antenna*?

25. An amplifier has a spectrally pure sine-wave input of 50 mV. It has a voltage gain of 60. A spectrum analyzer shows harmonics of 0.035 V, 0.027 V, 0.019 V, 0.011 V, and 0.005 V. Calculate the total harmonic distortion (THD) and the relative harmonic distortion. (2.864%, 38.66 dB)

26. An additional harmonic ($V_6 = 0.01$ V) was neglected in the THD calculation shown in Example 5-1. Determine the percentage error introduced by this omission. (0.91%)

SECTION 5-3

27. Determine the carrier frequency for the transmitter shown in Figure 5-11. (It is *not* 3 MHz).

28. List the advantages of the phase versus filter method of SSB generation. Why isn't the phase method more popular than the filter method?

29. Calculate a filter's required Q to convert DSB to SSB, given that the two sidebands are separated by 200 Hz. The suppressed carrier (40 dB) is 2.9 MHz. Explain how this required Q could be greatly reduced. (36,250)

30. An SSB signal is generated around a 200-kHz carrier. Before filtering, the upper and lower sidebands are separated by 200 Hz. Calculate the filter Q required to obtain 40-dB suppression. (2500)

31. Explain the operation of the phase-shift SSB generator illustrated in Figure 5-12. Why is the carrier phase shift of 90° not a problem, whereas that for the audio signal is?

SECTION 5-5

32. Draw a schematic diagram of a varactor diode FM generator and explain its operation.

*33. Draw a schematic diagram of a frequency-modulated oscillator using a reactance modulator. Explain its principle of operation.

34. Explain the principles of a Crosby-type modulator.

*35. How is good stability of a reactance modulator achieved?

*36. If an FM transmitter employs one doubler, one tripler, and one quadrupler, what is the carrier frequency swing when the oscillator frequency swing is 2 kHz? (48 kHz)

37. Draw a block diagram of a broadcast-band Crosby-type FM transmitter operating at 100 MHz, and label all frequencies in the diagram.

38. Explain the function of a discriminator.

*39. Draw a block diagram of an Armstrong-type FM broadcast transmitter complete from the microphone input to the antenna output. State the purpose of each stage, and explain briefly the overall operation of the transmitter.

40. Explain the difference in the amount of deviation when passing an FM signal through a mixer as compared to a multiplier.

41. What type of circuit is used to increase a narrowband FM transmission to a wideband one?

42. Explain the operation of the PLL FM transmitter shown in Figure 5-21.

43. Draw a block diagram of a stereo multiplex FM broadcast transmitter complete from the microphone inputs to the antenna output. State the purpose of each stage, and explain briefly the overall operation of the transmitter.

44. Explain how stereo FM can effectively transmit twice the information of a standard FM broadcast while still using the same bandwidth. How is the *S/N* at the receiver affected by a stereo transmission as opposed to monophonic?

45. Define *frequency-division multiplexing*.

46. Describe the type of modulation used in the L − R signal of Figure 5-23.

47. Explain the function of a matrix network as it relates to the generation of an FM stereo signal.

48. What difference in noise performance exists between FM stereo and mono broadcasts? Explain what might be done if an FM stereo signal is experiencing noise problems at the receiver.

CHAPTER 6

RECEIVERS

KEY TERMS

sensitivity
noise floor
selectivity
tuned radio frequency (TRF)
first detector
image frequency
double conversion
double conversion
up-conversion
preselector
auxiliary AGC diode
discriminator
automatic frequency control (AFC)
local oscillator
reradiation
cross-modulation
intermodulation distortion
dynamic range
limiter
sensitivity
quieting voltage

threshold voltage
limiting knee voltage
direct conversion
zero-IF receivers
envelope detector
synchronous detector
quadrature
subsidiary communication authorization (SCA)
noise floor
12-dB SINAD
dynamic range
intermod
third-order intercept point
input intercept
delayed AGC
variable bandwidth tuning (VBT)
electromagnetic interference (EMI)
S meter
squelch

The communications receiver extracts intelligence from the radio-frequency signal for subsequent use. The intelligence can take many forms. It may be analog, representing voice, music, or video, or it may take the form of data destined for processing in a digital environment. To do its job, the receiver must be able to select and extract the desired signal from all those present at its antenna. The receiver must also increase the gain of the received signal, often by a factor of many millions. Finally, the receiver must be able to perform in the presence of noise introduced in both the communication channel and in the receiver itself. These considerations pose a number of design challenges, which will be examined more fully in this chapter. Also, the process of demodulation or detection will be examined in more detail.

6-1 RECEIVER CHARACTERISTICS: SENSITIVITY AND SELECTIVITY

Two major characteristics of any receiver are its sensitivity and selectivity. A receiver's **sensitivity** is a function of its overall gain and may be defined in terms of its ability to drive the output transducer (e.g., speaker) to an acceptable level. A more technical definition is that sensitivity is the minimum input signal, usually expressed as a voltage, required to produce a specified output. Sometimes, sensitivity is defined in terms of the minimum discernible output. The range of sensitivities for communication receivers varies from the millivolt region for low-cost AM receivers down to the nanovolt region for ultrasophisticated units for more exacting applications. In essence, a receiver's sensitivity is determined by the amount of gain provided and, more important, by receiver noise characteristics. In general, the input signal must be somewhat greater than the noise at the receiver's input. This input noise is termed the **noise floor** of the receiver. It is not difficult to insert more gain in a radio, but getting noise figures below a certain level presents a more difficult challenge.

Selectivity may be defined as the extent to which a receiver is capable of differentiating between the desired signal and other frequencies (unwanted radio signals and noise). A receiver can also be overly selective. For instance, on commercial broadcast AM, the transmitted signal can include intelligence frequencies up to a maximum of 15 kHz, which subsequently generates upper and lower sidebands extending 15 kHz above and below the carrier frequency. The total signal has a 30-kHz bandwidth. Optimum receiver selectivity is thus 30 kHz, but if a 5-kHz bandwidth were selected, the upper and lower sidebands would extend only 2.5 kHz above and below the carrier. The radio's output would suffer from a lack of the full possible fidelity because the output would include intelligence up to a maximum of 2.5 kHz. On the other hand, an excessively wide bandwidth, say 50 kHz in this instance, results in the reception of adjacent, unwanted signals and additional noise. Recall that noise is directly proportional to the bandwidth selected.

As has been stated, broadcast AM can extend to about 30-kHz bandwidth. As a practical matter, however, many stations and receivers use a more limited bandwidth. The lost fidelity is often not detrimental because of the talk-show format of many AM stations. For instance, a receiver bandwidth of 10 kHz is sufficient for audio frequencies up to 5 kHz, which more than handles the human voice range. Musical reproduction with a 5-kHz maximum frequency is not high fidelity but is certainly adequate for casual listening.

6-2 THE TUNED RADIO-FREQUENCY RECEIVER

If you were to envision a block diagram for a radio receiver, you would probably go through the following logical thought process:

1. The signal from the antenna is usually very small—therefore, amplification is necessary. This amplifier should have low-noise characteristics and should be tuned to

accept only the desired carrier and sideband frequencies to avoid interference from other stations and to minimize the received noise. Recall that noise is proportional to bandwidth.

2. After sufficient amplification, a circuit to detect the intelligence from the radio-frequency carrier is required.

3. Following the detection of the intelligence, further amplification is necessary to give it sufficient power to drive a loudspeaker.

This logical train of thought leads to the block diagram shown in Figure 6-1. It consists of an RF amplifier, detector, and audio amplifier. The first radio receivers for broadcast AM took this form and are called **tuned radio frequency (TRF)** receivers. These receivers generally had three stages of RF amplification, with each stage preceded by a separate variable-tuned circuit. You can imagine the frustration experienced by the user when tuning to a new station. The three tuned circuits were each adjusted by individual variable capacitors. To receive a station required proper adjustment of all three, and a good deal of time and practice was necessary.

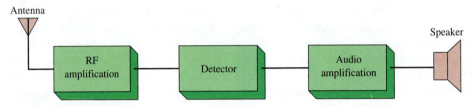

FIGURE 6-1   **Simple tuned-radio-frequency receiver block diagram.**

TRF Selectivity

Another problem with the TRF design is that it exhibits variable selectivity over its intended tuning range. Consider a standard AM broadcast-band receiver that spans the frequency range from 550 to 1550 kHz. If the approximate center of 1000 kHz is considered, we can use Equation (4-8) from Chapter 4 to find that, for a desired 10-kHz BW, a Q of 100 is required.

$$Q = \frac{f_r}{\text{BW}} \tag{4-8}$$

$$= \frac{1000 \text{ kHz}}{10 \text{ kHz}}$$

$$= 100$$

Now, since the Q of a tuned circuit remains fairly constant as its capacitance is varied, a change to 1550 kHz will change the BW to 15.5 kHz.

$$Q = \frac{f_r}{\text{BW}} \tag{4-8}$$

Therefore,

$$\text{BW} = \frac{f_r}{Q}$$

$$= \frac{1550 \text{ kHz}}{100}$$

$$= 15.5 \text{ kHz}$$

The receiver's BW is now too large, and the received signal will suffer from increased noise. On the other hand, the opposite problem is encountered at the lower end of the frequency range. At 550 kHz, the BW is 5.5 kHz.

$$\text{BW} = \frac{f_r}{Q}$$

$$= \frac{550 \text{ kHz}}{100}$$

$$= 5.5 \text{ kHz}$$

The fidelity of reception is now impaired. The maximum intelligence frequency possible is 5.5 kHz/2, or 2.75 kHz, instead of the full 5 kHz transmitted. This selectivity problem led to the general use of the superheterodyne receiver in place of TRF designs.

EXAMPLE 6-1

A TRF receiver is to be designed with a single tuned circuit using a 10-μH inductor.

(a) Calculate the capacitance range of the variable capacitor required to tune from 550 to 1550 kHz.

(b) The ideal 10-kHz BW is to occur at 1100 kHz. Determine the required Q.

(c) Calculate the BW of this receiver at 550 kHz and 1550 kHz.

SOLUTION

(a) At 550 kHz, calculate C using Equation (4-1) from Chapter 4.

$$f_r = \frac{1}{2\pi\sqrt{LC}} \qquad \text{(4-1)}$$

$$550 \text{ kHz} = \frac{1}{2\pi\sqrt{10 \ \mu\text{H} \times C}}$$

$$C = 8.37 \text{ nF}$$

At 1550 kHz,

$$1550 \text{ kHz} = \frac{1}{2\pi\sqrt{10 \ \mu\text{H} \times C}}$$

$$C = 1.06 \text{ nF}$$

Therefore, the required range of capacitance is from

1.06 to 8.37 nF

(b)
$$Q = \frac{f_r}{\text{BW}} \qquad \text{(4-8)}$$

$$= \frac{1100 \text{ kHz}}{10 \text{ kHz}}$$

$$= 110$$

(c) At 1550 kHz,

$$\text{BW} = \frac{f_r}{Q}$$

$$= \frac{1550 \text{ kHz}}{110}$$

$$= 14.1 \text{ kHz}$$

At 550 kHz,

$$\text{BW} = \frac{550 \text{ kHz}}{110} = 5 \text{ kHz}$$

The variable-selectivity problem in TRF systems led to the development and general usage of the superheterodyne receiver beginning in the early 1930s. This basic receiver configuration is still dominant after all these years, an indication of its utility, and is used in many types of receivers with all possible forms of modulation. Figure 6-2 shows the block diagram of an AM superheterodyne receiver. The SSB and FM receivers will be shown shortly to have a very similar layout. The first receiver stage, the RF amplifier, may or may not be required depending on factors to be discussed later. The next stage is the mixer, which accepts two inputs, the output of the RF amplifier (or antenna input when an RF amplifier is omitted) and a constant-amplitude sine wave from the local oscillator (LO). Recall that a mixer is a nonlinear device or circuit. Its function in the receiver is to mix the received signal with a sine wave to generate a new set of sum and difference frequencies. The mixer output will be a modulated signal that is usually centered around the difference frequency. The mixer output is called the intermediate frequency (IF), and it is a constant, usually lower frequency than that of the received signal. (The sum of the LO and RF frequencies can also be used, and there are advantages to doing so, but so-called *up-conversion* receiver architectures that employ IF frequencies higher than the received RF are somewhat more difficult to design and are less frequently encountered. The essential point is that the IF is different than the RF carrier frequency of the transmitter.) Because the IF does not change even as the receiver is tuned, receiver bandwidth is constant over the entire tuning range. Constant bandwidth is the key to the superior selectivity of the superheterodyne receiver. Additionally, since the IF frequency is usually lower than the RF, voltage gain of the received signal is more easily attained at the IF frequency.

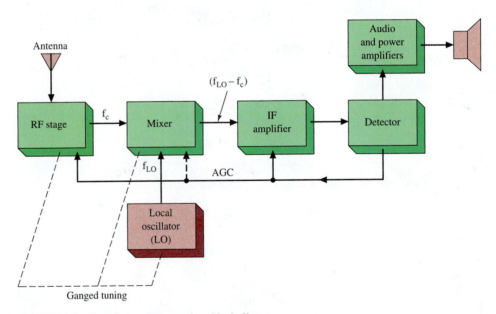

FIGURE 6-2   Superheterodyne receiver block diagram.

    The block diagram of Figure 6-2 shows that the mixer output, which is the IF, is then amplified by one or more IF amplifier stages, which provide the bulk of radio-frequency signal amplification, hence overall receiver gain; again this amplification occurs at a *fixed* frequency. Following the IF amplifiers is the detector, which extracts the intelligence from the radio signal for subsequent amplification by the audio amplifiers to a level that is sufficient to drive a speaker. A dc level proportional to received signal strength is extracted from the detector stage and fed back to the IF amplifiers and sometimes to the mixer and/or the RF amplifier. This is the automatic gain control (AGC) level, which allows relatively constant receiver output for widely variable received signals. AGC subsystems will be described more fully in Section 6-8.

Frequency Conversion

The mixer performs a frequency conversion process. Consider the situation shown in Figure 6-3 for a broadcast AM receiver. The AM signal into the mixer is a 1000-kHz carrier that has been modulated by a 1-kHz sine wave, thus producing side frequencies at 999 kHz and 1001 kHz. The LO input is a 1455-kHz sine wave. The mixer, being a nonlinear device, will generate the following components:

1. Frequencies at the original inputs: 999 kHz, 1000 kHz, 1001 kHz, and 1455 kHz.
2. Sum and difference components of all the original inputs: 1455 kHz ± (999 kHz, 1000 kHz, and 1001 kHz). This means outputs at 2454 kHz, 2455 kHz, 2456 kHz, 454 kHz, 455 kHz, and 456 kHz.
3. Harmonics of all the frequency components listed in 1 and 2 and a dc component.

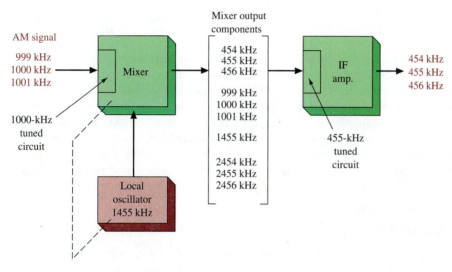

FIGURE 6-3 **Frequency conversion process.**

These results are as predicted from the description of the power series described in Chapter 4 and are the consequence of any nonlinear mixing action.

The IF amplifier has a tuned circuit that accepts components only near 455 kHz, in this case 454 kHz, 455 kHz, and 456 kHz. Since the mixer maintains the same amplitude proportion that existed with the original AM signal input at 999 kHz, 1000 kHz, and 1001 kHz, the signal now passing through the IF amplifiers is a replica of the original AM signal. The only difference is that now its carrier frequency is 455 kHz. Its envelope is identical to that of the original AM signal. A frequency conversion has occurred that has translated the carrier from 1000 kHz to 455 kHz, a frequency that is in between the carrier and intelligence frequencies, hence the term intermediate-frequency amplifier. Since the mixer and detector both have nonlinear characteristics, the mixer is often referred to as the **first detector**.

Tuned-Circuit Adjustment

Now consider the effect of changing the resonant frequency of the tuned circuit at the front end of the mixer to accept a station at 1600 kHz. Either the inductance or the capacitance (usually the latter) of the tank-circuit components must be reduced to change the resonant frequency from 1000 kHz to 1600 kHz. If the capacitance in the local oscillator's tuned circuit were simultaneously reduced so that its frequency of oscillation went up by 600 kHz, the situation shown in Figure 6-4 would now exist. The mixer output still contains a component at 455 kHz (among others), as in the previous case when the receiver was tuned to 1000 kHz. The frequency-selective circuits in the IF amplifiers reject all the other frequency components produced by the mixer.

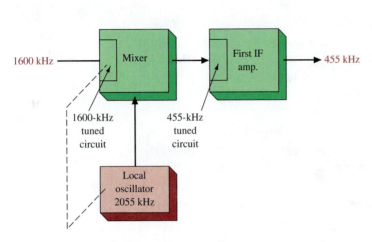

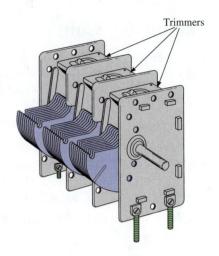

FIGURE 6-4   **Frequency conversion.**

FIGURE 6-5   **Variable ganged capacitor.**

Thus, the key to superheterodyne operation is to make the LO frequency track with the circuit or circuits that are tuning the incoming radio signal so that their difference is a constant frequency (the IF). For a 455-kHz IF frequency, the most common case for broadcast AM receivers, this means the LO should always be at a frequency 455 kHz above the incoming carrier frequency. The receiver's front-end tuned circuits are usually made to track together either by mechanically linking (ganging) the capacitors in these circuits on a common variable rotor assembly, as shown in Figure 6-5, or by using frequency-synthesis techniques based on the phase-locked loop principles discussed previously. Note that the ganged capacitor of Figure 6-5 has three separate capacitor elements.

Image Frequency

The great advantage of the superheterodyne receiver over the TRF is that the superhet shows constant selectivity over a wide range of received frequencies. Most of the gain in a superheterodyne receiver occurs in the IF amplifiers at a fixed frequency, and this configuration allows for relatively simple and yet highly effective frequency-selective circuits. A disadvantage does exist, however, other than the obvious increase in complexity. The frequency conversion process performed by the mixer–oscillator combination sometimes will allow a station other than the desired one to be fed into the IF. Consider a receiver tuned to receive a 20-MHz station that uses a 1-MHz IF. The LO would, in this case, be at 21 MHz to generate a 1-MHz frequency component at the mixer output. This situation is illustrated in Figure 6-6. If an undesired station at 22 MHz were also on the air, it would, however, be possible for the undesired signal also to become a mixer input. Even though the tuned circuit at the mixer's front end is "selecting" a center frequency of 20 MHz, a look at its

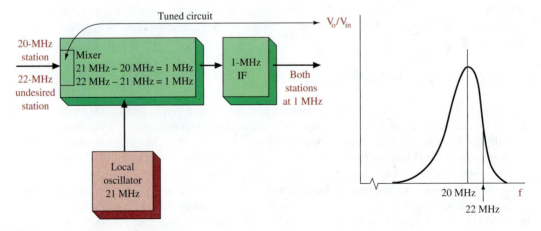

FIGURE 6-6   **Image frequency illustration.**

response curve in Figure 6-6 shows that it will not fully attenuate the undesired station at 22 MHz. As soon as the 22-MHz signal is fed into the mixer, we have a problem. It mixes with the 21-MHz LO signal and one of the components produced is 22 MHz − 21 MHz = 1 MHz—the IF frequency! Thus, we now have a desired 20-MHz station and an undesired 22-MHz station, which both look correct to the IF. Depending on the strength of the undesired station, it can either interfere with or even completely override the desired station.

EXAMPLE 6-2

Determine the image frequency for a standard broadcast band receiver using a 455-kHz IF and tuned to a station at 620 kHz.

SOLUTION

The first step is to determine the frequency of the LO. The LO frequency minus the desired station's frequency of 620 kHz should equal the IF of 455 kHz. Hence,

$$LO - 620\,\text{kHz} = 455\,\text{kHz}$$
$$LO = 620\,\text{kHz} + 455\,\text{kHz}$$
$$= 1075\,\text{kHz}$$

Now determine what other frequency, when mixed with 1075 kHz, yields an output component at 455 kHz.

$$X - 1075\,\text{kHz} = 455\,\text{kHz}$$
$$X = 1075\,\text{kHz} + 455\,\text{kHz}$$
$$= 1530\,\text{kHz}$$

Thus, 1530 kHz is the image frequency in this situation.

In the preceding discussion, the undesired received signal is called the **image frequency.** Designing superheterodyne receivers with a high degree of image frequency rejection is an important consideration.

Image frequency rejection on the standard (AM radio) broadcast band is not a major problem. A glance at Figure 6-7 serves to illustrate this point. In this case, the tuned circuit at the mixer input comes fairly close to attenuating the image frequency fully because 1530 kHz is far from the 620-kHz center frequency of the tuned circuit. Unfortunately, this situation is not so easy to attain at the higher frequencies used by many communication receivers. In these cases, a technique known as **double conversion** is employed to solve the image frequency problems.

The use of an RF amplifier with its own input tuned circuit also helps to minimize this problem. Now the image frequency must pass through two tuned circuits, both of which have been tuned to the desired frequency, before the image frequency is mixed. These tuned circuits at the input of the RF and mixer stages attenuate the image frequency to a greater extent than can the single-tuned circuit in receivers without an RF stage.

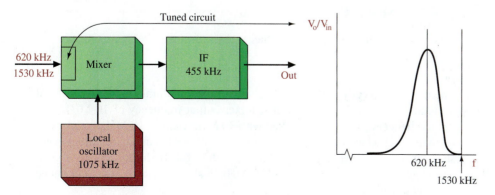

FIGURE 6-7 Image frequency not a problem.

Double Conversion

Many communications systems are designed for purposes other than mass communication to a general audience. Such applications may be "mission critical," such as for national defense or for the support of the business goals of the system operator, or systems may be designed for life-safety applications, including police and fire protection or emergency medical response. Because of their vital nature, communications receivers intended for use in mission-critical or life-safety applications are much more sophisticated than those designed solely for the reception of stations broadcasting to the public at large. One of the most obvious areas of change from broadcast receivers to communications receivers is in the mixing process. The two major differences are the widespread use of *double conversion* and the increasing popularity of *up-conversion* in communications equipment. Both refinements have as a major goal the minimization of image frequencies.

Double conversion is the process of stepping down the RF signal to a first, relatively high IF frequency and then mixing down again to a second, lower, final IF frequency. Figure 6-8 provides a block diagram for a typical double-conversion system. Notice that the first local oscillator is variable to allow a constant 10-MHz frequency for the first IF amplifier. Now the input into the second mixer is a constant 10 MHz, which allows the second local oscillator to be a fixed 11-MHz crystal oscillator. The difference component (11 MHz − 10 MHz = 1 MHz) out of the second mixer is accepted by the second IF amplifier, which is operating at 1 MHz. The following example illustrates the ability of double conversion to eliminate image frequency problems. Example 6-3 shows that, in this case, the image frequency is twice the desired signal frequency (40 MHz versus 20 MHz), and even the relatively broadband tuned circuits of the RF and mixer stages will almost totally suppress the image frequency. On the other hand, if this receiver uses a single conversion directly to the final 1-MHz IF frequency, the image frequency would most likely not be fully suppressed.

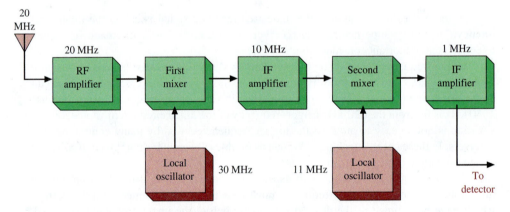

FIGURE 6-8   **Block diagram of a double-conversion receiver.**

E X A M P L E   6 - 3

Determine the image frequency for the receiver illustrated in Figure 6-8.

SOLUTION

The image frequency is the one that, when mixed with the 30-MHz first local oscillator signal, will produce a first mixer output frequency of 10 MHz. The desired frequency of 20 MHz mixed with 30 MHz yields a 10-MHz component, of course, but what *other* frequency provides a 10-MHz output? A little thought shows that if a 40-MHz input signal mixes with a 30-MHz local oscillator signal, an output of 40 MHz − 30 MHz = 10 MHz is also produced. Thus, the image frequency is 40 MHz.

The 22-MHz image frequency of Example 6-4 is very close to the desired 20-MHz signal. The RF and mixer tuned circuits will certainly provide attenuation to the 22-MHz image, but a high-level signal at the image frequency would almost certainly appear at the input to the IF stages and would not be removed from that point on. The graph of RF and mixer tuned circuit response in Figure 6-9 serves to illustrate the tremendous image-frequency rejection provided by the double-conversion scheme.

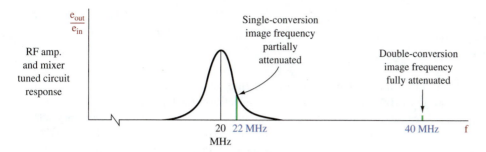

FIGURE 6-9   **Image frequency rejection.**

EXAMPLE 6-4

Determine the image frequency for the receiver illustrated in Figure 6-10.

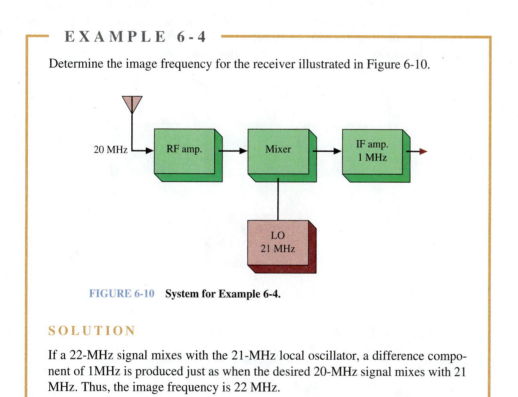

FIGURE 6-10   **System for Example 6-4.**

SOLUTION

If a 22-MHz signal mixes with the 21-MHz local oscillator, a difference component of 1MHz is produced just as when the desired 20-MHz signal mixes with 21 MHz. Thus, the image frequency is 22 MHz.

Image frequencies are not a major problem for low-frequency carriers, say, for frequencies below 4 MHz. For example, a single-conversion setup for a 4-MHz carrier and a 1-MHz IF means that a 5-MHz local oscillator will be used. The image frequency is 6 MHz, which is far enough away from the 4-MHz carrier that it will not present a problem. At higher frequencies, where images are a problem, the enormous number of transmissions taking place in our crowded communications bands aggravates the situation.

EXAMPLE 6-5

Why do you suppose that images tend to be somewhat less of a problem in FM versus AM or SSB communications?

SOLUTION

Recall the concept of the *capture* effect in FM systems (Chapter 3): It was shown that if a desired and undesired station are picked up simultaneously, the stronger one tends to be "captured" by inherent suppression of the weaker signal. Thus, a 2:1 signal-to-undesired-signal ratio may result in a 10:1 ratio at the output. This contrasts with AM systems (SSB included), where the 2:1 ratio is carried through to the output.

Up-Conversion

Until recently, the double-conversion scheme, with the lower IF frequency (often the familiar 455 kHz) providing most of the selectivity for the receiver, has been standard practice because the components available made it easy to achieve the necessary selectivity at low IF frequencies. However, now that VHF crystal filters (30 to 120 MHz) are available for IF circuitry, conversion to a higher IF than RF frequency is popular in sophisticated communications receivers. As an example, consider a receiver tuned to a 30-MHz station and using a 40-MHz IF frequency, as illustrated in Figure 6-11. This represents an **up-conversion** system because the IF is at a *higher* frequency than the received signal. The 70-MHz local oscillator mixes with the 30-MHz signal to produce the desired 40-MHz IF. Sufficient IF selectivity at 40 MHz is possible with a crystal filter.

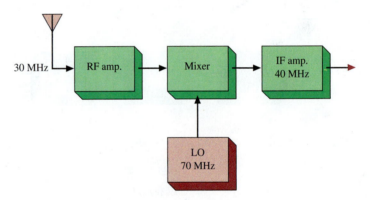

FIGURE 6-11   **Up-conversion system.**

EXAMPLE 6-6

Determine the image frequency for the system of Figure 6-11.

SOLUTION

If a 110-MHz signal mixes with the 70-MHz local oscillator, a 40-MHz output component results. The image frequency is therefore 110 MHz.

Example 6-6 shows the superiority of up-conversion. It is highly unlikely that the 110-MHz image could get through the RF amplifier tuned to 30 MHz. There is no need for double conversion and all its necessary extra circuitry. The only disadvantage to up-conversion is the need for a higher-Q IF filter and better high-frequency response IF transistors. The current state-of-the-art in these areas now makes up-conversion economically attractive. Additional advantages over double conversion include better image suppression and a less-extensive tuning range requirement for the oscillator. The smaller tuning range for up-conversion is illustrated in the following example and minimizes the tracking difficulties of a widely variable local oscillator.

┌─ **EXAMPLE 6-7** ─────────────────────────────────────┐

Determine the local oscillator tuning range for the systems illustrated in Figures 6-8 and 6-11 if the receivers must tune from 20 to 30 MHz.

SOLUTION

The double-conversion local oscillator in Figure 6-8 is at 30 MHz for a received 20-MHz signal. Providing the same 10-MHz IF frequency for a 30-MHz signal means that the local oscillator must be at 40 MHz. Its tuning range is from 30 to 40 MHz or 40 MHz/30 MHz = 1.33. The up-conversion scheme of Figure 6-11 has a 70-MHz local oscillator for a 30-MHz input and requires a 60-MHz oscillator for a 20-MHz input. Its tuning ratio is then 70 MHz/ 60 MHz, or a very low 1.17.

└───┘

The tuned circuit(s) prior to the mixer is often referred to as the **preselector.** The preselector is responsible for the image frequency rejection characteristics of the receiver. If an image frequency rejection is the result of a single tuned circuit of known Q, the amount of image frequency rejection can be calculated. The following equation predicts the amount of suppression in decibels. If more than a single tuned circuit causes suppression, the suppression contributed by each stage, in decibels, is calculated individually and the results added to provide the total suppression:

$$\text{image rejection (dB)} \cong 20 \log\left[\left(\frac{f_i}{f_s} - \frac{f_s}{f_i}\right)Q\right], \qquad (6\text{-}1)$$

where f_i = image frequency

f_s = desired signal frequency

Q = tuned circuit's Q

┌─ **EXAMPLE 6-8** ─────────────────────────────────────┐

An AM broadcast receiver has two identical tuned circuits prior to the IF stage. The Q of these circuits is 60 and the IF frequency is 455 kHz, and the receiver is tuned to a station at 680 kHz. Calculate the amount of image frequency rejection.

SOLUTION

Using Equation (7-1), the amount of image frequency rejection per stage is calculated:

$$\text{image rejection (dB)} \cong 20 \log\left[\left(\frac{f_i}{f_s} - \frac{f_s}{f_i}\right)Q\right]. \qquad (6\text{-}1)$$

The image frequency is 680 kHz + (2 × 455 kHz) = 1590 kHz. Thus,

$$20 \log\left[\left(\frac{1590\text{ kHz}}{680\text{ kHz}} - \frac{680\text{ kHz}}{1590\text{ kHz}}\right)60\right] = 20 \log 114.6$$
$$= 41\text{ dB}$$

Thus, the total suppression is 41 dB plus 41 dB, or 82 dB. This is more than enough to provide excellent image frequency rejection.

└───┘

A Complete AM Receiver

We have thus far examined the various sections of AM receivers. It is now time to look at the complete system as it would be built from individual, discrete components. Figure 6-12 shows the schematic of a widely used circuit for a low-cost AM receiver. In the schematic

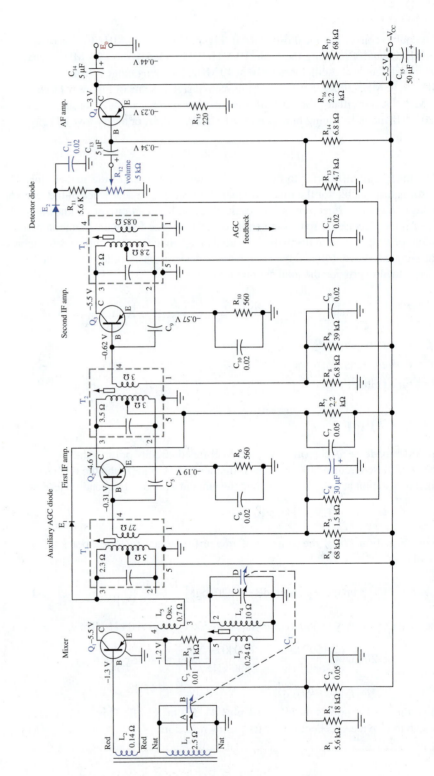

FIGURE 6-12 AM broadcast superheterodyne receiver schematic.

shown in Figure 6-12, the push-pull audio power amp, which requires two more transistors, has been omitted.

The $L_1 - L_2$ inductor combination is wound on a powdered-iron (ferrite) core and functions as an antenna as well as an input coupling stage. Ferrite-core loop-stick antennas offer extremely good signal pickup, considering their small size, and are adequate for the high-level signals found in urban areas. The RF signal is then fed into Q_1, which functions as the mixer and local oscillator (self-excited). The ganged tuning capacitor, C_1, tunes to the desired incoming station (the B section) and adjusts the LO (the D section) to its appropriate frequency. The output of Q_1 contains the IF components, which are tuned and coupled to Q_2 by the T_1 package. The IF amplification of Q_2 is coupled via the T_2 IF "can" to the second IF stage, Q_3, whose output is subsequently coupled via T_3 to the diode detector E_2. Of course, T_1, T_2, and T_3 are all providing the very good superheterodyne selectivity characteristics at the standard 455-kHz IF frequency. The E_2 detector diode's output is filtered by C_{11} so that just the intelligence envelope is fed via the R_{12} volume control potentiometer into the Q_4 audio amplifier. The AGC filter, C_4, then allows for a fed-back control level into the base of Q_2.

This receiver also illustrates the use of an **auxiliary AGC diode** (E_1). Under normal signal conditions, E_1 is reverse biased and has no effect on the operation. At some predetermined high signal level, the regular AGC action causes the dc level at E_1's cathode to decrease to the point where E_1 starts to conduct (forward bias), and it loads down the T_1 tank circuit, thus reducing the signal coupled into Q_2. The auxiliary AGC diode thus furnishes additional gain control for strong signals and enhances the range of signals that can be compensated for by the receiver.

SSB Receivers

The block diagram of Figure 6-13 shows the typical layout of an SSB receiver. The receiver is similar in many respects to the AM superheterodyne receiver just discussed; that is, it has RF and IF amplifiers, a mixer, a detector, and audio amplifiers. To permit satisfactory SSB reception, however, an additional mixer (the product detector introduced in Chapter 4, which, recall, is functionally identical to the balanced modulator) and oscillator, called the beat-frequency oscillator (BFO), must replace the rather simple detection scheme employed by an AM receiver designed for the reception of full-carrier, double-sideband transmissions.

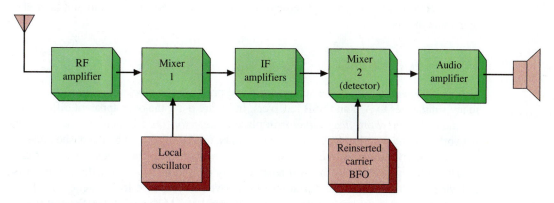

FIGURE 6-13 **SSB receiver block diagram.**

EXAMPLE 6-9

The SSB receiver in Figure 6-13 has outputs at 1 kHz and 3 kHz. The carrier used and suppressed at the transmitter was 2 MHz, and the upper sideband was utilized. Determine the exact frequencies at all stages for a 455-kHz IF frequency.

SOLUTION

RF amp and first mixer input $\}$	2000 kHz + 1 kHz = 2001 kHz 2000 kHz + 3 kHz = 2003 kHz
Local oscillator	2000 kHz + 455 kHz = 2455 kHz
First mixer output: IF amp and second mixer input (the other components attenuated by tuned circuits)	2455 kHz − 2001 kHz = 454 kHz 2455 kHz − 2003 kHz = 452 kHz
BFO	455 kHz
Second mixer output and audio amp $\}$	455 kHz − 454 kHz = 1 kHz 455 kHz − 452 kHz = 3 kHz

A second mixer stage is needed to demodulate an SSB signal properly because, for reasons to be developed more fully in the next section, recovery of the information contained within the intelligence is based on the nonlinear mixing of two signals, the carrier and the intelligence. Recall from chapters 2 and 5 that in any suppressed-carrier scheme, including SSB, the balanced modulator in the transmitter suppresses the carrier. The carrier is not needed to propagate the radio-frequency signal through the communications channel; its purpose in the transmitter is to convert the modulating signal to the frequency range of the carrier through the production of side frequencies. For proper intelligence detection, however, the process must essentially be reversed at the receiver, and this process requires that a carrier of the same frequency and phase as that used to create the sidebands in the transmitter be generated at the receiver and combined with the received sideband in the second mixer stage shown in Figure 6-13. The purpose of the BFO is to recreate the carrier. The BFO in the receiver illustrated in Figure 6-13 inserts a carrier frequency into the detector, and therefore would have a frequency equal to the intermediate frequency. In principle, however, the carrier or intermediate frequency may be inserted at any point in the receiver before demodulation.

The RF amplifier increases the gain of the SSB signal received at the antenna and applies it to the first mixer. The local oscillator output, when mixed (heterodyned) with the RF input signal, is the IF, which is then amplified by one or more stages. Up to this point the SSB receiver is identical to an AM superheterodyne receiver. The IF output is applied to the second mixer (detector). The detector output is applied to the audio amplifier and then on to the speaker.

FM Receivers

The FM receiver also uses the superheterodyne principle. In block diagram form, it has many similarities to AM and SSB receivers. In Figure 6-14, the only apparent differences are the use of the word *discriminator* in place of *detector,* the addition of a deemphasis network, and the fact that AGC may or may not be used as indicated by the dashed lines.

The **discriminator** extracts the intelligence from the high-frequency carrier and can also be called the detector, as in AM receivers. By definition, however, a discriminator is a device in which amplitude variations are derived in response to frequency or phase variations, and it is the preferred term for describing an FM demodulator. Several methods are in use to accomplish FM detection, and their operation will be described in more detail shortly.

The deemphasis network following demodulation is required to bring the high-frequency components of the intelligence back to the proper amplitude relationship with the lower frequencies. Recall that the high frequencies were preemphasized at the transmitter to provide them with greater noise immunity, as explained in Chapter 3.

The fact that AGC is optional in an FM receiver may come as a surprise. The AGC function is essential for satisfactory operation of the AM receiver. Without it, the wide

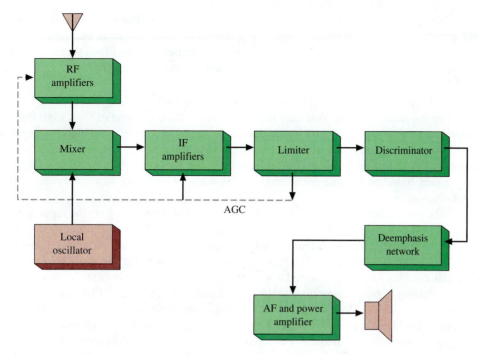

FIGURE 6-14   **FM receiver block diagram.**

variation in incoming signal strengths would produce a large volume change in the received audio signal, leading to a potentially unpleasant listening experience. However, the use of limiters in FM receivers essentially provides an AGC function, as will be explained shortly. Many older FM receivers also included an **automatic frequency control (AFC)** function. This circuit provides a slight automatic control over the local oscillator frequency. It compensates for drift in LO frequency that would otherwise cause a station to become detuned. It was necessary because it had not yet been figured out how to make an economical *LC* oscillator at 100 MHz with sufficient frequency stability. The AFC system is not needed in new designs.

The mixer, local oscillator, and IF amplifiers are basically similar to those discussed for AM receivers and do not require further elaboration. Higher frequencies are usually involved, however, because the bandwidth requirements of wideband FM systems necessitate their operation in the VHF or higher-frequency bands. The universally standard IF frequency for FM is 10.7 MHz, as opposed to 455 kHz for AM. Because of significant differences in all the other portions of the block diagram shown in Figure 6-14, they are discussed in the following sections.

RF AMPLIFIERS   Broadcast AM receivers normally operate quite satisfactorily without any RF amplifier. This is rarely the case with FM receivers, however, except for frequencies in excess of 1000 MHz (1 GHz), when it becomes preferable to omit the RF amplifier. The essence of the problem is that FM receivers can function with weaker received signals than AM or SSB receivers because of their inherent noise-reduction capability. FM receivers are called upon to deal with input signals of 1 μV or less as compared with perhaps a 30-μV minimum input for AM. If a 1-μV signal is fed directly into a mixer, the inherently high noise factor of an active mixer stage would make the received signal unintelligible. Therefore, the received signal must be boosted to 10 to 20 μV before mixing occurs. The FM system can tolerate 1 μV of noise from a mixer on a 20-μV signal but obviously cannot cope with 1 μV of noise with a 1-μV signal.

This reasoning also explains the abandonment of RF stages for the ever-increasing number of analog and digital systems employing FM and PM at frequencies of 1 GHz and above. At these frequencies, transistor noise increases while gain decreases, and a frequency will be reached where it is advantageous to feed the incoming FM signal directly into a diode mixer to step it down immediately to a lower frequency for subsequent amplification. Diode (passive) mixers are less noisy than active mixers.

Receiver designs with RF amplifiers do have advantages. The first is that use of an RF amplifier reduces the image frequency problem. Another benefit is the reduction in **local oscillator reradiation** effects. Without an RF amp, the local oscillator signal can get coupled back more easily into the receiving antenna and transmit interference.

FET RF AMPLIFIERS   Almost all RF amplifiers used in quality FM receivers use FETs as the active device. You may think that this is done because of their high input impedance, but this is *not* the reason. In fact, their input impedance at the high frequency of FM signals is greatly reduced because of their input capacitance. The fact that FETs do not offer any significant impedance advantage over other devices at high frequencies is not a deterrent, however, because the antenna impedance seen by the RF stage is several hundred ohms or less.

The major advantage is that FETs have a square-law relationship from the input to the output, while vacuum tubes have a $\frac{3}{2}$-power relationship and bipolar-junction transistors have a diode-type exponential characteristic. A square-law device has an output signal at the input frequency and a smaller distortion component at twice the input frequency, whereas the other devices mentioned have many more distortion components, with some of them occurring at frequencies close to the desired signal. The use of an FET at the critical small-signal level in a receiver means that distortion components produced by device nonlinearities are filtered out easily by the receiver tuned circuits because the closest distortion component is at twice the frequency of the desired signal. This consideration becomes significant when a weak station is in the immediate vicinity of a very strong adjacent signal. If the high-level adjacent signal gets through the input tuned circuit, even if it is greatly attenuated, a nonsquare-law device would most likely generate distortion components at the desired signal frequency, and the result would be audible noise in the speaker. This form of receiver noise is called **cross-modulation** and is similar to **intermodulation distortion,** which is characterized by the mixing of *two* undesired signals, resulting in an output, also undesired, at the same frequency as that of the desired signal. The possibility of intermodulation distortion is also greatly minimized by the use of FET RF amplifiers. Additional discussion of intermodulation distortion is included in Section 6-6.

MOSFET RF AMPLIFIERS   A dual-gate, common-source MOSFET RF amplifier is shown in Figure 6-15. The use of a dual-gate device allows a convenient isolated input for an AGC level to control device gain. Metal-oxide semiconductor field-effect transistor (MOSFET) offer the advantage of increased **dynamic range** over junction field-effect

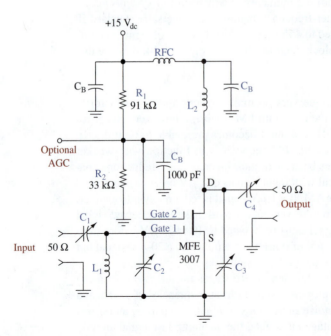

VHF Amplifier
The following component values are used for the different frequencies:

Component Values	100 MHz	400 MHz
C_1	8.4 pF	4.5 pF
C_2	2.5 pF	1.5 pF
C_3	1.9 pF	2.8 pF
C_4	4.2 pF	1.2 pF
L_1	150 nH	16 nH
L_2	280 nH	22 nH
C_B	1000 pF	250 pF

FIGURE 6-15   **MOSFET RF amplifier.**

transistor (JFET). That is, a wider range of input signal can be tolerated by the MOSFET while still offering the desired square-law input/output relationship. A similar arrangement is often used in mixers because the extra gate allows for a convenient injection point for the local oscillator signal. The accompanying chart in Figure 6-15 provides component values for operation at 100-MHz and 400-MHz center frequencies. The antenna input signal is coupled into gate 1 via the coupling/tuning network comprised of C_1, L_1, and C_2. The output signal is taken at the drain, which is coupled to the next stage by the $L_2 - C_3 - C_4$ combination. The bypass capacitor C_B next to L_2 and the radio-frequency choke (RFC) ensure that the signal frequency is not applied to the dc power supply. The RFC acts as an open to the signal while appearing as a short to dc, and the bypass capacitor acts in the inverse fashion. These precautions are necessary at RF frequencies because, while power-supply impedance is very low at low frequencies and dc, the power supply presents a high impedance to RF and can cause appreciable signal power loss. The bypass capacitor from gate 2 to ground provides a short to any high-frequency signal that may get to that point. It is necessary to maintain the bias stability set up by R_1 and R_2. The MFE 3007 MOSFET used in this circuit provides a minimum power gain of 18 dB at 200 MHz.

LIMITERS   A **limiter** is a circuit whose output is a constant amplitude for all inputs above a critical value. Its function in an FM receiver is to remove any residual (unwanted) amplitude modulation and the amplitude variations resulting from noise. Both these variations would have an undesirable effect if carried through to the speaker. In addition, the limiting function also provides AGC action because signals from the critical minimum value up to some maximum value provide a constant input level to the detector. By definition, the discriminator (detector) ideally would not respond to amplitude variations anyway because the information is contained in the amount of frequency deviation and the rate at which it deviates back and forth around its center frequency.

A transistor limiter is shown in Figure 6-16. Notice the dropping resistor, R_C, which limits the dc collector supply voltage. This provides a low dc collector voltage, which makes this stage very easily overdriven. This is the desired result. As soon as the input is large enough to cause clipping at both extremes of collector current, the critical limiting voltage has been attained and limiting action has started.

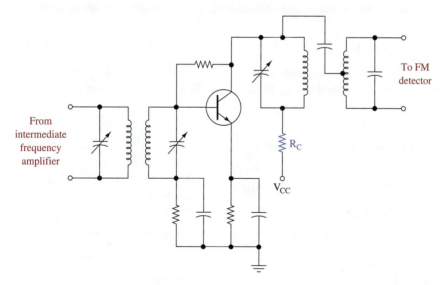

FIGURE 6-16   **Transistor limiting circuit.**

The input/output characteristic for the limiter is shown in Figure 6-17, and it shows the desired clipping action and the effects of feeding the limited (clipped) signal into an *LC* tank circuit tuned to the signal's center frequency. The natural flywheel effect of the tank removes all frequencies not near the center frequency and thus provides a sinusoidal output signal as shown. The omission of an *LC* circuit at the limiter output is desirable for some demodulator circuits. The quadrature detector (Section 6-4) uses the square-wave-like waveform that results.

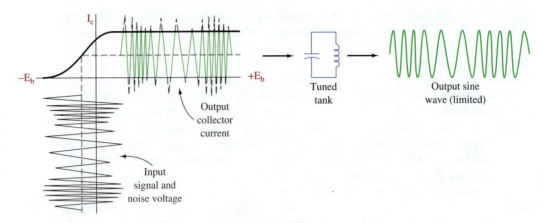

FIGURE 6-17 Limiter input/output and flywheel effects.

LIMITING AND SENSITIVITY A limiter, such as the one shown in Figure 6-16, requires about 1 V of signal to begin limiting. Much amplification of the received signal is, therefore, needed prior to limiting, which explains its position following the IF stages. When enough signal arrives at the receiver to start limiting action, the set *quiets*, which means that background noise disappears. The **sensitivity** of an FM receiver is defined in terms of how much input signal is required to produce a specific level of quieting, normally 30 dB. This specification means that the background noise will be reduced by 30 dB from a given input signal level. A good-quality wideband FM receiver will demonstrate quieting with input signals in the neighborhood of 1.5-μV. A narrowband FM receiver, such as one used in two-way radio service, should quiet with input signals of less than 1 μV.

The minimum required voltage for limiting is called the **quieting voltage**, **threshold voltage**, or **limiting knee voltage.** The limiter then provides a constant-amplitude output up to some maximum value that prescribes the limiting range. Going above the maximum value results either in a reduced or a distorted output. It is possible that a single-stage limiter will not allow for adequate range, thereby requiring a double limiter or the development of AGC control on the RF and IF amplifiers to minimize the possible limiter input range.

EXAMPLE 6-10

A certain FM receiver provides a voltage gain of 200,000 (106 dB) prior to its limiter. The limiter's quieting voltage is 200 mV. Determine the receiver's sensitivity.

SOLUTION

To reach quieting, the input must be

$$\frac{200 \text{ mV}}{200,000} = 1 \ \mu\text{V}$$

The receiver's sensitivity is therefore 1 μV.

Discrete Component FM Receiver

Though the vast majority of modern receiver designs rely heavily on large-scale integrated circuits and digital signal processing techniques, it is instructive to review the layout of a superheterodyne receiver composed of discrete analog components to gain an understanding of how the individual subsystem blocks fit together to form a complete system. In addition, some recent-vintage, high-fidelity FM receivers are partly or fully implemented with discrete components. A typical older-style FM receiver involves use of discrete MOSFET RF and mixer stages with a separately excited bipolar transistor local oscillator, as shown in Figure 6-18. The antenna input signal is applied through the tuning circuit L_1, C_{1A} to the

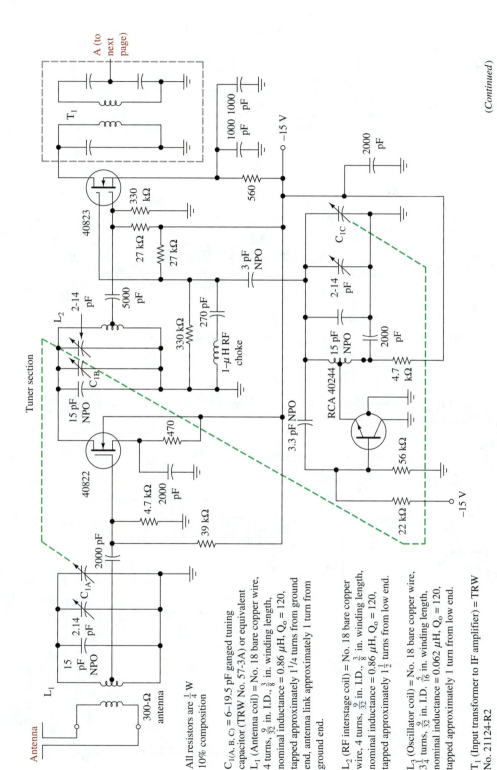

Antenna

L_1

15 pF NPO

C_{1A} 2.14 pF

2000 pF

300-Ω antenna

40822

4.7 kΩ

2000 pF

39 kΩ

470

Tuner section

L_2

15 pF NPO

C_{1B}

2-14 pF

5000 pF

330 kΩ

270 pF

1-μH RF choke

3.3 pF NPO

40823

330 kΩ

27 kΩ

27 kΩ

3 pF NPO

C_{1C}

2-14 pF

15 pF NPO

RCA 40244

2000 pF

4.7 kΩ

56 kΩ

22 kΩ

−15 V

560

1000 pF 1000 pF

−15 V

2000 pF

T_1

A (to next page)

All resistors are $\frac{1}{4}$ W 10% composition

$C_{1(A, B, C)}$ = 6–19.5 pF ganged tuning capacitor (TRW No. 57-3A) or equivalent
L_1 (Antenna coil) = No. 18 bare copper wire, 4 turns, $\frac{9}{32}$ in. I.D., $\frac{3}{8}$ in. winding length, nominal inductance = 0.86 μH, Q_o = 120, tapped approximately $1\frac{1}{4}$ turns from ground end, antenna link approximately 1 turn from ground end.

L_2 (RF interstage coil) = No. 18 bare copper wire, 4 turns, $\frac{9}{32}$ in. I.D., $\frac{3}{8}$ in. winding length, nominal inductance = 0.86 μH, Q_o = 120, tapped approximately $1\frac{1}{2}$ turns from low end.

L_3 (Oscillator coil) = No. 18 bare copper wire, $3\frac{1}{4}$ turns, $\frac{9}{32}$ in. I.D. $\frac{5}{16}$ in. winding length, nominal inductance = 0.062 μH, Q_o = 120, tapped approximately 1 turn from low end.

T_1 (Input transformer to IF amplifier) = TRW No. 21124-R2

FIGURE 6-18 **Complete 88- to 108-MHz stereo receiver.**

(Continued)

175

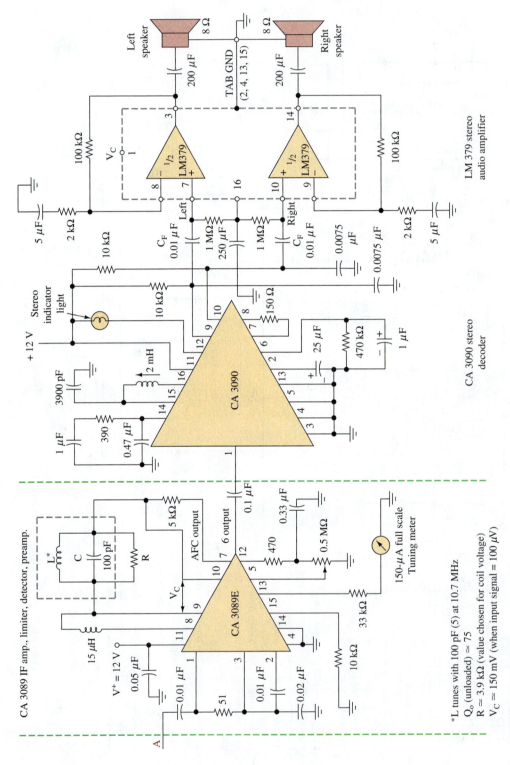

FIGURE 6-18 (Continued)

176

gate of the 40822 MOSFET RF amplifier. Its output at the drain is coupled to the lower gate of the 40823 mixer MOSFET through the $C_{1B} - L_2$ tuned circuit. The 40244 BJT oscillator signal is applied to the upper gate of the mixer stage. The local oscillator tuned circuit that includes C_{1C} uses a tapped inductor indicating a Hartley oscillator configuration. The tuning capacitor, C_1, has three separate ganged capacitors that vary the tuning range of the RF amp and mixer tuned circuits from 88 to 108 MHz while varying the local oscillator frequency from 98.7 to 118.7 MHz to generate a 10.7-MHz IF signal at the output of the mixer. The mixer output is applied to the commercially available 10.7-MHz double-tuned circuit T_1.

MOSFET receiver front ends offer superior cross-modulation and intermodulation performance as compared to other types. The Institute of High Fidelity Manufacturers (IHFM) sensitivity for this front end is about 1.75 μV. It is defined as the minimum 100% modulated input signal that reduces the total receiver noise and distortion to 30 dB below the output signal. In other words, a 1.75-μV input signal produces 30-dB quieting.

The front-end output through T_1 in Figure 6-18 is applied to a CA3089 IC. The CA3089 provides three stages of IF amplification—limiting, demodulation, and audio preamplification. It provides demodulation with an analog quadrature detector circuit (Section 6-4). It also provides a signal to drive a tuning meter and an AFC output for direct control of a varactor tuner. Its audio output includes the 30-Hz to 15-kHz (L + R) signal, 19-kHz pilot carrier, and 23- to 53-kHz (L − R) signal, which are then applied to the FM stereo decoder IC.

The audio outputs are then applied to a ganged volume control potentiometer (not shown) and then to an LM379 dual 6-W audio amplifier. It has two separate audio amplifiers in one 16-lead IC and has a minimum input impedance of 2 MΩ per channel. It typically provides a voltage gain of 34 dB, total harmonic distortion (THD) of 0.07% at 1-W output, and 70 dB of channel separation.

6-4 DIRECT CONVERSION RECEIVERS

In the superheterodyne receiver, the mixing action occurs first when the intermediate frequency is produced, and, second, when the intelligence is extracted from the down-converted IF. In contrast, the **direct conversion** receiver uses a single mixer stage to down-convert the incoming radio-frequency signal directly to the baseband frequency range of the intelligence. Put another way, direct-conversion receivers perform the frequency-conversion and demodulation functions in one step, rather than two.

Direct-conversion receivers are also known as **zero-IF receivers** because the mixer difference-frequency output is the intelligence rather than a higher-frequency IF. A typical block diagram is shown as Figure 6-19. The local-oscillator (LO) frequency, which in modern designs is most likely determined by a PLL frequency synthesizer, is made equal to that of the received signal. Sum and difference frequencies will be created as the result of mixing. The sum will be at a frequency twice that of the received signal, and the difference frequency in the absence of modulation will be 0 Hz. That is, there will be no output without modulation. In the presence of DSB or SSB AM, the sidebands mix with the LO output to recreate the original intelligence. A low-pass filter (LPF) removes the high-frequency (sum) mixing products without affecting the intelligence output.

The direct-conversion architecture has a number of advantages. One is relative simplicity, an important consideration in portable and low-power designs. The direct-conversion receiver requires no IF filter, which (particularly in up-conversion designs), is often a relatively expensive and bulky crystal, ceramic, or surface-acoustic wave (SAW) device. Because it will be called upon to pass only low-frequency baseband signals, the LPF can be

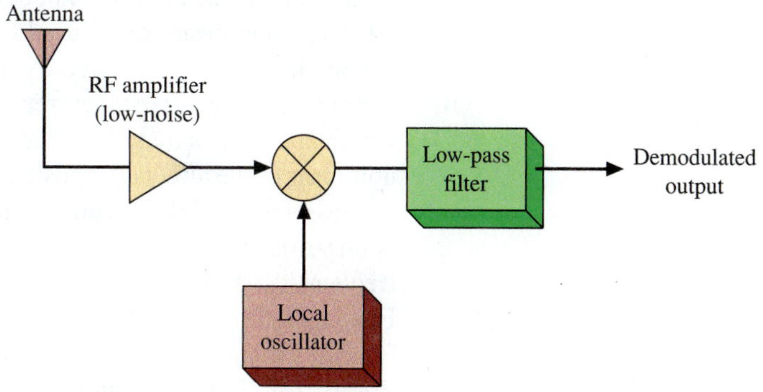

FIGURE 6-19 Direct-conversion receiver.

implemented with low-cost RC or active filters. There is also no need for a separate demodulator stage, contributing further to cost and space savings. Finally, the direct-conversion design is immune to the image-frequency problem, further simplifying matters by eliminating the need for multiple stages of filtering and elaborate tuning mechanisms.

One drawback to the zero-IF architecture is that it is susceptible to local-oscillator reradiation effects, whereby the LO output leaks back through the mixer to be radiated by the antenna. Such radiation causes undesired interference to other devices in the immediate vicinity and must be minimized through careful design. Placing a low-noise amplifier (LNA) before the mixer can help minimize, but not completely eliminate, undesired radiation. Another concern is the possible introduction of unintended dc offsets, which can adversely affect bias voltages and cause amplifier stages to saturate, thus preventing proper (linear) operation.

Finally, the zero-IF receiver shown in Figure 6-19 is capable of demodulating only AM, DSB, or SSB signals. It is incapable of recognizing either frequency or phase variations. For the direct-conversion architecture to be used as an FM, PM, or digital receiver, two mixers must be used along with a quadrature-detection arrangement. Though more complex than their AM counterparts, direct-conversion FM and digital receivers are very common, and, in fact, constitute the great majority of designs for cellular phones. The direct-conversion FM/PM receiver will be discussed in more detail in Chapter 8.

6-5 DEMODULATION AND DETECTORS

The purpose of the detector is to recover the low-frequency intelligence or baseband signal from the much higher intermediate- or radio-frequency carrier, at which point the carrier is no longer needed and can be discarded. The type of detector device or circuit used depends on the type of modulation employed, the presence or absence of the carrier, and on the fidelity requirements of the demodulated intelligence, among other factors. In an AM receiver, detection will be seen shortly to be the result of nonlinear mixing, and the sophistication of the detector circuit will depend on whether the carrier is present as part of the received signal (i.e., DSBFC AM from Chapter 2) or whether the carrier must be regenerated in the receiver (i.e., SSB). FM detection can be achieved in a number of ways, but in all cases the FM detector converts variations in frequency or phase to voltages that represent the intelligence.

AM Diode Detector

Passing the received AM signal, consisting of the carrier and sum and difference frequencies, through a nonlinear device will provide detection by means of the same mixing process that generated the frequency components of the modulated signal at the transmitter. The difference frequencies created by the AM receiver detector represent the intelligence. This result can be confirmed in both the time and frequency domains.

The nonlinear device used for detection may be nothing more than an ordinary silicon diode. The diode will rectify the incoming signal, thus distorting it. Recall that a diode exhibits a voltage-versus-current characteristic curve (*V-I* curve) shown in Figure 6-20, which approaches that of the ideal curve shown in the upper left of Figure 6-20. An ideal nonlinear curve affects the positive half-cycles of the modulated wave differently than the negative half-cycles. The curve shown in Figure 6-20(a) is called an *ideal curve* because it is linear on each side of the operating point *P* and, therefore, does not introduce harmonic frequencies. The average voltage of the modulated input waveform, shown at the left of Figure 6-20, is zero volts because it is symmetrical on either side of the operating point, *P*. (Recall the appearance of the modulated AM carrier in the time domain, as shown in Figure 2-8. The upper and lower contours of the modulation envelope represent equal amplitudes but have opposite polarities. The symmetry about the center point so implied demonstrates that the average voltage is zero.) Rectification by the diode distorts the

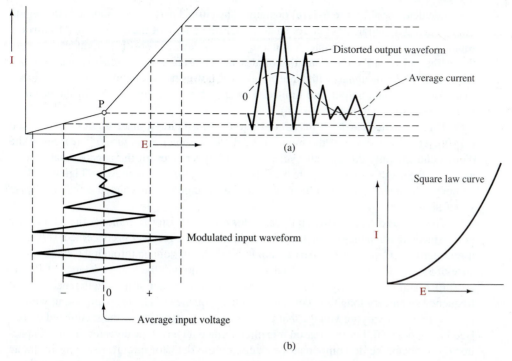

FIGURE 6-20   **Nonlinear device used as a detector.**

modulated waveform by causing the average voltage to have a net positive value and the average resultant current to vary with the intelligence signal amplitude.

One of the simplest and most effective types of AM detector circuits, and one with nearly an ideal nonlinear resistance characteristic, is the diode detector circuit shown in Figure 6-21(a). Notice the *V–I* curve in Figure 6-21(b). This is the type of curve on which

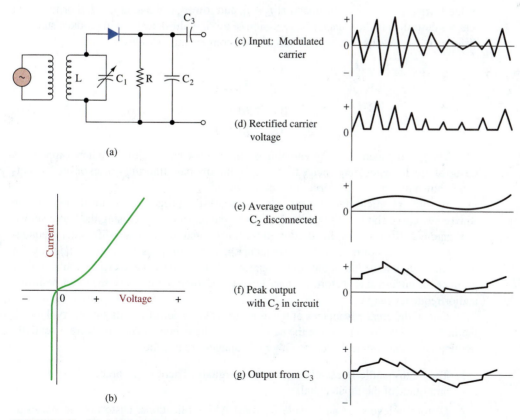

FIGURE 6-21   **Diode detector.**

the diode detector at Figure 6-21(a) operates. The curved part of its response is the region of low current and indicates that for small signals the output of the detector will follow the square law. (Recall from Chapter 4 that a square-law device is one that produces an output that is the square of the input and that diodes and field-effect transistors exhibit this characteristic.) For input signals with large amplitudes, however, the output is essentially linear. Therefore, harmonic outputs are limited. The abrupt nonlinearity occurs for the negative half-cycle as shown in Figure 6-21(b).

The modulated carrier is introduced into the tuned circuit made up of LC_1 in Figure 6-21(a). The wave shape of the input to the diode is shown in Figure 6-21(c). Since the diode conducts only during half-cycles, this circuit removes all the negative half-cycles and gives the result shown in Figure 6-21(d). The average output is shown at Figure 6-21(e). Although the average input voltage is zero, the average output voltage across R always varies above zero.

The low-pass filter, made up of capacitor C_2 and resistor R, removes the RF (carrier or intermediate frequency), which, so far as the rest of the receiver is concerned, serves no useful purpose. Capacitor C_2 charges rapidly to the peak voltage through the small resistance of the conducting diode, but discharges slowly through the high resistance of R. The sizes of R and C_2 normally form a rather short time constant at the intelligence (audio) frequency and a very long time constant at radio frequencies. The resultant output with C_2 in the circuit is a varying voltage that follows the peak variation of the modulated carrier [see Figure 6-21(f)]. For this reason it is often termed an **envelope detector** circuit. Capacitor C_3 removes the dc component produced by the detector circuit, resulting in the ac voltage wave shape seen in Figure 6-21(g). In communications receivers, the dc component is often used for providing automatic volume (gain) control.

Viewed in the time domain, we see that the detector reproduces the signal frequency by producing distortion of a desirable kind in its output. When the detector output is impressed upon a low-pass filter, the radio frequencies are suppressed and only the low-frequency intelligence signal and dc components are left. This result is shown as the dashed average current curve in Figure 6-20(a).

This result can be confirmed in the frequency domain to show that the output of the diode detector consists of multiple frequency components. The detector inputs will consist of the carrier, f_c, the upper sideband at $f_c + f_i$, and lower sideband at $f_c - f_i$. For the same reasons described in the Chapter 4 discussion of mixing, the detector will produce sum and difference frequencies (cross-products) of each component, as follows:

$$f_c + (f_c + f_i) = 2f_c + f_i$$
$$f_c - (f_c + f_i) = -f_i$$
$$f_c + (f_c - f_i) = 2f_c - f_i$$
$$f_c - (f_c - f_i) = f_i$$

The f_i component is the modulating or intelligence signal. All the other components reside at much higher frequencies than f_i, and the low-pass filtering action of the RC network shown in Figure 6-21 serves to remove them.

In some practical detector circuits, the nearest approach to the ideal curve is the square-law curve shown in Figure 6-20(b). The output of a device using the curve shown in Figure 6-20(b) contains, in addition to the carrier and the sum and difference frequencies, the harmonics of each of these frequencies. The harmonics of radio frequencies can be filtered out, but the harmonics of the sum and difference frequencies, even though they produce an undesirable distortion, may have to be tolerated because they can be in the audio-frequency range.

One of the great advantages of the diode (envelope) detector is its simplicity. Indeed, the first crystal radios, created in the early years of the twentieth century, were essentially nothing more than diode detectors. Other advantages are as follows:

1. They can handle relatively high power signals. There is no practical limit to the amplitude of the input signal.

2. Distortion levels are acceptable for most AM applications. Distortion decreases as the amplitude increases.

3. They are highly efficient. When properly designed, 90% efficiency is obtainable.

4. They develop a readily usable dc voltage for the automatic gain control circuits.

There are two disadvantages of diode detectors:

1. Power is absorbed from the tuned circuit by the diode circuit. This reduces the Q and selectivity of the tuned input circuit.

2. No amplification occurs in a diode detector circuit.

Detection of Suppressed-Carrier Signals

As already mentioned, demodulation of the SSB signal requires a mixer stage and a beat-frequency oscillator. The mixer serving as the demodulator is also known as the second detector. A simple diode detector cannot be used to demodulate an SSB signal because one of the frequencies, the carrier, is absent.

A time-domain representation of the suppressed-carrier DSB waveform reveals that the contours formed when the positive and negative peaks of the RF signal are joined together no longer represent the intelligence. In other words, the modulation envelope created does not retain the shape of the modulating signal as it did for full-carrier DSB AM. Figure 6-22(a) shows three different sine-wave intelligence signals; in Figure 6-22(b), the resulting AM waveforms are shown, and Figure 6-22(c) shows the DSB (no carrier) waveform. Notice that the DSB envelope (drawn in for illustrative purposes) looks like a full-wave rectification of the envelope of the corresponding AM waveform. The envelope is also at twice the frequency of the AM envelope. In Figure 6-22(d), the SSB waveforms are

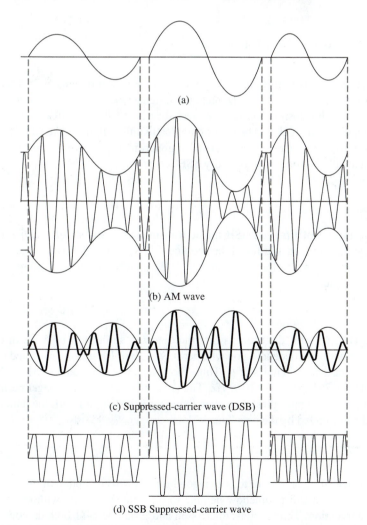

(a)

(b) AM wave

(c) Suppressed-carrier wave (DSB)

(d) SSB Suppressed-carrier wave

FIGURE 6-22 **AM, DSB, and SSB waves from sinusoidal modulating signals.**

simply pure sine waves. This is precisely what is transmitted in the case of a sine-wave modulating signal. These waveforms are either at the carrier plus the intelligence frequency (usb) or carrier minus intelligence frequency (lsb). If a diode detector were used in this situation, it would rectify the received sideband and produce a constant dc voltage output. For the SSB receiver to demodulate the received sideband, the carrier would have to be reinserted at the receiver.

A simple way to form an SSB detector is to use a mixer stage. When used in this fashion, the mixer is called a product detector, and it is identical in form and function to the balanced modulator found in the transmitter. The mixer is a nonlinear device, and the local oscillator input must be equivalent to the desired carrier frequency.

Figure 6-23 shows this situation pictorially. Consider a 500-kHz carrier frequency that has been modulated by a 1-kHz sine wave. If the upper sideband were transmitted, the receiver's demodulator would see a 501-kHz sine wave at its input. Therefore, a 500-kHz oscillator input will result in a mixer output frequency component of 1 kHz, which is the desired result. If the 500-kHz oscillator is not exactly 500 kHz, the recovered intelligence will not be exactly 1 kHz.

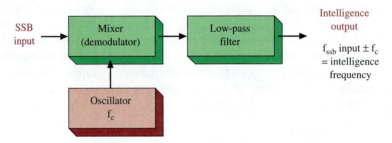

FIGURE 6-23 **Mixer used as SSB demodulator.**

Even minor drifts in BFO frequency can cause serious problems in SSB reception. If the oscillator drifts ± 100 Hz, a 1-kHz intelligence signal would be detected either as 1100 Hz or 900 Hz. Speech transmission requires less than a ± 100-Hz shift or the talker starts sounding like Donald Duck and becomes completely unintelligible. The BFO, therefore, is usually a variable-frequency oscillator, and for noncritical applications, the listener adjusts the BFO manually to obtain good-quality SSB reception.

EXAMPLE 6-11

At one instant of time, an SSB music transmission consists of a 256-Hz sine wave and its second and fourth harmonics, 512 Hz and 1024 Hz. If the receiver's demodulator oscillator has drifted 5 Hz, determine the resulting speaker output frequencies.

SOLUTION

The 5-Hz oscillator drift means that the detected audio will be 5 Hz in error, either up or down, depending on whether it is a usb or lsb transmission and on the direction of the oscillator's drift. Thus, the output would be either 251, 507, and 1019 Hz or 261, 517, and 1029 Hz. The speaker's output is no longer harmonic (exact frequency multiples), and even though it is just slightly off, the human ear would be offended by the new "music."

For critical applications, or to obtain good-quality SSB reception of music and digital signals, the receiver must somehow be able to recreate the carrier without the need for operator intervention. The previous discussion and Example 6-11 both demonstrate why. Tuning the sideband receiver is more difficult than tuning a conventional AM receiver because the BFO must be adjusted precisely to simulate the carrier frequency at all times. Any tendency for the oscillator to drift will cause the output intelligence to be distorted. For this reason, some forms of single-sideband transmission do not fully suppress the carrier but rather reduce it to a low level and transmit it along with the desired sideband. The reduced-power carrier is known as a pilot carrier, and, when received, can be used to recreate the intelligence. This process is known as coherent carrier recovery, and, among its other applications, coherent carrier recovery is used extensively in digital communications systems, whose transmitted data will be shown in Chapter 8 to share the characteristics of a DSB suppressed-carrier signal.

A **synchronous detector** regenerates the carrier from the received signal. It is a subsystem consisting of a product detector combined with additional circuitry to regenerate the original carrier, both in frequency and phase. The synchronous detector relies on the principle of the phase-locked loop discussed in Chapter 4 to regenerate the carrier in the receiver. In addition to the product detector previously discussed, the synchronous detector consists of a phase detector and voltage-controlled oscillator. The phase-locked loop locks into the reduced-amplitude incoming carrier, which serves as a reference to keep the receiver oscillator locked to the transmitter carrier.

Diode detectors are used in the vast majority of AM detection schemes where the carrier and sidebands are present. Since high fidelity is usually not an important aspect in AM, the distortion levels of several percentage points or more from a diode detector can be tolerated easily. In applications demanding greater performance, the use of a synchronous detector offers the following advantages:

1. Low distortion—well under 1%

2. Greater ability to follow fast-modulation waveforms, as in pulse-modulation or high-fidelity applications

3. The ability to provide gain instead of attenuation, as in diode detectors

A National Instruments Multisim implementation of a synchronous AM detector is provided in Figure 6-24(a). The AM signal feeds the Y input to the mixer circuit (A_1), and it is also fed through the high-gain limiter stage A_2. The high-gain limiter is used to limit the amplitude variations, leaving only the 900-kHz carrier signal. The mixer circuit outputs the sum and difference in the two input frequencies appearing on the X and Y input. The sum will be twice the AM carrier frequency plus the 1-kHz component, whereas the difference will be simply the intelligence frequency (1 kHz). Resistor R_1, Inductor L_1, and Capacitor C_1 form a low-pass filter with a cutoff frequency of about 1 kHz that is

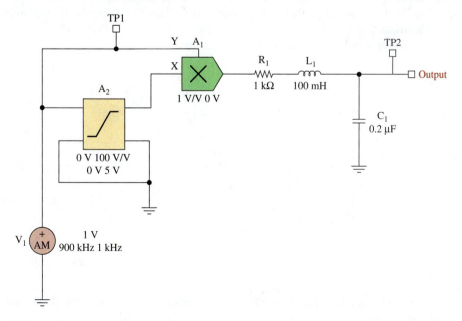

FIGURE 6-24 (a) A synchronous AM detector circuit as implemented in National Instruments Multisim.

(*Continued*)

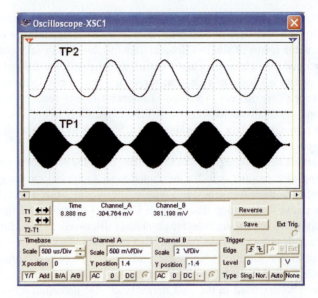

FIGURE 6-24 (*Continued*) **(b) The waveforms obtained from tP1 and TP2 for the synchronous AM detector provided in Figure 6-24(a).**

used to filter out the high-frequency components so only the intelligence frequency appears on the output.

The waveform of the AM input signal at Test Point 1 (TP1) is shown in Figure 6-24(b). This example is showing a 100% modulated AM waveform. The recovered 1-kHz signal is shown at TP2 in Figure 6-24(b).

Demodulation of FM and PM

The FM discriminator (detector) extracts the intelligence that has been modulated onto the carrier in the form of frequency variations. The detector should provide an intelligence signal whose amplitude is dependent on instantaneous carrier frequency deviation and whose frequency is dependent on the rate of carrier frequency deviation. A desired output amplitude versus input frequency characteristic for a broadcast FM discriminator is provided in Figure 6-25. Notice that the response is linear in the allowed area of frequency deviation and that the output amplitude is directly proportional to carrier frequency deviation. Keep in mind, however, that FM detection takes place following the IF amplifiers, which means that, in the case of broadcast FM, the $\pm$75-kHz deviation is intact but that carrier frequency translation (usually to 10.7 MHz) has occurred.

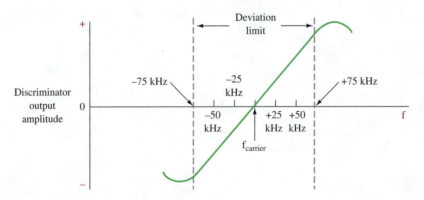

FIGURE 6-25 **FM discriminator characteristic.**

SLOPE DETECTOR The easiest FM discriminator to understand is the slope detector in Figure 6-26.

The *LC* tank circuit that follows the IF amplifiers and limiter is detuned from the carrier frequency so that f_c falls in the middle of the most linear region of the response curve.

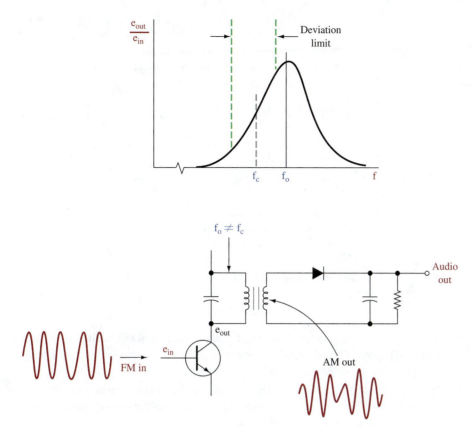

FIGURE 6-26 **Slope detection.**

When the FM signal rises in frequency above f_c, the output amplitude increases while deviations below f_c cause a smaller output. The slope detector thereby changes FM into AM, and a simple diode detector then recovers the intelligence contained in the AM waveform's envelope. In an emergency, an AM receiver can be used to receive FM by detuning the tank circuit feeding the diode detector. Slope detection is not widely used in FM receivers because the slope characteristic of a tank circuit is not very linear, especially for the large-frequency deviations of wideband FM.

FOSTER–SEELEY DISCRIMINATOR   The two classical means of FM detection are the Foster–Seeley discriminator and the ratio detector. While their once widespread use is now diminishing because of new techniques afforded by ICs, they remain a popular means of discrimination using a minimum of circuitry. Also, these circuits show how frequency or phase shifts can be converted to voltage changes using ordinary passive components. Because the need for phase-shift detection arises quite often in a variety of communications contexts, and because the concept of phase-shifting may be intuitively difficult to visualize, we will describe the operation of both types of circuits here.

A typical Foster–Seeley discriminator circuit is shown in Figure 6-27. In it, the two tank circuits [L_1C_1 and $(L_2+L_3)C_2$] are tuned exactly to the carrier frequency. Capacitors C_c, C_4, and C_5 are shorts to the carrier frequency. The following analysis applies to an unmodulated carrier input:

1. The carrier voltage e_1 appears directly across L_4 because C_c and C_4 are shorts to the carrier frequency.

2. The voltage e_s across the transformer secondary (L_2 in series with L_3) is 180° out of phase with e_1 by transformer action, as shown in Figure 6-28(a). The circulating $L_2L_3C_2$ tank current, i_s, is in phase with e_s because the tank is resonant.

3. The current i_s, flowing through inductance L_2L_3, produces a voltage drop that lags i_s by 90°. The individual components of this voltage, e_2 and e_3, are thus displaced by 90° from i_s, as shown in Figure 6-28(a), and are 180° out of phase with each other because they are the voltages from the ends of a center-tapped winding.

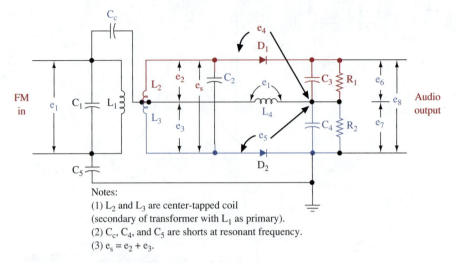

Notes:
(1) L_2 and L_3 are center-tapped coil (secondary of transformer with L_1 as primary).
(2) C_c, C_4, and C_5 are shorts at resonant frequency.
(3) $e_s = e_2 + e_3$.

FIGURE 6-27 **Foster–Seeley discriminator.**

4. The voltage e_4 applied to the diode D_1, C_3, and R_1 network will be the vector sum of e_1 and e_2 [Figure 6-28(a)]. Similarly, the voltage e_5 is the sum of e_1 and e_3. The magnitude of e_6 is proportional to e_4 while e_7 is proportional to e_5.

5. The output voltage, e_8, is equal to the sum of e_6 and e_7 and is zero because the diodes D_1 and D_2 will be conducting current equally (because $e_4 = e_5$) but in opposite directions through the R_1C_3 and R_2C_4 networks.

The discriminator output is zero with no modulation (zero frequency deviation), as is desired. The following discussion now considers circuit action at some instant when the input signal e_1 is above the carrier frequency. The phasor diagram of Figure 6-28(b) is used to illustrate this condition:

1. Voltages e_1 and e_s are as before, but e_s now sees an inductive reactance because the tank circuit is above resonance. Therefore, the circulating tank current, i_s, lags e_s.

2. The voltages e_2 and e_3 must remain 90° out of phase with i_s, as shown in Figure 6-28(b). The new vector sums of $e_2 + e_1$ and $e_3 + e_1$ are no longer equal, so e_4 causes a heavier conduction of D_1 than for D_2.

3. The output, e_8, which is the sum of e_6 and e_7, will go positive because the current down through R_1C_3 is greater than the current up through R_2C_4 (e_4 is greater than e_5).

The output for frequencies above resonance (f_c) is therefore positive, while the phasor diagram in Figure 6-28(c) shows that at frequencies below resonance the output goes negative. The amount of output is determined by the amount of frequency deviation, while the frequency of the output is determined by the rate at which the FM input signal varies around its carrier or center value.

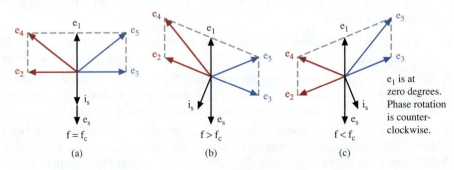

FIGURE 6-28 **Phasor diagram for Foster–Seeley discriminator.**

RATIO DETECTOR   While the Foster–Seeley discriminator just described offers excellent linear response to wideband FM signals, it also responds to any undesired input amplitude variations. The *ratio detector* does not respond to amplitude variations and thereby minimizes the required limiting before detection.

The ratio detector, shown in Figure 6-29, is a circuit designed to respond only to frequency changes of the input signal. Amplitude changes in the input have no effect upon the output. The input circuit of the ratio detector is identical to that of the Foster–Seeley discriminator circuit. The most immediately obvious difference is the reversal of one of the diodes.

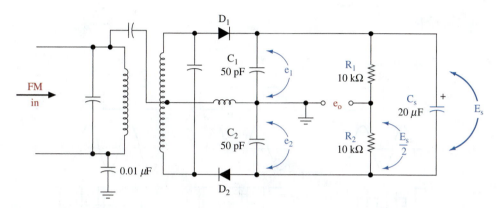

FIGURE 6-29   **Ratio detector.**

The ratio detector circuit operation is similar to the Foster–Seeley. A detailed analysis will therefore not be given. Notice the large electrolytic capacitor, C_5, across the R_1–R_2 combination. This maintains a constant voltage that is equal to the peak voltage across the diode input. This feature eliminates variations in the FM signal, thus providing amplitude limiting. The sudden changes in input-signal amplitude are suppressed by the large capacitor. The Foster–Seeley discriminator does not provide amplitude limiting. The voltage E_s is

$$E_s = e_1 + e_2$$

and

$$e_0 = \frac{E_s}{2} - e_2 = \frac{e_1 + e_2}{2} - e_2$$
$$= \frac{e_1 - e_2}{2}$$

When $f_{in} = f_c$, $e_1 = e_2$ and hence the desired zero output occurs. When $f_{in} > f_c$, $e_1 > e_2$, and when $f_{in} < f_c$, $e_1 < e_2$. The desired frequency-dependent output characteristic results.

The component values shown in Figure 6-29 are typical for a 10.7-MHz IF FM input signal. The output level of the ratio detector is one-half that for the Foster–Seeley circuit.

QUADRATURE DETECTOR   The Foster–Seeley and ratio detector circuits do not lend themselves to integration on a single chip because of the transformer required. This has led to increased usage of the quadrature detector and phase-locked loop (PLL).

Quadrature detectors derive their name from use of the FM signal in phase and 90° out of phase. The two signals are said to be in **quadrature**—at a 90° angle. The circuit in Figure 6-30 shows an FM quadrature detector using an exclusive-OR (XOR) gate. The limited IF output is applied directly to one input and the phase-shifted signal to the other. Notice that this circuit uses the limited signal that has not been changed back to a sine wave. The L, C, and R values used at the circuit's input are chosen to provide a 90° phase shift at the carrier frequency to the signal 2 input. The signal 2 input is a sine wave created by the LC circuit. The upward and downward frequency deviation of the FM signal results

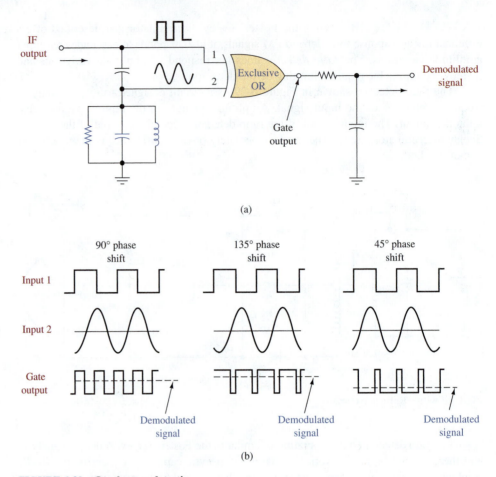

FIGURE 6-30 **Quadrature detection**

in a corresponding higher or lower phase shift. With one input to the gate shifted, the gate output will be a series of pulses with a width proportional to the phase difference. The low-pass *RC* filter at the gate output sums the output, giving an average value that is the intelligence signal. The gate output for three different phase conditions is shown at Figure 6-30(b). The *RC* circuit output level for each case is shown with dashed lines. This corresponds to the intelligence level at those particular conditions.

An analog quadrature detector is possible using a differential amplifier configuration, as shown in Figure 6-31. A limited FM signal switches the transistor current source (Q_1) of the differential pair $Q_2 + Q_3$. L_1 and C_2 should be resonant at the IF frequency. The

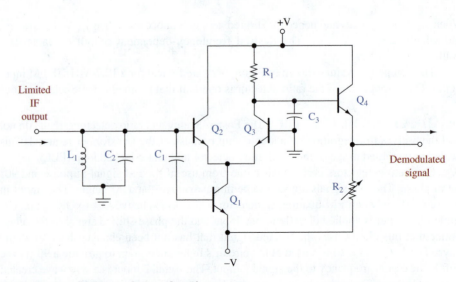

FIGURE 6-31 **Analog quadrature detection.**

L_1–C_2–C_1 combination causes the desired frequency-dependent phase shift between the two signals applied to Q_2 and Q_1. The conduction through Q_3 depends on the coincident phase relationships of these two signals. The pulses generated at the collector of Q_3 are summed by the R_1–C_3 low-pass filter, and the resulting intelligence signal is taken at the emitter of Q_4. R_2 is adjusted to yield the desired zero-volt output when an undeviating FM carrier is the circuit's input signal.

PLL FM DEMODULATOR   If the PLL input is an FM signal, the low-pass filter output (error voltage) is the demodulated signal. The modulated FM carrier changes frequency according to the modulating signal. The function of the phase-locked loop is to hold the VCO frequency in step with this changing carrier. If the carrier frequency increases, for example, the error voltage developed by the phase comparator and the low-pass filter rises to make the VCO frequency rise. Let the carrier frequency fall and the error voltage output drops to decrease the VCO frequency. Thus, we see that the error voltage matches the modulating signal back at the transmitter; the error signal is the demodulated output.

The VCO input control signal (demodulated FM) causes the VCO output to match the FM signal applied to the PLL (comparator input). If the FM carrier (center) frequency drifts because of local oscillator drift, the PLL readjusts itself and no realignment is necessary. In a conventional FM discriminator, any shift in the FM carrier frequency results in a distorted output because the *LC* detector circuits are then untuned. The PLL FM discriminator requires neither tuned circuits nor their associated adjustments and adjusts itself to any carrier frequency drifts caused by LO or transmitted carrier drift. In addition, the PLL normally has large amounts of internal amplification, which allows the input signal to vary from the microvolt region up to several volts. Since the phase comparator responds only to phase changes and not to amplitudes, the PLL is seen to provide a limiting function of extremely wide range. The use of PLL FM detectors is widespread in current designs.

6-6 STEREO DEMODULATION

FM stereo receivers are identical to monaural receivers up to the discriminator output. At this point, however, the discriminator output contains the 30-Hz to 15-kHz (L + R) signal *and* the 19-kHz subcarrier *and* the 23- to 53-kHz (L − R) signal. If a nonstereo (monaural) receiver is tuned to a stereo station, its discriminator output may contain the additional frequencies, but even the 19-kHz subcarrier is above the normal audible range, and its audio amplifiers and speaker would probably not pass it anyway. Thus, the monaural receiver reproduces the 30-Hz to 15-kHz (L + R) signal (a full monophonic broadcast) and is not affected by the other frequencies. This effect is illustrated in Figure 6-32.

The stereo receiver block diagram becomes more complex after the discriminator. At this point, filtering action separates the three signals. The (L + R) signal is obtained through a low-pass filter and given a delay so that it reaches the matrix network in step with the (L − R) signal. A 23- to 53-kHz bandpass filter selects the (L − R) double sideband signal. A 19-kHz bandpass filter takes the pilot carrier and is multiplied by 2 to 38 kHz, which is the precise carrier frequency of the DSB suppressed carrier 23- to 53-kHz (L − R) signal. Combining the 38-kHz and (L − R) signals through the nonlinear device of an AM detector generates sum and difference outputs of which the 30-Hz to 15-kHz (L − R) components are selected by a low-pass filter. The (L − R) signal is thereby retranslated back down to the audio range and it and the (L + R) signal are applied to the matrix network for further processing.

Figure 6-33 illustrates the matrix function and completes the stereo receiver block diagram of Figure 6-31. The (L + R) and (L − R) signals are combined in an adder that cancels R because (L + R) + (L − R) = 2L. The (L − R) signal is also applied to an inverter, providing −(L − R) = (−L + R), which is subsequently applied to another adder along with (L + R), which produces (−L + R) + (L + R) = 2R. The two individual signals for the right and left channels are then deemphasized and individually amplified to their own speakers.

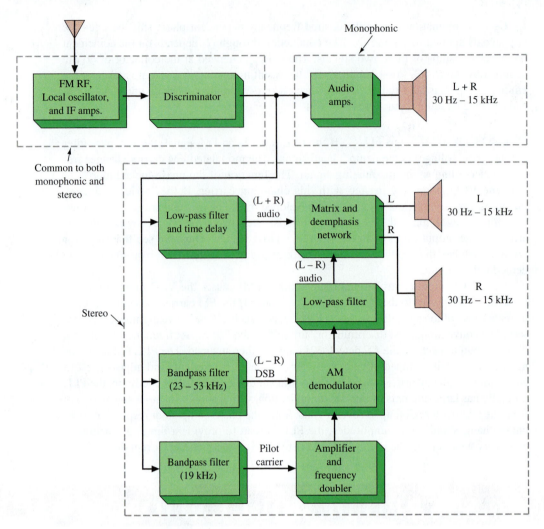

FIGURE 6-32   **Monaural and stereo receivers.**

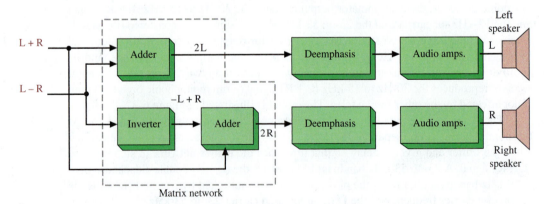

FIGURE 6-33   **Stereo signal processing.**

The process of FM stereo is ingenious in its relative simplicity and effectiveness in providing complete compatibility and doubling the amount of transmitted information through the use of multiplexing.

SCA Decoder

The FCC has also authorized FM stations to broadcast an additional signal on their carrier. It may be a voice communication or other signal for any nonbroadcast-type use. It is often used to transmit music programming that is usually commercial-free but paid for by subscription of department stores, supermarkets, and the like. It is termed the **subsidiary communication authorization (SCA)**. It is frequency-multiplexed on the FM modulating signal, usually with a 67-kHz carrier and ± 7.5-kHz (narrowband) deviation, as shown in Figure 6-34.

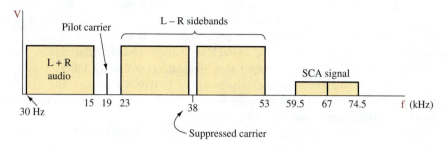

FIGURE 6-34   **Composite stereo and SCA modulating signal.**

6-7 RECEIVER NOISE, SENSITIVITY, AND DYNAMIC RANGE RELATIONSHIPS

Now that you have become more knowledgeable about receivers, it is appropriate to expand on the noise considerations discussed in Chapter 1. As you will see, there are various tradeoffs and relationships among noise figure, sensitivity, and dynamic range when dealing with high-quality receiver systems.

To understand fully these relationships for a receiver, it is first necessary to recognize the factors limiting sensitivity. In one word, the factor most directly limiting sensitivity is noise. Without noise, it would be necessary only to provide enough amplification to receive any signal, no matter how small. Unfortunately, noise is always present and must be understood and controlled as much as possible.

As explained in Chapter 1, there are many sources of noise. The overwhelming effect in a receiver is thermal noise caused by electron activity in a resistance. From Chapter 1, the noise power is

$$P_n = kT \, \Delta f \qquad (1\text{-}8)$$

For a 1-Hz bandwidth (Δf) and at 290 K

$$P = 1.38 \times 10^{-23} \, \text{J/K} \times 290 \, \text{K} \times 1 \, \text{Hz}$$
$$= 4 \times 10^{-21} \, \text{W} = -174 \, \text{dBm}$$

For a 1-Hz, 1-K system

$$P = 1.38 \times 10^{-23} \, \text{W} = -198 \, \text{dBm}$$

The preceding shows the temperature variable is of interest because it is possible to lower the circuit temperature and decrease noise without changing other system parameters. At 0 K no noise is generated. It is very expensive and difficult, however, to operate systems anywhere near 0 K. Most receiving systems are operated at ambient temperature. The other possible means to lower thermal noise is to reduce the bandwidth. However, the designer has limited capability in this regard.

Noise and Receiver Sensitivity

What is the sensitivity of a receiver? This question cannot be answered directly without making certain assumptions or knowing certain facts that will have an effect on the result. Examination of the following formula illustrates the dependent factors in determining sensitivity.

$$S = \text{sensitivity} = -174\ \text{dBm} + \text{NF} + 10\log_{10}\Delta f + \text{desired } S/N \qquad (6\text{-}2)$$

where −174 dBm is the thermal noise power at room temperature (290 K) in a 1-Hz bandwidth. It is the performance obtainable at room temperature if no other degrading factors are involved. The $10\log_{10}\Delta f$ factor in Equation (6-2) represents the change in noise power resulting from an increase in bandwidth above 1 Hz. The wider the bandwidth, the greater the noise power and the higher the noise floor. *S/N* is the desired signal-to-noise ratio in dB. It can be determined for the signal level, which is barely detectable, or it may be regarded as the level allowing an output at various ratings of fidelity. Often, a 0-dB *S/N* is used, which means that the signal and noise power at the output are equal. The signal can therefore also be said to be equal to the **noise floor** of the receiver. The receiver noise floor and the receiver output noise are one and the same thing.

Consider a receiver that has a 1-MHz bandwidth and a 20-dB noise figure. If an *S/N* of 10 dB is desired, the sensitivity (*S*) is

$$S = -174 + 20 + 10\log(1{,}000{,}000) + 10$$
$$= -84\ \text{dBm}.$$

You can see from this computation that if a lower *S/N* is required, better receiver sensitivity is necessary. If a 0-dB *S/N* is used, the sensitivity would become −94 dBm. The −94-dBm figure is the level at which the signal power equals noise power in the receiver's bandwidth. If the bandwidth were reduced to 100 kHz while maintaining the same input signal level, the output *S/N* would be increased to 10dB because of the reduction in noise power.

SINAD   As previously established, receiver sensitivity can be defined in terms of the minimum input signal required to produce a discernible (intelligible) output. Intelligibility is affected not only by internal and external noise but also by distortion components introduced by nonlinearities within the receiver itself. Distortion introduced by a receiver is not random like noise, but the effects of distortion on output intelligibility are similar to the effects of noise. Also, and as a practical matter, separating the effects of noise from those of distortion is difficult if not impossible to accomplish when receiver sensitivity measurements are made. For both these reasons, the sensitivity specification for receivers, particularly FM receivers, is often expressed in terms of an acronym known as SINAD (*SI*gnal plus *N*oise and *D*istortion), which quantifies how the received signal is impaired by both noise voltages and distortion (harmonic-frequency) components. SINAD is expressed in decibel terms and is defined as the ratio of the total power contributed by the received signal along with noise and distortion to the power of the noise and distortion components alone:

$$\text{SINAD} = 10\log\frac{S+N+D}{N+D} \qquad (6\text{-}3)$$

where *S* = signal power out

   N = noise power out

   D = distortion power out.

For narrowband FM receivers such as those found in two-way radios, the most commonly used sensitivity specification is **12-dB SINAD,** which defines the point where the amplitude of the received audio signal is 12 dB, or four times that of the noise and distortion present at the receiver output. At this level, the audio signal is considered to be barely intelligible under ideal reception conditions. The 12-dB SINAD level is by no means a full-quieting (i.e., no background noise) level. Instead, so much background noise would be audible that listening to the received audio over an extended time would be an unpleasant experience; however, the received signal would be sufficiently intelligible to be usable under emergency conditions. The 12-dB SINAD measurement has become an industry standard for quantifying minimum discernible signal level. This level also is relatively easy to determine using test instruments designed to measure audio levels only.

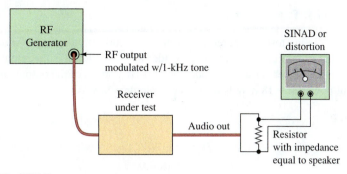

FIGURE 6-35   SINAD measurement setup.

Figure 6-35 shows a typical test setup for a 12-dB SINAD measurement. The RF generator applies a modulated signal at the carrier frequency to the antenna input of the receiver. The modulation consists of a constant, single-frequency tone at either 400 Hz or (more commonly for FM tests) 1 kHz. A distortion analyzer is connected across the audio output of the receiver and the volume increased to the manufacturer's maximum specified audio level, frequently called the "rated audio" level. Because the maximum audio level may be on the order of several watts, producing an unreasonably high volume in the speaker, the speaker may be replaced by an equal-impedance load resistor for the duration of the test, as shown in the figure. An ac voltmeter placed across the load resistor can be used to determine the rated audio level. Recall that $P = V^2/R$; therefore, $V = \sqrt{(PR)}$. As an example, a 5-W rated audio level across an 8-Ω load would be attained at 6.3 Vrms, so the volume would be increased until this voltage is indicated by the voltmeter.

As shown in Figure 6-36 the distortion analyzer is essentially an ac voltmeter with a very sharp notch filter tuned to the modulating signal frequency (either 400 Hz or 1 kHz) and that can be switched in and out of the test setup as needed. The notch filter removes the modulating signal (the S of Equation 6-33), leaving behind only the noise and distortion components, N + D, for measurement by the analyzer. The test is initially conducted with the notch filter switched out of the received signal path. In this condition the 1-kHz intelligence is demodulated and used to establish the rated output level. Then the notch filter is switched in, removing the modulating signal but leaving the noise and distortion components undisturbed. The RF generator output is then reduced until the distortion analyzer indicates that the noise and any distortion levels are 12 dB below the signal voltages previously established. For narrowband FM receivers, the RF level necessary to achieve 12-dB SINAD sensitivity is often less than 0.5 μV. Purpose-built distortion analyzers with meters calibrated in dB SINAD are widely available and are often incorporated into test instruments known variously as service monitors or communication test sets. These instruments have RF signal generators, spectrum analyzers, and other measurement instruments needed by field-service technicians in a single unit.

By making the 12-dB SINAD measurement at rated audio level, one can determine the overall health of the receiver with a single test. Excessive distortion in the audio amplifier stages or elsewhere will cause the receiver to fail its 12-dB SINAD test regardless of the receiver gain characteristics. Likewise, inadequate gain will cause a test failure for sensitivity reasons because excessive noise will appear at the receiver output. Therefore, all stages of a receiver that achieves its 12-dB SINAD specification can be verified to be in proper working order with a single test.

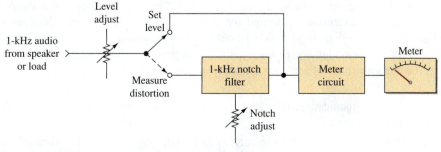

FIGURE 6-36   Distortion analyzer concept.

EXAMPLE 6-12

A receiver is being tested to determine SINAD. A 400-Hz audio signal modulates a carrier that is applied to the receiver. Under these conditions, the output power is 7 mW. Next a filter is used to cancel the 400-Hz portion of the output, and then an output power of 0.18 mW is measured. Calculate SINAD.

SOLUTION

$$S + N + D = 7 \text{ mW} \tag{6-3}$$
$$N + D = 0.18 \text{ mW}$$
$$\text{SINAD} = 10 \log\frac{S + N + D}{N + D}$$
$$= 10 \log\frac{7 \text{ mW}}{0.18 \text{ mW}}$$
$$= 15.9 \text{ dB}$$

Dynamic Range

The **dynamic range** of an amplifier or receiver is the input power range over which it provides a useful output. A receiver's dynamic and AGC ranges are usually two different quantities. The low-power limit is essentially the sensitivity specification discussed in the preceding paragraphs and is a function of the noise. The upper limit has to do with the point at which the system no longer provides the same linear increase as related to the input increase. It also has to do with certain distortion components and their degree of effect.

When testing a receiver (or amplifier) for the upper dynamic range limit, it is common to apply a single test frequency and determine the *1-dB compression point*. As shown in Figure 6-37, this is the point in the input/output relationship where the output has just reached a level where it is 1 dB down from the ideal linear response. The input power at that point is then specified as the upper power limit determination of dynamic range.

When two frequencies (f_1 and f_2) are amplified, the second-order distortion products are generally out of the system passband and are therefore not a problem. They occur at $2f_1$, $2f_2$, $f_1 + f_2$, and $f_1 - f_2$. Unfortunately, the third-order products at $2f_1 + f_2$, $2f_1 - f_2$, $2f_2 - f_1$, and $2f_2 + f_1$ usually have components in the system bandwidth. The distortion thereby introduced is known as *intermodulation distortion* (IMD). It is often referred to simply as **intermod**. Recall that the use of MOSFETs at the critical RF and mixer stages is helpful in minimizing these third-order effects. Intermodulation effects have such a major influence on the upper dynamic range of a receiver (or amplifier) that they are often specified via the **third-order intercept point** (or **input intercept**). This concept is illustrated in Figure 6-37. It is the input power at the point where straight-line extensions of desired and third-order input/output relationships meet. It is about 20 dBm in Figure 6-37. It is used only as a figure of merit. The better a system is with respect to intermodulation distortion, the higher will be its input intercept.

The dynamic range of a system is usually approximated as

$$\text{dynamic range (dB)} \approx \frac{2}{3}(\text{input intercept} - \text{noise floor}) \tag{6-4}$$

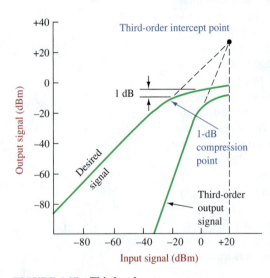

FIGURE 6-37 **Third-order intercept and compression point illustration.**

Poor dynamic range causes problems, such as undesired interference and distortion, when a strong signal is received. The current state of the art is a dynamic range of about 100 dB.

EXAMPLE 6-13

A receiver has a 20-dB noise figure (NF), a 1-MHz bandwidth, a +5-dBm third-order intercept point, and a 0-dB S/N. Determine its sensitivity and dynamic range.

SOLUTION

$$S = -174 \text{ dBm} + \text{NF} + 10 \log_{10} \Delta f + \frac{S}{N} \qquad (6\text{-}2)$$

$$= -174 \text{ dBm} + 20 \text{ dB} + 10 \log_{10} 10^6 + 0 = -94 \text{ dBm}$$

$$\text{dynamic range} \simeq \frac{2}{3}(\text{input intercept} - \text{noise floor}) \qquad (6\text{-}4)$$

$$= \frac{2}{3}[5 \text{ dBm} - (-94 \text{ dBm})]$$

$$= 66 \text{dB}$$

EXAMPLE 6-14

The receiver from Example 6-13 has a preamplifier at its input. The preamp has a 24-dB gain and a 5-dB NF. Calculate the new sensitivity and dynamic range.

SOLUTION

The first step is to determine the overall system noise ratio (NR). Recall from Chapter 1 that

$$\text{NR} = \log^{-1} \frac{\text{NF}}{10}.$$

Letting NR_1 represent the preamp and NR_2 the receiver, we have

$$\text{NR}_1 = \log^{-1} \frac{5 \text{ dB}}{10} = 3.16$$

$$\text{NR}_2 = \log^{-1} \frac{20 \text{ dB}}{10} = 100.$$

The overall NR is

$$\text{NR} = \text{NR}_1 + \frac{\text{NR}_2 - 1}{P_{G_1}} \qquad (1\text{-}14)$$

and

$$P_{G1} = \log^{-1} \frac{24 \text{ dB}}{10} = 251$$

$$\text{NR} = 3.16 + \frac{100 - 1}{251} = 3.55$$

$$\text{NF} = 10 \log_{10} 3.55 = 5.5 \text{ dB}$$

$$= \text{total system NF}$$

$$S = -174 \text{ dBm} + 5.5 \text{ dB} + 60 \text{ dB} = -108.5 \text{ dBm}$$

The third-order intercept point of the receiver alone had been +5 dBm but is now preceded by the preamp with 24-dB gain. Assuming that the preamp can deliver 5 dBm to the receiver without any appreciable intermodulation distortion, the system's third-order intercept point is +5 dBm − 24 dB = −19 dBm. Thus,

$$\text{dynamic range} \simeq \frac{2}{3}[-19 \text{ dBm} - (-108.5 \text{ dBm})]$$

$$= 59.7 \text{ dB}$$

EXAMPLE 6-15

The 24-dB gain preamp in Example 6-14 is replaced with a 10-dB gain preamp with the same 5-dB NF. What are the system's sensitivity and dynamic range?

SOLUTION

$$NR = 3.16 + \frac{100 - 1}{10} = 13.1$$

$$NF = 10 \log_{10} 13.1 = 11.2 \text{ dB}$$

$$S = -174 \text{ dBm} + 11.2 \text{ dB} + 60 \text{ dB} = -102.8 \text{ dBm}$$

$$\text{dynamic range} \simeq \frac{2}{3}[-5 \text{ dBm} - (-102.8 \text{ dB})] = 65.2 \text{ dB}$$

The results of Examples 6-13 to 6-15 are summarized as follows:

	RECEIVER ONLY	RECEIVER AND 10-dB PREAMP	RECEIVER AND 24-dB PREAMP
NF (dB)	20	11.2	5.5
Sensitivity (dBm)	−94	−102.8	−108.5
Third-order intercept point (dBm)	+5	−5	−19
Dynamic range (dB)	66	65.2	59.7

An analysis of the examples and these data shows that the greatest sensitivity can be realized by using a preamplifier with the lowest noise figure and highest available gain to mask the higher NF of the receiver. Remember that as gain increases, so does the chance of spurious signals and intermodulation distortion components. A preamplifier used prior to a receiver input has the effect of decreasing the third-order intercept proportionally to the gain of the amplifier, while the increase in sensitivity is less than the gain of the amplifier. Therefore, to maintain a high dynamic range, it is best to use only the amplification needed to obtain the desired noise figure. It is not helpful in an overall sense to use excessive gain. The data in the chart show that adding the 10-dB gain preamplifier improved sensitivity by 8.8 dB and decreased dynamic range by only 0.8 dB. The 24-dB gain preamp improved sensitivity by 14.5 dB but decreased the dynamic range by 6.3 dB.

Intermodulation Distortion Testing

It is common to test an amplifier for its IMD by comparing two test frequencies to the level of a specific IMD product. As previously mentioned, the second-order products are usually outside the frequency range of concern. This is generally true for all the even-order products and is illustrated in Figure 6-38. It shows some second-order products that would be outside the bandwidth of interest for most systems.

The odd products are of interest because some of them can be quite close to the test frequencies f_1 and f_2 shown in Figure 6-38. The third-order products shown ($2f_2 - f_1$ and $2f_1 - f_2$) have the most effect, but even the fifth-order products ($3f_2 - 2f_1$ and $3f_1 - 2f_2$) can be troublesome. Figure 6-39(a) shows a typical spectrum analyzer display when two test signals are applied to a mixer or small-signal amplifier. Notice that the third-order products are shown 80 dB down from the test signals, while the fifth-order products are more than 90 dB down.

Figure 6-39(b) shows the result of an IMD test when two frequencies are applied to a typical Class AB linear power amplifier. The higher odd-order products (up to the eleventh in this case) are significant for the power amplifier. Fortunately, these effects are less critical in power amplifiers than in the sensitive front end of a radio receiver.

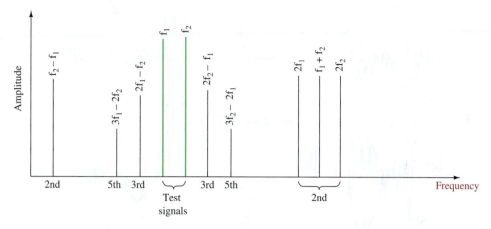

FIGURE 6-38   **IMD products (second-, third-, and fifth-order for two test signals).**

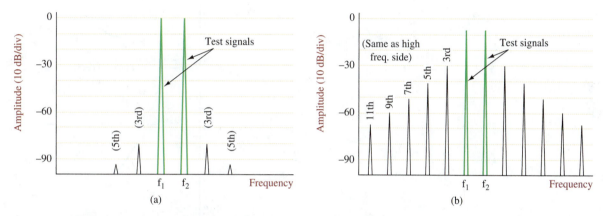

FIGURE 6-39   **IMD testing: (a) mixer; (b) Class AB linear power amplifier.**

6-8 AUTOMATIC GAIN CONTROL AND SQUELCH

An essential function within any receiver is some form of AGC. These systems can be relatively simple or somewhat more involved, but without AGC the receiver's usefulness would be seriously impaired. For example, simply tuning the receiver would be a nightmare. To avoid missing weak stations, the volume control on a non-AGC equipped set would have to be turned way up. A strong station would probably blow out the speaker, whereas a weak station might not be audible. In addition, received signals are constantly changing as a result of changing weather and ionospheric conditions, a particular concern to listeners of stations in the conventional AM broadcast as well as the short-wave (high-frequency) bands. The AGC allows for listening without the need to monitor the volume control constantly. Finally, many receivers are mobile. For instance, a standard broadcast AM car radio would be almost unusable without a good AGC to compensate for the signal variation in different locations.

Obtaining the AGC Level

Most AGC systems obtain the AGC level just following the detector. Recall that an *RC* filter following the detector diode removes the higher, intermediate frequency but leaves the low-frequency envelope intact. A slowly varying dc level can be obtained by increasing the *RC* time constant. The dc level changes with variations in the strength of the overall received signal.

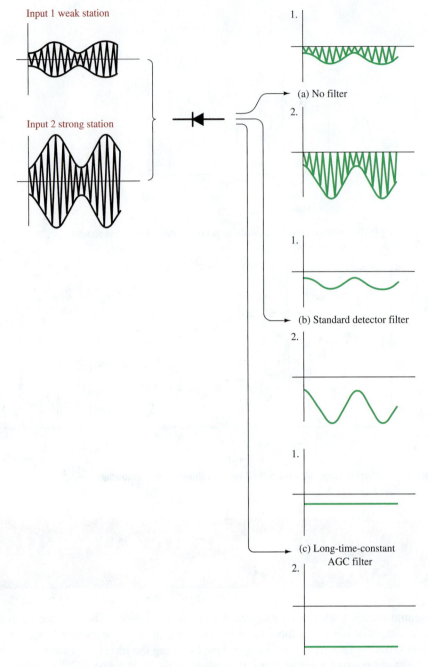

FIGURE 6-40   Development of AGC voltage.

Figure 6-40(a) shows the output from a diode detector with no filtering. In this case, the output is simply the AM waveform with the positive portion rectified out for two different levels of received signal into the diode. At (b), the addition of a filter has provided the two different envelope levels but without the high-frequency content. These signals correspond to an undesired change in volume of two different received stations. At (c), a filter with a much longer time constant has filtered the output into a dc level. Note that the dc level changes, however, with the two different levels of input signal. This is a typical AGC level that is subsequently fed back to control the gain of IF stages and/or the mixer and RF stages.

In this case, the larger negative dc level at C_2 would cause the receiver's gain to be decreased so that the ultimate speaker output is roughly the same for either the weak or strong station. It is important that the AGC time constant be long enough so that desired radio signal level changes that constantly occur do *not* cause a change in receiver gain. The AGC should respond only to average signal strength changes, and as such usually has a time constant of about a full second.

Controlling the Gain of a Transistor

Figure 6-41 illustrates a method whereby the variable dc AGC level can be used to control the gain of a common-emitter transistor amplifier stage. In the case of a strong received station, the AGC voltage developed across the AGC filter capacitor (C_{AGC}) is a large negative value that subsequently lowers the forward bias on Q_1. It causes more dc current to be drawn through R_2, and hence less is available for the base of Q_1, since R_1, which supplies current for both, can supply only a relatively constant amount. The voltage gain of a CE stage with an emitter bypass capacitor (C_E) is nearly directly proportional to dc bias current, and therefore the strong station reduces the gain of Q_1. The reception of very weak stations would reduce the gain of Q_1 very slightly, if at all. The introduction of AGC in the 1920s marked the first major use of an electronic feedback control system. The AGC feedback path is called the AGC bus because in a full receiver it is usually "bused" back into a number of stages to control gain over a large signal range.

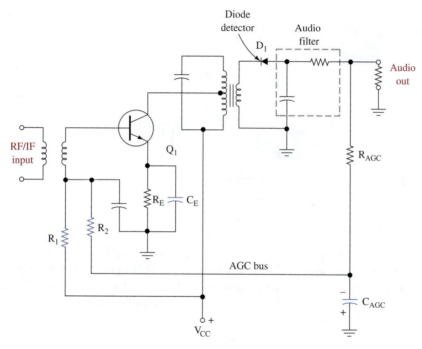

FIGURE 6-41   **AGC circuit illustration.**

Delayed AGC

The simple automatic gain control (AGC) just discussed has a minor disadvantage. It provides some gain reduction even to very weak signals. This is illustrated in Figure 6-42. As soon as even a weak received signal is tuned, simple AGC provides some gain reduction. Because communications equipment is often dealing with marginal (weak) signals, it is usually advantageous to add some additional circuitry to provide a **delayed AGC,** that is, an AGC that does not provide any gain reduction until some arbitrary signal level is attained. Delayed AGC, therefore, presents no reduction in gain for weak signals. This characteristic is also shown in Figure 6-42. It is important for you to understand that delayed AGC does not mean delayed in time.

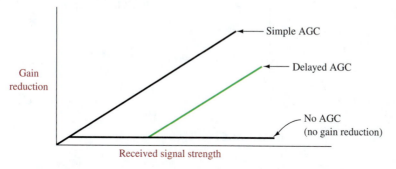

FIGURE 6-42   **AGC characteristics.**

A simple means of providing delayed AGC is shown in Figure 6-43. A reverse bias is applied to the cathode of D_1. Thus, the diode looks like an open circuit to the ac signal from the last IF amplifier unless that signal reaches some predetermined instantaneous level. For small IF outputs when D_1 is open, the capacitor C_1 sees a pure ac signal, and thus no dc AGC level is sent back to previous stages to reduce their gain. If the IF output increases, eventually a point is reached where D_1 will conduct on its peak positive levels. This will effectively *short* out the positive peaks of IF output, and C_1 will therefore see a more negative than positive signal and filter it into a relatively constant negative level used to reduce the gain of previous stages. The amplitude of IF output required to start feedback of the "delayed" AGC signal is adjustable by the delayed AGC control potentiometer of Figure 6-43. This may be an external control so that the user can adjust the amount of delay to suit conditions. For instance, if mostly weak signals are being received, the control might be set so that no AGC signal is developed except for very strong stations. This means the delay interval shown in Figure 6-42 is increased.

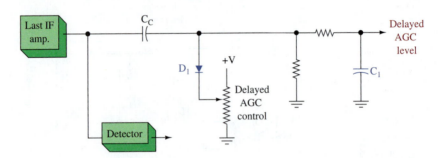

FIGURE 6-43 **Delayed AGC configuration.**

Auxiliary AGC

Auxiliary AGC is used (even on some broadcast receivers) to cause a step reduction in receiver gain at some arbitrarily high value of received signal. It then has the effect of preventing very strong signals from overloading a receiver and thereby causing a distorted, unintelligible output. A simple means of accomplishing the auxiliary AGC function is illustrated in Figure 6-44(a).

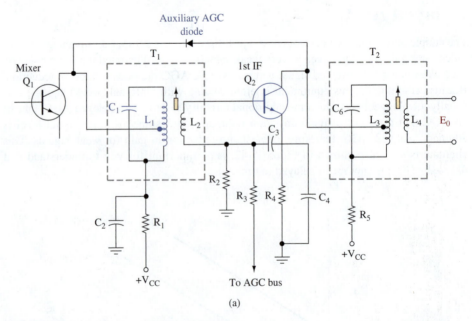

(a)

FIGURE 6-44 **(a) Auxiliary AGC.**

(*Continued*)

(b)

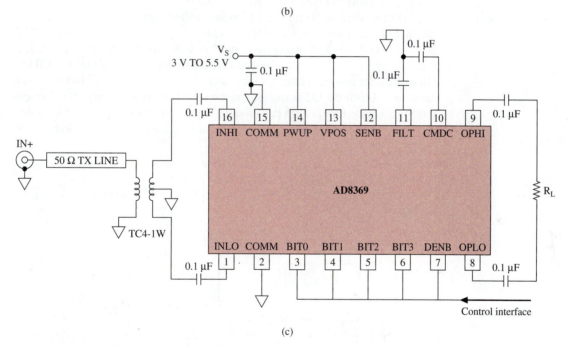

(c)

FIGURE 6-44 (*Continued*) **(b) the Analog Devices AD8369 variable-gain amplifier IC (Copyright Analog Devices. Reprinted with permission. All rights reserved.); (c) basic connections.**

Notice the auxiliary AGC diode connected between the collectors of the mixer and first IF transistors. Under normal signal conditions, the dc level at each collector is such that the diode is reverse-biased. In this condition the diode has a very high resistance and has no effect on circuit action. The potential at the mixer's collector is constant because it is not controlled by the normal AGC. However, the AGC control on the first IF transistor, for very strong signals, causes its dc base current to decrease, and hence the collector current also decreases. Thus, its collector voltage becomes more positive, and the diode starts to conduct. The diode resistance goes low, and it loads down the mixer tank ($L_1 C_1$) and thereby produces a step reduction of the signal coupled into the first IF stage. The dynamic AGC range has thereby been substantially increased.

An example of a modern digitally controlled variable gain amplifier, the Analog Devices AD8369, is shown in Figure 6-44(b). It provides a −5 dB to +40 dB digitally adjustable gain in 3-dB increments. The AD8369 is specified for use in RF receive-path AGC loops in cellular receivers. The minimum basic connections for the AD8369 are also shown in Figure 6-44(c). A balanced RF input connects to pins 16 and 1. Gain control of the device is provided via pins 3, 4, 5, 6, and 7. A 3 V to +5.5 V connects to pins 12, 13, and 14. The balanced differential output appears at pins 8 and 9.

Variable Sensitivity

Despite the increased dynamic range provided by delayed AGC and auxiliary AGC, it is often advantageous for a receiver also to include a variable sensitivity control. This is a manual AGC control because the user controls the receiver gain (and thus sensitivity) to suit the requirement. A communications receiver may be called upon to deal with signals over a 100,000:1 ratio, and even the most advanced AGC system does not afford that amount of range. Receivers that are designed to provide high sensitivity and that can handle very large input signals incorporate a manual sensitivity control that controls the RF and/or IF gain.

Variable Selectivity

Many communications receivers provide detection to more than one kind of transmission. They may detect Morse code transmissions, SSB, AM, and FM all in one receiver. The required bandwidth to avoid picking up adjacent channels may well vary from 1kHz for Morse code up to 30 kHz for narrowband FM.

Most modern receivers use a technique called **variable bandwidth tuning (VBT)** to obtain variable selectivity. Consider the block diagram shown in Figure 6-45. The input signal at 500 Hz has half-power frequencies at 400 and 600 Hz, a 200-Hz bandwidth. This is mixed with a 2500-Hz local oscillator output to develop signals from 1900 to 2100 Hz. These drive the bandpass filter, which passes signals from 1900 to 2100 Hz. The filter output is mixed with the LO output producing signals from 400 to 600 Hz. These are fed to another bandpass filter, which passes signals between 400 and 600 Hz. The system output, therefore, is the original 400- to 600-Hz range with the original bandwidth of 200 Hz.

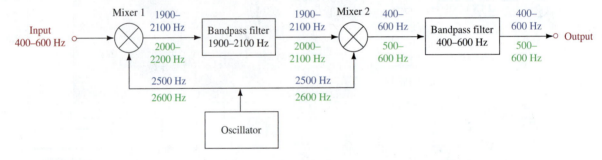

Blue frequencies when oscillator is 2500 Hz
Green frequencies when oscillator is 2600 Hz

FIGURE 6-45   **Variable-bandwidth tuning (VBT).**

Keeping the same 500-Hz input signal, let us now increase the LO frequency to 2600 Hz. The first mixer output now covers the range from 2000 to 2200 Hz, but only that portion from 2000 to 2100 Hz is passed by the first BPF. These are mixed to produce signals from 500 to 600 Hz, which are passed by the second BPF. The output bandwidth has been reduced to 100 Hz. Increase the LO to 2650 Hz and the system bandwidth drops to 50 Hz. In other words, the bandwidth is now a function of the variable LO frequency.

Noise Limiter

Manufactured sources of external noise are extremely troublesome to highly sensitive communications receivers as well as to any other electronics equipment that is dealing with signals in the microvolt region or less. The interference created by these human-made sources, such as ignition systems, motor communications systems, and switching of high current loads, is a form of **electromagnetic interference (EMI)**. Reception of EMI by a receiver creates undesired amplitude modulation, sometimes of such large magnitude as to affect FM reception adversely, to say nothing of the complete havoc created in AM systems. While these noise impulses are usually of short duration, it is not uncommon for them to have amplitudes up to 1000 times that of the desired signal. A noise limiter circuit is employed to silence the receiver for the duration of a noise pulse, which is preferable to a very loud crash from the speaker. These circuits are sometimes referred to as automatic noise limiter (ANL) circuits.

A common type of circuit for providing noise limiting is shown in Figure 6-46. It uses a diode, D_2, that conducts the detected signal to the audio amplifier as long as it is not greater than some prescribed limit. Greater amplitudes cause D_2 to stop conducting until the noise impulse has decreased or ended. The varying audio signal from the diode detector, D_1, is developed across the two 100-kΩ resistors. If the received carrier is producing a -10-V level at the AGC takeoff, the anode of D_2 is at -5 V and the cathode is at -10 V. The diode is on and conducts the audio into the audio amplifier. Impulse noise will cause the AGC takeoff voltage to increase instantaneously, which means the anode of D_2 also does. However, its cathode potential does not change instantaneously since the voltage from cathode to ground is across a 0.001-μF capacitor. Remember that the voltage across a capacitance cannot change instantaneously. Therefore, the cathode stays at -10 V, and as the anode approaches -10 V, D_2 turns off and the detected audio is blocked from entering the audio amplifier. The receiver is silenced for the duration of the noise pulse.

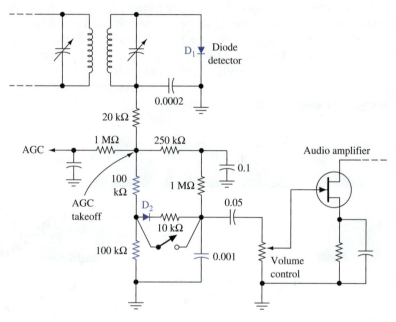

FIGURE 6-46   Automatic noise limiter.

The switch across D_2 allows the noise limiter action to be disabled by the user. This may be necessary when a marginal (noisy) signal is being received and the set output is being turned off excessively by the ANL.

Metering

Many communications receivers are equipped with a meter that provides a visual indication of received signal strength. It is known as the **S meter** and often is found in the emitter leg of an AGC controlled amplifier stage (RF or IF). It reads dc current, which is usually inversely proportional to received signal strength. With no received signal, there is no AGC bias level, which causes maximum dc emitter current flow and therefore maximum stage voltage gain. As the AGC level increases, indicating an increasing received signal, the dc emitter current goes down, which thereby reduces gain. The S meter can thus be used as an aid to accurate tuning as well as providing a relative guide to signal strength. Modern designs use LED bar graphs instead of an electrical meter. This reduces cost and improves reliability.

In some receivers, the S meter can be electrically switched into different areas to aid in troubleshooting a malfunction. In those cases, the operator's manual provides a troubleshooting guide on the basis of meter readings obtained in different areas of the receiver.

Squelch

Most two-way radio transmissions, including those in public safety and emergency services applications, consist of relatively brief-duration messages interspersed with long periods of inactivity. Because the messages are often of a critical nature, the user must continuously monitor the receiver for transmissions. In the absence of a received carrier, however, the automatic gain control circuits cause maximum system gain, which results in a great deal of noise in the speaker. The noise sounds like a hissing or rushing sound. Without a means to cut off the receiver audio during transmission lulls, the noise in the speaker would cause the user severe aggravation. For this reason, communications receivers have **squelch** circuits designed to silence the audio amplifier stages until a carrier is detected, at which time the received audio will be amplified and passed on to the speaker.

Squelch subsystems in FM narrowband receivers take two forms: noise squelch (sometimes called carrier squelch), and the somewhat more sophisticated tone squelch. Figure 6-47 is a block diagram of a noise squelch subsystem. Noise squelch combines the effects of FM limiting with the broadband nature of noise. Recall from Section 6-5 that the limiter is an easily saturated amplifier whose output voltage never exceeds a predetermined level. In the absence of a received carrier, the limiter output consists of noise voltages at random amplitudes and spanning all frequencies. When a signal is received, the signal amplitude will reduce the noise voltage. In other words, the limiter output can consist of any combination of signal or noise voltages, but these will appear in an inverse relationship to each other. A sufficiently high-level received signal will completely displace the noise voltages at the limiter output. This inverse relationship is the principle upon which noise squelch systems operate.

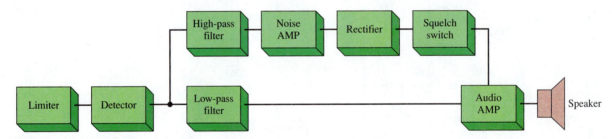

FIGURE 6-47   Noise squelch subsystem block diagram.

Referring again to Figure 6-47, note that the input voltages to the squelch switch are first passed through a 6-kHz *high*-pass filter. The purpose of the high-pass filter is to reject the 300-Hz to 3-kHz voice-band frequencies that would be present in the demodulated voice signal, leaving only noise. Because noise is broadband, a limiter output consisting only of noise will have voltages present with frequency components above 6 kHz. These high frequencies will pass through the filter to a rectifier, which in turn produces a dc bias voltage for the squelch switch transistor. The squelch switch is configured so that, when the transistor is biased on, the audio amplifiers are cut off, preventing the noise from being heard in the receiver speaker. When a carrier is received (with or without modulation), the carrier voltages will "crowd out" the noise voltages. The intelligence frequency components, being well below 6 kHz, are not present at the output of the high-pass filter and, therefore, will not trigger the squelch switch. With no rectified noise to produce a bias voltage, the squelch switch transistor will be turned off, thus allowing the audio amplifiers to pass their output to the speaker.

An enhancement to noise squelch is the Continuous Tone Coded Squelch System, or CTCSS. Originally developed by Motorola, CTCSS is often referred to by its original trade name, PL (for "Private Line"), even though the system is not truly private and has been implemented under different names by different manufacturers for a number of years. CTCSS allows multiple user groups to share a common communications channel without having to listen to each other's transmissions. It works by linearly summing one of 52 possible low-frequency audio tones with each voice transmission. The deviation of the tone squelch signal is most often set for a maximum of ±750 Hz. Because the maximum deviation in narrowband FM is ±5 kHz, the voice deviation must be reduced accordingly so that the maximum is not exceeded. A receiver with CTCSS enabled will not unsquelch unless its tone decoder picks up the predetermined CTCSS tone, thus the receiver would remain muted even when other audio signals with different CTCSS tones are present at the limiter output. The CTCSS tone is said to be subaudible because a low-pass filter in the receiver removes the tone before the audio is amplified and sent to the speaker.

A CTCSS-equipped receiver will be muted even in the presence of a transmission if it does not pick up the correct subaudible tone. Therefore, transmissions can interfere with each other if the user is unable to determine whether the channel is busy. For this reason, CTCSS-enabled radios are equipped with a monitor function, either in the form of a front-panel switch or as a switch that is activated when the microphone is removed from its holder. The monitor switch defeats CTCSS squelch and allows the user to determine if the channel is in use. Good operating practice requires the tone squelch function to be turned off before the push-to-talk switch on the microphone is pressed.

A still more-advanced squelch system is known as Digital Coded Squelch or DCS. Here, squelch action is enhanced by serially transmitting one of 106 possible binary codes along with the audio. The receiver decodes the serial code; if it is recognized as the correct code, the squelch switch passes the audio.

SUMMARY

The receiver converts the radio frequencies applied to its antenna down to baseband and extracts the intelligence from the received signal. Two important characteristics of any receiver are its sensitivity and its selectivity. Sensitivity relates to the minimum signal level that can be extracted from noise, and selectivity refers to the ability of a receiver to select a desired signal from all those around it.

The ability of the superheterodyne receiver to provide high gain at constant selectivity makes it among the most widely used designs. In the superhet, mixing action occurs in two stages. The mixer, or first detector, stage converts the incoming radio frequency to a (generally) lower, intermediate frequency by combining the incoming radio frequency signal with a sine wave generated by a local oscillator. The local oscillator frequency is changed when the radio is tuned; however, the intermediate frequency remains constant over the entire tuning range of the receiver. The bulk of receiver gain and selectivity occur at the intermediate frequency.

A second detector stage performs the mixing action a second time to extract the intelligence. The type of second detector used depends on the modulation and whether the transmitted carrier was present in the received signal. The most basic detector, the envelope detector, is used with full-carrier AM signals. The envelope detector uses a nonlinear device to produce distortion of a desirable kind, from which the intelligence is extracted from the intermediate frequency. For suppressed-carrier signals, such as SSB, a mixer termed a product detector recovers intelligence from the difference frequency created when the received sideband is mixed with a locally reinserted carrier generated by a beat-frequency oscillator. The beat-frequency oscillator may either be manually controlled or, in more sophisticated designs, may form part of a synchronous detector that, along with a phase-locked loop, recreates an exact replica of the transmitted carrier from recovered characteristics of the received signal.

The greatest advantage of the superheterodyne design is that it has a constant bandwidth over a wide tuning range. Its largest disadvantage other than complexity is that it is susceptible to image frequencies—undesired signals that, when mixed with the local-oscillator, produce an output at the intermediate frequency. To minimize images, preselector stages are employed to reject strong undesired signals before they can reach the mixer. Also, double conversion and, increasingly in modern designs, up-conversion architectures are used to ensure that the image frequencies are as far from the intended range as possible, thereby maximizing the possibility that the initial preselector stages will fully reject the unwanted signals.

Many of the newest receiver designs are of the direct conversion, or zero-IF variety, where the incoming radio signal is down-converted to the baseband range and the intelligence extracted in one step, rather than two. Direct conversion designs have a number of advantages, among them simplicity, power conservation, and immunity to image frequencies. For this reason, the zero IF architecture is used in a number of portable applications, including the receiver sections of cellular phones.

Frequency- and phase-modulated intelligence signals may be detected in a number of ways. The classical approaches are by means of discriminators, where frequency variations are converted to voltages as a result of the phase shifts among voltages and currents in *LC* circuits, both in and out of resonance. Though not widely used in modern designs because they do not lend themselves to fabrication on integrated circuits, discriminators are well suited to linear demodulation of wideband signals, and they serve to illustrate how instantaneous frequency and phase shifts can be detected using ordinary, passive components. More contemporary designs for FM demodulation use either the quadrature or, more commonly, the phase-locked loop detector because these approaches are well suited to large-scale integration.

Stereo FM is a form of multiplexing, where two separate channels of information are carried simultaneously over a common medium. Stereo broadcasting is designed to be fully compatible with monaural receivers and represents an application of frequency-division multiplexing. The stereo receiver uses synchronous detection to recover the suppressed stereo subcarrier and to recover the intelligence. The system relies on the mathematical recombination of phase-shifted left- and right-channel signals in the stereo receiver.

In addition to sensitivity and selectivity, receivers must be designed to perform adequately over a wide dynamic range. One of the most important measures in this regard is the third-order intercept point, or IP3 point, for this is a measure of how immune the receiver is to undesired mixing effects in the presence of strong adjacent signals. Other important tests are sensitivity tests, among them the SINAD test for FM receivers, as well as intermodulation distortion tests.

Receivers must have some form of automatic gain control and muting (squelch) capability to make them useful over a range of signal conditions. Automatic gain control can range from simple to sophisticated, and is used in conjunction with variable sensitivity and variable-bandwidth techniques to maximize effective under adverse signal conditions. Squelch systems allow for continuous monitoring of communications channels, and in FM systems rely on the inverse relationship between signal and noise levels that exists because of limiting action. As with AGC, squelch systems range in complexity, depending on application.

QUESTIONS AND PROBLEMS

SECTION 6-1

*1. Explain the following: sensitivity of a receiver; selectivity of a receiver. Why are these important characteristics? In what units are they usually expressed?

2. Explain why a receiver can be overly selective.

SECTION 6-2

*3. Draw a diagram of a tuned radio-frequency (TRF) radio receiver.

4. A TRF receiver is to be tuned over the range 550 to 1550 kHz with a 25-μH inductor. Calculate the required capacitance range. Determine the tuned circuit's necessary Q if a 10-kHz bandwidth is desired at 1000 kHz. Calculate the receiver's selectivity at 550 and 1550 kHz. (0.422 to 3.35 nF, 100, 5.5 kHz, 15.5 kHz)

SECTION 6-3

*5. Draw a block diagram of a superheterodyne AM receiver. Assume an incident signal, and explain briefly what occurs in each stage.

*6. What type of radio receiver contains intermediate-frequency transformers?

7. The AM signal into a mixer is a 1.1-MHz carrier that was modulated by a 2-kHz sine wave. The local oscillator is at 1.555 MHz. List all mixer output components and indicate those accepted by the IF amplifier stage.

*8. Explain the purpose and operation of the first detector in a superhet receiver.

9. Explain how the variable tuned circuits in a superheterodyne receiver are adjusted with a single control.

*10. If a superheterodyne receiver is tuned to a desired signal at 1000 kHz and its conversion (local) oscillator is operating at 1300 kHz, what would be the frequency of an incoming signal that would possibly cause *image* reception? (1600 kHz)

---

*An asterisk preceding a number indicates a question that has been provided by the FCC as a study aid for licensing examinations.

11. A receiver tunes from 20 to 30 MHz using a 10.7-MHz IF. Calculate the required range of oscillator frequencies and the range of image frequencies.

12. Show why image frequency rejection is not a major problem for the standard AM broadcast band.

*13. What are the advantages to be obtained from adding a tuned radio-frequency amplifier stage ahead of the first detector (converter) stage of a superheterodyne receiver?

*14. If a transistor in the only radio-frequency stage of your receiver shorted out, how could temporary repairs or modifications be made?

15. What advantages do dual-gate MOSFETs have over BJTs for use as RF amplifiers?

*16. What is the *mixer* in a superheterodyne receiver?

17. Why is the bulk of a receiver's gain and selectivity obtained in the IF amplifier stages?

18. A superhet receiver tuned to 1 MHz has the following specifications:

    RF amplifier: $P_G = 6.5$ dB, $R_{in} = 50$ Ω *Detector:* 4-dB attenuation
    Mixer: $P_G = 3$ dB   *Audio amplifier:* $P_G = 13$ dB
    3 IFs: $P_G = 24$ dB each at 455 kHz

    The antenna delivers a 21-μ V signal to the RF amplifier. Calculate the receiver's image frequency and input/output power in watts and dBm. Draw a block diagram of the receiver and label dBm power throughout. (1.91 MHz, 8.82 pW, −80.5 dBm, 10 mW, 10 dBm)

19. A receiver has a dynamic range of 81 dB. It has 0.55 nW sensitivity. Determine the maximum allowable input signal. (0.0692 W)

20. Define *dynamic range.*

21. List the components of an AM signal at 1 MHz when modulated by a 1-kHz sine wave. What is (are) the component(s) if it is (they are) converted to a usb transmission? If the carrier is redundant, why must it be "reinserted" at the receiver?

22. Explain why the BFO in an SSB demodulator has such stringent accuracy requirements.

23. Suppose the modulated signal of an SSBSC transmitter is 5 kHz and the carrier is 400 kHz. At what frequency must the BFO be set?

24. What is a product detector? Explain the need for a low-pass filter at the output of a balanced modulator used as a product detector.

25. Calculate the frequency of a product detector that is fed an SSB signal modulated by 400 Hz and 2 kHz sine waves. The BFO is 1 MHz.

*26. What is the purpose of a discriminator in an FM broadcast receiver?

27. Explain why the AFC function is usually not necessary in today's FM receivers.

*28. Draw a block diagram of a superheterodyne receiver designed for reception of FM signals.

29. The local FM stereo rock station is at 96.5 MHz. Calculate the local oscillator frequency and the image frequency for a 10.7-MHz IF receiver. (107.2 MHz, 117.9 MHz)

30. Explain the desirability of an RF amplifier stage in FM receivers as compared to AM receivers. Why is this not generally true at frequencies over 1 GHz?

31. Describe the meaning of *local oscillator reradiation,* and explain how an RF stage helps to prevent it.

32. Why is a square-law device preferred over other devices as elements in an RF amplifier?

33. Why are FETs preferred over other devices as the active elements for RF amplifiers?

34. Explain the need for the RFC in the RF amplifier shown in Figure 6-15.

*35. What is the purpose of a limiter stage in an FM broadcast receiver?

*36. Draw a diagram of a limiter stage in an FM broadcast receiver.

37. Explain fully the circuit operation of the limiter shown in Figure 6-16.

38. What is the relationship among limiting, sensitivity, and quieting for an FM receiver?

39. An FM receiver provides 100 dB of voltage gain prior to the limiter. Calculate the receiver's sensitivity if the limiter's quieting voltage is 300 mV. (3 μV)

40. Draw a block diagram for a double-conversion receiver when tuned to a 27-MHz broadcast using a 10.7-MHz first IF and 1-MHz second IF. List all pertinent frequencies for each block. Explain the superior image frequency characteristics as compared to a single-conversion receiver with a 1-MHz IF, and provide the image frequency in both cases.

---

*An asterisk preceding a number indicates a question that has been provided by the FCC as a study aid for licensing examinations.

41. Draw block diagrams and label pertinent frequencies for a double-conversion *and* up-conversion system for receiving a 40-MHz signal. Discuss the economic merits of each system and the effectiveness of image frequency rejection.

42. A receiver tunes the HF band (3 to 30 MHz), utilizes up-conversion with an intermediate frequency of 40.525 MHz, and uses high-side injection. Calculate the required range of local oscillator frequencies. (43.5 to 70.5 MHz)

43. An AM broadcast receiver's preselector has a total effective Q of 90 to a received signal at 1180 kHz and uses an IF of 455 kHz. Calculate the image frequency and its dB of suppression. (2090 kHz, 40.7 dB)

SECTION 6-5

*44. Explain the operation of a diode detector.

45. Describe the advantages and disadvantages of a diode detector.

46. Provide the advantages of a synchronous detector compared to a diode detector. Explain its principle of operation.

47. Draw a schematic of an FM slope detector and explain its operation. Why is this method not often used in practice?

48. Draw a schematic of a Foster–Seeley discriminator, and provide a step-by-step explanation of what happens when the input frequency is below the carrier frequency. Include a phase diagram in your explanation.

*49. Draw a diagram of an FM broadcast receiver detector circuit.

*50. Draw a diagram of a ratio detector and explain its operation.

51. Explain the relative merits of the Foster–Seeley and ratio detector circuits.

*52. Draw a schematic diagram of each of the following stages of a superheterodyne FM receiver:
    (a) Mixer with injected oscillator frequency.
    (b) IF amplifier.
    (c) Limiter.
    (d) Discriminator.

Explain the principles of operation. Label adjacent stages.

53. Describe the process of quadrature detection.

54. Explain in detail how a PLL is used as an FM demodulator.

SECTION 6-6

55. Explain how separate left and right channels are obtained from the (L + R) and (L − R) signals.

*56. What is SCA? What are some possible uses of SCA?

57. Determine the maximum reproduced audio signal frequency in an SCA system. Why does SCA cause less FM carrier deviation, and why is it thus less noise resistant than standard FM? (*Hint:* Refer to Figure 6-33.) (7.5 kHz)

SECTION 6-7

58. We want to operate a receiver with NF = 8 dB at S/N = 15 dB over a 200-kHz bandwidth at ambient temperature. Calculate the receiver's sensitivity. (−98 dBm)

59. Explain the significance of a receiver's 1-dB compression point. For the receiver represented in Figure 6-37, determine the 1-dB compression point. ($\cong$10 dBm)

60. Determine the third-order intercept for the receiver illustrated in Figure 6-37. ($\cong$+20 dBm)

61. The receiver described in Problem 58 has the input/output relationship shown in Figure 6-37. Calculate its dynamic range. (78.7 dB)

62. A receiver with a 10-MHz bandwidth has an S/N of 5 dB and a sensitivity of −96 dBm. Find the required NF. (3 dB)

63. A high-quality FM receiver is to be tested for SINAD. When its output contains just the noise and distortion components, 0.015 mW is measured. When the desired signal and noise and distortion components are measured together, the output is 15.7 mW. Calculate SINAD. (30.2 dB)

64. Explain SINAD.

SECTION 7-3

65. Discuss the advantages of delayed AGC over normal AGC and explain how it may be attained.

66. Explain the function of auxiliary AGC and give a means of providing it.

67. Explain the need for variable sensitivity and show with a schematic how it could be provided.

68. Explain the need for variable selectivity. Describe how VBT is accomplished if the oscillator in Figure 6-45 is changed to 2650 Hz.

69. What is the need for a noise limiter circuit? Explain the circuit operation of the noise limiter shown in Figure 6-46.

70. List some possible applications for *metering* on a communications transceiver.

*71. What is the purpose of a squelch circuit in a radio communications receiver?

72. List two other names for a squelch circuit. Provide a schematic of a squelch circuit and explain its operation. List five different squelch methods.

73. Describe the effects of EMI on a receiver.

74. Describe the operation of an ANL.

QUESTIONS FOR CRITICAL THINKING

75. Which of the factors that determine a receiver's sensitivity is more important? Defend your judgment.

76. Would passing an AM signal through a nonlinear device allow recovery of the low-frequency intelligence signal when the AM signal contains only high frequencies? Why or why not?

77. Justify in detail the choice of a superheterodyne receiver in an application that requires constant selectivity for received frequencies.

78. A superheterodyne receiver tunes the band of frequencies from 4 to 10 MHz with an IF of 1.8 MHz. The double-ganged capacitor used has a 325 pF maximum capacitance per section. The tuning capacitors are at the maximum value (325 pF) when the RF frequency is 4 MHz. Calculate the required RF and local oscillator coil inductance and the required tuning capacitor values when the receiver is tuned to receive 4 MHz and 10 MHz. (4.87 μ H, 2.32 μH, 52 pF, 78.5 pF)

79. If you were concerned with the sensitivity rating of a communications system, would noise reduction capability be a major factor in your decision-making? Why or why not?

80. Explain why a limiter minimizes or eliminates the need for the AGC function.

81. Draw a block diagram of an FM stereo demodulator. Explain the function of the AM demodulator and the matrix network so nontechnical users can understand. Add a circuit that energizes a light to indicate reception of a stereo station.

82. Describe the process of up-conversion. Explain its advantages and disadvantages compared to double conversion.

83. You have been asked to extend the dynamic range of a receiver. Can this be done? What factors determine the limits of dynamic range? Can they be changed? Explain.

84. In evaluating a receiver, how important is its ability to handle intermodulation distortion? Explain the process you would use to analyze a receiver's ability to handle this distortion. Include the concept of third-order intercept point in your explanation.

85. The receiver in Problem 61 has a 6-dB NF preamp (gain = 20 dB) added to its input. Calculate the system's sensitivity and dynamic range. (−99.94 dBm, 66.96 dB)

CHAPTER 7

DIGITAL COMMUNICATIONS TECHNIQUES

CHAPTER OUTLINE

KEY TERMS

regeneration

pulse modulation

time-division multiplexing (TDM)

pulse-amplitude modulation (PAM)

pulse-width modulation (PWM)

pulse-position modulation (PPM)

pulse-code modulation (PCM)

pulse-time modulation (PTM)

pulse-duration modulation (PDM)

pulse-length modulation (PLM)

aliasing

foldover distortion

Nyquist rate

codec

acquisition time

aperture time

natural sampling

flat-top sampling

quantization

quantile

quantile interval

quantization levels

quantizing error

quantizing noise

dynamic range (DR)

uniform

linear quantization levels

nonlinear

nonuniform coding

idle channel noise

amplitude companding

flash ADC

successive approximation

Hamming distance

minimum distance

(D_{min})

symbol substitution

block check character (BCC)

longitudinal redundancy check (LRC)

cyclic redundancy check(CRC)

systematic code

(n, k) cyclic codes

generating polynomial

syndrome

forward error-correcting (FEC)

Hamming code

interleaving

software-defined radio (SDR)

recursive

iterative

difference equation

7-1 INTRODUCTION TO DIGITAL COMMUNICATIONS

Changes in the physical world are often continuous and are, therefore, inherently analog in nature. In an analog communications system, transducers such as microphones or television cameras convert variations in sound-pressure level or light intensity to varying voltage levels representing audio or video information. Receiver transducers such as speakers or picture tubes essentially reverse the process by converting voltages back to their physical equivalents. Communications has historically been the purview of analog electronics because transmitted and received information (intelligence) signals were quite literally analogous to the physical quantities they represented. Certainly, much development of electronics as a technological discipline has revolved around the creation of circuits and devices that accurately represent phenomena of interest by faithfully preserving the essential characteristics of the voltages to be transmitted.

The all-analog approach has its advantages. One is relative simplicity. Transducers at the transmitter and receiver are straightforward in their operation, and it is not necessary to convert signals away from and then back to the analog domain. Also, the bandwidth requirements of all-analog transmissions can be shown to be lower than those of the alternative approaches in the absence of sophisticated modulation and compression techniques, which themselves add layers of complexity. The chief disadvantage of analog communication is susceptibility to noise. Undesired signals of sufficient amplitude will corrupt the transmitted information such that it cannot be recovered. The static heard in a radio broadcast or snow seen in an analog television picture represents information lost forever. Nothing in the transmitted analog signal allows for recovery of lost information. For this and other reasons, the communications field has, like many other applications of electronics technology, entered the digital realm, where information is encoded in the discrete, binary language of computers.

Digital communications systems have many advantages. Among the most important is the ability to recover lost information through various forms of error correction. Also extremely important is that noise and other impairments in the communications channel or equipment have much less effect for digital signals than for analog signals. In other words, digital systems can often perform at lower signal-to-noise ratios than their analog counterparts. Yet another advantage is embodied in the concept of **regeneration**, in which noise-corrupted signals are periodically restored to their original values, thereby recreating an ideal, perfect representation of data at regular intervals and enabling transmissions over long distances. These advantages, coupled with the great strides made in high-speed computing and the use of application-specific integrated circuits, make possible the software-based implementation of many functions that have traditionally been carried out in the form of discrete-component circuits. This development has led to a widespread embrace of digital signal processing and software-defined radio techniques, which, taken together, represent perhaps the most dramatic changes seen in the communications art in the past two decades.

For communications to take place digitally, analog signals must be converted to discrete samples for transmission as data. The data must also be "coded," or prepared for transmission over a communications channel that, in all likelihood, does not provide for direct electrical continuity between transmitter and receiver. In addition, one or more error detection and error correction techniques will often be deployed to enhance performance in the presence of noise. The data, which is often binary in nature, can then be sent over a communications channel, either in baseband form, or the data can form the intelligence that subsequently modulates a high-frequency carrier in the ways illustrated in previous chapters. We will look at the modulation of a carrier with data, as well as the recovery of digital information, in more detail in Chapter 8. The concept of coding will be examined in Chapter 9. This chapter, meanwhile, focuses on the preparatory steps required for effective digital communication in both the wired and wireless contexts as well as the rapidly expanding use of digital signal processing in communications applications.

7-2 PULSE MODULATION AND MULTIPLEXING

You have undoubtedly drawn graphs of *continuous* curves many times during your education. To do that, you took data at some finite number of discrete points, plotted each point, and then drew the curve. Drawing the curve may have resulted in a very accurate replica of the desired function even though you did not look at every possible point. In effect, you took *samples* and guessed where the curve went between the samples. If the samples had sufficiently close spacing, the result is adequately described. It is possible to apply this line of thought to the transmission of an electrical signal, that is, to transmit only the samples and let the receiver reconstruct the total signal with a high degree of accuracy. This concept is termed **pulse modulation**. The techniques used to produce pulse modulation, including sampling and the conversion from an analog signal to a digital bit stream and back again, form key underlying concepts for both wired and wireless digital communications systems.

The most important distinction between pulse modulation and the forms of amplitude modulation (AM) and frequency modulation (FM) studied in previous chapters is that in AM or FM some parameter of the modulated wave (formed from combining the low-frequency intelligence or information signal with the high-frequency carrier) varies continuously with the message, whereas in pulse modulation some parameter of a sample pulse is varied by each sample value of the message. The pulses are usually of very short duration so that a pulse-modulated wave is "off" most of the time. Short-duration pulses are desirable because transmitters operate on a very low duty cycle ("off" more than "on"), which is often desirable, particularly for certain microwave devices and lasers.

Another advantage of pulse-modulation schemes is that the time intervals between pulses can be filled with samples of other messages. In other words, a form of channel sharing known as **time-division multiplexing (TDM)** can be deployed. In general terms, *multiplexing* involves conveying two or more information signals over a single transmission channel. With *frequency-division multiplexing,* of which radio and television broadcasting are examples, each information signal modulates a carrier assigned to a specific frequency, and the modulated carriers occupy their respective portions of the available channel bandwidth simultaneously. Conversely, with *time-division multiplexing,* each information signal accesses the entire channel bandwidth but for only a small part of the available time. TDM is analogous to computer time sharing, where several users make use of a computer simultaneously. TDM is extensively used in both wired and wireless communication systems, especially telephone networks, and so will be studied in more detail in Chapter 9.

In the strictest sense, pulse modulation is not modulation as the term has been used up to this point but is, rather, a *message-processing* technique. The term "modulation" is appropriate, however, because, as will be shown shortly, multiplication takes place between the information signal and the pulses to produce sum and difference frequencies as well as harmonics. What is most important to understand at this juncture is that the message or information signal is sampled by pulses occurring at regular intervals and that an encoding process takes place whereby the pulse is somehow modified in response to characteristics of the information signal. The sampled (i.e., pulse-modulated) signals may reside at baseband frequencies, or the sampled pulses may subsequently amplitude- or angle-modulate a high-frequency carrier in the same manner described in previous chapters. Even though the term "modulation" is used in both contexts, keep in mind that these are two distinct processes.

The three basic forms of pulse modulation are **pulse-amplitude modulation (PAM), pulse-width modulation (PWM),** and **pulse-position modulation (PPM)**. Each is illustrated in Figure 7-1. A fourth form, **pulse-code modulation (PCM)**, has PAM encoding as an initial step. PCM is a waveform coding technique that permits analog signals to be represented as digital words. PCM is a

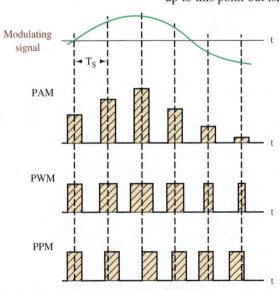

FIGURE 7-1 **Types of pulse modulation.**

fundamental building block of many digital systems and will be discussed more fully in the next section. First, however, let us look at the basic forms of pulse modulation.

For the sake of clarity, the pulse widths shown in Figure 7-1 are greatly exaggerated. Because pulse modulation is often used in conjunction with TDM, shorter pulse durations are desirable to leave room for more multiplexed signals. As shown in Figure 7-1, the pulse parameter that is varied in step with the analog signal is varied in direct step with the signal's value at each sampling interval. Notice that the pulse amplitude in PAM and pulse width in PWM are not zero when the signal is minimum. This is done to allow a constant pulse rate and is important in maintaining synchronization in TDM systems.

Pulse-Amplitude Modulation

In pulse-amplitude modulation (PAM), the pulse amplitude is made proportional to the amplitude of the modulating signal. This is the simplest form of pulse modulation to create because a simple sampling of the modulating signal at a periodic rate can be used to generate the varying-amplitude pulses. As will be seen shortly, the pulses can either be processed further to create digital words in a PCM system, or PAM pulses can directly modulate a high-frequency carrier. Further, the pulses can be time-division multiplexed over a single communication channel, as illustrated in Figure 7-2. The figure shows an eight-channel TDM system whose PAM pulses are applied as the modulating signals (i.e., intelligence) to an FM transmitter. At the transmitter, the eight signals to be transmitted are periodically sampled. The sampler illustrated is a rotating machine making periodic brush contact with each signal. A similar rotating machine at the receiver is used to distribute the eight separate signals, and it must be synchronized to the transmitter. It should be emphasized that the synchronized rotating switches shown in the figure are there mainly to illustrate the TDM concept. A mechanical sampling system such as this may be suitable for the low sampling rates like those encountered in some telemetry systems, but it would not be adequate for the rates required for voice transmissions. In that case, an electronic switching system would be incorporated.

At the transmitter, the variable amplitude pulses are used to frequency-modulate a carrier. A rather standard FM receiver recreates the pulses, which are then applied to the

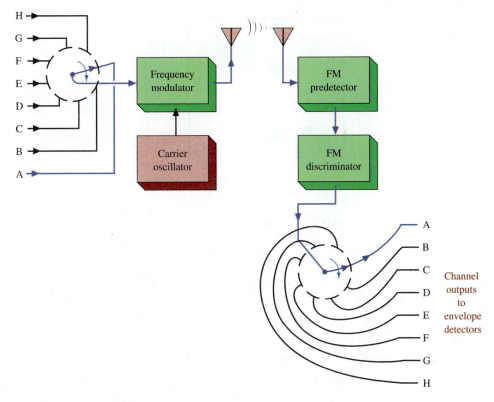

FIGURE 7-2   Eight-channel TDM PAM system.

electromechanical *distributor* going to the eight individual channels. This distributor is virtually analogous to the distributor in a car that delivers high voltages to spark plugs in a periodic fashion. The pulses applied to each line go into an envelope detector that serves to recreate the original signal. This can be a simple low-pass *RC* filter such as that used following the detection diode in a standard AM receiver.

The above description shows a PAM system that ultimately modulates a radio-frequency carrier. It should be emphasized again that PAM, as well as the other forms of pulse modulation discussed in this chapter, may also be used at baseband frequencies; that is, the pulse modulation may be an intermediate step in another process but one that does not necessarily involve modulating a radio-frequency carrier directly. Although PAM finds some use because of its simplicity, PWM and PPM provide superior noise performance because they use constant-amplitude pulses. The PWM and PPM systems fall into a general category termed **pulse-time modulation (PTM)** because their timing, and not amplitude, is the varied parameter.

Pulse-Width Modulation

Pulse-width modulation (PWM), a form of PTM, is also known as **pulse-duration modulation (PDM)** or **pulse-length modulation (PLM)**. A simple means of PWM generation is provided in Figure 7-3 using a 565 PLL. It actually creates PPM at the VCO output (pin 4), but by applying it and the input pulses to an exclusive-OR gate, PWM is also created. For the phase-locked loop (PLL) to remain locked, its VCO input (pin 7) must remain constant. The presence of an external modulating signal upsets the equilibrium. This causes the phase detector output to go up or down to maintain the VCO input (control) voltage. However, a change in the phase detector output also means a change in phase difference between the input signal and the VCO signal. Thus, the VCO output has a phase shift proportional to the modulating signal amplitude. This PPM output is amplified by Q_1 in Figure 7-3 just prior to the output. The exclusive-OR circuit provides a high output only when just

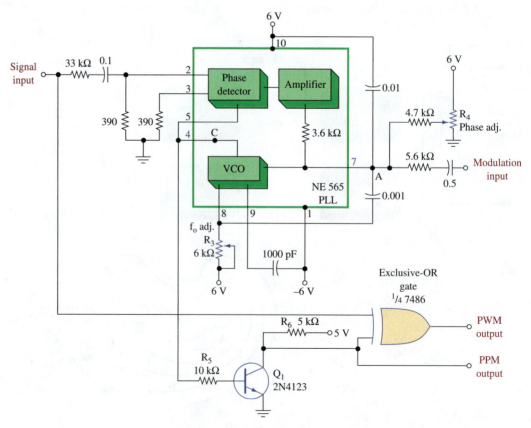

FIGURE 7-3 **PLL generation of PWM and PPM.**

one of its two inputs is high. Any other input condition produces a low output. By comparing the PPM signal and the original pulse input signal as inputs to the exclusive-OR circuit, the output is a PWM signal at twice the frequency of the original input pulses.

Adjustment of R_3 varies the center frequency of the VCO. The R_4 potentiometer may be adjusted to set up the quiescent PWM duty cycle. The outputs (PPM or PWM) of this circuit may then be used to modulate a carrier for subsequent transmission.

CLASS D AMPLIFIER AND PWM GENERATOR   Recall from Chapter 4 that class D amplifiers approach 100% efficiency because the active devices are alternately switched between cutoff and saturation rather than spending any time in the linear region of operation, where power is dissipated by the active device itself. PWM forms the basis for class D power amplification. The circuit in Figure 7-4 is a class D amplifier because the actual power amplification is provided to the PWM signal, and because it is of constant amplitude, the transistors alternate between cutoff and saturation. This allows for maximum efficiency (in excess of 90%) and is the reason for the increasing popularity of class D amplifiers as a means of amplifying any analog signal.

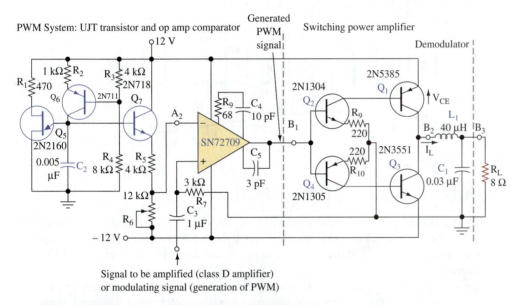

FIGURE 7-4   PWM generator and class D power amplifier.

The circuit of Figure 7-4 illustrates another common method for generation of PWM and also illustrates class D amplification. The Q_6 transistor generates a constant current to provide a linear charging rate to capacitor C_2. The unijunction transistor, Q_5, discharges C_2 when its voltage reaches Q_5's firing voltage. At this time C_2 starts to charge again. Thus, the signal applied to Q_7's base is a linear saw tooth as shown at A in Figure 7-5. That sawtooth following amplification by the Q_7 emitter-follower in Figure 7-4 is applied to the op amp's inverting input.

The modulating signal or signal to be amplified is applied to its noninverting input, which causes the op amp to act as a comparator. When the sawtooth waveform at A in Figure 7-5 is less than the modulating signal B, the comparator's output (C) is high. At the instant A becomes greater than B, C goes low. The comparator (op amp) output is therefore a PWM signal. It is applied to a push-pull amplifier (Q_1, Q_2, Q_3, Q_4) in Figure 7-4, which is a highly efficient switching amplifier. The output of this power amp is then applied to a low-pass LC circuit (L_1, C_1) that converts back to the original signal (B) by integrating the PWM signal at C, as shown at D in Figure 7-5. The output of the op amp in Figure 7-4 would be used to modulate a carrier in a communications system, while a simple integrating filter would be used at the receiver as the detector to convert from pulses to the original analog modulating signal.

FIGURE 7-5   PWM generation waveforms.

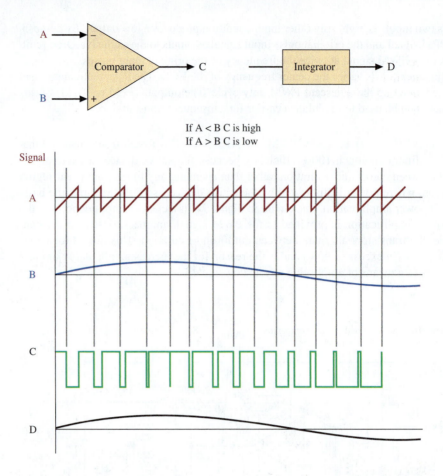

Pulse-Position Modulation

PWM and pulse-position modulation (PPM) are very similar, a fact that is underscored in Figure 7-6, which shows PPM being generated from PWM. Because PPM has superior noise characteristics, it turns out that the major use for PWM is to generate PPM. By inverting the PWM pulses in Figure 7-6 and then differentiating them, the positive and negative spikes shown are created. By applying them to a Schmitt trigger sensitive only to positive levels, a constant-amplitude and constant-pulse-width signal is formed. However, the position of these pulses is variable and is now proportional to the original modulating signal, and the desired PPM signal has been generated. The information content is *not* contained in either the pulse amplitude or width as in PAM and PWM, which means the signal now has a greater resistance to any error caused by noise. In addition, when PPM modulation is used to amplitude-modulate a carrier, a power savings results because the pulse width can be made very small.

At the receiver, the detected PPM pulses are usually converted to PWM first and then converted to the original analog signal by integrating, as previously described. Conversion from PPM to PWM can be accomplished by feeding the PPM signal into the base of one transistor in a flip-flop. The other base is fed from synchronizing pulses at the original

FIGURE 7-6   PPM generation.

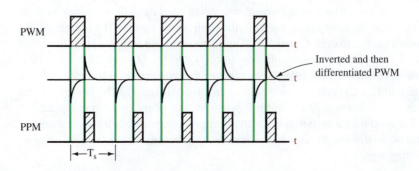

(transmitter) sampling rate. The period of time that the PPM-fed transistor's collector is low depends on the difference in the two inputs, and it is therefore the desired PWM signal.

This detection process illustrates the one disadvantage of PPM compared to PAM and PWM. It requires a pulse generator synchronized from the transmitter. However, its improved noise and transmitted power characteristics make it the most desirable pulse modulation scheme.

Demodulation

The process of demodulation is one of reproducing the original analog signal. We have noticed that the PAM signals contained harmonics of higher frequencies. Reshaping the original information signal would necessarily require removal of the higher frequencies. The use of a low-pass filter will accomplish this task. The upper cutoff frequency is selected to eliminate the highest frequency contained in the information. Figure 7-7 shows a block diagram of a PAM demodulator.

FIGURE 7-7 **Block diagram of a PAM demodulator.**

Demodulating a PWM signal is simple. The approach is similar to demodulation of the PAM signal. A low-pass filter can be used along with some wave-shaping circuit. By using an RS flip-flop, the PPM signal can be converted to PWM and then demodulated using the technique for demodulating PWM.

7-3 SAMPLE RATE AND NYQUIST FREQUENCY

Each pulse-modulation technique just described relies on producing a series of pulses to sample, at regular intervals, the amplitude of a continuously varying information or intelligence signal and then to represent the intelligence amplitude in terms of a pulse characteristic, such as pulse amplitude, width (duty cycle), or position. Only those parameters captured at the sample times are used. In other words, the continuously varying analog wave is broken down into a series of discrete sample points, and any information residing in between the sample points is simply not used or transmitted. Though it may seem counterintuitive at first, the "lost" information does not prevent the original, analog intelligence to be recreated provided that the sampling occurs often enough. The question naturally arises: just how fast does the sample rate need to be? The answer can be found by recalling that the essence of modulation is the multiplication of two or more frequencies and that additional frequencies are created as a result. As we shall see shortly, the act of sampling an analog waveform is a form of modulation because of the interaction that takes place between the sampled pulses and the information signal.

Recall from Chapter 1 that Fourier analysis allows complex, periodic waveforms to be broken down into a series of sine and cosine terms. The discussion in that chapter focused on the spectral (frequency-domain) content of square waves, which were shown at that time to consist of the fundamental and odd harmonics. The Fourier series for pulses can be established as well and, from Table 1-3(d), can be shown to take the following form:

$$v_s = \frac{\tau}{T} + 2\frac{\tau}{T}\left(\frac{\sin \pi\tau/T}{\pi\tau/T}\cos \omega_s t + \frac{\sin 2\pi\tau/T}{2\pi\tau/T}\cos 2\omega_s t + \frac{\sin 3\pi\tau/T}{3\pi\tau/T}\cos 3\omega_s t + \ldots\right),$$

where v_s = instantaneous voltage of the sampling pulse

τ = pulse duration

T = period of pulse train

ω_s = radian frequency of pulse train

It is not really necessary to understand the trigonometry to predict the result. Just be aware that the preceding equation represents the sampling signal, which is a series of narrow pulses (a *pulse train*) at a relatively high-frequency sample rate. The sample rate is roughly analogous to a high-frequency carrier in the context of how the term "modulation" has been introduced in previous chapters.

If the high-frequency sample rate (i.e., carrier) shown in the previous equation is used to digitize a single-tone (i.e., sine-wave) information signal, f_i, of the form $V_i\sin(2\omega_i t)$, where V_i is the peak voltage of the modulating signal, the result is multiplication of the information signal by the sample rate:

$$v(t) = V_i\frac{\tau}{T}\sin\omega_i t + 2V_i\frac{\tau}{T}\left(\frac{\sin \pi\tau/T}{\pi\tau/T}\sin\omega_i t\cos\omega_s t + V_i\frac{\sin 2\pi\tau/T}{2\pi\tau/T}\sin\omega_i t\cos 2\omega_s t\right.$$
$$\left. + V_i\frac{\sin 3\pi\tau/T}{3\pi\tau/T}\sin\omega_i t\cos 3\omega_s t + \ldots\right).$$

Though the above equation may look difficult to decipher at first, the result turns out to be the same as what we have seen many times before: multiplication created as the result of mixing. The first term is the sine-wave information signal. The presence of this term is significant because it predicts that the original intelligence information can be recovered from the sampled signal provided that a low-pass filter is used to remove the higher-order terms. The second and following terms represent the product of sine and cosine waves representing sum and difference frequencies of the sample rate and the information signal as well as sum and difference frequencies at the harmonics of the sampling frequency.

Figure 7-8 shows the result in the frequency domain. Sidebands are created above and below the sample frequency, f_s, and these occupy the region from $f_s - f_i$ to $f_s + f_i$. In other words, these are sum and difference frequencies. Sidebands also form around the harmonics of f_s. The harmonics and their sidebands are usually removed with a low-pass filter, but the presence of the sidebands around f_s, the fundamental term, gives a clue as to the relationship that must be maintained between the sample rate and the frequency of the signal to be sampled.

FIGURE 7-8 **Frequency-domain representation of sampled signal.**

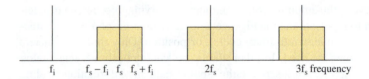

The preceding expression predicts that the original information signal, f_i, can be recovered provided that no other signals close in frequency to f_i are present when the intelligence signal is reconstructed. From the previous expression as well as Equation (2-7), we see that if the sample rate, f_s, is at least twice the frequency of the information signal, f_i, the difference frequency that is inevitably created will be above f_i, and if it is far enough above f_i, the low-pass filter can do its job. As an example, assume an intelligence frequency of 1 kHz. If the sample rate is 2 kHz or more, then the difference frequency would always be greater than the intelligence frequency, and a low-pass filter would be able to filter out all the harmonic components created as the result of sampling.

What would happen, though, if the sample rate were less than twice the intelligence frequency? In the preceding example, if the information signal remained at 1 kHz but the sample rate were reduced to, say, 1.5 kHz, the difference frequency created would then be 500 Hz—less than the frequency of the information signal and, therefore, impossible to remove with a low-pass filter. This undesired byproduct of mixing is known as an *alias*, and the phenomenon just described is called **aliasing** or **foldover distortion**. The undesired 500-Hz alias, rather than the expected 1-kHz information signal, would be recreated by digital-to-analog converters when the sampled signal is restored to analog form. This undesirable outcome can be avoided by placing a low-pass filter called an *anti-aliasing filter* between the modulating signal and the circuits used to convert the information signal to digital form. The purpose of the anti-aliasing filter is to ensure that the highest frequency component of the analog signal to be digitized is no more than one half the sample rate.

The analysis just given confirms that the sample rate must be *at least* twice the highest frequency of the intelligence or information signal to be sampled. Stated as an equation,

$$f_s \geq 2f_i. \tag{7-1}$$

The minimum sample rate is known as the **Nyquist rate,** named for Harry Nyquist, who in 1928 first demonstrated mathematically that an analog signal can be reconstructed from periodic samples. The Nyquist theorem applies to all forms of pulse modulation, including, perhaps most importantly, the PCM technique to be discussed shortly. PCM is used in virtually all telephone systems, and linear PCM (also to be discussed shortly) binary words make up the digitized music on compact discs. A voice channel on a telephone system is band-limited to a maximum of 4 kHz. The sample rate for the telephone system is 8 kHz, twice the highest input frequency. The highest-frequency audio signals for compact-disc recordings extend to 20 kHz, and these are sampled at a rate of 44.1 kHz, slightly more than twice the highest audio frequency to be preserved.

7-4 PULSE-CODE MODULATION

As already mentioned, PCM is a means of representing an analog signal in a digital format. PCM is used in many applications, including digital audio recording (DAT or digital audio tape), CD laser discs, digitized video special effects, voice mail, and digital video. PCM-formatted and multiplexed binary words also form the backbone of the global telephone system. Because PCM techniques and applications are a primary building block for many of today's advanced communications systems, we will take a look at the steps involved in some detail.

Pulse-code modulation takes the pulse-modulation techniques studied in Section 7-2 one step further. Instead of simply changing the amplitude, width, or position of a pulse, as with PAM or PTM, a PCM system uses a three-step process to convert a continuously varying waveform into a set of discrete values and then represents these values as digital words. The PCM encoding steps are

- Sampling—reading the amplitude of a continuously varying analog signal at regular, discrete intervals and storing the peak voltage of each resulting pulse.
- Quantizing—assigning one of a fixed number of numerical values to each pulse.
- Coding—expressing each quantized value as a binary word.

The process is essentially reversed when PCM words are converted back to their analog equivalents. The PCM decoding steps are

- Regeneration—receipt of binary pulses and recreation of the PCM words as square waves;
- Decoding—interpretation of PCM words in digital-to-analog decoder and translation into quantized amplitude values; and
- Reconstruction—conversion of quantized waveform into a stepped analog signal.

After it is low-pass filtered, the reconstructed analog signal obtained from a PCM decoder very closely resembles the original analog signal applied to the input of the encoder. PCM thus allows analog quantities to be processed in the digital domain with all the advantages in error correction, bandwidth conservation, and other signal processing considerations that digital communications systems can offer.

A block diagram of the encoding process is shown in Figure 7-9. The PCM architecture consists of a sample-and-hold (S/H) circuit and a system for converting the sampled signal into a representative binary format. First, the analog signal is input into an S/H circuit. At fixed time intervals, the analog signal is sampled and held at a fixed voltage level

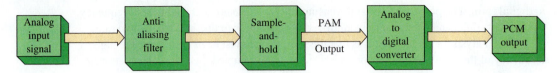

FIGURE 7-9　Block diagram of the PCM process.

until the circuitry inside an analog-to-digital (ADC) converter has time to complete the conversion process of generating a binary value. The binary value, made up of multiple bits, forms the digital word that is either stored (such as on a compact disc or computer hard drive) or transmitted through a communications channel.

The ADC circuitry in PCM systems is often referred to as the encoder. The digital-to-analog (DAC) circuitry at the receiver is correspondingly termed the decoder. These functions are often combined in a single LSI chip termed a **codec** (coder-decoder). The block diagram for a typical codec is provided in Figure 7-10. These devices are widely used in the telephone industry to allow voice transmission to be accomplished in digital form. The following sections describe the functional blocks as well as of the steps required for encoding and decoding.

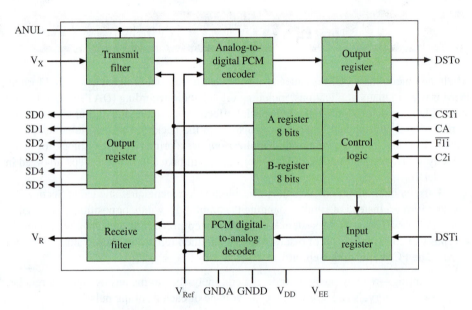

FIGURE 7-10　Codec block diagram.

The Sample-and-Hold Circuit

Most ADC integrated circuits have S/H circuits integrated into the system, but it is still important for the user to have a good understanding of the capabilities and the limitations of the S/H circuit. The S/H effectively functions as the modulator in a PCM system because it is here that the information signal will be sampled. A typical S/H circuit is shown in Figure 7-11. The analog signal is the input to a buffer circuit whose purpose is to isolate the input signal from the S/H circuit and to provide proper impedance matching as well as drive capability to the hold circuit. Many times the buffer circuit is also used as a current source to charge the hold capacitor. The output of the buffer is fed to an analog switch, which is typically the drain of a junction field-effect transistor (JFET) or a metal-oxide semiconductor field-effect transistor (MOSFET). The JFET or MOSFET is wired as an analog switch, which is controlled at the gate by a sample pulse generated by the sample clock. When the JFET or MOSFET transistor is turned on, the switch will short the analog signal from drain to source. This connects the buffered input signal to a hold capacitor. The capacitor begins to charge to the input voltage level at a time constant determined by the hold capacitor's capacitance and the analog switch's and buffer circuit's "on" channel resistance.

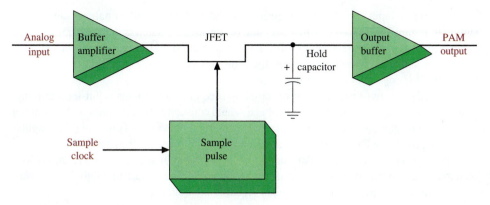

FIGURE 7-11   A sample-and-hold (S/H) circuit.

When the analog switch is turned off, the sampled analog signal voltage level is held by the hold capacitor. Figure 7-12(a) shows a picture of a sinusoid on the input of the S/H circuit. The sample times are indicated by the vertical green lines. In Figure 7-12(b) the sinusoid is redrawn as a sampled signal. Note that the sampled signal maintains a fixed voltage level between samples. The region where the voltage level remains relatively constant is called the *hold time*. The resulting waveform, shown in Figure 7-12(b), is a PAM signal. The S/H circuit is designed so that the sampled signal is held long enough for the signal to be converted by the ADC circuitry into a binary representation.

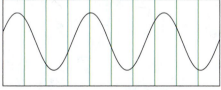 

(a)   Sample intervals for an input sinusoid.   (b) A pulse-amplitude-modulated (PAM) signal.

FIGURE 7-12   Generation of PAM.

The time required for an S/H circuit to complete a sample is based partly on the acquisition and aperture time. The **acquisition time** is the amount of time it takes for the hold circuit to reach its final value. (During this time the analog switch connects the input signal to the hold capacitor.)

The sample pulse controls the acquisition time. The **aperture time** is the time that the S/H circuit must hold the sampled voltage. The aperture and acquisition times limit the maximum input frequency that the S/H circuit can accurately process.

To provide a good-quality S/H circuit, a couple of design considerations must be met. The analog switch "on" resistance must be small. The output impedance of the input buffer must also be small. By keeping the input resistance minimal, the overall time constant for sampling the analog signal can be controlled by the selection of an appropriate hold capacitor. Ideally a minimal-size hold capacitor should be selected so that a fast charging time is possible, but a small capacitor will have trouble holding a charge for a very long period. A 1-nF hold capacitor is a popular choice. The hold capacitor must also be of high quality. High-quality capacitors have polyethylene, polycarbonate, or Teflon dielectrics. These types of dielectrics minimize voltage variations resulting from capacitor characteristics.

Natural and Flat-top Sampling

The concept of pulse-amplitude modulation (PAM) has already been introduced in this chapter, but there are a few specifics regarding the creation of a pulse-amplitude-modulated signal at the output of a sample-and-hold circuit that necessitate discussion.

Two basic sampling techniques are used to create a PAM signal. The first is called **natural sampling**. Natural sampling occurs when the tops of the sampled waveform (the sampled analog input signal) retain their natural shape. An example of natural sampling is shown in Figure 7-13(a). Notice that one side of the analog switch is connected to ground. When the transistor is turned on, the JFET will short the signal to ground, but it will pass the unaltered signal to the output when the transistor is off. Note, too, that there is not a hold capacitor present in the circuit.

Probably the most popular type of sampling used in PCM systems is **flat-top sampling**. In flat-top sampling, the sample signal voltage is held constant between samples. The method of sampling creates a staircase that tracks the changing input signal. This method is popular because it provides a constant voltage during a window of time for the binary conversion of the input signal to be completed. An example of flat-top sampling is shown in Figure 7-13(b). Note that this is the same type of waveform as shown in Figure 7-12(b). With flat-top sampling, the analog switch connects the input signal to the hold capacitor.

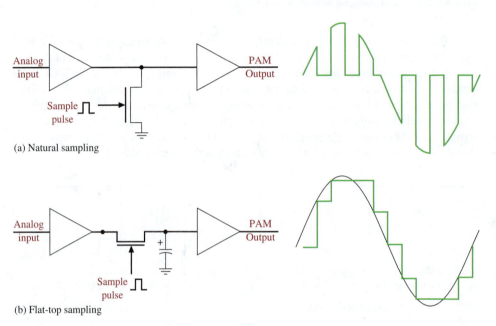

(a) Natural sampling

(b) Flat-top sampling

FIGURE 7-13   (a) Natural sampling; (b) flat-top sampling.

Quantization

Section 7-3 established that the Nyquist rate is the minimum sample rate, or number of samples that must be taken per second, for the analog (information or modulating) signal to be recreated fully. Each sample taken will be converted to a digital word consisting of multiple binary digits, or bits. In PCM systems the sampled signal is segmented into one of a predetermined number of voltage levels, with each level corresponding to a different binary number. This process is called **quantization**. Each quantization level step-size is called a **quantile**, or **quantile interval**.

The quantization levels also determine the resolution of the digitizing system. Each sampled amplitude value must be represented as a digital word with a fixed number of bits. Therefore, not every possible amplitude value can be represented by a digital equivalent. The number of bits making up the word (word size) determines the number of quantization levels. Additionally, the number of points at which the analog signal is sampled determines the number of digital words created. There is a tradeoff—increasing either the number of words created (increasing the sample rate) or the number of bits per word improves the ability of the system to represent the original signal faithfully but at the expense of increased bandwidth and overall system complexity.

Analog signals are quantized to the closest binary value provided in the digitizing system. This is an approximation process. For example, if our numbering system is the set of whole numbers 1, 2, 3, . . . , and the number 1.4 must be converted (rounded off) to the closest whole number, then 1.4 is translated to 1. If the input number is 1.6, then the number is translated to a 2. If the number is 1.5, then we have the same error if the number is rounded off to a 1 or a 2.

Figure 7-14 shows how the electrical representation of voice is converted from analog form to a PCM digital bit stream. The amplitude levels and sampling times are predetermined characteristics of the encoder. The amplitude levels are termed **quantization levels**, and 12 such levels are shown. At each sampling interval, the analog amplitude is quantized into the closest available quantization level, and the analog-to-digital converter (ADC) puts out a series of pulses representing that level in the binary code.

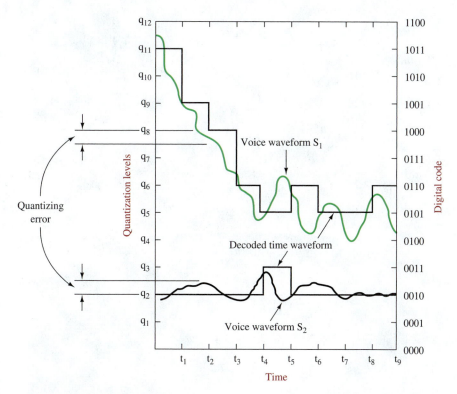

FIGURE 7-14   PCM encoding.

For example, at time t_2 in Figure 7-14, voice waveform S_1 is closest to level q_8, and thus the coded output at that time is the binary code 1000, which represents 8 in binary code. Note that the quantizing process resulted in an error, which is termed the **quantizing error**, or **quantizing noise**. The maximum voltage of the quantization error is one-half the voltage of the minimum step-size $V_{LSB/2}$. Voice waveform S_2 provides a 0010 code at time t_2, and its quantizing error is also shown in Figure 7-14. The amount of this error can be minimized by increasing the number of quantizing levels, which of course lessens the space between each one. The 4-bit code shown in Figure 7-14 allows for a maximum of 16 levels because $2^4 = 16$. The use of a higher-bit code decreases the error at the expense of transmission time and/or bandwidth because, for example, a 5-bit code (32 levels) means transmitting 5 high or low pulses instead of 4 for each sampled point. The sampling rate is also critical and must be greater than twice the highest significant frequency, as previously described. It should be noted that the sampling rate in Figure 7-14 is lower than the highest-frequency component of the information. This is not a practical situation but was done for illustrative purposes.

While a 4- or 5-bit code may be adequate for voice transmission, it is not adequate for transmission of television signals. Figure 7-15 provides an example of TV pictures for 5-bit and 8-bit (256 levels) PCM transmissions, each with 10-MHz sampling rates. In the first (5-bit) picture, contouring in the forehead and cheek areas is very pronounced. The 8-bit resolution results in an excellent-fidelity TV signal that is not discernibly different from a standard continuous modulation transmission.

FIGURE 7-15   PCM TV
transmission: **(a) 5-bit**
resolution; (b) 8-bit resolution.

Notice in Figure 7-16 that at the sample intervals, the closest quantization level is selected for representing the sine-wave signal. The resulting waveform has poor resolution with respect to the sine-wave input. *Resolution* with respect to a digitizing system refers to the accuracy of the digitizing system in representing a sampled signal. It is the smallest analog voltage change that can be distinguished by the converter. For example, the analog input to our PCM system has a minimum voltage of 0.0 V and a maximum of 1.0 V. Then

$$q = \frac{V_{max}}{2^n} = \frac{V_{FS}}{2^n}$$

where q = the resolution

n = number of bits

V_{FS} = full-scale voltage

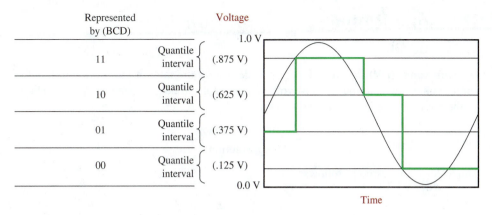

FIGURE 7-16   Voltage levels for a quantized signal.

If a 2-bit system is used for quantizing a signal, then 2^2, or 4, quantized levels are used. Referring to Figure 7-16 we see that the quantized levels (quantile intervals) are each 0.25 V in magnitude. Typically it is stated that this system has 2-bit resolution. This follows from the equation just presented.

To increase the resolution of a digitizing system requires that the number of quantization levels be increased. Increasing the number of quantization levels requires that the number of binary bits representing each voltage level be increased. If the resolution of the example in Figure 7-16 is increased to 3 bits, then the input signal will be converted to 1 of 8 possible values. The 3-bit example with improved resolution is shown in Figure 7-17.

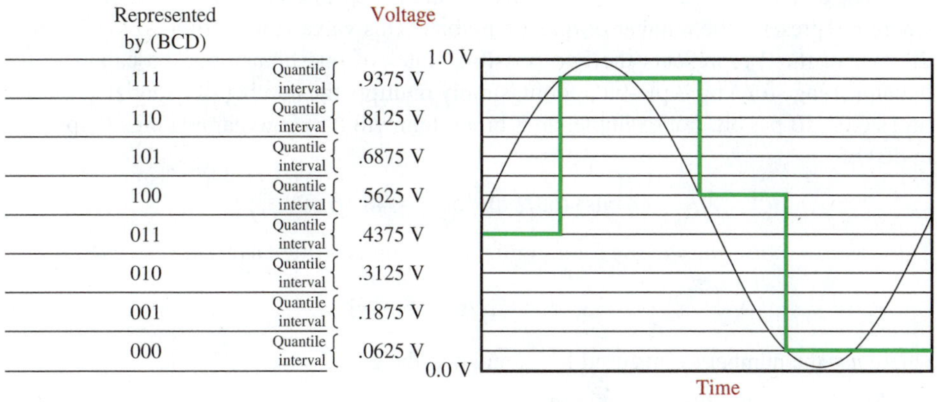

FIGURE 7-17   An example of 3-bit quantization.

Another way of improving the accuracy of the quantized signal is to increase the sample rate. Figure 7-18 shows the sample rate doubled but still using a 3-bit system. The resultant signal shown in Figure 7-18 is dramatically improved compared to the quantized waveform shown in Figure 7-17 by this change in sampling rate.

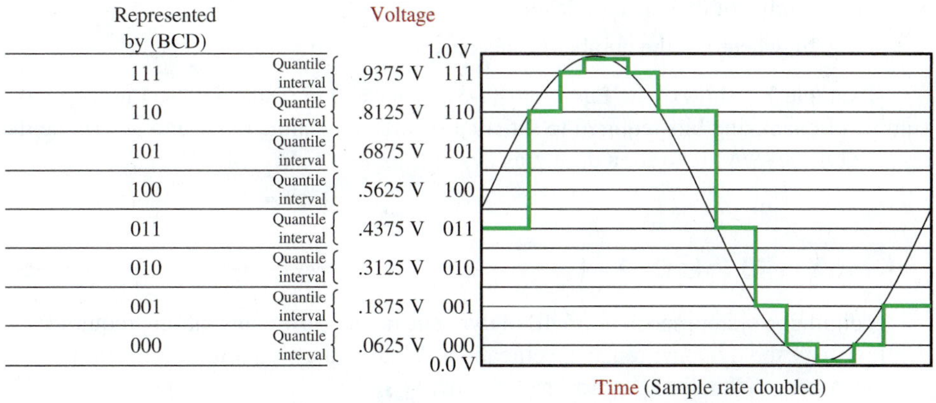

FIGURE 7-18   An example of 3-bit quantization with increased sample rate.

Dynamic Range and Signal-to-Noise Calculations

Dynamic range (DR) for a PCM system is defined as the ratio of the maximum input or output voltage level to the smallest voltage level that can be quantized and/or reproduced by the converters. It is the same as the converter's parameters:

$$\frac{V_{FS}}{q}, \frac{\text{full-scale voltage}}{\text{resolution}}.$$

This value is expressed as follows:

$$DR = \frac{V_{max}}{V_{min}} = 2^n. \tag{7-2}$$

Dynamic range is typically expressed in terms of decibels. For a binary system, each bit can have two logic levels, either a logical low or logical high. Therefore the dynamic range for a single-bit binary system can be expressed logarithmically, in terms of dB, by the expression

$$DR_{dB} = 20 \log\frac{V_{max}}{V_{min}}$$

$$DR_{dB} = 20 \log 2^n. \tag{7-3}$$

where n = number of bits in the digital word.

The dynamic range (DR) for a binary system is expressed as 6.02 dB/bit or $6.02 \times n$, where n represents the number of quantizing bits. This value comes from 20 log 2 = 6.02 dB, where the 2 represents the two possible states of one binary bit. To calculate the dynamic range for a multiple-bit system, simply multiply the number of quantizing bits (n) times 6.02 dB per bit. For example, an 8-bit system will have a dynamic range (expressed in dB) of

$$(8 \text{ bits}) (6.02 \text{ dB/bit}) = 48.16 \text{ dB}.$$

The *signal-to-noise ratio (S/N)* for a digitizing system is written as

$$S/N = [1.76 + 6.02n]. \tag{7-4}$$

where n = the number of bits used for quantizing the signal

  S/N = the signal-to-noise ratio in dB.

This relationship is based on the ratio of the rms quantity of the maximum input signal to the rms quantization noise.

Another way of measuring digitized or quantized signals is the *signal-to-quantization-noise level* $(S/N)_q$. This relationship is expressed mathematically, in dB, as

$$(S/N)_{q(dB)} = 10 \log 3L^2. \tag{7-5}$$

where L = number of quantization levels

  $L = 2n$, where n = the number of bits used for sampling

Example 7-1 shows how Equations (7-3), (7-4), and (7-5) can be used to obtain the number of quantizing bits required to satisfy a specified dynamic range and determine the *signal-to-noise (S/N)* ratio for a digitizing system.

EXAMPLE 7-1

A digitizing system specifies 55 dB of dynamic range. How many bits are required to satisfy the dynamic range specification? What is the signal-to-noise ratio for the system? What is $(S/N)_q$ for the system?

SOLUTION

First solve for the number of bits required to satisfy a dynamic range (DR) of 55 dB.

$$DR = 6.02 \text{ dB/bit } (n)$$
$$55 \text{ dB} = 6.02 \text{ dB/bit } (n)$$
$$n = \frac{55}{6.02} = 9.136. \tag{7-3}$$

Therefore, 10 bits are required to achieve 55 dB of dynamic range. Nine bits will provide a dynamic range of only 54.18 dB. The tenth bit is required to meet the 55 dB of required dynamic range. Ten bits provides a dynamic range of 60.2 dB. To determine the signal-to-noise (S/N) ratio for the digitizing system:

$$S/N = [1.76 + 6.02n] \text{ dB}$$
$$S/N = [1.76 + (6.02)10] \text{ dB}$$
$$S/N = 61.96 \text{ dB}. \tag{7-4}$$

Therefore, the system will have a *(S/N)* ratio of 61.96 dB. For this example, 10 sample bits are required; therefore, $L = 2^{10} = 1024$ and

$$(S/N)_{q(\text{dB})} = 10 \log 3L^2 = 10 \log 3(1024)^2 = 64.97 \text{ dB}. \tag{7-5}$$

For Example 7-1, the dynamic range is 60.2 dB, $S/N = 61.96$ dB, and $(S/N)q = 64.97$ dB. The differences result from the assumptions made about the sampled signal and the quantization process. For practical purposes, the 60.2-dB value is a good estimate, and it is easy to remember that each quantizing bit provides about 6 dB of dynamic range.

Companding

Up to this point our discussion and analysis of PCM systems have been developed around **uniform** or **linear quantization levels**. In linear (uniform) quantization systems each quantile interval is the same step-size. An alternative to linear PCM systems is **nonlinear** or **nonuniform coding**, in which each quantile interval step-size may vary in magnitude.

It is quite possible for the amplitude of an analog signal to vary throughout its full range. In fact, this is expected for systems exhibiting a wide dynamic range. The signal will change from a very strong signal (maximum amplitude) to a weak signal (minimum amplitude—V_{1sb} for quantized systems). For the system to exhibit good S/N characteristics, either the input amplitude must be increased with reference to the quantizing error or the quantizing error must be reduced.

The justification for the use of a nonuniform quantization system will be presented, but let us discuss some general considerations before proceeding. How can the quantization error be modified in a nonuniform PCM system so that an improved S/N results? The answer can be obtained by first examining a waveform that has uniform quantile intervals, such as that shown in Figure 7-19. Notice that poor resolution is present in the weak signal regions, yet the strong signal regions exhibit a reasonable facsimile of the original signal. Figure 7-19 also shows how the quantile intervals can be changed to provide smaller step-sizes within the area of the weak signal. This will result in an improved S/N ratio for the weak signal.

What is the price paid for incorporating a change such as this in a PCM system? The answer is that the large amplitude signals will have a slightly degraded S/N, but this is an acceptable situation if the goal is improving the weak signal's S/N.

Idle Channel Noise

Digital communications systems will typically have some noise in the electronics and the transmission systems. This is true of even the most sophisticated technologies currently

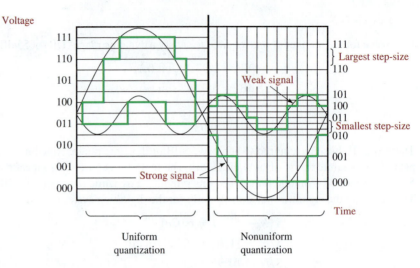

FIGURE 7-19   **Uniform (left) and nonuniform (right) quantization.**

available. One of the noise signals present is called **idle channel noise**. Simply put, this is a noise source of small amplitude that exists in the channel independent of the analog input signal and that can be quantized by the ADC converter. One method of eliminating the noise source in the quantization process is to incorporate a quantization procedure that does not recognize the idle channel noise as large enough to be quantized. This usually involves increasing the quantile interval step-size in the noise region to a large-enough value so that the noise signal can no longer be quantized.

Amplitude Companding

The other form of companding is called **amplitude companding**. Amplitude companding involves volume compression before transmission and expansion after detection. This is illustrated in Figure 7-20. Notice how the weak portion of the input is made nearly equal to the strong portion by the compressor but is restored to the proper level by the expander. Companding is essential to quality transmission using PCM.

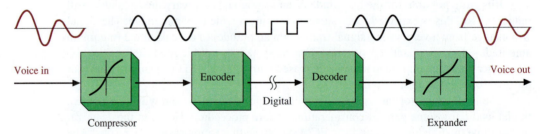

FIGURE 7-20   **Companding process.**

The use of time-division-multiplexed (TDM) PCM transmission for telephone transmissions has proven its ability to cram more messages into short-haul cables than frequency-division-multiplexed (FDM) analog transmission. This concept is explored fully in Chapter 9. The TDM PCM methods were started by Bell Telephone in 1962 and are now the only methods used in new designs. Once digitized, these voice signals can be electronically switched and restored without degradation. The standard PCM system in U.S. and Japanese telephony uses μ-*law* companding. In Europe, the Consultative Committee on International Telephone & Telegraph (CCITT) specifies A-law companding. The μ-*law* companded signal is predicted by

$$V_{out} = \frac{V_{max} \times \ln(1 + \mu V_{in}/V_{max})}{\ln(1 + \mu)}$$

(7-6)

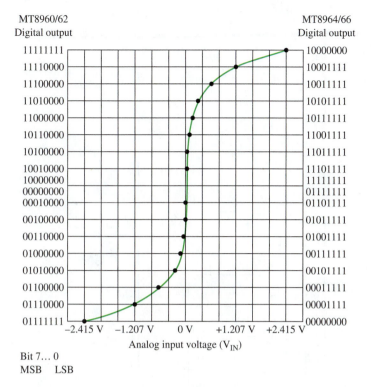

MT8960/62
Digital output

MT8964/66
Digital output

Bit 7... 0
MSB   LSB

FIGURE 7-21 μ-law encoder transfer characteristic.

The μ parameter defines the amount of compression. For example, $\mu = 0$ indicates no compression and the voltage gain curve is linear. Higher values of μ yield nonlinear curves. The early Bell systems have $\mu = 100$ and a 7-bit PCM code. An example of μ-*law* companding is provided in Figure 7-21. This figure shows the encoder transfer characteristic.

Coding and Analog-to-Digital Conversion

The final step in the creation of PCM words is *coding,* or the expressing of each quantized value as a binary word. Coding is largely a function of analog-to-digital (ADC) conversion, and ADC converters are found either as standalone integrated circuits or as part of PCM codec ICs. Many different methods of ADC conversion have been developed. One type is known as the **flash ADC**, and, although not widely used in PCM codecs for reasons that will become clear shortly, its operation is straightforward and illustrates how ADC conversion works as a general concept.

Figure 7-22 shows a flash ADC that produces a 3-bit binary word with each sample. (This example is simplified for the purposes of illustration; recall that practical PCM systems for voice or video would produce 8-bit words.) Each of the seven operational amplifiers (op-amps) is configured as a *comparator,* whose output will go to a logic high level whenever the analog voltage, V_{in}, applied to the noninverting (+) input exceeds the reference voltage for that comparator established by the resistive voltage divider and applied to the inverting (−) input. Again for the purposes of illustration, assume that the voltage divider is designed to produce a reference voltage that increases in 1-V steps from the bottom comparator shown in the figure to the top. Therefore, the bottom comparator has a reference voltage of 1 V, the one above it has a 2-V reference, and so on up to 7 V for the top comparator. This example comparator would, therefore, be capable of sampling an analog signal from 0 to 8 V in 1-V quantizing steps. Each comparator output becomes the input for a *priority encoder,* a digital logic circuit that produces a three-bit binary number representative of the *highest* input asserted.

The analog voltage V_{in} shown in the figure is the voltage to be sampled and digitized. The sample pulses, again at the Nyquist rate or higher, are applied to the enable input of the priority encoder. A 3-bit digital word, corresponding to the voltage level present when

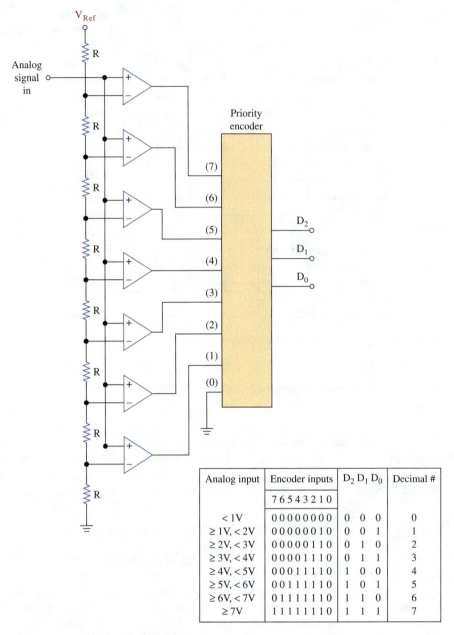

Analog input	Encoder inputs	D_2 D_1 D_0	Decimal #
	7 6 5 4 3 2 1 0		
< 1V	0 0 0 0 0 0 0 0	0 0 0	0
≥ 1V, < 2V	0 0 0 0 0 0 1 0	0 0 1	1
≥ 2V, < 3V	0 0 0 0 0 1 1 0	0 1 0	2
≥ 3V, < 4V	0 0 0 0 1 1 1 0	0 1 1	3
≥ 4V, < 5V	0 0 0 1 1 1 1 0	1 0 0	4
≥ 5V, < 6V	0 0 1 1 1 1 1 0	1 0 1	5
≥ 6V, < 7V	0 1 1 1 1 1 1 0	1 1 0	6
≥ 7V	1 1 1 1 1 1 1 0	1 1 1	7

FIGURE 7-22 **Flash ADC block diagram.**

each sample is taken, is produced at each sample point. The result is shown in Figure 7-23. The sine wave in (a) is sampled, and the priority encoder output produces the words (expressed as square waves) in (b).

The primary advantage of the flash ADC is that it is very fast and, therefore, ideal for capturing analog signals with a great deal of high-frequency information. For applications such as video, where the digitizing process needs to take place at speeds faster than 100 ns to capture high-frequency content adequately, flash ADCs capable of producing 6-, 8-, or 10-bit words are available. Their primary disadvantages are cost, complexity, and power consumption. The number of comparators required in a flash ADC is given by the expression $2^n - 1$. Thus, to produce an 8-bit word, such as would be required for telephone voice and some video systems, a flash ADC would require 255 comparators. Also, op-amp comparators are linear devices, implying high power consumption, making flash ADCs impractical for portable operation, and the sheer number of them required for high-bit-rate operation implies that integrated-circuit implementations would be difficult and expensive. They are the technology of choice, however, for high-speed applications.

Another type of ADC converter, and perhaps the most widely used because of its compactness and power economy, is the **successive approximation** ADC. This implementation is

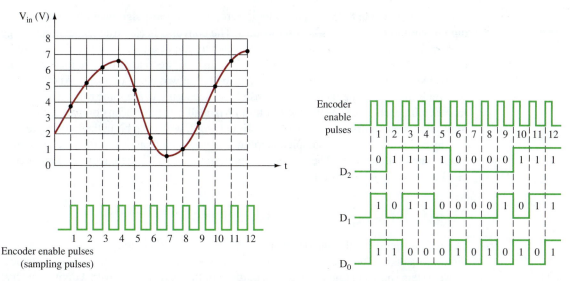

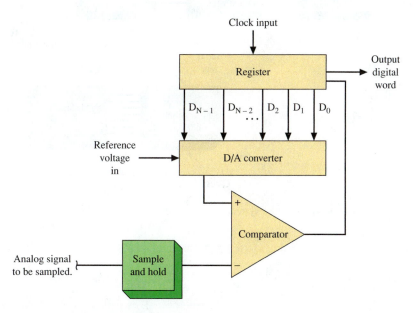

FIGURE 7-23   **Creation of 3-bit words from sampled analog signal.**   Adapted from Figures 13-17 and 13-18 in *Electronic Devices, 9th ed.* ©2012 Pearson Education, Inc.

seen in microcontrollers as well as PCM codecs. A conceptual block diagram is shown in Figure 7-24. In addition to a shift register and single comparator, the successive approximation ADC uses a digital-to-analog converter (DAC), whose operation will be described in more detail shortly. The DAC produces a voltage that is proportional to the digital word applied to its input; in other words, it performs the complement of the ADC conversion just described.

FIGURE 7-24   **Successive-approximation ADC.**

The successive approximation ADC is a sequential device, in which the output, a digital word representing the analog signal to be digitized, is created in a series of steps rather than all at once. This result is in contrast to the operation of the flash ADC just described, in which a complete digital word is produced with each sample. The basic idea of the successive approximation ADC is as follows: the DAC produces an analog output voltage that becomes the reference input to the comparator. The analog input signal to be sampled is the other comparator input. Control logic within the DAC initially sets its output to one-half the conversion range (the total possible range of analog voltages to be sampled), which forms the first reference voltage for comparison. The analog voltage to be sampled is compared with this reference. If the reference is greater than the analog input value, then the comparator outputs a logic zero, and the most significant bit (msb) of the digital word to be created is also set to zero. The converse is also true: an analog voltage greater than the reference causes the comparator to output a logic one and the msb to be one as well. The DAC output then either increases or decreases by one-quarter of the allowable range depending on whether the msb just set was a one or a zero. Another comparison is made, and the bit next to the msb is set according to the same rules. The process repeats with each subsequent bit set in turn. When all bits have been compared, the complete code is output from the register as a digital word, with each bit in the word representing the condition where the sampled amplitude was greater than or less than the amplitude of the output of the DAC. The successive approximation ADC, then, requires a number of clock cycles to produce its output.

Digital-to-Analog Converters

The reconstruction of the analog signal from a PCM representation is largely a process of digital-to-analog (DAC) conversion. The DAC is in the receiver section of the complete PCM system shown in Figure 7-25. The purpose of the DAC is to convert a digital (binary) bit stream to an analog signal. The DAC accepts a parallel bit stream and converts it to its analog equivalent. Figure 7-26 illustrates this point.

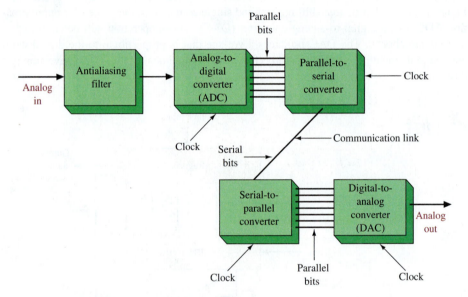

FIGURE 7-25 **PCM communication system.**

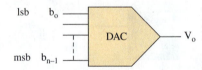

FIGURE 7-26 **DAC converter input and output.**

The least significant bit (lsb) is called b_0 and the most significant bit (msb) is called b_{n-1}. The resolution of a DAC is the smallest change in the output that can be caused by a change of the input. This is the step-size of the converter. It is determined by the lsb. The full-scale voltage (V_{FS}) is the largest voltage the converter can produce. In a DAC converter, the step-size or resolution q is given as

$$q = \frac{V_{FS}}{2^n} \tag{7-7}$$

where n is the number of binary digits.

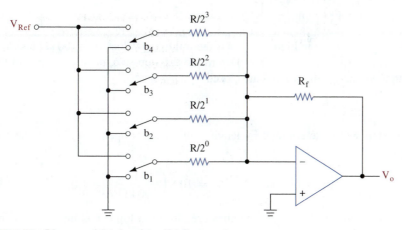

FIGURE 7-27   Binary-weighted resistor DAC converter.

A binary-weighted resistor DAC is shown in Figure 7-27. It is one of the more simple DACs to analyze. For simplicity we have used four bits of data. Note that the value of the resistor is divided by the binary weight for that bit position. For example, in bit position 2^0, which has a value of 1, the entire value of R is used. This is also the lsb. Because this is a summing amp, the voltages are added to give the output voltage.

The output voltage is given as

$$V_o = -V_{\text{Ref}}\left(\frac{b_1 R_f}{R/2^0} + \frac{b_2 R_f}{R/2^1} + \cdots + \frac{b_{n-1} R_f}{R/2^{n-1}}\right). \tag{7-8}$$

An *R-2R* ladder-type DAC is shown in Figure 7-28. This is one of the more popular DACs and is widely used. Note that each switch is activated by a parallel data stream and is summed by the amp. We show a 4-bit *R-2R* circuit for simplicity.

The output voltage is given as

$$V_o = V_{\text{Ref}}\left(1 + \frac{R_f}{R}\right)\left(\frac{b_n}{2^1} + \frac{b_{n-1}}{2^2} + \cdots + \frac{b_1}{2^n}\right). \tag{7-9}$$

where b is either 0 or 1, depending on the digital word being decoded.

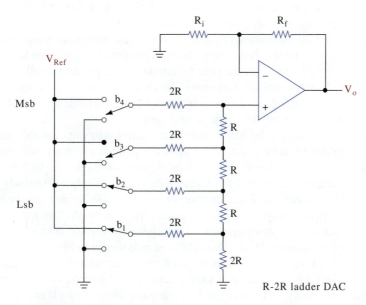

R-2R ladder DAC

FIGURE 7-28   R-2R ladder DAC converter.

EXAMPLE 7-2

Assume the circuit in Figure 7-27 has the following values: $R = 100 \ k\Omega$ and $R_f = 10 \ k\Omega$. Assume $V_{Ref} = -10$ V. Determine the step-size, or resolution, and the output voltage if all switches are closed.

SOLUTION

The step-size is determined by leaving all switches open and closing the lsb. Thus,

$$V_o = -(-10 \text{ V})(R_f/R) = 10 \text{ V}\left(\frac{10 \ k\Omega}{100 \ k\Omega}\right) = 1.0.$$

The resolution is 1.0. If all switches are closed, a logic 1 is input. So, using Equation (7-8), we have

$$V_o = -(-10 \text{ V})\left(\frac{10 \ k\Omega}{100 \ k\Omega} + \frac{10 \ k\Omega}{50 \ k\Omega} + \frac{10 \ k\Omega}{25 \ k\Omega} + \frac{10 \ k\Omega}{12.5 \ k\Omega}\right)$$

$$= (10 \text{ V})(0.1 + 0.2 + 0.4 + 0.8)$$

$$= (10 \text{ V})(1.5) = 15 \text{ V}.$$

7-5   CODING PRINCIPLES

An ideal digital communications system is error-free. Unfortunately, a digital transmission will occasionally have an error. Modifications to the data can provide an increase in the receive system's capability to detect and possibly correct the error. Suppose that we transmit a single zero or a single one. If either data value changes, then we have an error. How can the chance of an error be decreased? The process of decreasing an error depends on the transmission system being used and the digital encoding and modulation techniques employed. Even if the chances of detecting and correcting the error are increased, there is still some probability of receiving a data-bit error in the received message, but if the error can be corrected, then our message is still usable. Methods for improving the likelihood that the data-bit error can be both detected and corrected are presented next.

Our discussion on coding principles begins with a fundamental look at the basic coding techniques and establishing some rules for correcting errors. Let us first look at a very simple data system that has only two possible states. For this example we assume that the transmission consists of a single binary value with status indications for zero (0) and one (1). The system also requires that data-bit errors be corrected at the receiver without the need for retransmitting the data. If only a single zero and a single one are transmitted for each state, then the receiver will not be able to distinguish a correct bit from an error because all possible data values map directly to a valid state (0 or 1).

What can be done to the representative data value for each state so that an error can be detected? What if the number of binary bits representing each state is altered so that a logical zero is defined to be (00) and a logical one is defined to be (11)? Adding a data bit to each state effectively increases the distance between each code word to two. The distance can be illustrated by listing the possible binary states for the 2-bit words. The *distance* between each defined state is called the **Hamming distance**, also known as the **minimum distance (D_{min})**. This relationship is shown in Figure 7-29.

If either 01 or 10 is received, then a data-bit error has occurred. The receive system has no trouble detecting a 1-bit error. As for correcting the bit error, each possibility, (01) and (10), can represent an error in a 0 or a 1. Therefore, increasing the number of binary bits representing each state to a minimum distance, D_{min},

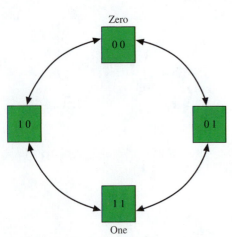

FIGURE 7-29 Coding for a zero (00) and a one (11) for a minimum distance, D_{min}, of 2.

of 2 improves the capability of the receive system to detect the error but does not improve the system's capability to correct the error.

Let's once again increase the number of binary bits for each state so that a logical zero is now defined to be 000 and a logical one is 111. The minimum distance between each code word is now 3. This relationship is depicted in Figure 7-30.

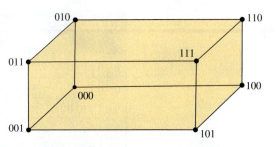

FIGURE 7-30   **Code message for a zero (000) and a one (111) with a minimum distance, D_{min}, of 3.**

If an error does occur in a data bit, can a coding system with $D_{min} = 3$ correct the error? This question can be answered by looking at an example. Assume that the data word 011 is received. Because this data word is neither a 111 nor a 000, it can be assumed that a data bit error has occurred. Look at Figure 7-30 and see if you can determine which code, 000 or 111, that the received error code, 011, is *most likely* to belong to. The answer is the 111 code because the distance from 011 to 111 is 1 and the distance from 011 to 000 is 2. Therefore, it can be stated that, for a coding system with a minimum distance between each code of 3 or greater, the errors in the received code can be detected and corrected. The minimum distance between each code will determine the number of data-bit errors that can be corrected. The relationships for the detection and correction of data-bit errors to minimum distance is provided in Table 7-1.

TABLE 7-1 • Error Detection and Correction Based on D_{min}

Error Detection

    For a minimum distance, D_{min}, between code words, $(D_{min} - 1)$ errors can be detected.

Error Correction

    If D_{min} is even, then $[(D_{min}/2) - 1]$ errors can be corrected.
    If D_{min} is odd, then $\frac{1}{2}(D_{min} - 1)$ errors can be corrected.

EXAMPLE 7-3

Determine the number of errors that can be detected and corrected for the distances

(a) 2.

(b) 3.

(c) 4.

SOLUTION

(a) $D_{min} = 2$; the number of errors detected is $(D_{min} - 1) = 2 - 1 = 1$.

    D_{min} is *even*; therefore, the number of errors corrected equals

$$(D_{min}/2) - 1 = \left(\tfrac{2}{2}\right) - 1 = 0$$

(b) $D_{min} = 3$; the number of errors detected equals $(D_{min} - 1) = 3 - 1 = 2$.

    D_{min} is *odd*; therefore, the number of errors corrected equals

$$\tfrac{1}{2}(D_{min} - 1) = \tfrac{1}{2}(3 - 1) = 1$$

(c) $D_{min} = 4$; the number of errors detected equals $(D_{min} - 1) = 4 - 1 = 3$.

    D_{min} is *even*; therefore, the number of errors corrected equals

$$(D_{min}/2) - 1 = \left(\tfrac{4}{2}\right) - 1 = 1$$

What would have to be done if all eight possible states shown in Figure 7-30 were transmitted and a minimum distance of 2 were required? To create an eight-level code with a minimum distance of 2 requires 4 bits [3 bits to provide eight levels (2^3) and 1 bit to provide a D_{min} of 2 (2^1)]. This is shown in Table 7-2.

TABLE 7-2 • Eight-Level Code with a Distance of 2	
0 0 0 0 (0 0 0)	1 1 0 0 (1 0 0)
0 0 0 1	1 1 0 1
0 0 1 1 (0 0 1)	1 1 1 1 (1 0 1)
0 0 1 0	1 1 1 0
0 1 1 0 (0 1 0)	1 0 1 0 (1 1 0)
0 1 1 1	1 0 1 1
0 1 0 1 (0 1 1)	1 0 0 1 (1 1 1)
0 1 0 0	1 0 0 0

TABLE 7-3 • Eight-Level Code with Distance 3	
0 0 0 0 0 (0 0 0)	0 1 0 1 1 (1 0 0)
0 0 0 0 1	0 1 0 1 0
0 0 0 1 1	0 1 0 0 0
0 0 0 1 0 (0 0 1)	1 1 0 0 0 (1 0 1)
0 0 1 1 0	1 1 0 0 1
0 0 1 1 1	1 1 0 1 1
0 0 1 0 1 (0 1 0)	1 1 1 1 1 (1 1 0)
0 0 1 0 0	1 1 1 0 1
0 1 1 0 0	1 1 1 1 0
0 1 1 0 1 (0 1 1)	1 0 1 1 0 (1 1 1)
0 1 1 1 1	
0 1 1 1 0	

A code with D_{min} equal to 2 cannot correct an error. For example, what if a (1 1 0 1) is received? Is the correct word (1 1 0 0) or is it (1 1 1 1)? Without sufficient overhead bits creating the required D_{min}, the error cannot be corrected. If the eight-level code is changed so that a 1-bit error can be corrected, 5 bits are now required to represent each word [3 bits to provide the eight levels (2^3) and 2 bits to provide a D_{min} of 3]. The 5-bit code is shown in Table 7-3.

To see how the code can correct, let us assume that the received code word is (0 0 1 1 1). To determine the distance of this code word to any of the possible receive codes, simply XOR the received code word with any of the eight possible valid codes. The XOR operation that yields the smallest result tells us which is the correct code. This process is demonstrated in Example 7-4.

EXAMPLE 7-4

Determine the distance for a received code of (0 0 1 1 1), shown in Table 7-2, to all the possible correct codes by XORing the received code with all the possible correct codes. The result with the least number of bit-position differences is most likely the correct code. Then state which is most likely the correct code based on the minimum distance.

SOLUTION

```
00111        00111        00111
00000        00010        00101
00111 (3)    00101 (2)    00010 (1)

00111        00111        00111
01101        01011        11000
01010 (2)    01100 (2)    11111 (5)

00111        00111
11111        10110
11000 (2)    10001 (2)
```

Therefore, based on comparing the received code with all the possible correct codes, the most likely code is (0 0 1 0 1), which is the code for (0 1 0).

7-6 CODE ERROR DETECTION AND CORRECTION

The amount of data transmitted increases every year. Needless to say, transmission reliability and data integrity are both of paramount importance. Without some means of error detection, it is not possible to know when errors have occurred, either as the result of noise or transmission system impairments. In contrast, impaired voice transmissions caused by noise or equipment problems produce an obviously audible noise in the receiver speaker.

With codes and data, error detection is accomplished through some form of redundancy. A basic redundancy system transmits everything twice to ensure an exact correlation between bits of the transmitted message. Transmitting redundant data uses bandwidth, which slows transfer. Fortunately, schemes have been developed that do not require such a high degree of redundancy and provide for a more efficient use of available bandwidth.

Parity

The most common method of error detection is the use of parity. A single bit called the *parity bit* is added to each code representation. If the added bit makes the total number of 1s even, it is termed *even parity,* and an odd number of 1s is *odd parity.* For example, if a standard numeric code for a capital letter "A" is to be generated, the code would take the form P1000001, where P is the parity bit. Odd parity would be 11000001 because the number of 1s is now 3 (see Figure 7-31). The receiver checks for parity. In an odd-parity system, if an even number of 1s occurs in a received character grouping (that is, if the digital word that was received has an even number of 1s), an error is indicated. In such instances the receiver usually requests a retransmission. Unfortunately, if two errors occur, parity systems will not indicate an error. In many systems a burst of noise causes two or more errors, so more elaborate error-detection schemes may be required.

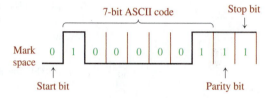

FIGURE 7-31   ASCII code for A with odd parity. Note that lsb b_1 is the first bit of the digital word transmitted.

Many circuits are used as parity generators and/or checkers. A simple technique is shown in Figure 7-32. If there are *n* bits per word, $n - 1$ exclusive-OR (XOR) gates are needed. The first 2 bits are applied to the first gate and the remaining individual bits to each subsequent gate. The output of this circuit will always be a 1 if there is an odd number of 1s and a 0 for an even number of 1s. If odd parity is desired, the output is fed through an inverter. When used as a parity checker, the word and parity bit is applied to the inputs. If no errors are detected, the output is low for even parity and high for odd parity.

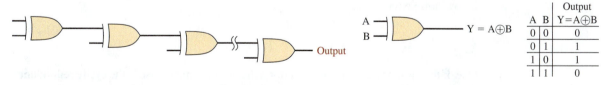

FIGURE 7-32   Serial parity generator and/or checker.

When an error is detected there are two basic system alternatives:

1. An automatic request for retransmission (ARQ)
2. Display of an unused symbol for the character with a parity error (called **symbol substitution**)

Most systems use a request for retransmission. If a block of data is transmitted and no error is detected, a positive acknowledgment (ACK) is sent back to the transmitter. If a parity error is detected, a negative acknowledgment (NAK) is made and the transmitter repeats that block of data.

Block Check Character

A more sophisticated method of error detection than simple parity is needed in higher data-rate systems. At higher data speeds, telephone data transmission is usually synchronous and blocked. A block is a group of characters transmitted with no time gap between them. It is followed by an *end-of-message* (EOM) indicator and then a **block check character (BCC)**. A block size is typically 256 characters. The transmitter uses a predefined algorithm to compute the BCC. The same algorithm is used at the receiver based on the block of data received. The two BCCs are compared and, if identical, the next block of data is transmitted.

Many algorithms are used to generate a BCC. The most elementary one is an extension of parity into two dimensions, called the **longitudinal redundancy check (LRC)**. This method is illustrated with the help of Figure 7-33. Shown is a small block of 4-bit characters using odd parity. The BCC is formed as an odd-parity bit for each vertical column. Now suppose that a double error occurred in character 2, as shown in Figure 7-33(b). With odd parity, the third and fourth bits from the left in the BCC should be zeros; instead they are 1s. As described previously, simple parity would not pick up this error. With the BCC, however, the error is detected. If a single error occurs [Figure 7-33(c)], the erroneous bit can be pinpointed as the intersection of the row and column containing the error. Correction is achieved by inverting the bad bit. If a double error occurs in a column, the scheme is defeated.

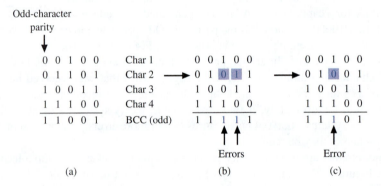

FIGURE 7-33   LRC error detection.

The error location process just described is not usually used. Rather, the receiver checks for character and LRC errors and if either (or both) occur, a retransmission is requested. Occasionally, an error will occur in the BCC. This is unavoidable, but the only negative consequence is an occasional unnecessary retransmission. This scheme is useful in low-noise environments.

Cyclic Redundancy Check

One of the most powerful error-detection schemes in common use is the **cyclic redundancy check (CRC)**. The CRC is a mathematical technique that is used in synchronous data transmission. It can effectively catch 99.95% of transmission errors.

In the CRC technique, each string of bits is represented by a polynomial function. CRC is a division technique, illustrated as follows:

$$\frac{M(x)}{G(x)} = Q(x) + R(x). \tag{7-10}$$

where $M(x)$ is the binary data, called the message function, and $G(x)$ is a special code by which the message function is divided, called the generating function. The process yields a quotient function, $Q(x)$, and a remainder function, $R(x)$. The quotient is not used, and the remainder, which is the CRC block check code (BCC), is attached to the end of the message. This is called a **systematic code**, where the BCC and the message are transmitted as separate parts within the transmitted code. At the receiver, the message and CRC check character pass through its block check register BCR. If the register's content is zero, then the message contains no errors.

Cyclic codes are popular not only because of their capability to detect errors but also because they can be used in high-speed systems. They are easy to implement, requiring only the use of shift registers, XOR gates, and feedback paths. Cyclic block codes are expressed as **(n, k) cyclic codes**, where

$$n = \text{length of the transmitted code}$$

$$k = \text{length of the message.}$$

For example, a (7, 4) cyclic code simply means that the bit length of the transmitted code is 7 bits ($n = 7$) and the message length is 4 bits ($k = 4$). This information also tells us the length or number of bits in the BCC, which is the code transmitted with each word. The number of bits in the BCC is ($n - k$). This relationship is expressed as

$$\text{BCC length} = n - k. \qquad (7\text{-}11)$$

This code is combined with the message to generate the complete transmit binary code and is used at the receiver to determine if the transmitted message contains an error. The general form for systematic (n, k) CRC code generation is shown in Figure 7-34.

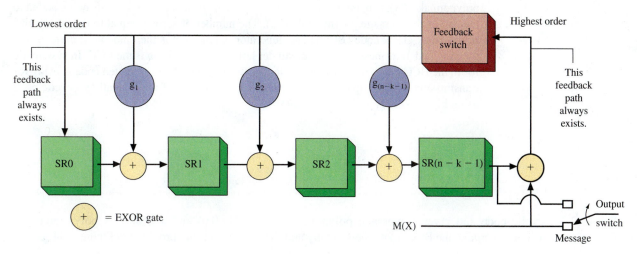

FIGURE 7-34   **CRC code generator for an (n, k) cyclic code.**

The number of shift registers required in the CRC generating circuit is the length of the block check code (BCC), which is ($n - k$). For a (7, 4) cyclic code, the number of shift registers required to generate the BCC is $7 - 4 = 3$. Note that the highest order of the generating polynomial described next (3 in this case, from x^3) is also the number of shift registers required.

Construction of a CRC generating circuit is guided by the **generating polynomial**, $G(x)$. The feedback paths to each XOR gate are determined by the coefficients for the generating polynomial, $G(x)$. If the coefficient for a variable is 1, then a feedback path exists. If the coefficient is 0, then there is no feedback path. The general form for the CRC generator with respect to $G(x)$ is

$$G(x) = 1 + g_1 x + g_2 x^2 + \cdots + g_{n-k-1} x^{n-k-1} + x^{n-k}. \qquad (7\text{-}12)$$

The lowest-order value is $x^0 = 1$; therefore, the feedback always exists. This is indicated in Figure 7-34.

The highest-order value, which varies, also has a coefficient of 1. In CRC circuits, the lowest- and highest-order polynomial values have a coefficient of 1. This feedback must always exist because of the cyclic nature of the circuit. For example, if the generating polynomial expression is $G(x) = 1 + x + x^3$, then the feedback paths are provided by 1 (x^0), x, and x^3. The coefficient for x^2 is 0; therefore, no feedback is specified.

In addition to the XOR gates and shift registers for the circuit shown in Figure 7-34, the CRC generator contains two switches for controlling the shifting of the message and code data to the output. The procedure for CRC code generation using this circuit is as follows:

CRC Code Generation: Procedure for Using Figure 7-34

1. Load the message bits serially into the shift registers. This requires:

   Output switch connected to the message input $M(x)$,

   Feedback switch closed.

   Note: k shifts are performed; k is obtained from the (n, k) specification. The k shifts load the message into the shift registers and, at the same time, the message data are serially sent to the output and the BCC is generated.

2. After completing the transmission of the kth bit, the feedback switch is opened and the output switch is changed to select the shift registers.

3. The contents of the shift registers containing the BCC are shifted out using $(n - k)$ shifts. *Note:* The total number of shifts to the CRC generator circuit is n, which is the length of the transmit code.

The operation of this circuit can be treated mathematically as follows. The message polynomial, $M(x)$, is being multiplied by $x^{(n-k)}$. This results in $(n - k)$ zeros being added to the end of the message polynomial, $M(x)$. The number of zeros is equal to the binary size of the block check code (BCC). The generator polynomial is then modulo 2 divided into the modified $M(x)$ message. The remainder left from the division is the BCC. In systematic form, the BCC is appended to the original message, $M(x)$. The completed code is ready for transmission. An example of this process and an example of implementing a cyclic code using Figure 7-34 is given in Example 7-5.

EXAMPLE 7-5

For a $(7, 4)$ cyclic code and given a message polynomial $M(x) = (1\,1\,0\,0)$ and a generator polynomial $G(x) = x^3 + x + 1$, determine the BCC (a) mathematically and (b) using the circuit provided in Figure 7-34.

SOLUTION

(a) The code message $M(x)$ defines the length of the message (4 bits). The number of shift registers required to generate the block check code is determined from the highest order in the generating polynomial $G(x)$, which is x^3. This indicates that three shift registers will be required, so we will pad the message $(1\,1\,0\,0)$ with three zeros to get $(1\,1\,0\,0\,0\,0\,0)$. Remember, a $(7, 4)$ code indicates that the total length of the transmit CRC code is 7 and the

$$\text{BCC length} = n - k. \tag{7-11}$$

The modified message is next divided by the generator polynomial, $G(x)$. This is called modulo-2 division.

$$
\begin{array}{r}
G(x)\overline{)M(x) \cdot x^{n-k}} \\
\end{array}
$$

```
              1110
      1011)1100000
           1011
           ----
           1110
           1011
           ----
            1010
            1011
            ----
             010
```

The remainder 010 is attached to $M(x)$ to complete the transmit code, $C(x)$. The transmit code word is 1100010.

(b) Next, we use the circuit in Figure 7-34 to generate the same cyclic code for the given $M(x)$ and $G(x)$. The CRC generating circuit can be produced from the general form provided in Figure 7-35. $G(x) = x^3 + x + 1$ for the (7, 4) cyclic code. This information tells us the feedback paths (determined by the coefficients of the generating polynomial expression equal to 1) as well as the number of shift registers required, $(n - k)$. The CRC generating circuit is provided in Figure 7-35. The serial output sequence and the shift register contents for each shift are shown in Table 7-4.

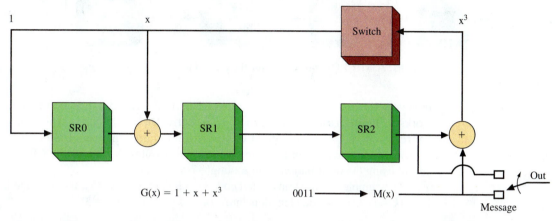

FIGURE 7-35   CRC code generator for a (7, 4) cyclic code using a generator polynomial $G(x) = x^2 + x + 1$.

TABLE 7-4 • The Shift Sequence for Example 7-5								
DATA [$M(x)$]					**REGISTER CONTENTS**			
X^0	X^1	X^2	X^3	**SHIFT NUMBER**	**SR0**	**SR1**	**SR2**	**OUTPUT**
0	0	1	1	0	0	0	0	
	0	0	1	1	1	1	0	1
		0	0	2	1	0	1	1
			0	3	1	0	0	0
			–	4	0	1	0	0

The result in both cases for Example 7-5 is a transmitted code of 1100010, where 1100 is the message $M(x)$ and 010 is the BCC. The division taking place in part (a) of Example 7-5 is modulo-2 division. Remember, modulo-2 is just an XORing of the values. The division of $G(x)$ into $M(x)$ requires only that the divisor has the same number of binary positions as the value being divided (the most significant position must be a 1). The value being divided does not have to be larger than the divisor.

CRC Code-Dividing Circuit

At the receive side, the received code is verified by feeding the received serial CRC code into a *CRC dividing circuit*. The dividing circuit has $(n - k)$ shift registers. The general form for a CRC divide circuit is given in Figure 7-36.

The arrangement of the feedback circuit and the shift registers depends on the generator polynomial, $G(x)$, and the coefficients for each expression. If the coefficient for X is 1, then a feedback path to an XOR is provided for that shift register. If the coefficient is 0, then there is not a feedback path to the input of the respective shift register. The number of shift registers required in the circuit is still $(n - k)$. The circuit requires k shifts to check the received data for errors. The result (or remainder) of the shifting (division) should be all 0s. The remainder is called the **syndrome**. If the syndrome contains all 0s, then it is assumed that the received data are correct. If the syndrome is a nonzero value, then bit error(s) have been detected. In most cases the receive system will

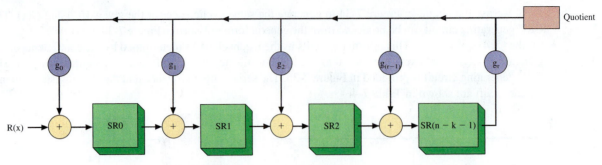

FIGURE 7-36 **CRC divide circuit for the general form of** $G(x) = g_0 + g_1x + g_2x^2 + \cdots + g_rx^r$.

request a retransmission of the message. This is the case for Ethernet computer networks. In limited cases, the code *can* then be corrected by adding the syndrome vector to the received vector. This method of correction requires the use of a look-up table to determine the correct word. Example 7-6 demonstrates the use of the CRC detection circuit using the information from Example 7-5.

In Example 7-6 the transmitted code length is seven, ($n = 7$); the message length is four, ($k = 4$); and the code length is three, ($n - k$) = (7 − 4) = 3.

EXAMPLE 7-6

The serial data stream 0 1 0 0 0 1 1 was generated by a (7, 4) cyclic coding system using the generator polynomial $G(x) = 1 + x + x^3$ (see Example 7-5). Develop a circuit that will check the received data stream for bit errors. Verify your answer both (a) mathematically and (b) by shifting the data through a CRC divide circuit.

SOLUTION

(a) To verify the data mathematically requires that the *received data* be divided by the coefficients defining generator polynomial, $G(x)$. This is similar to the procedure used in Example 7-5 except the data being divided contains the complete code.

$$
\begin{array}{r}
1110 \\
1011\overline{)1100010} \\
\underline{1011} \\
1110 \\
\underline{1011} \\
1011 \\
\underline{1011} \\
1011 \\
\underline{1011} \\
000
\end{array}
$$

The syndrome (remainder) is zero; therefore, the received data do not contain any errors.

(b) The CRC divide circuit is generated from the CRC generator polynomial expression, $G(x)$. $G(x)$ for this system is $1 + x + x^3$. The circuit is created using the form shown in Figure 7-37. The shift register contents are first reset to zero. Next, the received data are input to $R(x)$. The data are shifted serially into the circuit. The results of the shifting are provided in Table 7-5. A total of n shifts is required.

The result of the shifting in the data produced all 0s remaining in the shift registers. Therefore, the syndrome is 0, which indicates that the received data do not contain any bit errors.

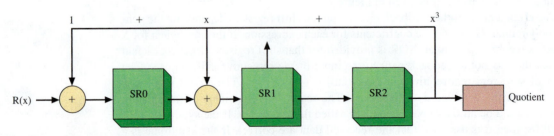

FIGURE 7-37 **CRC divide circuit for** $G(x) = 1 + x + x^3$.

TABLE 7-5 • The Shift Sequence for Example 7-6

CODE				DATA			SHIFT NUMBER	SR0	SR1	SR2	OUTPUT
LSB			MSB	LSB		MSB					
0	1	0	0	0	1	1		0	0	0	
	0	1	0	0	0	1	1	1	0	0	1
		0	1	0	0	0	2	1	1	0	1
			0	1	0	0	3	0	1	1	0
				0	1	0	4	1	1	1	0
					0	1	5	1	0	1	0
						0	6	0	0	0	1
						–	7	0	0	0	0

Hamming Code

The error-detection schemes thus far presented require retransmission if errors occur. Techniques that allow correction at the receiver are called **forward error-correcting (FEC)** codes. The basic requirement of such codes is for sufficient redundancy to allow error correction without further input from the transmitter. The **Hamming code** is an example of an FEC code named for R. W. Hamming, an early developer of error-detection/correction systems.

Figure 7-38 illustrates in simplified form the operation of a Hamming code. If m represents the number of bits in a data string and n represents the number of bits in the Hamming code, n must be the smallest number such that

$$2^n \geq m + n + 1. \qquad (7\text{-}13)$$

m = number of data bits
n = number of error-correction bits (Hamming bits)
Code length must be $2^n \geq m + n + 1$
Therefore, for 4-bit data code, n = 3 ($2^3 \geq 4 + 3 + 1$)
We will send the 4-bit data word 1 1 0 1 with data bits in places 3, 5, 6 & 7
3 parity bits are to be used; assume an even-parity system

	P_1	P_2		P_3				
Bit #	1	2	3	4	5	6	7	
Sent:	1	0	1	0	1	0	1	
Received:	1	0	1	0	0	0	1	

P_1 checks places 3, 5, 7
P_2 checks places 3, 6, 7
P_3 checks places 5, 6, 7

Parity error check: P_3 P_2 P_1
                      1   0   1

Parity error check result:
1 = parity wrong
0 = parity OK

Parity error check result spells out location of error: 1 0 1 = 5

FIGURE 7-38 Hamming code example of forward-error correction.

Consider a 4-bit data word 1101. The minimum number of parity bits to be used is 3 when Equation (7-13) is referenced. A possible setup, then, is

$$\begin{array}{ccccccc} P_1 & P_2 & 1 & P_3 & 1 & 0 & 1 \\ 1 & 2 & 3 & 4 & 5 & 6 & 7 \end{array} \quad \text{bit location.}$$

We'll let the first parity bit, P_1, provide even parity for bit locations 3, 5, and 7. P_2 does the same for 3, 6, and 7, while P_3 checks 5, 6, and 7. The resulting word, then, is

$$\begin{array}{ccccccc} 1 & 0 & 1 & 0 & 1 & 0 & 1 \\ 1 & 2 & 3 & 4 & 5 & 6 & 7 \\ P_1 & P_2 & & P_3 & & & \end{array} \quad \text{bit location.}$$

When checked, a 1 is assigned to incorrect parity bits, while a 0 represents a correct parity bit. If an error occurs so that bit location 5 becomes a 0, the following process takes place. P_1 is a 1 and indicates an error. It is given a value of 1 at the receiver. P_2 is not concerned with bit location 5 and is correct and therefore given a value of 0. P_3 is incorrect and is therefore assigned a value of 1. These three values result in the binary word 101. Its decimal value is 5, and this means that bit location 5 has the wrong value and the receiver has pinpointed the error without a retransmission. It then changes the value of bit location 5 and transmission continues. The Hamming code is not able to detect multiple errors in a single data block. More complex (and more redundant) codes are available if necessary.

Reed–Solomon Codes

Reed–Solomon (RS) codes are also forward error-correcting codes (FEC) like the Hamming code. They belong to the family of BCH (Bose–Chaudhuri–Hocquenghem) codes. Unlike the Hamming code, which can detect only a single error, RS codes can detect multiple errors, which makes RS codes very attractive for use in applications where bursts of errors are expected. Examples are compact disc (CD) players and mobile communications systems (including digital television broadcasting). A scratched CD can create multiple errors, and in mobile communications, error bursts are expected as the result of noise or system impairments.

RS codes often employ a technique called **interleaving** to enable the system to correct multiple data-bit errors. Interleaving is a technique used to rearrange the data into a nonlinear ordering scheme to improve the chance of correcting data errors. The benefit of this technique is that burst errors can be corrected because the likelihood is that the data message will not be completely destroyed. For example, a large scratch on a CD disc will cause the loss of large amounts of data in a track; however, if the data bits representing the message are distributed over several tracks in different blocks, then the likelihood that they can be recovered is increased. The same is also true for a message transmitted over a communications channel. Burst errors will cause the loss of only a portion of the message data, and if the Hamming distance is large enough, the message can be recovered.

The primary disadvantage of implementing the RS coding scheme is the time delay for encoding (interleaving the data) at the transmitter and decoding (deinterleaving the data) at the receiver. In applications such as a CD ROM, data size and the delay are not problematic, but implementing RS codes in mobile communications could create significant delay problems in real-time applications and possibly a highly inefficient use of data bandwidth.

7-7 DIGITAL SIGNAL PROCESSING

Certainly one of the most significant developments in the communications field in recent years has been the widespread adoption of digital signal processing (DSP) techniques. Nowhere is this trend more apparent than in portable devices such as next-generation smart phones. Thanks to advances in large-scale integration and increases in processor speed at reasonable cost, functions that had traditionally been the purview of analog circuits and discrete components can now often be implemented digitally. The operations performed by transmitter and receiver subsystems, such as modulation, demodulation, and filtering, can be carried out in the digital domain by means of mathematical operations implemented in software and executed by special-purpose DSP processor ICs. Developments in DSP implementation have also given rise to the idea of **software-defined radio (SDR)**, which permit the characteristics of entire radios to be defined in software and implemented in integrated circuits. Taken together, DSP and SDR represent a truly systems-level approach to communications, and a basic introduction to DSP is beneficial for acquiring a deeper understanding of the next-generation digital systems to be covered in the following chapters.

DSP involves the use of mathematical functions such as addition and multiplication to control the characteristics of a digitized signal. Often, the objective is to extract information from the processed signal, usually in the presence of noise. For example, an electrocardiogram (EKG) machine may need to extract and display relevant information about the heartbeat of a developing fetus from interfering signals representing the mother's heartbeat. DSP is used in many fields. In addition to medical applications such as EKG and imaging, DSP is found in areas ranging from seismology, nuclear engineering, and acoustics to more communications-related areas such as audio and video processing as well as speech and data communication. DSP is also extensively used in modern test equipment, such as digital storage oscilloscopes and spectrum analyzers.

Figure 7-39 is a block diagram of a typical DSP system. Conceptually, it is very straightforward. The first step is to use an ADC converter to convert the analog signal into a digital data stream. The data stream is fed into the DSP, which can be a computer, microprocessor, or dedicated DSP device. The DSP executes algorithms representing the function, such as filtering, to be carried out. The output stream of the DSP, representing the filtered or otherwise processed result, is then applied to a DAC converter to transform the digital signal back to its analog equivalent.

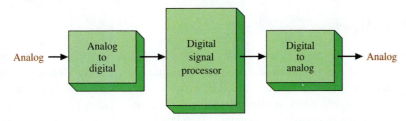

FIGURE 7-39   **DSP system.**

The primary distinction between DSP and other computational operations is that DSP is usually considered to occur in real time, or nearly in real time, whereas other forms of computer processing involve operations that may be similar in substance to real-time DSP but that occur with collected and stored data. Though any microprocessor can, in principle, be used to perform DSP, the need for near-real-time results means that dedicated IC DSP processors are often used. These chips are available from many vendors, including Altera, Analog Devices, and Texas Instruments. The DSP chip is essentially a special-purpose microprocessor whose internal architecture has been optimized for signal processing computations. As an example, a dedicated DSP chip may have separate memory spaces for program and data. This so-called *Harvard architecture* allows the processor to work with the next program instruction as it is reading or writing data for the previous instruction, thus speeding up the processor. This is in contrast with the *Von Neumann architecture* used by the majority of conventional microprocessors, in which the program and data share the same memory space. Also, because the dedicated DSP generally works with fairly small files, the processor has little or no need for external memory or for the interfacing and memory-addressing functions associated with conventional microprocessors. Memory required by the DSP is often built into the chip itself, as are other hardware resources such as multipliers, summers, and delay elements.

Like other microprocessors, DSPs can be easily reprogrammed, making them more flexible. Thus, functional changes in DSP operation can be implemented through changes in software. Also, DSPs, like all microprocessors, can perform different tasks at different points in its program. To prevent bottlenecks, multiple computational units or multiple *cores*, essentially copies of the entire processor, are used in the most powerful DSP ICs.

DSPs can emulate almost all communications system functions, often with better results than their analog counterparts. Nowhere is this truer than with filters. Recall from Chapter 4 that analog passive filters are formed from inductors, capacitors, and resistors. Filters have well-defined characteristics such as cutoff frequency and roll-off rate, bandwidth, shape factor, and Q. Key limitations of passive filters are the physical size of the components and the tendency of the output response to change with temperature and component aging. Active filters, where inductors are replaced with operational amplifiers, are more compact than their passive counterparts, but they are still subject to changes in output response because of changes in the surrounding temperature and aging of components. Digital filters, on the other hand, are very stable and are not subject to drift. They are often programmable, enabling the filter characteristics to be changed without requiring physical circuit components to be unsoldered and replaced. Finally, digital filters are extremely frequency versatile. They can be used at low frequencies, such as for audio applications, as well as at radio frequencies for various purposes.

DSP processing and filtering are used today in almost every area of electronic communication, including mobile phones, digital television, radio, and test equipment. The mathematics behind digital signal processing is quite complex. Though an understanding of the underlying mathematics would be necessary for DSP designers and engineers, it is not as necessary to have a deep understanding of DSP theory to understand DSP from a maintenance perspective. However, a conceptual understanding of how DSP works will be useful because DSP-based implementations of communications systems will become ever more prevalent. To illustrate the concept, here we will look at the implementation of two specific types of DSP filtering operations. Digital modulation and demodulation using DSP techniques will be covered in Chapter 8.

DSP Filtering

Chapter 1 introduced the idea that signals can be viewed in either the time or frequency domains. Fourier analysis allows for time-domain signals to be analyzed in the frequency domain and vice versa. In this way, a square wave can be seen to consist of a fundamental-frequency sine wave and its odd harmonics, for example. Other periodic waveforms, too, can be shown to consist of harmonically related combinations of sine waves with varying phase angles and with amplitudes that, in general, decrease as the harmonic multiple is increased.

Fourier analysis extends also to nonperiodic waveforms, such as pulses. Even single pulses are composed of sine waves, but, in contrast to periodic waveforms, the sine waves for single pulses are not harmonically related. Instead, a single pulse of time τ is made up of an infinite number of sine waves spaced infinitely closely together and with amplitudes that fall off on either side of the maximum-amplitude impulse in a sine wave–like fashion. In a time-domain representation, such as an oscilloscope display, a pulse of time τ (if it could be seen) would appear narrow, or compressed. However, because it consists of a number of frequency components represented by the large number of sine waves of which it is composed, the frequency-domain representation of that same pulse would appear wide as viewed on a spectrum analyzer; that is, it would have a wide bandwidth.

The relationship between the pulse widths in the time and frequency domains for certain types of pulses can be formalized with what is known as a sinc function. Pronounced "sink," the term is a shortened expression of the Latin term *sinus cardinalis,* or "cardinal sine." It has a number of interesting properties, particularly in describing filter behavior. Returning to our pulse of time τ, its sinc function expression would appear as

$$\operatorname{sinc}(f\tau) = \frac{\sin(\pi f\tau)}{(\pi f\tau)}.$$

What is important to recognize is that if the time interval represented by τ decreases, f must increase to maintain the equality. This result comports with our understanding of the relationship between frequency, period, and wavelength for repeating signals: the higher the frequency, the shorter the period and the smaller the wavelength. As viewed on an oscilloscope, increasing the frequency of a repeating signal causes the individual cycles to appear compressed, an indication that the period of each cycle (as well as the wavelength) has been shortened. The above expression allows us to extend what we know to be true for repetitive waveforms to impulse-like signals.

Continuing this line of thought, if our pulse continues to be narrowed, then its frequency spectrum continues to spread out. At the extreme, an infinitely narrow pulse in the time domain would have an infinitely wide frequency spectrum; that is, the spectrum would consist of a flat response from zero to infinity. Such an infinitely narrow pulse is called an *impulse,* and the fact that it has a flat frequency spectrum allows us to envision how to design filters that exhibit any desired frequency response.

Filters that exhibit the properties of the sinc function are sometimes referred to as $\sin(x)/x$ filters, because this is the form of the unnormalized sinc function. This function, shown graphically in Figure 7-40, is important in DSP filter implementations because it exhibits an ideal low-pass filter response. That is, it transitions instantly from the *passband* (the frequency range of minimum attenuation) to the *stopband* (the frequency range of maximum attenuation). Such filters would be impossible to build with discrete components. The digital implementation comes much closer to the ideal, however. Just how close it gets depends on the filter complexity and architecture.

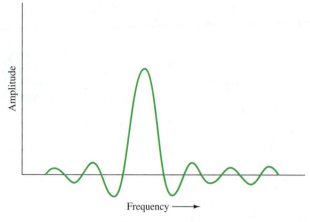

FIGURE 7-40   **Sin(x)/x (sinc) function.**

Since an infinitely narrow pulse (i.e., an impulse) has an infinitely wide and flat frequency spectrum, then a filter to which an impulse is applied at its input will output a signal whose impulse response is equal to that of the filter frequency response. This principle provides a starting point to approach the filter design: Determine what impulse response will give the desired spectrum (i.e., passband/stopband characteristics) and then design a filter to have that impulse response. The first type of DSP filter we will consider is the finite-impulse response (FIR) filter, which is one whose impulse response becomes zero at a specific point in time after receiving an impulse. This behavior is in contrast to that of the infinite impulse response (IIR) filter, whose output behavior (even when implemented digitally) is closer to that of the analog filters studied in Chapter 4. The output of any analog filter, or of a digital IIR filter, "rings," or decays exponentially toward zero but theoretically never quite gets there. Another contrast between the FIR and IIR filters has to do with whether the filtering algorithm makes use of previous output values as inputs for subsequent calculations. If it does, it is said to be **recursive** or **iterative**. The IIR filter is recursive, whereas the FIR filter is nonrecursive. We will look first at the DSP implementation of a nonrecursive (that is, FIR) filter first because it is perhaps conceptually easier to understand.

The essential idea behind a FIR DSP filter is that it creates the desired response (low-pass, high-pass, etc.) and characteristics (such as for the Butterworth or Chebyshev filters mentioned in Chapter 4) by repeatedly multiplying the amplitudes of input pulses by predetermined numerical values called *coefficients.* The coefficients can be thought of as being like horizontally arranged numerical entries in a spreadsheet. The amplitude of a given sample pulse, which is the filter input, is multiplied by the first coefficient. Then, that same pulse is shifted to the right, figuratively speaking, and multiplied by the second coefficient, then it is shifted and multiplied again, and the process repeats at the sample rate until the pulse has been multiplied by each available coefficient in sequence. The sample rate must be at least twice the highest frequency of the analog signal being sampled to conform to the Nyquist criterion. Each multiplication operation is associated with its own coefficient. The result of each multiplication is applied to a summing circuit as it is produced, and the summing circuit output represents the filtered signal sample. This last operation of multiplication and summing is known as a *dot product.* Designing DSP filters rests in no small part on determining the correct coefficients.

Figure 7-41 illustrates the concept just described. Whether implemented in hardware or software, the FIR filter is built from blocks that function as multipliers, adders, and delay elements. FIR filters can be series, parallel, or combination configurations. Shown in the figure is a parallel configuration because it is the easiest to understand conceptually. The figure shows a filter configuration with five "taps," each of which consists of a filter coefficient, a shift register, and a multiplier. An actual FIR filter implemented in DSP may have hundreds of taps. The number of taps equals the number of multipliers in a parallel implementation. The delay elements are shift registers, which cause the data to be shifted by one bit for each sample pulse.

As already established, the filter input is a data stream whose pulses will be represented as x_k. The output is a series of filtered data pulses, y_k. The filter coefficients are

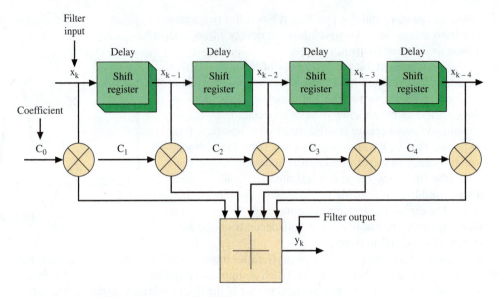

FIGURE 7-41 **FIR filter conceptual block diagram.**

designated C_m. The subscript k identifies the data sequence (i.e., where a pulse has moved as it is passed from one tap to the next). Thus, x_{k+1} follows the pulse x_k, and x_{k-1} precedes x_k. For simplicity, let us assume that the data stream is infinitely long. The coefficients are usually static (they do not change over time). Recall that the coefficients determine the filter frequency response. The output for a five-tap filter, therefore, can be represented by the equation

$$y_k = C_0 x_k + C_1 x_{k-1} + C_2 x_{k-2} + C_3 x_{k-3} + C_4 x_{k-4}.$$

Expressed as a general equation, a FIR filter of any length takes the form

$$y_k = \sum_{i=0 \ to \ N-1} C_i x_{k-i}.$$

The idea embodied by the above equation can be extended to a stream of pulses. The pulse stream represents data moving sequentially past an array of static (fixed, or location-specific) coefficients. With each clock cycle, the data values then in position are cross-multiplied with their associated coefficients, and the results of all the multiplications are summed. Then, the operations are repeated with the next clock cycle after the pulses have been shifted by one position with respect to the coefficients. This process illustrates the mathematical operation of *convolution,* in which the input signal and the impulse response act in concert to produce an output signal. Convolution can be shown to be frequency-domain multiplication, where the frequency spectrum of the output is the input frequency spectrum multiplied by the filter frequency spectrum.

The nonrecursive (no-feedback) design of the FIR filter makes its implementation in hardware or software relatively straightforward. The FIR filter is also naturally suited to sampled pulses. The IIR, or recursive filter, on the other hand, is a filter with feedback. IIR concepts are based on analog filter circuits, which are recursive in nature. IIR filters tend to be harder to design than FIR filters, and the underlying mathematics is more involved. One advantage of the IIR filter is that, with careful design, it can provide a much more sharply defined frequency response (i.e., sharp cutoff) than would be possible with FIR filters employing the same number of taps. Thus, IIR designs can often be implemented with fewer multiplier stages, hence fewer computations, for the same filter response. Another potential advantage has to do with phase response. FIR filters have a linear phase response, so no phase distortion occurs as a result of the filtering operation. IIR filters, on the other hand, emulate analog filters in many respects, including nonlinear phase response. This aspect can be advantageous if analog filter characteristics are to be preserved in a digital implementation. An example might be filters found in musical instrument applications such as guitar amplifiers or audio equipment.

Although it is beyond the scope of this text to delve deeply into DSP design, a worked-out example will serve to illustrate the computational process employed in the implementation of an IIR low-pass, Butterworth filter. Recall that a key feature of an IIR filter is the notion of feedback. Inside the DSP unit is the computational algorithm called the **difference equation**. The difference equation makes use of the present digital sample value of the input signal along with a number of previous input values, and possibly previous output values, to generate the output signal. Algorithms employing previous output values are said to be **recursive** or **iterative**.

A typical form for a second-order recursive algorithm is

$$y_0 = a_0 x_0 + a_1 x_1 + a_2 x_2 + b_1 y_1 + b_2 y_2.$$

where y_0 is the present output value; x_0 the present input value; and x_1, x_2, y_1, and y_2 are previous input and output values. This algorithm is said to be of second order because it uses up to two past values saved in memory. Contrast this equation with the one for the FIR filter presented earlier. Note in particular that the coefficients for the FIR filter are multiplied by normalized amplitudes of sampled pulses that are sequentially shifted with each sample time, whereas the IIR filter coefficients are multiplied by values that have been modified by the results of prior calculations. Put another way, the difference equation for the FIR filter does not include the previous output values (y_1, y_2, …). For both types of filters, the a and b quantities are the coefficients of the difference equation, and their values determine the type of filter and the cutoff frequency or frequencies. Finding these values requires tedious calculations when done by hand or calculator. The process is beyond the scope of this book, but there are computer programs that determine the coefficients based on specifications entered by the user, such as type of filter, cutoff frequencies, and attenuation requirements.

The following equation is an example of a second-order, recursive, low-pass Butterworth filter. The cutoff frequency is 1 kHz.

$$y_0 = 0.008685\, x_0 + 0.01737\, x_1 + 0.008685\, x_2 + 1.737\, y_1 - 0.7544\, y_2$$

To better illustrate the computation process of a recursive DSP algorithm, we will arrange the input values (x_0, x_1, and x_2) and the calculation results (y) as entries in a spreadsheet for three different input frequencies, 500 Hz, 1 kHz, and 2 kHz, at consecutive sample times (n), as shown in Table 7-6:

TABLE 7-6 • An Example of the Computational Process for the DSP Difference Equation

\multicolumn{3}{c}{$f = 500$ Hz}			\multicolumn{3}{c}{$f = 1000$ Hz}			\multicolumn{3}{c}{$f = 2000$ Hz}		
n	x	y	n	x	y	n	x	Y
−2	0	0	−2	0	0	−2	0	0
−1	0	0	−1	0	0	−1	0	0
0	0	0	0	0	0	0	0	0
1	0.09983	0.00087	1	0.19866	0.00173	1	0.38941	0.00338
2	0.19866	0.00495	2	0.38941	0.00980	2	0.71734	0.01881
3	0.29551	0.01475	3	0.56463	0.02895	3	0.93203	0.05374
4	0.38941	0.03187	4	0.71734	0.06182	4	0.99957	0.10934
5	0.47941	0.05718	5	0.84146	0.10916	5	0.90932	0.18088
6	0.56463	0.09093	6	0.93203	0.17006	6	0.67551	0.25897

For each of the three frequencies under consideration, entries are shown for

(a) The sample number n.

(b) The input sequence x, which is a sine wave sampled at a given sampling rate. In this case, a sample rate 10π times the cutoff frequency is being used to ensure that the Nyquist criterion is satisfied.

(c) The output sequence y, generated by the DSP algorithm.

The numbers shown for x and y are the normalized amplitudes of the signal at each sample time, n. (Normalization in this case means that the maximum amplitude is scaled to a value of 1 V and that all other values are scaled appropriately.) Specifically, the x values in the table represent the normalized amplitudes of the pulses created (the samples) when the input sine wave was digitized. The y values are the amplitudes of the output sine wave created as the result of filtering. The three different input frequencies used—500, 1000, and 2000 Hz—are, respectively, at one-half the cutoff frequency, at cutoff (which, by definition, is the frequency at which the voltage gain is reduced by 3 dB from maximum amplitude), and at twice the cutoff frequency. Because this is a low-pass filter, the maximum amplitudes will be found at 500 Hz, with roll-off expected to become increasingly pronounced as the frequency is increased from its initial starting point at an octave below the cutoff frequency up to cutoff, and then, finally, to an octave above cutoff.

Table 7-6 demonstrates the computational process for the first few samples. Recall that the difference equation describes a second-order, low-pass, Butterworth filter with a cutoff frequency of 1000 Hz. The recursive nature of the calculations can be seen from the normalized amplitudes in the table. For $f = 1000$, at sample number 3, for example, $x_0 = 0.56463$, $x_1 = 0.38941$, and $x_2 = 0.19866$, $y_0 = 0.02895$, $y_1 = 0.00980$, and $y_2 = 0.00173$. The present values are x_0 and y_0, whereas x_1, y_1, x_2, and y_2 are the previous values. This is repeated for all samples to create the output sequence (y). The computationally intensive operation of an algorithm such as this shows why dedicated processors are required to implement real-time digital filtering and signal processing.

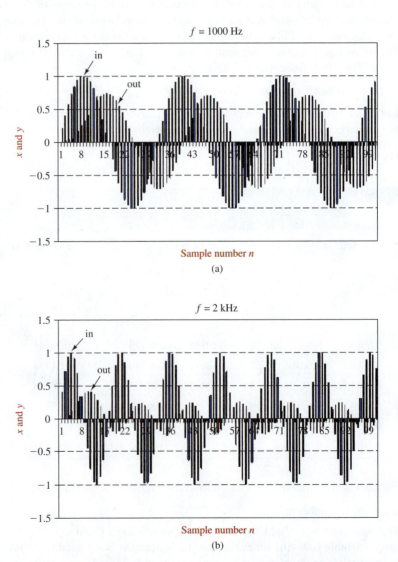

FIGURE 7-42 Input and output sequences for a second-order low-pass Butterworth filter at three frequencies: (a) 1000 Hz, (b) 2000 Hz.

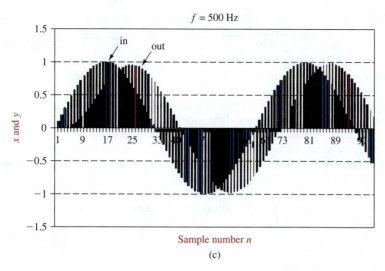

FIGURE 7-42   (c) 500 Hz.

Plotting the values from the three processes for the first hundred samples yields the sequences shown in Figure 7-42(a–c). As expected, the output amplitude at 1000 Hz (see Figure 7-42[a]), which is the cutoff frequency, is about 0.7, which is a 3-dB attenuation. The phase shift is clearly 90°, as it should be for this filter at the cutoff frequency. At 2000 Hz, twice the cutoff frequency (see Figure 7-42[b]), the nominal output should be 0.243, which is about what we see on the graph. At 500 Hz, an octave below the cutoff frequency (see Figure 7-42[c]), the output level is practically the same as the input level, which is in the filter passband, as expected. The phase shift at this frequency has a theoretical value of 43° (lagging). The phase shift shown on the plot is consistent with that value.

In addition to filtering, DSP techniques increasingly form the heart of digital implementations of all types, among them spread-spectrum-based mobile systems, radar, and other systems heavily reliant on compression and error-correction techniques. Also, and as already mentioned, many functions traditionally carried out with analog circuits are increasingly being implemented digitally because DSP implementations often produce results that are at least equivalent, if not superior, to those obtained with analog components. Despite its many advantages, DSP may not always be the best choice. High-power and high-frequency (microwave) applications are still largely the domain of analog circuits. Also, relatively simple devices (such as garage-door openers or radio-frequency identification devices, for example) do not require the computational power or performance characteristics of a DSP implementation. It is safe to predict, however, that as computational power continues to increase and costs continue to drop, the trend toward incorporating DSP into myriad applications will only continue to accelerate.

SUMMARY

The focus of this chapter has been on the preparatory stages needed for digital communications. Signals representing real-world phenomena are generally continuous, or analog in nature. Analog signals must be converted to a digital representation, which involves sampling the signal at regular intervals and preserving its essential characteristics in a digital format. Several formats are in use, but they all have in common some form of pulse modulation.

The three basic forms of pulse modulation are pulse-amplitude modulation, pulse-width modulation, and pulse-position modulation. With each form, the analog signal being sampled varies one or more parameters of the sample pulse. The sample rate must conform to the Nyquist criterion, where the sample rate must be at least twice as high as the highest frequency component of the signal to be sampled. Otherwise, an undesired form of mixing will occur between the sampling pulses and the analog signal. The difference frequencies created as a result are called alias frequencies because they are not present in the information signal. Because the alias frequencies are less than the highest analog signal frequency, the alias frequencies cannot be removed by low-pass filtering.

Pulse-amplitude modulation (PAM) is an intermediate step in pulse-code modulation systems. Pulse-code modulation (PCM) is a waveform-coding technique that converts analog information to digital words for subsequent transmission or storage. PCM is a vital building block in many digital communication applications and consists of three steps: sampling, quantizing, and coding. Conversion of the digital word to analog form also consists of three steps: regeneration, decoding, and reconstruction.

Digital communications systems have a number of important advantages over their analog counterparts. One of the most important is the ability to detect and, in some cases, to correct errors introduced as the result of noise or system impairments. Error detection systems work on the principle of parity, and correction is predicated on the notion that the correct logic state is the one residing the shortest distance from the received logic state. More sophisticated error-detection schemes involve use of the longitudinal redundancy check, which extends the concept of parity to two dimensions. A very powerful error-detection scheme, capable of catching 99.95% of transmission errors, is the cyclic redundancy check, which involves performing repetitive binary division on each block of data and checking the remainders. A remainder of zero indicates that the transmission was received correctly. A key advantage of the cyclic redundancy check is that it is easy to implement in hardware and can perform error detection in real time. When errors are detected, a retransmission is generally requested.

Forward error-correction schemes can also be implemented in which errors are corrected at the receiver without the need for retransmission. These techniques work on the principle of partial redundancy, in which sufficient additional bits are transmitted to pinpoint the exact location of erroneously received data. Codes also exist that can detect multiple errors. Among these are Reed–Solomon codes, which rely on the idea of interleaving data in a manner that allows for burst errors to be corrected.

Like many aspects of digital communications, error detection and correction techniques rely heavily on digital signal processing, which brings the power of computers to the communications realm. Many functions traditionally carried out with analog circuits, among them modulation, demodulation, and filtering, are implemented digitally in modern devices. Digital implementations often produce results that are at least equal, if not superior, to their analog counterparts. Filters, for example, often produce superior results when implemented digitally because of the lack of analog components that can change value over time or with changes in operating temperature. Digital signal processing is also an essential element of advanced communications techniques, among them spread spectrum and orthogonal frequency-division multiplexing, that form the core of next-generation wireless communications systems. These techniques, along with those used to modulate a carrier with data and, subsequently, to recover that data, will be the focus of the next chapter.

QUESTIONS AND PROBLEMS

SECTION 7-1

1. Explain what is meant by a baseband transmission.
3. Provide some possible reasons why an analog signal is digitized when an analog output is desired.
4. How are analog signals converted to digital?

SECTION 7-2

5. List two advantages of pulse modulation and explain their significance.
6. With a sketch similar to Figure 7-1, explain the basics of pulse-amplitude modulation (PAM), pulse-width modulation (PWM), and pulse-position modulation (PPM).
7. Describe a means of generating and detecting PWM.
8. Describe a means of generating and detecting PPM.
9. Draw a diagram showing demodulation of a PWM signal.

SECTION 7-4

10. Define *acquisition time* for a sample-and-hold (S/H) circuit.
11. Define *aperture time* for a S/H circuit.

12. What is the typical capacitance value for a S/H circuit?

13. Draw the PAM signal for a sinusoid using
    (a) Natural sampling.
    (b) Flat-top sampling.

14. An audio signal is band-limited to 15 kHz. What is the minimum sample frequency if this signal is to be digitized?

15. A sample circuit behaves like what other circuit used in radio-frequency communications?

16. What is the dynamic range (in dB) for a 12-bit pulse-code modulation (PCM) system?

17. Define the resolution of a PCM system. Provide two ways that resolution can be improved.

18. Calculate the number of bits required to satisfy a dynamic range of 48 dB.

19. What is meant by quantization?

20. Explain the differences between linear PCM and nonlinear PCM.

21. Explain the process of companding and the benefit it provides.

22. A μ-law companding system with $\mu = 100$ is used to compand a 0- to 10-V signal. Calculate the system output for inputs of 0, 0.1, 1, 2.5, 5, 7.5, and 10 V. (0, 1.5, 5.2, 7.06, 8.52, 9.38, 10)

SECTION 7-5

23. What is D_{min}?

24. How can the minimum distance between two data values be increased?

25. Determine the number of errors that can be detected and corrected for data values with a Hamming distance of
    (a) 2.
    (b) 5.

26. Determine the distance between the following two digital values:

$$1\ 1\ 0\ 0\ 0\ 1\ 0\ 1\ 0$$

$$0\ 1\ 0\ 0\ 0\ 0\ 0\ 1\ 0$$

27. What is the minimum distance between data words if
    (a) Five errors are to be detected?
    (b) Eight errors are to be detected?

SECTION 7-6

28. Cyclic redundancy check (CRC) codes are commonly used in what computer networking protocol?

29. Provide a definition of systematic codes.

30. With regard to (n, k) cyclic codes, define n and k.

31. What is a block check code?

32. How are the feedback paths determined in a CRC code–generating circuit?

33. Draw the CRC generating circuit if $G(x) = x^4 + x^2 + x + 1$.

34. Given that the message value is 1 0 1 0 0 1 and $G(x) = 1 1 0 1$, perform modulo-2 division to determine the block check code (BCC). (111)

35. What is a syndrome?

36. What does it mean if the syndrome value is
    (a) All zeros?
    (b) Not equal to zero?

37. What does it mean for a code to be forward error-correcting?

38. What popular commercial application uses Reed–Solomon (RS) coding and why?

SECTION 7-7

39. What is the type of difference equation that makes use of past output values?

40. What is the type of difference equation that makes use of only the present and past input values?

41. What is the order of a difference equation that requires four past input and output values?

42. What is the order of the following difference equation? Is this a recursive or a nonrecursive algorithm?

$$y_0 = x_0 - x_{(1)}$$

43. Given the following difference equations, state the order of the filter and identify the values of the coefficients.

    (a) $y = 0.9408\, x_0 - 0.5827\, x_1 + 0.9408\, x_2 + 0.5827\, y_1 - 0.8817\, y_2$

    (b) $y_0 = 0.02008\, x_0 - 0.04016\, x_2 + 0.02008\, x_4 + 2.5495\, y_1 - 3.2021\, y_2 + 2.0359\, y_3 - 0.64137\, y_4$

QUESTIONS FOR CRITICAL THINKING

44. Explain why a technique such as CRC coding is used. Can you achieve the same results using parity checking?

45. A PCM system requires 72 dB of dynamic range. The input frequency is 10 kHz. Determine the number of sample bits required to meet the dynamic range requirement and specify the minimum sample frequency to satisfy the Nyquist sampling frequency. (12 bits, 20 kHz)

46. The voltage range being input to a PCM system is 0 to 1 V. A 3-bit analog-to-digital (ADC) converter is used to convert the analog signal to digital values. How many quantization levels are provided? What is the resolution of each level? What is the value of the quantization error for this system? (8, 0.125, 0.0625)

CHAPTER 8

DIGITAL MODULATION AND DEMODULATION

CHAPTER OUTLINE

KEY TERMS

continuous wave (CW)
frequency-shift keying (FSK)
phase-shift keying (PSK)
binary phase-shift keying (BPSK)
differential phase-shift keying (DPSK)
quadrature phase-shift keying (QPSK)
minimum-shift keying (MSK)
Gaussian Minimum Shift Keying (GMSK)
bps
symbol
baud
M-ary modulation
quadrature amplitude modulation (QAM)
dibits
data bandwidth compression
orthogonal
composite
constellation pattern
spectral regrowth
offset QPSK (OQPSK)
pi-over-four differential QPSK ($\pi/4$ DQPSK)

bit error rate (BER)
energy per bit (E_b)
bit energy
E_b/N_o
raised-cosine
spread-spectrum
industrial, scientific, and medical (ISM)
Part 15
pseudonoise (PN)
spread
PN sequence length
maximal length
frequency-hopping spread spectrum
dwell time
direct-sequence spread spectrum (DSSS)
code-division multiple access (CDMA)
orthogonal frequency-division multiplexing (OFDM)
cyclic prefix
in-band on-channel (IBOC)
flash OFDM
OFDM with channel coding (COFDM)

Digital communications systems have a number of important advantages over their analog counterparts. The last chapter touched on some of these. Among the most important are immunity to noise and the ability to detect and correct transmission errors. In addition, the power of digital signal processing (DSP) and other computer-based techniques can be brought to bear to enable implementation of advanced features. Because of DSP, inexpensive and portable devices such as smart phones with features and data-transfer capabilities that would have been unimaginable only a few years ago are now commonplace.

Once an analog signal has been sampled and coded, the digital representation can be transmitted through either a wired or a wireless communications channel. In wired systems, the data may be applied directly to the transmission medium, in which case the communication is baseband in nature. In other cases, and especially in wireless systems, the data form the intelligence that modulates the carrier. With modulation, the transmitted signal is broadband because it occupies a wider bandwidth than it would without modulation. The modulation principles studied in previous chapters apply just as much to data as to analog signals, as do the considerations of bandwidth, power distribution, and performance in the presence of noise that were brought up earlier.

Though digital communication systems have a number of important advantages, such systems potentially bear a significant cost: a possible need for increased bandwidth over that required by an analog transmission with equivalent information content. As will be shown, for the simplest forms of digital modulation, and assuming ideal filtering, the bandwidth required for digital transmissions would be at least equal to the data rate in bits per second. Thus, a low-speed (by modern standards) computer network connection of 10 Mb/s (megabits per second) would require a minimum bandwidth, when modulated, of 10 MHz. Such a situation would be untenable in any bandwidth-limited medium, and many techniques have been developed to improve the spectral efficiency of digital transmissions. These techniques are the focus of this chapter.

The discussion of digital modulation schemes may appear daunting at first because of the many variations in use and the veritable alphabet soup of acronyms that have been adopted to name them. However, keep in mind that the principles established in Chapter 1 still apply and that the modulation techniques to be discussed shortly are largely an extension of the basic forms of amplitude and angle modulation covered previously. The variety of digital modulation schemes in use is a response to the bandwidth requirements and performance challenges arising in the digital context. These, too, will be considered in the following discussion.

Latest-generation, high-capacity digital systems make use of a number of newly deployed technologies to carry large amounts of data even as they co-exist with other systems occupying the same frequency bands. Two technologies that enable sharing and frequency reuse, spread spectrum and orthogonal frequency-division multiplexing, form the basis of wireless local-area networks and so-called fourth-generation (4G) mobile communications systems. Both technologies, which rely heavily on digital signal processing in their implementation, will be introduced in this chapter, and systems employing them will be examined in Chapter 10.

Finally, modulated signals must be demodulated to recover the original intelligence. As will be demonstrated, the modulated digital signal takes on many characteristics of the suppressed-carrier AM transmissions described in Chapter 2. The absence of a carrier implies that a carrier-regeneration process must take place at the receiver. Also, many of the receivers used in digital systems are of the direct-conversion type. These concepts will be explored in general terms in this chapter, with examples of wireless systems to follow in Chapter 10.

8-1 DIGITAL MODULATION TECHNIQUES

Digital modulation involves modifying the amplitude, frequency, or phase characteristics of a carrier with a data stream representing the ones and zeroes of computer language. The simplest digital modulation schemes change the carrier state with each bit of data. For example, in frequency-shift keying (FSK), the logic states (zero or one) would be represented by shifting the carrier above and below its center frequency by a predetermined amount. FSK is, therefore, a form of frequency modulation, and it has the same characteristics of sideband creation, bandwidth, and power distribution that analog FM does.

Other simple forms of digital modulation are amplitude-shift keying (ASK) and phase-shift keying (PSK). These, too, share many of the same characteristics as their analog counterparts. ASK is much like conventional amplitude modulation (AM). Like analog AM, the modulation envelope of ASK represents the intelligence, and since information is encoded as an amplitude level, ASK is susceptible to noise and requires linear amplification stages after modulation. Because of its susceptibility to noise, ASK is not usually used alone in digital communications schemes, particularly wireless systems. However, it is often used in conjunction with PSK. Both FSK and PSK are forms of angle modulation. Like analog frequency modulation (FM) and phase modulation (PM), both digital angle-modulation methods are *constant-envelope* schemes in that the amplitude of the modulated carrier stays constant; all information is encoded in the form of frequency or phase shifts. Also like their analog counterparts, FSK and PSK both produce one or more significant sideband pairs based on the amount of carrier deviation created by the modulating signal.

Amplitude-Shift Keying

The term "keying" is a holdover from the Morse code era, when a carrier was literally turned on and off with a type of paddle switch called a code key. Morse code transmissions represented the first type of digital communications and date back to the earliest days of radio. The Morse code is not a true binary code because it not only includes marks and spaces but also differentiates between the duration of these conditions. The Morse code is still used in amateur radio-telegraphic communications. A human skilled at code reception can provide highly accurate decoding. The International Morse code is shown in Figure 8-1. It consists of dots (short mark), dashes (long mark), and spaces. A *dot* is made by pressing the telegraph key down and allowing it to spring back rapidly. The length of a dot is one basic time unit. The *dash* is made by holding the key down (keying) for three basic time units. The spacing between dots and dashes in one letter is one basic time unit and between letters is three units. The spacing between words is seven units.

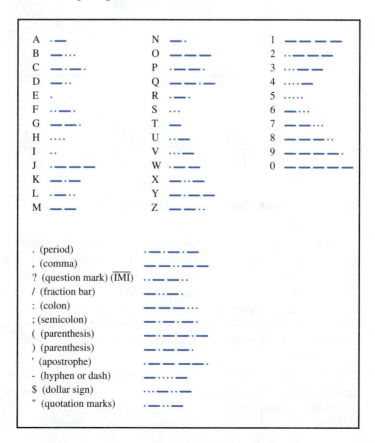

FIGURE 8-1   **International Morse code.**

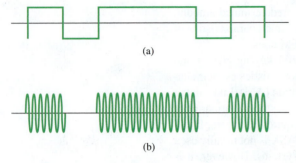

(a)

(b)

FIGURE 8-2 **CW waveforms.**

The most elementary form of transmitting highs and lows is simply to key a transmitter's carrier on and off. In fact, this form of amplitude-shift keying (ASK) is sometimes referred to as *on-off keying* (OOK). Figure 8-2(a) shows a dot-dash-dot waveform, while Figure 8-2(b) shows the resulting transmitter output if the mark allows the carrier to be transmitted and space cuts off transmission.

Thus, the carrier is conveying intelligence by simply turning it on or off according to a prearranged code. This type of transmission is called **continuous wave (CW)**; however, because the wave is periodically interrupted, it might more appropriately be called an interrupted continuous wave (ICW).

Whether the CW shown in Figure 8-2(b) is created by a hand-operated key, a remote-controlled relay, or an automatic system such as a computer, the rapid rise and fall of the carrier presents a problem. Notice that the modulation envelope shown in the figure is effectively the same as that for an overmodulated (modulation index greater than 100%) AM transmission. For the reasons described initially in Chapter 2, overmodulation must generally be avoided. The steep sides of the waveform are rich in harmonic content, which means the channel bandwidth for transmission would have to be extremely wide or else adjacent channel interference would occur. This is a severe problem because a major advantage of coded transmission versus direct voice transmission is that coded transmissions occupy much narrower bandwidths. The situation is remedied by use of an *LC* filter, as shown in Figure 8-3. The inductor L_3 slows down the rise time of the carrier, while the capacitor C_2 slows down the decay. This filter is known as a *keying filter* and is also effective in blocking the radio-frequency interference (RFI), created by arcing of the key contacts, from being transmitted. This is accomplished by the L_1, L_2 RF chokes and capacitor C_1 that form a low-pass filter.

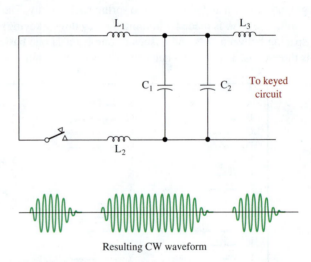

Resulting CW waveform

FIGURE 8-3 **Keying filter and resulting waveform.**

CW is a form of AM and therefore suffers from noise effects to a much greater extent than do angle-based systems. The space condition (no carrier) is also troublesome to a receiver because at that time the receiver's gain is increased by automatic gain control (AGC) action to make received noise a problem. Manual receiver gain control helps but not if the received signal is fading between high and low levels, as is often the case. The simplicity and narrow bandwidth of CW make it attractive to radio amateurs, but its major value today is to show the historical development of data transmission.

TWO-TONE MODULATION Two-tone modulation is a form of AM, but in it the carrier is always transmitted. Instead of simply turning the carrier on and off, the carrier is amplitude-modulated by two different frequencies representing either a one or zero. The two frequencies are usually separated by 170 Hz. An example of such a telegraphy system

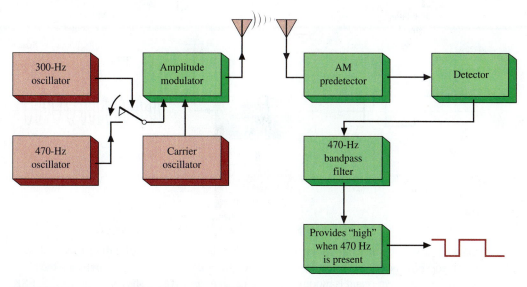

FIGURE 8-4   Two-tone modulation system (AM).

is provided in Figure 8-4. When the transmitter is keyed, the carrier is modulated by a 470-Hz signal (1 condition); it is modulated by a 300-Hz signal for the 0 condition. At the receiver, after detection, either 300- or 470-Hz signals are present. A 470-Hz bandpass filter provides an output for the 1 condition that makes the output high whenever 470 Hz is present and low otherwise.

E X A M P L E   8 - 1

The two-tone modulation system shown in Figure 8-4 operates with a 10-MHz carrier. Determine all possible transmitted frequencies and the required bandwidth for this system.

SOLUTION

This is an amplitude modulation system; therefore, when the carrier is modulated by 300 Hz, the output frequencies will be 10 MHz and 10 MHz ± 300 Hz. Similarly, when modulated by 470 Hz, the output frequencies will be 10 MHz and 10 MHz ± 470 Hz. Those are all possible outputs for this system. The bandwidth required is therefore 470 Hz × 2 = 940 Hz, which means that a 1-kHz channel would be adequate.

Example 8-1 shows that two-tone modulation systems use bandwidth very effectively. One hundred 1-kHz channels could be sandwiched in the frequency spectrum from 10 MHz to 10.1 MHz. The fact that a carrier is always transmitted eliminates the receiver gain control problems previously mentioned, and the fact that three different frequencies (a carrier and two side frequencies) are always being transmitted is another advantage over CW systems. In CW either one frequency, the carrier, or none is transmitted. Single-frequency transmissions are much more subject to ionospheric fading conditions than multifrequency transmissions, a phenomenon that will be discussed more fully in Chapter 14.

Frequency-shift Keying

Frequency-shift keying (FSK) is a form of frequency modulation in which the modulating wave shifts the output between two predetermined frequencies—usually termed the mark and space frequencies. In another instance of terminology from the days of telegraphy being used in modern digital systems, a logic one (high voltage) is usually termed a *mark* and a logic zero (low voltage) a *space*. (In early telegraph systems, a pen would place a

FIGURE 8-5 **FSK transmitter.**

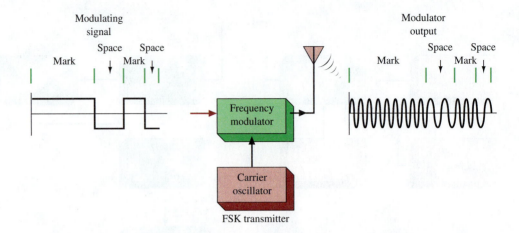

FIGURE 8-5 FSK transmitter.

mark on a strip of moving paper whenever a voltage level corresponding to a dash was received.) FSK is an FM system in which the carrier frequency is midway between the mark and space frequencies and is modulated by a rectangular wave, as shown in Figure 8-5. FSK systems can be narrowband or wideband. For a narrowband system, the mark condition causes the carrier frequency to increase by 42.5 Hz, while the space condition results in a 42.5-Hz downward shift. Thus, the transmitter frequency is constantly changing by 85 Hz as it is keyed. This 85-Hz shift is the standard for narrowband FSK, while an 850-Hz shift is the standard for wideband FSK systems.

EXAMPLE 8-2

Determine the channel bandwidth required for the narrowband and wideband FSK systems.

SOLUTION

The fact that narrowband FSK shifts a total of 85 Hz does not mean the bandwidth is 85 Hz. While shifting 85 Hz, it creates an infinite number of sidebands, with the number of significant sidebands determined by the modulation index. If this is difficult for you to accept, it would be wise to review the basics of FM in Chapter 3.

The terms "narrowband" and "wideband" are used in the same context as defined for FM in Chapter 3. A narrowband FM or FSK transmission is one occupying a bandwidth no greater than that of an equivalent AM transmission, which means that only one significant sideband pair can be generated. In practice, most narrowband FSK systems use a channel with a bandwidth of several kilohertz, while wideband FSK uses 10 to 20 kHz. Because of the narrow bandwidths involved, FSK systems offer only slightly improved noise performance over the AM two-tone modulation scheme. However, the greater number of sidebands transmitted in FSK allows better ionospheric fading characteristics than do two-tone AM modulation schemes.

FSK GENERATION   The generation of FSK can be accomplished easily by switching an additional capacitor into the tank circuit of an oscillator when the transmitter is keyed. In narrowband FSK it is often possible to get the required frequency shift by shunting the capacitance directly across a crystal, especially if frequency multipliers follow the oscillator, as is usually the case. (This technique is sometimes referred to as "warping" the crystal.) FSK can also be generated by applying the rectangular wave-modulating signal to a voltage-controlled oscillator (VCO). Such a system is shown in Figure 8-6. The VCO output is the desired FSK signal, which is then transmitted to an FM receiver. The receiver is a standard unit up through the IF amps. At that point a 565 phase-locked loop (PLL) is used for detecting the original modulating signal. As the IF output signal appears at the PLL input, the loop locks to the input frequency and tracks it between the two frequencies

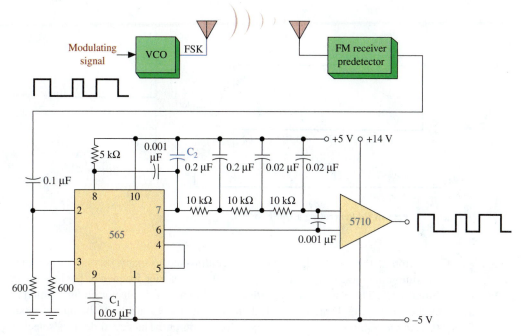

FIGURE 8-6   **Complete FSK system.**

with a corresponding dc shift at its output, pin 7. The loop filter capacitor C_2 is chosen to set the proper overshoot on the output, and the three-stage ladder filter is used to remove the sum frequency component. The PLL output signal is a rounded-off version of the original binary modulating signal and is therefore applied to the comparator circuit (the 5710 in Figure 8-6) to make it logic compatible.

Phase-shift Keying

The third type of data modulation, and one of the most efficient, is phase shift keying. **Phase-shift keying (PSK)** systems provide a low probability of error. In a PSK system, the transmitted binary data cause the phase of the sine-wave carrier to shift by a defined amount. The modulating signal may cause the carrier to shift to one or more possible phases with respect to its unmodulated (reference) phase. In the simplest form of PSK, known as **binary phase-shift keying (BPSK)**, the carrier can either stay at its reference phase (i.e., no phase shift) or be shifted in phase by 180°. Expressing these phase shifts in terms of radians, as shown in Figure 8-7, we see that the $+\sin(\omega_c t)$ vector provides the logical "1" and the $-\sin(\omega_c t)$ vector provides the logical "0." The BPSK signal does not require that the frequency of the carrier be shifted, as with the FSK system. Instead, the carrier is directly phase modulated, meaning that the phase of the carrier is shifted by the incoming binary data. This relationship is shown in Figure 8-8 for an alternating pattern of 1s and 0s.

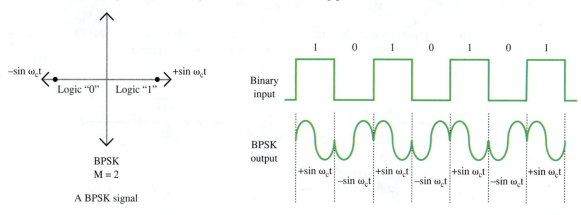

FIGURE 8-7   **Illustration of a BPSK constellation.**

FIGURE 8-8   **The output of a BPSK modulator circuit for a 1010101 input.**

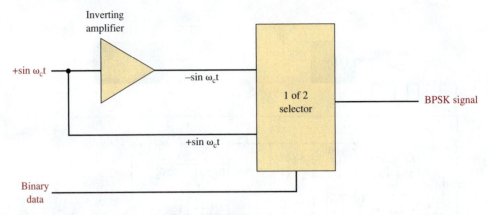

FIGURE 8-9 **Circuit for generating a BPSK signal.**

Generation of the BPSK signal can be accomplished in many ways. A block diagram of a simple method is shown in Figure 8-9. The carrier frequency $[+\sin(\omega_c t)]$ is phase-shifted 180°. The + and − values are then fed to a 1 of 2 selector circuit, which is driven by the binary data. If the bit is a 1, then the output is $+\sin(\omega_c t)$. If the binary-input bit is a 0, then the $-\sin(\omega_c t)$ signal is selected for the output. The actual devices selected for performing this operation are dependent on the binary input data rate and the transmit carrier frequency.

A BPSK receiver detects the phase shift in the received signal. One possible way of constructing a BPSK receiver is by using a mixer circuit. The received BPSK signal is fed into the mixer circuit. The other input to the mixer circuit is driven by a reference oscillator synchronized to the carrier, $\sin(\omega_c t)$. This is referred to as *coherent carrier recovery*. The recovered carrier frequency is mixed with the BPSK input signal to provide the demodulated binary output data. A block diagram of the receive circuit is provided in Figure 8-10. Mathematically, the BPSK receiver shown in Figure 8-10 can provide a 1 and a 0 as follows:

$$\text{“1” output} = [\sin(\omega_c t)][\sin(\omega_c t)] = \sin^2(\omega_c t)$$

and by making use of the trigonometric identity $\sin^2 A = \frac{1}{2}[1 - \cos 2A]$, we obtain

$$\text{“1” output} = \tfrac{1}{2} - \tfrac{1}{2}[\cos(2\omega_c t)].$$

The $\frac{1}{2}[\cos(2\omega_c t)]$ term is filtered out by the low-pass filter shown in Figure 8-10. This leaves

$$\text{“1” output} = \tfrac{1}{2}.$$

Similar analysis will show

$$\text{“0” output} = [-\sin(\omega_c t)][\sin(\omega_c t)]$$
$$= -\sin^2(\omega_c t)$$
$$= -\tfrac{1}{2}.$$

The $\pm\frac{1}{2}$ represent dc values that correspond to the 1 and 0 binary values. The $\pm\frac{1}{2}$ values can be conditioned for the appropriate input level for the receive digital system.

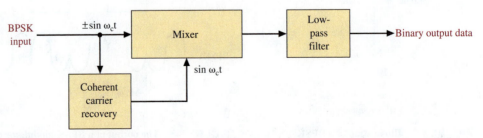

FIGURE 8-10 **A BPSK receiver using coherent carrier recovery and a mixer circuit.**

Synchronization and Carrier Reinsertion

An alternate way of looking at the preceding result is to consider the modulated digital signal as a form of suppressed-carrier transmission, in which the modulated data form the sidebands. Recall from Chapter 2 that suppressed-carrier systems require the carrier to be reinserted at the receiver so that the demodulator (which, because it is a product detector, is a mixer circuit) can recover the modulating signal. Recall also that the difference frequency created as the result of mixing the carrier with the side frequencies produces the original intelligence. To do this in a digital system we can make use of a synchronous detector, which, as described in Chapter 6, is a subsystem consisting of a product detector combined with additional circuitry to regenerate the original carrier, both in frequency and phase.

Figure 8-11 shows the coherent carrier recovery system for the BPSK receiver of Figure 8-10. Within the system, the *carrier synchronizer* is responsible for producing a reference carrier signal for the product detector. The heart of the carrier synchronizer is the PLL. Recall that the phase detector inputs in a basic PLL are the reference frequency (which, in this case, will be the PSK-modulated signal) and the output of the VCO, a portion of which is fed back to the phase detector. The phase detector produces an error voltage that corrects any drift in the VCO output frequency, causing the VCO to become locked in frequency and phase to the incoming signal. The PLL carrier synchronizer is used for all forms of digital modulation (ASK, FSK, or PSK), but in the case of PSK the received signal must be applied to a frequency doubler before being input to the PLL; the signal applied to the PLL reference input is, therefore, twice the frequency of the incoming PSK signal. The frequency doubler is functionally similar to a full-wave rectifier similar to that found in a power supply. One of the characteristics of a full-wave rectifier is that the output frequency is twice the input frequency. The received PSK signal is a double-sideband suppressed carrier signal, similar to analog single sideband in its appearance in the time domain. That is, the PSK signal does not contain a carrier, only upper and lower sidebands. However, the output of the frequency doubler will contain a signal to which the PLL can be locked. This frequency doubler output signal will be equal to twice the carrier frequency and will be unaffected by phase changes to the modulated signal. Because the full-wave rectifier serving as the frequency doubler is a nonlinear device, it mixes the frequency components present at the input and provides sum and difference frequencies (upper and lower sidebands) at its output plus other harmonics, which are removed by a bandpass filter. The upper sideband is equal to the carrier frequency, f_o, plus the data rate R $(f_o + R)$, and the lower sideband is equal to the carrier frequency, f_o, minus the data rate R $(f_o - R)$. When these two sidebands are mixed in the frequency doubler, the result is the sum and difference of the sidebands: $[(f_o + R) + (f_o - R)] = 2f_o$ and $[(f_o + R) - (f_o - R)] = 2R$. The $2f_o$ signal is recognized as twice the carrier frequency and can be sent to a divide-by-2 stage to produce the carrier reference that is stable in both frequency and phase. The result $2R$ is a frequency twice the data rate; this frequency can be rejected by a post-detection filter.

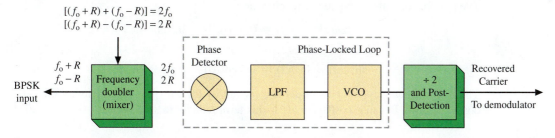

FIGURE 8-11   Coherent carrier recovery subsystem block diagram.

Differential Phase-shift Keying

An important and widely used variation on the basic form of BPSK just described is called **differential phase-shift keying (DPSK)**. DPSK uses the BPSK vector relationship for generating an output. However, a DPSK receiver is simpler to implement because it does not require a coherent carrier recovery subsystem. The logical "0" and "1" are determined by comparing the phase of two successive data bits. In BPSK, $\pm \sin \omega_c t$ represent a 1 and 0, respectively. This is also true for a DPSK system, but in DPSK generation of the 1 and 0 outputs is accomplished by initially comparing the first bit (1) transmitted with a reference value (0). The two values are XNORed to generate the DPSK output (0). The "0" output is used to drive a PSK circuit shown in Figure 8-9. A 0 generates $-\sin \omega_c t$, and a 1 generates $+\sin \omega_c t$.

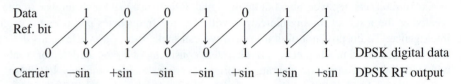

Data	1	0	0	1	0	1	1	
Ref. bit								
	0	0	1	0	0	1	1	1	DPSK digital data
Carrier	−sin	+sin	−sin	−sin	+sin	+sin	+sin	DPSK RF output

A simplified block diagram of a DPSK transmitter is shown in Figure 8-12.

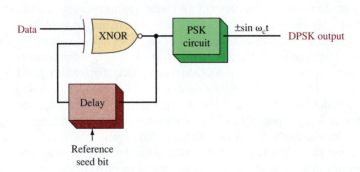

FIGURE 8-12   Simplified block diagram of a DPSK transmitter.

Data — XNOR — PSK circuit — $\pm\sin \omega_c t$ — DPSK output

Delay

Reference seed bit

At the DPSK receiver, the incoming data is XNORed with a 1-bit delay of itself. For the previous example, the data will be shifted as shown:

Rx carrier	−sin	+sin	−sin	−sin	+sin	+sin	+sin
(ref)							
Shifted Rx carrier	−sin	−sin	+sin	−sin	−sin	+sin	+sin	+sin
Recovered data	1	0	0	1	0	1	1

The advantage of a DPSK system is that carrier recovery circuitry is not required. The disadvantage of DPSK is that it requires a good signal-to-noise (S/N) ratio for it to achieve a bit error rate equivalent to **quadrature phase-shift keying (QPSK)**, a technique that permits one of four possible phase states, each separated by 90° degrees. In such a system two bit streams are applied to a modulator in a quadrature relationship, implying that there is a 90° phase relationship between them.

A DPSK receiver is shown in Figure 8-13. The delay circuit provides a 1-bit delay of the DPSK $\pm \sin \omega_c t$ signal. The delayed signal and the DPSK signal are mixed together. The possible outputs of the DPSK receiver are

$$(\sin \omega_c t)(\sin \omega_c t) = 0.5 \cos(0) - 0.5 \cos 2w_c t$$
$$(\sin \omega_c t)(-\sin \omega_c t) = -0.5 \cos(0) + 0.5 \cos 2w_c t$$
$$(-\sin \omega_c t)(-\sin \omega_c t) = 0.5 \cos(0) - 0.5 \cos 2w_c t.$$

The higher-frequency component 2ω is removed by the low-pass filter, leaving the ± 0.5 cos(0) dc term. Because cos(0) equals 1, this term reduces to ± 0.5, where $+0.5$ represents a logical 1 and -0.5 represents a logical 0.

FIGURE 8-13   A DPSK receiver.

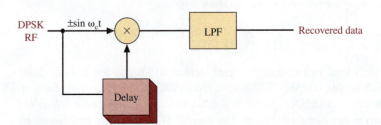

DPSK RF — $\pm\sin \omega_c t$ — × — LPF — Recovered data

Delay

Minimum Shift Keying

Another important and widely used variation on FSK/PSK is known as **minimum-shift keying (MSK)**. It is widely used in cordless phones, and a variant called Gaussian MSK (GMSK) (explained later in this section) is the modulation technique employed in the GSM (Global System for Mobile Communications) cellular standard. Recall from your study of FM in Chapter 3 that phase shifts result in instantaneous frequency changes and that a frequency change, likewise, results in a phase shift. Phase modulation in that context was described as indirect FM. With PSK, phase modulation is created directly because the modulating signal (the binary data input) is directly applied to the carrier in the modulator stage, thus affecting the carrier phase. With reference to Figure 8-8, which is a time-domain representation of the input to and output from the BPSK modulator, note that the output signal undergoes an abrupt reversal in direction at the edges of the input pulses. (In the next section, this will be shown to involve the concept of negative frequency, where the vector representation of the sine-wave carrier abruptly shifts from counter-clockwise rotation to clockwise rotation.) The sharp points seen as the sine waves abruptly change phase represent high-frequency components that contribute to sidebands and, hence, increased bandwidths. Recall also from Chapter 3 that the occupied bandwidth of frequency- or phase-modulated signals is determined by the modulation index, m_f, which in turn is defined by the intelligence signal frequency and amount of carrier deviation: $m_f = \delta/f_i$. In a bandwidth-limited system, then, some form of filtering will need to be employed to "smooth out" the sharp transitions and remove the high-frequency content from the modulated signal.

The abrupt, 180° phase shift represents an extreme case arising from direct PSK. A similar but less drastic form of the same phenomenon can occur in FSK as well, as shown in Figure 8-14(a). Note that a phase discontinuity can occur during the space-to-mark (0 to 1) or mark-to-space (1 to 0) transitions, and these discontinuities too create high-frequency components that contribute to the bandwidth characteristics of the modulated signal. Minimum-shift keying, on the other hand, is a form of what is known as *continuous-phase* FSK, in which the periods of the mark and space frequencies are such that cycles of the sine-wave carrier cross zero right at the edges of the modulating-signal pulse transitions. Thus, with careful selection of the mark and space frequencies it is possible to produce a modulated signal without phase discontinuities and thus a lower bandwidth requirement.

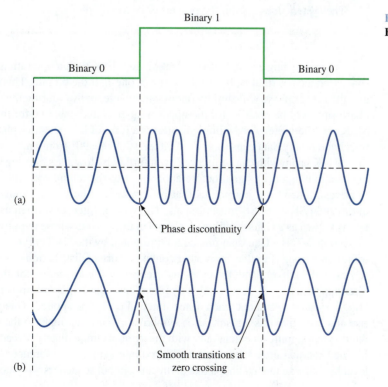

FIGURE 8-14 **Comparison of FSK and MSK.**

Such a signal is also said to be *coherent* because the cycles associated with each bit state start and stop at the zero crossing points. In addition to a lower bandwidth requirement, coherent signals have better noise performance characteristics, an important advantage in mobile wireless applications.

MSK modulation takes the concept one step further by using a low modulation index, generally 0.5. Like continuous-phase FSK, MSK minimizes bandwidth by ensuring a smooth transition between data states, as shown in Figure 8-14(b). This result occurs because the mark and space frequencies are integer multiples of the bit clock frequency. The modulation index for an FSK system is determined by deviation, δ (as for analog FM) as well as the bit time (T) of the data rate:

$$m_f = \delta(T). \tag{8-1}$$

In an FSK system the deviation is defined as the frequency shift between the mark and space frequencies. The bit time is the reciprocal of the data rate.

EXAMPLE 8-3

An FSK system has a bit clock frequency of 1800 bps with mark and space frequencies set at 1800 Hz and 2700 Hz, respectively. Does this system exhibit the characteristics of MSK modulation?

SOLUTION

The characteristics of MSK are that the modulation index equals 0.5 and that the mark and space frequencies are integer multiples of the bit clock frequency. The modulation index is given as

$$m_f = \delta(T), \tag{8-1}$$

and δ is the difference between the mark and space frequencies, or 2700 Hz − 1800 Hz = 900 Hz.

The period is the reciprocal of the bit rate in bits per second,

or $\dfrac{1}{1800bps} = 0.0005555$ s.

The modulation index is, therefore, $900(0.0005555) = 0.5$.

The system does exhibit the properties of MSK.

The preceding example shows that MSK is FSK with a separation between mark and space frequencies (that is, frequency shift) of one-half the bit rate. This relationship means that the carrier phase will shift by one-quarter cycle, or 90°, during the symbol period. The phase shift will be positive for the upper frequency and negative for the lower frequency. MSK can, therefore, also be implemented as a form of DPSK with a phase shift of ±90°, in which the binary data cause a shift of either +90° or −90° from one symbol to the next.

MSK bandwidth can be reduced even further by low-pass filtering the binary modulating (i.e., intelligence) signal before applying it to the modulator. This additional filtering removes higher-level harmonics that in turn create the wider bandwidths seen in non-optimized FSK and PSK transmissions. The type of filter most often used for this application is known as a Gaussian filter. A Gaussian filter is one whose impulse response (covered in Section 7-7) is a Gaussian function, represented by the "bell curve" shape of normal distributions perhaps familiar from other contexts (normal distribution of peoples' heights, test scores, and the like). Both the time- and frequency-domain representations of the Gaussian filter appear as a bell curve, and such filters have the important property that they do not "ring" or cause overshoot when a step function (pulse) is applied. This property minimizes rise and fall times. As a result, each symbol applied as an input to the Gaussian filter will interact to a significant extent only with the symbols immediately preceding and succeeding it. The Gaussian filter keeps such undesired intersymbol interference (ISI) to a reasonable level by reducing the likelihood that particular symbol sequences will interact in a destructive manner. The advantage in system implementations is that amplifiers are easier to build and more efficient than for alternative designs. MSK systems (either FSK or PSK) that use Gaussian filters to shape the digital bit stream before modulation produce **Gaussian Minimum Shift Keying (GMSK)**, a spectrally efficient form of modulation. An extension of the idea presented in this section, in which PSK is extended to two dimensions, is the basis of the modulation format used in GSM mobile communications systems deployed throughout the world. Such quadrature modulation techniques will be examined in more detail shortly, but first let us consider why the additional complexity implied therein is necessary.

8-2 BANDWIDTH CONSIDERATIONS OF MODULATED SIGNALS

The digital modulation schemes presented to this point modulate the carrier with each bit of data. That is, each change in logic state causes the amplitude, frequency, or phase of the carrier to be changed. Let us now consider the bandwidth implications of such systems.

Bandwidth considerations for digital systems are much like those of analog systems. Technical specifications such as frequency response and available frequency spectrum are still of great importance. The available bandwidth dictates the amount of data that can be transmitted over time (i.e., bits per second, **bps**) and the bandwidth (in Hertz) influences the choice of the digital modulation techniques.

The first consideration is that the bandwidth of our transmission channel will not be infinite and, therefore, the transmitted pulse will not be a perfect rectangle. Remember from Chapter 1 that the square wave is made up of a summation of sinusoids of harmonically related frequencies. The bandwidth-limiting effects of a channel will attenuate the higher-frequency harmonics and will cause an applied pulse stream to look increasingly like a sine wave at the fundamental frequency of the pulse stream. It is the fundamental frequency that determines the minimum bandwidth of the channel.

The bandwidth requirements for a data stream depend on the encoding method used to transmit the data. The various encoding methods and the reasons for their use will be examined in more detail in Chapter 9, but for right now let us assume that the encoding method used causes an alternating 1 0 1 0 pattern as shown in the left portion of Figure 8-15. This pattern represents the highest baseband frequency and, therefore, the minimum bandwidth. The period of an individual bit, T_b, is shown in the figure to be one half the period of a full cycle of the sinusoidal component. Put another way, the period of one sine-wave cycle is associated with two bits—one bit (a logic 1) for the positive half, and the second bit (a logic 0) for the negative half of the sine wave. The minimum bandwidth, BW_{min}, is determined by the sinusoidal component contained in the data stream with the shortest period of T (in seconds), which is the alternating 1 0 pattern. Therefore, at baseband, the minimum bandwidth is

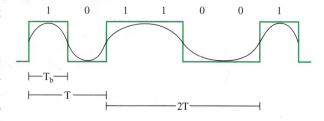

FIGURE 8-15 **A digital pulse stream of frequency 1/T and a pulse width of T_b.**

$$BW_{min} = \frac{1}{2T_b} = \frac{1}{T} \text{ [Hz]}. \tag{8-2}$$

Equation (8-2) predicts that the required channel bandwidth will be one-half the bit rate. Put another way, the channel capacity (in bits per second) is twice the bandwidth. Again, this statement assumes that the coding format produces an alternating 0 1 0 1 pattern, because this pattern produces a fundamental frequency with the shortest period and, therefore, the highest frequency. Contrast this result with the sinusoid produced by a sequence of 1 1 0 0, for example, also shown in Figure 8-15. The 1 1 0 0 sequence has a period twice that of the 1 0 pattern. Therefore, the 1 1 0 0 sequence will not be a limiting factor for the minimum bandwidth.

EXAMPLE 8-4

An 8-kbps data stream with a repeating 1 0 1 0 pattern is used in a digital communication link. Determine the minimum bandwidth required for the communications link.

SOLUTION

Using Equation (8-2), the minimum bandwidth is

$$T_b = \frac{1}{8 \text{ kbps}} = 125 \, \mu s$$

$$\text{BW}_{min} = \frac{1}{2 \, T_b} = \frac{1}{2 \cdot 125 \, \mu s} = 4 \text{ kHz}. \tag{8-2}$$

The discussion up to this point has focused on baseband signals, that is, signals containing low-frequency components applied to a communications channel without modulation. Examples include telephone lines, twisted-pair copper-wire networks, or coaxial cable networks. In all these cases, the bit rate is seen to be twice the minimum bandwidth. However, if the baseband signal modulates a carrier, as will often be the case (particularly in wireless systems), then a minimum of one pair of sidebands will be produced. As always, these sidebands represent sum and difference frequencies, and the bandwidth of the modulated signal will be twice that of the highest modulating signal frequency. For a digital system in which the carrier state changes with each bit, the occupied channel bandwidth in Hertz will be at least equal to the data rate in bits per second. We say "at least" because the bandwidth is also dependent on the type and effectiveness of filtering used. In the cases studied so far, where the carrier is modulated by each bit, *and with ideal filtering*, a data rate of, say, 10 Mb/s (megabits per second) would produce a modulated signal bandwidth of 10 MHz. Contrast this required bandwidth with that allocated to AM radio (10 kHz), FM radio (200 kHz), or broadcast television (6 MHz) and the problem becomes immediately apparent: The bandwidth requirements of high-capacity digital systems potentially become unmanageable in situations where bandwidth is scarce. To provide high information-carrying capacities in such cases, more spectrally efficient forms of modulation must be employed. These involve the use of so-called *M-ary* systems, where two or more data bits are first grouped together, and it is these groups of bits, rather than the bits individually, that modulate the carrier. A grouping of two or more bits is called a **symbol**, and the symbol rate is sometimes referred to as the **baud** rate. By grouping bits into symbols, we will see the bandwidth of the modulated signal will be associated with the symbol rate, not the bit rate.

8-3 M-ARY MODULATION TECHNIQUES

The term "M-ary" in **M-ary modulation** derives from the word "binary," or two-state. BPSK is a binary modulation scheme because the carrier can shift to one of two predetermined phase states, usually separated by 180°. A technique that permits one of four possible phase states, each separated by 90° degrees, would be a quarternary PSK or QPSK system. In such a system two bit streams are applied to a modulator in a quadrature relationship, implying that there is a 90° phase relationship between them. Hence another term for QPSK is quadrature phase-shift keying. Higher-level PSK systems are possible, up to a practical limit. If eight phase states are possible, then the modulation is called 8PSK, 16 levels, 16PSK, and so on.

Although more than sixteen phase states are possible in theory, 16PSK is the practical limit for systems that depend strictly on phase modulation because phase states established too close together will be difficult if not impossible for the receiver to detect reliably. This close proximity would detract from one of the biggest advantages of PSK, which is the ability to transmit data with a low probability of error. More states are achievable by combining phase shifts with amplitude shifts. This combination gives rise to **quadrature amplitude modulation (QAM)** schemes, in which a given grouping of bits is represented by a particular combination of phase and amplitude. Effectively, then, both amplitude and angle modulation are in use at the same time. Both PSK and QAM are widely used, and the following section describes their implementation in more detail.

M-Ary PSK

As already established, in a PSK system the incoming data cause the phase of the carrier to shift by a defined amount. The most basic form, BPSK, with two possible states, is suitable for low-data-rate applications where reliability of information transfer is of paramount concern. An example is a deep-space probe. The principal drawback, however, is that BPSK is bandwidth-intensive. For higher-data-rate applications, the carrier phase must be able to shift to a number of allowable phase states. In general, the output voltage can be expressed as

$$V_o(t) = V \sin\left[\omega_c(t) + \frac{2\pi(i-1)}{M}\right],$$ (8-3)

where $i = 1, 2, \ldots, M$

$M = 2^n$, number of allowable phase states

$n =$ the number of data bits needed to specify the phase state

$\omega_c =$ angular velocity of carrier.

The three most common versions of the PSK signal are shown in Table 8-1. When viewed in the time domain, the symbol locations produced as the result of amplitude and/or phase shifts can be represented as a *constellation* of points in a two-dimensional space. Each point represents a symbol; the number of possible bit combinations determines the number of possible points. In a BPSK signal, the phase of the carrier will shift by 180° (i.e., $\pm \sin \omega_c t$). A diagram showing a BPSK constellation was shown in Figure 8-7. For a QPSK signal, the phase changes 90° for each possible state.

TABLE 8-1 • Common PSK Systems		
Binary phase shift keying—BPSK	$M = 2$	$n = 1$
Quadrature phase shift keying—QPSK	$M = 4$	$n = 2$
8PSK	$M = 8$	$n = 3$

In the QPSK constellation, binary data are first grouped into two-bit symbols (sometimes called **dibits**). M-ary systems employ two data channels, one of which is phase-shifted by 90° with respect to the other. The non-phase-shifted channel is identified as the *I* channel (for "in-phase"), and the phase-shifted channel the *Q* (for "quadrature") channel. ("Non-phase-shifted" means that the instantaneous carrier phase is the same as that of the unmodulated carrier, which serves as the reference, or starting point.) Each channel contributes to the direction of the vector within the phase constellation, as shown in Figure 8-16(a). The two-bit symbols produce four possible locations, each with a magnitude that is the vector sum of the voltages applied to the I and Q channels, and each with a phase angle that is offset from the I channel by one of four possible angles (45°, 135°, 225°, and 315°). The combination of four possible amplitudes and four possible phase shifts gives four possible data points, each associated with a pair of bits (0 0, 0 1, 1 0, or 1 1) as shown in Figure 8-16(b).

FIGURE 8-16   The QPSK phase constellation: (a) vector representation; (b) the data points.

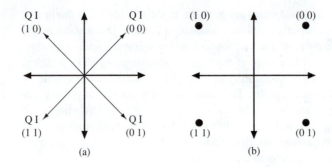

(a)   (b)

As already established, the BPSK signal requires a minimum bandwidth equal to the bit rate, or $BW_{min} = f_b$. The QPSK signal will require only $f_b/2$ for each channel. The quantity f_b refers to the frequency of each of the original bits in the data. This relationship is shown pictorially in Figure 8-17. The figure shows that grouping bits into pairs causes the period of the fundamental sine-wave component (i.e., what would remain after ideal filtering) to be doubled and its frequency to be cut in half. Therefore, the required bandwidth will ideally be reduced in half as well. This result is similar to what would happen with a repeating 1 1 0 0 transmission (Figure 8-15): Now four bit states can be contained within each cycle of the sine wave, rather than two. QPSK transmissions realize a form of **data bandwidth compression**, where more data are being compressed into the same available bandwidth.

Returning to Figure 8-17, observe the relationship between the logic states of the data stream forming the modulator input (represented as a square wave with a positive peak voltage for a logic 1 and a negative peak for a logic zero) and the sine-wave output of the modulator, which represents the information signal after filtering to remove all higher-order harmonics. Note in particular how the output represents a logic 0 as a sine wave with a positive peak voltage and a logic 1 as a voltage with a negative peak. Comparing Figure 8-17 to the constellation diagram of Figure 8-16, we see that the symbol (1 0), with an I bit of 1 and a Q bit of 0, is formed as the result of the vector sum of

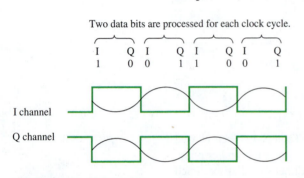

FIGURE 8-17   The I and Q data channels for a QPSK signal.

voltages representing a negative excursion on the I axis (−I, facing West) and a positive excursion on the Q axis (+Q, facing North). Likewise, the combination (0 1) results from the vector sum of +I and −Q. Not shown in Figure 8-17 are the combinations (0 0) and (1 1); however, these can be shown to be the result of simultaneous positive or negative excursions along both the I and Q axes, respectively.

In modern designs QPSK and other forms of M-ary modulation are carried out with DSP-based chipsets. However, a conceptual block diagram of an I/Q modulator is shown in Figure 8-18. The incoming data stream is first split into two paths, each of half the

FIGURE 8-18   Block diagram of a QPSK modulator.

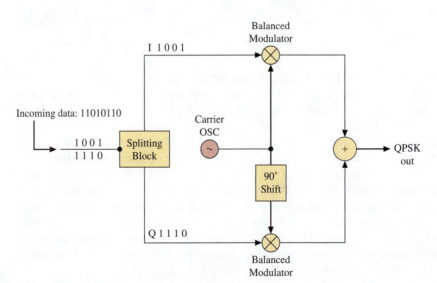

incoming data rate. The upper path in the figure is the I channel, and the lower path the Q channel. Note also from the figure that the data stream is directed to the I and Q channels in alternating fashion—the first bit goes to the I channel, the next to Q, back to I, and so on. Each stream is applied as one input to its respective mixer (multiplier). The other mixer input is a sine-wave carrier, which is either applied directly to the I-channel mixer or shifted by 90 degrees for application to the Q-channel mixer. Sum and difference frequencies from each mixer are created in the usual fashion, and these are applied to a summer (Σ) to form the complete modulated signal.

For reasons that will be developed more fully in Section 8-5, signals that are in quadrature are said to be **orthogonal** because they do not interfere with each other. The I and Q bit streams can be considered to be two independent components of the modulated signal that, when combined, form a **composite** signal. However, the two channels maintain their independent characteristics, and their orthogonal relationship makes it possible for the receiver to split the received signal into its independent parts.

A QPSK receiver is essentially the mirror image of the modulator just described. A block diagram of a QPSK demodulating circuit is shown in Figure 8-19. It also makes use of the principle of coherent carrier recovery described earlier. A carrier recovery circuit is used to generate a local clock frequency, which is locked to the QPSK input carrier frequency ($\sin \omega_c t$). This frequency is then used to drive the phase-detector circuits for recovering the I and Q data. The phase detector is basically a mixer circuit where $\sin \omega_c t$ (the recovered carrier frequency) is mixed with the QPSK input (expressed as $\sin \omega_c t + \phi_d$). The ϕ_d term indicates that the carrier frequency has been shifted in phase, as expected. The phase detector output (v_{pd}) is then determined as follows:

$$v_{pd} = (\sin \omega_c t)(\sin \omega_c t + \phi_d)$$
$$= 0.5A \cos (0 + \phi_d) - 0.5A \cos 2\omega_c t.$$

A low-pass filter (LPF) removes the $2\omega_c$ high-frequency component, leaving

$$v_{pd} = 0.5A \cos \phi_d.$$

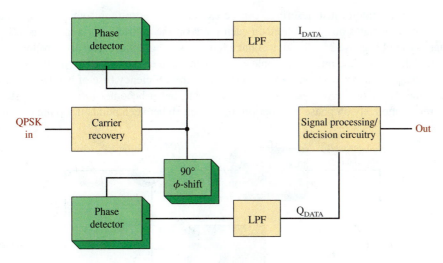

LPF: Low-Pass Filter

FIGURE 8-19   **Block diagram of a QPSK demodulator.**

The remaining ϕ_d value is the dc voltage representing the direction of the data vector along the I-axis. The Q data are recovered in the same way, and together the I and Q values define the direction of the data vector. For a data value of (0, 0), the data point is at a vector of 45°. Therefore, $v_{pd} = (0.5)(\cos 45°) = (0.5)(0.707) = 0.35$ V. The decoded data point is shown in Figure 8-20. The amplitude of the data and the phase, ϕ, will vary as a result of noise and imperfections (such as circuit nonlinearities) in data recovery. Each contributes to a noisy constellation, as shown in Figure 8-21. This is an example of what the decoded QPSK digital data can look like at the output of a digital receiver. The values in each quadrant are no longer single dots, as in the ideal case [Figure 8-16(b)]. During

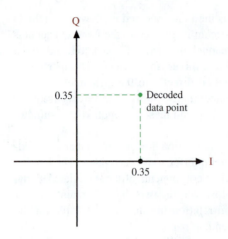

FIGURE 8-20   **The decoded data point.**

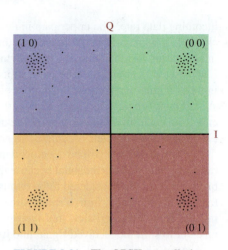

FIGURE 8-21   **The QPSK constellation with noise at the receiver. The decision boundaries are shown by the different shaded areas.**

transmission, noise is added to the data, and the result is a cluster of data in each quadrant of the QPSK constellation.

Once the I and Q data are decoded, signal processing and a decision circuit are used to analyze the resulting data and assign a digital value to it (e.g., 00, 10, 11, 01). The shaded areas in Figure 8-21 represent the decision boundaries in each quadrant. The data points, defined by the I and Q signals, will fall in one of the boundary regions and are assigned a digital value based on the quadrant within which the point resides.

Quadrature Amplitude Modulation

The QPSK technique just described can be extended to more than four possible phase states. If eight states are possible, then each point on the constellation represents a 3-bit binary symbol because $2^3 = 8$. What about a four-bit symbol? Since $2^4 = 16$, it follows that the modulation method must be capable of producing sixteen distinct states for the receiver to decode. Such a system would exhibit better spectral efficiency (use of bandwidth) than either the BPSK or QPSK systems described so far, but it becomes difficult for systems that rely completely on phase modulation to distinguish between points if too many points

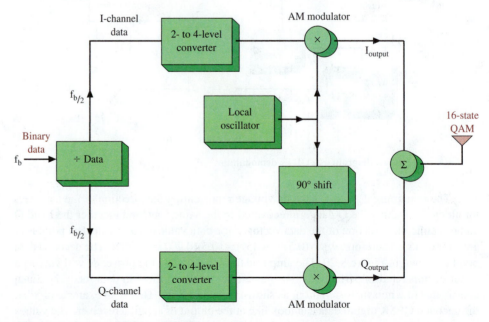

FIGURE 8-22   **16-QAM transmitter.**

are represented in the constellation. For higher-throughput systems achieving even higher spectral efficiencies than QPSK, both ASK and PSK are used together to produce QAM.

The block diagram in Figure 8-22 shows a 16-QAM transmitter. It will produce a constellation of sixteen points, each associated with a 4-bit symbol. Just as with the QPSK system just described, the binary data are first fed to a data divider block (labeled ÷ data in the figure) that essentially produces two data signals at half the original bit rate by applying the data bits alternately to the two outputs. Each distinct amplitude level, in conjunction with phase, represents a possible combination of data bits. The resultant four-level symbol streams are then applied to the I and Q modulators, as shown in Figure 8-22. Like its QPSK counterpart, the QAM demodulator reverses the process of modulation, thereby recovering the original binary data signal.

The constellation diagrams of M-ary systems can be viewed in the time domain with a dual-channel oscilloscope set to "X-Y" mode, which causes the instrument time base to be turned off and the beam to be deflected horizontally and vertically based on the voltages applied to the vertical inputs. On most oscilloscopes, channel 1 is designated the X channel, whereby a voltage input deflects the beam horizontally, and channel 2 is the Y channel, deflecting the beam vertically. Applying the demodulated I signal to channel 1 and the Q signal to channel 2 provides insight into system health. The gain and position of each channel must be properly adjusted and a signal must also be applied to the scope's Z-axis (intensity modulation) input to kill the scope intensity during the digital state transition times. The result (shown in Figure 8-23) is called a **constellation pattern** because of its resemblance to twinkling stars in the sky.

The constellation pattern gives a visual indication of system linearity and noise performance. Because QAM involves AM, linearity of the transmitter's power amplifiers can be a cause of system error. Unequal spacing in the constellation pattern indicates linearity problems. Excessive blurring and spreading out of the points indicate noise problems.

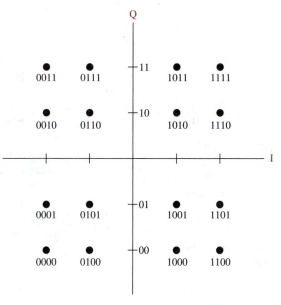

FIGURE 8-23   16-QAM (4 × 4) constellation pattern.

By grouping bits into ever-larger symbols and by combining amplitude modulation with quadrature phase shifts, we gain the ability to transmit large amounts of digital data through limited bandwidth channels. QAM digital transmission systems have become dominant in recent years. Besides the 4 × 4 system introduced here, systems up to 16 × 16 are also commonly used. The 16 × 16 system produces a constellation of 256 symbol points, each associated with an 8-bit symbol. So-called 256-QAM systems represent the current state of the art and are used in digital cable television and some wireless broadband applications.

Offset Modulation

Because QAM schemes are based in part on amplitude modulation, they suffer from some drawbacks associated with any form of AM. In addition to susceptibility to noise, the level shifters in the modulator must be extremely linear to preserve the amplitude levels represented in the transmitted signal. Another problem has to do with the signal trajectory from one symbol to the next: If the transition between symbols passes through the origin, which is the center point in the I/Q representation, the signal amplitude goes to zero for a short time. This condition implies that the RF power amplifiers in the transmitter must be extremely linear. Recall from Chapter 4 that linear (i.e., class A) amplifiers are inefficient. The power consumption required for linearity in such cases is a serious drawback, particularly in portable applications where battery life is an important design consideration. In addition, even the most linear of amplifiers, as a practical matter, exhibit less-than-ideal linearity characteristics at close to zero power output. The more nonlinearities that are present, and the larger the amplitude excursion demanded of the transmitted signal, the more likely the transmitted signal will display unwanted, wide modulation sidebands that must subsequently be eliminated. This undesired phenomenon is closely related to intermodulation distortion (Chapter 6) and is termed **spectral regrowth**. Therefore, it is desirable to have the symbol trajectory (the path taken from one symbol to the next) avoid the origin if at all possible. One way to do this is to offset the symbol transitions in the I and Q channels by one-half symbol. Thus, in an **offset QPSK (OQPSK)** modulation scheme, for a given symbol, the I channel will change state, and then the Q channel will change state half a symbol time later. This arrangement has the effect of causing the trajectories to bypass the origin, thus allowing for the use of amplifiers exhibiting less-than-perfect linearity. OQPSK modulation is used in cellular phones based on the code-division multiple access (CDMA) standard (to be described more fully in Chapter 10).

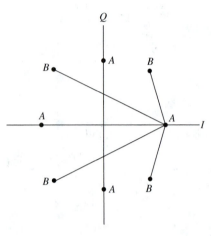

FIGURE 8-24 **Permissible transition states for $\pi/4$ DQPSK modulation.**

Another approach to avoiding the origin is embodied in a variant of 8-PSK known as **pi-over-four differential QPSK ($\pi/4$ DQPSK)**. The $\pi/4$ designation derives from the location of the symbols every 45° (hence, $\pi/4$ radians) around a circle in I/Q space. Because the circle is of constant amplitude, $\pi/4$ DQPSK, like other forms of PSK modulation, need not rely on linear circuits to produce its constellation. Rather, PSK is a constant-envelope form of modulation like conventional FM, allowing for the use of more efficient amplifiers. To avoid the zero-crossing problem, with any given symbol only four of the eight possible locations are used, and the symbol location always changes by an odd multiple of 45°. With reference to Figure 8-24, note that a symbol at any of the four locations marked "A" can transition only to one of the points marked "B." Put another way, if a symbol is on one of the four points located along the I or Q axes, then the next symbol must reside off the axes, and vice versa. This arrangement ensures that none of the possible symbol trajectories (represented in the figure by lines between the symbols) takes a path through the origin.

Another advantage to $\pi/4$ DQPSK is that it is a differential modulation scheme. Like with the DPSK method covered earlier, the differential modulation approach means that the receiver does not need to have a system of coherent carrier recovery for the determination of absolute phase. Rather, the information is derived from the difference in state represented by one symbol with respect to the one immediately preceding it. The simplified receiver architecture that results is ideally suited to large-scale integration and digital signal processing techniques.

8-4 SPECTRAL EFFICIENCY, NOISE PERFORMANCE, AND FILTERING

The widely used forms of digital modulation are FSK, PSK, and QAM, with DPSK, OQPSK, and $\pi/4$ DQPSK representing refinements that address particular shortcomings of the basic formats. As has been the case throughout the study of communications systems, the fundamental constraints are bandwidth and noise. We are now in a position to see how the various approaches stack up in both areas as well as to examine how filtering optimizes the bandwidth characteristics of the transmitted signal.

Spectral Efficiency

Combining bits into multiple-bit symbols improves bandwidth efficiency. Bandwidth efficiency is a measure of how well the modulation scheme handles data within a defined bandwidth. The measurement unit is b/s/Hz (bits per second per Hertz). The simplest modulation schemes, BPSK and MSK, which modulate the carrier with each bit, have a theoretical efficiency of 1 b/s/Hz. As we have seen, this means that the data transmission rate is, at best, equal to the bandwidth. The other modulation formats we have studied trade increased complexity for better bandwidth efficiency. Table 8-2 provides the theoretical bandwidth efficiencies for several modulation schemes shown so far.

TABLE 8-2 • Spectral Efficiencies of Digital Modulation Formats.	
MODULATION FORMAT	**THEORETICAL BANDWIDTH EFFICIENCY (B/S/HZ)**
FSK	<1
BPSK, MSK	1
GMSK	1.35
QPSK	2
8PSK	3
16QAM	4
64QAM	6
256QAM	8

An important caveat is that the bandwidth efficiencies shown in Table 8-2 would only be attainable if the radio had a perfect filter. A perfect filter is one that produces a rectangular response in the frequency domain (i.e., instantaneous transition from passband to stopband). With perfect filtering, the occupied band would be equal to the symbol rate (composed of multiple-bit groups), rather than the number of bits per second. We will see shortly that the choice of filters involves tradeoffs as well, particularly involving bandwidth, output power requirements, and implementation complexity, but for right now it is sufficient to note that the appropriate choice of filtering is one means for a practical system to approach the theoretical efficiencies shown in the table. The presence of the Gaussian modulation prefilter, for example, accounts for a modulation efficiency of 1.35 for GMSK as opposed to 1.0 for unfiltered MSK. Other approaches to maximizing spectral efficiency have also been covered. One is to relate the data rate to the frequency shift, which was done to produce MSK. Another is to restrict the types of permitted transitions, as is seen in OQPSK and $\pi/4$ DQPSK, which also helps to minimize the potential for spectral regrowth because the transmitter amplifiers are least likely to operate in their nonlinear regions.

Noise Performance

Another very important consideration in the selection of a modulation scheme is its performance in the presence of noise. Recall from Figure 8-21 that, in any practical system, the amplitude and/or phase characteristics of the received signal will be affected by noise as well as by nonlinearities and other imperfections of the transmitter and receiver stages. The result is a noisy constellation with the individual symbol states represented on an I/Q diagram by fuzzy "clouds" rather than sharply defined "stars." A decision-making circuit in the receiver is charged with assigning the received signal to the appropriate state. For a QPSK signal, this decision is relatively clear-cut because it largely involves choosing the right quadrant; there is only one symbol point per quadrant. For higher-level modulation schemes where the symbol points are closer together, the decision circuit must be able to distinguish from among several possible amplitude and phase states within each quadrant, a somewhat harder task because the clouds may overlap. Intuitively, then, it would seem that the modulation methods having the fewest symbols (and symbol changes) or the lower spectral efficiencies (that is, smaller number of bits per hertz) would have the best performance in the presence of noise. Largely, this is indeed the case. Let us try to put some numbers to these ideas.

The first issue is how to define the minimum acceptable signal level in a digital system. Recall from Chapter 6 that, in the analog context, receiver sensitivity is a measure of the minimum discernable signal. An example is the 12-dB SINAD specification for an FM receiver, which is an industry-standardized expression of the lowest level at which a received signal would be intelligible in the presence of background noise. What is the minimum acceptable level for a digital system, and how is it measured? The answer to the first part depends on the application, but the measurement is most often expressed in terms of **bit error rate (BER)**. The BER is a comparison of the number of bits received in error for a given number of bits transmitted. (The "given number of bits transmitted" may or may not be the same as the *total* number of bits transmitted. BER is commonly given in terms of the number of bits transmitted per unit of time [which is usually one second] for a type of digital service. The number of errors per total number of bits received is properly expressed as the *error probability, P_e*.) For example, if 1 error bit occurs for 100,000 bits transmitted over the course of 1 s, the BER is 1/100,000 or 10^{-5}. The acceptable BER is determined by the type of service: Voice-grade PCM telephone systems may function well with a BER of 10^{-6}, whereas more demanding applications may require a BER of 10^{-9} or even lower.

The second issue is to relate the energy contained within the signal to the noise energy. Recall that the signal-to-noise (*S/N*) ratio is defined as the ratio of signal power to noise power. Expressed as a ratio, SNR = P_s/P_n. A high SNR is necessary to minimize errors in the detected signal. If the SNR is 1 the signal cannot be recognized. An SNR of 2 means that the noise power is one-half that of the signal. With an SNR of 5 the noise power is only one-fifth the signal power. As the signal increases it is less affected by noise. More commonly for digital systems, the ratio is expressed in terms of the total power in the carrier and sidebands (C) to the noise power (N). The noise power of interest is usually thermal noise, which, recall, is directly proportional to bandwidth. The ratio of carrier-and-sideband power to noise power is called the carrier-to-noise (C/N) ratio and is usually expressed in decibels. A C/N of 20 dB means that the carrier energy is 20 dB higher (100 times the power or 10 times the rms amplitude, assuming constant impedance) than that of the noise.

Figure 8-25 shows how different modulation methods compare in terms of BER and C/N ratio. The results are pretty much as expected. Recall that the system requirements

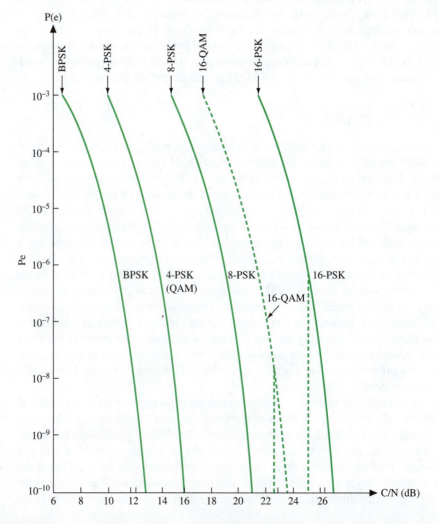

FIGURE 8-25 BER vs C/N ratio for digital modulation formats. Adapted from *Electronic Communication Systems: Fundamentals Through Advanced, 4th ed.,* by Wayne Tomasi, 2001.

determine the maximum number of acceptable errors per second, hence the BER. That is to say, a BER of 10^{-6} may be acceptable for one system but not for another. Using a BER of 10^{-6} as an example, we see from the figure that a BPSK transmission, the required C/N is 11 dB, whereas for 16QAM the C/N would have to be at least 20 dB to achieve the same BER. This result should not be too surprising upon reflection. QAM, like any form of amplitude modulation, is more susceptible to noise in the communications channel than any form of angle modulation, so QAM would have to contain more power to achieve the desired C/N ratio and BER.

Another way to look at the issue is to look at the amount of energy within each bit of data and to compare this energy to the noise power density. The **energy per bit (E_b)**, or **bit energy**, is the amount of power in a digital bit for a given amount of time:

$$E_b = P_t \cdot T_b, \qquad (8\text{-}4)$$

where E_b = energy per bit (in joules/bit)

$\quad P_t$ = total carrier power (watts)

$\quad T_b$ = 1/(bit frequency, f_b).

EXAMPLE 8-5

The transmit power for a cellular phone is 0.8 W. The phone is transferring digital data at a rate of 9600 bps. Determine the energy per bit, E_b, for the data transmission.

SOLUTION

$$\text{bit frequency} = 1/9600 = 1.042 \times 10^{-4}$$

$$E_b = P_t \cdot T_b = 0.8 \cdot 1.042 \times 10^{-4} = 8.336 \ 10^{-5} \text{ J (or } 83.36 \ \mu\text{J)}. \qquad (8\text{-}4)$$

The value for the bit energy (E_b) is typically provided as a relative measure to the total system noise (N_o). Recall from Chapter 1 that the electronics in a communications system generate noise and that the primary noise contribution is thermal noise. The thermal noise can be expressed in terms of noise density, which is the same as N_o, the noise power normalized to a 1-Hz bandwidth. Recall from Chapter 1, Equation (1-8), that noise power is related to Boltzmann's constant, equivalent noise temperature, and bandwidth. In Equation (1-8) the bandwidth was shown as Δf, but this is the same as bandwidth, B. Therefore, N_o can be expressed in terms of noise power and bandwidth:

$$N_o = \frac{N}{B} = \frac{kTB}{B} = kT.$$

Because T_b in Equation (8-4) is equal to $1/f_b$, then

$$E_b = \frac{P_t}{f_b}.$$

The average wideband carrier power, including sideband power, was defined earlier as C. If we reference C to a 1-Hz normalized bandwidth, then C in this case is equal to the total power, P_t. Therefore,

$$E_b = \frac{C}{f_b}.$$

The relationship for the bit-energy-to-noise value is stated as E_b/N_o. From the definitions just given, then,

$$\frac{E_b}{N_o} = \frac{C/f_b}{N/B} = \frac{CB}{Nf_b}.$$

Separating the terms on the right-hand-most side of the preceding equation, we have the following expression:

$$\frac{E_b}{N_o} = \left(\frac{C}{N}\right) \times \left(\frac{B}{f_b}\right). \tag{8-5}$$

This expression simply says that the E_b/N_o ratio is the product of the C/N ratio and the noise bandwidth-to-bit-rate ratio. Its utility lies in the fact that with it the bandwidth occupied by the various schemes does not need to be taken into account, whereas with the previous analysis the bandwidth would need to be taken into account to get a full picture of system behavior. Thus, the same data from Figure 8-25 can be expressed in terms of E_b/N_o, as shown in Figure 8-26. Graphs like that of Figure 8-26 allow for a more direct comparison and contrast of the behavior of various modulation schemes because E_b/N_o for each of the schemes has already been normalized to a noise bandwidth of 1 Hz.

Figure 8-26 also confirms that in digital communication systems, the probability of a bit error (P_e) is a function of two factors, the bit-energy-to-noise ratio (E_b/N_o) and the method used to modulate the data digitally. Basically, if the bit energy increases (within operational limits), then the probability of a bit error, P_e, decreases, and if the bit energy is low, then the probability of a bit error will be high.

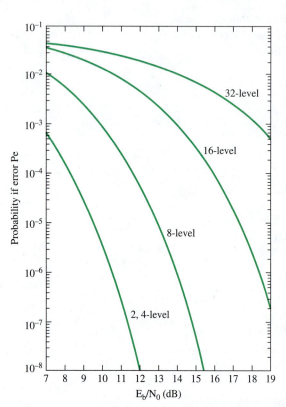

FIGURE 8-26 E_b vs N_o for digital modulation formats.
Adapted from *Electronic Communication Systems: Fundamentals Through Advanced, 4th ed.*, by Wayne Tomasi, 2001.

Filtering

As we have already seen, filtering plays a prominent role in all forms of communications systems. The choice of filter involves a number of tradeoffs, principally pitting spectral efficiency against complexity and, sometimes, the need for additional transmit power. Several filter types have already been introduced. In addition to the analog filters covered in Chapter 4, the other types of filters covered to this point include the sin(x)/x and Gaussian filters, both of which are used extensively in digital communications contexts. Several other filter types are also in use. This section covers these as well as some of the design considerations that must be made.

In digital systems the ultimate goal of filtering is to reduce transmitted bandwidth as much as possible without removing the information content of the data to be transmitted. Fast transitions, such as those represented by square waves or sharp impulses, create high-frequency information that require wide bandwidths. Narrowing the bandwidth involves smoothing out or slowing down the transitions. In addition, filtering is used to prevent signals from interfering with each other, whether these signals emanate from within the system itself or are generated by other systems in the same frequency range.

One important issue in filtering for digital communications systems is that of "overshoot," which is the tendency of signals to go beyond, rather than directly to, the intended target when transitioning from one symbol state to the next. This situation is undesirable because it represents an inefficiency: Additional output power from the transmitter would be required than would be necessary to transmit the symbol without overshoot. Any attempt to reduce or limit power to prevent overshoot would cause the transmitted spectrum to spread out again, defeating the purpose of filtering. Therefore, one tradeoff is determining how much and what type of filtering to use to produce the spectral efficiencies needed without creating excessive overshoot.

Another issue with filtering in digital communications systems is the possibility of ISI, which is the tendency of the symbols to blur together to the point where they are affected by surrounding symbols. Excessive ISI would mean that a receiver could not differentiate one symbol from another. We have seen that one advantage of the Gaussian filter used in GMSK modulation is that it affects only the immediately preceding and subsequent symbols—that is, it holds ISI to a minimum. ISI is a function of the impulse response of the filter, which is an expression of how the filter responds in the time domain.

A final set of tradeoffs has to do with more prosaic concerns. Filters make radios more complex, and, particularly if implemented as analog circuits, can make them larger. For this reason, much if not most filtering in digital systems is implemented through DSP techniques, a topic to which we will return.

RAISED-COSINE FILTER   One widely used type of filter, and one that is often implemented in DSP, is called the **raised-cosine** filter. This type of filter has the important property that its impulse response rings at the symbol rate, which means that it is capable of minimizing ISI.

With reference to Figure 8-27, note that the filter output amplitude periodically goes to zero at the symbol rate, τ. An amplitude of zero at the symbol rate is desirable because it enhances the receiver's ability to detect the symbol state without error. Part of the receiver's job is to sample the detected signal at appropriate intervals and then to decide which binary condition each sampled signal state represents. To do this with minimum probability of error, the receiver must be able to take pulse samples without interference from previous or subsequent pulses. This objective can be achieved in part by ensuring that each pulse shape exhibits a zero crossing at sample points associated with any other pulse (that is, any sample point not associated with the pulse being sampled sees zero energy from all but the intended pulse). If energy from adjacent pulses is present, then errors can possibly be introduced into the receiver's decision-making process. Another important consideration is that the decay outside the pulse interval be very rapid. Rapid decay is desirable because any real system will exhibit some timing jitter, in which the sample point will move from its optimal, zero-crossing position and be affected by adjacent pulses. Therefore, the more rapidly a pulse decays to zero, the less likely it is that timing jitter will cause sampling errors to occur when adjacent pulses are sampled. These considerations are in addition to that of the overall objective: minimizing transmitted bandwidth.

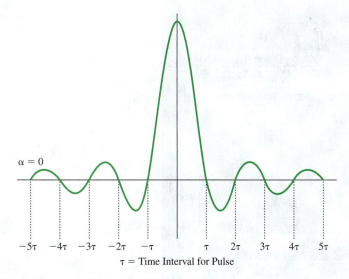

τ = Time Interval for Pulse

FIGURE 8-27   **Spectrum of raised-cosine filter.**

The raised-cosine filter is so widely used because it exhibits the properties just described. In fact, the time-domain impulse response of the filter follows very closely the $\sin(\pi x)/\pi x$ (sinc) function describing rectangular pulses (first introduced in Chapter 7). The sinc function is part of the equation defining the time-domain response of the raised cosine filter. The bandwidth (i.e., sharpness in the frequency domain) of the raised cosine filter is adjustable. A parameter represented as alpha (α), which is the rate at which the tails decay, defines the occupied bandwidth of the system. The relationship is

$$\text{occupied BW} = \text{symbol rate} \times (1 + \alpha). \qquad \text{(8-6)}$$

Alpha varies from 0 to 1. For $\alpha = 0$, the bandwidth equals the symbol rate (the theoretical limit) if the filter exhibits sharp transitions from passband to stopband (sometimes referred to as a *brick-wall* response). For $\alpha = 1$, the occupied bandwidth is twice the symbol rate; therefore, twice as much bandwidth is used for $\alpha = 1$ as for $\alpha = 0$. Note that as alpha becomes greater than zero, the occupied bandwidth increases over the theoretical minimum. For this reason, alpha is sometimes called the *excess bandwidth factor* because it describes the additional bandwidth required as the result of less-than-perfect filtering.

An alpha of zero is impossible to achieve, as is a perfect brick-wall filter, so all practical raised-cosine filters will have an excess bandwidth factor; just how much is a design choice involving a tradeoff between data rate, rate of decay in the tails, and power requirements. A low α implies a higher data rate for a given bandwidth but at the expense of a much slower rate of decay in the pulse tails. A higher α implies an increased bandwidth (or lower data rate, if bandwidth cannot increase) but with rapidly decaying tails in the time domain. For systems with relatively high jitter at the receiver, a higher α may be desirable, even at the expense of transmission throughput.

As for power tradeoff, lower α values create a need for additional power from the transmitter. The reason is shown in Figure 8-28, which shows alpha settings of 1, 0.75, and 0.375, respectively. Overshoot is greater with the lower alpha setting, even as the bandwidth is improved. Note in particular how some trajectories between symbol states loop beyond the edges of the constellation. These circuitous paths, which are a result of

FIGURE 8-28 **Raised cosine filter with alpha settings of 1, 0.75, and 0.375.**

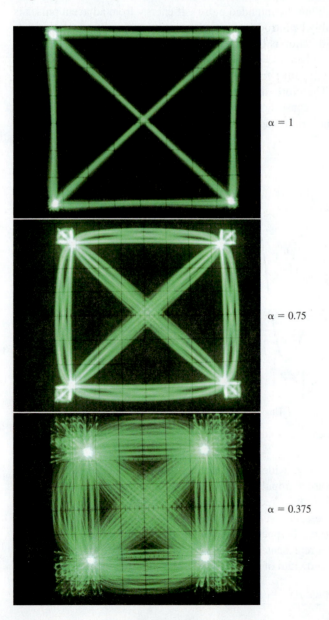

$\alpha = 1$

$\alpha = 0.75$

$\alpha = 0.375$

overshoot in the filter's step response, translate into higher peak powers required of the transmitter amplifiers.

Typical values for alpha range from 0.35 to 0.5, though it is possible to implement an alpha below 0.2, and some video system use alphas near 0.1. Lower alphas reduce required transmitted bandwidth, but they increase ISI, and receivers must compensate by operating in a low-jitter environment (with consequences for specifications such as clock accuracy). Lower alphas also produce more overshoot, thus requiring higher peak carrier powers. The larger amplifiers needed to deliver such powers may be impractical in portable applications for reasons of size, heat generation, power consumption, and electromagnetic interference. For applications such as portable cellular phones, these tradeoffs manifest themselves in characteristics such as portability versus talk time: Smaller and lighter handsets may be limited in their operating time as a result of system constraints, including tradeoffs made in filter design.

Many, if not most, filtering operations for digital communications applications are integrated into large-scale integrated circuits and are implemented as DSP algorithms, particularly for latest-generation and portable implementations. In fact, the operations of I/Q modulation and demodulation are also ideally suited for DSP, as are the wideband modulation schemes of spread spectrum and orthogonal frequency-division multiplexing to be introduced shortly. We will next look at an alternate means of describing the behavior of signals in quadrature, for the model to be developed in the next section will help describe how the functions described up to this point can be implemented as software routines carrying out mathematical operations.

8-5 THE COMPLEX EXPONENTIAL AND ANALYTIC SIGNALS

Another way to look at the I/Q diagrams introduced in the previous section is to view them through the perspective of the complex exponential. With it we develop the idea of an analytic signal, which is a signal with both "real" and "imaginary" components that, taken together, form a complex signal. Though the material presented in this section is not necessary for an understanding of the basics of digital modulation and demodulation, it is enlightening because among other things, it helps to illuminate why a 90° phase shift between the I and Q data streams is important. This discussion will also be beneficial to the understanding of digital signal processing approaches to digital modulation and demodulation. Finally, textbooks and other literature, particularly literature pertaining to DSP, often describe the cyclical behavior of signals in complex exponential form.

Complex Numbers

You may already be familiar with the underlying mathematical concepts of the complex exponential from your study of the properties of impedance. Recall in particular that impedance is an AC quantity (Z) composed of both resistance (R) and reactance (X), and that the reactive components are 90° out of phase (either leading or lagging, depending on whether the reactance is inductive or capacitive) with the resistive component. For impedance-related calculations to be valid, the phase shift between voltage and current occurring in a circuit must be taken into account, and we can use the notion of an imaginary number to do so. To review, an imaginary number, i, is defined as the square root of negative one: $i = \sqrt{-1}$. The term "imaginary" applies in a mathematical sense because *no* real number produces a negative square root. Any negative number, when squared, produces a positive result. However, the ideas embodied by the concept of imaginary numbers can and do apply to very real, physically realizable things such as reactances and out-of-phase signals. The underlying mathematical concept is very useful, though, because it allows us to set up a number line at right angles to the "real" number line—the line associated with numbers we think of in our everyday experience—to represent electrical quantities that are phase-shifted with respect to other quantities. Put another way, if the ordinary number line used to represent real numbers faces East–West, then another number line facing North–South can be established to represent imaginary numbers. It bears repeating that nothing is imaginary about the electrical quantities represented along the North–South number line, but the mathematical tool works because quantities falling along the imaginary line can be shown largely to follow the same rules of arithmetic as do real numbers. Because the letter i is used to represent instantaneous current, in electronics the letter j (the "j operator") is used to represent the square root of negative one, and arithmetic operations involving both real and imaginary quantities are known as *complex* operations. Returning to the impedance example, we see that it is a complex quantity because it can be composed of both resistive and reactive elements. The complex impedance, therefore, takes the form

$$Z = R \pm jX,$$

where R is the resistive (real) quantity, X is the reactive (imaginary) quantity, and $j = \sqrt{-1}$. The j operator—that is, any quantity prefixed with the letter j—indicates quantities along the imaginary (North–South) number line, implying a phase shift from those on the real number line. The j operator, then, can be thought of as a bookkeeping device to allow us to identify and keep separate the sets of quantities and their locations along their respective number lines.

What does this discussion have to do with modulation? Complex impedance represents a familiar instance of how to handle quantities that are related but that are not simply additive. We will see shortly that I/Q modulation can be modeled in much the same way, but in the case of digital modulation the quantities of interest are streams of data that have been phase-shifted by 90° rather than resistances and reactances. The number lines at right angles to each other establish a two-dimensional, or *Cartesian*, coordinate system. Any two-dimensional quantity can thus be represented by a point established within the Cartesian coordinate space. This point is the intersection formed by two lines extending at right angles from the tips of vectors representing the distance travelled along the horizontal and vertical axes, respectively. Returning to the impedance example for a moment, the magnitude of the impedance, Z, is established by the length of a vector originating where the number lines intersect (the origin) and extending to the point just defined. The magnitude is also the hypotenuse of a right triangle whose sides are formed from the vectors established along the real (representing resistance) and imaginary (representing reactance) axes. From the Pythagorean theorem, the magnitude of impedance is equal to the square root of the sum of the squares of resistance and reactance:

$$Z = \sqrt{(R^2 + X^2)}.$$

This idea can be extended to represent the magnitude of any complex quantity. In general, if X represents the quantity along the real axis and Y represents the quantity along the imaginary axis, then the resultant, Z is

$$Z = \sqrt{(X^2 + Y^2)}.$$

The location of any point in Cartesian space can also be described in terms of magnitude coupled with an angle, ω, formed from the positive real axis (the East-facing line) to the hypotenuse. The angle ω is related to the ratio of the lengths of the two sides of the right triangle. The ratio of the side opposite ω to the side adjacent to ω is the *tangent* function in trigonometry, that is,

$$\tan \omega = \frac{Y}{X}.$$

To determine the angle from the ratio, we take the inverse tangent, or *arctangent* ("arctan") function, which is usually represented on scientific calculators as $\tan^{-1}$. Thus,

$$\omega = \tan^{-1}\left(\frac{Y}{X}\right)$$

Therefore, an equivalent way to represent a point described by the function $Z = X \pm jY$, also known as rectangular form, is to specify the point in terms of magnitude and phase angle. This so-called polar notation takes the form $Z \angle \omega$, where the symbol $\angle$ represents the angle ω, usually in degrees. The polar and rectangular forms are equivalent, and conversion between the two, either by hand or with a calculator, is straightforward. The advantage to polar form is that multiplication and division operations involving complex quantities are much easier to carry out than they would be if left in rectangular form.

To see how these ideas apply to digital modulation schemes, let us look at the polar form representation of complex numbers. To do so, we will take a "unit circle," that is, a circle with a radius, R, of 1. As before, any point Z on the unit circle can be expressed either as $\angle\omega$ or in terms of what is known as the complex exponential, which takes the form $e^{j\omega}$, where e is Euler's number (equal to 2.71828 . . .), the base of the natural logarithm. By extension, any point on the complex plane (not just those on the unit circle) can be expressed either as $R\angle\omega$ or in complex exponential form as $Re^{j\omega}$. Again, it is worth recalling that the j operator functions primarily as an aid to organizing two distinct sets of numbers: Whenever you see the j, it reminds you that the quantities associated with it are associated with the imaginary (vertical, or North–South) number line, as distinct from the real quantities residing on the horizontal number line.

What does this complex exponential representation tell us? Let us look at some specific examples, again using the unit-circle representation, $e^{j\omega}$. If $\omega = 0°$, then this corresponds to a point 1 on the positive real axis because anything raised to the zero power equals one. If $\omega = 90°$, then the complex exponential expression appears as e^{j90}. This expression can be evaluated by invoking what is known as Euler's relation, which, in this case, takes the form

$$e^{j\omega} = \cos\omega + j\sin\omega;$$

therefore,

$$e^{j90} = \cos(90) + j\sin(90).$$

In trigonometry, the cosine of a right triangle is the ratio of the adjacent side to the hypotenuse, and, as established previously, the sine is the ratio of the opposite side to the hypotenuse. The cosine of a 90° angle is 0, and the sine of 90° is 1. Therefore, for the preceding expression,

$$e^{j90} = \cos(90) + j\sin(90) = 0 + j \cdot 1 = j.$$

This result means that $\omega = 90°$ places the point Z on the positive imaginary axis (facing North). The point has effectively rotated 90° in a counter-clockwise direction. A similar argument can be made for any other angle. At 180°, the sine expression is 0, but the cosine of 180° is −1, which is on the negative real-number axis (i.e., facing West), so the point has rotated counter-clockwise by yet another 90°. At 270°,

$$e^{j270} = \cos(270) + j\sin(270) = 0 + j(-1) = -j,$$

so this point would be on the negative imaginary axis, or facing South. As ω continues to increase, the point Z continues to move counter-clockwise about the unit circle, and at 360° it has returned to where it started, only to have the process repeat.

What we have just described is a cyclic waveform (sine or cosine wave) expressed as a phasor rotating about the unit circle at an angular velocity expressed in radians per second, where each full revolution about the unit circle is equal to 2π radians. Thus, the angle ω expresses the frequency in terms of angular velocity and, as previously established, is equal to 2π times the cyclic frequency in Hertz, or $2\pi f$.

Analytic Frequency

The complex exponential allows us to picture the notion of analytic or "complex" frequency by showing that any waveform is composed of both sine and cosine waves, and the presence of the j operator implies that there is a 90° phase shift between them. Thus, the cosine wave can be thought of as being on the "real" number plane, and the sine wave on the "imaginary" number plane. As an example, the complex representation of a cosine wave (which looks like a sine wave but shifted in phase by 90°) is

$$x(t) = \cos\omega t = \frac{1}{2}[(\cos\omega t + j\sin\omega t) + (\cos\omega t - j\sin\omega t)].$$

The preceding result implies that the cosine wave is the sum of two complex signals, where the left-hand term produces a positive result and the right-hand term produces a negative result. In other words, the signal is composed of both positive and negative frequencies. The expression $\omega(t)$ is associated with a positive frequency, that is, one in which the phasor is rotating counter-clockwise, as just established. What about $-\omega(t)$? The expression $-\omega(t)$ is equal to $\omega(-t)$, so, in a manner of speaking, the expression is telling us that time is running backward. Also, $\cos(-\omega t) = \cos(\omega t)$. A negative frequency, expressed as $-\omega t$, is associated with a *clockwise* rotation around the unit circle.

The cosine wave, then, can be thought of as being composed of two phasors rotating in opposite directions. Two equal-amplitude signals, one of positive frequency (counter-clockwise rotation) and the other of negative frequency (clockwise rotation), when combined, will produce a cosine wave of twice the amplitude of either signal individually because the sine portions cancel:

$$x(t) = [\cos(\omega t) + \sin(\omega t)] + [\cos(-\omega t) + \sin(-\omega t)] = 2\cos(\omega t).$$

The preceding result predicts that a cosine-wave signal in the I/Q plane will always lie on the I axis and will have twice the amplitude of either signal alone; that is, it will oscillate sinusoidally between $+2$ and -2. This result may not be immediately apparent from the preceding equation, but remember that the sine and cosine waves are always $90°$ out of phase with each other. Therefore, when $\cos(\omega t) = 1$, $\sin(\omega t) = 0$. Ninety degrees or one quarter-cycle later $\cos(\omega t) = 0$ and $\sin(\omega t) = 1$. You can imagine the I signal cosine wave moving back and forth in a sinusoidal fashion along the I plane, and the Q signal sine wave doing likewise along the Q plane.

The preceding discussion also helps us to develop the notion of an *analytic signal,* one in which the amplitude and phase (that is, I and Q) are considered to be separate from the portion that oscillates (usually the carrier). The oscillating portion can be expressed by the complex exponential $e^{-j\omega t}$, and, as before, the I/Q modulation can be expressed as a complex number, x, of the form $I + jQ$. As always, modulation is a multiplication process, so the complete modulated signal, $x(t)$, takes the following form:

$$x(t) = x \cdot e^{-j\omega t} = (I + jQ)(\cos(\omega t) - j\sin(\omega t)),$$

which, when multiplied, becomes

$$x(t) = [(I\cos(\omega t) + Q\sin(\omega t)) + j(Q\cos(\omega t) - I\sin(\omega t))]. \tag{8-7}$$

The significance of the preceding expression is in the first two terms: $I\cos(\omega t) + Q\sin(\omega t)$. These represent the "real" (i.e., without phase shift) or scalar part of the modulated signal. The scalar part is what would be viewable on an oscilloscope.

Though the concept of negative frequency may seem like something of a mathematical artifice, it does help explain a number of results with which we are already acquainted, among them the production of side frequencies as the result of modulation. As has been established, a single sinusoidal signal, $\cos(\omega t)$ can be considered to contain two frequencies, $+\omega$ and $-\omega$. These two frequencies, when "unfolded," will reside above and below the carrier; they do so because the upper and lower frequency components are already present in the modulating signal. To demonstrate this point, we can express a cosine-wave modulating signal as a complex exponential using Euler's identity:

$$\cos\omega t = \frac{1}{2}(e^{j\omega t} + e^{-j\omega t}).$$

When mixed with a carrier of the form $\cos\omega_0 t$, the result is the product of the two inputs:

$$(\cos\omega_0 t)(\cos\omega_0 t) = \left[\frac{(e^{j\omega_0 t} + e^{-j\omega_0 t})}{2}\right]\left[\frac{(e^{j\omega t} + e^{-j\omega t})}{2}\right]$$

$$= \frac{[e^{j(\omega_0+\omega)t} + e^{-j(\omega_0+\omega)t}] + [e^{j(\omega_0-\omega)t} + e^{-j(\omega_0-\omega)t}]}{4}$$

$$= \frac{1}{2}[\cos(\omega_0 + \omega)t + \cos(\omega_0 - \omega)t].$$

Notice the result. The multiplication of the two cosine waves produces sum and difference frequencies. In other words, the positive and negative frequencies, when "translated up," produce two positive frequencies separated in the frequency domain by twice the original signal frequency. In addition, the amplitude of each is one-half that of the original amplitude. Recall from Chapter 4 that this is precisely the result predicted by a balanced mixer.

The concept of analytic frequency is somewhat abstract, especially when the idea of negative frequency is invoked, but the analytic frequency concept is particularly useful in seeing how digital modulation and demodulation can be carried out in DSP. Here are the most important points so far:

- A sine or cosine wave can be thought of as consisting of both positive and negative frequencies.
- The complex exponential is another way of representing a signal in phasor form.
- Digital modulation relies heavily on the notion of phase-shifted data, and this can be modeled in the form of signals on the real and imaginary (I/Q) planes.
- Digital signal processing techniques make use of the complex exponential representation of frequency to carry out the modulation and demodulation functions required by digital communications systems.

DSP Modulation and Demodulation

The modulation of an I/Q signal, like analog modulation, is a process of frequency up-conversion, which involves moving the baseband frequency range to the higher frequency range of the carrier. The signal must ultimately be converted to the analog domain for transmission, but in modern systems the digital-to-analog (DAC) conversion can happen after up-conversion thanks to the availability of high-speed DAC converters. The complex baseband signal (I/Q data streams in quadrature) is multiplied by a carrier of the desired frequency, which is represented by a complex exponential, $e^{j\omega t}$, as shown in Figure 8-29.

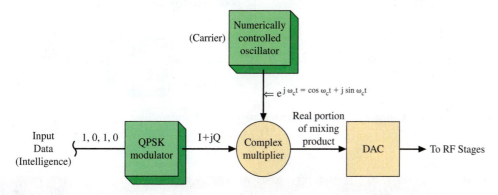

FIGURE 8-29   Complex exponential representation of I/Q modulation.

The carrier can be created with a numerically controlled oscillator (NCO), which was first introduced in the Section 4-5 discussion of direct digital synthesis. The NCO will produce a complex carrier—that is, one with two sampled sinusoidal signals that are offset by 90°. This complex signal takes the form $\cos \omega_c t + j\sin \omega_c t$, as established previously through the use of Euler's identity to resolve the signal $e^{j\omega t}$. If the complex carrier is multiplied by a complex (I/Q), sinusoidal modulating (intelligence) signal of the form $\cos \omega_i t + j\sin \omega_i t$, then the result, after simplification, can be shown to be

$$\cos[(\omega_c + \omega_i)t] + j\sin[(\omega_c + \omega_i)t]. \qquad (8\text{-}8)$$

Note that this is the same result predicted by Equation (8-7). The first part in brackets is the real (scalar) portion. The imaginary portion ($j\sin$ . . .) of this result is not needed because, once the signal has been shifted to the carrier frequency range, the baseband frequency components appear above and below the carrier. The imaginary portion can, therefore, be discarded. The scalar portion is sent to a DAC, which produces the final output for transmission.

Looking at this result from another point of view, remember that the baseband signal has both positive and negative frequency components. This can be visualized for a QPSK signal by remembering that the modulator output vectors rotate both clockwise and counterclockwise, depending on the input data. The result of modulation is both to up-convert the signal and to "unfold" it so that the positive and negative baseband components lie on either side of the carrier rather than overlie each other. The total bandwidth is doubled, as expected, but it is in the form of quadrature signals, with the I and Q signals each occupying the frequency spectrum of their respective unmodulated baseband signals. In short, the I/Q modulator takes the analytic baseband signal (I + jQ) and converts it to a scalar signal at the carrier frequency. The positive and negative frequencies of the I/Q frequency spectrum are up-converted and centered on the carrier frequency, with counter-clockwise-rotating (positive) frequency components appearing above the carrier frequency and clockwise-rotating (negative) components appearing next.

Implementations of modulators and demodulators as DSP functions avoid all the problems of analog circuitry. Analog-to-digital (ADC) converters can convert analog I/Q streams to digital bit streams. Mixers, oscillators, and other circuit elements can be implemented digitally. For a demodulator, such as that shown in Figure 8-30, the modulation process is essentially reversed. In the most modern implementations the incoming radio-frequency signal will be immediately digitized with an ADC converter and applied to a complex multiplier, in which the same complex mixing process described for modulations is carried out to recover the scalar I/Q stream for subsequent delivery to a demodulator/pulse shaper subsystem.

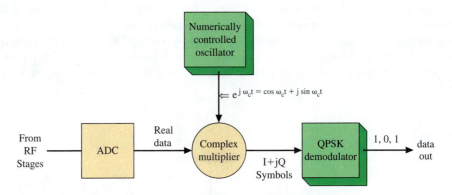

FIGURE 8-30   **Complex exponential representation of I/Q demodulation.**

8-6 WIDEBAND MODULATION

The emphasis in our analysis up to this point has been on conserving and using bandwidth wisely. Hartley's law allows us to predict the minimum bandwidth required for information transfer in a given time interval. Channel capacity considerations influence in no small way the type and complexity of modulation employed. Also, for all the modulation approaches studied so far, the transmitted signal power was seen to be concentrated within the spectrum immediately surrounding the carrier. It may seem, therefore, that any scheme occupying a wider bandwidth than the absolute minimum necessary would be inherently useless. In fact, though, quite the opposite is true: As counterintuitive as it may seem at first, definite advantages accrue in some cases when the rules for bandwidth conservation are deliberately flouted and transmissions occupy much *more* than the minimum bandwidth necessary for information transfer. Among other advantages, multiple users can make use of a given portion of the frequency spectrum at the same time without interfering with each other. It is also possible, in some cases, to extract signals from background noise; in other words, in some cases the receiver can perform adequately even with a negative *S/N* ratio. Such wideband modulation schemes are also the basis of latest-generation mobile communications standards that promise data capacities that would otherwise be impossible to envision, much less achieve. Thanks to advances in DSP and high-speed ADCs and DACs, both spread-spectrum communications and orthogonal frequency-division multiplexing (OFDM) are becoming more widely used even in low-cost, mass-market applications such as wireless access points and smart phones. In addition, wideband modulation is helping to revolutionize the way designers think about the limits posed by the fundamental constraints of bandwidth and noise.

Spread-Spectrum Techniques

A **spread-spectrum** scheme is one in which the occupied bandwidth is deliberately made to be much greater than that specified by Hartley's law. In other words, at the transmitter the signal energy is spread out from the carrier over a much wider bandwidth than it would otherwise need to be for the information to be transmitted without an excessive number of errors. The converse to the spreading operation happens at the receiver. The energy of the received signal is *de-spread*, or recollected and returned to its original, narrowband form for subsequent recovery of the information in the usual way.

The two basic forms of spread spectrum are *frequency hopping* and *direct sequence*. Both forms accomplish the goal of spreading out the transmitted energy over a wide bandwidth, but they do so in somewhat different ways. Though the two forms can be (and sometimes are) used together, they represent distinct approaches to performing the spreading task. In addition to the advantages of multiple simultaneous use and perform- ance in the presence of noise mentioned earlier, another major attribute of spread spec- trum communications is secrecy and resistance to jamming. In fact, these important qualities, coupled with the relative complexity of spread spectrum transmitters and receivers, meant that, for many years, spread-spectrum technologies were largely reserved for military applications. Spreading the information over a wide bandwidth produces transmissions that are resistant to jamming (i.e., being rendered useless by a counter "noise" signal at the same frequency) and enemy detection, characteristics that are critical for defense-related applications.

Military applications are still widespread, but spread spectrum technology became mainstream for consumer use after 1985, when the Federal Communications Commis- sion (FCC) adopted rules allowing for the unlicensed operation of spread-spectrum- based devices at transmitter powers up to 1 W in three frequency bands: 902–928 MHz, 2400–2483.5 MHz, and 5725–5850 MHz. These are designated the **industrial, scien- tific, and medical (ISM)** bands, and devices that are authorized by law to emit radio- frequency energy without the requirement for a station or operator license are often required to operate within the ISM bands, whether they are based on spread-spectrum technologies or not. Very low-power emitters of radio-frequency energy such as garage- door openers, wireless headsets and microphones, and wireless computer networking equipment, are also sometimes referred to as **Part 15** devices, so named after the sec- tion of the FCC regulations governing unlicensed operation of low-power devices. Part 15 of the FCC Rules now applies to "unintentional radiators," such as computers and other electronic devices that emit signals into the surrounding environment as a byprod- uct of their normal operation. Such undesired emissions fall under the general category of electromagnetic interference (EMI), and an extensive regulatory framework has been established to limit the maximum permissible levels of such emissions to prevent destructive interference. These rules are very strict, and electronic equipment intended for import or sale in the United States must often be subjected to a series of regulatory compliance tests to ensure compliance with Part 15 before they can be legally offered for sale, even if the transmission of radio-frequency energy is not their primary purpose or application.

Spread spectrum has become the technology of choice for many wireless applica- tions, including cellular telephones, mobile Internet, wireless local-area networks (WLANs), and automated data collection systems using portable scanners of universal product code (UPC) codes. Other applications include remote heart monitoring, industrial security systems, and very small aperture satellite terminals (VSAT). Multiple spread-spectrum users can coexist in the same bandwidth if each user is assigned a different "spreading code," which will be explained subsequently.

Regular modulation schemes tend to concentrate all the transmitted energy within a relatively narrow frequency band. Noise will degrade the signal, and the concentration of energy makes the signal vulnerable to jamming. This single-frequency band also allows detection by undesired recipients, who may use direction-finding techniques to track down the signal source. Both forms of the spread-spectrum solution to these problems, frequency hopping and direct sequence, make use of what is known as a **pseudonoise (PN)** spreading code to distribute energy over a wide frequency spectrum in an apparent random fashion. At the same time the spreading code allows for multiple users to share the same frequency spectrum without causing destructive interference because transmissions associated with all spreading codes except the one to which a receiver is programmed to respond are ignored and appear as background noise. The background noise level can simply be disregarded provided it is not too high.

The operation of the spreading code is perhaps best described by an analogy to everyday experience. Imagine you are at a business meeting or social gathering with a number of other people. You strike up a conversation with an associate near you, and all other attendees do likewise with those nearest them. The conversations all happen in the same geographic space and over the same frequency spectrum (i.e., voice frequencies). The conversations can all co-exist as long as nobody gets too loud or the gathering doesn't get so crowded that the background noise level—that is, the sum total of all the other conversations going on at the same time—climbs to the point that nobody can make out what anyone else is saying. The unique "frequency signature" of each person's voice (along with visual cues) allows participants to focus on those with whom they are interacting and to ignore what is going on in the background. In this case, the frequency characteristics that make each speaker's voice unique among all the others in the room serve as a sort of spreading code that makes it possible for a listener to identify and concentrate on what one person has to say to the exclusion of all the others.

Imagine a similar get-together at the United Nations, where pairs of conference participants carry on conversations in different languages. Now (and assuming you aren't distracted by the novelty of the languages being spoken around you) you can easily envision how *language* acts as a spreading code because you understand, and will only listen to, those who speak your language. You simply "tune out" all the other languages in the room. You might even be able to do this if the background noise were high enough to prevent intelligible communications in a single language because you seek out and concentrate on the unique characteristics of your own language, even if there is a great deal of background noise in the form of languages you don't understand. The spreading code in a spread-spectrum system is the unique language that allows receivers to listen only to the intended message and to ignore all others, even if there are many others occupying the same frequency spectrum at the same time.

CODE GENERATION Spreading codes are used in both types of spread spectrum to give the transmitted information a distinct language. For frequency hopping spread spectrum, the spreading code defines how the carrier is switched from one frequency to the next. For direct sequence spread spectrum, the code performs the carrier spreading operation directly by first modulating the carrier with an apparently random and high-speed bit pattern and then by modulating the bit pattern with the much-lower speed information signal. We will examine both types of spreading operation in more detail shortly, but because the spreading code idea is common to both forms of spread spectrum, we will first look at how to go about generating it.

Spreading codes in spread-spectrum systems are usually based on the use of a stream of data bits displaying pseudonoise (PN) characteristics, which means that the bits appear random (and hence noiselike) but form a repeating pattern, albeit one that may take hours or days to repeat. The structure and pattern are such that the receiver can recognize and reconstruct the transmitted data stream. Recall from Chapter 1 that one of the characteristics of noise is that it is broadband; that is, it is present at all frequencies. By having the random data streams **spread** the RF signal, it appears as noise to an outside observer (such as a potentially eavesdropping receiver) because the carrier is spread over a range of frequencies in an apparently random fashion. How often the pattern repeats depends on the PN sequence generating circuit.

The method for implementing PN codes is quite simple. It requires the use of shift registers, feedback paths, and XOR gates. A random pattern of 1s and 0s are output as the circuit is clocked. The data pattern will repeat if the circuit is clocked a sufficient number of times. For example, the circuit for a 7-bit PN sequence generator is shown in Figure 8-31. The 7-bit PN sequence generator circuit contains three shift registers, an XOR gate, and feedback paths. The circuit structure and the number of shift registers in the circuit can be used to determine how many times the circuit must be clocked before repeating. This is called the **PN sequence length**. The equation for calculating the length of the PN sequence is provided in Equation (8-9).

$$\text{PN sequence length} = 2^n - 1 \qquad \text{(8-9)}$$

where $n =$ the number of shift registers in the circuit. A PN sequence that is $2^n - 1$ in length is said to be of **maximal length**. For example, the PN sequence generator shown in Figure 8-31 contains three shift registers ($n = 3$) and a maximal length of $2^3 - 1 = 7$. An example of using Equation (8-9) to determine the PN sequence length is provided in Example 8-6.

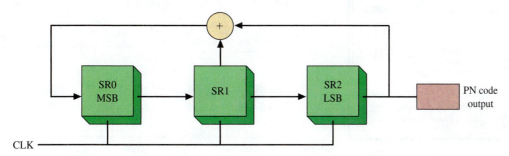

CLK

FIGURE 8-31   **A 7-bit sequence generator.**

EXAMPLE 8-6

Determine the sequence length of a properly connected PN sequence generator containing

(a) 3 shift registers ($n = 3$).
(b) 7 shift registers ($n = 7$).

SOLUTION

(a) $n = 3$, PN sequence length $= 2^3 - 1 - 8 - 1 = 7$ ˘ (8-9)

(b) $n = 7$, PN sequence length $= 2^7 - 1 = 127$ (8-9)

An implementation of a 7-bit PN sequence generator circuit using National Instruments Multisim is shown in Figure 8-32. The circuit consists of three shift registers using 74LS74 D-type flip-flops (U1A, U1B, and U2A). The system clock drives each shift register. The Q outputs of U1B and U2A are fed to the input of the XOR gate U3A (74LS86). The circuit is clocked $2^n - 1$ times (seven times) before repeating. The data is output serially, as shown in Figure 8-33. Notice that the PN code output data stream repeats after seven cycles. The reason the PN circuit repeats after $2^n - 1$ times is that an all zero condition is not allowed for the shift register because such a condition will not propagate any changes in the shift-register states when the system is clocked.

FIGURE 8-32
Implementation of a 7-bit PN sequence generator using National Instruments Multisim.

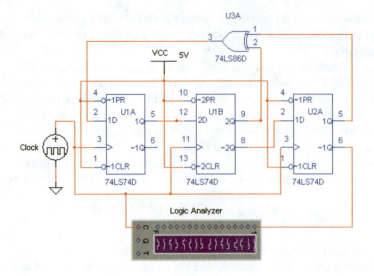

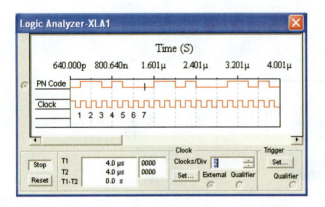

FIGURE 8-33 Serial data output stream for the 7-bit PN sequence generator.

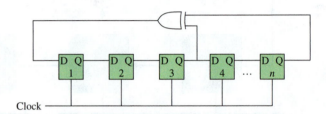

Clock

FIGURE 8-34 PN sequence generator circuit structure for the connections listed in Table 8-3. The circuit shown is connected with five shift registers ($n = 5$) and a maximal length of 31.

The circuits required to implement PN sequences are well known. Modern spread-spectrum circuits have the PN sequence generators integrated into the system IC. Table 8-3 lists the connections required to create different sequence lengths for the circuit structure provided in Figure 8-34. The table shows only a partial listing of some of the shorter PN sequence lengths. Many spread-spectrum systems use a PN sequence length that is much longer. The PN circuit shown in Figure 8-34 contains five shift registers ($n = 5$). Table 8-3 shows the XOR inputs for the circuit that come from the Q outputs of the shift registers.

TABLE 8-3 • Connections for Creating PN Sequence Generators		
NUMBER OF SHIFT REGISTERS (*N*)	**SEQUENCE LENGTH**	**XOR INPUTS**
2	3	1, 2
3	7	2, 3
4	15	3, 4
5	31	3, 5
6	63	5, 6
7	127	6, 7
9	511	5, 9
25	33,554,431	22, 25
31	2,147,483,647	28, 31

FREQUENCY-HOPPING SPREAD SPECTRUM   The first type of spread-spectrum system developed, and probably the easier of the types to understand conceptually, is frequency-hopping spread spectrum. Its genesis dates back to World War II, though its first practical implementations were in military communications systems starting in the late 1950s. In **frequency-hopping spread spectrum** the information signal modulates a carrier that is switched in frequency in a pseudorandom fashion. *Pseudorandom* implies a sequence that can be re-created (e. g., at the receiver) but has the properties of randomness. The time of each carrier block is called **dwell time**. Dwell times are usually less than 10 ms. The receiver knows the order of the frequency switching, picks up the successive blocks, and assembles them into the original message. Spreading occurs because the carrier is rapidly moved to different frequencies within a wide bandwidth rather than being kept at a single frequency.

A block diagram for a frequency hopping system is provided in Figure 8-35. Essentially identical programmable frequency synthesizers and hopping sequence generators are the basis for these systems. The receiver must synchronize itself to the transmitter's hopping sequence, which is done via the synchronization logic.

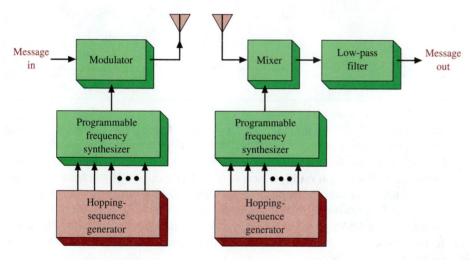

FIGURE 8-35   **Frequency-hopping spread spectrum.**

A simplified picture of the RF spectrum for a frequency hopping signal is provided in Figure 8-36. This image shows the shifting of the carrier frequency over a 600 kHz range. Note that seven frequencies are displayed. A PN code sequence generator generates the pattern for the shifting of the frequencies. In this case, the 3-bit contents of the PN sequence generator shown in Figure 8-31 were used to generate the frequency-hopping sequence. The current state of the shift registers (SR0–SR2) randomly select each frequency, which requires that the PN sequence generator be initiated with a seed value.

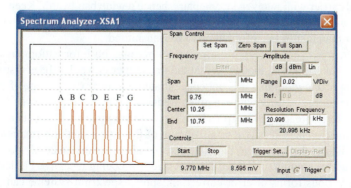

FIGURE 8-36   **Simplified RF spectrum for a frequency-hopping spread-spectrum signal.**

The first step for configuring the PN sequence generator is to initialize the shift registers with a known value. In this case, the sequence generator is seeded with a 1 0 0 value. This step is shown in Table 8-4, which lists the shift register contents as the generator is clocked. At shift 0, the 1 0 0 contents select frequency D. At shift 1, the shift register contents of 0 1 0 select frequency B. This repeats for the length of the PN sequence generator and produces the seven unique states and frequencies A–G. The frequencies are randomly selected until the sequence begins repeating on shift 7, as shown with the shift register contents of 1 0 0 and the selection of frequency D.

TABLE 8-4 • The Shift Register Contents for a PN Sequence Generator with $N = 3$ and a Seed Value of 1 0 0

SHIFT NUMBER	SR0	SR1	SR2	FREQENCY
0	1	0	0	D
1	0	1	0	B
2	1	0	1	E
3	1	1	0	F
4	1	1	1	G
5	0	1	1	C
6	0	0	1	A
7	1	0	0	D

DIRECT-SEQUENCE SPREAD SPECTRUM   The second major form of spread spectrum is **direct-sequence spread spectrum (DSSS)**. DSSS is widely used in many familiar applications, among them wireless access points, and is the basis for cellular systems employing the **code-division multiple access (CDMA)** standard. If you have ever wondered how so many coffeehouse patrons can use their laptops simultaneously to access the store's wireless Internet connection without interfering with each other, part of the secret is the large role that DSSS plays in the various wireless networking standards. The standards themselves will be covered in more detail in Chapter 10, but the basic concepts of DSSS will be described here.

The radio-frequency carrier does not hop around to different frequencies in a DSSS system as it does for frequency hopping. Instead, the fixed-frequency carrier is modulated with a high-data-rate, pseudorandom spreading code, and it is the spreading code that is modulated with the intelligence, which is at a much lower data rate. The spreading code is *not*, in and of itself, the information signal. Rather, and returning to the United Nations analogy for a moment, spreading codes are the languages the receiver would be able to recognize to distinguish one "conversation" from another. Recall the Hartley's law statement that bandwidth is directly related to data rate. Higher data rates imply higher bandwidth signals. DSSS works on the principle that an RF carrier modulated with a high-bit-rate spreading code will occupy a much wider bandwidth than it would if it were modulated by the lower-bit-rate information signal directly. Put another way, the spreading code creates a modulated carrier with many sideband pairs, hence a wide bandwidth. Keep in mind that the spreading code consists of a known and repeating sequence of bits and with a bit rate much higher than that of the information to be transmitted.

At first, this exercise may seem pointless because no information has been transferred. What has happened, though, is that the transmitter power has been distributed over a wide enough bandwidth that it is not concentrated within any one range of frequencies; therefore, the energy density within a given frequency range is much lower than it would have been before spreading. The spreading has implications for security and resistance to destructive interference: If only a small amount of the transmitted information is resident within any slice of the spectrum, then interference affecting that slice of spectrum would not be destructive enough to prevent the message from being received. Particularly when paired with various forms of error correction, spread spectrum systems can provide for reliable communication in high-noise environments.

The binary data representing the intelligence (the information signal), like all binary data, can take one of two possible states: 1 or 0. These data are combined with the PN spreading code. The usual technique is to invert all the spreading-code bits when the intelligence bit is a 1 and to leave them uninverted when the intelligence bit is a 0. This task can be accomplished with an XOR gate into which both the spreading code and the serial data stream are inputs, as shown in Figures 8-37 and 8-38. Note that, if the information-bit input to the XOR gate is a 1, then the output of the XOR is an inverted form of the spreading-code bits (i.e., all the 1 bits of the spreading code are changed to 0, and vice versa). If the information-bit input to the XOR is a 0, the spreading code is passed unchanged to the XOR output. Effectively, then, the intelligence data bits are multiplied by the much higher-bit-rate spreading code bits, which causes the data to be chopped into the much shorter increments of time associated with each bit of the spreading code. These smaller time increments are called *chips,* and the bit rate associated with the spreading sequence is termed the *chip rate* or *chipping rate.* The output of the XOR gate, which is a bit stream at the chipping rate because it is composed of the data bits combined with the chips, then becomes the input to a PSK modulator. Typically, a balanced modulator (mixer) is used to produce a BPSK signal, though other forms of modulation can be used as well. The modulator output is a sine wave with 180° phase shifts occurring at the pulse edges of the input bit stream. These phase shifts are evident in the time domain, as shown in Figure 8-38. Also recall that a balanced modulator produces sidebands without a carrier, which has implications for recovery of the information in the receiver.

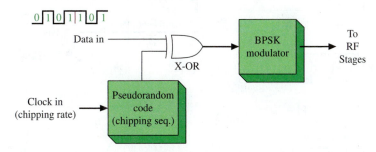

FIGURE 8-37 **Direct-sequence spread-spectrum transmitter.**

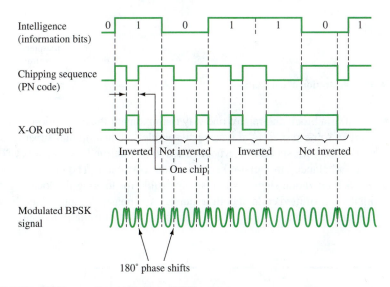

FIGURE 8-38 **XOR operation to produce DSSS output.**

At the receiver the wideband signal must be de-spread or returned its original, narrowband state and the information signal recovered. The receiver generates the same PN code used as a spreading code at the transmitter to de-spread the signal. To do so in a DSSS system the receiver must undertake a comparison process called *correlation* to match the locally generated PN code with the transmitted code. Correlation takes place in a balanced modulator (a mixer), whereby the received signal is multiplied with the receiver-generated PN code. The output of the correlation mixer will be large when the codes match and very small or zero at all other times.

Figure 8-39 illustrates conceptually why the correlator produces the highest output only when the two PN codes are equal. Assume for the purposes of illustration that a bit that has not been phase shifted has an amplitude of +1 and that a bit with a 180° phase shift has an amplitude of −1. Assume also that the DSSS system produces a 7-bit PN sequence for each information bit. From the previous discussion, then, the group of 7 PN bits was inverted if the single data bit was a 1 and left uninverted if the bit was a zero. If both the received bit sequence and the locally generated sequence are fed to a multiplier, the maximum output (all +1 values) will only occur when all the +1 values and the −1 values are aligned (remember that two negative numbers multiplied together produce a positive result). The result will be less than maximum if one bit stream is offset from the other so that at least some positive values are multiplied by negative values, or vice versa.

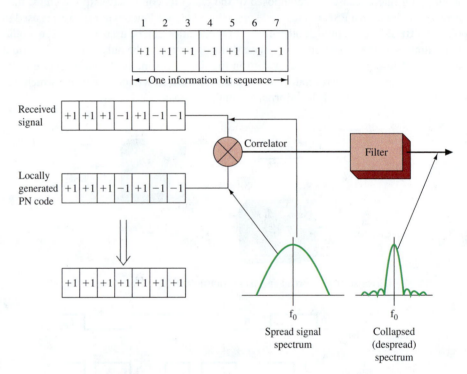

FIGURE 8-39   Illustration of correlation.

Figure 8-40 is a block diagram of one way to implement the correlation function in a DSSS receiver. In some respects the receiver resembles a conventional superheterodyne, in which the incoming RF signal is downconverted to an intermediate frequency (IF). The IF is applied to a balanced modulator serving as the correlator. The correlator output is applied to a synchronization stage, which is responsible for locking the locally generated PN code with the transmitted code in both frequency and phase. Although the PN codes of the transmitter and receiver are equal, the correlator would not initially recognize them as

FIGURE 8-40

Autocorrelation in DSSS receiver.

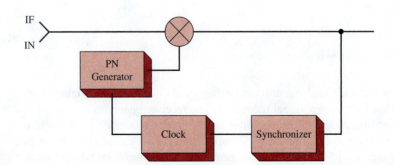

such because they are initially out of phase. The synchronizer works by varying the clock frequency of the local PN code generator, thus causing the code frequency to vary. By slightly speeding up and slowing down the clock driving the receiver PN generator, the synchronizer finds the point where the two codes are exactly equal. At this point, the correlator output is maximum for the reason illustrated in Figure 8-39. You can think of the synchronizer's job as being that of "sliding" the locally generated PN code back and forth until each bit polarity is in perfect alignment.

For DSSS systems employing BPSK modulation, the *spread bandwidth,* which is the bandwidth occupied by the modulated RF carrier, is approximately equal to the chip rate. (Recall from the earlier discussion that BPSK has a theoretical spectral efficiency of 1 b/s/Hz.) The unfiltered spectrum of the spread signal has a sinc or $\sin(x)/x$ function, as shown in Figure 8-41. In a practical BPSK transmitter, a bandpass filter is generally employed to remove all but the main lobe of the transmitted signal, which causes the transmitted bandwidth to approach that of the theoretical limit.

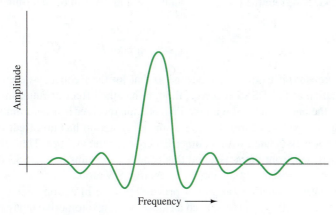

FIGURE 8-41   Sinc function.

The power of DSSS systems is largely in the spreading and dispreading of the received signal because it gives rise to a phenomenon known as *processing gain,* which helps to improve the *S/N* ratio at the receiver. Both types of spread spectrum exhibit processing gain, but it is a particularly powerful concept for DSSS systems because, with it, the receiver can perform even in the presence of a negative signal-to-noise ratio. That is, when spread, the overall signal level can reside below the noise floor, but the receiver can detect the desired PN code and still demodulate the intelligence, even if the overall noise level is greater than the level of the signal after spreading, a truly counterintuitive result. In effect, the DSSS receiver is doing what participants in that noisy United Nations conference do when they pick out and listen for the unique characteristics of the languages they are attuned to understand. By "correlating" their understanding of the characteristics of the language they are concerned with to all the languages they are hearing, and then by focusing only on their desired language when a match is found, the conference participants extract the desired message content from the din of the background noise.

In a frequency-hopping spread-spectrum system, the processing gain is defined in decibels as

$$G = 10 \times \log\left(\frac{\text{spread bandwidth}}{\text{unspread bandwidth}}\right). \qquad (8\text{-}10)$$

The unspread bandwidth is generally equal to the data rate—that is, a BPSK transmission of 20 kbps before spreading would have a modulated bandwidth of around 20 kHz. The spread bandwidth for a frequency-hopping system would be the available range of carrier frequencies. However, recall that noise is directly proportional to bandwidth, so the increase in *S/N* ratio predicted as a result of spreading gain is exactly offset by the reduction in *S/N* resulting from the increase in bandwidth. There is no net advantage in *S/N* ratio for the spread versus the unspread versions of the same signal. Because the frequency-hopping receiver moves from one carrier frequency to the next in synchronization with the transmitter, the receiver will have a narrow bandwidth in its front-end stages, as would a conventional receiver. The frequency-hopping receiver does not need to pick up signals from a broad range of frequencies at the same time, as would its DSSS counterpart. For the frequency-hopping receiver, even as the receiver frequency is made to change as the result of the frequency-hopping spreading code, the bandwidth of its front end is made narrow enough to pass only the carrier and sidebands associated with that frequency position at that instant. For a frequency-hopping system, then, the preceding statement leads to the conclusion that the S/N ratio of the received signal must still be positive; that is, the signal level must be above the noise floor at each carrier frequency. The PN codes for other frequency-hopping transmitters would still appear as random noise to the receiver, to which it would be insensitive.

In contrast with the frequency-hopping receiver, the DSSS receiver must have a front-end bandwidth equal to that of the fully spread signal, consisting of upper and lower sidebands, and this characteristic has implications for performance in the presence of high noise levels. As established, the modulated signal has a bandwidth at least equal to the chip rate, so for a DSSS system the processing gain, again in dB, can be determined as

$$G = 10 \times \log\left(\frac{\text{chip rate}}{\text{bit rate}}\right). \tag{8-11}$$

This gain relationship is essentially the same as that for the frequency-hopping case. The dispreading action at the DSSS receiver, though, has the effect of returning the desired signal (that is, the one associated with the PN code the receiver is programmed to receive) to its original, narrowband form, but the dispreading action has no effect on the noise, including the noise associated with all other PN codes in the vicinity. Thus, the signal can be lifted or extracted from the noise, and this behavior explains how DSSS receivers can perform even in the presence of an apparent negative *S/N* ratio.

A particularly dramatic example of processing gain in action occurs with Global Positioning System (GPS) receivers, such as those used in automotive navigation systems and smart-phone location applications. The GPS system consists of a "constellation" of at least twenty-four satellites in constant orbit, and from any spot on Earth a minimum of four satellites is in view. All satellites transmit their location data on the same radio frequency, which, for nonmilitary applications, is 1575.42 MHz. So that the receiver can distinguish one satellite from another, GPS employs DSSS with each satellite assigned its own unique PN spreading code. The chip rate for civilian applications is 1.023 Mcps (million chips per second). BPSK is the modulation format, so the transmitted bandwidth (hence, the bandwidth required of the receiver) is about double the chip rate, or 2.048 MHz, because upper and lower sidebands are produced. The average receive signal level on the Earth's surface from any satellite is about −130 dBm, equal to 70 nV or 0.07 μV for a 50-Ω system.

To get a perspective on just how small this level really is, we can compare it to the expected noise floor at the receiver front end. Recall from Equation (1-8) that noise power is the product of Boltzmann's constant (1.38×10^{-23} J/K), ambient temperature (in Kelvin), and bandwidth (in Hertz) and that noise is proportional to bandwidth. The bandwidth of interest is 2.048 MHz, the receiver input bandwidth, and room temperature is approximately 290 K; therefore,

$$P_n = kTB$$
$$= (1.38 \times 10^{-23})(290)(2.048 \times 10^6) = 8.2 \times 10^{-15} \text{ W or 8.2 fW.}$$

Converting this level to dBm, we obtain

$$\text{dBm} = 10 \log\left(\frac{8.2 \times 10^{-15}}{1 \times 10^{-3}}\right) \tag{8-11}$$

$$= -110.9 \text{ dBm.}$$

This level just calculated represents the "noise floor," and it is seen to be nearly 20 dB *above* the received signal level from the GPS satellites. How is it possible for the receiver to perform with an apparent −20-dB *S/N* ratio? Here is where process gain comes to the rescue.

To determine process gain, we need to know that the bit rate of the GPS positioning data stream is 50 bps (bits per second, not megabits). Therefore, the process gain in dB is

$$G = 10 \log\left(\frac{1.023 \times 10^6}{50}\right)$$
$$= 43.1 \text{ dB}.$$

Because the gain is expressed in decibels, it is directly additive to the received signal level. Thus, the de-spread signal is at -130 dBm $+ 43.1$ dB $= -86.9$ dBm, almost 24 dB above the receiver noise floor after dispreading, well within the operating capabilities of any well-designed receiver. This example also shows why it is advantageous to have a high chip rate in comparison to the information signal rate: the processing gain is substantially improved, even as the receiver front-end bandwidth needs to be made larger.

Orthogonal Frequency-Division Multiplexing (OFDM)

Another very important form of wideband modulation is **orthogonal frequency-division multiplexing (OFDM)**. OFDM belongs to a class called multi-carrier modulation and is sometimes called *multi-tone modulation*. Applications where OFDM is used include 802.11a and 802.11g wireless local-area networks, some DSL and cable modems, a wide-area wireless networking standard called WiMax, and North American digital radio. OFDM is also a crucial aspect of so-called LTE ("long-term evolution") 4G cellular and wireless video standards. The OFDM concept originated in the 1960s but has only recently been implemented because of the availability of affordable digital signal processing integrated circuits. OFDM is considered to be a form of spread spectrum because the transmitted data are spread over a wide bandwidth; however, rather than using a single carrier and high-speed data to create bandwidth spreading, as with DSSS, or moving a single carrier rapidly over a wide bandwidth, as with frequency-hopping, OFDM transmits data at relatively slow rates over multiple carriers simultaneously. OFDM achieves high spectral efficiencies because the spacing of each carrier is as close as theoretically possible.

With conventional frequency-division multiplexing (such as ordinary broadcasting or narrowband communications, for example) each carrier must be far enough away in frequency from its nearest neighbor so that the sidebands created by modulation do not overlap to any significant degree. Also, "guard bands" are often established to minimize the potential for overlap. High data rates create wide-bandwidth sidebands in conventional systems, which in turn would require the carriers to be spread even farther apart than they would be otherwise. All of this leads to a relatively inefficient use of spectrum. The low data rates associated with OFDM carriers cause them to have narrow bandwidths and a square-shaped frequency spectrum. This characteristic allows the carriers to be spaced close together, contributing to a high degree of spectral efficiency. High data rates can be achieved through the use of many closely spaced carriers arranged in a spectrally efficient fashion rather than a single carrier modulated with fast data.

How closely can the carriers be spaced without causing interference to signals associated with adjacent-frequency carriers? The answer again lies in the $\sin(x)/x$ or sinc function, which describes the shape of the unfiltered spectrum with random data:

$$\text{sinc}(f/f_s) = \frac{\sin(\pi f/f_s)}{\pi f/f_s}.$$

The f_s term in the preceding equation is the symbol rate. From the relevant Fourier expressions, random or pseudorandom data can be shown to contain energy at all frequencies from zero to half the symbol rate as well as all the odd harmonics of those frequencies. The harmonics can be filtered out because they are not necessary for demodulation. The spectrum of the sinc function, first shown earlier, has been redrawn as Figure 8-42 with the horizontal axis relabeled in terms of the symbol rate. The center, where the energy levels are highest, corresponds to the RF carrier frequency. The energy falls off symmetrically on either side of the carrier frequency and is minimum at the symbol rate and at points where the frequency (f) is an integer multiple of the symbol rate. Thus, if the OFDM carrier spacing is

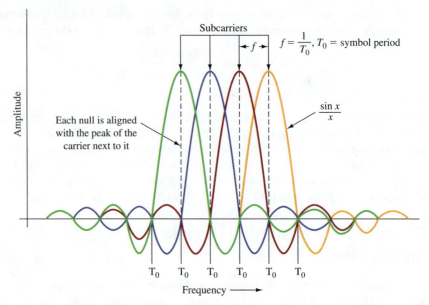

FIGURE 8-42 **OFDM carrier spacing.**

equal to the symbol rate, the necessary filters can be designed relatively easily to be insensitive to adjacent-carrier energy. The peak of each adjacent carrier is centered on the symbol rate, which is the null of the carrier next to it. Thus, the carriers are said to be *orthogonal* to each other because each theoretically has zero correlation with any other.

In OFDM, the information is divided into multiple data streams and is transmitted using multiple subcarriers at fixed frequencies over a common wired or wireless communications channel. If implemented with discrete components, an OFDM transmitter would be composed of multiple carrier-generating oscillators, modulators, and associated filtering and combining circuits. Likewise, an OFDM receiver would consist of all the functional blocks involved with conventional receivers, including local oscillators, mixing and demodulation circuits, and the like. Implementing OFDM with analog building blocks would clearly be impractical, particularly when hundreds if not thousands of individual carriers are involved, but it is a technique ideally suited to I/Q modulation and DSP.

Recall from Section 8-5 that an analytic signal is of the form I + jQ, meaning that it consists of both a "real," or in-phase cosine wave, and an "imaginary," or phase-shifted sine wave. Section 8-5 also showed that these two waveforms are orthogonal to each other: When the cosine-wave component was maximum the sine-wave component was zero, and vice versa. The analytic signal can just as well be thought of as having an amplitude magnitude and phase angle. The underlying idea is similar to how impedance was described in polar form as a quantity consisting of both a magnitude in ohms and a phase angle representing the separation between voltage and current seen by the source. Polar form can also be used to describe an I/Q signal; the magnitude was shown in Section 8-5 to be the amplitude as represented by length of the I/Q vector and predicted by the Pythagorean theorem, and the phase angle was defined with respect to the +I axis. Therefore, the full analytic signal consists of radian frequency (ω) and phase (φ) and can be expressed as

$$x(t) = Ae^{-j(\varphi + \omega t)},$$

which, for the reasons described earlier, is equal to

$$A[\cos(\varphi + \omega t) + j\sin(\varphi + \omega t)].$$

The scalar portion of the signal (i.e., the part viewable on an oscilloscope-based constellation diagram or on a spectrum analyzer) is the "real" portion, which is the constant A multiplied by the cosine term.

These results imply that signals can be described fully in I/Q space in terms of amplitude and phase characteristics, which lend themselves to DSP implementations and the fast Fourier transform. The basic idea of an OFDM receiver implemented in DSP is shown in Figure 8-43(a). Recall from Chapter 1 that a *fast Fourier transform* (FFT) is an algorithm

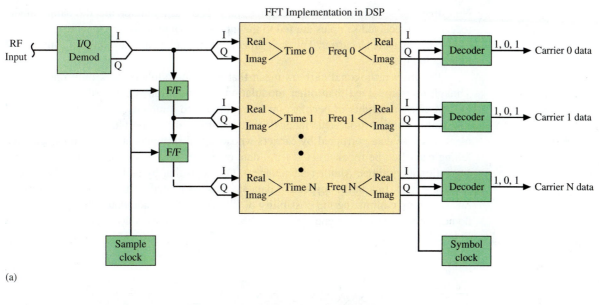

(a)

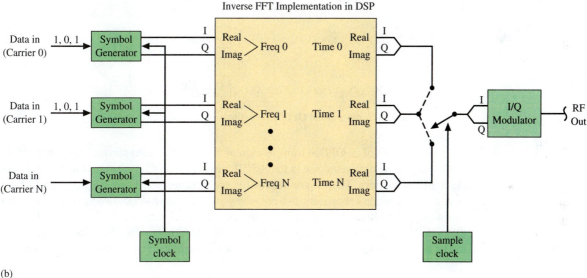

(b)

FIGURE 8-43   **Conceptual block diagram of (a) OFDM receiver and (b) OFDM modulator.**

that allows for conversion between time and frequency-domain representations. This algorithm is essentially a means of implementing a hardware spectrum analyzer in software. If samples are gathered at regular time intervals, the FFT output will be a series of samples representing the frequency components of that signal. For an OFDM system the sample rate is selected so that each FFT output is associated with one carrier. As shown in the figure, the demodulated I/Q symbols in the form of sampled pulses are applied to the FFT block after being stored in a series of shift registers. The inputs to the FFT algorithm represent symbols in the time domain. Each FFT output is a complex number representing the real and imaginary parts (amplitude and phase, which map to I and Q components) of the modulated carrier. The amplitude and phase at each carrier frequency contains the information needed to determine where the demodulated symbol should appear in the I/Q representation. Once determined, the I/Q data can then be decoded to their respective symbol equivalents.

OFDM modulation, shown in Figure 8-43(b), is essentially the converse of demodulation. Each data symbol is routed to an available carrier, which when modulated is represented in the frequency domain as a vector containing the amplitude and phase information of the symbol mapped to it. The encoder performs the inverse FFT (IFFT) to convert each carrier amplitude and phase from its frequency-domain representation to a series of time samples in I/Q space. These then become the inputs, sequentially applied, to an I/Q modulator. The sample clock is arranged to apply the I/Q outputs serially to the I/Q modulator.

Note that the sample clocks for the modulator and demodulator are the same; at the receiver, the received samples are fed to the inputs shown in sequence to recreate the data used to modulate each carrier. The data streams can modulate the carriers using any digital modulation method.

The fully orthogonal carriers mean that each demodulator sees only one carrier. Though more involved than other modulation/demodulation schemes described so far, OFDM offers a number of benefits, including improved immunity to RF interference and low multipath distortion. This latter benefit is particularly important in mobile environments because data corrupted by carriers whose signal strengths are reduced by selective fading can often be reconstructed from data obtained from other carriers through error-correction algorithms. Another advantage to OFDM is that by making the symbol length relatively long, bits in the respective data streams associated with each carrier are separated, thereby minimizing the possibility of ISI. Thus, the demodulators for each channel do not see an interfering signal from an adjacent channel. This concept is demonstrated in Figure 8-44.

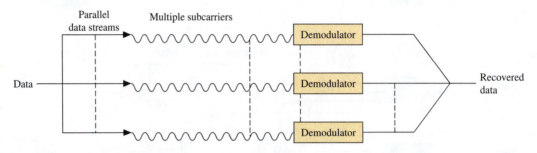

FIGURE 8-44   **Block diagram of OFDM reception.**

An example of an OFDM transmission of a data sequence is shown in Figure 8-45. In this example, the data sequence is the ASCII message "BUY" transmitted in parallel over eight BPSK carriers. The binary values for each carrier are listed. Subcarriers 1–4 are cosine waves, while subcarriers 5–8 are sine waves. Figure 8-45 shows three segments for the multiple BPSK carriers. Segment 1 shows the ASCII code for "B." Segment 2 shows the ASCII code for "U," and segment 3 shows the ASCII code for "Y." The ASCII values for BUY are provided in Table 8-5. Note that each subcarrier is being used to carry just

FIGURE 8-45   **Example of the OFDM transmission of the ASCII characters B, U, and Y.**

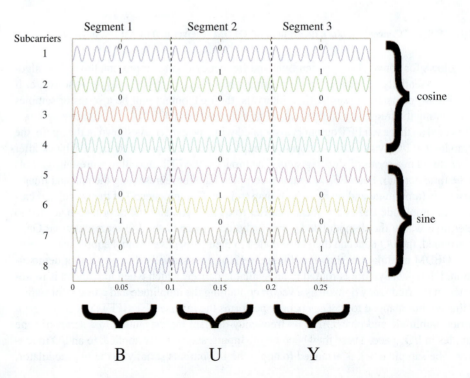

LETTER	8-BIT ASCII VALUE
B	0 1 0 0 0 0 1 0
U	0 1 0 1 0 1 0 1
Y	0 1 0 1 1 0 0 1

TABLE 8-5 • The ASCII Values for the Letters B, U, and Y

one bit of information per segment. To prevent intersymbol interference in OFDM systems, a small guard time is usually inserted (not shown in Figure 8-45). This guard time may be left as a gap in transmission or it may be filled with a **cyclic prefix**, which means that the end of a symbol is copied to the beginning of the data stream, thus leaving no gap. This modification makes the symbol easier to demodulate.

HD RADIO   Another technology that uses OFDM for transporting the digital data is HD radio, which is a digital radio technology that operates in the same frequency bands as broadcast AM (530–1705 kHZ) and FM (88–108 MHz). Another name for HD radio technology is **in-band on-channel (IBOC)** and was developed in 1991 by iBiquity Digital. The FCC officially approved the HD radio technology in 2002.

In AM HD radio, the digital signal is placed above and below the analog AM carrier as shown in Figure 8-46. This is called a "hybrid" AM signal because both the analog and digital signal share the same bandwidth. The normal spectral bandwidth of an analog AM transmission is 10 kHz (±5 kHz for the upper and lower sidebands). In an HD radio hybrid transmission, an additional 10 kHz is added to the spectrum on each side of the carrier, resulting in a total spectral bandwidth for the AM HD radio transmission of 30 kHz. AM HD radio transmission uses eighty-one OFDM carriers with a 181.7-Hz spacing between carriers. The digital data rate for AM HD radio is 36 kbps, which produces an audio signal with comparable quality to the current analog FM.

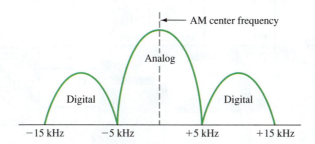

FIGURE 8-46   The RF spectrum for a hybrid analog and digital AM transmission.

The frequency spectrum for a hybrid FM transmission is shown in Figure 8-47. The analog transmitted FM signal occupies ±130 kHz of bandwidth. The digital data is located in the 130–199-kHz range above and below the center frequency of the FM signal. The total spectral bandwidth used by the hybrid FM system is ±200 kHz. The digital data rate for the primary audio channel is 96 kbps, which produces a near CD quality audio signal.

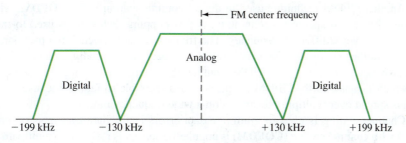

FIGURE 8-47   The RF spectrum for a hybrid (analog and digital) FM transmission.

HD RECEIVERS The simultaneous transmission of the analog and digital signals in HD radio does not affect analog only radios (non HD radios). The analog radios will receive the entire RF signal but will demodulate only the analog signal. However, an HD radio receiver will first try to lock to the analog signal. The HD radio will next try to lock to the FM stereo signal (for FM transmission) if present and will finally lock to the digital signal.

Philips Semiconductor has a high-performance AM/FM radio chip set that supports HD radio. A simplified block diagram of the HD radio is provided in Figure 8-48. The chip set includes a single-chip radio tuner (TEF 6721) for the reception of AM, FM, and FM IBOC RF signals. The TEF 6721 uses a 10/7 MHz IF for both AM and FM signals. The IF output of the TEF 6721 is fed to the IF A/D input of the SAF 7730. The SAF 7730 is a programmable DSP chip that performs digital audio processing. Additionally, this device incorporates multipath cancellation of RF signals. The IF output of the SAF 7730 feeds the IF input of the SAF 3550. The SAF 3550 is an HD radio processor that supports both hybrid and all-digital modes. The SAF 3550 outputs a signal to the SAF 7730 blend input for additional audio processing.

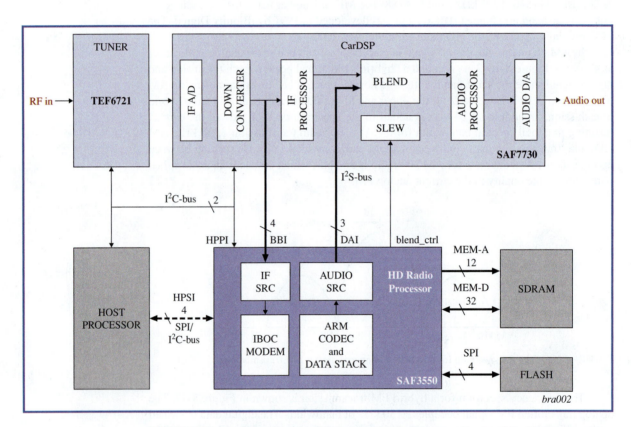

FIGURE 8-48 **An HD radio—simplified system block diagram.** (Courtesy Philips Semiconductor.)

Another OFDM technique used in digital communication is **flash OFDM**, which is considered a spread-spectrum technology. A fast hopping technique is used to transmit each symbol over a different frequency. The frequencies are selected in a pseudorandom manner. In other words, the frequency selected appears to be totally random. This technique provides the advantages of CDMA but has the additional benefit of frequency diversity, which means that the OFDM signal is less susceptible to fading because the entire signal is spread over multiple subcarriers on a wide frequency range.

Channel coding is used to minimize data errors in the data transmission process. **OFDM with channel coding (COFDM)** is popular because it is resistant to multipath signal effects. CODFM is the format for digital video broadcasting in Europe.

SUMMARY

Modern communications systems employ various forms of digital modulation because advances in high-speed computing on low-cost integrated circuits make these modulation schemes practical. The fundamental constraints of bandwidth and noise are still present, and these constraints have given rise to a number of enhancements to the basic formats, but technologies that had been ideas on paper for many years are now a practical reality.

Digital modulation approaches are derived from the fundamental forms of amplitude and angle modulation. The basic forms of amplitude-, frequency-, or phase-shift keying are employed, either singly or in combination to produce methods of digital modulation that overcome the biggest disadvantage of any digital modulation scheme: The fact that, in their simplest forms, digital modulation formats occupy more bandwidth than the analog alternatives. However, with M-ary modulation approaches, where individual data bits are grouped into multiple-bit symbols and then amplitude- and/or phase-shift keyed, advances in spectral efficiency can be attained because bandwidth is related to symbol rate rather than data rate. Each form of digital modulation has implications in terms of bandwidth, spectral efficiency, and power distribution. For example, phase-shift keying approaches (not involving various forms of amplitude modulation) are "constant envelope" schemes, implying that efficient class C amplifiers can be used in the power-output stages of the transmitters. Approaches that use both phase- and amplitude-shift keying may employ a form of offset modulation to minimize the likelihood that the power amplifier output will go to zero, thus allowing for less linear amplifiers to be used in transmitters. Differential keying methods, in which the information is encoded in terms of a state change from the previously received bit, do not require receivers with coherent carrier recovery capability, thus making for more straightforward implementations.

An alternative model for achieving high spectral efficiency is to spread the transmitted energy over an exceptionally wide spectrum, much wider than that mandated by Hartley's law. Such spread-spectrum schemes appear to upend the sacred rules of communications theory; bandwidth appears to be wasted, and in some cases the desired signal resides below the noise floor. However, advantages accrue in information security and resistance to interference, as well as the ability for multiple users to share a given portion of the frequency spectrum simultaneously. Spectral efficiency is also a key characteristic of orthogonal frequency-division multiplexing, where multiple carriers are modulated even as they are spaced as closely together as theoretically possible. These forms of wideband modulation are the underlying technologies that make high-speed wireless communications systems a practical reality.

QUESTIONS AND PROBLEMS

SECTION 8-1

1. Define *continuous-wave transmission*. In what ways is this name appropriate?

2. Explain the automatic gain control (AGC) difficulties encountered in the reception of continuous wave (CW). What is two-tone modulation, and how does it remedy the receiver AGCC problem of CW?

3. Calculate all possible transmitted frequencies for a two-tone modulation system using a 21-MHz carrier with 300- and 700-Hz modulating signals to represent mark and space. Calculate the channel bandwidth required.

4. What is a frequency-shift-keying (FSK) system? Describe two methods of generating FSK.

5. Explain how the FSK signal is detected.

6. Describe the phase-shift keying (PSK) process.

7. What do the *M* and *n* represent in Table 8-1?

8. Explain a method used to generate binary phase-shift keying (BPSK) using Figure 8-7 as a basis.

9. Describe a method of detecting the binary output for a BPSK signal using Figure 8-10 as a basis.

10. What is coherent carrier recovery?

11. Describe the recovery of quadrature phase-shifting keying (QPSK) using Figure 8-19 as a basis.

12. Provide a brief description of the quadrature amplitude modulation (QAM) system. Explain why it is an efficient user of frequency spectrum.

13. Describe how to generate a constellation.

14. What is the purpose of any eye pattern?

15. The input data to a differential phase-shift keying (DPSK) transmitter is 1 1 0 1. Determine the DPSK digital data stream and the DPSK radio-frequency (RF) output.

SECTION 8-6

16. What is a pseudonoise (PN) code and why is it noiselike?

17. What does it mean to spread the RF signal?

18. What is the PN sequence length of a PN sequence generator with each of the following?
    (a) four shift registers
    (b) nine shift registers
    (c) twenty-three shift registers

19. What does it mean for a PN sequence to be of maximal length?

20. Draw the circuit for a PN sequence generator with $n = 5$.

21. Explain the concept of frequency hopping spread spectrum.

22. Define *dwell time*.

23. Generate the shift-register contents for a PN sequence generator with $n = 3$ and a seed value of 1 0 1.

24. Define *code division multiple access*.

25. Define the term *hit* relative to spread spectrum.

26. What is a signature sequence?

27. Draw the block diagram of a direct-sequence spread spectrum (DSSS) transmit-receive system.

28. Determine the spreading of a DSSS signal if the chip rate is 1 Mbps and the modulation rate is 56 kbps.

29. Draw the block diagram of an orthogonal frequency-division multiplexing (OFDM) transmission.

30. How does flash OFDM differ from OFDM?

31. Determine the received OFDM signal for the traces shown in Figure 8-49.

FIGURE 8-49

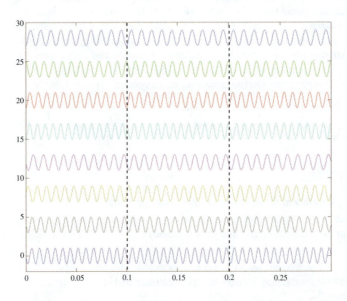

32. What is a hybrid amplitude modulation (AM) or frequency modulation (FM) signal?

33. What is the digital data rate for AM digital and what is its comparable quality?

34. What is the digital data rate for FM digital and what is its comparable quality?

35. Draw a picture of the RF spectrum for (a) hybrid AM and (b) hybrid FM.

36. What technology does HD radio use to transport the digital signal?

QUESTIONS FOR CRITICAL THINKING

37. Why is OFDM not considered a true spread-spectrum system?

38. Determine the 7-bit code generated for the PN generator shown in Figure 8-31 if a seed value of 1 1 1 is used. What happens if a seed value of 0 0 0 is used?

39. Explain how it is possible for multiple code-division multiple access (CDMA) communications devices to share the same RF spectrum.

CHAPTER 9

TELEPHONE NETWORKS

CHAPTER OUTLINE

KEY TERMS

tip
ring
local loop
trunk
T3
OC-1
loaded cable
Attenuation distortion
delay distortion
delay equalizer
synchronous
asynchronous
handshaking
protocols
framing
line control
multipoint circuits
flow control
sequence control
character insertion
character stuffing
bit stuffing
transparency
multilevel binary
time-division multiple access (TDMA)
time slot
demultiplexer (DMUX)
guard times
intersymbol interference (ISI)
fractional T1 (FT1)
point of presence
channel service unit/ data service unit (CSU/ DSU)
D4 framing
extended superframe framing (ESF)
loopback
AMI
bipolar coding
bipolar 8 zero substitution (B8ZS)
minimum ones density
bipolar violations
Shannon-Hartley theorem
packets
statistical concentration
frame relay
public data network (PDN)
X.25
committed information rate (CIR)
bursty
committed burst information rate (CBIR)
asynchronous transfer mode (ATM)
packet switching
payload
virtual path connection (VPC)
virtual channel connection (VCC)
switched virtual circuits (SVCs)
virtual path identifier (VPI)
virtual circuit identifier (VCI)
signaling systems
public switched telephone network (PSTN)
in-band
out-of-band
SS7
eye patterns

9-1 INTRODUCTION

Chapter 8 focused on the various methods developed to modulate data for effective transmission over a bandwidth-limited medium. The medium can be and often is wireless, for it is here that bandwidth availability is generally most constrained. However, many ideas introduced in the previous two chapters apply to wired networks as well. Certainly the largest wired network is the worldwide telephone system. This chapter focuses specifically on the wired ("land-line") telephone network because of its universal presence and because of the frequent need to interconnect other, privately owned ("customer premises") equipment to it. Certainly, mobile wireless systems for both voice and computer applications have become nearly as universal in recent years, and these will be the focus of Chapter 10. Finally, keep in mind that many of the ideas and methods used in computer networks have their basis in telephone technology. Telephone and computer networks have blended together in their functionality. Computer networks now have their own telephone systems (IP telephony) and wireless telephones now provide Internet access and browsing capabilities. For this reason, we will see that some of the topics first brought up in the context of wired telephone networks bear repetition, and will be seen again in the Chapter 11 discussion of computer networks.

A network is an interconnection of users that allows communication among them. As mentioned, the most extensive existing network is the worldwide telephone grid. It allows direct connection between two users simply by dialing an access code. Behind that apparent simplicity is an extremely complex system. Because the telephone network is so convenient and inexpensive, it is often used to allow one computer to "speak" with another. Therefore, the distinction between telephone and wired computer networks is becoming blurred, and many of the underlying principles used for one network can be found to apply to the other. The primary focus of this chapter is wired telephone systems. However, keep in mind that many of the underlying principles, among them the importance of line coding and the distinction between synchronous and asynchronous communication, apply as well to computer-based networks.

9-2 BASIC TELEPHONE OPERATION

The Greek word *tele* means "far" and *phone* means "sound." The telephone system represents a worldwide grid of connections that enables point-to-point communications between the many subscribers. The early systems used mechanical switches to provide routing of a call. Initially, the *Strowger stepping switch* was used, followed by *crossbar switching*. Today's systems utilize solid-state electronics for switching under computer control to determine and select the best routing possibilities. The possible routes for local calls include hard-wired paths or fiber-optic systems. Long-distance calls are routed using these same paths, but they can also use radio transmission via satellite or microwave transmission paths. It is possible for a call to use all these paths in getting from its source to its destination. These multipath transmissions are especially tough on digital transmissions, as we will later see.

The telephone company (telco) provides two-wire service to each subscriber. One wire is designated the **tip** and the other the **ring**. The telco provides −48 V dc on the ring and grounds the tip, as shown in Figure 9-1. The telephone circuitry must work with three signal levels: the received voice signal, which could be as low as a few millivolts; the transmitted voice signal of 1 to 2 V rms; and an incoming ringing signal of 90 V rms. They also accept the dc power of −48 V at 15 to 80 mA. Until the phone is removed from the hook, only the ring circuits are connected to the line. The subscriber's telephone line is usually either AWG 22, 24, or 26 twisted-pair wire, which, because of filters installed to limit bandwidth to the range of voice frequencies, handles audio signals within the 300-Hz to 3-kHz voice range. Without filtering, the copper wires can handle frequencies up to

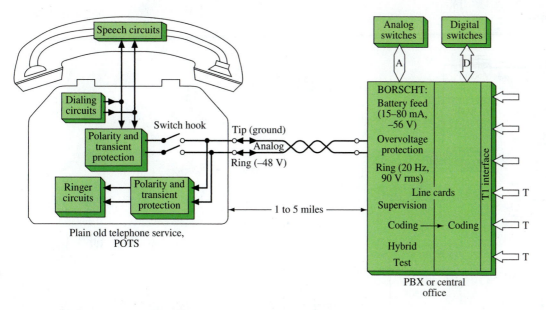

FIGURE 9-1   **Telephone representation.**

several megahertz. The twisted-pair line, or **local loop**, runs up to a few miles to a central phone office, or in a business setting to a PBX (private branch exchange). When a subscriber lifts the handset, a switch is closed to complete a dc loop circuit between the tip and the ring line through the phone's microphone. The handset earpiece is transformer or electronically coupled into this circuit also. The telco senses the off-hook condition and responds with an audible dial tone. The central office/PBX function shown in Figure 9-1 will be detailed later in this section.

At this point the subscriber dials or keys in the desired number. Dial pulsing is the interruption of the dc loop circuit according to the number dialed. Dialing 2 interrupts the circuit twice and an 8 interrupts it eight times. Tone-dialing systems use a dual-tone multifrequency (DTMF) electronic oscillator to provide this information. Figure 9-2 shows the arrangement of this system. When selecting the digit 8, a dual-frequency tone of 852 Hz and 1336 Hz is transmitted. When the telco receives the entire number selected, its central computer makes a path selection. It then either sends the destination a 90-V ac ringer signal or sends the originator a busy signal if the destination is already in use.

The connection paths for telephone service were all initially designed for voice transmission. As such, the band of frequencies of interest were about 300 to 3000 Hz. To meet the increasing demands of data transmission, the telco now provides special dedicated lines with enhanced performance and bandwidths up to 30 MHz for high-speed applications. The characteristics of these lines have a major effect on their usefulness for data transmission applications. Line-quality considerations are explored later in this section.

	1209	1336	1477
697	1	2	3
770	4	5	6
852	7	8	9
941	*	0	#
Frequency (Hz)	1209	1336	1477

FIGURE 9-2   **DTMF dialing.**

Telephone Systems

A complete telephone system block diagram is shown in Figure 9-3. On the left, three subscribers are shown. The top one is an office with three phones and a computer (PC). The digital signal of the PC is converted to analog by the modem. The office system is internally tied together through its private branch exchange (PBX). The PBX also connects it to the outside world, as shown. The next subscriber in Figure 9-3 is a home with two phones, while the third subscriber is a home with a phone and a PC. Notice the switch used for voice/data communications.

The primary function of the PBX and central office is the same: switching one telephone line to another. In addition, most central offices multiplex many conversations onto one line. The multiplexing may be based on time or frequency division, and

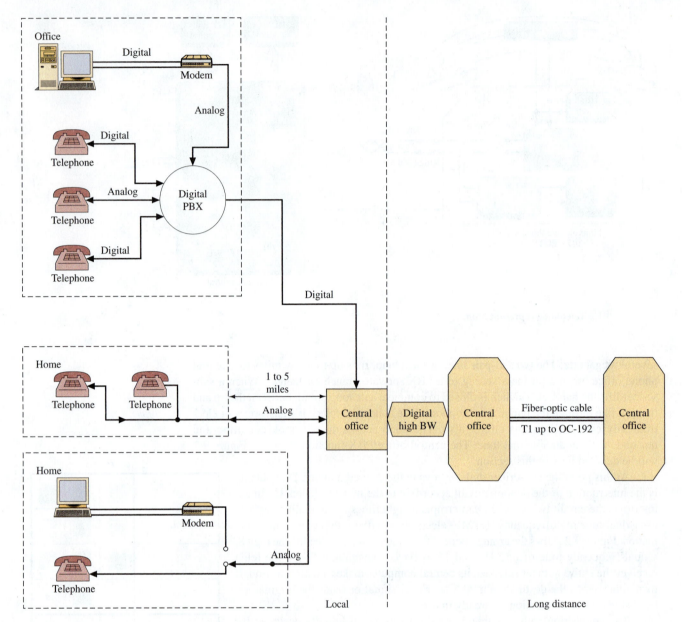

FIGURE 9-3 Telephone system block diagram.

the transmitted signals may be analog or digital; however, in current transmission practice, multiplexing is almost universally PCM digital.

Before switching at the PBX or central office, circuitry residing on the line cards handles the so-called *BORSCHT function* for their line. These interface circuits are also called the *subscriber loop interface circuit* (SLIC). BORSCHT is an acronym generated as follows:

Battery feeding: supplying the −48 V at 15- to 80-mA power required by the telephone service. Batteries under continuous charge at the central office allow phone service even during power failures.

Overvoltage protection: guarding against induced lightning strike transients and other electrical pickups.

Ringing: producing the 90-V rms signal shared by all lines; a relay or high-voltage solid-state switch connects the ring generator to the line.

Supervision: alerting the central office to on- and off-hook conditions (dial tone, ringing, operator requests, busy signals); also the office's way of auditing the line for billing.

Coding: if the central office uses digital switching or is connected to TDM/PCM digital lines, there are codes and their associated filters on the line card.

Hybrid: separating the two-wire subscriber loops into two-wire pairs for transmitting and receiving. The phone system is obviously a full-duplex system. Other than the local loop, four wires are used by the phone company: two for transmitting and two for receiving. This arrangement is shown in Figure 9-4. The hybrid was a transformer-based circuit, but today an electronic circuit contained within an IC provides the BORSCHT functions.

Testing: enabling the central office to test the subscriber's line.

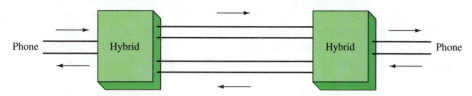

FIGURE 9-4   **Two- to four-wire conversion.**

Historically, analog switches ranged from stepping switches and banks of relays to crossbar switches and crosspoint arrays of solid-state switches. They were slow, had limited bandwidth, ate up space, consumed lots of power, and were difficult to control with microprocessors or even CPUs. To overcome the limitations inherent in analog switching and to enable central offices to use TDM/PCM links in urban areas, telephone companies have converted to digital transmission using 8-bit words. The conversions were almost complete by the late 1990s.

On the other hand, if the digital words are multiplexed onto a T1 carrier en route to another office, the serial bit stream is converted to the equivalent of 24 voice channels plus supervisory, signaling, and framing bits. Frames are transmitted every 125 μs (an 8-kHz rate). The circuit connecting one central office to another is called the **trunk** line. The T-carrier concept is central to the telephone system and will be examined in more detail in Section 9-5.

Bundles of T1 lines carry most voice channels between central offices in densely populated areas. If the lines stretch more than 6000 ft or so, a repeater amplifies the signal and regenerates its timing. When the number of T1s becomes large, they are further combined into a **T3** signal, which contains a total of 28 T1s and operates at a data rate of 44.736 Mbps. At higher channel densities, it is necessary to convert to optical signals. The T3 then becomes known as an **OC-1** or optical carrier level 1, operating at 51.84 Mbps. On very high-density routes, levels of OC-12 and OC-192 are common, and on long domestic transcontinental routes, OC-48 and OC-192 are in service. Optical fibers have replaced most of the copper wire used in telephone systems, especially in urban areas. Refer to Chapter 16 for details on these fiber-optic systems.

Line Quality Considerations

An ideal telephone line would transmit a perfect replica of a signal to the receiver. Ideally, this would be true for a basic analog voice signal, an analog version of a digital signal, or a pure digital signal. Unfortunately, the ideal does not occur. In many cases, the existing cable infrastructure in the United States is more than fifty years old and will be in use for many more years. We now look at the various reasons that a signal received via telephone lines is less than perfect (i.e., is distorted).

Attenuation Distortion

The local loop for almost all telephone transmissions is a two-wire twisted-pair cable. This rudimentary transmission line is usually made up of copper conductors surrounded by polyethylene plastic for insulation. The transmission characteristics of this line are dependent on the wire diameter, conductor spacing, and dielectric constant of the insulation. The resistance of the copper causes signal attenuation. Unfortunately, as explained in Chapter 12, transmission lines have inductance and capacitance that have a frequency-dependent effect. A frequency versus attenuation curve for a typical twisted-pair cable is shown in Figure 9-5. The higher frequencies are obviously attenuated much more than the lower ones. This distortion can be very troublesome to digital signals because the pulses become very rounded and data errors (is it a 1 or a 0?) occur. This distortion is also troublesome to analog signals.

This higher-frequency attenuation can be greatly curtailed by adding inductance in series with the cable. The dashed curve in Figure 9-5 is a typical frequency versus attenuation response for a cable that has had 88 mH added in series every 6000 ft. Notice that the response is nearly flat below 2 kHz. This type of cable, termed **loaded cable**, is universally used by the phone company to extend the range of a connection. Loaded cable is denoted by the letters H, D, or B to indicate added inductance every 6000, 4500, or 3000 ft, respectively. Standard values of added inductance are 44, 66, or 88 mH. A twisted-pair cable with a 26 D 88 label indicates a 26-gauge wire with 88 mH added every 4500 ft.

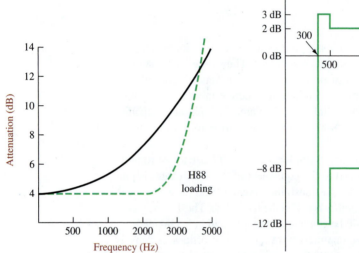

FIGURE 9-5   Attenuation for 12,000 ft of 26-gauge wire.

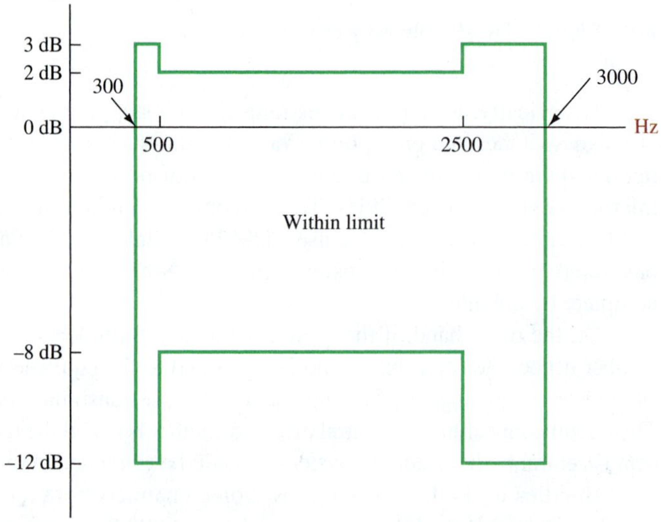

FIGURE 9-6   Attenuation distortion limit for 3002 channel.

Attenuation distortion is the difference in gain or loss at some frequency with respect to a reference tone of 1004 Hz. The basic telephone line (called a 3002 channel) specification is illustrated graphically in Figure 9-6. As can be seen, from 500 to 2500 Hz, the signal cannot be more than 2 dB above the 1004-Hz level or 8 dB below the 1004-Hz level. From 300 to 500 Hz and 2500 to 3000 Hz, the allowable limits are +3 dB and −12 dB. A subscriber can sometimes lease a better line if necessary. A commonly encountered "improved" line is designated as a C2 line. It has limits of +1 dB and −3 dB between 500 and 2800 Hz. From 300 to 500 Hz and 2800 to 3000 Hz, the C2 limits are +2 dB and −6 dB.

Delay Distortion

A signal traveling down a transmission line experiences some delay from input to output. That is not normally a problem. Unfortunately, not all frequencies experience the same amount of delay. The **delay distortion** can be quite troublesome to data transmissions. The basic 3002 channel is given an *envelope delay* specification by the Federal Communications Commission (FCC) of 1750 μs between 800 and 2600 Hz. This specifies that the delay between any two frequencies cannot exceed 1750 μs. The improved C2 channel is specified to be better than 500 μs from 1000 to 2600 Hz, 1500 μs from 600 to 1000 Hz, and 3000 μs from 500 to 600 and 2600 to 2800 Hz.

The delay versus frequency characteristic for a typical phone line is shown with dashed lines in Figure 9-7, while the characteristics after delay equalization are also provided. The **delay equalizer** is a complex *LC* filter that provides increased delay to those frequencies least delayed by the phone line, so that all frequencies arrive at nearly the same time. Delay, or phase, equalizers typically have several sections, one for each small group of frequencies across the band. Also, they may be fixed or adjustable.

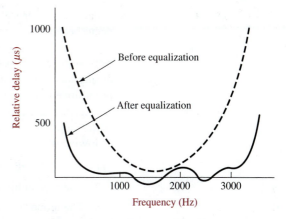

FIGURE 9-7   **Delay equalization.**

Telephone Traffic

Let us take a typical workday. During rush-hour traffic, plenty of vehicles are lined up in every available lane of traffic: in the morning before office hours, at lunch, and after office hours. Motorists take into account delays caused by rush-hour traffic to arrive at their destinations in a timely manner. Those who are unfamiliar with traffic-flow patterns are likely to miss appointments (how they wish there were always enough lanes and no intersections on the way to their destinations).

Normal telephone traffic is not so different from vehicular traffic. There are two times in a typical workday when telephone traffic intensifies: between 9:00 and 11:00 in the morning and between 2:00 and 4:00 in the afternoon. The morning traffic is the heaviest, because of the volume of business calls, followed by a lesser volume of calls in the afternoon. That period of one hour when traffic is heaviest is referred to as the busy hour. An example of a busy hour period is from 9:45 a.m. to 10:45 a.m. The busy hour traffic intensity on Mondays is mostly the highest, and it tapers off toward Wednesdays. It starts to pick up again on Thursdays and even more on Fridays (especially on payday Fridays).

The Unit of Traffic

In telephony, one way to define traffic is the average number of calls in progress during a specific period of time, often one hour. A circuit or path that carries its usage for one traffic call at a time is referred to as a trunk. Telephone traffic, although it is a dimensionless quantity, is expressed either in erlang (named after Agner Krarup Erlang, a Danish pioneer of traffic theory) or in hundred-call-seconds (CCS), the latter being commonly used in North America. One erlang equals 36 CCS. A simple way to understand this unit is to look at a bicycle. In one hour, the most that you can ride this bicycle is 36 hundred seconds. Two or more people can ride, or occupy, or hold this bicycle one at a time, but they cannot exceed a total of 36 hundred seconds (or 36 hundred-ride-seconds) in a period of one hour. A group of ten bicycles for rent, while it has a theoretical capacity of 360 hundred-ride-seconds, may sometimes be used only for an average of 36 hundred-ride-seconds on a Monday morning, but on a Sunday this same group of ten bicycles may not be enough, so "congestion" occurs. Some lose interest, others wait for the next available bicycle, and some try again later.

Telephone trunks are arranged in groups. The traffic capacity of a group of trunks depends on the nature or distribution of call durations or holding time. Widely distributed call holding times tend to reduce traffic handling capacity. Traffic carried by a group of trunks may, therefore, be stated as follows:

$$A = C \times \frac{H}{T},$$

where A = traffic in erlangs

C = average number of calls in progress during a period of time

H = average holding time of each call

T = 3,600 seconds (1 hour).

Congestion

Congestion occurs when calls are unable to reach their destination as a result of excess demand for system capacity. Getting a busy tone instead of a ringback tone because the called subscriber station is busy is not congestion by definition. Rather, congestion can occur because, while the central office has sufficient equipment installed to provide terminal access for all subscribers, most likely it does not have sufficient facilities to handle the peak volume of calls encountered during the busy hour. Some calls are allowed to be lost during this time to meet the economic objectives of providing service. It is highly cost-prohibitive to provide sufficient trunks to carry all traffic offered to a system with no calls lost at all. The measure of calls lost during a busy hour period is known as grade of service. In the dimensioning of a telephone network, traffic engineers look at grade of service in the following ways:

1. The probability that a call will be lost as a result of congestion
2. The probability of congestion
3. The proportion of time during which congestion occurs

Grade of service, B, is designed into the system by traffic engineers. After the system is put into service, it is observed and verified using traffic scanning devices. Traffic observation and measurement show how many calls are offered, carried, and lost in the system. Grade of service, B, is then determined as:

$$B = \frac{\text{number of calls lost}}{\text{number of calls offered}}$$

or

$$B = \frac{\text{traffic lost}}{\text{traffic offered}}.$$

The lower this number, the higher the grade of service.

Traffic Observation and Measurement

Just as the traffic control center of a city would like to ensure a smooth flow of vehicular traffic, telephone-operating companies consider traffic management to be the most important function in providing reliable and efficient public telephone service. Continuous traffic measurement is done to detect and resolve potential sources of congestion. Calls that are either delayed or lost because of congestion problems mean customer dissatisfaction and, ultimately, lost revenues. Traffic measurement studies are made to determine customer calling patterns, which serve as a basis for discounted toll rates. Although operating companies provide contingencies for their network failures, telephone systems are never engineered to handle all calls without congestion. It is also based on studying traffic that operating companies forecast future demands and project capital expenditures for their expansion programs.

9-3 DIGITAL WIRED NETWORKS

The previous discussion hinted that, although the "subscriber loop" from the central office to a customer's phone generally handles analog signals, virtually all the rest of the telephone network is fully digital. This is indeed so, and many issues pertaining to digital networks of all types, including computer networks, also apply in the context of telephone systems and will be introduced here. Two issues of particular significance involve protocols and the need for line coding. The choice of line code has bandwidth implications, and these will be examined further as well.

Communication Links and Protocols

Transmission of information in a communications link is defined by three basic protocol techniques: simplex, half duplex, and full duplex. An example of simplex operation is a radio station transmitter. The ratio station is sending transmissions to you, but there is typically not a return communications link. Most walkie-talkies operate in the half-duplex mode. When one unit is transmitting, the other unit must be in the receive mode. A cellular telephone is an example of a communications link that is operating full duplex. In full-duplex mode, the units can be transmitting and receiving at the same time. The following terms apply to both analog and digital communications links.

Simplex   Communication is in one direction only.

Half duplex   Communication is in both directions but only one can talk at a time.

Full duplex   Both parties can talk at the same time.

The communications link can also be classified according to whether it is a **synchronous** or an **asynchronous** channel. Synchronous operation implies that the transmit- and receive-data clocks are locked together. In many communications systems, this requires that the data contain clocking information (called *self-clocking data*). The other option for transmitting data is to operate asynchronously, which means that the clocks on the transmitter and receiver are not locked together. That is, the data do not contain clocking information and typically contain start and stop bits to lock the systems together temporarily. We will see shortly that one of the primary functions of line codes is to provide clocking information, either explicitly or in a form where the clock can be recreated at the receiver.

PROTOCOLS   The vast number of digital data facilities now in existence require complex networks to allow equipment to "talk" to one another. To maintain order during the interchange of data, rules to control the process are necessary. Initially, procedures allowing orderly interchange between a central computer and remote sites were needed. These rules and procedures were called **handshaking**. As the complexity of data communications systems increased, the need grew for something more than a "handshake." Thus, sets of rules and regulations called **protocols** were developed. A protocol may be defined as a set of rules designed to force the devices sharing a channel to observe orderly communications procedures.

Protocols have four major functions:

1. **Framing:** Data are normally transmitted in blocks. The **framing** function deals with the separation of blocks into the information (text) and control sections. A maximum block size is dictated by the protocol. Each block will normally contain control information such as an *address field* to indicate the intended recipient(s) of the data and the block check character (BCC) for error detection. The protocol also prescribes the process for error correction when one is detected.

2. **Line control:** **Line control** is the procedure used to decide which device has permission to transmit at any given time. In a simple full-duplex system with just two devices, line control is not necessary. However, systems with three or more devices (**multipoint circuits**) require line control.

3. **Flow control:** Often there is a limit to the rate at which a receiving device can accept data. A computer printer is a prime example of this condition. **Flow control** is the protocol process used to monitor and control these rates.

4. **Sequence control:** This is necessary for complex systems where a message must pass through numerous links before it reaches its final destination. **Sequence control** keeps message blocks from being lost or duplicated and ensures that they are received in the proper sequence. This has become an especially important consideration in packet-switching systems. They are introduced later in this chapter.

Protocols are responsible for integration of control characters within the data stream. Control characters are indicated by specific bit patterns that can also occur in the data stream. To circumvent this problem, the protocol can use a process termed **character insertion**, also called **character stuffing** or **bit stuffing**. If a control character sequence is detected in the data, the protocol causes an insertion of a bit or character that allows the receiving device to view the preceding sequence as valid data. This method of control character recognition is called **transparency**.

Protocols are classified according to their framing technique. Character-oriented protocols (COPs) use specific binary characters to separate segments of the transmitted information frame. These protocols are slow and bandwidth-inefficient and are seldom used any more. Bit-oriented protocols (BOPs) use frames made up of well-defined fields between 8-bit start and stop flags. A flag is fixed in both length and pattern. The BOPs include high-level data link control (HDLC) and synchronous data link control (SDLC). Synchronous data link control is considered a subset of HDLC. The byte-count protocol used in the digital data communications message protocol (DDCMP) uses a header followed by a count of the characters that will follow. They all have similar frame structures and are therefore independent of codes, line configurations, and peripheral devices. There are many protocol variations.

To illustrate the use of BOPs, we will discuss SDLC in more detail. SDLC was developed by IBM. Data link control information is transferred on a bit-by-bit basis. Figure 9-8 illustrates the SDLC BOP format.

Start flag	Address	Control	Message	Frame check sequence	Stop flag
8 bits	8 bits	8 bits	Multiples of 8 bits	8 bits	8 bits

FIGURE 9-8 **Bit-oriented protocol format, SDLC frame format.**

The start and stop flag for the SDLC frame is 01111110 ($7E_H$). The SDLC protocol requires that there be six consecutive ones (1s) within the flag. The transmitter will insert a zero after any five consecutive ones. The receiver must remove the zero when it detects five consecutive 1s followed by a zero. The communication equipment synchronizes on the flag pattern.

Line Codes

Line coding refers to the format of the pulses sent over the communications link (copper wire, optical fiber, or radio-frequency channel). Regardless of whether the signals are modulated or not, data must often first be *coded* or prepared for transmission. Line coding serves three purposes. The first is to eliminate the need for data states to be represented in terms of absolute voltage levels. In many cases, there is no dc continuity between devices within a network. Transformer coupling is often used in wired networks to provide isolation between elements or subsystems, or blocking capacitors may be installed to remove dc voltages. Therefore, reliance on absolute voltage levels (such as 0 V for a logic zero and 5 V for a logic 1) to convey binary data would be unworkable in such cases. Another problem arises even in situations where dc continuity can be maintained. Over distance, voltages applied to parallel conductors will cause them to have either a positive or negative *dc offset,* or average voltage that is either above or below 0 V. Effectively, the conductors behave as charged capacitors (remember that a capacitor is simply two conductive surfaces separated by a dielectric or insulating layer, and that is what two wires with insulation essentially are). With an average voltage present, what had started out as discrete pulses at, say, 0 and 5 V at the transmit end would converge toward the average value at the receive end—the logic low would no longer be near 0 V and the logic high would probably be less than 5 V. These complications, coupled with the likelihood that network subsystems located in different areas have their own power sources that do not share a common ground reference, mean that any coding method dependent on absolute voltage levels would be unworkable over more than the shortest distances.

Line codes are also important for maintaining synchronization between transmitter and receiver clocks. Recall that synchronous systems are those in which the transmitter and receiver data clocks are somehow locked together. Perhaps the most intuitive way to go about achieving synchronization is to send the clock signal over the communications channel; however, doing so uses bandwidth and potentially reduces data capacity. Instead, characteristics of the transmitted data can be used to "embed" the clocking information within them. The receiver can make use of these characteristics to derive a signal from which it locally recreates clocking information. The degree to which this is done, and ultimately the stability of the recovered clock information, is partly determined by the characteristics of the line code. In concept the idea is very similar to that of coherent carrier recovery, introduced in the context of suppressed-carrier systems like analog SSB in Chapter 6 and digital PSK in Chapter 8.

Finally, line codes enable a form of error detection. Certain coding formats, such as the alternate-mark inversion (AMI) scheme widely used in telephone systems, have certain "ground rules" that must be followed. If errors are introduced, either deliberately or as the result of noise or other transmission impairments, then the receiving subsystem is immediately alerted to their presence if the error causes the ground rules to be violated. The system responds in a manner determined by the underlying protocols governing its operation.

Line codes can be grouped either in terms of the number of possible states (either binary or multi-level) and in terms of how the logic high and low levels are represented. Three of the most important, because of their use in a wide variety of digital systems, are known as nonreturn-to-zero (NRZ), return-to-zero (RZ), and Manchester. The distinguishing feature of each code is how it represents a logic high or logic low within a given bit time.

NRZ CODES   The NRZ group of codes for encoding binary data describe, for example, how pulses are represented within a piece of equipment. NRZ codes are also one of the easiest to implement. NRZ codes get their name from the fact that the data signal does not return to the zero-volt state during a given bit-time interval. With reference to the NRZ-L code shown in Figure 9-9, note that the clock frequency determines the bit time, which is the time allotted for each binary data bit to be transmitted. The vertical orange bands in the figure denote the bit times. In the NRZ format, the voltage level for each bit is fixed for the entire bit time; that is, the binary zero and binary one remain at their respective voltages for the duration of the bit time defined by the clock frequency. Notice also that the code can remain constant for several clock cycles for a series of zeros or ones. Because of this

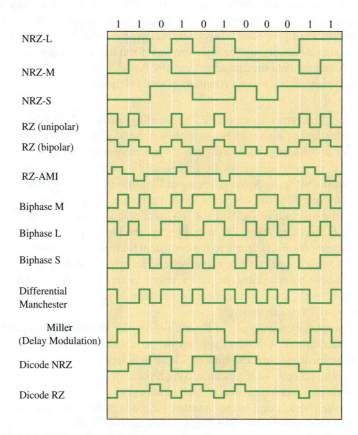

FIGURE 9-9   **Digital signal encoding formats.**

feature, the code has a dc component in the waveform. For example, a data stream containing a chain of 1s or 0s will appear as a dc signal at the receive side.

Another important factor to consider is that the NRZ codes do not contain any self-synchronizing capability. NRZ codes will require the use of start bits or some kind of synchronizing data pattern to keep the transmitted binary data synchronized. There are three coding schemes in the NRZ group: NRZ-L (level), NRZ-M (mark), and NRZ-S (space). The waveforms for these formats are provided in Figure 9-9. The NRZ code descriptions are provided in Table 9-1.

TABLE 9-1 • NRZ Codes	
NRZ-L	(nonreturn-to-zero—level)
	1 (hi)—high level
	0 (low)—low level
NRZ-M	(nonreturn-to-zero—mark)
	1 (hi)—transition at the beginning of the interval
	0 (low)—no transition
NRZ-S	(nonreturn-to-zero—space)
	1 (hi)—no transition
	0 (low)—transition at the beginning of the interval

In addition to describing how pulses are represented within a piece of equipment, NRZ codes may be used for communication over very short distances, such as between components on a board, or between circuit boards on a common chassis, or perhaps between computers and peripherals. However, because the NRZ code does not contain any self-synchronizing capability and because the waveform has a dc component, NRZ codes are not well suited for data transmission over any distance or where the transmitting and receiving equipment do not share a common ground reference.

RZ CODES   The return-to-zero (RZ) line-coding formats are better suited to communication over longer distances because they generally include some resynchronization capability and because some versions do not have a dc component. With the RZ code, logic zero or logic one states are represented by the waveform amplitude during the *first half* of the bit time. Again referring to Figure 9-9, note that for the RZ-unipolar code, the binary one is at the one level (positive voltage) and the binary zero is at the zero level for the first half of the bit time. The second half of the bit time is always represented by a binary zero level. Put another way, the logic zero levels stay at the zero state and the logic one levels return to the zero state within each bit time. In an RZ bipolar format, the logic 1 goes to a positive voltage and the logic zero goes to a negative voltage, but in both binary states the pulses return to zero for the second half of the bit time. As will be seen, the extra transitions between logic one and logic zero provide some resynchronization capability.

The *RZ-unipolar* code shown in Figure 9-9 has the same limitations and disadvantages as the NRZ group. A dc level appears on the data stream for a series of 1s or 0s. Synchronizing capabilities are also limited. These deficiencies are overcome by modifications in the coding scheme, which include using bipolar signals and alternating pulses. The *RZ-bipolar* code provides a transition at each clock cycle, and a bipolar pulse technique is used to minimize the dc component. Another RZ code is *RZ-AMI*. The alternate-mark-inversion (AMI) format provides alternating pulses for the 1s. (Recall from the Chapter 8 discussion of frequency-shift keying that a logic 1 is often referred to as a "mark" and a logic zero a "space.") This technique almost completely removes the dc component from the data stream, but since a data value of 0 is 0 V, the system can have poor synchronizing capabilities if a series of 0s is transmitted. This deficiency can also be overcome by transmission of the appropriate start, synchronizing, and stop bits. Table 9-2 provides descriptions of the RZ codes.

TABLE 9-2 • RZ Codes
RZ (unipolar) (return-to-zero)
1 (hi)—transition at the beginning of the interval
0 (low)—no transition
RZ (bipolar) (return-to-zero)
1 (hi)—positive transition in the first half of the clock interval
0 (low)—negative transition in the first half of the clock interval
RZ-AMI (return-to-zero—alternate-mark inversion)
1 (hi)—transition within the clock interval alternating in direction
0 (low)—no transition

BIPHASE AND MILLER CODES   The popular names for phase-encoded and delay-modulated codes are biphase and Miller codes. Biphase codes are popular for use in optical systems, satellite telemetry links, and magnetic recording systems. Biphase M is used for encoding Society of Motion Picture and Television Engineers (SMPTE) time-code data for recording on videotapes. The biphase code is an excellent choice for this type of media because the code does not have a dc component to it. Another important benefit is that the code is self-synchronizing, or self-clocking. This feature allows the data stream speed to vary (tape shuttle in fast and slow search modes) while still providing the receiver with clocking information.

One particularly important code used in computer networks is the *biphase L* or *Manchester* code. This code is used on the Ethernet standard IEEE 802.3 for local-area networks (LANs). In contrast to the NRZ and RZ codes the Manchester format requires a transition between amplitudes at the center of *each* bit time. The *transition* between amplitudes, not the amplitudes themselves, represents the logic state of the transmitted data. A binary one is defined as a change in amplitude from a high to a low level, whereas a binary zero is a low-to-high amplitude change. The transition occurs at the center of each bit time no matter which logic level is represented. Thus, even long strings of ones and zeros would still result in the amplitude of the transmitted signal changing from high to low or vice versa within each bit time. This fact, coupled with the bipolar or biphase nature of Manchester codes, in which a positive amplitude is offset by a corresponding negative amplitude within each bit time, means that there is no dc component within the transmitted data stream. The Manchester code is thus well suited for communications between remotely located pieces of equipment or in situations where dc continuity and ground references cannot be maintained between transmitter and receiver.

Figure 9-9 provides examples of these codes and Table 9-3 summarizes their characteristics.

MULTILEVEL BINARY CODES   Codes that have more than two levels representing the data are called **multilevel binary** codes. In many cases the codes will have three levels. We have already examined two of these codes in the RZ group: RZ (bipolar) and RZ-AMI. Also included in this group are *dicode NRZ* and *dicode RZ*. Table 9-4 summarizes the multilevel binary codes.

TABLE 9-3 • Phase-Encoded and Delay-Modulation (Miller) Codes	
Biphase	M (biphase-mark)
	1 (hi)—transition in the middle of the clock interval
	0 (low)—no transition in the middle of the clock interval
	Note: There is always a transition at the beginning of the clock interval.
Biphase	L (biphase-level/Manchester)
	1 (hi)—transition from high-to-low in the middle of the clock interval
	0 (low)—transition from low-to-high in the middle of the clock interval
Biphase	S (biphase-space)
	1 (hi)—no transition in the middle of the clock interval
	0 (low)—transition in the middle of the clock interval
	Note: There is always a transition at the beginning of the clock interval.
Differential Manchester	1 (hi)—transition in the middle of the clock interval
	0 (low)—transition at the beginning of the clock interval
Miller/delay modulation	1 (hi)—transition in the middle of the clock interval
	0 (low)—no transition at the end of the clock interval unless followed by a zero

TABLE 9-4 • Multilevel Binary Codes	
Dicode	NRZ
	One-to-zero and zero-to-one data transitions change the signal polarity.
	If the data remain constant, then a zero-level is output.
Dicode	RZ
	One-to-zero and zero-to-one data transitions change the signal polarity in half-step voltage increments. If the data don't change, then a zero-voltage level is output.

SYNCHRONIZATION AND BANDWIDTH   Careful examination of Figure 9-9 reveals that waveforms produced by all the coding formats discussed up to this point share one common characteristic: The transition between logic states results in an "edge," or fast change in the rise or fall times of the transmitted pulses. This edge, created by the amplitude transitions of the transmitted data, is the embedded feature used by the receiver to recover the transmitter clock. The underlying principle is similar to that of coherent carrier recovery described in the context of digital modulation formats in Chapter 8. An edge-detector circuit within the receiver produces a pulse train that is, in turn, applied to one input of a phase-locked loop (PLL); the other input of the loop is the receiver clock oscillator. The PLL uses the pulses created by the edge detector as a frequency reference to synchronize the receiver clock oscillator with the transmitter oscillator that originally generated the data pulses. This resynchronization capability is essential for the operation of any synchronous digital communications system, and the ability of the receiver PLL to acquire and retain lock is essential to maintaining synchronization. It is also directly affected by how frequently the amplitude of the transmitted data stream changes state. In the RZ case, the number of logic ones determines the transition frequency, whereas Manchester-coded data will have the maximum number of transitions because the coded data signal changes amplitude within each bit time. For this reason, Manchester-encoded data are said to be *fully reclocking* because the maximum number of transitions occurs, thus producing the highest number of recovered clock pulses and hence the most stable frequency reference for the receiver PLL. The RZ code is *partially reclocking* because it produces fewer transitions than does the Manchester code, but the transitions are made to occur often enough for the receiver to maintain synchronization. This also means that the bandwidth requirements for partially reclocking codes are lower than those for their fully reclocking brethren. We will see shortly that partially reclocking codes used in telephone systems employ additional circuitry designed to guarantee a certain "ones density"; that is, circuits will insert patterns with logic ones into a data stream that has many zeros in sequence to force amplitude transitions, thus helping to guarantee that the receiver detects enough transitions to maintain synchronization. The NRZ code is nonreclocking because amplitude transitions only occur at the beginning of each bit interval. For all the schemes, the logic level of each data bit is determined within the bit time rather than at the beginning or end to minimize ambiguities caused by changing amplitudes. Fully reclocking schemes such as Manchester encoding do not require the insertion of synchronization bits but will occupy the widest bandwidth because the transmitted signal amplitude is changing with every data bit, thereby producing the highest frequency baseband signal. Partially reclocking schemes such as RZ or its close relative, AMI, operate at a lower baseband frequency, hence reducing the transmitted bandwidth, because amplitude transitions in the modulated signal are governed only the number of logic ones in the data stream. The coding scheme ultimately chosen is thus determined largely by the available bandwidth and the need for the transmitter and receiver to maintain synchronization.

9-4 THE T-CARRIER SYSTEM AND MULTIPLEXING

The backbone of the global network of interconnected telephone systems, as well as interconnections within the systems themselves, is built up from multiplexed streams of pulse-code modulated (PCM) representations of voice and other data in the form of digital words. This backbone is composed of a hierarchical data structure that is time-division multiplexed and aggregated into larger and larger streams. In North America, the arrangement is known as the T-carrier system (the "T" is for telephone), whereas in the rest of the world a similar multiplexed data hierarchy is dubbed the E-Carrier system ("E" for Europe).

Time-Division Multiplexing

The concept of *multiplexing* was introduced as part of the discussion of pulse modulation in Chapter 7. In general terms, *multiplexing* involves conveying two or more information signals over a single transmission channel. Several types of multiplexing are possible, but the type of multiplexing used in T- and E-carrier systems is *time-division multiplexing* (TDM), in which each information signal accesses the entire channel bandwidth but for only a small part of the available time. An extension of the basic idea of TDM is **time-division multiple access (TDMA)**, which is a technique used to transport data from multiple sources over the same serial data channel. TDMA techniques are used for transporting data over wired systems and are also used for wireless communication (e.g., some cellular telephones). An example of generating a TDMA output is provided in Figure 9-10. A1–A4, B1–B4, and C1–C4 represent digital data from three different users. A multiplexer (MUX) is used to combine the source data into one serial data stream. A key to successful data multiplexing is that the multiplexing frequency (f_m) must be fast enough so that the throughputs of the A, B, and C data are output fast enough, so that the system does not become congested and none of the data are lost.

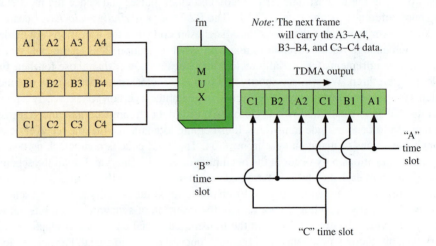

FIGURE 9-10   An example of generating a TDMA output.

The TDMA output data stream (Figure 9-10) shows the placement of the A, B, and C data in the TDMA frame. Each block represents a **time slot** within the frame. The time slot provides a fixed location (relative in time to the start of a data frame) for each group of data so at the receiver, the data can be recovered easily. The process of recovering the data is shown in Figure 9-11. The serial data is input into a **demultiplexer (DMUX)**, which recovers the A, B, and C group data. If only the information in the A time slots is to be recovered, then the data in the B and C time slots are ignored.

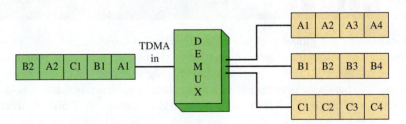

FIGURE 9-11   An example of recovering TDMA data.

In wireless systems the arrival of the TDMA data can be an issue because of potential multipath RF propagation problems with stationary systems and the added RF propagation path problems introduced by mobile communications. To compensate for the variation in data arrival times, **guard times** are added to the TDMA frame. If the data arrive too closely together, then potential **intersymbol interference (ISI)** from overlapping data can occur, resulting in an increase in the bit error rate (BER). The guard times provide an additional margin of error, thereby minimizing ISI and bit errors.

T-CARRIER MULTIPLEXING   Telephone systems make extensive use of TDM because the PCM representations of many different telephone calls can be simultaneously transmitted and received over the same line. The basic concept was first illustrated in Figure 7-2 in the context of a pulse-amplitude modulated signal being prepared for transmission over a wireless channel. The only difference between the system shown in that figure and the present discussion is that T-carrier multiplexing involves PCM words. To review, the multiplexing circuit in the transmitter and demultiplexing circuit in the receiver operate as synchronized rotating switches. The transmitter multiplexer causes a portion of the PCM signal from each input to be applied to the communication channel in sequence; the receiver demultiplexer outputs the received PCM signal on the appropriate line. In a T-carrier TDM system PCM data from channel 1 are transmitted first, then data from channel 2, followed by data from channels 3 and 4 in sequence before the process repeats. Data from each PCM input are interleaved for transmission on the communications channel, that is, portions of each input signal are arranged with portions of all other signals in a way that preserves the identity of each part.

Industry standards govern PCM data rates, synchronization process, and number of channels used. In telephone systems, PCM coded data are multiplexed into ever-larger groups for transport between central offices and long-distance switches. The groups are arranged in a hierarchy, with the smallest number of multiplexed channels at the bottom of the hierarchy and the largest at the top. In North America, the lowest level of the PCM hierarchy is the T1 and the highest the T4, whereas in Europe, PCM data are arranged into E carriers, which are similar to T carriers but with more channels and a higher overall data rate.

Tables 9-5 and 9-6 show the data rates for T and E carriers, respectively. The "DS" shown in Table 9-5 means "digital signal." Although the terms T1 and DS-1 (or T2 and DS-2, etc.) are often used interchangeably, there is, strictly speaking, a distinction. The DS designation, as the name implies, refers to the characteristics of the electrical signal itself, such as data rate and amplitude. This designation also describes the bottom or physical layer of the seven-layer open systems interconnection (OSI) reference model (to be covered in the context of computer networks in Chapter 11). The T designation refers to the physical components such as wires, plugs, and repeaters. These devices, when put together, form the T1 line on which the DS1 pulses are transported. This distinction is often not observed in practice, however; just be aware that both terms will be used to refer to the same thing.

TABLE 9-5 • Data Rates for the T and DS Carriers	
DESIGNATION	**DATA RATE**
T1 (DS-1)	1.544 Mbps
T2 (DS-2)	6.312 Mbps
T3 (DS-3)	44.736 Mbps
T4 (DS-4)	274.176 Mbps

TABLE 9-6 • E1 and E3 Data-Transmission Rates	
DESIGNATION	**DATE RATE**
E1	2.048 Mbps
E3	34.368 Mbps

THE T1 SIGNAL   Recall from the Chapter 7 introduction to PCM that a codec converts each analog input sample into a digital word at a rate determined by the Nyquist criterion, which requires the sample rate to be at least twice that of the highest-frequency component of the audio being digitized. The number of permissible quantum steps determines the number of bits in the digital word. Also remember that the T-carrier system was originally developed to handle telephone voice traffic, so the digital signal specifications were established to allow adequate voice reproduction without the need for excessive bandwidth. Voice frequencies range up to about 3.5 kHz or so, and filters are placed on the subscriber loop to remove energy at higher frequencies. The sample rate was set to be 8 kHz, somewhat above the Nyquist minimum. An 8-bit word is created with each sample, meaning that there are $2^8 = 256$ possible quantum levels. Therefore, each phone call requires 64 kbps:

$$8 \text{ bits/sample} \times 8000 \text{ samples/second} = 64,000 \text{ bits per second (64 kbps)}$$

The T1 line has capacity for 24 individual, 64-kbps, time-division-multiplexed telephone calls. Framing bits are added to the multiplexed voice data at a rate of 8 kbps to maintain data flow. This brings the total data rate for a T1 channel to:

$$24 \text{ channels} \times 64 \text{ kbps/channel} = 1.536 \text{ Mbps}$$
$$+ \quad \underline{\text{framing bits} = 8 \text{ kbps}}$$
$$\text{total bit rate} = 1.544 \text{ Mbps}$$

Multiplexing takes place when one 8-bit word from channel 1 is interleaved with one word from channel 2, and so on through channel 24, at which time the process repeats. From these interleaved words the codec produces *frames*, or sets of consecutive time slots, for eventual transmission on the communications medium. Frame synchronization is maintained by adding one bit after the data from channel 24 are output. Each frame consists of 24 time slots of 8 bits each, plus the synchronization bit, or 193 bits. Each second, 8000 total frames are created. The total bit rate, therefore, is 193 bits × 8000 frames per second = 1.544 Mb/s, the bit rate for the T1 carrier. Figure 9-12 shows the relationship between the time slots, frames, and synchronization bits. Systems using the E-carrier standard use the same sample rate and number of quantum steps but produce frames consisting of 32 channels for an E1 data rate of 2.048 Mb/s. The hierarchical structure is also borne out by the data rates shown in Table 9-5. The T3/DS-3 data rate, for example, consists of multiplexed data from 28 DS-1s.

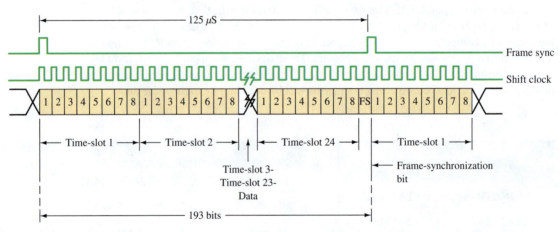

FIGURE 9-12 **T1 multiplexing and framing.**

Although the T-carrier system was originally developed for channelized voice traffic, the data need not be voice and do not necessarily need to be channelized. T-carrier systems instead act as pipelines through which any kind of data can be transported, and the capacity may be divided up and data formatted in a number of ways. The only requirements are that the data rates and signal specifications conform to those established by the T-carrier standard. The data lines can be leased from a communications carrier for carrying any type of data, including voice and video. It is important to note that when you lease a T1 (or higher capacity) line for providing a data connection from point A to point B, the communications carrier does not provide you with your own private physical connection. What they are providing is sufficient bandwidth in their system to carry your data as well as a guarantee of reliability. Your data will most likely be multiplexed with hundreds or even thousands of other T1 data channels.

A data facility may need only a portion of the data capability of the T1 bandwidth. **Fractional T1 (FT1)** is the term used to indicate that only a portion of the T1 bandwidth is being used. An individual 64-kb portion, equivalent to one voice channel, is sometimes designated as a DS-0. Fractional T1 service consisting of any number of DS-0's is available.

When data capacity is leased from a common carrier such as a wireline telephone service provider ("telco"), the carrier will install copper wire to the customer location. The point where the communication carrier brings in service to a facility is called the

point of presence. The point of presence also usually serves as the "demarc" or demarcation point, where responsibility for maintenance shifts from one party to the other: Everything facing back to the customer side from the demarc (the "customer premises equipment") is the customer's responsibility, whereas all equipment and connections facing back toward the carrier side is the carrier's responsibility. The point of presence is the point where users connect their data to the communications carrier's facilities. The connection can be by means of copper wire, fiber-optic lines, digital microwave systems, or digital satellite links.

Communications carriers also require that the connection be through a unit known as a **channel service unit/data service unit (CSU/DSU)**. The CSU/DSU provides the data interface to the carrier side, which includes adding the framing information for maintaining the data flow, storing performance data, and providing management of the line. An example of inserting the CSU/DSU in the connection to the Telco cloud is shown in Figure 9-13. The CSU/DSU also has three alarm modes for advising the user about problems on the link. The alarms are red, yellow, and blue. The conditions for each alarm are defined in Table 9-7. Also, in some cases, the carrier will install a network interface (NI), which provides many of the same functions as the CSU/DSU. The primary distinction is that the NI is the carrier's property and is under its control, while the CSU/DSU is usually customer property and under customer control.

FIGURE 9-13   **Insertion of the CSU/DSU in the connection to the Telco cloud.**

TABLE 9-7 • The CSU/DSU Alarms	
Red alarm	A local equipment alarm that indicates that the incoming signal has been corrupted.
Yellow alarm	Indicates that a failure in the link has been detected.
Blue alarm	Indicates a total loss of incoming signal.

T1 FRAMING   As already established, multiplexed T1 voice data is time-division multiplexed into 24 time slots plus framing bits. There is one framing bit appended to each frame. The state of each bit (0 or 1) depends on how the T1 circuit is provisioned. The original framing for the data in T1 circuits, and one still used in voice telephone circuits, is based on **D4 framing**. The D4 framing sequence consists of 12 bits forming the following pattern:

$$1\ 0\ 0\ 0\ 1\ 1\ 0\ 1\ 1\ 1\ 0\ 0$$

The purpose of framing, recall, is to maintain synchronization of the receiving equipment. The preceding pattern represents the logic states of the individual framing bits appended to the end of each frame in a group of 12 consecutive D4 frames.

Extended superframe framing (ESF) is an improvement in data performance over D4 framing. ESF extends the bit sequence to 24 frames. The extended frame length creates 24 ESF framing bits with several purposes, as shown in Table 9-8.

ESF uses only 6 bits for frame synchronization compared to the 12 synchronizing bits used in D4 framing. ESF uses 6 bits for computing an error check code using the cyclic redundancy check (CRC) methodology described in Chapter 7. The code is used to verify the data transmission was received without errors. Twelve bits of the ESF frame are used for maintenance and control of the communications link. Examples of the use of the maintenance and control bits include obtaining performance data from the link and configuring loopbacks for testing the link.

TABLE 9-8 • The Function of the 24 ESF Framing Bits	
6 bits	Frame synchronization
6 bits	Error detection
12 bits	Communications link control and maintenance

LOOPBACKS   The ability to perform loopbacks is very useful in system maintenance. Many digital systems include a **loopback** capability, which causes transmitted data to be routed back to the originating location by the receiver or, in telephone and wireline data systems, by the CSU/DSU or NI. Often, the equipment can be commanded remotely to perform the loopback, allowing testing to be done by one person without the need to travel to a remote site. Whether done in the receiver or by an external piece of equipment, when looped back the received data are sent back to the transmitter. (For voice traffic, this means you would be talking to yourself.) The data are then compared with the originally transmitted data to provide an indication of system performance. Bit errors can occur both in the original transmission and in the loopback, so the source of the errors could not be pinpointed with a single test because it will not be known where the error occurred. In that case, either an "end-to-end" test or a series of loopbacks from different locations in the system would have to be performed. Nonetheless, the loopback test is very helpful in diagnosing the basic system performance.

Three loopback tests are shown in Figure 9-14. The loopback test marked A is used to test the cable connecting the router to the CSU/DSU. Loopback test B is used to test the link through the CSU/DSU. Loopback test C tests the CSU and the link to telco.

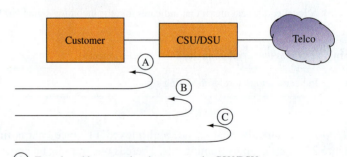

Ⓐ Tests the cable connecting the router to the CSU/DSU

Ⓑ Tests the link through the CSU/DSU

Ⓒ Tests the CSU/DSU and the link to Telco

FIGURE 9-14   Three loopback configurations used to test the serial communications link.

T1 LINE CODING   Recall from the earlier discussion that the purpose of line coding in digital systems is threefold: Line codes allow for synchronization, they eliminate the need for data states to be represented in terms of absolute voltage levels, and they provide some error detection ability. Two line codes of particular importance for T1 data are AMI and B8ZS. Though these two codes serve the same functions as the codes described earlier, they were developed specifically to maintain synchronization in the bandwidth-constrained environment of the telephone system.

A fundamental coding scheme that was developed for transmission over T1 circuits, and one that is still in use for voice-grade circuits in telephone systems, is **AMI**. The AMI code provides for alternating voltage level pulses (+V and −V) for representing the ones. This technique removes the dc component of the data stream almost completely, which helps to maintain synchronization. An example of the AMI coded waveform is shown in Figure 9-15. Notice that successive 1s are represented by pulses in the opposite direction (+V and −V). This is called **bipolar coding** and gives the AMI format an error-detection capability because the ground rule established by the protocol is that any two consecutive, valid logic 1 states *must* have opposite polarities. If, as the result of a burst of noise or equipment malfunction, two pulses of the same polarity are received, then a *bipolar violation* (BPV) is deemed to have occurred, and the system handles this occurrence in accordance with the terms of its underlying protocols. Usually, bit-error-rate testing equipment or CSU/DSUs will have a visual indication that BPVs are occurring and will sometimes keep count of the number of occurrences per unit time. In some cases, code patterns are used in which BPVs have been introduced deliberately. We will see an example of this shortly.

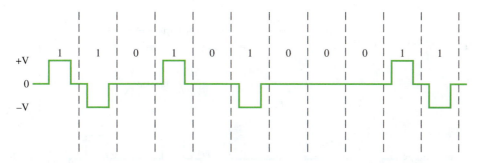

FIGURE 9-15   An example of the AMI data encoding format.

The 0s have a voltage level of 0 V. Notice that when there are successive 0s a straight line at 0 V is generated. A flat line generated by a long string of 0s can produce a loss of timing and synchronization. This deficiency can be overcome by the transmission of the appropriate start, stop, and synchronizing bits, but this comes at the price of adding overhead bits to the data transmission and consuming a portion of the data communication channel's bandwidth.

The **bipolar 8 zero substitution (B8ZS)** data encoding format was developed to address the synchronization issue created by long strings of zeros. Recall from the earlier discussion that synchronous data systems such as T1 circuits derive their timing information from the edges created when a pulse transitions from one logic state to the other. However, T1 circuits only produce pulse transitions when logic ones are present; long strings of zeros produce periods of no activity. T1 circuits require that a minimum *ones density* level be met so that the timing and synchronization of the data link are maintained. Maintaining a **minimum ones density** means that a pulse is intentionally sent in the data stream, even if the data being transmitted is a series of 0s. Intentionally inserting the pulses in the data stream helps to maintain the timing and synchronization of the data stream. The inserted data pulses include two **bipolar violations**, which means that the pulse is in the same voltage direction as the previous pulse in the data stream.

In B8ZS encoding, eight consecutive 0s are replaced with an 8-bit sequence that contains two intentional bipolar violations. An example of B8ZS encoding is shown in Figure 9-16 for both cases of bipolar swing prior to the transmission of eight consecutive 0s. Figure 9-16(a) shows the bipolar swing starting with a +V, whereas Figure 9-16(b) shows the bipolar swing starting with −V. The receiver detects the bipolar violations in the data stream and replaces the inserted byte (8 bits) with all 0s to recover the original data stream. The result is that timing is maintained without corrupting the data. The advantage of using the B8ZS encoded fromat is that the bipolar violations enable the timing of the data transmission to remain synchronized without the need for overhead bits. This arrangement provides for use of the full channel data capacity.

BANDWIDTH CONSIDERATIONS   The discussion just concluded demonstrated that B8ZS coding was developed to improve the synchronization capabilities of line codes

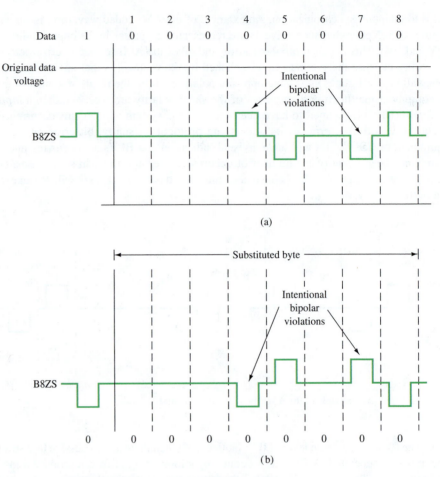

FIGURE 9-16   **Intentional bipolar violations in B8ZS encoding.**

used in T carrier networks. Synchronization was improved by ensuring a sufficient number of pulse transitions, but these transitions only occurred in the presence of logic one levels. No transitions occurred with zeros. Manchester codes, however, force a transition with each bit, zero or one, making it fully reclocking. Why, then, aren't Manchester codes used? The answer has to do with bandwidth and is related to the ideas introduced in the context of modulated signals from Section 8-2. Recall in particular from that discussion and from Example 8-4 that the highest-bandwidth baseband signal occurs with a recurring 1 0 1 0 pattern. Not stated there was the fact that bandwidth is also dependent on the form of line coding used. Thus, an alternating NRZ-L pattern would produce the highest bandwidth signal when the recurring 1 0 1 0 pattern is transmitted, but the Manchester coded signal will always have the highest bandwidth.

We can now formalize the relationship between bit rate in a digital system and channel bandwidth. This relationship holds true for both baseband and modulated data and demonstrates that system capacity and bit rate are not unlimited. Instead, the **Shannon-Hartley theorem** allows the channel capacity to be determined when its bandwidth and noise are known:

$$C = \mathrm{BW} \log_2 (1 + S/N), \tag{9-1}$$

where C = channel capacity (bps)

   BW = bandwidth of the system

   S/N = signal-to-noise (S/N) ratio.

An example using Equation (9-1) is provided in Example 9-1.

EXAMPLE 9-1

Calculate the capacity of a telephone channel that has an S/N of 1023 (60 dB).

SOLUTION

The telephone channel has a bandwidth of about 3 kHz. Thus,

$$\begin{aligned}
C &= BW \log_2(1 + S/N) \\
&= 3 \times 10^3 \log_2(1 + 1023) \\
&= 3 \times 10^3 \log_2(1024) \\
&= 3 \times 10^3 \times 10 \\
&= 30,000 \text{ bits per second.}
\end{aligned} \tag{9-1}$$

Example 9-1 shows that a telephone channel could theoretically handle 30,000 b/s if an *S/N* power ratio of 1023 (60 dB) exists on the line. The Shannon–Hartley theorem represents a fundamental limitation. The consequence of exceeding it is a very high bit error rate. Generally, an acceptable bit error rate (BER) of 10^{-5} or better requires significant reductions from the Shannon–Hartley theorem prediction.

9-5 PACKET-SWITCHED NETWORKS

Since its inception, the telephone system has been based on a "circuit-switching" topology. The circuit can be a direct path supporting current flow, such as the path established between a subscriber phone and the central office, or a virtual one involving many connections until it reaches the phone of the destination party. The distinctive feature is that, once it has been set up, the physical or virtual circuit remains established for the duration of the call. An alternative technique, and one that is seeing greater prominence because of the blurring of the boundaries between telephone and computer networks, is the idea of packet switching. In this data-handling technique, data are divided into small segments called **packets**. A typical packet size is 1000 bits. The individual packets of an overall message do not necessarily take the same path to a destination. They are held for very short periods of time at switching centers and then retransmitted. The process happens in near real time, but the packets do not necessarily take the same path to their destination. Processors at the switching centers monitor the packets continuously from the standpoint of source, destination, and priority. The processors then direct each packet so the network is used most efficiently. The process is termed **statistical concentration**. The price paid for the very high efficiency afforded and near-real-time transmission is the need for very complex protocols and switching arrangements. As packet switching technology progresses, the user may not need to choose in advance between packet and circuit switching; the techniques may merge. This could be the result of "fast" packet switches. The development of packet switch systems operating at millions of packets per second could eliminate the need for circuit switching.

Frame Relay

A **frame relay** is a packet switching network designed to carry data traffic over a **public data network (PDN)** such as the local telephone company. Frame relay is an extension of the X.25 packet switching system; however, frame relay does not provide the error-checking and data-flow control of X.25. **X.25** was designed for data transmission over analog lines, and frame relay is operated over higher-quality, more reliable digital data lines.

Frame relay operates on the premise that the data channels will not introduce bit errors or, at worst-case, minimal bit errors. Without the overhead bits for error checking and data-flow control, the transfer of data in a frame relay system is greatly improved. If an error is detected, then the receiver system corrects the error. The frame relay protocol enables calls or connections to be made within a data network. Each data frame contains a connection number that identifies the source and destination addresses.

The commercial carrier (telco) provides the switch for the frame relay network. Telco provides a guaranteed data rate, or **committed information rate (CIR)**, based on the service and bandwidth requested by the user. For example, the user may request a T1 data service with a CIR of 768 kbps. A T1 connection allows a maximum data rate of 1.544 Mbps. The telco allows for bursty data transmissions and will allow the data transfer to go up to the T1 connection carrier bandwidth, even though the CIR is 768 kbps. *Bursty* means that the data rates can momentarily exceed the leased CIR data rate of the service. Communication carriers use what is called a **committed burst information rate (CBIR)**, which enables subscribers to exceed the committed information rate (CIR) during times of heavy traffic. Note that the bursty data transmission rate can never exceed the data rate of the physical connection. For example, the maximum data rate for a T1 data service is 1.544 Mbps, and the bursty data rate can never exceed this rate.

Asynchronous Transfer Mode

The **asynchronous transfer mode (ATM)** is a cell relay technique designed for voice, data, and video traffic. Cell relay is considered to be an evolution of **packet switching** because packets or cells are processed at switching centers and directed to the best network for delivery. The stations connected to an ATM network transmit octets (8 bits of data) in a cell that is 53 octets (bytes) long. Forty-eight bytes of the cell are used for data (or **payload**) and 5 bytes are used for the cell header. The ATM cell header contains the data bits used for error checking, virtual circuit identification, and payload type.

All ATM stations are always transmitting cells, but the empty cells are discarded at the ATM switch. This technique provides more efficient use of the available bandwidth and allows for bursty traffic. The station is also guaranteed access to the network with a specified data frame size. This is not true for IP networks, where heavy traffic can bring the system to a crawl. The ATM protocol was designed for use in high-speed multimedia networking, including operation in high-speed data transmission from T1 up to T3, E3, and SONET. (SONET is the synchronous optical network and is covered in Chapter 16.) The standard data rate for ATM is 155 Mbps, although the data rates for ATM are continually evolving.

ATM is connection oriented and uses two different types of connections, a **virtual path connection (VPC)** and a **virtual channel connection (VCC)**. A virtual channel connection is used to carry the ATM cell data from user to user. The virtual channels are combined to create a virtual path connection that is used to connect the end users. Virtual circuits can be configured as permanent virtual connections (PVCs) or they can be configured as **switched virtual circuits (SVCs)**.

Five classes of services are available with ATM. These classes are based on the needs of the user. In some applications, the user needs a constant bit rate for applications such as teleconferencing. In another application, the user may need only limited periods of higher bandwidth to handle bursty data traffic. The five ATM service classes are provided in Table 9-9.

TABLE 9-9 • The Five ATM Service Classes

ATM SERVICE CLASS	ABBREVIATION	DESCRIPTION	TYPICAL USE
Constant bit rate	CBR	Cell rate is constant	Telephone, videoconferencing, television
Variable bit rate/not real time	VBR-NRT	Cell rate is variable	E-mail
Variable bit rate/real time	VBR-RT	Cell rate is variable but can be constant on demand	Voice traffic
Available bit rate	ABR	Users are allowed to specify a minimum cell rate	File transfers/e-mail
Unspecified bit rate	UBR		TCP/IP

ATM uses an 8-bit **virtual path identifier (VPI)** to identify the virtual circuits used to deliver cells in the ATM network. A 16-bit **virtual circuit identifier (VCI)** is used to identify the connection between the two ATM stations. The VPI and VCI numbers are provided by telco. Together, the numbers are used to create an ATM PVC through the ATM cloud.

9-6 SIGNALING SYSTEM 7

Over the last 125 years, telephone service providers have used many different types of **signaling systems** to administer phone calls in various ways. Signaling systems are responsible for many functions, including setting up and removing each call from the telephone network and routing calls from their origin to their destination. Signaling systems are also what allow the various elements of the **public switched telephone network (PSTN)** to communicate with each other. PSTN is a general term describing a shared system used by all telephone service providers throughout the United States.

Telephone systems were originally set up to handle only voice traffic through analog networks. The use of T1 lines (1.544 Mbps data rate links) to TDM calls provided more efficient utilization of each physical connection. In 1984 telephone service providers started using a system called integrated services digital network (ISDN) to administer calls from private business locations having customer premise equipment (CPE).

ISDN systems are physically "**in-band**" and logically "**out-of-band**." In-band means that the same physical wires are used to multiplex both the voice traffic and the data traffic required to administer the system. Out-of-band means that various increments of time are dedicated for signaling and are not available for voice traffic.

Beginning in 1980, and by the end of 1999, almost all telephone service providers in the United States implemented a signaling system called **SS7** to administer the PSTN. Today, most telephone systems throughout the world use some version of SS7. In contrast with ISDN, SS7 uses physical out-of-band signaling. This means that the network that "sets up" and "tears down" the phone calls is totally separate from the network that provides the actual voice traffic. "Setting up" a call means that a customer decides to make a call, dials the call and is connected to the number he/she wanted to call. "Tearing down" the call means that the call has ended and one party hangs up, making the trunk (telephone circuit) available for another call.

The functions of ISDN and SS7 follow the guidelines provided by the OSI (open systems interconnection) model. (The layers of the OSI model will be discussed in Chapter 11.) This model has seven layers. ISDN uses layers 1–3 of the OSI model and SS7 uses layers 1–7. The SS7 levels are illustrated in Figure 9-17.

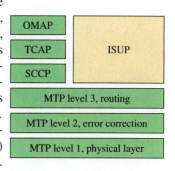

FIGURE 9-17   SS7 levels.

The SS7 layers: OMAP (operations, maintenance, and admistration part), TCAP (transactions capabilities application) part, SCCP (signaling connection control part), MTP (message transfer part), and ISUP (ISDN user part).

One characteristic of the OSI model is that each layer "provides a service" to the next higher layer. The following is a description for each SS7 layer that will illustrate the services provided.

1. The first SS7 layer (shown in Figure 9-17) is called the "physical layer" and is also called "MTP 1." This layer includes the physical hardware as well as the T1 multiplexing method.

2. The second SS7 layer, MTP 2, provides error checking. It consists of three types of message units: MSU (message signaling unit), LSSU (link status signaling unit), and FISU (fill-in signaling unit). The MSU carries signals to level MTP 3 and other upper layers. The LSSU message is used to connect and disconnect links. In an operational network, these messages can indicate serious problems. FISUs occupy the link when there is no traffic. In the analogy of a railroad train, the physical layer would be the railroad track. A FISU would be like an empty railroad car.

3. The third SS7 layer, MTP 3, is involved in routing signaling messages. Key information in this layer is the OPC (origination point code) and the DPC (destination point code). This corresponds to identifying the switch closest to the person making the call (OPC) and the person receiving the call (DPC).

4. The fourth SS7 layer, SCCP, provides a variety of translating functions. One type of translation would be converting "800 prefix" numbers to standard area codes.

5. The fifth SS7 layer, TCAP, provides a method for different telephone service providers to communicate with each other.

6. The sixth SS7 layer, OMAP, provides various tests to make sure that the network is operating correctly.

7. The seventh SS7 layer, ISUP, communicates directly with MTP 3, SCCP, TCAP, and OMAP. This is the level that SS7 users would normally use first in order to study the performance of the network and troubleshoot problems.

There are fifty-nine types of ISUP messages, but we will discuss only the following five types: IAM, ACM, ANM, REL, and RLC.

1. **IAM (initial address message):** This message starts the call; that is, a user has dialed a number.

2. **ACM (address complete message):** This indicates that the called party has been found and that the phone is ringing.

3. **ANM (answer message):** This shows that the called party (or answering machine) has picked up the phone.

4. **REL (release message):** This occurs when either the calling or called party hangs up the phone.

5. **RLC (release complete message):** This acknowledges the REL message and makes the voice trunk available for another call.

Troubleshooting SS7 Networks

A typical telephone network will receive thousands of signaling messages per second. Technicians use instruments called *protocol analyzers* to sort through these messages to find "a needle in a haystack" that can identify a problem. Figure 9-18 is a display from a Tektronix K15 protocol analyzer. The columns show the following information:

1. The number of the messages received.

2. The time the message was recorded. Using the first two columns you can see that 30 messages (14891–14921) were received within 1 second (1:53:35 p.m. to 1:53:36 p.m.) on just the one monitored link.

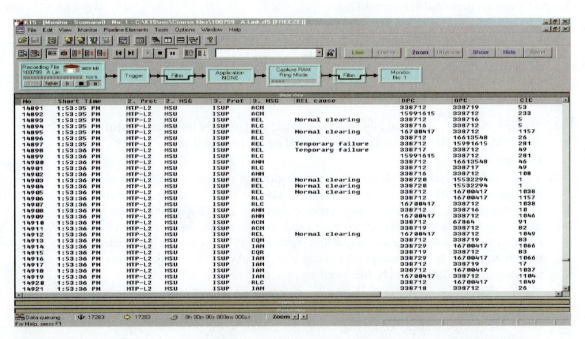

FIGURE 9-18   An example of captured telephone signaling messages as displayed using a Tektronix K15 protocol analyzer.

3. The SS7 layer type. This is MPT-L2, the "error-checking layer."

4. The type of message unit. This is "MSU" (message-signaling unit).

5. These are all "ISUP" messages, which are "seventh layer messages."

6. These are types of ISUP messages. Note: The REL are the messages that will be studied in this example.

7. "REL cause." There are thirty-five different reasons for releasing a call. The most common is "normal clearing," which means that one party hangs up the phone.

8. The DPC (destination point code). This identifies the switch closest to the called party.

9. The OPC (origination point code). This identifies the switch closest to the person making the call.

10. The CIC (circuit identification code). This is the trunk identifier.

The information displayed in Figure 9-19 is the result of applying a filter to look only at a specific type of release message called *temporary failure*. Note that 23 calls could not be completed during a 6-minute period. That's 230 per hour, or 5,520 per day, or more than 2,000,000 lost calls per year! This would likely be true for the other 30-to-40 links (average switching center). The total could therefore equal between 60 and 80 million lost calls annually and a loss in revenue of about $725,000!

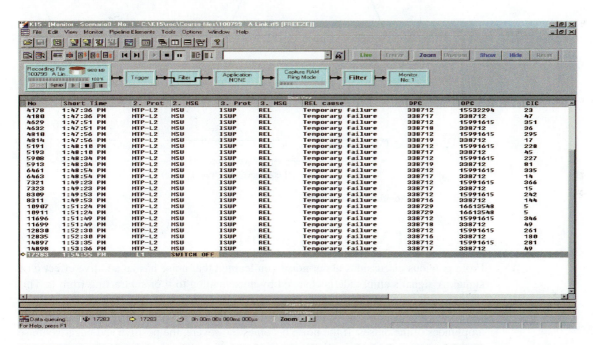

FIGURE 9-19 Configuration on the protocol analyzer set to display only the "temporary failure" messages.

9-7 TROUBLESHOOTING

Digital data are used in every aspect of electronics in one form or another. In this section we will look at digital pulses and the effects that noise, impedance, and frequency have on them. Digital communications troubleshooting requires that the technician be able to recognize digital pulse distortion and to identify what causes it.

After completing this section you should be able to

• Identify a good pulse waveform;

• Identify frequency distortion;

• Describe effects of incorrect impedance on the square wave; and

• Identify noise on a digital waveform.

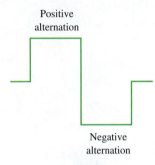

Positive alternation

Negative alternation

FIGURE 9-20 **Ideal square wave with 50% duty cycle.**

The Digital Waveform

A square-wave signal is a digital waveform and is illustrated in Figure 9-20. The square-wave shown is a periodic wave that continually repeats itself. It is made up of a positive alternation and a negative alternation. The ideal square-wave will have sides that are vertical. These sides represent the high-frequency components. The flat-top and bottom lines represent the low-frequency components. A square wave is composed of a fundamental frequency and an infinite number of odd harmonics, as described in Chapter 1.

Figure 9-21 shows this same square wave stretched out to illustrate a more true representation of it. Notice the sides are not ideally vertical but have a slight slope to them. The edges are rounded off because transition time is required for the low pulse to go high and back to low again. Figure 9-21 shows the positive alternation as a positive pulse and the negative alternation as a negative pulse. Rise time refers to the pulse's low-to-high transition and is normally measured from the 10% point to the 90% point on the waveform. Fall time represents the high-to-low transition and is measured from the 90% and 10% points. From the pulse's maximum low point to its maximum high point is the amplitude measured in volts. By observing the pulse waveform in response to circuit conditions that it may encounter, much can be determined about the circuit.

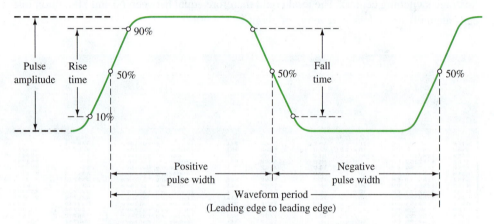

FIGURE 9-21 **Nonideal square-wave.**

Effects of Noise on the Pulse

From previous discussions about noise you learned that noise has an additive effect on a signal. A signal's amplitude is changed by noise adding to it or subtracting from it. This concept is depicted in Figure 9-22. The positive pulse and the negative pulse have changed from the ideal as a direct result of encountering noise. The noise has changed the true amplitudes of the pulses. If the noise becomes too severe, the positive and negative noise excursions might be mistaken by logic circuits as high and low pulses. Proper noise compensating techniques, shielding, and proper grounds help to reduce noise. If the digital waveforms under test show signs of deterioration due to noise, troubleshoot by checking compensation circuits, shielding, and ensuring that proper grounds are made.

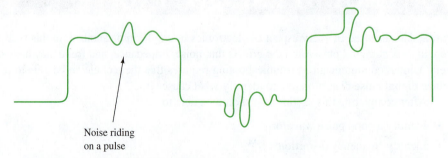

Noise riding on a pulse

FIGURE 9-22 **Illustration of noise on a pulse.**

Effects of Impedance on the Pulse

The square wave pulse can show the effects of impedance mismatches, as seen in Figure 9-23. The pulse in Figure 9-23(a) is severely distorted by an impedance that is below that required. For example, if RG-58/AU coaxial cable, carrying data pulses, became shorted or if one of its connectors developed a low-resistance leakage path to ground, then the data pulses would suffer low-impedance distortion. Figure 9-23(b) shows a type of distortion on the top and bottom of the square-wave called ringing. Ringing is the result of the effects of high impedance on the pulse. If a transmission line is improperly terminated or develops a high resistance for some reason, then ringing can occur. A tank circuit with too high of a Q will cause ringing. By adjusting the tank's Q or by adding proper termination to a transmission line, the waveform can be brought back to normal, as shown in Figure 9-23(c). When working with data lines, ensure that proper terminations are made. The effect of impedance loading is reduced in communications equipment like transmitters, receivers, and data handling circuits when proper repair and alignment maintenance techniques are used.

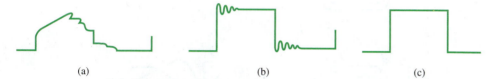

(a)   (b)   (c)

FIGURE 9-23   **Effects of impedance mismatches: (a) impedance is too low; (b) impedance is too high (ringing); (c) impedance is matched.**

Effects of Frequency on the Pulse

Digital pulses will not be distorted when passing through an amplifier with a sufficient bandwidth or a transmission line with a sufficient bandwidth. Figure 9-24(a) shows that the pulse does not become distorted in any fashion. The dip at the top and bottom of the waveform in Figure 9-24(b) represents low-frequency attenuation. The high-frequency harmonic components pass without being distorted, but the low-frequency components are attenuated. Coupling capacitors in communications circuits can cause low-frequency distortion. Low-frequency compensating network malfunctions will also attenuate the low-frequency components of the pulse waveform.

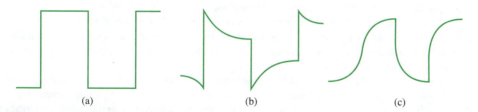

(a)   (b)   (c)

FIGURE 9-24   **Effects of frequency on data pulses: (a) good pulses; (b) low-frequency attenuation; (c) high-frequency attenuation.**

High-frequency distortion occurs when the high-frequency harmonic components of a digital square wave are lost. The edges become rounded off, as seen in Figure 9-24(c). The high harmonic frequencies are lost when the media bandwidth becomes too narrow to pass the pulse in its entirety. A circuit's bandwidth can change when the value of circuit components such as coils, capacitors, and resistors increases or decreases. Open and shorted capacitors found in tank circuits can also cause high-frequency losses.

Eye Patterns

Another technique that is extremely helpful in diagnosing the performance of a digital system is the generation of **eye patterns**. They are generated by "overlaying" on an oscilloscope all the digital bit signals received. Ideally, this would result in a rectangular display because of the persistence of the CRT phosphor. The effects of transmission cause various rounding effects, which result in a display resembling an eye.

Refer to Figure 9-25 for the various possible patterns. The opening of the eye represents the voltage difference between a 1 or a 0. A large variation indicates noise problems, while nonsymmetrical shapes indicate system distortion. Jitter and undesired phase shifting are also discernible on the eye pattern. The eye pattern can be viewed while making adjustments to the system. This allows for the immediate observation of the effects of filter, circuit, or antenna adjustments.

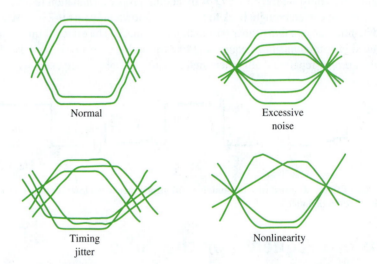

Normal

Excessive noise

Timing jitter

Nonlinearity

FIGURE 9-25   **Eye patterns.**

SUMMARY

The focus of this chapter was on the wired telephone system. Though the lines between telephone and computer networks are becoming increasingly blurred, and many of the principles introduced here in the context of telephone networks find applicability elsewhere, the worldwide reach of telephone systems, and the fact that interfacing with them is often a job requirement, merits their study as a separate topic from other forms of communication.

Communications network designers are concerned with efficient use of resources, so much attention is paid to usage patterns and deployment of resources. Carriers are concerned with traffic measurement and congestion management, and telephone traffic is measured either in units of erlangs or hundred-call-seconds.

The interface from the central office to the subscriber's home is analog, but from the central office through the system to its ultimate destination, any calls made will have been converted to a digital format. This fact requires the development of communications links and protocols for handshaking, line control, and other essential system functions.

A crucial aspect of any digital communications system is embodied in the concept of line coding. Line codes permit synchronization of transmitter and receiver clocks, obviate the need for continuity between system elements, and provide a degree of error detection. A number of these codes is in use; the choice largely boils down to the bandwidth available and the amount of reclocking, hence frequency stability, required. Two line codes used in telephone networks in particular are the alternate-mark inversion and bipolar 8 zero substitution (B8ZS) codes, with the latter becoming increasingly dominant. The advantage to B8ZS is that it increases synchronization capability by preventing long strings of logic low (zero) levels from being transmitted.

Time-division-multiplexed voice traffic and other data is organized into time slots governed by the specifications of the T-carrier system, a hierarchical arrangement with 64-kbps voice channels at the bottom of the hierarchy and ever-increasing groupings of data multiplexed together at the top. The T carrier system is standard in North America. A similar arrangement, E carriers, is used throughout much of the rest of the world. The base data rate 1.544 Mbps is defined by the DS1 framed format of 24 voice channels plus 8 kbps framing bits. Two types of framing, D4 and ESF, are used. DS1 data rates can be further multiplexed into higher data rates such as DS3.

Telephone systems are primarily circuit-switched networks. However, the blurring of the lines between telephone and computer networks has given rise to packet-switched network topologies as well. Among these are frame relay networks, which carry data traffic over reliable digital lines. Another packet-switching technique is based on ATM networking, in which packets of data are processed at switching centers and directed to the best network for delivery. ATM operates with either virtual path or virtual channel connections and is capable of five classes of service.

The world's public switched telephone networks, both wireline and mobile, communicate with each other by means of a signaling protocol known as Signaling System 7. Signaling systems are responsible for all aspects of call management, among them call setup and termination, routing, and billing ("toll ticketing"). SS7 uses physical out-of-band signaling, which means that the signaling network is completely separate from the voice network.

QUESTIONS AND PROBLEMS

SECTION 9-1

1. Describe the basic limitation of using the telephone system in computer communication.

SECTION 9-2

2. List the three signal levels that telephone circuitry must work with.

3. Describe the sequence of events taking place when a telephone call is initiated through to its completion.

4. Transcribe your phone number into the two possible electrical signals commonly used in the phone system.

5. What is a private branch exchange (PBX), and what is its function?

6. List the BORSCHT functions.

7. Explain the causes of *attenuation distortion*.

8. Define *loaded cable*.

9. What is a C2 line? List its specification.

10. What are repeaters and when are they used on a T1 line?

11. Define *attenuation distortion*.

12. Define *delay distortion* and explain its causes.

13. Describe the envelope delay specification for a 3002 channel.

14. Why are telephone lines widely used for transmission of digital data? Explain the problems involved with their use.

15. What is telephone traffic? What is busy-hour telephone traffic?

16. What are the two units used to express telephone traffic?

17. What is congestion?

18. What is grade of service? How do traffic engineers interpret or look at grade of service?

19. How is grade of service measured? What does it mean when this figure is very low?

20. What is the significance of continuously observing and measuring traffic?

SECTION 9-3

21. Describe the differences between an asynchronous and a synchronous communications channel. Provide examples for each.

22. Describe the purpose of a communication protocol.

23. In what ways is coded voice transmission advantageous over direct transmission? What are some possible disadvantages?

24. Define *coding* and *bit*.

25. Describe the characteristics of the four basic encoding groups: NRZ, RZ, phase-encode binary, and multilevel binary.

26. Sketch the data waveforms for 1 1 0 1 0 using NRZ-L, biphase M, differential Manchester, and dicode RZ.

27. What are the differences between bipolar codes and unipolar codes?

28. Which coding format is considered to be self-clocking? Explain the process.

SECTION 9-4

29. Describe the time-division multiple access (TDMA) technique for transmitting data over a serial communications channel.

30. What is the purpose of guard time and why is it necessary to provide this in a TDMA system?

31. Draw a block diagram of a TDMA system. Are there any special requirements for the multiplexing frequency? Describe what is required.

32. List the data rates for the T and DS carriers.

33. How many telephone calls can be carried over a T1 line? How is the total of 1.544 Mbps obtained for a T1 line?

34. Define *fractional T1*.

35. Define *point of presence*.

36. What is the purpose of a CSU/DSU?

SECTION 9-5

37. Describe how packet switching works. What is statistical concentration?

38. Explain the operation of a frame relay network. What technique does it use to increase data throughput and why is this a viable approach?

39. What is ATM and how are data moved quickly with this technique?

SECTION 9-6

40. Compare the logic and physical configurations of SS7 and integrated services digital network (ISDN). Relate your answer to "in-band" and "out-of-band" signaling.

41. Which layers of the open systems interconnection (OSI) model are used by ISDN?

42. Which layers of the OSI model are used by SS7?

43. Which layer of the OSI model communicates directly with the four other layers? What are the other four layers?

44. Name and describe five types of ISUP messages.

45. What type of test equipment is commonly used for troubleshooting SS7 networks?

46. What is an OPC (origination point code)?

47. What is a DPC (destination point code)?

48. What is a CIC (circuit identification code)?

49. How many different reasons are there for sending an REL message? What is the most common reason?

CHAPTER 10

WIRELESS COMMUNICATIONS SYSTEMS

CHAPTER OUTLINE

KEY TERMS

wireless local-area networks (WLANs)
pseudorandom
unlicensed national information infrastructure (U-NII)
multiple-input multiple-output (MIMO)
carrier sense multiple access with collision avoidance (CSMA/CA)
worldwide interoperability for microwave access (WiMAX)
broadband wireless access (BWA)
nonline-of-sight (NLOS)
inquiry procedure
paging procedure
piconet
personal-area network (PAN)
radio frequency identification (RFID)
backscatter
slotted aloha
Global System for Mobile Communications (GSM)

code-division multiple access (CDMA)
cell sites
base stations
mobile telephone switching office (MTSO)
switch
frequency reuse
handoff
Rayleigh fading
wireless application protocol (WAP)
latency
wireless markup language (WML)
WMLScript
microbrowser
passive attack
active attack,
external attack
internal attack
countermeasures
key
data encryption standard (DES)
specialized land-mobile radio (SMR)

Though wireless communications was dominated for many years by analog radio and television broadcasting and two-way radio applications, today the field is crowded with digital wireless systems of all types. Certainly among the most visible are cellular phones, which are treated as nearly throwaway commodities even as their utility and functionality have expanded to the point that they are considered indispensable. Other applications are equally widespread. Short-range systems, among them wireless local-area networks and Bluetooth-enabled electronic products, rely heavily on wideband-modulation techniques to perform in crowded frequency bands. Without a doubt the development and widespread adoption of low-cost digital signal processing integrated circuits has been the impetus behind the explosion of short- and long-range wireless applications. This chapter surveys a number of these systems and applications and presents examples of the most significant trend in communications system design to appear in recent years: the wholesale migration to software-defined radio.

10-1 WIRELESS COMPUTER NETWORKS

Wireless Local-Area Networks

A typical computer network uses twisted-pair and fiber-optic cable for the data links. Wired networks will be described in more detail in Chapter 11, but at this point we will look at the radio portion of the most widely used wireless networking standards. These are based on the IEEE 802.11 wireless standard.

The advantages of wireless accessibility are probably quite apparent by now, probably so much so that you take them for granted. The biggest advantages include user mobility and cost-effectiveness. The concept of user mobility provides for workplace flexibility. Workers can potentially access the network from almost any location within the workplace. Accessing information from the network is as easy as if the information is on a disk. Implementing wireless networks is often much more cost-effective than outfitting an office or home with the wired infrastructure needed for permanent, fixed networks. The **wireless local-area networks (WLANs)** in wide use today derive from standards initially created and ratified in 1996 by working groups operating under the auspices of the Institute of Electrical and Electronics Engineers (IEEE), the professional association of electrical engineers and other technology professionals that, among its many activities, establishes technology standards.

The IEEE 802.11 wireless LAN standard defines the physical and media access control (MAC) layers of the open systems interconnection (OSI) reference model as well as the MAC management protocols and services. The OSI reference model will be described in more detail in Chapter 11. For now, it is sufficient to know that the physical layer defines the data transmission method (radio or infrared frequencies) and that the MAC layer defines three aspects of the transmitted data. These are data service reliability, access control to the wireless medium, and data privacy. Finally, the standard specifies wireless management protocols and services pertaining to authentication, association, data delivery, and privacy. A large number of 802.11 standards has been defined, each with its own letter suffix and associated specifications for frequency bands, maximum data rates, access methods (i.e., direct-sequence or frequency-hopping spread spectrum, and/or orthogonal frequency-division multiplexing), and maximum operating range. The most fully developed and widely used standards for wireless networking are designated with the suffixes a, b, g, and n. The RF, frequency access, and modulation characteristics will be surveyed here; configuration topologies and other topics pertaining to the setup and operation of both wireless and wired computer networks will be covered in Chapter 11.

802.11B   The first widely adopted standard for wireless networking, and still one of the most used, is IEEE 802.11b. It has a maximum data rate of 11 Mbps with provision for reduction to 5.5, 2, or 1 Mbps in the presence of channel impairments or when operated at the specified maximum range of 100 meters. The frequency spectrum used is the 2.4-GHz unlicensed industrial, scientific, and medical (ISM) band first described in the Chapter 8 introduction to spread-spectrum communications. Eleven channels, each spaced 5 MHz apart, are available, for a total bandwidth of 83.5 MHz. The frequency channels used in North America are listed in Table 10-1. The maximum transmit power of 802.11b wireless devices is 1000 mW; however, the nominal transmit power level is 100 mW.

At the two highest data rates the modulation scheme is differential quadrature phase-shift keying (DQPSK). Recall from Chapter 8 that an advantage of differential keying schemes is that coherent carrier recovery is not needed to demodulate the data, thus making the receiver simpler to implement. Also recall that QPSK has a best-case spectral efficiency of 2 b/s/Hz, implying that at all but the lowest data rates the occupied bandwidth is much greater than the 5-MHz channel spacing shown in Table 10-1. Indeed this is true: Each channel is 22 MHz wide, implying a significant degree of overlap. Therefore, for multiple signals to coexist on the same band, direct-sequence spread spectrum (DSSS), also first described in Chapter 8, is specified for channel access.

TABLE 10-1 • The DSSS Channels

CHANNEL NUMBER	FREQUENCY (GHz)
1	2.412
2	2.417
3	2.422
4	2.427
5	2.432
6	2.437
7	2.442
8	2.447
9	4.452
10	2.457
11	2.462

From the previous discussion of DSSS we saw that the modulated signal bandwidth is increased by a pseudorandom spreading code with a much higher bit rate than that of the information to be transmitted. **Pseudorandom** means that the high-bit-rate sequence appears to an outside observer to be random but that it does repeat, typically after some lengthy period of time. The high bit rate is called the *chipping rate*, and the individual bits are called *chips*. Each binary data bit modifies groups of chips, usually by inverting their states if the data bit is a 1 and by not inverting the chips if the data bit is a 0. The modified chips then modulate the RF carrier using any of the modulation schemes discussed in Chapter 8. When it was first implemented, the 802.11b standard accommodated bit rates up to a maximum of 2 Mbps; the higher speeds were a result of enhancements to the spreading code that permitted a higher data throughput within the same occupied bandwidth.

The 802.11b rates of 5.5 and 11 Mbps both use DQPSK modulation with a variant of the chip-modification scheme just discussed called *complementary code keying*, or CCK, which produces a spreading code with a length of 8 chips and a chipping rate of 11 Mcps (millions of chips per second). Recall that groupings of bits are referred to as symbols, so a single symbol will consist of 8 chips. The resulting symbol rate is 11 Mcps ÷ 8 chips/symbol = 1.375 MS/s, which produces a modulated signal occupying roughly the same bandwidth as that produced by a 2-Mbps data rate. Recall also from Section 8-5 that a complex signal can be defined in terms of magnitude and phase and represented in terms of real and imaginary components in I/Q space, where I is the in-phase or real component and

Q represents the quadrature or imaginary component. Each CCK code word is complex and consists of 8 bits, with each bit composed of a specific combination of from one to four phase parameters. These in turn map to locations for each pair of bits (dibits) in I/Q space when the symbol is modulated. The 8-bit CCK code word, then, acts as a spreading code to modify the serial data signal (information signal) at the full, 11-Mbps bit rate. There are a total of 64 possible 8-bit spreading codes available to modify the serial data signal at the highest bit rate. At the lower bit rate, 5.5 Mb, the 8-bit code is replaced by a 4-bit CCK spreading code. CCK has the advantage of performing well in the presence of noise, and the large number of available codes means that receivers have an easier time decoding them under adverse conditions.

As channel or reception conditions continue to degrade, the 802.11b standard provides for an automatic reduction in data rate. At the lower rates (1 or 2 Mbps) DSSS is still used but it is created with a fixed, 11-bit spreading sequence known as a Barker code rather than the more involved CCK code. Each data bit is XORed (exclusively ORed) with the 11-bit Barker code (10110011000). The result is a spread signal with minimum likelihood of other signals occupying the same frequency spectrum. At 2 Mbps the modulation is DQPSK and at 1 Mbps it is DBPSK. With reference to Figure 8-26, note that BPSK modulation can perform at the lowest signal-to-noise ratio, which explains why the most straightforward modulation techniques are employed under the most adverse signal conditions.

802.11A   The 2.4 GHz frequency range used by 802.11b is shared by many technologies and devices, including Bluetooth networks (to be described shortly), cordless telephones, WLANs, and microwave ovens. Radio-frequency emissions from these devices contribute noise and can affect wireless data reception. A significant improvement in wireless performance is available with the IEEE 802.11a standard because equipment operates in the much-less-crowded 5-GHz range, where there is significantly less interference.

The 802.11a standard uses orthogonal frequency-division multiplexing (OFDM) to transport the data over 12 possible radio-frequency channels, each 20 MHz wide. These channels reside within what is known as the **unlicensed national information infrastructure (U-NII)**, which is composed of bands of frequencies set aside by the Federal Communications Commission (FCC) to support short-range, high-speed, wireless data communications. The operating frequencies for 802.11a are listed in Table 10-2. The transmit power levels for 802.11a are provided in Table 10-3.

TABLE 10-2 • The IEEE 802.11a Channels and Operating Frequencies		
	CHANNEL	CENTER FREQUENCY (GHz)
Lower Band	36	5.180
	40	5.20
	44	5.22
	48	5.24
Middle Band	52	5.26
	56	5.28
	60	5.30
	64	5.32
Upper Band	149	5.745
	153	5.765
	157	5.785
	161	5.805

TABLE 10-3 • Maximum Transmit Power Levels for 802.11a with a 6 dBi Antenna Gain	
BAND	POWER LEVEL
Lower	40 mW
Middle	200 mW
Upper	800 mW

Recall from Chapter 8 that OFDM transmits data at relatively low rates over many simultaneous subcarriers spaced as closely as theoretically possible so as to be orthogonal to each other. Within each 20-MHz RF channel, the 802.11a device splits the data into 52 subcarriers, each of which is about 300 kHz wide. Of the subcarriers, 48 are for data and 4 are set aside for error-correction codes. The fastest supported data rate is 54 Mbps with provision for automatic reduction to any of a number of lower rates. Each data rate is associated with a different form of modulation. At the lowest rate, 6 Mbps, the modulation format is BPSK, whereas QPSK is used at 12 Mbps. Higher-order M-ary modulation formats are used to achieve still higher rates. For example, 16-QAM is used at 24 Mbps, and 64-QAM is employed at the highest rate of 54 Mbps.

IEEE 802.11a equipment is not compatible with 802.11b. One advantage is that the 802.11a equipment will not interfere with 802.11b. Therefore, 802.11a and 802.11b links can run next to each other without causing any interference. The downside of 802.11a is the increased cost of the equipment and the increased power consumption because of the OFDM technology. These factors are of particular concern with mobile users and the effect it can have on battery life. Also, the maximum usable distance (RF range) is about half that for the 802.11b.

802.11G   Another IEEE 802.11 wireless standard is the IEEE 802.11g. Like 802.11a, the 802.11g standard uses OFDM to support data transmission rates of up to 54 Mbps, but the 11g standard specifies operation in the same 2.4-GHz range as 802.11b. The 802.11g equipment is also backward compatible with 802.11b equipment, which means that 802.11b wireless clients can communicate with the 802.11g access points and the 802.11g wireless client equipment can communicate with the 802.11b access points.

The obvious advantage of this is that companies with an existing 802.11b wireless network can migrate to the higher data rates provided by 802.11g without sacrificing network compatibility. In fact, some manufacturers support both the 2.4 GHz and 5.8-GHz standards.

802.11N   The newest wireless standard, and the one capable of maintaining the highest data rates even under adverse conditions, is designated IEEE 802.11n. Like 802.11g, the 11n standard uses the 2.4-GHz band and OFDM. Unlike any of the other standards, however, it also uses multiple antennas at both the transmit and receive ends to set up a potentially very large number of wireless paths for data transfer. This arrangement is termed **multiple-input multiple-output (MIMO)**, and the multiple paths provide for much higher data rates than do single-path solutions. At the same time, signal-processing algorithms in the receiver use characteristics of each received signal to estimate the amount and type of signal degradation occurring and to compensate for it. These actions serve to optimize link quality even in the presence of significant impairments.

MIMO represents the marriage of a significantly enhanced variant of an age-old technique called *spatial diversity* with the latest advances in digital signal processing. Spatial diversity systems recognize that radio-frequency energy emanates from a transmit antenna in a more-or-less spherical fashion and will inevitably take more than one path from transmitter to receiver. In addition to the direct path, the transmitted signal will also most likely be reflected from one or more objects, thus causing it to reach the receive antenna some small amount of time after the line-of-sight signal has arrived. The direct and reflected signals combine at the receiver antenna, and the result may either be constructive (in-phase) or destructive (out-of-phase) addition, depending on the instantaneous amplitudes. In a diversity reception scheme, two or more receive antennas are spaced from each other, usually by multiples of a full wavelength. This way, signals received out of phase by one antenna will, in all likelihood, be received in phase by the other antenna because of the slightly longer time-to-arrival of the signal taking the longer path.

MIMO is a significant advance beyond simple spatial diversity because it does not rely on sending the same information along multiple paths and relying on the receiver to reassemble the data stream. Rather, each transmission path consists of different data, and because the antennas are spatially separated by at least one wavelength, the transmission paths are truly separate. Multiple antennas are used at both the transmitter and

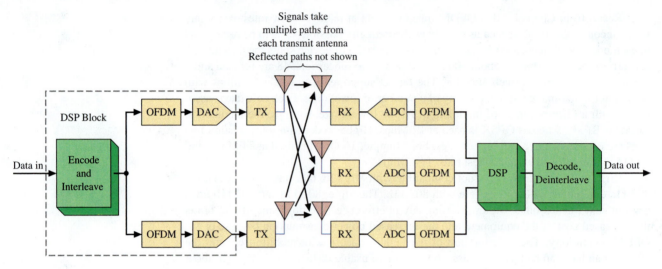

FIGURE 10-1   **2-by-3 MIMO system.**

receiver. A typical MIMO arrangement is shown in Figure 10-1, with two antennas at the transmitter and three at the receiver. Other arrangements can be posited also: two-by-two, two-by-four, and so on. The baseband data to be transmitted are, first, significantly processed before they are divided into separate data streams for subsequent application to DSP blocks, modulators, and antennas. The purpose of the processing is to create data streams that are substantially independent of each other but with characteristics that permit missing information to be recovered. Each receiver antenna is also separated by one wavelength, and every receiver antenna receives all the transmitted signals, both direct and reflected.

Essentially, MIMO relies on the fact that the DSP algorithms can recognize the unique characteristics of all the transmitted signals, among them the different phase characteristics resulting from the various paths taken from each transmitter to each receiver antenna. The receiver DSP algorithms do not simply choose the "best" (i.e., strongest) received signal, as a conventional diversity arrangement would, but instead they take characteristics of all the received signals and then combine them in a manner that compensates for signal degradation encountered in all the transmit-receive paths. Within the DSP block multiple equations are solved simultaneously and other statistical correlation and matrix-inversion algorithms are invoked to produce a result that compensates, or corrects for, errors within each receiver path. This manipulation serves to minimize multipath effects and recovers information that would have been lost had there been only one transmitter and one receiver. The functions just described have largely been integrated into chipsets where the "heavy lifting" is done by the DSP blocks. What would have been virtually impossible using multiple circuits is now relatively straightforward and economical. The truly amazing increases in data throughput and reliability brought about by MIMO mean that the technique is being explored not only for local-area networks but also for a number of other technologies, most notably next-generation (so-called 4G) mobile communications systems.

All the wireless local-area networks described use an access protocol called **carrier sense multiple access with collision avoidance (CSMA/CA)**. The CSMA/CA protocol avoids collisions by waiting for an acknowledgment (ACK) that a packet has arrived intact before initiating another transmission. If the sender does not receive an ACK, then it is assumed that the packet was not received intact and the sender retransmits the packet.

An example of an 802.11b wireless Ethernet office LAN is shown in Figure 10-2. Each PC has a module connected to it called a wireless LAN adapter (WLA). This unit connects to the Ethernet port on the PC's network interface card or is plugged into the motherboard. The wireless LAN adapters use a low-gain (2.2 dBi) dipole antenna. The wireless LAN adapters communicate directly with another wireless device called an *access point*. The access point serves as an interface between the wireless LAN and the wired network.

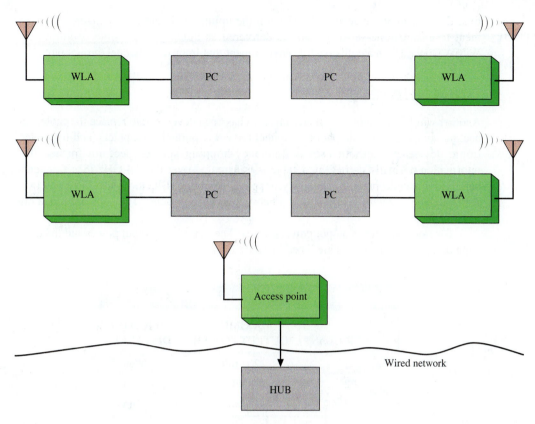

WLA: wireless LAN adapter

FIGURE 10-2   **Example of a wireless LAN and its interface to a wired network.**

WiMAX

Worldwide Interoperability for Microwave Access (WiMAX) is a broadband wireless system and has been developed for use as **broadband wireless access (BWA)** for fixed and mobile stations and will be able to provide a wireless alternative for last mile broadband access in the 2 GHz–66 GHz frequency range. BWA access for fixed stations can be up to 30 miles, whereas mobile BWA access is 3–10 miles. Internationally, the WiMAX frequency standard will be 3.5 GHz, whereas the United States will use both the unlicensed 5.8-GHz and the licensed 2.5-GHz spectrum. There are also investigations with adapting WiMAX for use in the 700-MHz frequency range. Information transmitted at this frequency is less susceptible to signal blockage from trees. The disadvantage of the lower frequency range is the reduction in the bandwidth.

WiMAX uses OFDM as its signaling format. This signaling format was selected for the WiMAX IEEE 802.16a standard because of its improved **nonline-of-sight (NLOS)** characteristics in the 2-GHz–11-GHz frequency range. An OFDM system uses multiple frequencies for transporting the data, which helps to minimize multipath interference problems. Some frequencies may experience interference problems, but the system can select the best frequencies for transporting the data.

WiMAX also provides flexible channel sizes (e.g., 3.5 MHz, 5 MHz, and 10 MHz), which provide adaptability to standards for WiMAX worldwide. This also helps to ensure that the maximum data transfer rate is being supported. For example, the allocated channel bandwidth could be 6 MHz, and the adaptability of the WiMAX channel size allows it to adjust to use the entire allocated bandwidth.

Additionally, the WiMAX media access control (MAC) layer differs from the IEEE 802.11 Wi-Fi MAC layer in that the WiMAX system only has to compete once to gain entry into the network. Once a WiMAX unit has gained access, it is allocated a time slot by the base station, thereby providing the WiMAX with scheduled access to the network. The WiMAX system uses time-division multiplexing (TDM) data streams on the downlink, time-division multiple access (TDMA) on the uplink, and centralized channel management to ensure time-sensitive data is delivered as quickly as possible. Additionally, WiMAX operates in a collision-free environment that improves channel throughput.

Bluetooth

Another wireless technology called *Bluetooth* has been developed to replace the cable connecting computers, mobile phones, handheld devices, portable computers, and fixed electronic devices. Bluetooth uses a frequency-hopping spread-spectrum format. The information normally carried by a cable is transmitted over the 2.4-GHz ISM frequency band via a pseudo-random frequency hopping technique. Bluetooth uses 79 hops from 2.402–2.480 GHz with 1-MHz spacing between hop frequencies. The hops occur at a rate of 1600 per second.

Bluetooth has three output power classes. The maximum output power and the operating distance for each class are listed in Table 10-4.

TABLE 10-4 • Bluetooth Output Power Classes		
POWER CLASS	**MAXIMUM OUTPUT POWER**	**OPERATING DISTANCE**
1	20 dBm	~100 m
2	4 dBm	~10 m
3	0 dBm	~1 m

When a Bluetooth device is enabled, it will use an **inquiry procedure** to determine whether any other Bluetooth devices are available and to allow itself to be discovered by other devices. Bluetooth uses a dedicated channel for inquiry requests and replies.

When a Bluetooth device is discovered, it sends an inquiry reply back to the Bluetooth device initiating the inquiry. Next, the Bluetooth devices enter the paging procedure.

The **paging procedure** is used to establish and synchronize a connection between two Bluetooth devices. Once the procedure for establishing the connection has been completed, the Bluetooth devices establish a **piconet**. A piconet is an ad-hoc network of up to eight Bluetooth devices, such as a computer, mouse, headset, earpiece, and so forth. In a piconet, one Bluetooth device (the master) is responsible for providing the synchronization clock reference. All other Bluetooth devices are called *slaves*.

Two modulation schemes are in use, depending on the Bluetooth version. Up to version 1.2, Gaussian-filtered frequency-shift keying (FSK) with a ± 160-kHz frequency shift between binary data levels is used. This format allows for a maximum data rate of 1 Mbps, of which 723.2 kbps is for one-way (simplex) transmissions of user data, with an additional 276.8 kbps allotted to overhead functions such as error correction and detection. For two-way (duplex) applications, such as a wireless headset for a cellular phone, the data rate is reduced to 433.9 kbps.

A newer Bluetooth version called Enhanced Data Rate (EDR) achieves a raw data rate of 3 Mbps and a user rate of more than 2.1 Mbps by using a different modulation format. To maintain compatibility with earlier versions, the initial setup, paging, and link-establishment functions are performed with Gaussian FSK. Once established, the Bluetooth device switches over to a variant of QPSK called pi-over-four differential quadrature phase-shift keying (π/4 DQPSK) for user data. Recall from Chapter 8 that π/4 DQPSK is a form of offset modulation designed to minimize excursions through the origin in I/Q space from one symbol to the next. This arrangement allows for the use of less linear but more efficient power amplifiers in the Bluetooth device. Efficiency is a particularly important consideration because many Bluetooth devices are designed for portable operation. Also, like the IEEE 802.11b format, the differential modulation scheme used in EDR Bluetooth obviates the need for coherent carrier recovery, thus allowing for implementation of a simpler receiver in the device.

ZigBee

The Bluetooth protocol just covered is an example of a **personal-area network (PAN)**, and the ad-hoc networks formed when Bluetooth-enabled products form links are called piconets. The network topology is that of a star, where the Bluetooth device providing the primary timing reference interacts directly with secondary devices. Another PAN gaining widespread popularity for a number of applications requiring relatively low data rates with extremely low power-consumption requirements is known as ZigBee. It is based on the IEEE 802.15.4 wireless standard, which specifies the physical and media-access control requirements of wireless devices. ZigBee is the trade name applied to devices based upon this standard and produced by members of the ZigBee Alliance, consisting of hardware and software vendors that also specify higher-order functions of ZigBee-enabled products, including networking and data security capabilities.

Unlike the relatively high data rates carried by Bluetooth devices, ZigBee rates are much lower. It is designed for applications such as building automation and remote process monitoring and control, applications where high network reliability is needed but that generally do not require high data throughput. The achievable data rates depend on which of the unlicensed ISM frequency bands is being used: 868 MHz (Europe only), 915 MHz, or 2.4 GHz. Table 10-5 shows the modulation schemes and data rates available for each band.

TABLE 10-5 • ZigBee Data Rates and Modulation Formats by Frequency Band.

FREQUENCY BAND	NUMBER OF CHANNELS	DATA RATE (Kbps)	TYPE OF MODULATION
868 MHz	1	20	BPSK
915 MHz	1	40	BPSK
2.4 GHz	16	250	OQPSK

As expected, the lowest frequencies have the lowest data rates because these have the smallest available bandwidths. Direct-sequence spread spectrum is used in all frequency bands. As with 802.11 devices the access method for ZigBee is CSMA/CA.

The maximum range achievable varies inversely with frequency and data rate. At the highest frequency, 2.4 GHz, the range is from about 30 m for indoor environments to a maximum of about 400 m for outdoor applications with a direct line of sight. At the lower frequencies, the outdoor range (again assuming a clear line of sight) can extend up to 1000 m. At present the 2.4-GHz variant of ZigBee is the most widely used.

One distinguishing feature of ZigBee is that it can be used in a number of network topologies; however, the most intriguing of these is the *mesh* network because of its implications for reliability and range enhancement. ZigBee networks are made up of a number of devices, each with different levels of functionality. The devices are the ZigBee coordinator (ZC), of which there is one per network, the ZigBee router (ZR), and the ZigBee end device (ZED). There are potentially many ZRs and ZEDs in the network; the practical limit is 65,536 nodes, with each identified by a unique 16-bit address. The coordinator establishes and serves as the network "master control," and it stores network information. The ZR runs applications to, for example, observe a remote sensor or to control a device. Importantly, it also acts as a router: The ZR can receive data from other devices and retransmit it, including to devices in other nodes. The ZED is the most basic device, and potentially the one requiring the least amount of power because it can spend most of its time in a "sleep" state. The ZED can talk to a coordinator or router but cannot relay information from other devices, as can a ZR.

The mesh topology shown in Figure 10-3 improves reliability and increases range. The ZRs act as relay devices by routing data from one node to another. The ZRs also act as monitor and control points by gathering data for transmission and responding to commands. Thus, if the intended destination for data from a given node is out of range, data can be passed through alternate paths involving multiple nodes in its quest to the final

FIGURE 10-3 **ZigBee mesh network.**

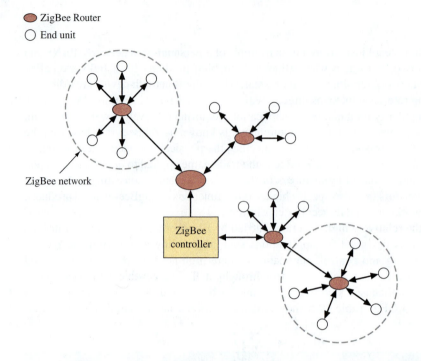

ZigBee network

destination. In addition, the mesh topology provides for built-in redundancy. The network is robust in that if any one node is disabled, data can still travel through alternate paths to any of the remaining nodes. The mesh topology thus achieves a much higher level of reliability than the alternatives because data can reach its destination even if paths are interrupted because of equipment failure or unfavorable path conditions.

The relatively low data rates supported by the ZigBee specification are not a drawback for many intended applications. These applications are principally in the areas of building automation and control and remote monitoring—that is, the control of functions such as lighting and the heating, ventilating, and air conditioning (HVAC) utilities in large buildings. Also possible are industrial process control and monitoring functions in manufacturing operations, refineries, chemical plants, and the like, as well as applications in the medical field, medical sensors, and home automation. All these applications share in common the need for high reliability but not necessarily high data throughput: Much of the control involves turning devices on or off or controlling them in relatively straightforward ways. The monitoring generally involves converting a relatively slowly changing analog signal representing some physical process into a digital equivalent for transmission to a control center or other monitoring point. Another major area being deployed is automatic utility (gas and electric) meter reading. The hallmarks of ZigBee systems are that they are low cost and have extremely low power consumption requirements. With these advantages, ZigBee is bound to find application in a number of as-yet-undefined ways in coming years.

Radio-Frequency Identification

Radio frequency identification (RFID) is a technique that uses radio waves to track and identify people, animals, objects, and shipments. This is done by the principle of modulated **backscatter**. The term *backscatter* refers to the reflection of the radio waves striking the RFID tag and reflecting stored unique identification information back to the transmitter source.

The basic block diagram for an RFID system is provided in Figure 10-4. The RFID system consists of two things:

- *RFID tag* (also called the *RF transponder*), which includes an integrated antenna and radio electronics

- *Reader* (also called a transceiver), which consists of a transceiver and an antenna. A transceiver is the combination of a transmitter and receiver

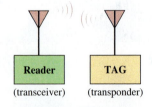

FIGURE 10-4 **Basic block diagram of an RFID system.**

The reader (transceiver) transmits radio waves that activate (turn on) an RFID tag. The tag then transmits modulated data, containing its unique identification information stored in the tag, back to the reader. The reader then extracts the data stored on the RFID tag.

The RFID idea dates back to 1948 when the concept of using reflected power as a means of communication was first proposed. The 1970s saw further development in RFID technology, in particular, a UHF scheme that incorporates rectification of the RF signal for providing power to the tag. Development of RFID technology significantly increased in the 1990s. Applications included tollbooths that allow vehicles to pass through at highway speeds while still recording data from the tag.

Today, RFID technology is being used to track inventory shipments for major commercial retailers, the transportation industries, and the Department of Defense. Additionally, RFID applications are being used in Homeland Security applications such as tracking container shipments at border crossings.

Three parameters define an RFID system. These include the following:

- Means of powering the tag
- Frequency of operation
- Communications protocol (also called the air *interface protocol*)

POWERING THE TAG   RFID tags are classified in three ways based on how they obtain their operating power. The three different classifications are passive, semiactive, and active.

- **Passive:** Power is provided to the tag by rectifying the RF energy transmitted from the reader that strikes the RF tag antenna. The rectified power level is sufficient to power the ICs on the tags and also provides sufficient power for the tag to transmit a signal back to the reader. An example of two passive RFID tags (also called *inlays*) is shown in Figure 10-5.

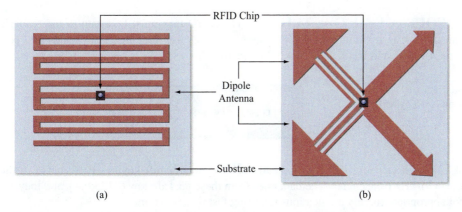

FIGURE 10-5   **Examples of (a) single-dipole and (b) dual-dipole RFID inlays.**

The tag inlays include both the RFID chip and antenna mounted on a substrate. The antenna for the RF inlay shown in Figure 10-5(a) is a single-dipole antenna, and the inlay shown in Figure 10-5(b) incorporates a dual-dipole antenna. The single-dipole antenna configuration for the RFID tag works well if the orientation of the inlay is properly aligned to the reader's antenna. RFID readers can use a circularly polarized antenna to minimize the effect of tag orientation. However, there is an issue of limited read distance with circularly polarized antennas because of the distribution of the receive power between the horizontal and vertical elements.

The advantage of the dual-dipole (shown in Figure 10-5[b]) is the improvement in the independence of the tag relative to the reader's antenna. The two dipoles of the dual-dipole antenna (Figure 10-5[b]) are oriented at 90° angles to each other. This arrangement means the tag orientation is not critical for the dual-dipole inlay.

- **Semiactive:** The tags use a battery to power the electronics on the tag but use the property of backscatter to transmit information back to the reader.
- **Active:** The tags use battery power to transmit a signal back to the reader. Basically, the device is a radio transmitter. New active RFID tags are incorporating Ethernet 802.11b/g connectivity. An example is the RN-171 "WiFly" module and asset tag from Microchip Technologies shown in Figure 10-6. The power consumption of the RN-171 is 4 μA in the sleep mode, and it uses a 3.6-Vdc power source with an expected lifetime of many years. The RN-171 also has location capability, making it suitable for use as a tag for tracking the location of materials within a facility, for example expensive or controlled medications on a hospital cart. Location capability is accomplished by making RSSI (receive signal strength indication) measurements from three separate wireless access points. The three measurements provide sufficient information to make a triangulation measurement for use in locating the object.

(a) (b)

FIGURE 10-6 **(a) The RN-171 "WiFly" module and (b) asset tag from Microchip Technologies.** (Courtesy of Microchip Technologies.)

FREQUENCY OF OPERATION The RFID tags must be tuned to the reader's transmit frequency to turn on. RFID systems typically use three frequency bands for operation, LF, HF, and UHF, as shown in Figure 10-7.

FIGURE 10-7 **Frequency bands used by RFID tags.**

LF	HF	UHF
125/134 kHz	13.56 MHz	860–960 MHz
		2.4 GHz

Low-frequency (LF) tags typically use frequency-shift keying (FSK) between the 125–134-kHz frequencies. The data rates from these tags are low (~12 kbps), and they are not appropriate for any applications requiring fast data transfers.

However, the low-frequency tags are suitable for animal identification, such as dairy cattle and other livestock. The RFID tag information is typically obtained when the livestock are being fed. The read range for low-frequency tags is approximately 0.33 m.

High-frequency (HF) tags operate in the 13.56-MHz industrial band. High-frequency tags have been available commercially since 1995. It is known that the longer wavelengths of the HF radio signal are less susceptible to absorption by water or other liquids. Therefore, these tags are better suited for tagging liquids. The read range for high-frequency tags is approximately 1 m. The short read range provides for better-defined read ranges. The applications for tags in this frequency range include access control, smart cards, and shelf inventory.

Ultra-high-frequency (UHF) tags work at 860–960 MHz and at 2.4 GHz. The data rates for these tags can be from 50–150 kbps and greater. These tags are popular for tracking inventory. The read range for passive UHF tags is 10–20 ft., which makes them a better choice for reading pallet tags. However, if an active tag is used, a read range of up to 100 m is possible.

COMMUNICATIONS (AIR INTERFACE) PROTOCOL   The air interface protocol adopted for RFID tags is **slotted aloha**, a network communications protocol technique similar to the Ethernet protocol. In a slotted aloha protocol, the tags are only allowed to transmit at predetermined times after being energized. This technique reduces the chance of data collisions between RFID tag transmissions and allows for the reading of up to 1000 tags per second. (Note, this is for HF tags). The operating range for RFID tags can be up to 30 m. This means that multiple tags can be energized at the same time, and a possible RF data collision can occur. If a collision occurs, the tag will transmit again after a random back-off time. The readers transmit continuously until there is no tag collision.

10-2 CELLULAR PHONE VOICE SYSTEMS

It is no exaggeration to state that cellular phones are everywhere. In early 2010, the number of cellular subscribers worldwide passed 4.6 *billion*, and that number has certainly grown since then. The number is truly mind-boggling when you stop to think it represents over half the people on the face of the Earth, all using technologies that have been commercially available for less than thirty years.

Though mobile telephone service originated in the late 1940s, the systems were never widely used because frequency allocations were limited and equipment was expensive. Early systems also used high-power transmitters. Any time a channel was in use, it would then become unavailable anywhere over a wide geographic area, which in turn severely limited overall system capacity. The modern cellular concept came about as a result of several developments. First, in the mid-1970s the FCC reallocated frequency spectrum in the 800- to 900-MHz band away from UHF television broadcasting (the little-used channels 70 through 83) toward what was then termed the Advanced Mobile Phone Service (AMPS), the analog precursor to today's all-digital wireless services. This additional capacity allowed for the hundreds of analog voice channels required for large-scale mobile service. At the same time, advances in semiconductor technology and computer-aided engineering design provided for the development of reasonably priced portable transceivers that worked reliably at the intended frequency ranges (not a trivial problem in itself). Perhaps most fundamental was a shift in the way wireless networks were conceived. Rather than relying on a single, high-power transmitter providing service to a small number of users dispersed over a wide area, networks would be composed of a large number of relatively low-power base stations, each intended to serve subscribers within the immediate vicinity—often a radius of a mile or less. The new networks relied on the concept of *frequency reuse,* whereby access to the scarcest resource—the radio-frequency spectrum— would be actively managed by locating the base stations fairly close together such that a large number of subscribers could be served simultaneously without interfering with each other. The base stations and subscriber units would all be under the control of a central computer at the mobile telephone switching office (MTSO), which would constantly allocate radio frequencies and manage power levels with the aim of maximizing efficient spectrum use even as subscribers experience uninterrupted service.

The first-generation AMPS cellular networks were analog and were designed strictly as mobile telephone systems, somewhat akin to two-way radio. They were definitely not the high-speed voice and data networks that are commonplace today. The AMPS system was designed to operate in the 800-to-900-MHz frequency band and used frequency modulation with a peak deviation of 12 kHz and with 30-kHz spacing between channels. A duplex phone conversation required two 30-kHz channels, one for transmitting and one for receiving. A typical metropolitan system with 666 duplex channels, therefore, required a 40-MHz (30 kHz $\times$ 2 $\times$ 666) spectrum allocation. System capacity quickly became constrained by the amount of spectrum allocated, the need for 60 kHz of bandwidth per call, and the fact that only one call per channel was supported. These constraints caused carriers to develop and, over time, to adopt digital technologies that made more efficient use of available spectrum. The second-generation technology, designated IS-136, relied on TDMA multiplexing techniques to enable three calls to be handled per RF channel, leading to a large increase in system capacity without the need for additional spectrum. The IS-136 TDMA standard, now also largely obsolete, was intended to work along with the analog infrastructure and to provide an evolutionary upgrade path for the network carrier. The evolutionary upgrade path meant that many aspects of the second-generation digital standard, such as channel bandwidth and the method for handling call transfer as the user changed location, were directly derived from existing analog systems.

Today, most mobile telephone service providers use one of two digital technologies to administer mobile phone calls. Other technologies, falling under the general heading *3G* (for third-generation), have been established for high-speed data transfer and advanced services such as wireless video. Third-generation standards and the technologies used in their implementation are distinct from the networks used for voice traffic. Voice calls are still handled by one of two second-generation (2G) digital standards, either the **Global System for Mobile Communications (GSM)**, which is also based on TDMA multiplexing but with wider-bandwidth channels and more users per channel than IS-136 TDMA, or a competing technology based on direct-sequence spread spectrum known as **code-division multiple access (CDMA)**. Notwithstanding the large number of network types and technologies in use, several fundamental concepts dating to the earliest cellular networks are common to all of them. Two of these ideas, *frequency reuse* and *cell splitting,* pertain directly to maximizing capacity.

The Cellular Concept

The cellular concept involves using an essentially regular arrangement of transmitter–receiver systems called **cell sites**. Figure 10-8 shows a cellular system with 21 "cells" being served by seven different channel groups. The cells cover the entire geographic area to be served by the system. Spacing between sites is a function of both geography and capacity. In sparsely populated areas with flat terrain and low traffic, the sites may be spaced on the order of 10 mi apart. Densely populated, high-traffic areas such as busy intersections, airports, stadiums, or big-city business districts may have several sites placed within a half-mile radius. The hexagon shape shown in the figure is an approximation of the coverage areas associated with each site. The most basic coverage patterns are *omnidirectional,* meaning they radiate RF energy equally in all directions, producing a circular coverage area. The cell-site antennas are located close enough to each other that they overlap to a degree. The point at which responsibility for a call transitions from one site to the next, called the handoff point, can be approximated with a straight line drawn between the points where the overlapping circles for each site's coverage pattern intersect. The hexagon shape that results was chosen after detailed studies showed it led to the most cost-efficient and easily managed system. One of the differences between GSM and CDMA systems pertains to how the handoff function is managed, a topic to which we shall return shortly.

FIGURE 10-8   **Cellular phone system layout.**

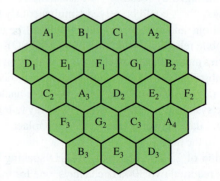

Within the cell sites are a number of **base stations**, which are transceivers for communicating with mobile phone users. The base station enables communication between the mobile and the rest of the interconnected phone system. In addition to the transceivers, each cell site contains transmitting and receiving antennas as well as combiners to permit multiple transmitter-receiver pairs to share antennas, along with signal level monitoring equipment and all the support functions needed for the base stations to communicate with the

mobile telephone switching office (MTSO). Within the MTSO is a central computer called the **switch**. It controls cell-site actions and many actions of the mobile units through commands relayed to them by the cell sites. Each switch communicates with a large number of base stations and cell sites (on the order of 200 or more), normally with T1 links but also with terrestrial microwave or, in some cases, fiber-optic connections. The switch also interconnects mobile users with the public switched telephone network (PSTN), which is the land-line (wired) telephone system. An 11-cell system is shown in Figure 10-9, with the MTSO providing a link with two mobile users.

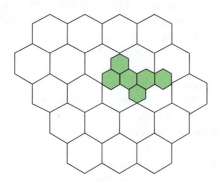

FIGURE 10-9   **MTSO linking two mobile users.**

Frequency Reuse

As mentioned, two fundamental cellular system concepts address the issue of capacity maximization, which pertains directly to efficient use of radio-frequency spectrum. **Frequency reuse** is the process of using the same carrier frequency (channel) in different, geographically separated cells. Power levels are kept low enough so that cochannel interference is not objectionable. Thus cell sites A1 and A2 in Figure 10-8 use channels at the same frequency but have enough separation that they do not interfere with each other. Actually, this process of *space-division multiplexing* is used in most radio services but not on the shrunken geographic scale of cellular telephone. Instead of covering an entire metropolitan area from a single transmitter site with high-power transceivers, the service provider uses moderate-power transmitters at each cell site. Through frequency reuse, a cellular system can handle several simultaneous calls, greatly exceeding the number of allocated channels. The multiplier by which the system capacity exceeds allocated channels depends primarily on the number of cell sites.

Cell Splitting

Several frequency channels are assigned to each cell in the system. This is called a *channel set*. In an analog cell phone system, one channel is required for each phone call taking place at any one instance of time. If all traffic in the cell increases beyond reasonable capacity, a process called *cell splitting* is employed. Figure 10-10 illustrates this process.

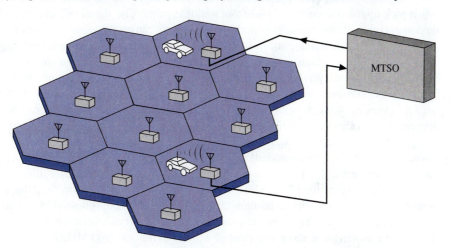

FIGURE 10-10   **Cell splitting.**

An area from Figure 10-9 is split into a number of cells that perhaps correspond to the city's downtown area, where phone traffic is heaviest. Successive stages of cell splitting would further increase the available call traffic, if necessary. The techniques of frequency reuse and cell splitting permit service to a large and growing area without expanding the frequency allocation.

An extension of the frequency reuse and cell splitting concepts is that of *sectorization,* which involves the use of directional antennas to focus energy in preferred directions at the expense of others. Sectorized cell cites have base stations that transmit and receive signals in specific directions called sectors. Most cell sites are split into three sectors (though more or fewer can be used) normally identified as "α," "β," and "γ" sectors (alpha, beta, gamma) with at least three sets of antennas, each covering an arc of about 120°. Each sector in effect acts as a separate cell. In the old AMPS as well as the present-day GSM networks, good frequency planning minimizes interference by avoiding the use of the same or adjacent channels by adjacent sectors or nearby cells.

As subscribers move about, the system must be able to transfer the call from one cell site to the next as needed. The means by which this transfer, known as a **handoff**, is accomplished depends on the underlying network access method, either TDMA (for GSM) or CDMA. For either type of access method the system periodically determines the strength (hence, quality) of the signal being received at the cell site, and, when the signal strength falls below a predetermined threshold, the system looks for another site to serve the call. In a TDMA-based system such as GSM, where specific channels (frequencies) are assigned to each site, the switch then sends a command to the user device to retune to a channel allocated to the new cell site. In CDMA, the handoff involves changing the spreading code used to differentiate one call from the next. Handoffs occur not only from cell to cell but also from sector to sector within the same cell. The procedure causes only a brief interruption of the conversation, typically 50 ms, and is not noticed by the user. A command signal from the base station causes the frequency synthesizer (under microprocessor control) in the phone to switch automatically to the carrier frequency of the new cell.

The power output of the mobile is controlled by a power up/down signal from the base station. This is done to reduce interference with other phones and to minimize overloading of the base station receiver. In GSM and CDMA, the power from the downlink (base station to mobile) is increased as the base-station-to-mobile distance is increased. This leads to "soft handoff," which reduces the probability of a call being "dropped" (disconnected) during handoff.

Rayleigh Fading

The FM capture effect is very helpful in minimizing cochannel interference effects in cellular systems. Unfortunately, this benefit is considerably degraded by **Rayleigh fading**, a rapid variation in signal strength received by mobile units in urban environments. To maintain adequate signal strength during the fades, transmitter power must be increased by up to 20 dB to provide an adequate fade margin.

Rayleigh fading comes about because, particularly in urban areas, the receiver will often pick up the transmitted signal from multiple paths. The signal reflects off buildings and many other obstructions. This multipath reception causes the signal to contain components from many different path links. The fading results because, in some relative positions, phases of the signals arriving from the various paths interfere constructively, while in other positions the phases add destructively. In a mobile communication environment, where subscriber units are moving at high speeds, Rayleigh fading can cause the received signal to vary by as much as 30 dB at a very rapid rate. The signal strengths vary from one extreme to the other every half-wavelength. That is, if there is maximum destructive interference between two signals at one point, because the received signals are 180° out of phase with each other, then one-half wavelength away the signals will be in phase, producing constructive interference, because the signal taking the longer path will have undergone a half-wavelength phase shift. Another half-wavelength away will be another out-of-phase condition with maximum destructive interference, followed by constructive interference, and so on. This phenomenon occurs over short distances for cellular systems because a wavelength is about one-third of a meter in the 800–900 MHz band.

Cellular Frequency Bands

The FCC assigned the frequencies shown in the "CELLULAR" section of Figure 10-11 for mobile phone operation. This is 824–829 MHz for "phone transmit" (also called "uplink") and 869–894 MHz for "base transmit" (also called "downlink"). As mobile phones became more widely used, the FCC assigned an additional set of frequencies called personal communication systems (PCS) for mobile phone service. These frequencies, from 1850–1990 MHz, are also shown in Figure 10-11.

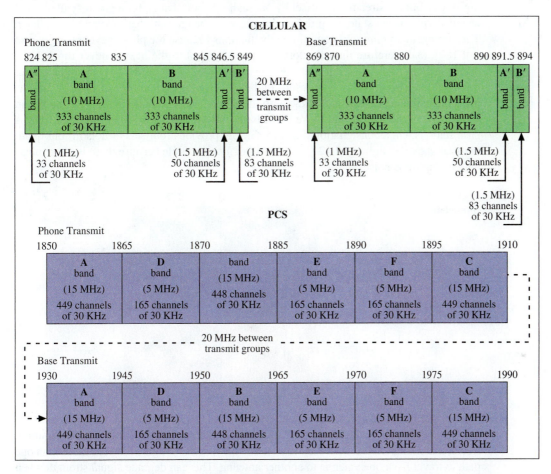

FIGURE 10-11   A comparison of the cellular and PCSA spectrums.   (Courtesy of the International Engineering Consortium and the IEC Web Pro Forums, http://www.iec.org/tutorials/) It should be pointed out that the only difference between "cellular" and "PCS" is the frequency assignment. There is a common misconception that there is "some other difference" between cellular and PCS. The term PCS is outdated, however the frequency assignments are still valid. These frequency bands are now used by all mobile phone services.

Global System for Mobile Communications

Several countries in Western Europe, starting around 1990, developed **Global System for Mobile Communications (GSM)**. GSM is the standard used by carriers throughout Europe. In North America, AT&T Mobility and T Mobile are the two largest GSM-based voice-system operators. The system is very well defined. The document provided by the European Telecommunications Standards Institute (ETSI) is 5000 pages. GSM uses 200-kHz channels, which are time-shared by eight users. It is, therefore, considered to be a TDMA-based system. The extra bandwidth provides even greater opportunities than does IS-136 for enhancing system performance. One enhancement is that the power sent by the base to the mobile user is increased as the distance between the two is increased. When a mobile is approaching a location at which it will be handed off to another base station, the possibility of the call being dropped is minimized. Since GSM "knows" how far the mobile is located from each base station, the mobile's precise location can be found as soon as the phone is turned on. GSM has sufficient timing accuracy so that it can use the velocity of light to calculate distance on the basis of delay time. Another advantage of GSM's wider bandwidth is that additional features can be added, such as photographs and e-mails.

GSM uses Gaussian minimum-shift keying (GMSK) as its modulation method. Recall from Chapter 8 that the most basic form of MSK is as a form of PSK that involves a minimum (smooth) phase shift between binary states. The term *Gaussian* refers to the shape produced by the filter used to keep the channel within its assigned bandwidth. GMSK as implemented in the GSM system is based on offset quadrature phase-shift keying (OQPSK). Like standard QPSK, the data streams are split and sent to I and Q modulating stages, but the streams are offset by one-half symbol; that is, for a given symbol, the I channel will change state, then the Q channel will change state half a symbol time later. This arrangement minimizes amplitude fluctuations because the phase never changes by a full 180°, and therefore the trajectory from one symbol to the next is never through the origin in I/Q space, meaning that the transmitter amplifiers never have to reduce their power all the way to zero. This result implies that a major advantage of OQPSK and its implementation as GMSK is that the amplifiers in a base station do not need to be linear. This is one of the factors that make GSM base stations much less expensive than CDMA base stations. On a constellation diagram, GMSK provides four sets of dots in four quadrants, as shown in Figure 10-12. These dots fall along a circle because there is phase noise but no amplitude noise.

FIGURE 10-12 **The GMSK constellation.**

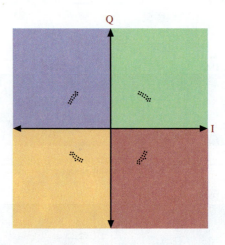

GSM uses an "equalizing filter" to minimize the effects of multipath propagation. Recall that multipath is a term used to describe the fact that a signal may use more than one path to travel from one antenna to another antenna. This can degrade signal strength when two (or more signals) arriving at the receiving antenna are out of phase with each other. In certain circumstances, this filter even enhances the received signal. Another feature of GSM is the subscriber identity module (SIM) card, which allows a user to carry his identity between different phones. In contrast with the CDMA platform to be discussed shortly, GSM uses a base station controller (BSC) between various base stations and the switch. Among other functions, the BSC is responsible for assigning operating frequencies to the mobiles. In CDMA, some of the functions similar to those in the BSC are located in either the base station or the switch.

The base station provides six types of control signals to the mobile. This link is called the *downlink*.

1. **FCCH:** frequency control channel; like a "lighthouse" for the base station.
2. **SCH:** synchronization.
3. **BCCH:** broadcast common control.
4. **AGCH:** access grant channel.
5. **PCH:** paging channel.
6. **CBCH:** cell broadcast channel.

The mobile provides a random access channel (RACH) to the base station. This link is called the *uplink*. There are three other control channels that are used for both uplink and downlink. These are the standalone dedicated control channel (SDCCH), slow associated control channel (SACCH), and fast associated control channel (FACCH). A few examples of the functions of the control channels are as follows:

1. The FCCH is the "lighthouse," which is always transmitting and is always looking for a mobile. As soon as a customer turns on his or her phone, the GSM system identifies the user, determines his or her home location, and determines such characteristics as account status (for example, whether the customer has been paying bills on time).

2. The PCH is a paging channel. This channel starts operating when someone places a call to the mobile. It basically sets up the call, including assigning a frequency.

3. The SCH is used for synchronization. It is used to maintain timing during the call.

4. The RACH is used only by the mobile. Its function is to get the attention of the base station.

Though GSM is primarily considered to be a TDMA-based system, it also employs frequency hopping on the speech channels to minimize the possibility of interference between adjacent cells, to avoid noisy channels, and to minimize the effects of multipath fading. Although the technique is related to frequency-hopping spread spectrum (FHSS), the form implemented in GSM is not considered to be spread spectrum because the same frequencies are not in use simultaneously by multiple users, as would be the case with FHSS. Instead, GSM uses "slow frequency hopping," which can be thought of as a form of frequency diversity intended to maximize transmission quality. The hop rate is set to be 217 hops per second, and a full data burst (on the order of 1200 bits per hop) is completed before the frequency is changed. The frequency-hopping algorithm is broadcast on the BCCH.

As mentioned, GSM is based on TDM. In addition, the total frequency spectrum allocated to each carrier is subdivided into a number of 200-kHz-wide channels, each with its own carrier frequency. Thus, frequency-division multiple access is also a key component of GSM multiplexing. GSM is perhaps best thought of as a "two-dimensional" access method in that it uses both TDMA and FDMA at the same time. Each carrier frequency carries one TDMA "frame," which in turn is subdivided into eight time slots. As shown in Figure 10-13, each frame has a total duration of 4.615 ms, within which a total of 1248 bits is transmitted, so each user time slot has a duration of 4.615 ms ÷ 8 = 0.577 ms, or 156 bits.

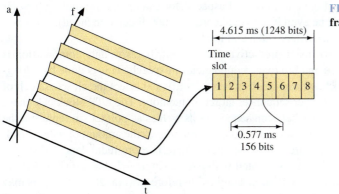

FIGURE 10-13   **The GSM frame.**

VOCODERS   The basic data rate per channel is 270 kbps, which must fit into a 200-kHz-wide channel. Recall that the most basic forms of digital modulation have a spectral efficiency of 1 b/s/Hz, so as a rule of thumb the modulated bandwidth will be approximately equal to the data rate. However, OQPSK (as part of MSK modulation) and Gaussian filtering together serve to reduce the occupied bandwidth, leading to a spectral efficiency of 270/200 = 1.35 b/s/Hz. Perhaps more striking, however, is the amount of data compression that must take place for voice to be represented with adequate fidelity within such a bandwidth-limited channel. This compression task is the function of the *vocoder*, which, along with the substantial amount of error correction and other signal-enhancement functions that must take place for reliable communication, is a DSP operation that is critical to the successful implementation of GSM systems.

The purpose of the vocoder is to compress the voice data for transmission. Recall from Chapter 7 that a pulse-code-modulated (PCM) voice channel consists of an 8-bit word produced at a sample rate of 8 kHz to represent a voice-grade telephone call with a maximum frequency of no more than 4 kHz (the Nyquist limit), for a total bit rate of 64 kbps. Multiply this rate by 8 time slots, the standard for GSM, and the total becomes 512 kbps. The required bandwidth would be, at a minimum, 512 kHz, and this number does not take into account the considerable "overhead" needed for call setup and reliable operation in the mobile environment. Clearly, the number of bits needed to send voice must be reduced but without making the voice quality unacceptably low.

A vocoder (short for "voice coder") is a compression technique for reducing the amount of data needed to represent human speech. All digital mobile communications systems use vocoders, of which many types have been developed and used. One is called *regular pulse excitation-linear prediction coding* (RPE-LPC), which produces "full-rate" coding because it compresses the 64-kbps PCM bit stream to the maximum bit rate of 13 kbps per call specified by the GSM standard. At this bit rate, the compression is sufficient to fit 8 TDMA channels within a 200-kHz bandwidth. In very basic terms, modern vocoders create a mathematical model of the human vocal tract for processing by a DSP algorithm. Both voiced (speech) and unvoiced sounds emanate from the mouth after having passed through the larynx. The larynx and mouth together effectively act as filters, giving both the vocal and nonvocal characteristics of the sounds a distinctive and frequently changing frequency response. The vocoder periodically captures the salient characteristics of these emanations by implementing the DSP equivalent of a multi-pole bandpass filter. Parameters representing the excitation (characteristics such as pitch and loudness) and filter response of the PCM-encoded speech are then transmitted and reconstructed at the receiver. The transmission interval is on the order of 20 ms. The mathematical process used to generate the filter parameters based on voice characteristics forms the basis for the class of vocoders known as *linear predictive coders.* What is transmitted, then, is not the digitized voice directly but rather the excitation and filter parameters, thus permitting the essential speech and non-speech characteristics to be preserved but at a much lower bit rate than that of the original, digitized voice-grade signal.

The input to the LPC-RPE vocoder block used in GSM consists of 2080 bits representing 20 ms of audio at a time. For each 20-ms chunk, the vocoder produces 8 filtering coefficients and an excitation signal. The coefficients represent specific aspects of the human voice, including audible but nonverbal aspects (for example, representing how air passes by the teeth and tongue even if the vocal cords are not directly engaged). The excitation signal encodes aspects such as pitch and loudness.

Another aspect of how linear predictive coding contributes to data compression is that it identifies correlations and removes redundancies in speech. This is done by a long-term prediction (LTP) block, whose function is to identify and select the closest match of a given data sequence with earlier sequences stored in memory. The LTP block calculates and then transmits the *difference* between sequences as well as a "pointer" to represent the earlier sequence used for comparison. This process compresses the amount of data considerably because redundant data are not transmitted. There is more redundancy in speech than you might think. The sounds associated with even a single syllable are considerably longer than the 20 ms associated with each sample. Therefore, several 20-ms samples may largely consist of the same data. By telling the receiver which data are redundant, the LPC function shifts responsibility for recreating the data to the receiver, rather than having the redundant data transmitted directly.

The amount of data compression accomplished is substantial. The LPC-RPE vocoder, along with LTP, reduces the 2080 bits input every 20 ms by a factor of eight, down to 260 bits. The net data rate of 13 kbps represents 260 bits every 20 ms. As mentioned, other vocoders are also in use, including a GSM "half-rate" vocoder with a bit rate of 5.6 kbps. With it, the number of calls carried by the network could in theory be doubled but at the expense of audio quality. All the decoders trade off audio quality for channel capacity, and newer vocoder implementations seek to make the compressed voice sound as natural as possible without requiring excessive bandwidth. Also, vocoders are optimized for human speech. They do not do well with nonspeech audio. If you have ever wondered why "music on hold" sometimes sounds so awful while you wait on your cell phone for a customer service representative to take your call (which, after all, is very important to the firm, as you were no doubt reassured repeatedly while waiting), at least some of the blame rests with the vocoder.

Code-Division Multiple Access

The second major class of second-generation, digital cellular voice network is based on the multiple-access principles embodied by direct sequence spread spectrum (DSSS). In code-division multiple access (CDMA)–based cellular systems, the multiplexing is based not on time or frequency but on a variant of the spreading-code idea discussed in Chapter 8. Recall from that discussion that the function of the spreading code is to assign a unique "signature" or identity to each transmission even as the bandwidth is deliberately made much wider than the minimum necessary to send the message. The receiver knows the possible spreading sequences, and when its locally generated sequence is correlated with that of the received signal—that is, when a spreading sequences exactly match—the information signal can be extracted, even as the frequency spectrum is shared with other users at the same time.

CDMA-based cellular systems operate on a variation of the basic ideas behind DSSS: Multiple calls are handled over the same frequency range at the same time and in the same vicinity. In fact, multiple cell sites share the same frequencies in the same geographic area, a distinct departure from the principles of frequency reuse and active channel management embodied with analog and TDMA-based systems. However, CDMA systems do not use pseudorandom (PN) bit streams exclusively as spreading codes. Although PN codes are used (for reasons to be discussed shortly), the spreading codes used to differentiate individual users within a given area are known as *Walsh codes.* There are 64 such codes in the basic CDMA implementation (known as IS-95), and each code is 64 bits long. The Walsh codes have the mathematical property of orthogonality, which means that each code is completely independent of every other code. With no correlation between any of the Walsh codes, the receiver will not mistake one code for another because the receiver correlator will only produce an output when the codes are exactly equal. The base stations assign unique Walsh codes to every call, which serves to keep the calls separated, even as all traffic occupies the same frequency spectrum.

As mentioned, the Walsh codes are repeating sequences of 64 bits, with each Walsh code bit forming a single chip. The Walsh codes are clocked at the chipping rate, which is 1.2288 Mcps. At this high rate the RF signal will occupy a 1.25-MHz bandwidth, so spreading has definitely occurred. The data rate, which consists of the voice signal after vocoder compression as well as bits allocated to error-correction and other housekeeping functions, is 19.2 kbps per Walsh code. The low-rate information bits are combined with the much higher-rate chips in the manner described previously. Effectively, the combination is an XOR operation in which the phase of the chips is inverted or not depending on whether the data bit is a 0 or a 1. Each data bit is applied to and modifies one Walsh sequence of 64 bits. If the data bit is a 0, there is no phase shift of the Walsh sequence (that is, all 64 Walsh bits are left as they were), but if the data bit is a 1, then all 64 bits of the Walsh sequence are inverted—the ones become zeroes and the zeroes become ones, representing a 180° phase shift. The Walsh codes so modified are then applied to the modulator, and the modulated signal produces a constellation that is a form of QPSK shown in Figure 8-21.

The preceding discussion illustrates two interesting properties of CDMA systems. First is the central role played by the Walsh codes and the receiver correlator. For the correlation to work, all Walsh codes must be transmitted with exactly the same start and end timing. Thus, the base stations and subscriber units (phones) must be very precisely synchronized. Synchronization among the base stations is maintained with Global Positioning System (GPS) receivers at each base station. CDMA base stations derive their frequency reference from GPS receivers for satellite signals. They also derive their timing reference, called *even second clock,* from these receivers. TDMA-based systems do not require such precise synchronization because the phones reacquire the timing associated with each base station at each handoff. This need for precision timing has implications for the mobile units (to be discussed shortly) as well as for system maintenance. Interference with the

GPS received signal is about −140 dBm at 1.575 GHz. This means that even an extremely small signal can cause problems at this frequency. This is a common cause of degradation of base station performance.

The second property has to do with handoffs between cells or among sectors within a cell site. Recall that CDMA systems use a common 1.25-MHz frequency band shared among sites and sectors and that separation is maintained through the use of Walsh codes. In contrast, TDMA systems actively manage frequency allocations to eliminate the possibility of overlap, and when a mobile user moves from one site to the next the call is momentarily disconnected until the user's phone changes frequency and reestablishes contact with the new site, whereupon the connection is continued. TDMA and analog systems, therefore, engage in a "hard handoff" as users move about within the coverage area of the system. For CDMA systems, nearby cell sites do use the same Walsh codes, but for reasons to be discussed shortly these can be differentiated as well. The phone, therefore, can be and often is served by more than one cell site simultaneously. Thus, as the phone moves to the edges of a site, it can maintain communication with the weaker base station even as it establishes contact with the stronger one. In fact, this simultaneous communication can occur with three or more sites. As the signal from one base station becomes too weak to provide reliable service, the mobile unit can be gracefully transitioned to another base station on the same frequency by simply changing Walsh codes. The phone does not need to disconnect from the base station to retune to a new carrier frequency. Because CDMA systems are capable of such "soft" handoffs the systems potentially provide for more reliable and higher quality voice service, particularly under weak signal conditions. The trade-off, however, is in terms of network capacity because a single user with connections to multiple base stations simultaneously consumes more network resources than would be the case if the connection were to a single base station. This capacity issue and the companion issue of how resources are allocated in general are variables under the control of the system operator.

An important aspect of CDMA system operation is embodied in the *pilot signal* transmitted by each base station. Up to a point, the pilot signal serves a role similar to the FCCH in GSM. The pilot signal is assigned Walsh code 0, which is a 64-bit word consisting of all zeroes. The pilot signal serves as a beacon, constantly transmitting an all-zero signal, which the mobiles use to recognize that they are in the vicinity of a base station. The pilot signals serve several purposes, including timing and base station identification, and employ pseudonoise (PN) sequences to differentiate themselves. The PN code used for this purpose is sometimes referred to as the "short code" because it repeats every 26.67 ms. Although GPS is used for base-station synchronization, it is not practical for the mobile units because GPS signals are easily blocked indoors or in the vicinity of tall buildings. Timing information embodied in the pilot signal provides a known reference for the mobile to know the start and end times of the Walsh codes, in the same way the GPS receivers did so for the base stations. The pilot signals also allow for coherent carrier regeneration in the receiver. Mobile receivers are not capable of resolving absolute phase information because phase changes occur whenever the mobile unit moves with respect to the base station; an increase in distance causes the signal to take longer to travel from base to mobile, thus introducing a phase delay. One way to get around this problem is to use differential PSK schemes, which is what is done in TDMA systems as well as some others discussed previously, but these are more susceptible to noise, a serious concern in a mobile environment. CDMA systems use coherent modulation, in which the carrier must be recreated at the receiver. One of the purposes of the pilot signal is to provide sufficient information to determine and recreate the received carrier phase.

The pilot signal is also responsible for differentiating among base stations. It does so by making use of a phase offset introduced into the short code, which, as we shall see shortly, will be applied to all the channels to be transmitted. Each base station or sector differentiates itself by sending the pilot sequence with a different offset to the short code. All base stations use the same short code; what differentiates them is the phase offset. There are 64 different offsets possible, each in increments of 512 chips, which is a significant enough delay that the phone will recognize the received short code as being from the pilot signal sent by a nearby base rather than the result of a transmission delay. The offset is similar to several racers running around a track, with each racer assigned his own starting position.

Base stations are identified with offset short codes from #1 to #64. Offset #1 could be considered the inside lane of the track. This CDMA "track" has 64 lanes with short-code offset #64 the outside lane. In addition to providing a timing signal and phase reference for the receiver, the pilot signal also assists with the soft handoff function. The phone assesses the strength of the pilot signal to determine whether a handoff is warranted. At the same time, the phone processes information regarding the strengths of pilot signals from adjacent cells and passes that information back to the network. This information, taken together with other parameters, assists in handoff decisions.

In addition to Walsh code 0 being reserved for the pilot signal, several other Walsh codes are defined for specific system functions:

- Codes 1–6 are reserved for paging. This function is similar to the role of the PCH in GSM systems. Code "1," consisting of a repeating sequence of all zeroes for bits 1–32 and all ones for bits 33–64, is normally the code chosen for paging. The phone monitors the paging channel for incoming calls. Other information, such as when to put up a "voicemail received" indication, is also carried on the paging channel.

- Code 32 is the synchronization signal. This is similar to the SCH for GSM. Walsh code 32 is the following: 101010101010 … (i.e., 32 sets of 10). The synchronization channel includes information about the short offset acquired by the mobile as well as system time, derived by the GPS receivers at the bases.

- The remaining Walsh codes are available for traffic from mobile phone subscribers. In practice, one channel can support only about twenty phone calls, even though there are more Walsh codes. The reason for this is that "other users" appear to the system as random noise, and the system noise level increases every time a user is added.

Figure 10-14 is a conceptual block diagram of various data streams within a CDMA base station. Data streams associated with the pilot, paging, and synchronization functions, along with their associated Walsh codes, are shown, as well as that for the user traffic. As established, the pilot channel consists of all zeroes, as does Walsh code 0. Because the inputs to the XOR function are the same logic state, the XOR output remains unchanged. To produce the QPSK modulation characteristic of CDMA, the XOR output for the pilot and all other channels is split into parallel paths and XOR-combined with the I and Q short-code PN sequences. There are actually two forms of the short code, one for data destined for what will become the I channel when the signal is modulated with the RF carrier, and another form of the short code for the phase-shifted Q channel data. The pilot channel is simply a QPSK signal that has been modified by the PN short code, which, among its other purposes, allows the phone to differentiate among base stations in the vicinity.

The same I and Q short-code PN sequences will be XOR-combined with all the other channels as well. However, these other channels undergo significantly more processing than does the pilot. First, data from the synch, paging, and traffic channels undergo mathematical operations, executed as part of digital signal processing (DSP) operations, designed to encode and interleave the data in such a way that it becomes more robust and is less likely to be affected by burst errors. Then, for the synch channel, the coded data are XOR-combined with Walsh code 32 before the I and Q PN operations. The purpose of the synchronization channel is to provide the GPS-based timing information needed for the Walsh-based autocorrelation to work properly in the phone. Recall that all these Walsh codes produce bits at the chipping rate of 1.2288 Mcps.

Both the paging and user channels undergo a similar process but with an additional XOR operation. In addition to undergoing convolutional encoding and interleaving, as did the sync channel, the paging and user channels also undergo an XOR combination with a 2^{42}-bit PN "long code" (which really is long—it repeats after about forty-two days). Along with spreading the signal, the long code associated with the user channels provides privacy by encrypting the signal. On the paging channel, a "public long code" is applied in part to encrypt the electronic serial number (ESN) of the phone. ESN information is exchanged between the base and mobile to create the random number used as an encryption key used to create the "private long code" associated with the user channels. The private long code changes with each call, but the public long code is a constant number used by the base to

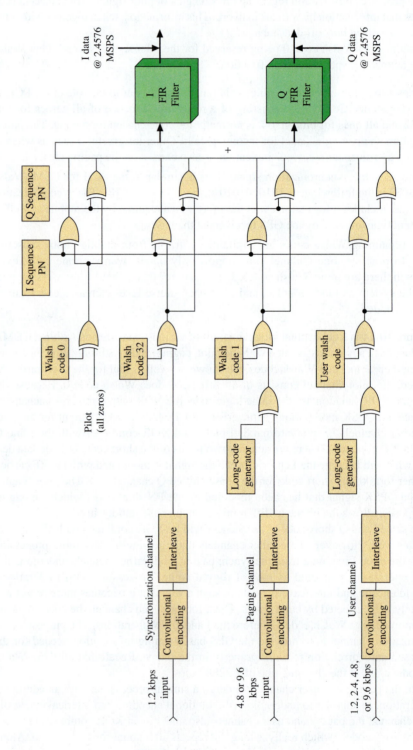

FIGURE 10-14 **Conceptual diagram of CDMA base station.**

reach all the phones. When a mobile responds to a call, the private long code will be generated and applied before the call is connected.

After XOR combination with the long code, the paging and user channels are XOR'd with their respective Walsh codes: code 1 for the paging channel, and any of the remaining fifty-five codes for the user channels. All these I/Q modulated channels are then summed together. The result is a signal that is up to 14 bits wide. This signal undergoes further processing to equalize gain and then a number of other steps (to be discussed in more detail in section 10-5) in preparation for transmission as an RF signal with QPSK modulation and an occupied bandwidth of 1.25 MHz.

The final aspect of CDMA systems to be considered has to do with the "uplink," which is the path from the subscriber unit (phone) to the base station. The discussion of CDMA up to this point has focused on the "downlink" path, the link from the base station to the phone. The CDMA multiplexing works well for the downlink because all the base stations are synchronized with GPS receivers. This level of precision ensures that all the Walsh codes are received synchronously and can be decoded properly. However, the uplink or return path presents a problem in the form of the propagation delay presented by the trip the signals take from base station to subscriber unit and back. Radio frequencies travel at the speed of light, which is about one mile per microsecond, which works out to about one chip. Because the phone may be constantly changing location with respect to the base, and because the base is receiving signals from many phones simultaneously at different locations, the uplink signals cannot possibly be perfectly aligned at the base station receiver. Therefore, the CDMA multiplexing process used for the downlink cannot also be directly used for the uplink.

The answer is in the long code. The phone also produces a long code using information from the synchronization and paging channel to generate its own cyclically shifted code. Like the base station long code, the code in the phone consists of a 2^{42}-bit-long sequence, and each user is assigned a different phase based on information sent from the base through the paging channel. Out-of-phase code sequences appear as noise at the base receiver, as would be the case if different codes were used. Walsh codes are also associated with the uplink, but not to separate the calls directly. Instead, each grouping of 6 data bits is associated with one of the 64 possible Walsh codes. Recall that Walsh codes are 64 bits long, so once the association has been made, all the Walsh code bits (the chips) are mixed with the selected phase of the long code. The base station receiver has the job of detecting the Walsh code. Once it has done so, then the original 6-bit data stream can be recreated. Although 64 bits can be combined in many possible ways, the only valid combinations are the 64 Walsh sequences. Essentially, the base-station receiver's task boils down to matching the received sequence with the most likely (i.e., closest) valid Walsh sequence.

Although this process seems roundabout, it does address the problem posed by the variations in phase delay seen by the base-station receiver. However, the tradeoff is that the correlation process used to match up the PN-code phase with the users is not perfect. While the use of Walsh codes in the downlink direction effectively eliminates the possibility that one user would have an effect on any other user, in the uplink direction there is some amount of interference between users because of the less-than-perfect nature of the correlation. The uplink is, therefore, the weaker of the two paths because it is interference limited. In contrast, the downlink tends to be power-limited because the power amplifiers in the base station must have sufficient linearity to provide adequate power to all users.

POWER CONTROL   As with every sophisticated system, the devil is in the details, and many details of system operation were left out of the previous discussion in the interest of clarity. However, it should be mentioned that satisfactory operation of CDMA systems is critically dependent on power control, both in the pilot signal to allow the correct handoff decisions to be made as well as in the uplinks and downlinks. For optimal capacity, each channel must be transmitted at the lowest power possible to provide reception with an acceptably low number of bit errors. In the uplink, the power levels from all the mobiles must be constantly adjusted so that the received signal strength at the base station antenna is roughly the same. Otherwise, the base will tend to "latch on" to the strongest signal and ignore all the others. This is sometimes known as the "near-far" problem.

Because the mobiles are different distances from the base, and are most likely in motion, the power emitted by each mobile must constantly be adjusted to maintain equal signal strength. This is done with a power control bit generated by the base and sent within each downlink data frame, which causes the mobile power to be incremented up or down as often as every 1.25 ms.

Other open- and closed-loop power control applications are running in the background. Many of these are proprietary to Qualcomm, the primary developer of IS-95 CDMA. The vulnerability of CDMA-based systems to power issues is perhaps not too much of a surprise if you remember the conference/party analogy introduced in Chapter 8 to describe the spreading-code concept. As long as everyone cooperates by talking at a reasonable volume, multiple conversations can take place in the same space at the same time and over the same frequency range. But what would happen if the slightly drunk married couple in the corner got into an argument and started shouting at each other? Most likely, all other conversations in the room would come to a screeching halt. The many power-control systems built into CDMA implementations are largely there to prevent anything in the system from becoming the out-of-control married couple in the room.

RAKE RECEIVER   A unique characteristic of CDMA-based systems is seen in the receiver architecture. The type of receiver used is known as a *rake receiver* because it has multiple correlators that can be programmed either for different Walsh codes or to compensate for time differences of signals arriving by different routes as a result of multipath effects. In a manner similar to how the tines of a garden rake gather leaves, the multiple correlators in the rake receiver gather energy from multiple signals to provide the best possible reception. The basic idea is illustrated in the block diagram of Figure 10-15. Each correlator is known as a "finger," and each finger essentially acts as a sub-receiver. The rake configuration is particularly well suited to CDMA systems because of the prevalence of multipath effects. Multipath is severe both because of the short wavelengths involved and because of the high chip rate used to spread the CDMA signal. The rake receiver is well suited to optimizing reception of signals that would be degraded as the result of multipath.

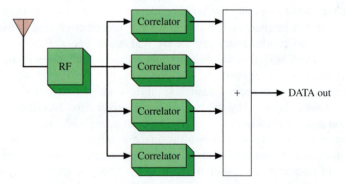

FIGURE 10-15   **Simplified block diagram of rake receiver.**

Troubleshooting and Interference Considerations

The "open architecture" of GSM assists technicians with troubleshooting. An open architecture means that the information about the system is available for all users. RF technicians can easily use the results obtained with the open architecture to identify RF problems, such as interference. If an area is identified that has strong signal strength and still has an excessive number of dropped calls, the area probably has interference problems. The use of the protocol analyzer instead of "drive testing" can be a more efficient use of a technician's time. In contrast, CDMA systems were developed by private manufacturers that have chosen to keep information about the links between the base stations and switches

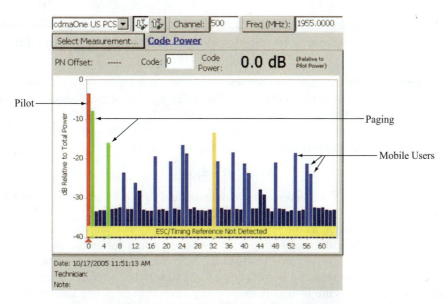

FIGURE 10-16 **Display photo from Tektronix YBT250, showing Walsh codes. Walsh code 0 (red) is the Pilot, codes 1–6 (green) are paging, code 32 (yellow) is synchronization, light blue codes are actual mobile users, and dark blue codes are Walsh codes not in use.** (*Note:* **This measurement was made over the air; "ESC/Timing Reference Not Detected" indicates that the YBT250 is not physically connected to a base station.**)

"proprietary." However, specialized test equipment has been developed to assist with the troubleshooting of CDMA networks, particularly with respect to system-quality issues and interference considerations. Interference is of particular concern in CDMA because some received signal levels are very low. Recall in particular the importance of GPS timing in maintaining base station coordination. Also remember from Chapter 8 just how small the received signal level is: around −130 dBm or so, which means that potential interference at the GPS frequency of 1.575 GHz can occur in the vicinity of −140 dBm, which is an extremely small signal.

An instrument called a *modulation analyzer* can be used to demodulate the CDMA signal and separate the signal into Walsh codes (See Figure 10-16). Various other measurements for base stations are as follows:

1. **Rho:** This is a comparison between the actual CDMA and a perfect undistorted signal. A perfect rho = 1.

2. **EVM:** Error vector magnitude. This is a measurement of the total noise, both AM and FM.

3. **Carrier feed through:** This is a measure of the amount of unmodulated carrier that "leaks through" to the output signal. It needs to be at least 25 dB less than the CDMA pilot signal.

Interference can degrade the performance of both GSM and CDMA systems. GSM systems can have interference with themselves in the form of *cochannel* and *adjacent-channel interference.* Updating the frequency planning can reduce this problem. The update is normally done empirically. The engineer makes initial frequency assignments on the basis of mathematical models. The technician identifies some "problem channels" that the engineer uses to update the frequency planning. In CDMA, all users are operating at the same frequency, and the use of orthogonal codes reduces interference with other users. Nevertheless, each user adds to the noise level at that frequency, and this limits the number of users to about twenty.

Interference can also be caused by many other sources. These fall into the following three categories:

1. Signals that interfere with the GPS receiver at 1.575 GHz. The GPS satellites provide signals that GSM and CDMA use for frequency and timing references.

2. Signals that interfere with the uplink, which is 821–849 MHz for the cellular band.

3. Signals that interfere with the downlink, which is 869–894 MHz for the cellular band.

One source of interference can be harmonics from other signals. One example is the fifth harmonic of VHF television channel 7. The fifth harmonic of the channel 7 frequencies are within the cellular downlink frequency band.

Another source of interference can be mixing products from other signals. The frequencies can be combinations calculated from Equation (10-1).

$$\text{Fout} = n\text{F}_1 \pm m\text{F}_2 \tag{10-1}$$

where n and m are integers representing harmonics, and F_1 and F_2 are separate potential interfering signals.

Interference created as the result of undesired mixing is commonplace and often presents some of the most challenging troubleshooting issues for communications system technicians. These issues are by no means limited to cellular installations. Rather, they can arise in any situation involving multiple transmitters and receivers, but the issue is particularly acute in installations where several high-power transmitters are located within close proximity to each other. This type of "co-location" arrangement is the rule rather than the exception. In many areas, the most desirable sites for communications installations of all types are relatively few and are usually found at high elevations such as mountaintops or the roofs of tall buildings. Also, zoning ordinances in many jurisdictions mandate that communications installations be co-located wherever possible to reduce the proliferation of antennas and towers. These sites are usually leased to many unrelated tenants, who share facilities including tower space for antennas. The challenge for communications engineers and technicians is similar to that faced by anyone moving into a new community: You can't choose your neighbors. As long as all the systems are operating within their licensed parameters (and they usually are), any interference issues that arise (and they inevitably will) must be resolved through the mutual cooperation of the parties involved. Interference can often be mitigated through the installation of various types of filters in either the transmit or receive signal paths, or sometimes both.

One example of interference resulting from undesired mixing would be a harmonic of an FM radio station (about 100 MHz) mixing with a paging signal (about 160 MHz). The second harmonic of the 99.9 MHz FM signal mixed with the third harmonic of the paging signal (160 MHz) produces a potentially interfering signal at 839.8 MHz, as shown in the following equation: using Equation (10-1).

$$F_1 = 100 \text{ MHz} \qquad F_2 = 160 \text{ MHz}$$
$$n = 2 \text{ (second harmonic)} \qquad n = 4 \text{ (fourth harmonic)}$$
$$\text{Fout} = 2 \times 99.9 \text{ MHz} + 4 \times 160 \text{ MHz} = 839.8 \text{ MHz}$$

This type of problem would be especially difficult to isolate because of the "bursting" nature of paging signals. An instrument called a *real-time spectrum analyzer* is useful for identifying this type of problem. This is illustrated in Figures 10-17 and 10-18. The display shown is called a *spectrogram*. Note that the horizontal axis is frequency, the vertical axis is time, and power is displayed by the color. The instrument stores eighty-seven separate data records, which can be viewed in order to determine the time at which interference occurred. The signal shown is FSK. At various times, the frequency is at one of three values.

The large amount of information contained in Figures 10-17 and 10-18 can be understood by making the following observations:

1. The horizontal scale of the upper and lower photos in Figure 10-17 is frequency, with a span of 15 MHz.

2. The center frequency is 2.445325 MHz, as shown at the bottom of both the upper and lower photos. (Figure 10-17)

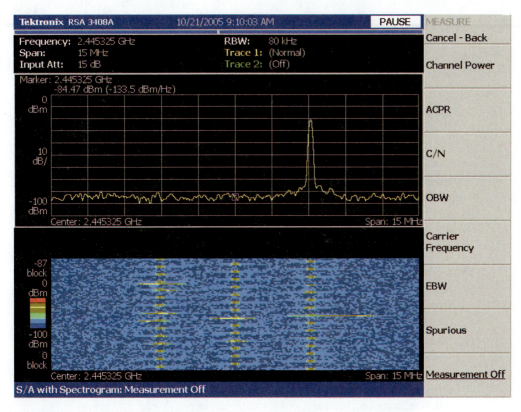

FIGURE 10-17   Display photo from Tektronix RSA 3408A, showing spectrogram.

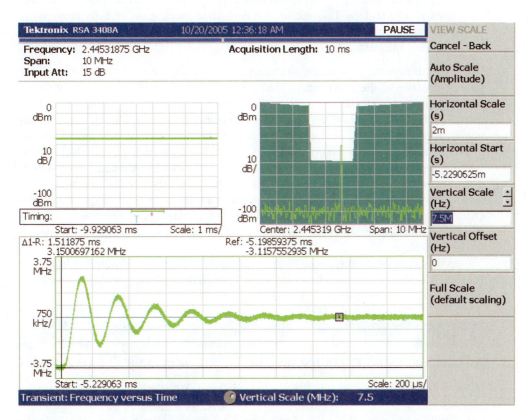

FIGURE 10-18   Display photo from Tektronix RSA 3408A, showing the playback of a single captured line from the spectrogram shown in Figure 11-19.   Note the settling time for the FSK phase lock loop.

3. Using a ruler, one can determine that the frequency between to the left row of measured "spots" and the right row of "spots" is 6 MHz. (Figure 10-17)

4. The vertical scale of the upper photo is power. The photo shows that the power level at the right row of "spots" is −20 dBm. (Figure 10-17)

5. Power in the lower photo is indicated by the color. The yellow color of the "spots" indicates −20 dBm, but this is not as accurate as the measurement in the upper photo. (Figure 10-17)

6. The vertical scale of the lower photo is "block number." The bottom of the photo is block 0 and the top of the photo is block −87. (Figure 10-17)

7. Each block represents an increment of time. Note from the lower photo in Figure 10-18 that the time scale is 200 μs per division. Because there are ten divisions, the time for a full span is 2 ms.

8. Because there are 88 blocks, the time elapsed between the bottom, lower border of Figure 10-17 and the upper border is 176 ms.

9. Using a ruler, one can determine that the approximate duration of each "spot" is 6 ms.

10. Summarizing what we have learned, we now know that "hopping" occurs between three different frequencies that are each separated by 3 MHz. The signal dwells at each frequency for about 6 ms. (Figure 10-17)

11. When this instrument is in use, the display in the lower photo of Figure 10-17 scrolls upward. Block −87 disappears and is lost (unless it is stored in an external device) and a new block 0 will appear. The time duration of each block can be adjusted. Each block can be set to measure any the following: maximum signal during the block duration, minimum, average, or maximum/minimum.

12. The lower photo in Figure 10-18 provides additional information about the settling time for the phase lock loop at each frequency. Note the initial overshoot of −3.75 MHz. About 150 μs later, the overshoot is +2.25 MHz.

13. Note that the term *real-time spectrum analyzer* applies to an instrument that captures all of the data. This means that each time block must be fully contiguous. Many spectrum analyzers that produce spectrograms are technically not "real-time" spectrum analyzers.

When troubleshooting an interference problem, it may be important to identify the physical location of the interference. The interference could result from intermodulation with a nearby transmitter. In this case, *intermodulation* refers to the mixing of signals from two properly working transmitters to produce an interfering signal. Mixing occurs in any nonlinear device, and the device does not need to be electronic. Interference problems could also be the result of mixing created by the nonlinear junction set up in bad antenna connections or even to a rusty fence or to rusty piles of metal located near a base station. The effectiveness of this type of "mixer" can be substantially changed by rain, making the problem more difficult to isolate. Tracking down interference problems is one of the most challenging troubleshooting assignments that communications and cell-site technicians will face. Successful resolution of some interference issues may require help from specialists in the field.

A list of potential interference from the harmonics of common powerful radio signals is listed in Table 10-6. These signals are strong enough to cause significant interference to the mobile radio frequencies.

TABLE 10-6 • A Partial List of Some Potential Interfering Signals			
CELLULAR BAND	**FREQUENCIES (MHz)**	**INTERFERER**	**HARMONIC**
US Cellular Base TX	869–894	VHF TV channel 7	5
US PCS Base TX	1930–1990	UHF TV channels 16–18	4
US PCS Base RX	1850–1910	UHF TV channels 14–15	4
GPS Receiver	1574.9–1575.9	UHF TV channel 23	3

10-3 MOBILE AND CELLULAR DATA NETWORKS

Third-Generation Systems

So-called *third-generation* or *3G* systems have been widely deployed, and it is now commonplace to expect a great deal of interconnectivity in the mobile environment. At the same time, the demand for ever higher data rates for applications such as streaming video seems to continue unchecked. The path to 3G has been a long one, however. The timing for the widespread deployment of 3G is closely tied to market conditions and the availability of spectrum. Because 3G services require more bandwidth, they also cost mobile phone users more money. All the major wireless service providers offer an interim type of service called *2.5G* or *2.75G*. These services provide much higher data speed than does 2G and at much lower cost to the mobile phone customer than does 3G.

2.5G has two basic paths, one which is "backwards compatible" with GSM and one which is backwards compatible with CDMA. The term *backwards-compatible* means that the customer's phone will work for both the new technology and the old technology. Basic characteristics of the two paths to 3G are listed in Table 10-7.

SYSTEM	GENERATION	NAME	DATA RATE	MODULATION TYPE	BANDWIDTH
GSM	2G	GSM	14 Kbps	GMSK	200 kHz
GSM	2.5G	GPRS	144 Kbps	GMSK	200 kHz
GSM	2.75G	GPRS/EDGE	288 Kbps	8 PSK	200 kHz
GSM	3G	WCDMA	2.4 Mbps	—	5 MHz
CDMA	2G	CDMAone	14 Kbps	QPSK	1.2 MHz
CDMA	2.5G	1XRTT	144 Kbps	QPSK	1.2 MHz
CDMA	2.75G	1XEVD0	288 Kbps	8 PSK	1.2 MHz
CDMA	3G	CDMA2000	2.4 Mbps	—	5 MHz
UMTS	4G	UMTS	2.4 Mbps	—	5 MHz

TABLE 10-7 • The Basic Characteristics of the Paths to 3G

Several of the technologies shown in the table bridge the gap from 2G voice to 3G data. One in particular is CDMA2000, which is an enhancement to the 2G CDMA voice-grade system previously discussed and which is also termed IS-95 CDMA. Like its predecessor, CDMA2000 was developed by Qualcomm, which also holds many of the patents on both forms of the technology. CDMA2000 is fully backwards-compatible with IS-95, which makes it attractive to system operators. In its earliest form, CDMA2000 was designated $1 \times$ RTT. The RTT stands for "radio transmission technology," and the $1 \times$ designation indicated that a single 1.25-MHz radio channel was in use, the same as for conventional CDMA. This form of CDMA was considered to be a "2.5G" technology because it offered a higher data rate than the 2G implementation. It did so by changing modulation formats and coding to double the voice handling capacity. At the same time, $1 \times$ RTT introduced packet-based data-handling to the wireless arena to provide a maximum data rate of 144 kbps. The packet approach is in contrast to the circuit-switched topology of voice cellular systems. (Packet switching was introduced in Chapter 9.) A subsequent enhancement called $3 \times$ RTT used three radio channels for a total bandwidth of 3.75 MHz and a higher chip rate to increase the data rate by a factor of 3 as well, to 432 kbps.

While $3 \times$ RTT is arguably a "2.75G" technology, the latest enhancement to CDMA2000, known as $1 \times$ EV-DO, brings the CDMA platform into the 3G realm. EV-DO stands for Evolution-Data Optimized. Again, a single 1.25-MHz channel is used, and several different forms of modulation, from QPSK to 16-QAM, are supported depending on the data rate. The maximum data rate is 2.4 Mbps for downloads using 16-QAM. The upload data rate in all cases is a maximum of 153.6 kbps.

UNIVERSAL MOBILE TELECOMMUNICATIONS SERVICE AND WIDEBAND CDMA   A competing 3G technology is a wideband form of CDMA called, appropriately enough, WCDMA. It is not the same as CDMA2000; however, it does have a number of characteristics in common with other CDMA implementations such as reliance on direct-sequence spread spectrum. WCDMA also goes by the name Universal Mobile Telecommunications Service (UMTS) and is the form of 3G recommended by the International Telecommunication Union (ITU), a United Nations organization charged with allocating global radio spectrum and coordinating affairs among nations pertaining to telecommunications standards. Originally, the ITU specified characteristics for 3G systems operating in the 2-GHz range and envisioned a worldwide standard and upgrade path. The idea was to achieve interoperability and enable worldwide roaming. The ITU did not endorse a particular technology, however, so several competing 3G approaches have been proposed and tried, of which CDMA2000 and WCDMA are two examples. At the same time, implementation delays for 3G systems and the passage of time have caused consumer expectations for downlink capacity to leapfrog the original ITU goal of 2 Mbps demand for high data capacities, causing carriers to play catch-up and to roll out a number of competing proposals.

As mentioned, the WCDMA format is based on direct-sequence spread spectrum (DSSS). It uses a 5-MHz-wide RF channel with a chipping rate of 3.84-Mcps and QPSK modulation. The data rate achievable with this configuration is 2 Mbps. For full-duplex operation (uplink and downlink) two RF channels are needed. The wide channel bandwidth required is a problem shared by all 3G systems, but it is particularly acute for WCDMA. As has been established, radio spectrum is scarce and expensive. When chunks become available, government bodies often auction it to the highest bidder. Auction prices can go into the hundreds of millions, if not billions, of dollars. Needless to say, this represents a significant investment and barrier to entry for any prospective carrier. It also means that different frequency bands are available in different regions of the world.

An attempt to address the bandwidth issue is embodied in a variant of WCDMA, also endorsed by the ITU, called TD-SCDMA (time-division, synchronized CDMA). The primary advantage of this implementation is that it requires a narrower-bandwidth channel: 1.6 MHz. It uses a 1.28-Mcps chipping rate. As implied by the name, time-division multiplexing is also used, and different time slots within the same frequency bandwidth can be assigned to the uplink and downlink. This assignment function is also dynamic. The system operator can assign more or fewer timeslots to activities based on the load experienced by the system at any given time. A hint to the biggest problem lies with the word *synchronized* in the system name. For it to work properly, the system must have extremely accurate timing and synchronization throughout. For this reason, TD-SCDMA systems are very complex to implement and maintain.

To address consumer demand for download data rates in excess of 2 Mbps, so-called 3.5G technologies compatible with WCDMA have been rolled out. One of these is called high-speed download packet access (HSDPA). It can be considered an upgrade to WCDMA. One of its features is that it automatically adapts to channel conditions. Data capacity is specified in terms of "categories," with Category 1, the lowest, at 1.2 Mbps. Higher categories offer higher rates: Category 10 sports 14.4 Mbps using 16-QAM, for example, and a Category 20 system in Australia claims 42 Mbps. Uplink capacities are considerably slower because achieving a high data rate reliably from a portable device to the base station is more problematic. However, a companion standard to HSDPA called high-speed uplink packet access (HSUPA), providing a maximum uplink data rate of 5.76 Mbps, is in the early stages of deployment.

Fourth-Generation Systems

Competitive pressures, along with the seemingly insatiable demand for wireless capacity and connectivity, have caused carriers to begin promoting so-called *4G* or *fourth-generation* wireless data networks. Whether these are truly 4G systems is a matter of some debate.

Whereas the definition of a 3G system is clear-cut, this is less so for 4G, and true 4G systems are still in the evolutionary stages of development. A 3G system is one that meets the ITU standard for what it termed an International Mobile Telecommunication 2000 (IMT-2000) phone: one that could achieve a data rate of up to 2.048 Mbps while stationary, 384 kbps while in slow motion (i.e., the subscriber is walking), and 144 kbps while in fast motion (such as in a vehicle). Many of the systems and protocols just discussed do truly meet or exceed the definition of 3G, and some fall into the category of 3.5G or even 3.9G with various upgrades to enhance data throughput.

Fourth-generation systems, however, have not been fully defined as yet. The ITU defines a 4G system as embodying architectures identified as *long-term evolution-advanced*, an umbrella term encompassing many technologies already implemented to a greater or lesser degree in the present generation of advanced 3G systems. The next generation of 4G envisions a 1-Gb/s upper limit to data rates and would require entirely new network infrastructure. What is being touted as 4G at the present time is largely an enhancement of existing infrastructure rather than a wholesale replacement.

The most advanced 3G systems (perhaps most accurately thought of as 3.9G) make use of the first generation of long-term evolution (LTE) standards. One significant difference between LTE and earlier 3G implementations is that LTE relies on orthogonal frequency-division multiplexing (OFDM) rather than CDMA for the downlink. Recall that OFDM is already in use in some wireless local-area networks and that the technique is predicated on modulating data onto a large number of closely spaced but orthogonal subcarriers. OFDM in wireless systems is essentially a "one-to-many," or point-to-multipoint, system. As implemented in wireless mobile communications, the basic form of OFDM must be enhanced with a multiple-access system to permit simultaneous use by many subscribers. Thus, OFDM coupled with some form of multiple access, as is the case in LTE implementations, is termed OFDMA.

Several advantages potentially accrue in the use of OFDMA as a multiple-access technique, even though system implementation is much more involved and would not be practical were it not for the heavy reliance on digital signal processing and other software-based implementations of functions traditionally carried out in hardware. One advantage was mentioned earlier: spectrum is used much more effectively than with the traditional form of frequency-division multiplexing because the need for guard bands between users is eliminated. In fact, the sidebands produced on either side of each subcarrier do overlap to a substantial degree. However, in an OFDM receiver all the subcarriers are demodulated simultaneously with fast Fourier transform calculations implemented in DSP. This idea implies that we do not have to worry about less-than-ideal filter responses (that is, the inability to produce a perfect "brick-wall" transition from passband to stopband) and the consequent need for additional frequency separation. The subcarriers in LTE systems are separated by 15 kHz, and orthogonality is maintained by separating the carriers by an integer multiple of the symbol rate associated with the modulating signal.

A second advantage also has to do with spectrum utilization. OFDMA systems can support multiple bandwidths, which can be adjusted to accommodate changes in demand for system capacity. In LTE the channel bandwidth (again, consisting of many subcarriers at 15-kHz intervals) can be adjusted from as little as 1.25 MHz up to 10 or 20 MHz, as conditions demand and as spectrum allocations and business considerations permit. In contrast, CDMA and WCDMA systems work with a fixed bandwidth per channel, which, as shown previously, is largely determined by the chip rate and choice of filtering. The only way for CDMA systems to adjust bandwidth is by using multiple channels. A third advantage pertains to adaptability in the presence of impairments in the communications channel, such as noise. OFDMA-capable transceivers will be capable of changing modulation formats in a dynamic fashion. A good-quality, noise-free link will enable higher-level QAM modulation techniques for maximum data throughput, whereas a noisy channel will spur the system to "kick-back" to a form of modulation such as QPSK that has better performance under noisy conditions but at the expense of fewer symbol locations to represent data, thus less data throughput. The connection will still be maintained, however, providing a more satisfactory user experience for the mobile subscriber.

A second noteworthy aspect of LTE systems is the reliance on multiple-input multiple-output (MIMO) configurations, first introduced in the discussion of IEEE 802.11n wireless

LANs (WLANs), to increase data throughput. Practical MIMO implementations are also largely made possible by the wholesale shift to DSP. MIMO involves the use of multiple antennas at both the transmitter and receiver to, in essence, set up multiple and independent transmission paths for data. This data independence, coupled with the ability of the receiver to take advantage of the additive phase effects of signals arriving at its several antennas by different paths, promises a large increase in data-handling capability. When paired with OFDM, LTE systems employing MIMO potentially sport best-case data rates of up to 100 Mb/s in the downlink direction (from base to mobile) and 50 Mb/s in the uplink direction. It is left as an exercise for the student to imagine why anyone would need such capacity in a mobile environment.

So, are the systems advertised as 4G really 4G? That is a matter of perspective. Strictly speaking, true "4G," as codified by international standards-setting bodies, is not yet a reality and probably will not be for some time, as the standards are still being finalized. In all likelihood 4G will involve wider bandwidth channels and more MIMO antennas at both the transmitter and receiver. Just be aware that even though the systems now being deployed may not strictly align with formal definitions of 4G, their capacities have surpassed the original 3G specification to such a degree that some recognition must be given to the tremendous ingenuity behind the capabilities offered by latest-generation networks, no matter how they are branded. One other point should be made as well: the 3G and 4G networks are data networks only. As of this writing, voice traffic is still handled by the 2G, circuit-switched CDMA and GSM networks discussed earlier. However, it is not too much to assume that the integration of voice traffic with data will no doubt continue as part of the migration to packet switching as mobile systems continue to evolve.

Wireless Application Protocol

The **wireless application protocol (WAP)** is a world standard that has been developed to bridge the gaps between mobile communications, the Internet, and corporate intranets. WAP was developed to address the issues and standardize the solutions for providing web-based services and wireless Internet access. The WAP standards address key delivery problems with wireless data networks. These include limited bandwidth, **latency**, connection stability, and availability. Latency is the time delay from the request for information until a response is obtained.

The WAP specification includes several key specifications:

Wireless markup language (WML), which optimizes hypertext methodologies for a wireless environment, including graphics, and **WMLScript**, which is the WML comparable version of Javascript.

A specification for a **microbrowser** that is web browser adapted for the wireless environment.

The framework for wireless telephony applications (WTA) within the WML environment.

Some features of WAP include:

A wireless transaction protocol (WTP), used to manage the data transfer in the WAP protocol. The WTP is analogous to the TCP layer in computer networks. WTP provides the minimum information required to handle each request/response transaction.

A wireless transport layer security (WTLS), which provides for a more secure wireless connection. This layer is comparable to the computer networking industry standard transport layer security (TLS) protocol, which used to be called the secure socket layer.

WAP provides enormous possibilities for growth in the wireless environment. The standard addresses the key limiting issues of wireless phone communications but provides the adaptability for supporting existing information sources such as the web and provides a methodology for incorporating a multitude of new uses. The majority of the world's mobile telephone companies now support WAP.

This chapter has addressed many issues dealing with both wired and wireless communication networks. This section addresses the important issue of securing the transmission of data over these networks. The availability of the Internet and the mobility provided by cell phones, wireless laptop computers, WiMAX, Bluetooth devices, and many other related wireless technologies provide the user with an easier way to share information (e.g., data, images). But with this flexibility, the threat of eavesdropping, jamming, or even theft of information over the communication channel has increased. This section provides an overview of the basic communication security concepts a person responsible for maintaining a communications link should know.

One of the biggest differences between wired and wireless communications channels is that with a wireless communication device, you are using a "shared" broadcast channel. This means that when you transmit a signal, it will go everywhere; to your intended recipients and unfortunately to the "bad guys." This means that the "bad guys" will not have much trouble getting access to the wireless channel. Remember, a wireless channel is transmitting an RF signal and a radio receiver can pick up the transmission. This is unlike a wired communications channel that requires the "bad guy" to gain physical access to the communications facility and to the communication cables carrying the information to intercept the transmission.

There are five aspects to consider when securing a communication link. These five aspects (*confidentiality*, *integrity*, *authentication*, *nonrepudiation*, and *availability of the network*) are first defined and then followed by a discussion on how these apply to wireless security.

1. **Confidentiality (privacy):** This means that you want to keep unauthorized people from gaining access to your information.

2. **Integrity:** Integrity means that you can rely on the data getting through the communication channel without modification. This doesn't mean that the "bad guys" can't modify the data; rather, it means that the unauthorized modification can be detected. Integrity is a guarantee that any modification to the data set will be detected. For example, a checksum can be applied to the data so that you can detect if the data has been changed. A checksum is a count of the number of bits in a transmitted message. If the number of bits in the received message equals the number in the checksum, it can be assumed that the received message is correct. But there can be a problem; if the bad guy can change the data, the checksum will change and therefore the change is not detected.

3. **Authentication:** This requires proving you are who you say you are. For example, when you do online banking, the bank wants to make sure an authorized person is establishing the connection and issuing the commands for a transaction. This typically requires that an authorized personal identification number (PIN) be entered before allowing the user to gain access to an account.

4. **Nonrepudiation:** The bank will insist on nonrepudiation, which means they can prove you issued any bank transaction commands that they accepted.

5. **Availability of the network:** Another area the "bad guy" can use to disrupt the communication link is to transmit an interfering signal over the same communication channel. An example is called *jamming*, in which the "bad guys" transmit an interfering signal over the same channel, thereby disrupting or slowing down communications.

People like wireless networks because such networks allow their communication nodes (e.g., laptop computers and cell phones) to be mobile. The mobility of communication devices requires the use of a fairly insecure operating environment, an RF link. For example, a user sitting in an airport with his or her laptop computer with a wireless connection established is susceptible to having his or her data intercepted or jammed. This set up results in a very insecure operating environment.

Another threat to wireless communication devices is the resource limitation of the nodes. The communications could be battery operated or may have limited memory. An adversary can degrade the quality of the network by consuming the user's resources by disrupting the wireless communications link (jamming). This can cause the communications device to have to retransmit continuously, running down the battery and making it necessary to reestablish communication channel connections, which can cause the internal memory to fill.

The attacks that run on a wireless network can be categorized as passive and active. **Passive attack** means that the "bad guy" is just listening and picking up what information can be obtained. It is hard to detect a passive attack because nothing is being transmitted, but the "bad guy" is receiving (listening) to the transmission. In an **active attack**, the bad guy is transmitting a signal, often a very powerful one, disrupting the communications link. Fortunately, this type of transmission can be detected.

The attacks can also be classified as external or internal. An **external attack** is perpetrated by someone that doesn't have access to the network. What problems can occur by an external attack? They can listen to the transmission or can transmit interfering signals (jamming). An **internal attack** is perpetrated by someone inside the network. These are often more subtle and more dangerous because measures are often put in place to protect against external, but not necessarily internal, attacks. The internal attack can do more damage and can hurt your organization. The attacker can masquerade as someone else within the organization, even though access to certain servers or switches might be restricted to members of the network. Gaining an internal IP address or access code can make it look as though the intruder is actually an authorized user of the network.

Countermeasures are the last point in this overview. Countermeasures often use cryptography to protect the transmitted data from eavesdropping. The use of cryptography means that the information being transmitted over the air is encrypted. This directly addresses any privacy concerns. If the bad guys can't decrypt the data being transmitted, they can't read your traffic.

Cryptography makes a mathematical transformation of the data that the good guys know how to undo but that the bad guys don't. The objective is to make data recovery easy for the good guys but to make it so that the bad guys have to do a tremendous amount of work to recover the same data. This means that if data is encrypted, the good guys will know the "key" for the encryption and the bad guys will have to obtain the key through a lot of guesswork. A **key** is the secret code used in the encryption algorithm to create cipher text and to decode (decrypt) the message. The length of the key is a factor to be considered when determining how long it will take for the bad guy to decrypt a message.

If the bad guy has to use a brute force method to guess the key, we want to make sure the number of guesses required is a huge number. The **Data Encryption Standard (DES)**, a method used to encrypt data in the United States, uses a 56-bit key, which is 2^{56} possible keys. This seems like a large number, but for today's computers, all the keys can be tried in less than one day.

There are a lot of overhead requirements once cryptography is used because the privacy of the whole system depends on keeping the key secret. How does the administrator distribute the key without requiring that an armed courier deliver the key to all authorized users in the network? Protocols are available that enable the transfer of the key over the air as long as everybody has some secret key to start with. Electronic commerce requires that everyone have a secret key and a public key. People can encrypt things with their public key, and only the intended recipient can decrypt it using his or her secret key.

All communication systems want confidentiality or privacy, and the usual way to achieve this is with cryptography. The first way our new generation of cell phones are more secure is that they are digital, which is not cryptographic. Some cell phones use a time-slotted system (TDMA) for transmitting the data that requires the receiver to be synchronized with the transmitted channel because not all bits belong to one conversation. Another system used by cell phones is code-division multiple access (CDMA). In CDMA, you are not talking about cryptography but rather a pseudorandom sequence of bits that effectively spread the signal over the entire spectrum. If the codes are set up so that they don't interfere with each other, then we only have a slight increase in the noise channel. Each spreading sequence is applied to the digital voice signal. This spreads the signal in a way so that when the signals are added together they don't interfere, except as noise. This means that the receiver has to know what spreading sequence to apply. This information is not encrypted or secret, but it is coded. Spreading sequences are well known. It takes a more sophisticated bad guy to be able to listen to and extract information from modern cell phone conversations.

The need for properly securing data transmitted over a communication link can easily be justified by examining how easy it is to decode simple encrypted messages (cipher text). Studies have shown that the frequency of occurrence of certain alphabetical characters and keystrokes is predictable. It is well-known that the frequency distribution of alphabetical characters in the English language follows a pattern similar to that shown in Figure 10-19. The "space" character has been added because this is the most common ASCII character found in a normal typed text message. Look carefully at the pattern generated by the frequency distribution. Notice that A, E, N, O, and T are the five most frequent letters. These letters are the most common in the English language, and this frequency distribution can be used to identify the shift in characters used in a simple cipher.

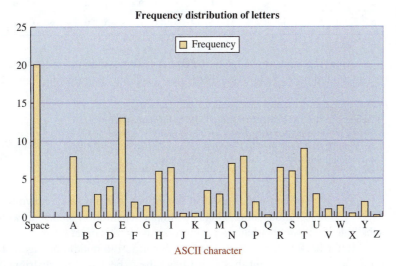

FIGURE 10-19 **A graph of the character frequency for the ASCII characters.**

An example of cipher text (an encrypted message) is provided in Figure 10-20. At first glance, this message appears to be unreadable and the characters totally random. However, the cipher-text message can be input into a software program that analyzes the frequency distribution of the characters. The objective of analyzing the frequency distribution of the characters is to determine whether the pattern is similar to that shown in Figure 10-19. The result of the frequency analysis of the cipher-text message is shown in Figure 10-21.

```
[ipgsqi$xs$xli$QSHIVR$IPIGXVSRMG$GSQQYRMGEXMSR$wigx
msr$sr$[mvipiww$Wigyvmx}2$$Mj$}sy$evi$viehmrk$xlmw$}sy$I
ezi$hixivqmrih$xlex
```

FIGURE 10-20 **An example of cipher text (an encrypted message).**

FIGURE 10-21 **The histogram of the cipher-text message shown in Figure 10-20.**

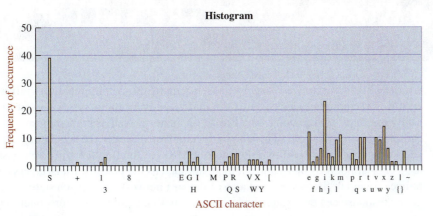

Look at the frequency pattern for the lowercase characters shown on the right side of Figure 10-21. There appears to be a distribution similar to the letter frequency distribution shown in Figure 10-19, except the characters don't match. In fact, the letters appear to be shifted by four, as shown in Table 10-8. The "e" in the cipher text might be an "a." The "x" in the cipher text might be a "t."

Also, the bar representing the frequency of occurrence for the "$" character shows that it was the most frequently used ASCII character in the message. The most common ASCII character in a normal message is the space (SP). Referring ahead to Figure 11-1 in Chapter 11, one can see that the $ is four ASCII letters away from the SP. Therefore, it is a reasonable assumption that this text has been shifted by four letters.

TABLE 10-8 • The Predicted Character Shift for Cipher-text Message Shown in Figure 10-20																									
a	b	c	d	e	f	g	h	I	j	K	l	M	n	o	p	q	r	s	t	u	v	w	x	Y	z
e	f	g	h	i	j	k	l	M	n	O	p	Q	r	s	t	u	v	w	x	y	z	a	b	C	d

Applying this knowledge to the cipher-text message in Figure 10-21, we can convert the encrypted message into one that is now readable, as shown in Figure 10-22.

FIGURE 10-22   The decrypted cipher-text message.

> **Welcome to the MODERN ELECTRONIC COMMUNICATION** section on Wireless Security. **If you are reading this you have determined that a Caesar-Shift of 4 has been used. If you are not reading this well then you aren't reading this.**

The type of encryption used in the cipher-text message shown in Figure 10-20 is called a simple *Caesar-shift*. This is a simple substitution cipher, in which each letter is substituted with another letter k positions away. In this case, $k = 4$.

Refer back to the group of letters shown in the middle of Figure 10-21. These are uppercase characters, but they don't have the expected letter distribution shown in the right side of Figure 10-21. The reason is that the sample size is small (thirty-four uppercase letters). The message could be reanalyzed by converting all characters to either uppercase or lowercase, and then the letter frequency distribution would be similar to the right side of Figure 10-21.

This section defined five important aspects that should be considered when securing a communication link. These are confidentiality, integrity, authentication, nonrepudiation, and availability. Additionally, an overview of cryptography and an example of why data should be encrypted was also presented. It should never be assumed that the "bad guys" can't see the information you are transmitting; therefore, you should always take the necessary steps to protect your information.

10-5 TWO-WAY AND TRUNKED RADIO SYSTEMS

A critically important but largely unseen part of the communications infrastructure within any community consists of the many two-way radio systems used by public-safety agencies such as police and fire departments as well as by numerous other public and private organizations. These systems largely fall under the regulatory designation known as **specialized land-mobile radio (SMR)**. In the public-safety arena, such systems consist not only of visible components such as the portable transceivers carried by police officers or firefighters, but also of the vast network of dispatch stations and emergency response centers established to support both the public and personnel in the field.

These centers communicate through a largely unseen "backbone" consisting of repeaters, computerized controllers of various sorts, and interconnection facilities that can include telephone lines, T1 or T3 data links, and, in many cases, terrestrial microwave or fiber-optic connections. Also, because the systems are often of a critical, life-safety nature—meaning that people can die if the message is not received—they are often designed to ensure that there is no single point at which a system can be rendered inoperable. In some cases the systems are fully redundant, with duplicate installations at secure, alternate locations.

SMR systems may be analog or digital. If analog, the radios use conventional, narrowband FM, mainly in the VHF (150–174-MHz) and UHF (421–512-MHz) bands. (In the United States, some Federal government users occupy allocations somewhat outside these ranges but within the generally accepted designations for the frequency bands shown in Table 1-1.) Most analog systems use a maximum channel bandwidth of 25 kHz, attained by limiting the deviation of the modulated signal to a maximum of ±5 kHz. In the interest of spectrum efficiency, however, regulatory authorities (the FCC in the United States) have mandated that the channels be narrowed to 12.5 kHz, potentially allowing for a doubling of capacity within each band. This regulatory initiative is known as *narrowbanding,* and it is achieved in analog systems by further reducing the permissible deviation. There is a further mandate to reduce occupied channel bandwidth to ±6.25 kHz, again to improve spectrum efficiency and to make room for more users. Since such narrow bandwidths cannot be attained in all-analog systems, some form of digital modulation is necessary. In addition, voice compression through the use of vocoders, such as those described for GSM cellular systems, would also be required with digital systems to reduce data rates to a level sufficient to fit within the narrow channels.

Both analog and digital land-mobile systems usually depend on repeaters to provide wide-area coverage. A repeater is a transceiver (combination transmitter and receiver) usually located atop a mountain, tall building, or other high-elevation point. Its purpose is to retransmit the relatively low-power signals from a portable or mobile radio at higher power (usually 100 W or more) on a different frequency over a wide area. Repeaters are often necessary because "simplex" operation directly from one unit to another is strictly line-of-sight at VHF and UHF frequencies. Repeaters are assigned *frequency splits,* in which the repeater receives a transmission from a mobile unit at one frequency and retransmits it at a different frequency. Frequency splits vary, but they are often several hundred kilohertz, if not several megahertz apart. In many cases, repeaters also have a direct connection by wire to one or more *base remote* units. Along with control signals that command various repeater functions such as *keying* (turning on the transmitter), frequency changes, and the like, the remote units send and receive voice audio by dedicated telephone lines to a monitoring point. In a public safety installation, the base remote is most likely part of a communications console staffed by trained emergency communications dispatchers. In addition to radio communications capabilities, the console consolidates multiple telephone lines as well as a myriad of other essential functions. Among these are caller position location (911 location services), location mapping, call recording, and computer-aided dispatch activities, including determining which emergency resources are available to respond to calls.

The simplest two-way radio systems consist in concept of a high-elevation repeater serving a number of remote units (portable and mobile radios) dispersed over a wide geographic area. Such systems are reliable and simple, but they suffer from a frequency-utilization and access problem: If you were to look at how often the radio-frequency channel is used over a given time interval, you would find that it is unused for much of the time. However, during times of heavy usage (such as during an emergency, in the case of public-safety systems), the channel becomes congested, with many users attempting to access it at the same time, and most being denied service. This is an inefficient use of spectrum from an allocation standpoint. For this reason, many public safety agencies as well as other users have installed *trunked* radio systems to make maximal use of scarce radio spectrum.

The concept of a trunked radio system is somewhat akin to that of cellular service, but trunked two-way radio systems use fewer radio frequencies than do cellular phone systems and employ sets of repeaters that cover a wide geographic area. Each repeater has its own set of frequencies, so trunking systems rely on frequency-division multiple-access (FDMA) methods of multiple-user access. Many trunked systems, particularly those used by public-safety agencies, operate within the 800-MHz UHF band at frequencies close to those occupied by cellular systems; some Federal government users operate trunked systems at frequencies near 400 MHz.

The term *trunking* has its roots in telephone technology. Local telephone offices route incoming calls to "trunk lines" for delivery to the destination office. The assigned trunk line is based on availability, call priority level, and toll charges. Each call to a particular destination may use a different trunk line. By continually switching traffic to the next available trunk line, the telephone network can cost-effectively accommodate considerably more customers than it has lines. Conceptually, trunked calls are similar to bank customers waiting in a single queue for the next available teller. Even when one window is closed, the queue order is undisturbed and customers continue to be served. Similarly for the telephone system, failure of one or several trunk lines will not disrupt service.

A trunked radio system applies the trunking concept to two-way radio traffic, which usually consists of many short-duration transmissions. A basic trunked radio system normally employs five or more radio repeaters, a system controller (computer) and management interface, a database of users, and mobile and portable radios. Each repeater uses a pair of frequencies to provide wide-area coverage, with the system controller directing and monitoring each repeater. The controller assigns one repeater to become the control channel, which provides a continuous, two-way data link to every unit enrolled in the system. The field units (mobile and portable radios) continually monitor the control channel for instructions as to frequency assignments, range limits, and incoming transmissions. Although the number of voice channels increases with larger systems, there is never more than one control channel. The controller can generally be reprogrammed to rotate the control channel between all the repeaters such that each shares this continuous-duty load.

Figure 10-23 is a block diagram of a basic trunked radio system. To place a call, the operator need only select the individual or user group with whom he or she wishes to communicate. When the operator presses the microphone push-to-talk button, the radio sends a request over the data channel for a talk path. The controller verifies that the caller's radio is authorized to use the system and selects an available channel. A

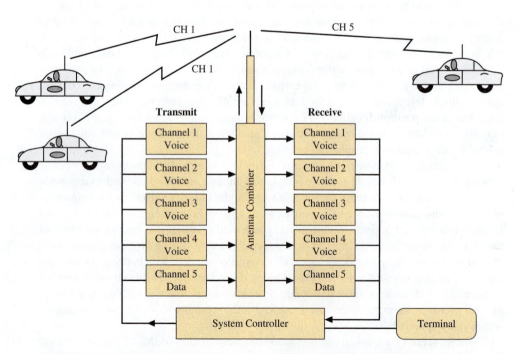

FIGURE 10-23 **Block diagram of a basic trunked radio system.**

command is then sent to the radio of the operator initiating the call and to the radios of the party or parties being called to move to the selected frequency. Communication can then take place. The entire process executes without operator intervention and takes less than one second.

Through the controller the system manager establishes *talk groups*, which effectively provide users with private channels because all radios outside the group using a channel at any given time are prevented from monitoring or transmitting on the occupied frequency. Different groups, such as fire, police, or medical services, can thus communicate over a common trunking backbone, while the controller keeps each user group distinct. Although radios are normally programmed to transmit and receive only within authorized talk groups, in an emergency the system manager or emergency dispatcher can remotely add to or delete units from the groups. Individuals can also identify a specific radio to make a private call. From the perspective of users in the field, each talk group or transmission operates on a private channel, but from a system perspective, transmissions are distributed among the assigned channels to make optimal use of available radio frequencies.

Trunking systems have become mainstream in recent years and are now commonplace, particularly in public-safety installations. One significant problem for both trunked and nontrunked systems, however, relates to the issue of *interoperability,* that is, the ability of different public-safety agencies to be able to talk to each other during times of crisis. This issue became apparent to the public after the terrorist attacks of September 11, 2001, but had been identified as an issue of critical concern to professionals in the public-safety field some time before. For this reason, efforts have been made to establish a common *air interface* to allow radio systems under the control of different agencies to be able to communicate with each other when necessary.

One of the longest-running interoperability efforts has been a project undertaken by members of the Association of Public-Safety Communications Officials (APCO), the professional organization of technical professionals and managers in the public-safety communications field. This effort is known as Project 25 (P25), and P25-compliant radios are available from a number of vendors. P25 is a digital LMR format relying on FDMA to divide the assigned spectrum to 12.5-kHz or (in the future) 6.25-kHz channels. The modulation format for 12.5-kHz P25 is four-level FSK to provide a spectral efficiency of 2 b/s/ Hz and a data rate of 9.6 kbps. With four-level FSK, the carrier is shifted on either side of the rest frequency to one of four specific frequencies, which in this case of P25 is ± 1.8 kHz or ± 600 Hz, to encode the four possible symbol states.

One reason for the selection of FSK is that it is a *constant-envelope* format, like other forms of FM. In contrast, amplitude modulation encodes information in terms of changing carrier amplitudes, which in turn requires the use of linear power amplifiers. The advantage of constant-envelope modulation is that efficient class C amplifiers can be used in both the base stations and mobile units. Particularly for portable operation, efficiency is a key design criterion because it impacts battery life and the size and weight of the radio. As an example, it is generally desirable for the radio batteries to have sufficient capacity to last at least for an 8-hour shift. Providing this level of capacity in a reasonably sized battery requires careful attention to all aspects of power use within the transceiver.

P25 also provides an upgrade path to 6.5-kHz channels without the need for wholesale replacement of the backbone repeaters. The so-called Phase 2 implementation envisions the use of QPSK modulation to permit the same data rates to be carried in half the bandwidth. The constellation produced will be similar to that seen in Figure 8-16. One key advantage is that only the modulators in the transmitter need to be upgraded to support Phase 2. The demodulators in P25 radios have been designed to detect either the four-level FSK or QPSK-encoded data.

Countries in Europe, Africa, Asia, and Latin America use a trunked radio standard called TETRA, for *Terrestrial Trunked Radio*. It is a TDMA format in which four voice or data channels are multiplexed onto time slots within a single 25-kHz-wide RF channel. This arrangement effectively provides capacity equivalent to four independent channels of 6.25 kHz each. The modulation format is pi-over-4 DQPSK with a bit rate per time slot of 7.2 kbps.

If there is one common theme to the elements of this chapter, it is the role that digital signal processing techniques and large-scale integrated circuits play in the implementation of advanced communications technologies. Many functions traditionally carried out in circuits with discrete components—such as modulation, demodulation, and filtering—are now being implemented as software routines running in DSP hardware. The migration to software-defined radio (SDR) represents one of the biggest shifts in communications technology in recent years. The trend to full system implementations in the form of SDR will no doubt accelerate as ever faster but lower-cost DSP chips, as well as ADC and DAC converters, continue to be introduced.

SDR can be loosely defined as radio in which at least some subsystem functions are carried out with software-based signal-processing routines. The ultimate SDR implementation would look something like the block diagram of Figure 10-24. The analog RF signal received by the antenna, after passing through a preselector, would immediately be digitized by the ADC converter and applied to a DSP. Algorithms in the DSP would handle all the functions necessary to produce a baseband signal, everything from demodulation and filtering to tasks such as error correction and decompression. The practicality of such an implementation depends on such factors as the RF bands in use, the speed capabilities of the ADC converter, and cost as well as portability and power consumption considerations.

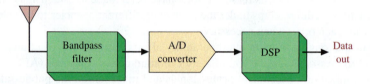

FIGURE 10-24 **Full software-defined radio implementation.**

Although they are becoming more widely available, ADC and DAC converters capable of working in the gigahertz range have been expensive and did not even exist until recently. For this reason, many currently used SDR implementations "push back" the conversion from analog to digital (or vice versa, in the case of SDR transmitters) to lower-frequency stages. The SDR receiver block diagram of Figure 10-25 is a typical configuration. First, note how much of this block diagram is similar to what we have seen before. In it, the received analog RF signal is first split into two paths and each fed to its respective mixer. The mixers are also supplied with a locally generated signal from an on-board frequency synthesizer acting as the local oscillator (LO). One of the two LO signals is phase-shifted by 90°, and the mixer outputs form orthogonal I and Q signals. From Section 8-5, recall that these represent cosine and sine components in quadrature:

$$V(t) = I(t) \cos(\omega_c t) + Q(t) \sin(\omega_c t).$$

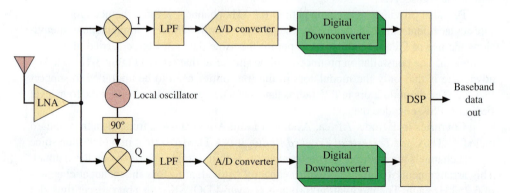

FIGURE 10-25 **Practical software-defined receiver.**

Using the notion of analytic frequency, in which the above expression represents the scalar (real) portion, we can separate out the amplitude and phase components from the oscillating portion of the carrier. The I and Q components in quadrature contain the amplitude and phase information for DSP algorithms to demodulate the information signal by detecting variations in amplitude, frequency, or phase.

After bandpass filters select the difference frequencies produced by mixing the LO with the incoming RF, the down-converted signal is applied to an ADC converter. Depending on the sample rates and DSPs involved, the digital words output by the ADC may be down-converted again to lower the sample rate. The I/Q data are then applied to the DSP block for demodulation and any other signal processing (such as decompression, equalization, error correction, etc.) required to recover the transmitted data.

An SDR transmitter is shown in Figure 10-26. In it, modulation takes place in the initial DSP block, where the I and Q signals are formed. These take parallel paths, often through digital up-converters (DUCs) to increase the sample rate, and then to DAC converters, whereupon the outputs are low-pass filtered to remove high-frequency components created as part of the sample-and-hold process. The low-pass-filtered outputs are then applied to mixers, into which the RF sine-wave carrier, either in phase or shifted by 90°, is also applied, this time to produce sum frequencies (up-conversion) at the final transmitted frequency. The up-converted signal is then filtered and amplified, often in more than one amplifier stage, before being applied to the antenna. Depending on the modulation format, the intermediate and power amplifiers may have to be linear (in the case of CDMA, extremely linear with sophisticated feedback and control systems to maintain linearity) so as not to distort amplitude variations in the transmitted signal. This need for linearity, and its consequent implications it has for efficiency and power consumption, would be present in any of the QAM-based modulation formats, for example.

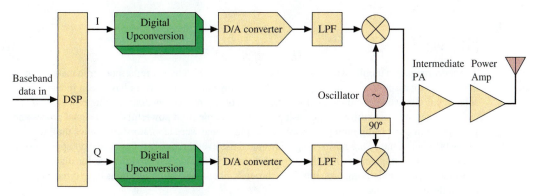

FIGURE 10-26   **Practical software-defined transmitter.**

SDRs have many of the advantages already discussed pertaining to DSP. Specifically, subsystems can perform much closer to the ideal. For example, filters can be designed to exhibit much sharper transitions from passband to stopband, and with extremely stable performance characteristics that are not affected by changes in ambient temperature or by component aging. SDRs are potentially simpler to implement in hardware, and with time and the tendency to implement the DAC and ADC conversion at stages closer to the antenna, fewer and fewer discrete components will be needed. Also, SDRs can be made reconfigurable through reprogramming. Thus, new features and modulation formats can be supported, and bug fixes or other software-related issues can be addressed without the need for hardware modifications. These advantages, along with the increasing availability of very high-performance and fast DSPs, DAC converters, and ADC converters, virtually guarantees that all significant future developments in wireless communications will involve a wholesale migration to SDR platforms.

An extension of SDR is what has been labeled *cognitive radio*. As the name implies, cognitive radios can "think," in the sense that they can take note of and adapt to their surroundings. Another appropriate term for such a device is *adaptive radio*. A cognitive radio would, for example, be able to change operating frequencies in the presence of interference and make use of unoccupied spectrum. An example where cognitive radio presents potentially significant opportunities is in the use of the so-called "white space" or unused spectrum surrounding broadcast UHF television stations. Recall from the previous discussions of frequency-division multiplexing (of which television broadcasting is certainly an example) that *guard bands* are assigned around licensed channels to minimize the possibility of interference. Also, in most areas much of the UHF spectrum allocated by law for television broadcasting sits unused, either because there is no station in a given area or because of legal requirements to maintain a minimum spacing between stations on nearby channels. Again, these requirements are in place to minimize the possibility of interference. The untapped spectrum represents an underused natural resource that can potentially be exploited for any number of applications, among the most intriguing of which are "open access" wide-area networks proposed by Google and other technology firms. Another potential application is expansion of existing 3G and 4G networks. As was pointed out earlier, these services require huge amounts of spectrum, and projections of growth in demand have network operators reaching the limits of capacity in the next few years. Expansion will depend in part on finding new spectrum opportunities along with even more creative approaches to using existing allocations efficiently. The cognitive radio concept is a creative approach to spectrum use in that it envisions the coexistence of both licensed and unlicensed services in the same frequency bands, a potentially revolutionary transformation of long-held notions of frequency management. Experimental cognitive radios have been developed and deployed, primarily for military applications, but the concept has not yet seen widespread application in the consumer realm as yet, partly because of resistance by incumbent users of the affected spectrum to relinquish or allow intrusions onto any of their exclusive allocations. How these developments play out will be a prime example of the sometimes politicized nature of spectrum-allocation decisions in the regulatory and legislative arenas.

SUMMARY

The short-range and mobile wireless networks surveyed in this chapter represent a true maturation of the system approach to communications studies. Entire radio transceivers have been reduced from subsystems consisting of hundreds if not thousands of discrete components to single integrated circuits. Further, thanks to the ever-increasing speeds available from powerful digital signal processing ICs, spectrum-conserving multiple-access technologies that were laboratory curiosities until the end of the twentieth century are now seeing widespread use even in low-cost, mass-market applications. Finally, whether implemented in the form of discrete subsystems or as software for digital signal processors, many underlying concepts from the study of analog communications have direct application in the digital realm.

Wireless computer networks make use of one or more unlicensed bands in the 2.4 or 5.8-GHz range to provide connectivity at rates ranging from 1 to 54 Mb/s. Several modulation schemes are in use, with those exhibiting the highest throughput also, as expected, having the shortest ranges. The various networks all belong to the IEEE 802.11 family, with suffixes denoting specific standards. The 802.11b is the original and still widely used wireless network, with data rates up to 11 Mb/s using direct-sequence spread spectrum with complementary-code keying to occupy a bandwidth equivalent to the original maximum rate of 2 Mb/s. The 802.11a and 11g standards add orthogonal frequency-division multiplexing (OFDM) to increase data throughput; the 11a uses the less-crowded 5.8-GHz band for maximum performance, while the 11g enjoys backward compatibility with installed 11b networks. Finally, the 802.11n standard promises the highest data throughputs by deploying multiple-input multiple-output configurations to provide independent signal paths between transmitter and receiver antennas. The WiMAX networking format takes the ideas deployed for local-area networks (LANs) and deploys them on a metropolitan scale. In particular, WiMAX also makes use of OFDM.

Other LANs operating on an even smaller scale are classified as *piconets.* Bluetooth accommodates a relatively high data rate and is suited for such applications as computer-to-peripheral interconnectivity and wireless handset use. Bluetooth devices form a star network with one device acting as the primary unit and all others deriving their timing signals and other important information from it. Another piconet, ZigBee, trades off a lower data rate for higher reliability through its mesh-network configuration. ZigBee is used for applications such as building automation, process control, remote-sensor monitoring, and automatic utility-meter reading. A final local-area communications application cannot really be considered a network, but some of the underlying principles are the same. Radio-frequency identification makes use of a number of frequency ranges and active or passive tags to track and identify shipments, pets, and so on.

Certainly no technology has become more widespread over the last quarter-century than the cellular phone. Nearly 70% of the world's population counts themselves as wireless subscribers. The underlying concepts of space diversity and frequency reuse are holdovers from the earliest cellular networks based on narrowband FM two-way radio. So-called *second-generation* digital networks are in use for voice traffic; these networks allowed carriers to accommodate more traffic within their existing frequency allocations by means of either time-division multiplexing (TDM) or code-division multiple-access (CDMA) techniques. The Global System for Mobile Communication (GSM) is widespread throughout the world and relies on time-division multiple access (TDMA) for capacity enhancement. Also important to all digital voice systems is the notion of voice compression in the form of vocoders.

The other principal multiple-access technique used for voice traffic is based on direct-sequence spread spectrum. The CDMA standard is based on a modification of DSSS in which wireless subscribers are identified with one of a limited number of Walsh codes. Walsh codes take the place of pseudonoise (PN) codes in CDMA systems, and it is this aspect that distinguishes them from other forms of direct-sequence spread spectrum.

GSM and CDMA are deployed worldwide for voice networks. Extensions to these networks are intended to provide high-speed data capabilities while at the same time acting as natural overlays to the existing network. So-called 3G networks employ variants of CDMA called CDMA2000 or WCDMA. The latter is also known as UMTS, for Universal Mobile Telephone Service, a standard proposed and promoted by the International Telecommunication Union (ITU). All the 3G networks in use far surpass the original benchmarks set by the ITU for 3G service and can be properly classified as 3.5G services, offering very high-speed data download capability and other features. These enhanced services employ wider bandwidths and higher-level modulation schemes to achieve their higher download data rates.

Fourth-generation wireless data networks are on the horizon. Already existing are 3.9G networks, which make use of the OFDM and MIMO enhancements first deployed in LANs to achieve even higher data rates than their 3G or 3.5G predecessors. True 4G networks are still in the future; these will make use of even more OFDM subcarriers and wider bandwidths as well as additional antennas at the transmitter and receiver to produce theoretical data rates approaching 1 Gb/s.

Though largely unseen, a very important part of the communications infrastructure is that represented by two-way radio. Many of these systems perform critical roles in the public-safety arena, where ultimate reliability is sometimes a matter of life and death. To promote spectrum efficiency, two developments are taking place. One is the drive to reduce channel bandwidths from 25 kHz, first to 12.5 kHz and then to 6.25 kHz. Analog systems can manage 12.5-kHz bandwidth by reducing permissible deviation, but narrower bandwidths are not possible in analog systems. Many two-way systems are making the move to digital modulation, and also in the interest of spectrum efficiency and effective channel utilization, many public-safety systems make use of some form of trunking, where individual users are assigned to talk groups that in turn share radio frequencies that are allocated under computer control. The most widely used trunking air interface in North America is FDMA based on the APCO Project 25 standard, whereas much of the rest of the world uses a TDMA implementation called TETRA.

Both LANs and mobile communications systems are becoming increasingly reliant on software-based radio implementations. Software-defined radio (SDR) makes use of high-speed DAC and ADC converters as well as powerful digital signal processing (DSP) chips to perform many functions of a radio transmitter or receiver in software. These architectures have many advantages, chief among them those of reconfigurability and reliability under a variety of operating conditions. An extension of software-defined radio is that of cognitive radio, in which adaptable radios can reconfigure themselves on the fly to respond to changes in their environment. Cognitive radios provide the promise that licensed and unlicensed services can coexist in the same spectrum, thereby promoting more efficient utilization of a scarce natural resource.

QUESTIONS AND PROBLEMS

SECTION 10-1

1. What is the current data rate for the IEEE 802.11 wireless LAN protocol? What protocol is used with wireless communications?

2. What is the WiMax frequency standard for the United States?

3. Why was OFDM selected for WiMax?

4. How does WiMax differ from Wi-Fi?

5. What transmission method does WiMax use on the uplink and downlink, and why do they differ?

6. In what frequency band does Bluetooth operate?

7. How many output power classes does Bluetooth have? List the power level and the operating range for each class.

8. What is a *piconet?*

9. What is the purpose of the inquiry procedure in Bluetooth?

10. What is the purpose of the paging procedure in Bluetooth?

11. Define the term *backscatter.*

12. What are the three parameters that define a radio frequency identification (RFID) system?

13. Explain how power is provided to a passive RFID tag.

14. Explain why some RFID tags (inlays) incorporate dual-dipole antennas.

15. Cite three advantages for using an active RFID tag.

16. What are the three frequency bands typically used for RFID tags?

SECTION 10-2

17. Redesign the cellular system in Figure 10-8 so that only six different channel groups (*A, B, C, D, E, F*) are used instead of the seven shown.

18. Describe the concepts of frequency reuse and cell splitting.

19. Design the split-cell system in Figure 10-10 so that the minimum number of channel groups are used.

20. Describe the sequence of events when a mobile user makes a call to a land-based phone. Include in your description the handoff once the call has been made.

21. The cellular system in Figure 10-10 requires splitting of the two cells surrounded by other cells. They need to be split into five cells (from the original two). Determine the minimum number of channel sets that can serve the original and new systems.

22. Describe Rayleigh fading and explain how to minimize its effects.

23. A cellular system operates at 840 MHz. Calculate the Rayleigh fading rate for a mobile user traveling at 40 mi/h (100 fades per second).

24. A mobile user is transmitting two steps above minimum power. Calculate its power output. (1.11 W)

25. Explain the functions of the base station, switch, and the public switched telephone network (PSTN).

26. What are the two systems that most mobile telephone service providers presently use?

27. What is the bandwidth for each channel of IS-136?

28. What modulation method is used by IS-136?

29. What is the name of the standards group that documented the Global System for Mobile Communications (GSM)?

30. What modulation method is used by GSM?

31. How many users are time-multiplexed into a single GSM frequency?

32. How many users are time-multiplexed into a single IS-136 frequency?

33. In GSM, what device is used to minimize the effects of multipath interference?

34. In code-division multiple access (CDMA), what device is used to minimize the effects of multipath interference?

35. In GSM, what is the function of the frequency control channel (FCCH) control signal?

36. In CDMA, what Walsh code has a function similar to the FCCH used in GSM?

37. In CDMA, what is the function of Walsh code 32?

38. In GSM, what control signal has a function similar to Walsh code 32, used by CDMA?

39. What is the name of the standards group that registered CDMA?

40. What is another name for the 2G version of CDMA?

41. In CDMA, what type of orthogonal code separates mobile users?

42. In CDMA, what type of orthogonal code identifies base stations?

43. In CDMA, how is the system performance affected as the number of users increase?

44. In CDMA where do the base stations get their frequency reference? What is the frequency and typical power level of this signal?

45. In CDMA what is carrier feed through? What is its specified power level?

46. In CDMA and GSM what are the three main categories of interference signals?

47. What equation is used to calculate the frequencies of interference due to mixing products?

48. Give two examples of unintentional "mixers" that can cause interference problems when located near base stations.

49. Refer to Figure 10-18 (display photo for Tektronix RSA 3408A).
   (a) What is the horizontal scale for the upper and lower photos?
   (b) What is the vertical scale for the upper and lower photos?
   (c) How far apart in frequency are the three frequency-shift keying (FSK) signals shown in the lower photo?
   (d) How many data records are included in the lower photo?

50. Refer to Figure 10-18 (display photo for Tektronix RSA 3408A).
   (a) What is the horizontal and vertical scale for the lower photo?
   (b) In the lower photo, what is the maximum deviation from the final frequency?
   (c) In the lower photo, what is the maximum deviation from the final frequency after a period of 200 μ sec.?

SECTION 10-4

51. Discuss the key issues of the wireless application protocol (WAP).

52. What are the five aspects to consider when securing a communication link?

53. Why is integrity important?

54. Define "jamming" as it is related to RF communication.

55. What are the two categories of attacks on a wireless network? Provide a description of each.

56. Which type of attack, external or internal, is the biggest threat and why?

57. Why is information encrypted?

58. What is a *key* or a *cipher key*?

CHAPTER 11

COMPUTER COMMUNICATION AND THE INTERNET

CHAPTER OUTLINE

KEY TERMS

American Standard Code for Information Interchange (ASCII)
parity
Extended Binary-Coded Decimal Interchange Code (EBCDIC)
Baudot code
Gray code
asynchronous system
start bit
stop bit
synchronous system
universal serial bus (USB)
hot-swappable
Type A
Type B
FireWire A (IEEE 1394a)
FireWire B (IEEE 1394b)
RS-232
null modem
data terminal equipment (DTE)
data communications equipment (DCE)
RS-422
RS-485
balanced mode
local-area network (LAN)
topology
protocol
carrier sense multiple access with collision detection (CSMA/CD),
network interface card (NIC)
media access control (MAC) address
broadcast address
100BaseT
RJ-45
metropolitan area networks (MANs)
wide-area network (WAN)
open systems interconnection (OSI)
Internet Assigned Numbers Authority (IANA)
IP telephony (voice-over IP)
quality of service (QoS)
V.44 (V.34)
V.92 (V.90)
asymmetric operation
Cable modems
ranging
xDSL
digital subscriber line
Asymmetric DSL (ADSL)
discrete multitone (DMT) modulation

This chapter focuses specifically on computer communication, initially that between computers and peripheral devices and then that between groups of computers, both locally situated and widely spaced. We first discuss the means by which alphanumeric codes can be expressed as binary data. Then, the various methods by which data are communicated to peripherals are outlined. Following that is a description of the ways by which computers can be organized into local-area networks (LANs) and interconnected. Finally, we turn our attention to what is perhaps the most significant societal shift caused by communications technology over the past generation: the vast network of computer networks called the internet. The chapter concludes with a discussion of the various methods by which computers can interface to the internet, as well as security and troubleshooting issues that can arise in computer networks.

11-1 ALPHANUMERIC CODES

The most common alphanumeric coding scheme for binary data is the American Standard Code for Information Interchange (ASCII). Another code, the Extended Binary-Coded Decimal Interchange Code (EBCDIC), is still used in large computing systems. EBCDIC and another code, Baudot, may largely be obsolete in modern implementations; however, they are presented for their historical importance as well as for their use in certain specialized applications.

The ASCII Code

The **American Standard Code for Information Interchange (ASCII)** is a 7-bit code used for representing alphanumeric symbols with a distinctive code word. A committee of the American National Standards Institute (ANSI) developed the ASCII code for the purpose of coding binary data. ASCII-77 is the adopted international standard. Figure 11-1 provides a list of the codes.

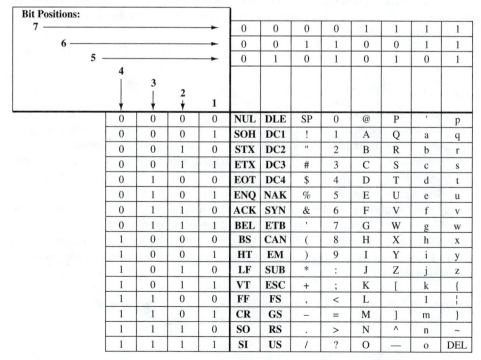

FIGURE 11-1 American Standard Code for Information Interchange (ASCII).

4	3	2	1	0 0 0	0 0 1	0 1 0	0 1 1	1 0 0	1 0 1	1 1 0	1 1 1
0	0	0	0	NUL	DLE	SP	0	@	P	'	p
0	0	0	1	SOH	DC1	!	1	A	Q	a	q
0	0	1	0	STX	DC2	"	2	B	R	b	r
0	0	1	1	ETX	DC3	#	3	C	S	c	s
0	1	0	0	EOT	DC4	$	4	D	T	d	t
0	1	0	1	ENQ	NAK	%	5	E	U	e	u
0	1	1	0	ACK	SYN	&	6	F	V	f	v
0	1	1	1	BEL	ETB	'	7	G	W	g	w
1	0	0	0	BS	CAN	(	8	H	X	h	x
1	0	0	1	HT	EM	)	9	I	Y	i	y
1	0	1	0	LF	SUB	*	:	J	Z	j	z
1	0	1	1	VT	ESC	+	;	K	[	k	{
1	1	0	0	FF	FS	,	<	L		l	¦
1	1	0	1	CR	GS	–	=	M	]	m	}
1	1	1	0	SO	RS	.	>	N	^	n	~
1	1	1	1	SI	US	/	?	O	—	o	DEL

Sample of Control Characters (Bold)

STX = Start of text
EOT = End of transmission
CR = Carriage return
HT = Horizontal tabulation

Examples:

1000011 = C
0110011 = 3
1010000 = P
0110000 = 0 (Zero)
0100000 = SP (space)

There are 2^7 (128) possible 7-bit code words available with an ASCII system. The binary codes are ordered sequentially, which simplifies the grouping and sorting of the characters. The 7-bit words are ordered with the least significant bit (lsb) given as bit 1 (b_1), while the most significant bit (msb) is bit 7 (b_7). Notice that a binary value is not specified by the code for bit 8 (b_8). Usually the bit 8 (b_8) position is used for parity checking. Recall from Chapter 7 that **parity** is an error detection scheme that identifies whether an even or odd number of logical ones is present in the code word. For ASCII data used in a serial transmission system b_1, the lsb bit, is transmitted first.

The ASCII system is based on the binary-coded-decimal (BCD) code in the last 4 bits. The first 3 bits indicate whether a number, letter, or character is being specified. Notice that 0110001 represents "1," while 1000001 represents "A" and 1100001 represents "a." It uses the standard binary progression (i.e., 0110010 represents "2"), and this makes mathematical operations possible. Because the letters are also represented with the binary progression, alphabetizing is also achieved via binary mathematical procedures. You should also be aware that analog waveform coding is accomplished simply by using the BCD code for pulse-code modulation (PCM) systems, also covered in Chapter 7.

In some systems the actual transmission of these codes includes an extra pulse at the beginning (start) and ending (stop) for each character. When start/stop pulses are used in the coding of signals, it is called an *asynchronous* (nonsynchronous) transmission. A synchronous transmission (without start/stop pulses) allows more characters to be transmitted within a given sequence of bits. The transmission of information between various computer installations may require the less efficient asynchronous transmitting mode depending on computer characteristics.

The EBCDIC Code

The **Extended Binary-Coded Decimal Interchange Code (EBCDIC)** is an 8-bit alphanumeric code. The term *binary-coded decimal* is used in the name because of the structure present in the coding scheme, which uses only the 0–9 positions. A list of the code words for the EBCDIC system is given in Figure 11-2, and the acronyms for the control characters are listed in Table 11-1.

The Baudot Code

Another interesting code presented for historical reasons is the **Baudot code.** The Baudot code was developed in the days of teletype machines such as the ASR-33 Teletype terminal. Baudot is an alphanumeric code based on five binary values. The Baudot code is not very powerful, but it does have its place in communications history. The Baudot code is provided in Figure 11-3.

The alphabet has twenty-six letters, and there is an almost equal number of commonly used symbols and numbers. The 5-bit Baudot code is capable of handling these possibilities. A 5-bit code can have only 2^5 or 32 bits of information but actually provides 26×2 bits by transmitting a 11111 to indicate all following items are "letters" until a 11011 transmission occurs, indicating "figures." Notice that no provision for lowercase letters is provided.

Figure 11-4(a) shows an example of the Baudot code to transmit "YANKEES 4 REDSOX 3." Be sure to work out the code in Figure 11-4(b) on your own; it is the only "X-rated" part of this book that we were allowed to include.

The Gray Code

The last alphanumeric code we will look at is the **Gray code.** The Gray code is a numeric code for representing the decimal values from 0 to 9. It is based on the relationship that only one bit in a binary word changes for each binary step. For example, the code for 7

FIGURE 11-2 The Extended Binary-Coded Decimal Interchange Code.

EBCDIC CODES

Bit Positions 4, 5, 6, 7 / Second Hexadecimal Digit (row labels); Bit Positions 0,1 (00, 01, 10, 11); Bit Positions 2,3 (00, 01, 10, 11); First Hexadecimal Digit (0–F)

Bin	Hex	0	1	2	3	4	5	6	7	8	9	A	B	C	D	E	F
0000	0	NUL	DLE	DS		SP	&	-						(	)	\	0
0001	1	SOH	DC1	SOS		RSP		/		a	j	~		A	J	NSP	1
0010	2	STX	DC2	FS	SYN					b	k	s		B	K	S	2
0011	3	ETX	DC3	WUS	IR					c	l	t		C	L	T	3
0100	4	SEL	RES/ENP	BYP/INP	PP					d	m	u		D	M	U	4
0101	5	HT	NL	LF	TRN					e	n	v		E	N	V	5
0110	6	RNL	BS	ETB	NBS					f	o	w		F	O	W	6
0111	7	DEL	POC	ESC	BOT					g	p	x		G	P	X	7
1000	8	GE	CAN	SA	SBS					h	q	y		H	Q	Y	8
1001	9	SPS	EM	SPE	IT				▲	i	r	z		I	R	Z	9
1010	A	RPT	UBS	SM/SW	RFF	¢	!			:					SHY			
1011	B	VT	CU1	CSP	CU3	.	$	,	#								
1100	C	FF	IFS	MFA	DC4	<	*	%	@								
1101	D	CR	IGS	ENQ	NAK	(	)	–	▲								
1110	E	SO	IRS	ACK		+	;	>	=								
1111	F	SI	IUS/ITB	BEL	SUB	¦	¬	?	"								BO

TABLE 11-1 • The EBCDIC Code—List of Abbreviations

ACK	Acknowledge	ETB	End of Transmission	RFF	Required Form Feed
BEL	Bell	ETX	End of Text	RNL	Required New Line
BS	Backspace	FF	Form Feed	RPT	Repeat
BYP/	Bypass/Inhibit	FS	Field Separator	SA	Set Attribute
INP	Presentation	GE	Graphic Escape	SBS	Subscript
CAN	Cancel	HT	Horizontal Tab	SEL	Select
CR	Carriage Return	IFS	Interchange File Sep.	SFE	Start Field Extend
CSP	Control Sequence Prefix	IGS	Interchange Group Sep.	SI	Shift In
CU1	Customer Use 1	IR	Index Return	SM/SW	Set Mode/Switch
CU3	Customer Use 3	IRS	Interchange Record Sep.	SO	Shift Out
DC1	Device Control 1	IT	Indent Tab	SOH	Start of Heading
DC2	Device Control 2	IUS/	Interchange Unit Sep./	SOS	Start of Significance
DC3	Device Control 3	ITB	Intermediate Text Block	SPS	Superscript
DC4	Device Control 4	LF	Line Feed	STX	Start of Text
DEL	Delete	MFA	Modify Field Attribute	SUB	Substitute
DLE	Data Link Escape	NAK	Negative Acknowledge	SYN	Synchronous Idle
DS	Digit Select	NBS	Numeric Backspace	TRN	Transparent
EM	End of Medium	NL	New Line	UBS	Unit Backspace
ENQ	Enquiry	NUL	Null	VT	Vertical Tab
EO	Eight Ones	POC	Program-Operator Comm.	WUS	Word Underscore
EOT	End of Transmission	PP	Presentation Position		
ESC	Escape	RES/NEP	Restore/Enable Presentation		

FIGURE 11-3 The Baudot code.

Character Shift		Binary Code
		BIT
Letter	Figure	4 3 2 1 0
A	–	1 1 0 0 0
B	?	1 0 0 1 1
C	:	0 1 1 1 0
D	$	1 0 0 1 0
E	3	1 0 0 0 0
F	!	1 0 1 1 0
G	&	0 1 0 1 1
H	#	0 0 1 0 1
I	8	0 1 1 0 0
J	.	1 1 0 1 0
K	(	1 1 1 1 0
L	)	0 1 0 0 1
M	.	0 0 1 1 1
N	,	0 0 1 1 0
O	9	0 0 0 1 1
P	0	0 1 1 0 1
Q	1	1 1 1 0 1
R	4	0 1 0 1 0
S	BEL	1 0 1 0 0
T	5	0 0 0 0 1
U	7	1 1 1 0 0
V	;	0 1 1 1 1
W	2	1 1 0 0 1
X	/	1 0 1 1 1
Y	6	1 0 1 0 1
Z	"	1 0 0 0 1
Figure Shift		1 1 1 1 1
Letter Shift		1 1 0 1 1
Space		0 0 1 0 0
Line Feed		0 1 0 0 0
Null		0 0 0 0 0

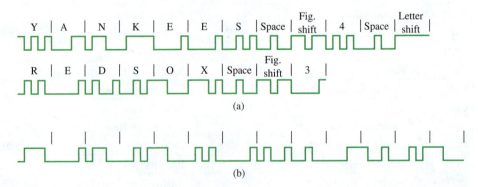

FIGURE 11-4 Baudot code examples.

Binary	#
0000	0
1000	1
1100	2
0100	3
0110	4
1110	5
1010	6
0010	7
0011	8
1011	9

1-bit change for each step value.

FIGURE 11-5 The Gray code.

is 0010 while the code for 8 is 0011. Notice that only one binary bit changes when the decimal value changes from 7 to 8. This is true for all of the numbers (0–9). The Gray code is shown in Figure 11-5.

The Gray code is used most commonly in telemetry systems that have slowly changing data or in communication links that have a low probability of bit error. This coding scheme works well for detecting errors in slowly changing outputs, such as data from a temperature sensor (thermocouple). If more than one change is detected when words are decoded, then the receiving circuitry assumes that an error is present.

The data communication that takes place between computers and peripheral equipment is of two basic types—serial and parallel—and uses ASCII format. In addition, the data that are sent in serial form (i.e., one bit after another on a single pair of wires) may be classified as either synchronous or asynchronous.

In an **asynchronous system,** the transmit and receive clocks free-run at approximately the same speed. Each computer word is preceded by a **start bit** and followed by at least one **stop bit** to frame the word. In a **synchronous system** both sender and receiver are exactly synchronized to the same clock frequency. This is most often accomplished by having the receiver derive its clock signal from the received data stream.

Many choices are available for selecting a serial communications interface. Most often the computer and the electronic equipment dictate the choice. For example, a spectrum analyzer might have to contain a GPIB and RS-232 interface, while a digital camera might contain a USB or IEEE 1394 interface. Industrial equipment might be interconnected using the RS422 or RS485 standard. This section addresses the key standards that the user may find on computer and electronic communications equipment, which include the following:

- USB
- IEEE 1394
- RS232
- RS485
- RS422
- GPIB

Universal Serial Bus

The **universal serial bus (USB)** port has become a nearly universal high-speed serial communications interface. The reasons are simple:

- Almost all computer peripherals (mouse, printers, scanners, etc.) are now available in a USB version.
- The USB devices are **hot-swappable,** which means the external devices can be plugged in or unplugged at any time.
- USB devices are detected automatically once they are connected to the computer.
- The USB 2 supports a maximum data rate of 480 Mbps; USB 1.1 supports 12 Mbps. An even newer form, USB 3.0 ("SuperSpeed"), has a 5 Gbps signaling rate.
- A total of 127 peripherals can be connected to one USB port.

The USB cable used to interconnect the peripheral to the computer consists of four wires inside a shielded jacket. The function of the four wires and their wire colors are listed in Table 11-2.

Two types of connectors are used with USB ports and cables, **Type A** and **Type B.** They are shown in Figure 11-6. The Type A connector is the upstream connection that connects to the computer. The Type B connector is for the downstream connection to the peripheral.

An example of a USB connection using the MAX3451 transceiver is shown in Figure 11-7. This diagram shows the MAX3451 being used to interface the peripheral device through the IC to the PC. The SPD input is used to select the data transfer rate of 1.5 Mbps (SPD = low) or 12 Mbps (SPD = +V). The D+ and D− pins are bidirectional bus connections. The OE pin is used to control the data flow. To transmit data from the peripheral to the USB side, bring OE low and SUS low. Receiving data requires OE to be high and SUS to be low. VP and VM terminals function as receiver outputs when OE is high (transmit mode) and duplicate D+ and D− when OE is low (receiver mode). The supply voltage range for the device is +1.65 V to +3.6 V.

TABLE 11-2 • The USB Wire Colors and Functions	
COLOR	**FUNCTION**
Red	+5 V
Brown	Ground
Yellow	Data
Blue	Data

FIGURE 11-6 The USB Type A and Type B connectors.

(a)   (b)

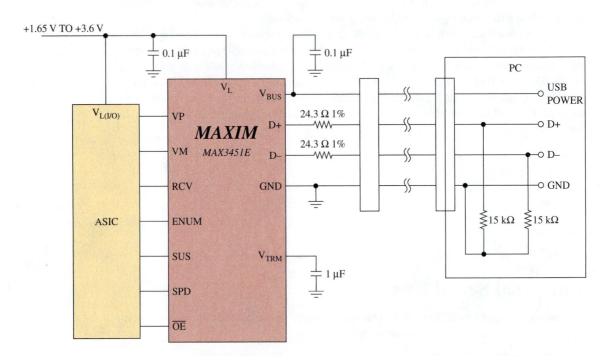

FIGURE 11-7 An example of using the MAX3451 transceiver for establishing a USB connection.

IEEE 1394

The IEEE 1394 connection is another high-speed serial connection available for computers and peripherals. Originally developed by Apple, the connection is usually referred to by its trademarked name, FireWire. **FireWire A (IEEE 1394a)** supports data transfers up to 400 Mbps, whereas **FireWire B (IEEE 1394b)** supports 800 Mbps. The IEEE 1394 connection specifies a shielded, twisted-pair cable with three pairs (six wires). Two of the wire pairs are used for communication and the third is used for power. The connector pin assignments are shown in Figure 11-8.

FIGURE 11-8 The IEEE 1394 connector and pin assignments.

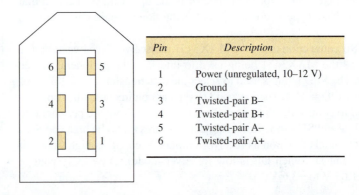

Pin	Description
1	Power (unregulated, 10–12 V)
2	Ground
3	Twisted-pair B–
4	Twisted-pair B+
5	Twisted-pair A–
6	Twisted-pair A+

RS-232 Standard

Older serial data communications follow a standard called **RS-232,** or more correctly, RS-232 C. Usually everyone is referring to the "C" version of RS-232 because that is what is currently in use, but often the "C" is omitted. Be aware that even though we may refer to the standard as RS-232, we really mean RS-232 C. The Electronics Industry Association (EIA) sets the RS-232 C standard.

In addition to setting a standard of voltages, timing, and so on, standard connectors have also been developed. This normally consists of a DB-25 connector, which is a connector with two rows of pins, arranged with thirteen pins in one row and twelve in the other. A diagram of this connector is provided in Figure 11-9.

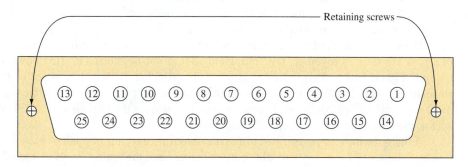

FIGURE 11-9 DB-25 connector.

RS-232 "D-type" connector—front view.

Pin	Name	Abbreviation
1	Frame ground	FG
2	Transmit data	TD
3	Receive data	RD
4	Request to send	RTS
5	Clear to send	CTS
6	Data set ready	DSR
7	Signal ground	SG
8	Data carrier detect	DCD
20	Data terminal ready	DTR

Most popular pins implemented in RS-232 connections.

It should be noted that even though the DB-25 connector is usually used, the RS-232 C standard specifications do not define the actual connector. Another connector is also used for RS-232. This is the DB-9 connector, which has become a sort of quasi-standard for use on IBM compatible personal computers. See Figure 11-10 for a diagram of the DB-9 connector. You may ask, "How can twenty-five pins from a DB-25 connector all fit into the nine pins of a DB-9 connector?" As we will see, all twenty-five pins of the DB-25 connector are not used, and, in fact, nine pins are enough to do the job.

The original purpose of RS-232 was to provide a means of interfacing a computer with a modem. The computer in this case was likely a mainframe computer because personal computers, at least as we know them today, had not yet been developed. Modems were always external, so it was necessary that some means of connection between the modem and the computer be made. It would be even nicer if this connection would be made standard so that all computers, and all modems, could be connected interchangeably. This was the original purpose of RS-232. However, it has evolved into being many other things.

Today an RS-232 interface is used to interface a mouse to a personal computer, to interface a printer to a personal computer, and probably to interface about as many other

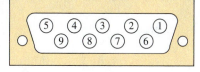

Front view

Pin			
	1	DCD	(8)
	2	RD	(3)
	3	TD	(2)
	4	DTR	(20)
	5	SG	(7)
	6	DSR	(6)
	7	RTS	(4)
	8	CTS	(5)
	9	RI	(22)

FIGURE 11-10 DB-9 connector.

things as one can think of to a personal computer. In many cases, it is used to interface instrumentation to a PC. This means that the standard has evolved, and in the process of this evolution has changed in terms of the real world.

The standard did not define the connector but did define signal levels and different lines that could be used. The signal levels that are defined are very broad. The voltage levels are to be between 3 and 25 V. A minus voltage indicates a "1" and a plus voltage indicates a "0." Although the definition covers 3 to 25 V, the real signal levels are usually a nominal 12 to 15 V. Many chips available today will not respond to the 3-V levels, so from a real-world practical point of view, the signal level is between 5 and 15 V. This is still a very broad range and can obviously allow for a lot of loss in a cable.

In addition to the signal levels, the RS-232 standard specifies that the maximum distance for a cable is 50 ft, and the capacitance of the cable cannot exceed 2500 pF. In reality, it is the capacitance of the cable that limits the distance. In fact, distances that far exceed 50 ft are commonly used today for serial transmission.

The RS-232 standard also contains another interesting statement. It says that if any two pins are shorted together, the equipment should not be damaged. This obviously requires a good buffer. This buffering is normally provided. It should be noted that the standard says only that the equipment will not be damaged. It does not say that the equipment will work in that condition. In other words, if you short the pins on an RS-232 connector, there should be no smoke, but it might not work!

Perhaps the most important part of the standard from a technical point of view is that it defines the way the computer should "talk" to the modem; the timing involved, including the sequence of signals; and how each is to respond.

RS-232 Line Descriptions

Now that we know a little about what RS-232 is designed to do, it is time to look at the actual signal lines involved and see what they do. As explained earlier, the DB-25 connector definition is not a part of the original standard, but because it has become a de facto standard, we will use it in our discussion. Refer to Figure 11-9 as a reference in this description. The complete signal description chart in Figure 11-11 will also be helpful.

1. **Ground pins:** Actually, there are two ground pins in RS-232. They are not the same, however, and serve very different purposes.

    Pin 1 is the *protective ground* (GND). It is connected to the chassis ground and is there simply to make certain that no potential difference exists between the chassis of the computer and the chassis of the peripheral equipment. This is *not* the signal ground. The protective ground functions much the same as the third prong of a 120-V ac three-prong outlet. The circuit will work without a connection to this pin, but the operator will also lose protection. In other words, make certain this pin is connected.

    Pin 7 is the *signal ground* (SG). This is the pin used for the ground return of all the other signal lines. Look at the location of the signal ground, pin 7. One of the problems often associated with RS-232 and especially DB-25 connectors is how to determine whether you are looking from the front or back and which end you should start counting from. Many connectors have the pin numbers printed on them, but this printing is usually so small that you cannot read it. Note that regardless of which end you count from, pin 7 always comes out in the same place!

2. **Data signal pins:** Data can be sent from both the computer and peripheral equipment. Therefore a bidirectional path is necessary.

    Pin 2 is *transmit data* (TD). In theory, this pin will contain the actual data flowing from the computer to the peripheral equipment.

    Pin 3 is *receive data* (RD). In theory, this pin has the actual data flowing from the peripheral equipment to the computer. Note that it is the same as pin 2 listed previously, but in the opposite direction.

    The problem is that theory and reality are not always the same. What if you want to link computers together? Which one is sending data, and which one is

PIN NO.	EIA CKT.	CCITT CKT.	Signal Description	Common Abbrev.	From DCE	To DCE
1	AA	101	Protective (chassis) ground	GND		
2	BA	103	Transmitted data	TD		X
3	BB	104	Received data	RD	X	
4	CA	105	Request to send	RTS		X
5	CB	106	Clear to send	CTS	X	
6	CC	107	Data set ready	DSR	X	
7	AB	102	Signal ground/common return	SG	X	X
8	CF	109	Received line signal detector	DCD	X	
9			Reserved			
10			Reserved			
11			Unassigned			
12	SCF	122	Secondary received line signal detector		X	
13	SCB	121	Secondary clear to send		X	
14	SBA	118	Secondary transmitted data			X
15	DB	114	Transmitter signal element timing (DCE)		X	
16	SBB	119	Secondary received data		X	
17	DD	115	Receiver signal element timing		X	
18			Unassigned			
19	SCA	120	Secondary request to send			X
20	CD	108/2	Data terminal ready	DTR		X
21	CG	110	Signal quality detector	SQ	X	
22	CE	125	Ring indicator	RI	X	
23	CH	111	Data signal rate selector (DTE)			X
23	CI	112	Data signal rate selector (DCE)		X	
24	DA	113	Transmitter signal element timing (DTE)			X
25			Unassigned			

FIGURE 11-11   Signal description for DB-25.

receiving data? It would seem that there should be an easy answer to this question, but such is not the case. Which end of the cable are you looking at? In this instance, when one is transmitting data, those same data become receive data to the other computer. What this really means is that it is not easy to define which is receive and which is transmit.

Because of this problem, what is usually called a **null modem** cable has been developed. It has pins 2 and 3 crisscrossed—that is, pin 2 at one end is connected to pin 3 at the other end, and vice versa. Of course, this problem does not exist if we were to use RS-232 as originally intended, that is, for the computer to talk to a modem.

Another way of completing an RS-232 connection, a null modem connection, is shown in Figure 11-12. The RTS-CTS and DSR-DTR pins are connected together as shown. This tricks the RS-232 device into believing a piece of data communications equipment is connected and enables the TX and RX lines to function properly.

3. **Handshaking pins:** Pins 4 and 5 are used for handshaking or flow control. More correctly, pin 4 is called the *request-to-send* (RTS) pin, while pin 5 is the *clear-to-send* pin (CTS). These two lines work together to determine that everything is fine for data to flow. Originally, this was used to turn on the modem's carrier, but today it is more often used to check for buffer overflow. Almost all modern modems (and other serial devices) have some sort of buffer that is used when receiving and sending data. It would not be satisfactory for that buffer to be sending more information than it could actually hold. Pins 4 and 5 may be used to handle this problem. Note that if one computer is talking to another one via a null modem cable, these pins need to be crisscrossed just like pins 2 and 3.

In many cases, software flow control is used, and pins 4 and 5 do not serve any useful purpose at all. In this case, pins 4 and 5 need to be jumpered at the DB-25 connector. You may wonder why it would be necessary to jumper these pins if they are

FIGURE 11-12 **The RS-232 null modem connection.**

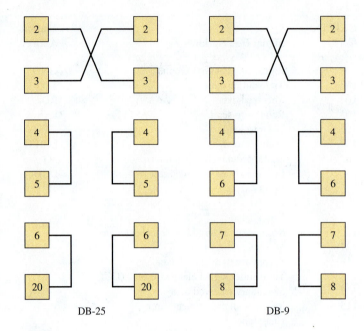

DB-25   DB-9

not actually used. The reason is that many serial ports will check these pins before they will send any data. Many serial interface cards are configured so that this signal is required for any data transfer to take place.

4. **Equipment ready pins:** Pins 6 and 20 are complementary pairs much as pins 4 and 5 are. Actually, pin 6 is called *data set ready* (DSR), while pin 20 is called *data terminal ready* (DTR). The original purpose of these pins was simply to make certain that the power of the external modem was turned on and that the modem was ready to go to work. In some instances, these pins were also used to indicate if the phone was off hook or on hook.

Today, these pins may be used for any number of purposes, such as paper out indicators on printers. The two pins work together and normally should be jumpered on the DB-25 connector like pins 4 and 5 if their use is not expected.

Note that, up to this point, the complementary pairs of pins have been adjacent to one another. Obviously, pins 6 and 20 are not in consecutive order. An inspection of the DB-25 connector will reveal, however, that pins 6 and 20 are almost directly above and below each other in their respective rows.

5. **Signal detect pin:** Pin 8 is usually called the *data carrier detect* (DCD) or sometimes simply the *carrier detect* (CD) line. In an actual modem, it is used to indicate that a carrier (or signal) is present. It may also indicate that the signal-to-noise (*S/N*) ratio is such that data transmission may take place.

Many computer interface cards require that this signal be present before they will communicate. If the RS-232 connection is not a modem requiring this signal, it is often tied together with pins 6 and 20 (DSR and DTR). As has been the case with other pins, this pin is probably not used for its original purpose. What it really is used for is highly dependent on the device in use.

6. **Ring indicator pin:** Pin 22 is known as the *ring indicator* (RI). Its original purpose was to do just what its name indicates—to let the computer know when the phone was ringing. Most modems are equipped with automatic answer capabilities. Obviously, the modem needed to tell the computer that someone with data to send was calling!

Pin 22 is forced true only when the ringing voltage is present. This means that this signal will go on and off in rhythm with the actual telephone ring. As is true with other signals, this pin may not be used. Often it is tied together with pins 6, 20, and 8. In other cases, it is simply ignored. Again, the actual use of this pin today will vary widely with the equipment connected.

7. **Other pins:** Up to this point, we have discussed the ten pins used most often. In reality, the chassis ground is usually accomplished by grounding each piece of equipment

separately and is not necessary with battery-powered computer equipment. These considerations have led to the 9-pin connector usage shown in Figure 11-10. When the DB-25 connector is used, the remaining pins are occasionally used for special situations, as warranted by the equipment connected.

It should be noted that the terms *computer, modem,* and *peripheral equipment* have been used in the discussion thus far. Two more technically correct terms may be used in literature, texts, and diagrams. The term **data terminal equipment (DTE)** is used to indicate a computer, computer terminal, personal computer, and so on. The term **data communications equipment (DCE)** is used to indicate the peripheral equipment (such as the modem, printer, mouse, etc.). Notice that the DCE label is used in Figure 11-11.

Many communication standards available on today's computers complement the use of RS-232. Table 11-3 provides a brief overview of many of the standards currently being used.

TABLE 11-3 • Overview of Current Serial Computer Communication Standards	
RS-232	The common serial data connection for computer modems and the mouse interface.
RS-422	A balanced serial communications link that can support data speeds up to 10 Mbps at a distance of 4000 ft.
RS-449	Intended as a replacement for the RS-232 standard, but the two standards are not completely compatible either mechanically or electrically. Data speeds are from 9600 bps to 10 Mbps, and they depend on the length and type of cable used.
RS-485	A balanced differential output that allows multiple data connections. Data speeds of 30 Mbps are supported.
RS-530	A standard replacing RS-449 and also complementing the RS-232 standard. This standard will also interface the balanced systems RS-422 and 423.

Source: www.blackbox.com

RS-422, RS-485

The RS-232 output is a single-ended signal, which means that a signal line and a ground line are used to carry the data. This type of arrangement can be susceptible to noise and ground bounce. The **RS-422** and **RS-485** use a differential technique that provides a significant improvement in performance and thus yields greater distances and the capacity to support higher data rates. In a differential technique, the signals on the wires are set up for a high (+) and low (−) signal line. The (+) indicates that the phase relationship of the signal on the wire is positive and the (−) indicates that the phase of the signal on the wire is negative; both signals are relative to a virtual ground. This is called a **balanced mode** of operation. In a balanced mode of operation, the balanced operation of the two wire pairs helps to maintain the required level of performance in terms of crosstalk and noise rejection.

RS-422 and RS-485 also support multidrop applications, which means that the standard supports multiple drivers and receivers connected to the same data line. RS-422 is not a true multidrop network because it can drive ten receivers but only allows one transmitter to be connected. RS-422 supports a data rate of 10 Mbps over a distance of 4000 feet. The RS-485 standard allows a true multiport connection. The standard supports thirty-two drivers and thirty-two receivers on a single two-wire bus. It supports data rates up to 30 Mbps over a distance of 4000 feet.

Computer interface cards are available that provide the functions for virtually any communications test set or gear. For example, spectrum analyzers and data and protocol analyzer interface cards are available on a PC interface card. Table 11-4 outlines some of the common computer bus interfaces available today. This information is helpful for the user in specifying the proper bus or an interface card for a computer. This list is not complete but it does provide some of the more common computer bus interfaces.

TABLE 11-4 • Standard Computer Bus Interfaces	
PCI (Peripheral Component Interconnect)	PCI is the best bus choice for current computers. PCI supports 32- and 64-bit implementations.
PCIe (PCI Express)	Used primarily for high-end video displays.
AGP (Accelerated Graphics Port)	Used in high-speed 3D graphics applications.
ISA (Industry Standard Architecture)	Allows for 16-bit data transfers between the motherboard and the expansion board.
EISA (Extended Industry Standard Architecture)	EISA extends the ISA bus to 32-bit data transfers.
MCA (Micro Channel Architecture)	
VLB [VESA (Video Electronics Standards Association) Local Bus]	Supports 32-bit data transfers at 50 MHz.
USB (universal serial bus)	Does not require that the computer be turned off or rebooted to activate the connection. Supports data rates up to 12 Mbps. Two type of USB connectors are shown in Figures 11-6(a) and (b).
IEEE 1394 (FireWire, i-Link)	A high-speed, low-cost interconnection standard that supports data speeds of 100 to 400 Mbps (see Figure 11-8).
SCSI (Small Computer System Interface)	Consists of an SCSI host adapter; SCSI devices such as hard drives; DVD; CD-ROM; and internal or external SCSI cables, terminators, and adapters. SCSI technology supports data transfer speeds up to 160 Mbps.
IDE (Integrated Drive Electronics)	Standard electronic interface between the computer's motherboard and storage devices such as hard drives. The data transfer speeds are slower than SCSI. Faster than SCSI, speeds up to 133 Mbps are supported.
Serial ATA (Advanced Technology Attachment)—SATA	Uses a high-speed serial cable to achieve transfer speeds up to 6 Gbps.

11-3 LOCAL-AREA NETWORKS

The dramatic decrease in computer system cost and increase in availability have led to an explosion in computer usage. Organizations such as corporations, colleges, and government agencies have acquired large numbers of single-user computer systems. They may be dedicated to word processing, scientific computation, process control, and so on. A need to interconnect these locally distributed computer networks soon became apparent. Interconnection allows the users to send messages to the other network members. It also allows resource sharing of expensive equipment such as high-quality graphics printers or access to a robust, dedicated server to run programs too complicated for the local computer. The local computer is usually a personal microcomputer-type system. The network used to accomplish this is called a **local-area network (LAN)**. LANs are typically limited to separations of a mile or two and to several hundred users, but are usually smaller in scope.

LANs are defined in terms of the **topology** (architecture) used to interconnect the networking equipment and the **protocol** used for accessing the network. The most common architectures for LANs are shown in Figure 11-13. Two of the networking protocols in common use are **carrier sense multiple access with collision detection (CSMA/CD),** which is associated with the bus and star topologies, and token passing, which is associated with the token-ring topology.

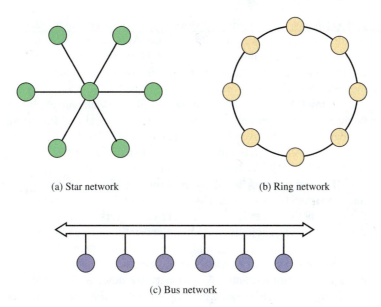

FIGURE 11-13 **Network topologies.**

(a) Star network

(b) Ring network

(c) Bus network

The token-ring topology is shown in Figure 11-14. The token-passing technique is well suited to the ring network topology. An electrical token is placed in the channel and circulates around the ring. If a user wishes to transmit, the station must wait until possession of the token exists. Each station is assured access for transmission of its messages. A disadvantage of this system is that if an error changes the token pattern, it causes the token

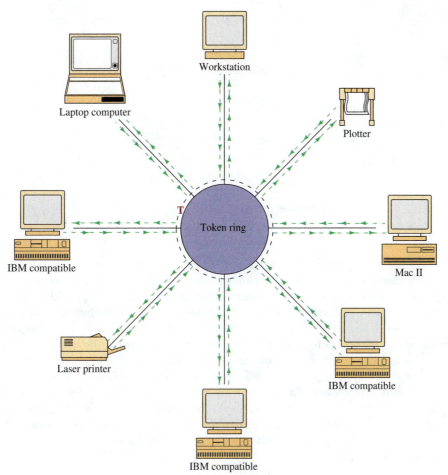

FIGURE 11-14 **The token-ring topology.**

Workstation

Plotter

Laptop computer

IBM compatible

Token ring

Mac II

Laser printer

IBM compatible

IBM compatible

– – ► – – ► The passing of the token

to stop circulating. Also, ring networks rely on each system to relay all data to the next user. A failed station causes data traffic to cease. Another approach for the token-ring technique is to attach all the computers to a central token-ring hub. Such a device manages the passing of the token rather than assigning this task to individual computers, which improves the reliability of the network.

Many LAN systems are currently available. Many are applicable to a specific manufacturer's equipment only. The Institute of Electrical and Electronics Engineers (IEEE) standards board approved LAN standards in 1983. The following IEEE 802 standards provided impetus for different manufacturers to use the same codes, signal levels, and so on: IEEE 802.3 CSMA/CD IEEE 802.5 Token-ring.

The bus network topology is shown in Figure 11-15. A bus network shares the media for data transmission. This means that while one computer is talking on the LAN, the other network devices (e.g., other computers) must wait until the transmission is complete. For example, if computer 1 is printing a large file, the line of communication will be between computer 1 and the printer, and this will tie up the network bus for a good portion of the time. All network devices on the bus see the data traffic from computer 1 to the printer, and the other network devices must wait for pauses in transmission or until the transmission is complete before they can assume control of the bus and initiate transmission. This means that bus topologies are not very efficient. This is one reason—but not the only reason—that bus topologies are seldom used in modern computer networks.

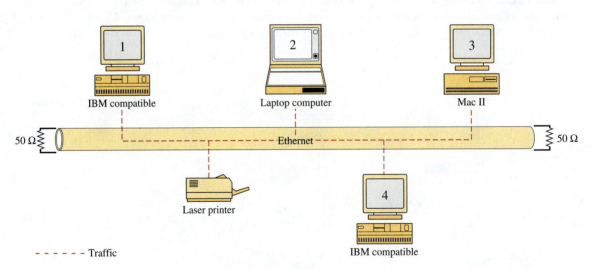

FIGURE 11-15   **The bus topology.**

The star topology, shown in Figure 11-16, is the most common in today's LANs. At the center of a star network is either a hub or a switch. The hub or the switch is used to connect the network devices together and facilitate the transfer of data. For example, if computer 1 wants to send data to the network printer, the hub or switch provides the network connection. Actually, either a switch or a hub can be used, but there is a significant advantage to using a switch. In a hub environment, the hub will rebroadcast the message to all computers connected to the star network. This is very similar to the bus topology because all computers are seeing all data traffic on the LAN. However, if a switch is used instead of a hub, then the message is transmitted directly from computer 1 to the printer. This greatly improves the efficiency of the available bandwidth. It also permits additional devices to communicate without tying up the network. For example, while computer 1 is printing a large file, computers 5 and 6 can communicate with each other.

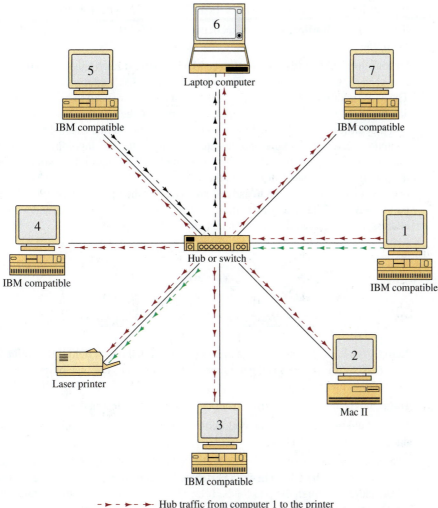

FIGURE 11-16 The star topology.

- –►- –►- –►- → Hub traffic from computer 1 to the printer
- –►- –►- –►- → Switch traffic from computer 1 to the printer
- –►- –►- –►- → Switch traffic from computer 5 to computer 6

Ethernet LAN

Ethernet is a baseband CSMA/CD protocol local area network system. It originated in 1972, and the full specification was provided via a joint effort among Xerox, Digital Equipment Corporation, and Intel in 1980.

Basically, for a computer to talk on the Ethernet network, it first listens to see if there is any data traffic (carrier sense). This means that any computer on the LAN can be listening for data traffic and any of the computers on the LAN can access the network (multiple access). There is a chance that two or more computers will attempt to broadcast a message at the same time; therefore, Ethernet systems have the capability for detecting data collisions (collision detection).

How is the destination for the data determined? The Ethernet protocol provides information regarding the source and destination addresses. The structure for the Ethernet frame is shown and described in Figure 11-17.

Preamble	Start frame delimiter	Destination MAC address	Source MAC address	Length type	Data	Pad	Frame check sequence

FIGURE 11-17 The data structure for the Ethernet frame.

Preamble: an alternating pattern of 1s and 0s used for synchronization.

Start frame delimiter: A binary sequence of 1 0 1 0 1 0 1 1 that indicates the start of the frame.

Destination MAC address and source MAC address: Each Ethernet **network interface card (NIC)** has a unique **media access control (MAC) address** associated with it. The MAC address is 6 bytes in length. The first 3 bytes are used to indicate the vendor, and the last 3 bytes are unique numbers assigned by the vendor. This is the information that ultimately enables the data to reach a destination in a LAN. This is also how computer 1 and the printer communicated directly in the star topology example using the switch (Figure 11-16). The switch used the MAC address information to redirect the data from computer 1 directly to the printer. *Note:* If the destination MAC address is all 1s, then this is called a **broadcast address,** and the message is sent to all stations on the network. The following codes are examples of a MAC address and a broadcast address; the addresses shown are in hexadecimal code (base 16).

	VENDOR	NIC CARD ID
MAC address	0 0 A A 0 0	B 6 7 A 5 7
Broadcast address	F F F F F F	F F F F F F

Length/type: an indication of the number of bytes in the data field if this value is less than 1500. If this number is greater than 1500, it indicates the type of data format, for example, IP and IPX.

Data: the data being transferred from the source to the destination.

Pad: a field used to bring the total number of bytes up to the minimum of 46 if the data file is less than 46 bytes.

Frame check sequence: a 4-byte cyclic redundancy check (CRC) value used for error detection. The CRC check is performed on the characters from the destination MAC address through the pad fields. If an error is detected, then the system requests a retransmission.

The minimum length of the Ethernet frame is 64 bytes from the destination MAC address through the frame check sequence. The maximum Ethernet frame length is 1522 bytes. The 0s and 1s in the Ethernet frame are formatted using Manchester encoding (biphase-L, see Section 9-3). An example of Manchester encoding is shown in Figure 11-18.

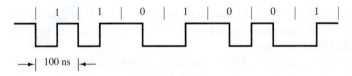

FIGURE 11-18   Manchester encoding.

11-4 ASSEMBLING A LAN

This section presents two examples of assembling LANs. The examples demonstrate a technique that can be used to assemble an office LAN and a building LAN. These examples are presented from the point of view of assembling the hardware necessary for establishing network communications between the computers and ancillary network devices. Many possible configurations can be used to solve these problems; this is one solution. Note that all computer networks require some type of networking software to run the LAN. Networking software is available with Microsoft Windows, Linux, UNIX, and the Macintosh operating system, to name a few.

The Office LAN Example

Our example of an office LAN consists of ten computers, two printers, and one server. The layout for the office LAN is shown in Figure 11-19. Each computer, the printers, and the server on the LAN are all connected to a common switch. The connection from each unit in the network to the switch is provided by a CAT6 (category 6) twisted-pair cable. CAT6 cables are capable of carrying 1000 Mbps of data up to a length of 100 meters. Twisted-pair cables and the various category specifications are discussed in Chapter 12, Section 12-2. If the network hardware and software are properly set up, all computers will be able to access the server, the printer, and other computers.

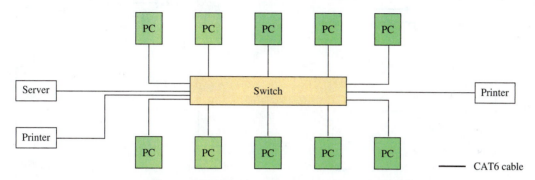

FIGURE 11-19   **An example of an office LAN.**

The media used for transporting data in a modern computer network are either twisted-pair or fiber cables. Fiber cables, optical LANs, and their numerics are discussed in Chapter 16. Table 11-5 lists the common numerics used to describe the data rates for the copper coaxial cable and twisted-pair media being used in a LAN.

TABLE 11-5 • Common Numerics for LAN Cabling	
NUMERIC	**DESCRIPTION**
10Base2	10 Mbps over coax up to 185 m, also called ThinNet (seldom used anymore)
10Base5	10 Mbps over coax up to 500 m, also called ThickNet (seldom used anymore)
10BaseT	10 Mbps over twisted pair
100BaseT	100 Mbps over twisted pair
100BaseFX	100 Mbps over fiber
1000BaseT	1 Gbps over twisted pair
1000BaseFX	1 Gbps over fiber
10GBase—	The family of fiber products supporting 10-gigabit Ethernet (10GbE)

Assembling a Building LAN

A building LAN describes a network where multiple LANs within a building are connected together. An example of assembling a building LAN is provided in Figure 11-20. For this example, three switches were required because the distance from the computers to a central switch exceeded the 100-m maximum distance for CAT6 twisted-pair cable. To meet the 100-m maximum-distance requirement, it was decided to place a switch inside a closet in each of the three wings. The connection from each device is **100BaseT.** This means that the data rate is 100-Mbps baseband, and CAT6 twisted-pair cable was used. All network devices in each wing were routed to their respective switches, located in either closet A, B, or C. Each closet has **RJ-45** patch panels for routing the cables. RJ-45 connectors are 8-pin modular types used to connectorize CAT6 twisted-pair cable. The patch panels were included to maximize the flexibility of the network, including future wiring changes needed to accommodate changes in the network. The number of computers, printers, workstations, and servers input to each switch is listed by the respective switch. The fiber optic feeds from switches A and B are combined with switch C and sent to a router. The router provides a connection to the internet and routes data traffic back to closets A, B, and C.

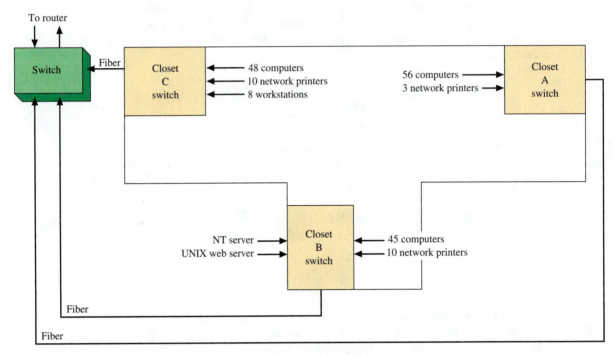

FIGURE 11-20 An example of a building LAN.

11-5 LAN INTERCONNECTION

The utility of LANs led to the desire to connect two (or more) networks together. For instance, a large corporation may have had separate networks for its research and engineering and for its manufacturing units. Typically, these two systems used totally different technologies, but it was deemed necessary to "tie" them together. This led to **metropolitan area networks (MANs)**—two or more LANs linked together within a limited geographical area. Once the techniques were in place to do this, it was decided that it would be helpful to link the MAN with the marketing division on the other side of the country. Now two or more LANs were linked together over a wide geographical area, resulting in a **wide-area network (WAN).**

To allow different types of networks to be linked together, an **open systems interconnection (OSI)** reference model was developed by the International Organization for Standardization (ISO). When introduced, the OSI model was competing with TCP/IP to become the default protocol for Ethernet networks. Though TCP/IP became the default protocol suite for modern networks, the OSI model is used to introduce the subject of networking because it is so thorough. The OSI reference model contains seven layers, as shown in Figure 11-21. It provides for everything from the actual physical network interface to software applications interfaces.

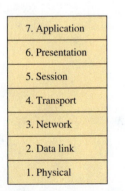

FIGURE 11-21 OSI reference model.

1. **Physical layer:** provides the electrical connection to the network. It doesn't speak to the modulation or physical medium used.

2. **Data link layer:** handles error recovery, flow control (synchronization), and sequencing (which terminals are sending and which are receiving). It is considered the "media access control layer."

3. **Network layer:** accepts outgoing messages and combines messages or segments into packets, adding a header that includes routing information. It acts as the network controller.

4. **Transport layer:** is concerned with message integrity between the source and destination. It also segments/reassembles the packets and handles flow control.

5. **Session layer:** provides the control functions necessary to establish, manage, and terminate the connections as required to satisfy the user request.

6. **Presentation layer:** accepts and structures the messages for the application. It translates the message from one code to another if necessary.

7. **Application layer:** logs the message in, interprets the request, and determines what information is needed to support the request.

Interconnecting LANs

The interconnection of two or more LANs (into a MAN or WAN) is accomplished in several ways, depending on the LAN similarities.

Switch: Layer 2 switches use only the bottom two OSI layers to link LANs that usually have identical protocols at the physical and data link layers. This is illustrated in Figure 11-22.

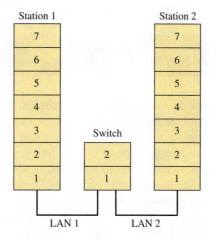

FIGURE 11-22 **Switch connecting two LANs.**

Routers: Routers interconnect LANs by using the bottom three OSI layers, as shown in Figure 11-23. They manage traffic congestion by employing a flow-control mechanism to direct traffic to alternative paths and can provide protocol conversion when needed.

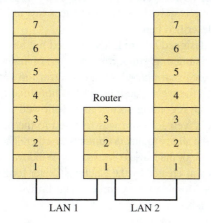

FIGURE 11-23 **Router connecting two LANs.**

Gateways: This is an older term used to describe a device that encompasses all seven OSI layers. It interconnects two networks that use different protocols and formats. In the capacity shown in Figure 11-24, the gateway is being used to perform protocol conversion at the applications layer. Today, routers manage all the functions previously handled by a gateway.

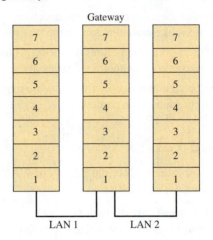

FIGURE 11-24   Gateway connecting two LANs.

11-6 INTERNET

The most exciting network development in recent times is the internet (commonly referred to simply as the Net). It allows for the interconnection of LANs and individuals. In a few short years it has become a pervasive force in our society. A system originally called ARPANET was developed in the early 1970s to link academic institutions and their researchers involved with military defense activities. It has evolved into the internet—a network that has worldwide broadcasting capability, a mechanism for information dissemination, and a medium for collaboration and interaction between individuals and their computers without regard to geographic location. The internet is a packet-switched, global network system that consists of millions of local area networks and computers (hosts).

A definition of the term *internet* has been provided by the Federal Networking Council (FNC): "Internet refers to the global information system that—(i) is logically linked together by a globally unique address space based on the Internet Protocol (IP) or its subsequent extensions/follow-ons; (ii) is able to support communications using the Transmission Control Protocol/Internet Protocol (TCP/IP) suite or its subsequent extensions/follow-ons, and/or other IP-compatible protocols; and (iii) provides, uses or makes accessible, either publicly or privately, high level services layered on the communications and related infrastructure described herein."[1]

The internet was designed before LANs existed but has accommodated this network technology. It was envisioned as supporting a range of functions including file sharing, remote login, and resource sharing/collaboration. Over the years it has also enabled electronic mail (e-mail) and the web. The internet has not finished evolving as evidenced by the internet telephone that is coming online, to be followed by internet television.

In 1992, Tim Lee's web software was released to the public. It is called a *hypertext system*—one that gives the ability to link documents together. The major breakthrough came in June 1993, with the release of the Mosaic browser for Windows, which dramatically improved its look and interface. It was created by the National Center for Supercomputing Applications. The initial versions of the Mosaic are very similar to the browsers we use today. Web popularity is shown by the fact that it became the dominant internet use one year later in 1994.

---

[1]"Definition of 'Internet,'" The Networking and Information Technology Research and Development (NITRD) Program home page. http://www.nitrd.gov/fnc/Internet_res.aspx (accessed November 8, 2012).

The web is a method (and system) that provides users the opportunity to create and disseminate information on a global basis. It unleashes the power of individual creativity and allows a cross-connection of all people of the world. The growth of the web has been rapid and has become very commercial as product marketing and pay-for-access sites become common. There is a risk that it will lose its diversity and democratic nature with increased commercialism. One of the great appeals of the internet is that ordinary people with limited resources can publish and/or gather material just as do large corporations and organizations. If this capability is lost, the internet may regress to just another passive medium like television.

Internet Protocol Addressing

Moving data across the country and even moving data through routers in LANs requires a better addressing scheme than the MAC address. The MAC address provides the physical address for the network interface card, but where is it located—on which LAN, which building, which city, or even which country? Internet protocol (IP) addressing provides a solution to worldwide addressing through incorporating a unique address that tells on which network the computer is located. The current addressing scheme is a 32-bit address ($2^{32} = 4.29 \times 10^9$) format known as IPv4, which is being phased out in favor of the more robust IPv6 with its 128-bit addressing scheme. This enhancement provides for a theoretical maximum of 3.4×10^{38} possible addresses—a number so huge it is almost inconceivable. In addition to providing an endless supply of IP addresses, IPv6 was designed with security as a foremost concern.

IP network numbers are assigned by **Internet Assigned Numbers Authority (IANA),** an agency that assigns IP addresses to computer networks and makes sure no two different networks are assigned the same IP address. IP addresses are issued based on the class of the network. Examples of the three classes of IP networks are provided in Table 11-6.

TABLE 11-6 • The Three Classes of IP Networks		
CLASS	**DESCRIPTION**	**IP NUMBER RANGE**
Class A	Governments, very large networks	0.0.0.0. − 127.255.255.255
Class B	Midsize companies, universities, etc.	128.0.0.0 − 191.255.255.255
Class C	Small networks	192.0.0.0 − 223.255.255.255

The network addresses shown in Table 11-6 indicate the network portion of the IP address for each class. This provides sufficient information for routing the data to the appropriate destination network. The destination network uses this information to direct the packet to the destination computer. The complete address is typically assigned by the local network system administrator or is dynamically assigned when users need access outside their local networks. For example, your internet service provider (ISP) dynamically assigns an IP address to your computer when you log on to the internet.

11-7 IP TELEPHONY

IP telephony (voice-over IP) is the telephone system for computer networks. It incorporates technologies comparable to private branch exchange (PBX) telephone systems while maintaining the flexibility of computer networks. In fact, the 3COM Corporation calls the IP telephone system *network branch exchange (NBX)*. The NBX easily enables the user to incorporate telephone systems within their facility. Features that you would expect of a traditional PBX are provided with the NBX, including internal calls, message forwarding, speed dialing, voice mail, and access to the local public switched telephone network

(PSTN). The access to the PSTN is provided through traditional telephone-line connections. Long-distance calls can be placed through the access to the PSTN or through internet IP delivery via the NBX. The computer networking staff can typically do installation and management of the IP telephone system. The cable requirements and terminations are the same as CAT5e/6 and RJ-45 cabling. 3COM's NBX 100 uses an internet browser, which has been set up for a direct connection to access its internal management features.

Each telephone in the NBX system is assigned an internal extension number, much in the same way as in a PBX phone system. In addition to the phone number, each telephone has its own MAC address, which is used to deliver the voice data traffic within the LAN. The telephone can also be assigned an IP number and a gateway address so that phone calls can be routed outside the immediate LAN for long-distance calling over the internet or over leased corporate computer data links.

The phones in the NBX system are connected in a star topology to a central switch so that the **quality of service (QoS)** for voice traffic is not affected by computer network usage. Traditional telephone systems are very reliable, and the public expects a high quality of service. IP telephone systems must adhere to this implied measure of quality for the public to accept them as a viable alternative to the traditional PSTN. The computer network can experience very heavy data traffic and the NBX system can experience heavy voice traffic without any loss in system performance because the network switch provides the direct connection between the parties participating in the telephone call.

An example of an IP telephone LAN is provided in Figure 11-25. This network looks very similar to the star topology network (see Figure 11-16). Integration into the computer network LAN is provided through a connection to a switch or hub.

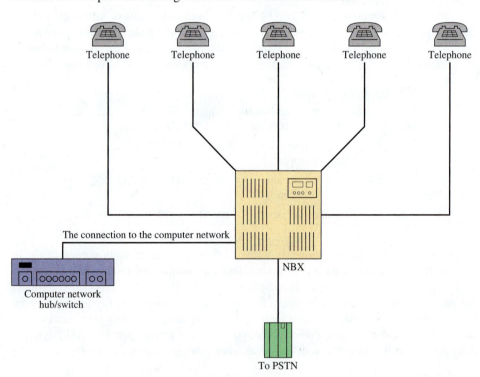

FIGURE 11-25 An IP telephone network.

11-8 INTERFACING THE NETWORKS

Chapter 9 and the previous sections of this chapter introduced the basics of both telephone and computer networks. Not too long ago, telephone and computer networks were considered totally separate technologies. Today, both sides of these networks are developing technologies and applications that will help integrate their technologies and capabilities into a total information network framework. This section addresses the current issues of interfacing the networks and discusses the latest in modem technologies and standards (including

V.92), cable modems, traditional data connections such as integrated services digital network (ISDN), and the latest in data connection (xDSL), as well as providing a summary of the new protocols being developed that help facilitate the integration of the networks.

Modem Technologies

The voice frequency channels of the PSTN are used extensively for the transmission of digital data. To use these channels, the data must be converted to an analog form that can be sent over the bandwidth-limited line. In voice-grade telephone lines, the transformers, carrier systems, and loading considerations attenuate all signals below 300 Hz and above 3400 Hz. While the bandwidth from 300 to 3400 Hz is suitable for voice transmission, it is not appropriate for digital data transmission because the digital pulse contains harmonics well outside this range. To transmit data via a phone requires the conversion of a signal totally within the 300- to 3400-Hz range. This conversion is provided by a modem.

There are currently two major standards for providing high-speed modem connections to an analog telephone line. These standards are **V.44 (V.34),** which is totally analog and which provides data rates up to 33.6 Kbps, and **V.92 (V.90),** which is a combination of digital and analog and provides data rates up to 56 Kbps. The V.92 (V.90) modem connection requires a V.92 (V.90) compatible modem and a service provider who has a digital line service back to the phone company. The data transfer with V.92 (V.90) is called **asymmetric operation** because the data-rate connection to the service provider is typically at V.44 (V.34) speeds, whereas the data rate connection from the service provider is at the V.92 (V.90) speed. The difference in the data rates in asymmetric operation is caused by the noise introduced by the analog-to-digital conversion process. The modem link from your computer to the PSTN (your telephone connection) is typically analog. This analog signal is then converted to digital at the phone company's central office. If the ISP has a digital connection to the phone company, then an analog-to-digital conversion is not required. The signal from the ISP through the phone company is converted back to analog for reception by your modem. However, the digital-to-analog process does not typically introduce enough noise to affect the data rate.

Cable Modems

Cable modems provide an alternative way of accessing a service provider. Cable modems capitalize on their high-bandwidth network to deliver high-speed, two-way data. Data rates range from 128 kbps to 10 Mbps upstream (computer to the cable-head end) and 10 to 30 Mbps downstream (cable-head end back to the computer). The cable modem connections can also be one-way when the television service implemented on the cable system precludes two-way communications. In this case, the subscriber connects to the service provider via the traditional telephone and receives the return data via the cable modem. The data service does not impair the delivery of the cable television programming. Currently the cable systems are using Ethernet protocol for transferring the data over the network. Many subscribers use the same upstream connection. This leads to potential collision problems, so a technique called **ranging** is used, where each cable modem determines the amount of time needed for its data to travel to the cable-head end. This technique minimizes collision rate, keeping it less than 25%.

Integrated Services Digital Network

The integrated services digital network (ISDN) is an established data communications link for both voice and data using a set of standardized interfaces. For business, the primary attractions will be increased capability, flexibility, and decreased cost. If one type of service—say, facsimile—is required in the morning and another in the afternoon—perhaps teleconferencing or computer links—it can easily shift back and forth. At present, the hookup for a given service might take a substantial amount of time to complete. With ISDN, new capacities will be available just by asking for them through a terminal.

The ISDN contains four major interface points as shown in Figure 11-26. The R, S, T, and U interface partitions allow for a variety of equipment to be connected into the system.

Type 1, or TE1, equipment includes digital telephones and terminals that comply with ISDN recommendations. Type 2, or TE2, gear is not compatible with ISDN specifications. It needs a terminal adapter to change the data to the ISDN's 64-kbps *B* channel rate. The TE2 equipment interfaces the network via the *R* reference point.

FIGURE 11-26 **ISDN setup illustration of R, S, T, and U interfaces.**

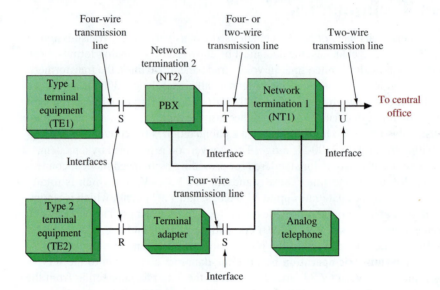

The ISDN standards also define two network termination (NT) points. NT$_1$ represents the telephone companies' network termination as viewed by the customer. NT$_2$ represents the termination of items such as local area networks and private branch exchanges (PBXs).

The customer ties in with the ISDN's NT$_1$ point with the *S* interface. If an NT$_2$ termination also exists, an additional *T* reference point linking both NT$_2$ and NT$_1$ terminations will act as an interface. Otherwise, the *S* and *T* reference points are identical. The ITU-T recommendations call for both *S* and *T* reference points to be four-wire synchronous interfaces that operate at a basic access rate of 192 kbps. They are called the local loop. Reference point *U* links NT$_1$ points on either side of a pair of users over a two-wire 192-kbps span. The two termination points are essentially the central office switches.

The ISDN specifications spell out a basic system as two *B* channels and one *D* channel (2*B* + D). The two *B* channels operate at 64 kbps each, while the *D* channel is at 16 kbps, for a total of 144 kbps. The 48-kbps difference between the basic 192-kbps access rate and the 2*B* + D rate of 144 kbps is mainly for containment of protocol signaling. The 2*B* + D channels are what the *S* and *T* four-wire reference points see. The *B* channels carry voice and data while the *D* channel handles signaling, low-rate packet data, and low-speed telemetry transmissions.

The ITU-T defines two types of communication channels from the ISDN central office to the user. They are shown in Figure 11-27. The basic access service is the 192-kbps channel already discussed and serves small installations. The primary access channel has a total overall data rate of 1.544 Mbps and serves installations with large data rates. This channel contains 23 64-kbps *B* channels plus a 64-kbps *D* channel. From any angle, the potential for ISDN is enormous. The capabilities of worldwide communications are now taking a quantum leap forward.

xDSL Modems

The **xDSL** modem is considered to be the next generation of high-speed internet access technology. DSL stands for **digital subscriber line,** and the "x" generically represents the various types of DSL technologies that are currently available. The DSL technology uses the existing copper telephone lines for carrying the data. Copper telephone lines can carry high-speed data over limited distances, and the DSL technologies use this trait to provide a high data-rate connection. However, the actual data rate depends on the quality of the copper cable, the wire gauge, the amount of crosstalk, the presence of load coils, the bridge taps, and the distance of the connection from the phone service's central office.

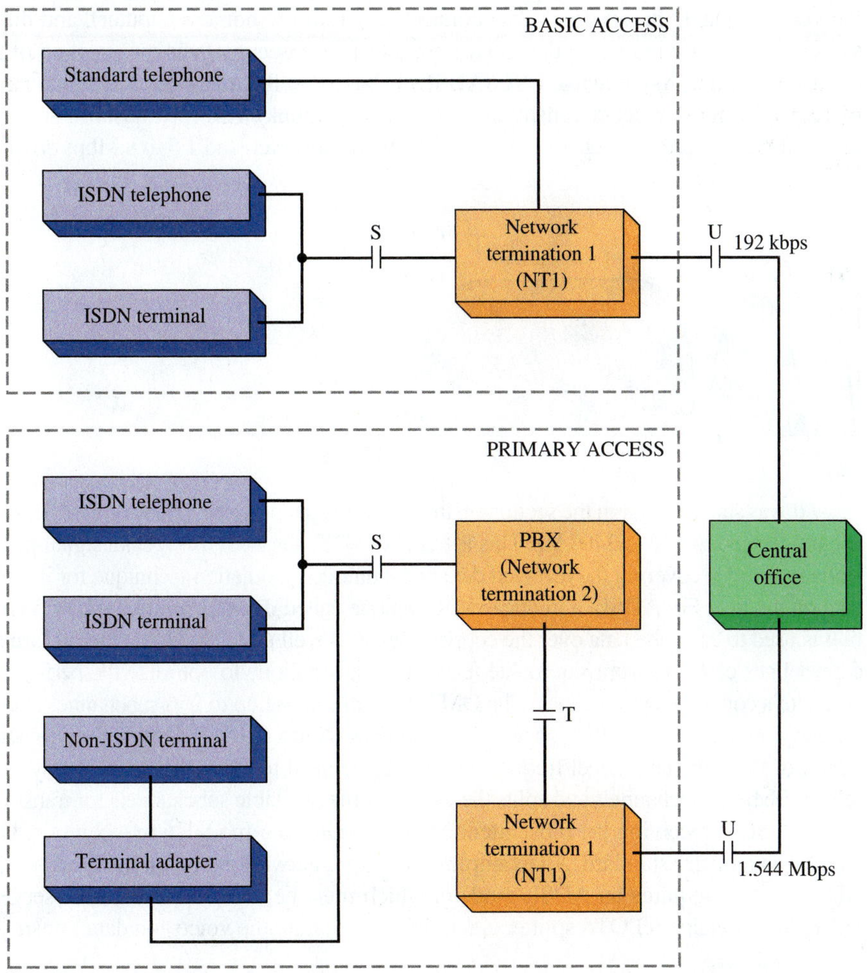

FIGURE 11-27 **Basic and primary access ISDN system.**

DSL is the base technology in the xDSL services. It is somewhat related to the ISDN service; however, the DSL technologies provide a significant increase in bandwidth and DSL is a point-to-point technology. ISDN is a switch technology and can experience traffic congestion at the phone service's central office. The available xDSL services and their projected data rates are provided in Table 11-7.

TABLE 11-7 • xDSL Services and Their Projected Data Rates

TECHNOLOGY	DATA RATE	DISTANCE LIMITATION
ADSL	1.5–8 Mbps downstream	18,000 ft
	Up to 1.544 Mbps upstream	
IDSL	Up to 144 kbps full-duplex	18,000 ft
HDSL	1.544 Mbps full-duplex	12,000 to 15,000 ft
SDSL	1.544 Mbps full-duplex	10,000 ft
VDSL	13–52 Mbps downstream	1,000 to 4,500 ft
	1.5–2.3 Mbps upstream	

Upstream: computer user to the service provider
Downstream: from the service provider back to the computer user
ADSL: Asymmetric digital subscriber line
IDSL: ISDN digital subscriber line
HDSL: High-bit-rate digital subscriber line
SDSL: Single-line digital subscriber line
VDSL: Very high-bit-rate digital subscriber line

Source: "xDSL Local Loop Access Technology, Delivering Broadband over Copper Wires," ©3COM Technical Paper. Reproduced with permission of 3COM Corporation.

DSL services use filtering techniques to enable the transport of data and voice traffic on the same cable. Figure 11-28 shows an example of the ADSL frequency spectrum. Note that the voice channel, the upstream data connection (from the home computer), and the downstream data connection (from the service provider) each occupy their own portion of the frequency spectrum. **Asymmetric DSL (ADSL)** is based on the assumption that the user needs more bandwidth to receive transmissions (downstream link) than for transmission (upstream link). ADSL can provide data rates up to 1.544 Mbps upstream and 1.5 to 8 Mbps downstream.

FIGURE 11-28   The ADSL frequency spectrum.

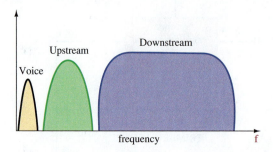

It was stated earlier in the section on interfacing to the network that a copper telephone line is band-limited to 300 to 3400 Hz. This is true, but xDSL services use special signal-processing techniques for recovering the received data and a unique modulation technique for inserting the data on the line. For ADSL, a multicarrier technique called **discrete multitone (DMT) modulation** is used to carry the data over the copper lines. It is well understood that the performance of copper lines can vary from site to site. DMT uses a technique to optimize the performance of each site's copper telephone lines. The DMT modem can use up to 256 subchannel frequencies for carrying the data over the copper telephone lines. A test is initiated at start-up to determine which of the 256 subchannel frequencies should be used to carry the data. The system then selects the best subchannels and splits the data over the available subchannels for transmission.

ADSL is receiving the most attention because its data-modulation technique, DMT, is already an industry standard. An example of an xDSL network is shown in Figure 11-29. The ADSL system requires an ADSL modem, which must be compatible with the service provider. Additionally, a POTS splitter is required to separate the voice and data transmission.

FIGURE 11-29   An xDSL connection to an ISP.

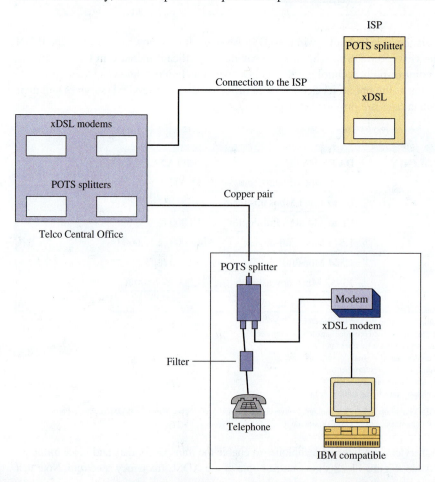

LANs are finding their way into every kind of business at an ever-increasing rate. From small offices to very large government agencies, LANs are becoming an indispensable part of business communications. Computer data and audio and video information are shared on LANs every day. In this section, you will be introduced to a typical LAN configuration and to some common LAN problems. Opportunities abound in LAN technology for those who are willing to prepare with specialized training. LAN seminars, community college classes, and hands-on training will help prepare you for this technology.

After completing this section you should be able to

- Define near-end crosstalk interference
- Describe two common problems when using twisted pair
- Name the two types of modular eight connectors and their proper use

Troubleshooting a LAN

The most common maintenance situation for larger LAN installations is to set up a help desk. LAN users who experience problems call the help desk. Usually a technician is dispatched if the problem can't be resolved over the phone. Let's take a look at some of the problems the technician may encounter when dispatched.

Some preliminary checks should be made first. Using the OSI model to troubleshoot is an excellent method. Start at the physical layer and work your way up. Ensure that the workstation is plugged into electrical power. Is it turned on? Is the CRT brightness turned down? These are obvious things that are easily overlooked and warrant a check. The network interface card should be checked for proper installation in the PC. Check the hub-to-workstation connection. Is the user's account set up properly on the server? Sometimes passwords and accounts get deleted accidentally. Check the user's LAN connection software for proper boot-up. This software can become corrupted and may need to be reinstalled on the workstation.

The following two commands are useful for troubleshooting computer networks. The first is ping, which provides a way to verify the operation of the network. The command structure is as follows:

```
Usage:        ping[-t][-a][-n count][-l size][-f][-i
              TTL][-v TOS]
              [-r count][-s count][[-j host-list]|[-k
              host-list]
              [-w timeout] destination-list
```

Options

```
-t              Ping the specified host until stoppedTo
                see statistics and continue,
                type Control-Break
                To stop, type Control-C
-a              Resolve addresses to host-names
-n count        Number of echo requests to send
-l size         Send buffer size
-f              Set Don't Fragment flag in packet
-i TTL          Time To Live
-v TOS          Type Of Service
-r count        Record route for count hops
-s count        Timestamp for count hops
-j host-list    Loose source route along host-list
-k host-list    Strict source route along host-list
-w timeout      Timeout in milliseconds to wait for each
                reply
```

The following is an example of pinging a network site using its IP address. You can try this if your computer is connected to the internet. The site is a computer called invincible.nmsu.edu and has an IP address of 128.123.24.123. This computer has been set up for outside users to experiment with, so feel free to ping this site.

```
Pinging 128.123.24.123 with 32 bytes of data:

Reply  from  128.123.24.123:  bytes=32  time=3ms  TTL=253
Reply  from  128.123.24.123:  bytes=32  time=2ms  TTL=253
Reply  from  128.123.24.123:  bytes=32  time=2ms  TTL=253
Reply  from  128.123.24.123:  bytes=32  time=3ms  TTL=253

Ping statistics for 128.123.24.123:
packets: sent=4, received=4, lost=0 (0% loss)
Approximate round trip times in milliseconds:
minimum=2 ms, maximum=3 ms, average=2 ms
```

You will get a timed-out message if the site does not respond.

Another command that is useful for troubleshooting networks is tracert, which traces the route of the data through the network

```
Usage:          tracert [-d] [-h maximum_hops][-j host-
                list] [-w timeout]
                target_name
```

Options

```
-d              Do not resolve addresses to hostnames.
-h maximum_     Maximum number of hops to search for
hops            target
-j host-list    Loose source route along host-list
-w timeout      Wait timeout milliseconds for each reply
```

The following is an example of tracing the route to 128.123.24.123.

```
Tracing route to pc-ee205b-8.NMSU.Edu [128.123.24.123]
              over a maximum of 30 hops:
1  1ms  1ms1ms  jett-gate-e2.NMSU.Edu [128.123.83.1]
2  3ms  2ms2ms  r101-2.NMSU.Edu [128.123.101.2]
3  2ms  2ms3ms  pc-ee205b-8.NMSU.Edu [128.123.24.123]
```

The trace is complete. The tracert command provides the path the data traveled.

TROUBLESHOOTING UNSHIELDED TWISTED-PAIR NETWORKS

Twisted-pair networks represent a different challenge to the troubleshooter than coaxial cable. Two common problems that occur with unshielded twisted-pair wiring are that the pairs become crossed or split up. Both conditions produce data signal degeneration. Near-end crosstalk (NEXT) is generated from split pairs. NEXT stems from interference between the twisted pairs. Let's use the example of a signal being transmitted from the workstation to the hub. The signal is smallest (maximum attenuation) at the hub. A transmitted signal originating at the hub—a strong signal—will feed over into the attenuated weak signal. Preventing crossed and split pairs is the best insurance against NEXT. Crossed pairs are not difficult to find but usually require a certain amount of wire tracing and continuity checking. Split pairs are more difficult to find, and special test instruments should be used to track down the splits. A LAN cable meter has several specialized test functions, and the miswired feature is one of them.

Another common problem that happens with unshielded twisted-pair wiring is that the wrong kind of connector is used. Stranded copper conductors need a piercing-type modular eight connector, and solid core conductors use a connector that straddles the wire. Both types of connectors look alike, so they can easily be mixed up and often are. When the wrong connector is used, the result is an open or an intermittent connection. Use care in replacing connectors. Keep these connectors separate in clearly labeled bins.

SOME CABLING TIPS   With a little extra precaution, many LAN problems can be eliminated. Be sure to keep wiring links short without stretching the wire tight. Twisted pair should never be placed near ac power lines or other noise sources. Install your cable base carefully. Wiring runs should be dry. Moisture causes corrosion over a period of time. Use only good-quality connectors. Never untwist more twists than necessary when making twisted-pair connections. The rule of thumb is to untwist a maximum of $\frac{1}{2}$ in. Finally, keep good-quality wiring diagrams of the network installation.

SUMMARY

The focus of this chapter has been on wired computer communication, including communication between computers and peripherals and communication among interconnected groups of computers. The line dividing telephone and computer networks continues to get blurrier, and many techniques introduced in the Chapter 9 discussion of wired telephone systems find application in wired computer networks as well. The most widely used format for representing alphanumeric characters in binary format is the American Standard Code for Information Interchange (ASCII). Other codes used for similar purposes, EBCDIC and Baudot, are largely obsolete but may find use in specialized applications.

Communication between computers and peripheral equipment can take place in either serial or parallel form. Parallel connections are generally used only between computers and closely located peripheral devices such as printers. Perhaps the most widely used format in modern equipment is the universal serial bus (USB), a four-wire, high-speed serial interface that has largely replaced the older RS-232 serial connector. Another high-speed serial connection is the IEEE 1394 FireWire, which is a 6-wire system capable of data rates up to 800 Mbps. Other serial formats, including RS-422 and RS-485, are balanced systems using differential signal techniques to afford performance improvements over longer distances and in high-noise environments.

Wired local-area computer networks are defined in terms of their topology or architecture and the protocol used for accessing the network. Topologies may take the form of a bus, star, or token ring. The two most widely used access methods are carrier sense multiple access (CSMA) and token-passing. The Ethernet protocol is the most widely used for local-area networks (LANs). Ethernet is a format using CSMA with collision detection for access. The Ethernet frame consists of a preamble, start frame delimiter, source and destination media access control (MAC) addresses, and a frame check sequence, as well as bytes indicating the amount and type of data to be transmitted.

Different types of networks are linked together through the open systems interconnection (OSI) model. The OSI model is used to introduce the subject of networking because it is so thorough. The seven-layer OSI model defines everything from the physical network to the software applications interface. The layers, from lowest to highest, are physical, data link, network, transport, session, presentation, and application.

The internet is the worldwide network of networks. Internet protocol (IP) addressing is used to distinguish between networks. IP addresses are assigned by the Internet Assigned Numbers Authority (IANA) based on the class of the network. The largest are Class A networks, followed by Class B, or midsize networks, and Class C (small) networks. Portions of the network address are either assigned by the local system administrator or are dynamically assigned when users need access outside their local networks.

IP addressing is increasingly finding its way to packet-switched telephone networks. Voice-over-IP networks maintain the flexibility of computer networks in environments similar to those of private branch exchange telephone systems. IP telephone networks are configured in a star topology similar to that of computer networks.

Networks can be interfaced in a number of ways. Traditionally, the voice-frequency channels of the public switched telephone network were used in conjunction with various modem technologies to provide data rates up to 56 kbps. These "dial-up" rates have largely been supplanted by higher-rate technologies, including cable and xDSL modems. Several types of digital subscriber line technologies are available. DSL is somewhat related to the older, switch-based ISDN technology but with significantly higher bandwidth and point-to-point connectivity. DSL modems use discrete multitone (DMT) modulation, a technique similar to orthogonal frequency-division multiplexing, with up to 256 subchannel frequencies to carry data over copper telephone lines. DSL modems adapt themselves to the characteristics of the copper wires to optimize the carrier frequencies selected for the highest-speed data transfer.

QUESTIONS AND PROBLEMS

SECTION 11-1

1. What do the abbreviations ASCII and EBCDIC stand for?

2. Provide the ASCII code for 5, a, A, and STX.

3. Provide the EBCDIC code for 5, a, A, and STX.

4. Describe the Gray code.

5. Provide an application of the Gray code.

SECTION 11-2

6. Describe the origin of the RS-232 C standard and discuss its current status as a standard.

7. What are the universal serial bus (USB) wire functions and colors?

8. The USB Type A connector is used for upstream or downstream connection?

9. What is handshaking?

10. What data rates do USB and FireWire support?

11. Describe the difference in USB Type A and B connectors.

SECTION 11-3

12. Provide a general description of a LAN.

13. List the basic topologies available for LANs and explain them.

14. Describe the operation of the Ethernet protocol.

15. Describe the Ethernet frame structure, including a description of media access control (MAC) level addressing.

16. Define the concept of a broadcast address in an Ethernet frame.

SECTION 11-4

17. Discuss the layout, interconnection of the devices, and issues in implementing an office LAN.

18. Discuss the layout, interconnection of the devices, and issues in implementing a building LAN.

19. List the common numerics used for LAN cabling.

SECTION 11-5

20. Describe the differences among local-area networks (LANs), metropolitan area networks (MANs), and wide-area networks (WANs).

21. Provide a brief description of the functions addressed by the OSI reference model.

22. Explain the functions of bridges and routers.

SECTION 11-6

23. Describe the evolution of the internet.

24. Describe the concept of IP addressing.

25. Find four internet sites that provide technical tutorials on cellular communications.

SECTION 11-7

26. Discuss the operation of IP telephony.

27. Discuss the issue of quality of service (QoS).

SECTION 11-8

28. Describe the operation of ADSL (asymmetric digital subscriber line).

29. Discuss the implementation of cable modems for interfacing the networks.

30. Describe how V.92 can achieve such high data rates over the analog phone lines.

31. Explain the objective of ISDN and briefly explain its organization with the help of Figures 11-26 and 11-27.

32. List five xDSL services and indicate their projected data rates.

33. Describe the use of DMT (discrete multitone) operation in ADSL systems.

QUESTIONS FOR CRITICAL THINKING

34. You are at a company where phone lines are being used for signal transmission without delay equalization. Predict the results of unequal delays to the different frequency components of a received signal and justify the need for delay equalization.

35. You hear someone refer to the "handshaking protocol." Is this an accurate use of terms? Why or why not?

36. Discuss the issues of connecting a building network (LAN) to a T1 connection that has been brought into the building. What information do you need to know to complete the job?

CHAPTER 12

TRANSMISSION LINES

CHAPTER OUTLINE

KEY TERMS

waveguides
transmission line
category 6 (CAT6)
category 5e (CAT5e)
RJ-45
attenuation
near-end crosstalk (NEXT)
Crosstalk
attenuation-to-crosstalk ratio (ACR)
delay skew
power-sum NEXT (PSNEXT)
return loss
unbalanced line
balanced line
common mode rejection (CMR)
baluns
characteristic impedance
surge impedance
skin effect

velocity of propagation
delay line
velocity constant
velocity factor
wavelength
nonresonant line
traveling waves
resonant line
reflection
standing wave
voltage standing wave ratio (VSWR)
standing wave ratio (SWR)
flat line
quarter-wavelength matching transformer
electrical length
Smith chart
normalized
single-stub tuner
double-stub tuner
slotted line

12-1 INTRODUCTION

In previous chapters we have been concerned with the generation and reception of communications signals. We now turn our attention to the means by which energy is transferred from a transmitter final amplifier to its antenna as well as the means by which received signals are coupled from antennas to the front-end stages of receivers. Energy may be coupled by means of two-conductor wire lines or by **waveguides,** which are hollow tubes used to conduct electromagnetic energy at microwave frequencies. Both types of energy-carrying media are types of **transmission line,** which may be defined as the conductive connections between system elements that carry signal power. You may be wondering why an entire chapter is devoted to the study of transmission lines if they are really nothing more than pairs of wires. It turns out that at radio frequencies, even simple wire lines and connections start behaving as circuits with inductance and capacitance as well as resistance. What appears at low frequencies to be a short or open circuit may no longer act that way at higher frequencies, particularly at UHF and above. Also, we will see that, at all frequencies, some or all the energy sent down the transmission line will be reflected back toward the source if there is any type of impedance mismatch between system elements. Therefore, determining impedance as well as determining how to resolve impedance-mismatch issues are both crucial concepts in the design of properly working communications systems. The ideas of impedance matching and wave propagation play a central role in transmission line theory and form the central themes of this chapter. A good understanding of how transmission lines interact with other system elements is critical not only for the discussion of antennas to follow in Chapter 14 but also for understanding the importance of the roles that impedance matching and signal integrity concepts play in the effective design of communications systems generally.

12-2 TYPES OF TRANSMISSION LINES

Two-Wire Open Line

One type of parallel line is the two-wire open line illustrated in Figure 12-1. This line consists of two wires that are generally spaced from $\frac{1}{4}$ to 6 in. apart. It is sometimes used as a transmission line between antenna and transmitter or antenna and receiver. An advantage of this type of line is its simple construction. Another type of parallel line is the twin lead or two-wire ribbon type. This line is illustrated in Figure 12-2. This line is essentially the same as the two-wire open line, except that uniform spacing is assured by embedding the two wires in a low-loss dielectric, usually polyethylene. The dielectric space between conductors is partly air and partly polyethylene.

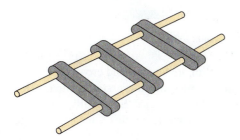

FIGURE 12-1   Parallel two-wire line.

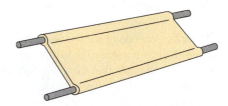

FIGURE 12-2   Two-wire ribbon-type lines.

FIGURE 12-3   **Twisted pair.**

Twisted Pair

The twisted-pair transmission line is illustrated in Figure 12-3. As the name implies, the line consists of two insulated wires twisted to form a flexible line without the use of spacers. It is not used for high frequencies because of the high losses that occur in the rubber insulation. When the line is wet, the losses increase greatly.

Unshielded Twisted Pair (UTP)

Unshielded twisted-pair (UTP) cable plays an important role in computer networking. Local-area networks (LANs) are often wired using unshielded twisted pair. The most common UTP categories used for computer networking are **category 6 (CAT6)** and **category 5e (CAT5e),** both of which are tested to provide data rates up to 1000 Mbps for a maximum length of 100 m. CAT6/5e cable consists of four color-coded pairs of 22- or 24-gauge wires terminated with an **RJ-45** connector. The precise way that the twist of the cables is maintained, even at the terminations, provides a significant increase in signal-transmission performance. CAT5e standards allow 0.5 in. of untwisted conductors at the termination; CAT6 recommends about 3/8 in. untwisted. The balanced operation of the two wires per pair help to maintain the required level of performance in terms of crosstalk and noise rejection.

The need for increased data rates is pushing the technology of twisted-pair cable to even greater performance requirements. The CAT6/5e designation is simply a minimum performance measurement of the cables. The cable must satisfy minimum **attenuation** loss and **near-end crosstalk (NEXT)** for a minimum frequency of 100 MHz. Attenuation loss defines the amount of loss in signal strength as it propagates down the wire. When current travels in a wire, an electromagnetic field is created. This field can induce a voltage in adjacent wires resulting in crosstalk.

Crosstalk is what you occasionally hear on the telephone when you can faintly hear another conversation. NEXT is a measure of the level of crosstalk, or signal coupling, within the cable. The measurement is called near-end testing because the receiver is more likely to pick up the crosstalk from the transmit to the receiver wire pairs at the ends. The transmit-signal levels at each end are strong, and the cable is more susceptible to crosstalk at this point. Additionally, the receive-signal levels have been attenuated by normal cable path loss and are significantly weaker than the transmit signal. A high NEXT (dB) value is desirable. Near-end crosstalk is graphically depicted in Figure 12-4.

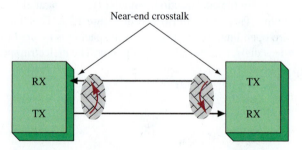

FIGURE 12-4   **A graphical illustration of near-end crosstalk (NEXT).**

Manufacturers combine the two measurements of attenuation and crosstalk on data sheets and list this combined measurement as the **attenuation-to-crosstalk ratio (ACR).** A larger ACR indicates that the cable has a greater data capacity. Essentially, ACR is a figure of merit for twisted-pair cable, where figure of merit implies a measure of the quality of the cable.

Wiring of the RJ-45 connector for RJ-45 CAT6/5e cable is defined by the Telecommunications Industry Association standard TIA568B. Within the TIA568B standard are the wiring guidelines T568A and T568B. These are shown in Table 12-1.

TABLE 12-1 • T568A/T568B Wiring

WIRE COLOR	PAIR	PIN NO. T568A	PIN NO. T568B
White/blue	1	5	5
Blue/white	1	4	4
White/orange	2	3	1
Orange/white	2	6	2
White/green	3	1	3
Green/white	3	2	6
White/brown	4	7	7
Brown/white	4	8	8

Table 12-2 lists the different categories, a description, and bandwidth for twisted-pair cable. Notice that CAT1, CAT2, and CAT4 are not listed. There never were CAT1 and CAT2 cable specifications, although you see them in many textbooks, handbooks, and online sources. The first CAT or category specification was for CAT3. The CAT4 specification was removed from the TIA568B standard because it is an obsolete specification. CAT3 is still listed and is still used to a limited extent in telephone installations; however, modern telephone installations use CAT6/5e.

Enhanced data capabilities of twisted-pair cable include new specifications for testing **delay skew.** It is critical in high-speed data transmission that the data on the wire pair arrive at the other end at the same time. If the wire lengths of different wire pairs are significantly different, then the data on one wire pair will take longer to propagate along the wire, hence arriving at the receiver at a different time and potentially creating distortion of the data. Therefore, delay skew is a measure of the difference in time between the fastest and the slowest wire pair in a UTP cable. Additionally, the enhanced twisted-pair cable must meet four-pair NEXT requirements, which is called **power-sum NEXT (PSNEXT)** testing. Basically, power-sum testing measures the total crosstalk of all cable pairs. This test ensures that the cable can carry data traffic on all four pairs at the same time with minimal interference.

TABLE 12-2 • Different Categories for Twisted-Pair Cable, Based on TIA568B

CATEGORY	DESCRIPTION	BANDWIDTH/DATA RATE
Category 3 (CAT3)	Telephone installations Class C	Up to 16 Mbps
Category 5 (CAT5)	Computer networks Class D	Up to 100 MHz/100 Mbps for a 100-m length
Enhanced CAT5 (CAT5e)	Computer networks	100-MHz/1000 Mbps applications with improved noise performance
Category 6 (CAT6)	Current standard for higher-speed computer networks Class E	Up to 250 MHz/ 1000 Mbps
Category 7 (CAT7)	Proposed standard for higher-speed computer networks Class F	Up to 600 MHz

It has also been indicated that twisted pair can handle gigabit Ethernet networks at 100 m. The gigabit data-rate capability of twisted pair requires the use of all four wire pairs in the cable, with each pair handling 250 Mbps of data. The total bit rate is 4×250 Mbps, or 1 Gbps.

An equally important twisted-pair cable measurement is **return loss.** This provides a measure of the ratio of power transmitted into a cable to the amount of power returned or reflected. The signal reflection results from impedance changes in the cable link and the impedance changes contributing to cable loss. Cables are not perfect, so there will always be some reflection. Examples of the causes for impedance changes are nonuniformity in impedance throughout the cable, the diameter of the copper, cable handling, and dielectric differences. Return-loss tests are now specified for qualifying CAT5/class D, CAT5e, and CAT6/class E twisted-pair links.

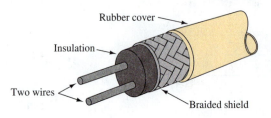

FIGURE 12-5 **Shielded pair.**

Shielded Pair

The shielded pair, shown in Figure 12-5, consists of parallel conductors separated from each other, and surrounded by a solid dielectric. The conductors are contained within a copper braid tubing that acts as a shield. The assembly is covered with a rubber or flexible composition coating to protect the line from moisture or mechanical damage.

The principal advantage of the shielded pair is that the conductors are balanced to ground; that is, the capacitance between the cables is uniform throughout the length of the line. This balance is a result of the grounded shield surrounding the conductors with a uniform spacing along their entire length. The copper braid shield isolates the conductors from external noise pickup. It also prevents the signal on the shielded-pair cable from radiating to and interfering with other systems.

Coaxial Lines

There are two types of coaxial lines: the rigid or air coaxial line, and the flexible or solid coaxial line. The electrical configuration of both types is the same; each contains two concentric conductors.

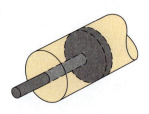

FIGURE 12-6 **Air coaxial: cable with washer insulator.**

The rigid air coaxial line consists of a wire mounted inside, and coaxially with, a tubular outer conductor. This line is shown in Figure 12-6. In some applications the inner conductor is also tubular. Insulating spacers, or beads, at regular intervals, insulate the inner conductor from the outer conductor. The spacers are made of Pyrex, polystyrene, or some other material possessing good insulating characteristics and low loss at high frequencies.

The chief advantage of this type of line is its ability to minimize radiation losses. The electric and magnetic fields in the two-wire parallel line extend into space for relatively great distances, and radiation losses occur. No electric or magnetic fields extend outside the outer (grounded) conductor in a coaxial line. The fields are confined to the space between the two conductors; thus, the coaxial line is a perfectly shielded line. It is important, however, to have the connectors properly installed to avoid leakage. Noise pickup from other lines is also prevented.

This line has several disadvantages: it is expensive to construct, it must be kept dry to prevent excessive leakage between the two conductors, and although high-frequency losses are somewhat less than in previously mentioned lines, they are still excessive enough to limit the practical length of the line.

The condensation of moisture is prevented in some applications by the use of an inert gas, such as nitrogen, helium, or argon, pumped into the line at a pressure of from 3 to 35 psi. The inert gas is used to dry the line when it is first installed, and a pressure is maintained to ensure that no moisture enters the line.

Concentric cables are also made, with the inner conductor consisting of flexible wire insulated from the outer conductor by a solid, continuous insulating material. Flexibility may be gained if the outer conductor is made of braided wire, although this reduces the level of shielding. Early attempts at obtaining flexibility employed the use of rubber insulators between the two conductors. The use of rubber insulators caused excessive losses at high frequencies and allowed moisture-carrying air to enter the line, resulting in high leakage current and arc-over when high voltages were applied. These problems were solved by the development of polyethylene plastic, a solid substance that remains flexible over a wide range of temperatures. A coaxial line with a polyethylene spacer is shown in

Figure 12-7. Polyethylene is unaffected by seawater, gasoline, oils, and liquids. High-frequency losses that result from the use of polyethylene, although greater than the losses would be if air were used, are lower than the losses resulting from the use of most other practical solid dielectric material. Solid flexible coaxial transmission lines are the most frequently used type of transmission line. Teflon is also commonly used as the dielectric in transmission lines.

Balanced/Unbalanced Lines

The amplitude of an electrical signal carried by the center conductor in a coaxial line is measured with respect to the grounded outer conductor. This is called an **unbalanced line.** Operation with the other types (two-wire open, twisted pair, shielded pair) is usually done with what is termed a **balanced line.** In it, the same current flows in each wire but 180° out of phase. Figure 12-8 shows the common technique for converting between unbalanced and balanced signals using a center-tapped transformer.

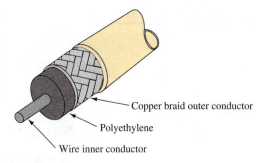

FIGURE 12-7   Flexible coaxial.

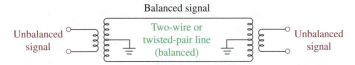

FIGURE 12-8   Balanced/unbalanced conversion.

Any noise or unwanted signal picked up by the balanced line is picked up by both wires. Because these signals are 180° out of phase, they ideally cancel each other at the output center-tapped transformer. This is called **common mode rejection (CMR).** Practical common mode rejection ratio (CMRR) figures are 40–70 dB. In other words, the undesired noise or signal picked up by a two-wire balanced line is attenuated by 40–70 dB.

Circuits that convert between balanced and unbalanced operation are called **baluns.** The center-tapped transformers in Figure 12-8 are baluns. In Section 12-9, a balun using just a section of transmission line will be introduced.

12-3 ELECTRICAL CHARACTERISTICS OF TRANSMISSION LINES

Two-Wire Transmission Line

The end of a two-wire transmission line that is connected to a source is ordinarily called the *generator end* or *input end.* The other end of the line, if connected to a load, is called the *load end* or *receiving end.*

The electrical characteristics of the two-wire transmission line depend primarily on the construction of the line. Because the two-wire line can be viewed as a long capacitor, the change of its capacitive reactance is noticeable as the frequency applied to it is changed. Because the long conductors have a magnetic field about them when electrical energy is being passed through them, the properties of inductance are also observed. The values of the inductance and capacitance present depend on various physical factors, and the effects of the line's associated reactances also depend on the frequency applied. No dielectric is perfect (electrons manage to move from one conductor to the other through the dielectric), so a value for conductance, representing current flow through the dielectric, is associated with each type of two-wire transmission line as well. If the line is uniform (all values equal at each unit length), one small section of the line, perhaps several feet long, may be represented as shown in Figure 12-9.

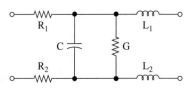

L_1 = inductance of top wire
L_2 = inductance of bottom wire
R_1 = resistance of top wire
R_2 = resistance of bottom wire
G = conductance between wires
C = capacitance between wires

FIGURE 12-9   Equivalent circuit for a two-wire transmission line.

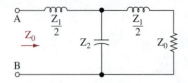

FIGURE 12-10 Simplified circuit terminated with its characteristic impedance.

In many applications, the values of conductance and resistance are insignificant and may be neglected. If they are neglected, the circuit appears as shown in Figure 12-10. Notice that this network is *terminated* with a resistance that represents the impedance of the infinite number of sections exactly like the section of line under consideration. The termination is considered to be a load connected to the line.

Characteristic Impedance

A line infinitely long can be represented by an infinite number of inductors and capacitors. If a voltage is applied to the input terminals of the line, current would begin to flow. Because there are an infinite number of these sections of line, the current would flow indefinitely. If the infinite line were uniform, the impedance of each section would be the same as the impedance offered to the circuit by any other section of line of the same unit length. Therefore, the current would be of some finite value. If the current flowing in the line and the voltage applied across it are known, the impedance of the infinite line could be determined by using Ohm's law. This impedance is called the **characteristic impedance** of the line. The symbol used to represent the characteristic impedance is Z_0. If the characteristic impedance of the line could be measured at any point on the line, it would be found to be the same. The characteristic impedance is sometimes called the **surge impedance.**

In Figure 12-10 the distributed inductance of the line is divided equally into two parts in the horizontal arms of the T. The distributed capacitance is lumped and shown connected in the central leg of the T. The line is terminated in a resistance equal to that of the characteristic impedance of the line as seen from terminals *AB*. The reasons for using this value of resistive termination will be fully explained in Section 12-6. Because the circuit in Figure 12-10 is nothing more than a series–parallel *LCR* circuit, the impedance of the network may be determined.

The impedance, Z_0, looking into terminals *AB* of Figure 12-10 is

$$Z_0 = \frac{Z_1}{2} + \frac{Z_2\left[(Z_1/2) + Z_0\right]}{Z_2 + (Z_1/2) + Z_0}. \tag{12-1}$$

Simplifying yields

$$Z_0 = \frac{Z_1}{2} + \frac{(Z_1 Z_2/2) + Z_0 Z_2}{Z_2 + (Z_1/2) + Z_0}. \tag{12-2}$$

Expressing the right-hand member in terms of the least common denominator, we obtain

$$Z_0 = \frac{Z_1 Z_2 + (Z_1^2/2) + Z_1 Z_0 + (2Z_1 Z_2/2) + 2Z_0 Z_2}{2\left[Z_2 + (Z_1/2) + Z_0\right]}. \tag{12-3}$$

If both sides of this equation are multiplied by the denominator of the right-hand member, the result is

$$2Z_2 Z_0 + \frac{2Z_1 Z_0}{2} + 2Z_0^2 = Z_1 Z_2 + \frac{Z_1^2}{2} + Z_1 Z_0 + \frac{2Z_1 Z_2}{2} + 2Z_0 Z_2. \tag{12-4}$$

Simplifying gives us

$$2Z_0^2 = 2Z_1 Z_2 + \frac{Z_1^2}{2} \tag{12-5}$$

or

$$Z_0^2 = Z_1 Z_2 + \left(\frac{Z_1}{2}\right)^2. \tag{12-6}$$

If the transmission is to be accurately represented by an equivalent network, the T-network section of Figure 12-10 must be replaced by an infinite number of similar sections. Thus, the distributed inductance in the line will be divided into n sections, instead of the number (2) as indicated in the last term of Equation (12-6). As the number of sections approaches infinity, the last term Z_1/n will approach zero. Therefore,

$$Z_0 = \sqrt{Z_1 Z_2}. \qquad (12\text{-}7)$$

Because the term Z_1 represents the inductive reactance and the term Z_2 represents the capacitive reactance,

$$Z_0 = \sqrt{2\pi f L \times \frac{1}{2\pi f C}}$$

and

$$Z_0 = \sqrt{\frac{L}{C}}. \qquad (12\text{-}8)$$

The derivation resulting in Equation (12-8) shows that the characteristic impedance of a line depends *only* on inductance and capacitance, not on length, applied frequency, or anything else.

EXAMPLE 12-1

A commonly used coaxial cable, RG-8A/U, has a capacitance of 29.5 pF/ft and inductance of 73.75 nH/ft. Determine its characteristic impedance for a 1-ft section and for a length of 1 mi.

SOLUTION

For the 1-ft section,

$$Z_0 = \sqrt{\frac{L}{C}}$$

$$= \sqrt{\frac{73.75 \times 10^{-9}}{29.5 \times 10^{-12}}} = \sqrt{2500} = 50\,\Omega. \qquad (12\text{-}8)$$

For the 1-mi section,

$$Z_0 = \sqrt{\frac{5280 \times 73.75 \times 10^{-9}}{5280 \times 29.5 \times 10^{-12}}} = \sqrt{\frac{5280}{5280} \times 2500} = 50\,\Omega.$$

Example 12-1 shows that the line's characteristic impedance is independent of length and is, in fact, a *characteristic* of the line. The value of Z_0 depends on the ratio of the distributed inductance and the capacitance in the line. An increase in the separation of the wires increases the inductance and decreases the capacitance. This effect takes place because the effective inductance is proportional to the flux established between the two wires. If the two wires carrying current in opposite directions are placed farther apart, more magnetic flux is included between them (they cannot cancel their magnetic effects as completely as if the wires were closer together), and the distributed inductance is increased. The capacitance is lowered if the plates of the capacitor (i.e., the two conducting wires) are more widely spaced.

Thus, the effect of increasing the spacing of the two wires is to increase the characteristic impedance because the L/C ratio is increased. Similarly, a reduction in the diameter of the wires also increases the characteristic impedance. The reduction in the size of the wire affects the capacitance more than the inductance because the effect is equivalent to decreasing the size of the plates of a capacitor to decrease the capacitance. Any change in the dielectric material between the two wires also changes the characteristic impedance. If a change in the dielectric material increases the capacitance between the wires, the characteristic impedance, by Equation (12-8), is reduced.

The characteristic impedance of a two-wire line may be obtained from the formula

$$Z_0 \cong \frac{276}{\sqrt{\varepsilon}} \log_{10} \frac{2D}{d}, \tag{12-9}$$

where D = spacing between the wires (center to center)

$\quad d$ = diameter of one of the conductors

$\quad \varepsilon$ = dielectric constant of the insulating material relative to air.

The characteristic impedance of a concentric or coaxial line also varies with L and C. However, because the difference in construction of the two lines causes L and C to vary in a slightly different manner, the following formula must be used to determine the characteristic impedance of the coaxial line:

$$Z_0 \cong \frac{138}{\sqrt{\varepsilon}} \log_{10} \frac{D}{d}, \tag{12-10}$$

where D = inner diameter of the outer conductor

$\quad d$ = outer diameter of the inner conductor

$\quad \varepsilon$ = dielectric constant of the insulating material relative to air.

The relative dielectric constant of air is 1; polyethylene, 2.3; and Teflon, 2.1.

EXAMPLE 12-2

Determine the characteristic impedance of

(a) A parallel wire line with $D/d = 2$ with air dielectric.

(b) An air dielectric coaxial line with $D/d = 2.35$.

(c) RG-8A/U coaxial cable with $D = 0.285$ in. and $d = 0.08$ in. It uses a polyethylene dielectric.

SOLUTION

(a)
$$Z_0 \cong \frac{276}{\sqrt{\varepsilon}} \log_{10} \frac{2D}{d}$$

$$\cong \frac{276}{1} \log_{10} 4$$

$$\cong 166\Omega \tag{12-9}$$

(b)
$$Z_0 \cong \frac{138}{\sqrt{\varepsilon}} \log_{10} \frac{D}{d}$$

$$\cong \frac{138}{1} \log_{10} 2.35$$

$$\cong 51.2\Omega \tag{12-10}$$

(c)
$$Z_0 \cong \frac{138}{\sqrt{2.3}} \log_{10} \frac{0.285}{0.08}$$

$$\cong 50\Omega$$

Coaxial cable must not be stepped on, crimped, or bent in too small a radius. Any of these conditions changes the D/d ratio and therefore changes Z_0. As you will see from the discussion to follow, unintended changes in characteristic impedance cause problems with respect to circuit operation.

Transmission Line Losses

Whenever the electrical characteristics of lines are explained, the lines are often thought of as being loss-free. Although this allows for simple and more readily understood explanations, the losses in practical lines cannot be ignored. Three major losses occur in transmission lines: copper losses, dielectric losses, and radiation or induction losses.

The resistance of any conductor is never zero ohms. When current flows through a transmission line, energy is dissipated in the form of I^2R losses. A reduction in resistance will minimize the power loss in the line. The resistance is indirectly proportional to the cross-sectional area. Keeping the line as short as possible will decrease the resistance and the I^2R loss. The use of a wire with a large cross-sectional area is also desirable; however, this method has its limitations, partly because of increased cost and weight.

At high frequencies the I^2R loss results mainly from the **skin effect.** When a dc current flows through a conductor, the movement of electrons through its cross section is uniform. The situation is somewhat different when ac is applied because the flux density is greater at the center of a conductor than it is at the outer edge. Therefore, the inductance and inductive reactances are greater, causing lower current in the center and more along the outer edge. This effect increases with frequency. Forcing current to the edge effectively reduces the cross-sectional area and, therefore, the area of the conductor through which current can flow. Resistance increases because it is inversely proportional to cross-sectional area. This increase in resistance is called skin effect. At frequencies in the high UHF and microwave regions, skin effect becomes great enough that wires can no longer be used to carry current. At these frequencies, single-conductor "pipes" called waveguides are used to couple or guide the energy from source to load. Waveguides will be covered in Chapter 15.

Dielectric losses are proportional to the voltage across the dielectric. They increase with frequency and, when coupled with skin-effect losses, limit most practical operation to a maximum frequency of about 18 GHz. These losses are lowest when air dielectric lines are used. In many cases the use of a solid dielectric is required, for example, in the flexible coaxial cable, and if losses are to be minimized, an insulation with a low dielectric constant is used. Polyethylene allows the construction of a flexible cable whose dielectric losses, though higher than air, are still much lower than the losses that occur with other types of low-cost dielectrics. Because I^2R losses and dielectric losses are proportional to length, they are usually lumped together and expressed in decibels of loss per meter. The loss versus frequency effects for some common lines are shown in Figure 12-11.

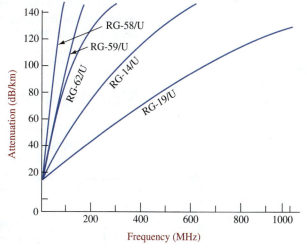

FIGURE 12-11   **Line attenuation characteristics.**

The electrostatic and electromagnetic fields that surround a conductor also cause losses in transmission lines. The action of the electrostatic fields is to charge neighboring objects, while the changing magnetic field induces an electromagnetic force (EMF) in nearby conductors. In either case, energy is lost.

Radiation and induction losses may be greatly reduced by terminating the line with a resistive load equal to the line's characteristic impedance and by properly shielding the line. Proper shielding can be accomplished by the use of coaxial cables with the outer conductor grounded. The problem of radiation loss is of consequence, therefore, only for parallel-wire transmission lines.

12-4 PROPAGATION OF DC VOLTAGE DOWN A LINE

Physical Explanation of Propagation

To understand the characteristics of a transmission line with an ac voltage applied, the infinitely long transmission line will first be analyzed with a dc voltage applied. This will be accomplished using a circuit like the one illustrated in Figure 12-12. In this circuit the resistance of the line is not shown. The line is assumed to be loss-free.

Considering only the capacitor C_1 and the inductor L_1 as a series circuit, when we apply voltage to the network, capacitor C_1 charges through inductor L_1. A characteristic of an inductor is that maximum voltage is developed across it at the first instant in time when voltage is applied; consequently, minimum current passes through it. At the same time, the capacitor has a minimum of voltage across it and is capable of passing maximum current (or, to be precise, capable of having maximum current flow into and out of it). The maximum current is not permitted to flow at the first instant, however, because of the inductor current-blocking action just described, which occurs because the inductor is in the charge path of the capacitor. At this instant the voltage across points c and d is zero. Because the remaining portion of the line is connected to points c and d, 0 V is developed across it at the first instant of time. The voltage across the rest of the line is dependent on the charging action of the capacitor, C_1. Some finite amount of time is required for capacitor C_1 to charge through inductor L_1. As capacitor C_1 is charging, the ammeter records the changing current. When C_1 charges to a voltage that is near the value of the applied voltage, capacitor C_2 begins to charge through inductors L_1 and L_2. The charging of capacitor C_2 again requires time. In fact, the time required for the voltage to reach points e and f from points c and d is the same time as was necessary for the original voltage to reach points c and d. This is true because the line is uniform, and the values of the reactive components are the same throughout its entire length. This action continues in the same manner until all the capacitors in the line are charged. Because the number of capacitors in an infinite line is infinite, the time required to charge the entire line would be an infinite amount of time. It is important to note that current is flowing continuously in the line and that it has some finite value.

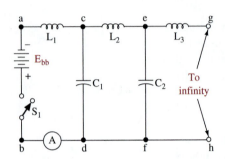

FIGURE 12-12   DC voltage applied to a transmission line.

Velocity of Propagation

When a current is moving down the line, its associated electric and magnetic fields are said to be *propagated* down the line. Time is required to charge each unit section of the line, and if the line were infinitely long, it would require an infinitely long time to charge. The time for a field to be propagated from one point on a line to another may be computed because if the time and the length of the line are known, the **velocity of propagation** may be determined. The network shown in Figure 12-13 is the circuit that will be used to compute the time required for the voltage wave-front to pass a section of line of specified length. The total charge (Q) in coulombs on capacitor C_1 is determined by the relationship

$$Q = Ce \tag{12-11}$$

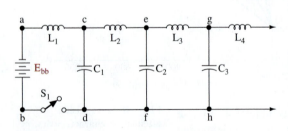

FIGURE 12-13   Circuit for computing time of travel.

Because the charge on the capacitor in the line had its source at the battery, the total amount of charge removed from the battery will be equal to

$$Q = it. \tag{12-12}$$

Because these charges are equal, they may be equated:

$$Ce = it. \tag{12-13}$$

As the capacitor C_1 charges, capacitor C_2 contains a zero charge. Because capacitor C_1's voltage is distributed across C_2 and L_2, at the same time the charge on C_2 is practically zero, the voltage across C_1 (points c and d) must be, by Kirchhoff's law, entirely across L_2. The value of the voltage across the inductor is given by

$$e = L\frac{\Delta i}{\Delta t}. \tag{12-14}$$

Because current and time start at zero, the change in time and the change in current are equal to the final current and the final time. Equation (12-14) becomes

$$et = Li. \tag{12-15}$$

Solving the equation for i, we have

$$i = \frac{et}{L}. \tag{12-16}$$

Solving the equation that was a statement of the equivalency of the charges for current [Equation (12-13)], we obtain

$$i = \frac{Ce}{t}. \tag{12-17}$$

Equating both of these expressions yields

$$\frac{et}{L} = \frac{Ce}{t}. \tag{12-18}$$

Solving the equation for t gives us

$$t^2 = LC$$

or

$$t = \sqrt{LC}. \tag{12-19}$$

Because velocity is a function of both time and distance ($V = d/t$), the formula for computing propagation velocity is

$$V_P = \frac{d}{\sqrt{LC}} \tag{12-20}$$

where V_p = velocity of propagation

d = distance of travel

$\sqrt{LC}$ = time(t).

It should again be noted that the time required for a wave to traverse a transmission line segment depends on the value of L and C and that these values will be different, depending on the type of the transmission line considered.

Delay Line

The velocity of electromagnetic waves through a vacuum is the speed of light, or 3×10^8 m/s. It is just slightly reduced for travel through air. It has just been shown that a transmission line decreases this velocity because of its inductance and capacitance. This property is put to practical use when we want to delay a signal by some specific amount of time. A transmission line used for this purpose is called a **delay line.**

EXAMPLE 12-3

Determine the amount of delay and the velocity of propagation introduced by a 1-ft section of RG-8A/U coaxial cable used as a delay line.

SOLUTION

From Example 12-1, we know that this cable has a capacitance of 29.5 pF/ft and inductance of 73.75 nH/ft. The delay introduced by 1 ft of this line is

$$
\begin{aligned}
t &= \sqrt{LC} \\
&= \sqrt{73.75 \times 10^{-9} \times 29.5 \times 10^{-12}} \\
&= 1.475 \times 10^{-9}\,\text{s} \quad \text{or} \quad 1.457\,\text{ns.}
\end{aligned} \tag{12-19}
$$

The velocity of propagation is

$$
\begin{aligned}
V_p &= \frac{d}{\sqrt{LC}} \\
&= \frac{1\,\text{ft}}{1.475\,\text{ns}} = 6.78 \times 10^8\,\text{ft/s} \quad \text{or} \quad 2.07 \times 10^8\,\text{m/s.}
\end{aligned} \tag{12-20}
$$

Example 12-3 showed that the energy velocity for RG-8A/U cable is roughly two-thirds the velocity of light. This ratio of actual velocity to the velocity in free space is termed the **velocity constant** or **velocity factor** of a line. It can range from about 0.55 up to 0.97, depending on the type of line, the D/d ratio, and the type of dielectric. As an approximation for non-air-dielectric coaxial lines, the velocity factor, v_f, is

$$
v_f \cong \frac{1}{\sqrt{\varepsilon_r}}, \tag{12-21}
$$

where v_f = velocity factor

ε_r = relative dielectric constant

$$
\varepsilon_r(\text{RG} - 8\text{A/U}) \cong 2.3.
$$

EXAMPLE 12-4

Determine the velocity factor for RG-8A/U cable by using the results of Example 12-3 and also by using Equation (12-21).

SOLUTION

From Example 12-3, the velocity was 2.07×10^8 m/s. Therefore,

$$
v_f = \frac{2.07 \times 10^8\,\text{m/s}}{3 \times 10^8\,\text{m/s}} = 0.69.
$$

Using Equation (12-21), we obtain

$$
v_f \cong \frac{1}{\sqrt{\varepsilon_r}} = \frac{1}{\sqrt{2.3}} = 0.66.
$$

Wavelength

A wave that is radiated through "free space" (i.e., a vacuum) travels at the speed of light, or about 186,000 mi/s (3×10^8 m/s). The velocity of this wave is constant regardless of frequency, so that the distance traveled by the wave during a period of one cycle (called one **wavelength**) can be found by the formula

$$
\lambda = \frac{c}{f} \tag{12-22}
$$

where λ (the Greek lowercase letter lambda, used to symbolize wavelength) is the distance in meters from the crest of one wave to the crest of the next, f is the frequency, and c is the velocity of the radio wave in meters per second. Keep in mind that the wave travels more slowly in a wire than it does in free space.

EXAMPLE 12-5

Determine the wavelength (λ) of a 100-MHz signal in free space and while traveling through an RG-8A/U coaxial cable.

SOLUTION

In free space, the wave's velocity (c) is 3×10^8 m/s. Therefore,

$$\lambda = \frac{c}{f}$$
$$= \frac{3 \times 10^8 \text{ m/s}}{1 \times 10^8 \text{Hz}} = 3 \text{ m.} \qquad \textbf{(12-22)}$$

Note: The free-space wavelength is typically labeled λ_0. In RG-8A/U cable we found in Example 12-3 that the velocity of propagation is 2.07×10^8 m/s. Therefore,

$$\lambda = \frac{c}{f}$$
$$= \frac{2.07 \times 10^8 \text{ m/s}}{1 \times 10^8 \text{ Hz}} = 2.07 \text{ m.} \qquad \textbf{(12-22)}$$

Example 12-5 shows that line wavelength is less than free-space wavelength for any given frequency signal.

12-5 NONRESONANT LINE

Traveling DC Waves

A **nonresonant line** is defined as a line of infinite length or as one terminated with a resistive load equal in ohmic value to the characteristic impedance of the line. In a nonresonant line, the load resistance and any inherent resistance in the line absorb all energy transferred or propagated down the line. The voltage and current waves are called **traveling waves** and move in phase with one another from the source to the load.

Because the nonresonant line may either be an infinite-length line or one of finite length but terminated in its characteristic impedance, the physical length of the line is not critical because its behavior does not change. In the resonant line, which will be discussed shortly, the physical length of the line is quite important.

The circuit in Figure 12-14 shows a line terminated with a resistance equal to its characteristic impedance. The charging process and the ultimate development of a voltage across the load resistance will now be described. At the instant switch S_1 is closed, the total applied voltage is felt across inductor L_1. After a very short time has elapsed, capacitor C_1 begins to assume a charge. C_2 cannot charge at this time because all the voltage felt between points c and d is developed across inductor L_2 in the same way the initial voltage was developed across inductor L_1. Capacitor C_2 is unable to charge until the charge on C_1 approaches the amplitude of the supply voltage. When this happens, the voltage charge on capacitor C_2 begins to rise. The voltage across capacitor C_2 will be felt between points e and f. Because the load resistor is also effectively connected between points e and f, the voltage across the resistor is equal to

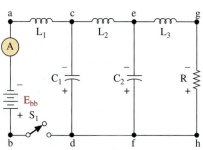

FIGURE 12-14 Charged nonresonant line.

the voltage appearing across C_2. The voltage input has been transferred from the input to the load resistor. While the capacitors were charging, the ammeter recorded a current flow. After all of the capacitors are charged, the ammeter will continue to indicate the load current that will be flowing through the dc resistance of the inductors and the load resistor. The current will continue to flow as long as switch S_1 is closed. When it is opened, the capacitors discharge through the load resistor in much the same way as filter capacitors discharge through a bleeder resistor.

Traveling AC Waves

There is little difference between the charging of the line when an ac voltage is applied to it and when a dc voltage is applied. The charging sequence of the line with an ac voltage applied will now be discussed. Refer to the circuit and waveform diagrams in Figure 12-15.

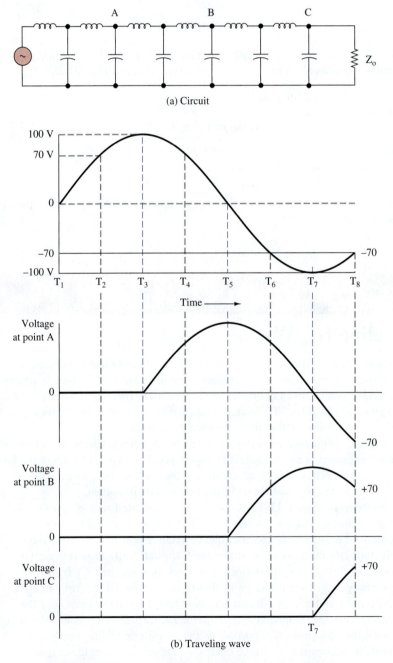

(a) Circuit

(b) Traveling wave

FIGURE 12-15 **Alternating-current charging analysis.**

As the applied voltage begins to go positive, the voltage wave begins traveling down the line. At time t_3, the first small change in voltage arrives at point A, and the voltage at that point starts increasing in the positive direction. At time t_5, the same voltage rise arrives at point B, and at time t_7, the same voltage rise arrives at the end of the line. The waveform has moved down the line as a wavefront. The time required for the voltage changes to move down the line is the same as the time required for the dc voltage to move down the same line. The time for both of these waves to move down the line for a specified length may be computed by using Equation (12-19). Therefore, the following conclusions can be drawn regarding the charging behavior of the nonresonant line:

- All of the instantaneous voltages produced by the generator travel down the line in the order in which they were produced.
- If the voltage waveform is plotted at any point along the line, the resulting waveform will be a duplicate of the generator waveform.
- Because the line is terminated with its characteristic impedance, all of the energy produced by the source is absorbed by the load impedance.

This last conclusion is particularly important. It states that a line terminated in its characteristic impedance will have all its applied energy fully absorbed by the load with none of it being returned to the source. If the load is a resistor, the applied energy will be converted to thermal (heat) energy. If the load is an antenna, the applied energy will be converted to electromagnetic energy for reasons to be discussed in Chapter 13. In any event, all the energy will be converted if the characteristic impedance of the line is matched to that of the load. If the impedances are not matched, then some of the energy will not be converted, and because energy cannot be destroyed, it will remain in the transmission line to be returned to the source. To understand why this is so, we must contrast the behavior of the terminated line just described with one that has not been terminated in its characteristic impedance. Such a line is known as a resonant line.

12-6 RESONANT TRANSMISSION LINE

A **resonant line** is defined as a transmission line that is terminated with an impedance that is *not* equal to its characteristic impedance. Unlike the nonresonant line, the length of the resonant line is critical. In some applications, the resonant line may be terminated in either an open or a short. When this occurs, some very interesting effects may be observed.

DC Applied to an Open-Circuited Line

A transmission line of finite length terminated in an open circuit is illustrated in Figure 12-16. The line characteristic impedance is assumed to be equal to the internal impedance of the source.

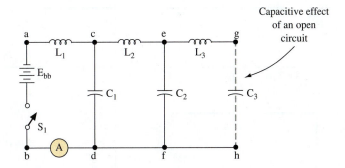

FIGURE 12-16   Open-ended transmission line.

Because the impedances are equal, the applied voltage is divided equally between the source and line. When switch S_1 is closed, current begins to flow as the capacitors begin to charge through the inductors. As each capacitor charges in turn, the voltage will move down the line. As the last capacitor is charged to the same voltage as every other capacitor, there will be no difference in potential between points e and g. This is true because the capacitors possess exactly the same charge. The inductor L_3 is also connected between points e and g. Because there is no difference in potential between points e and g, there can no longer be current flow through the inductor. This means that the magnetic field about the inductor is no longer sustained. The magnetic field must collapse. It is characteristic of the field about an inductor to tend to keep current flowing in the same direction when the magnetic field collapses. This additional current must flow into the capacitive circuit of the open circuit, C_3. Because the energy stored in the magnetic field is equal to that stored in the capacitor, the charge on capacitor C_3 doubles. The voltage on capacitor C_3 is equal to the value of the applied voltage. Because there is no difference of potential between points c and e, the magnetic field about inductor L_2 collapses, forcing the charge on capacitor C_2 to double its value. The field about inductor L_1 also collapses, doubling the voltage on C_1. The combined effect of the collapsing magnetic field about each inductor in turn causes a voltage twice the value of the original apparently to move back toward the source. This voltage movement in the opposite direction caused by the conditions just described is called **reflection.** The voltage reflection was of the same polarity as that of the original charge. This point bears repeating: a transmission line terminated in an open has a reflected voltage wave that is always of the same polarity and amplitude as the incident voltage wave. When this reflected voltage reaches the source, the action stops because of the cancellation of the voltages. The current, however, is reflected back with an opposite polarity because when the field about the inductor collapsed, the current dropped to zero. As each capacitor is charged, causing the reflection, the current flow in the inductor that caused the additional charge drops to zero. When capacitor C_1 is charged, current flow in the circuit stops, and the line is charged. It may also be said that the line now "sees" that the impedance at the receiving end is an open.

Incident and Reflected Waves

A specific example with numbers will help to illustrate the principles just described. The situation for a 100-V battery with a 50-Ω source resistance is illustrated in Figure 12-17. The battery is applied to an open-circuited 50-Ω characteristic impedance line at time $t = 0$.

FIGURE 12-17 Direct current applied to an open-circuited line.

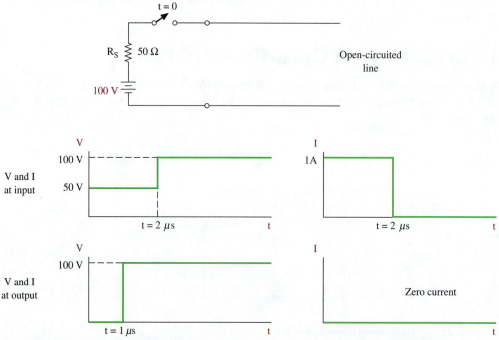

Initially, a 50-V level propagates down the line, and 1 A is drawn from the battery. Notice that 50 V is dropped across R_s at this point in time. The reflected voltage from the open circuit is also 50 V so that the resultant voltage along the line is 50 V + 50 V, or 100 V, once the reflection gets back to the battery. If it takes 1 µs for energy to travel the length of this transmission line, the voltage at the line's input will initially be 50 V until the reflection gets back to the battery at $t = 2$ µs. Remember, it takes 1 µs for the energy to get to the end of the line and then an additional 1 µs for the reflection to get back to the battery. Thus, the voltage versus time condition shown in Figure 12-17 for the line's input is 50 V until $t = 2$ µs, when it changes to 100 V. The voltage at the load is zero until $t = 1$ µs, as shown, when it jumps to 100 V as a result of the incident and reflected voltages being summed.

The current conditions on the line are also shown in Figure 12-17. The incident current is 1 A, which is $100V \div (R_s + Z_0)$. The reflected current for an open-circuited line is out of phase and, therefore, is −1 A with a resultant of 0 A. The current at the line's input is 1 A until $t = 2$ µs, when the reflected current reaches the source to cancel the incident current. After $t = 2$ µs, the current is, therefore, zero. The current at the load is always zero because, when the incident 1 A reaches the load, it is immediately canceled by the reflected −1 A.

Thus, after 1 µs at the load and after 2 µs at the source, the results are predicted by standard analysis; that is, the voltage on an open-circuited wire is equal to the source voltage, and the current is zero. The point to be aware of is that this end result, though preordained and perhaps intuitive by now, does not come about instantaneously. Some small but definite time interval passes before the open-circuit conditions are realized because of the propagation behavior of both the forward (incident) and reflected waves. This idea will be useful to keep in mind when we look at the somewhat more involved situation occurring when ac signals are applied. In that case, the results are not quite as simple to visualize because the incident and reflected signals occur continuously and repetitively. However, the underlying mechanism, as well as the conclusions to be drawn, are largely the same.

DC Applied to a Short-Circuited Line

The condition of applying a 100-V, 50-Ω source resistance battery to a shorted 50-Ω transmission line is illustrated in Figure 12-18. Once again, assume that it takes 1 µs for energy to travel the length of the line. A complete analysis as presented with the open-circuited line would now be repetitive. Instead, only the differences and final results will be provided.

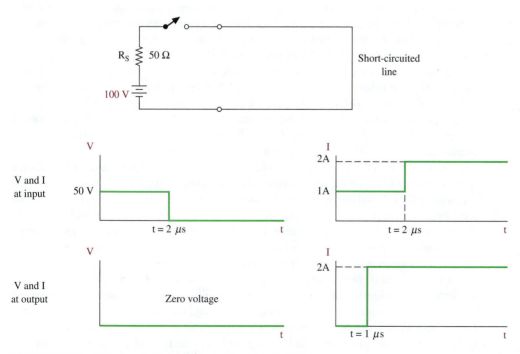

FIGURE 12-18 Direct current applied to a short-circuited line.

When the incident current of 1 A $\lfloor 100\ \text{V} \div (R_s + Z_0) \rfloor$ reaches the short-circuited load, the reflected current is 1 A and in phase so that the load current becomes $1\ \text{A} + 1\ \text{A} = 2\ \text{A}$. The incident voltage ($+50$ V) is reflected back out of phase (-50 V) so that the resultant voltage at the short circuit is zero, as it must be across a short. The essential differences here from the open-circuited line are:

1. The voltage reflection from an open circuit is in phase, while from a short circuit it is out of phase.

2. The current reflection from an open circuit is out of phase, while from a short circuit it is in phase.

The resultant load voltage is therefore always zero, as it must be across a short circuit, and is shown in Figure 12-18. However, the voltage at the line's input is initially $+50$ V until the reflected out-of-phase level (-50 V) gets back in 2 µs, which causes the resultant to be zero. The load current is zero until $t = 1$ µs, when the incident and reflected currents of $+1$ A combine to cause 2 A of current to flow. The current at the line's input is initially 1 A until the reflected current of 1 A arrives at $t = 2$ µs to cause the total current to be 2 A.

Standing Waves: Open Line

The open or shorted transmission lines represent somewhat extraordinary conditions in that they illustrate the "worst case" situations, where the maximum amount of incident-wave energy is reflected back to the source as the result of an impedance mismatch. Reflected waves are, on the whole, highly undesirable. Maximum power transfer occurs when the source and line impedances are equal and when the line and load are matched, that is, when the line is terminated in its characteristic impedance. In the matched condition the load completely absorbs the applied energy and no reflected waves are created. If the line is not terminated in its characteristic impedance, however, reflected waves will be present on the line, and the extent of the reflection as well as the phase relationship between voltage and current on the line will depend on the type and amount of mismatch.

When a mismatch occurs, the incident and reflected waves interact. When the applied signal is ac, this interaction results in the creation of a new kind of wave called a **standing wave,** so-called because it apparently remains in one position, varying only in amplitude. These waves, and the variations in amplitude, are illustrated in Figure 12-19.

The left-hand column in Figure 12-19 represents the in-phase reflection of the voltage wave on an open-circuited transmission line or, alternatively, the current wave on a shorted line. Note the dashed line in the top left-hand part of the diagram that extends past the vertical line labeled "termination." It is an extension of the incident traveling wave, shown in black and represented with the arrowhead facing right. The dashed-line extension is in-phase because it continues in the same direction past the termination as it was traveling when it reached it. By "folding" the dashed line back toward the source from the termination point, we obtain the reflected wave for in-phase reflection. All the diagrams of in-phase reflection in the left-hand column show the incident, reflected, and resultant waves on a line at various instants of time. Keep in mind that these are graphs of wave amplitude versus *position* at each instant of time and *not* wave amplitude versus time, as you are accustomed to seeing in a time-domain representation. Each wave pattern shown in the drawings arranged vertically along the left-hand column represents a snapshot of the incident, reflected, and combined wave patterns at points in time. These instances in time are labeled a through h. The resultant waves are shown with a green line and are simply the vector sum of the incident and reflected waves as they existed at each instant in time.

Note that at positions d_1 and d_3 in Figure 12-19 the resultant voltage (or current) on the line shown in green is always zero. If you stationed yourself at these points and could see the waves (or looked at them with an oscilloscope), you would see that the amplitude of the standing wave (the resultant) would always be zero at positions d_1 and d_3. In other words, no wave would ever be seen once the very first incident wave and then reflected wave had passed by. On the other hand, at position d_2, the resultant oscillates up and down between its positive and negative peaks. The important point to recognize is that the two wave minima are always at points d_1 and d_3 and that the maximum will always appear at position d_2. The resultant is truly a *standing* wave. If readings of the rms voltage and current waves for an open line were taken

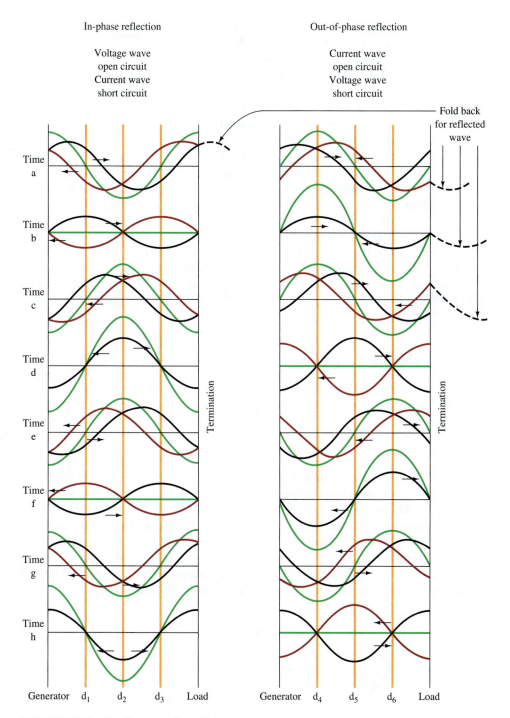

In-phase reflection

Voltage wave
open circuit
Current wave
short circuit

Out-of-phase reflection

Current wave
open circuit
Voltage wave
short circuit

Fold back
for reflected
wave

Time a

Time b

Time c

Time d

Time e

Time f

Time g

Time h

Termination

Termination

Generator d_1 d_2 d_3 Load

Generator d_4 d_5 d_6 Load

FIGURE 12-19   Development of standing waves.

along the line, the result would be as shown in Figure 12-20. This is a conventional picture of standing waves. The voltage at the open is maximum, while the current is zero, as it must be for an open circuit. The positions d_1 through d_6 correspond to those positions shown in Figure 12-19 with respect to summation of the absolute values of the resultants shown in Figure 12-19.

Standing Waves: Shorted Line

The right-hand column of Figure 12-19 shows the out-of-phase reflection that occurs for current on an open line and voltage on a shorted line. The first *graph* shows that the incident wave is extended (in dashed lines) past the load 180° *out of phase* from the incident wave and then folded back to provide the reflected wave. At *b* the reflected wave coincides with the incident wave, providing the maximum possible resultant. At *d* the

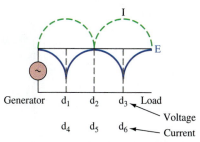

I

E

Generator d_1 d_2 d_3 Load

d_4 d_5 d_6 Voltage

Current

FIGURE 12-20   Conventional picture of standing waves—open line.

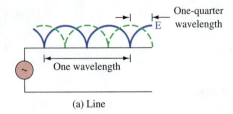

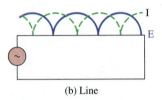

One-quarter
wavelength

(a) Line

(b) Line

FIGURE 12-21   Standing waves of voltage and current.

reflected wave cancels with the incident wave so that the resultant at that instant of time is zero at all points on the line.

At the end of a transmission line terminated in an open, the current is zero and the voltage is maximum. This relationship may be stated in terms of phase. The voltage and current at the end of an open-ended transmission line are 90° out of phase. At the end of a transmission line terminated in a short, the current is maximum and the voltage is zero. The voltage and current are again 90° out of phase.

The current–voltage phase relationships shown in Figure 12-21 are important because they indicate how the line acts at different points throughout its length. They also indicate something else: that impedance changes with position along a mismatched line and that there is a repeating pattern to these impedance changes that can be expressed in wavelength units. Remember that impedance is voltage divided by current. If the ratio of instantaneous voltage and current is constantly varying along the resonant line, then it follows that impedance varies along the line as well. A transmission line has points of maximum and minimum voltage as well as points of maximum and minimum current. The position of these points can be predicted accurately if the applied frequency and type of line termination are known.

EXAMPLE 12-6

An open-circuited line is 1.5 λ long. Sketch the incident, reflected, and resultant voltage and current waves for this line at the instant the generator is at its peak positive value.

SOLUTION

Recall that to obtain the reflected voltage from an open circuit, the incident wave should be continued past the open (in your mind) and then folded back toward the generator. This process is shown in Figure 12-22(a). Notice that the reflected wave coincides with the incident wave, making the resultant wave double the amplitude of the generator voltage. The current wave at an open circuit should be continued 180° out of phase, as shown in Figure 12-22(b), and then folded back to provide the reflected wave. In this case, the incident and reflected current waves cancel one another, leaving a zero resultant all along the line at this instant of time.

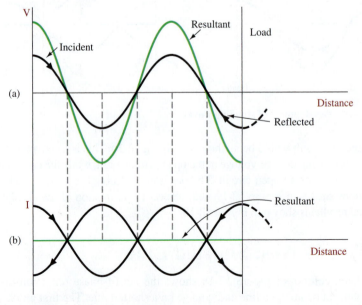

FIGURE 12-22   Diagram for Example 12-6.

Quarter-Wavelength Sections

Figure 12-21 also illustrates another critically important point. Although the voltage and current patterns *repeat* every half wavelength, they *invert* every quarter wavelength; that is, a voltage maximum and current minimum at one point along a transmission line become a voltage minimum and current maximum a quarter-wavelength away. Put yet another way, there is a phase inversion of voltage and current every quarter wavelength. This idea is the underlying principle behind many types of transmission-line-based filters. With reference to Figure 12-21(a), note that at the open-circuited end opposite the source the voltage is maximum (and equal to the source voltage) and the current is zero, as would be expected for an open circuit. However, at a point one-quarter wavelength closer to the generator, the voltage is minimum and the current maximum. A generator placed at that point would "see" a short circuit *at the frequency at which the length of transmission line is one-quarter wavelength.* At higher or lower frequencies, there would be little or no effect. The same relationship holds true for the shorted quarter-wave stub shown in Figure 12-21(b): The voltage-minimum/current-maximum relationship expected for a short circuit and present at the stub end become a voltage-maximum/current-minimum relationship one-quarter wavelength away. In other words, the shorted stub acts like an open circuit to a generator placed one-quarter wavelength away from the short, and again this effect is exhibited at the specific frequency at which the line is one-quarter wavelength long.

A setup such as that of Figure 12-23 shows an open-ended quarter-wave stub acting as a notch filter. With a tee connector the open-ended stub is connected between a tracking generator and the input of a spectrum analyzer. Many spectrum analyzers have built-in tracking generators installed either as standard or optional equipment. The tracking generator outputs an RF signal that sweeps across the frequency range displayed by the spectrum analyzer. In this way, the generator/analyzer combination can produce a frequency-response curve showing filter behavior in real time. Though RF spectrum analyzers with built-in tracking generators are quite expensive, they are almost indispensable for characterizing the behavior of filters in radio systems.

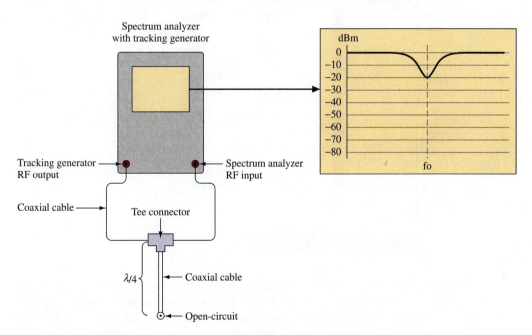

FIGURE 12-23   **Test setup and frequency-domain plot for notch filter.**

The frequency-domain plot shown in Figure 12-23 confirms the notch-filter behavior of the open-ended, quarter-wave stub. At the quarter-wavelength frequency, the output is attenuated by approximately 20 dB, whereas frequencies on either side are attenuated less severely, and those some number of kilohertz away are not affected at all.

Again, the notch effect occurs because the open circuit at the far end of the tee connection appears as a short circuit to the quarter-wavelength frequency at the tee connector itself, thereby notching it out.

A *cavity filter* or *selective cavity* extends the idea of the quarter-wavelength coaxial stub filter to high-power operation. These devices are often called "bottles" in the trade because they are mostly empty. A cutaway view of the cavity filter is shown in Figure 12-24. Its physical dimensions are such that it has a high Q even as it handles powers in the hundreds of watts. It is also contained within an enclosed space so that RF energy does not radiate outside its confines. (This is a problem for the open-ended stub just described; energy radiated from the open-circuited end will cause it to exhibit less-than-ideal behavior.) The cavity interior surface has a smooth metallic coating and forms one conductor, while the adjustable-length center rod forms the other conductor. The center rod is tuned to be one-quarter wavelength at the operating frequency. The conductor pair forms an open-circuited stub at the end opposite the connection point. The end with the open circuit is a point of maximum voltage and minimum current. At the top of the cavity, the phase inversion between current and voltage that is characteristic of quarter-wavelength sections produces maximum current and, with it, a maximum concentration of electromagnetic energy. A coupling loop placed into this "pool" of RF energy effectively acts as a small antenna and couples electromagnetic energy into and out of the cavity. If a single loop is placed at the top of the cavity, the device behaves as a notch filter, much like the transmission-line notch filter just described. The same cavity can be used as a bandpass filter by using a pair of current-sense loops. Because maximum magnetic energy is resident at the top of the cavity, nearest the loops, the loops couple the maximum signal into and out of the cavity at the quarter-wave frequency.

FIGURE 12-24 Cutaway view of cavity filter.

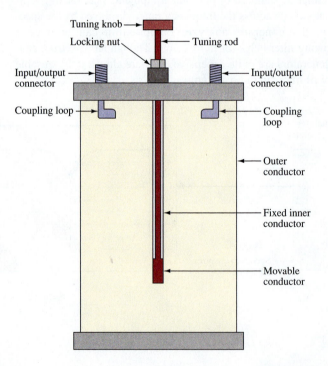

Quarter-wavelength transmission lines, either in the form of cavity filters, or (for low-power operation) as traces on circuit boards, can be used to build *duplexers,* which allow a transmitter and receiver to share a single antenna even when the transmitter is turned on, as well as *combiners,* which allow a single antenna to be used by multiple transmitters operating at the same time. Duplexers can be configured either with bandpass or notch filters, but the idea in either case is that the signal from the transmitter is "steered" to the antenna, bypassing the receiver, and signals received by the antenna are steered to the receiver input, leaving the transmitter uninvolved. A notch-filter duplexer works on the principle that one or (usually) more cavities placed in both the transmit and receive paths appear as open circuits (high impedances) to the undesired frequency. In the receive path, for example, the transmitter frequency is undesired, so a notch-filter cavity tuned to the *transmit* frequency is placed in the receive path. The filter looks like a high impedance (i.e., open circuit) to the transmit frequency, so the transmitter output is routed from the transmitter to the antenna rather than to the receiver. Likewise, a notch cavity in the transmit path is tuned to act as a high impedance to the receive frequency, routing it from the antenna to the receiver rather than to the transmitter. A combiner uses several cavity filters, sometimes of both notch and pass variety, to allow the outputs of multiple transmitters all to be steered to the antenna rather than to each other.

The frequency response of each notch or pass filter is quite broad. For this reason, multiple cavities are placed in series to improve the filtering action. To prevent one filter from loading down the others in series with it, each filter must be series-connected through a quarter-wavelength section of coaxial cable so that the neighboring filters appear as open circuits.

12-7 STANDING WAVE RATIO

A standing wave is created as the result of the instantaneous vector addition of the incident-wave amplitude with that of the reflected wave created as the result of impedance mismatch between the transmission line and the load. Reflections occur whenever there is an impedance mismatch, including one between the source and the line. However, because these conditions can usually be controlled, it is often assumed that the source impedance and the characteristic impedance of the transmission line are equal. Thus, any indication of a mismatch is usually taken to be the result of a discontinuity between line and load. Such a discontinuity can result from a maintenance issue, such as a dirty or loose connector, water in the transmission line, or a damaged or missing antenna. Alternatively, the mismatch can be the result of normal variations in load impedance, perhaps created as the result of changes in operating frequency. For reasons that will become clear in Chapter 14, antennas have impedances and act as loads as they convert electrical energy in the form of voltages and currents into electromagnetic energy in the form of electric and magnetic waves. Ideally, the antenna impedance is closely matched to that of the transmission line, but maintaining an impedance match is often not possible, particularly if the antenna is expected to operate over a wide frequency range. Thus, an indication of system operation and overall health can often be attained by ascertaining the degree of mismatch. The absence of a standing wave is an indication of the matched condition, which, as we have seen, is the condition of maximum power transfer and which is often the most-desired outcome. Expressions of the degree of mismatch can take several forms, all of which can be shown to be related to each other. These forms are the *reflection coefficient,* the *voltage standing-wave ratio,* and the *return loss.* If one of these is known, the others can be determined, but it is important to note that these terms are just different ways of expressing the same thing.

The ratio of reflected voltage to incident voltage is called the *reflection coefficient,* Γ. That is,

$$\Gamma = \frac{E_r}{E_i}, \tag{12-23}$$

where Γ is the reflection coefficient, E_r is the reflected voltage, and E_i is the incident voltage. When a line is terminated with a short circuit, an open circuit, or a purely reactive load, no energy can be absorbed by the load, so total reflection takes place. The reflected wave equals the incident wave so that $|\Gamma| = 1$. When a line is terminated with a resistance equal to the line's Z_0, no reflection occurs, and therefore $\Gamma = 0$. For all other cases of termination, the load absorbs some of the incident energy (but not all), and the reflection coefficient is between 0 and 1.

The reflection coefficient may also be expressed in terms of the load impedance:

$$\Gamma = \frac{Z_L - Z_0}{Z_L + Z_0}, \tag{12-24}$$

where Z_L is the load impedance. Expressing the impedance match in terms of the reflection coefficient is particularly useful in applications involving Smith chart calculations, which will be shown in the next section.

Another way of looking at the degree of mismatch, and the method that is often most useful from a maintenance point of view, is to determine the voltage standing wave ratio. On a lossless line, the standing-wave amplitude maxima and minima are constant. As the mismatch becomes more severe, the standing wave has a greater amplitude, which means that its maximum voltage, E_{max}, is higher and its minimum voltage, E_{min}, is lower. The ratio of the maximum voltage to minimum of the standing wave on a line is called the **voltage standing wave ratio (VSWR)**:

$$\text{VSWR} = \frac{E_{max}}{E_{min}}. \qquad (12\text{-}25)$$

With reference to Figure 12-25, note that the standing-wave amplitudes get larger as the VSWR gets larger, which is the result of a larger mismatch. Thus, VSWR becomes a direct proxy for the degree of mismatch in a system. This notion is particularly important because VSWR is a relatively easy parameter to measure. Simple instruments (essentially voltmeters) permanently placed in-line can be used to measure VSWR continuously, and malfunctions can be detected when the VSWR is seen to increase beyond some predetermined threshold. In a large system (such as a cellular communications system), operational parameters like VSWR are continuously relayed to a central monitoring location where technicians and automated equipment continuously assess overall system health. An abnormal VSWR at any cell site, for example, would trigger an alarm resulting in a maintenance technician being dispatched to determine the cause.

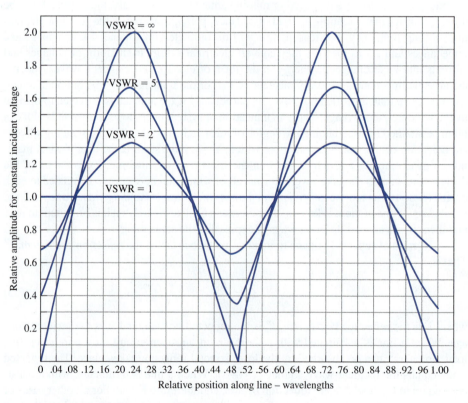

FIGURE 12-25   Amplitudes of standing waves.

In a more general sense, the VSWR is sometimes referred to as simply the **standing wave ratio (SWR)** because it is also equal to the ratio of maximum current to minimum current. However, there is a distinction between the two terms. In general, the VSWR is taken to be the voltage ratio directly, as determined by Equation (12-25). The SWR is the VSWR expressed in decibels. Therefore,

$$\text{SWR(dB)} = 20 \log \text{VSWR},\tag{12-26}$$

where VSWR is the ratio of Equation (12-25).

The VSWR can also be expressed in terms of Γ:

$$\text{VSWR} = \frac{1 + |\Gamma|}{1 - |\Gamma|}.\tag{12-27}$$

The VSWR is infinite when total reflection occurs because E_{min} is 0 and is equal to 1 when no reflection occurs. No reflection means no standing wave, so the rms voltage along the line is always the same (neglecting losses); thus E_{max} equals E_{min} and the VSWR is 1.

Yet another way of representing the degree of mismatch is by expressing it in terms of return loss. Return loss is the ratio of transmitted or incident power to reflected power:

$$RL = \frac{P_{in}}{P_{ref}},\tag{12-28}$$

where P_{in} is the incident power and P_{ref} is the reflected power. Like the relationship between VSWR and SWR, return loss can be expressed in decibel form as well. Because this is a power relationship,

$$RL_{(dB)} = 10 \log RL.\tag{12-29}$$

The previous expression shows that the least reflection occurs with the largest return loss. Therefore, a perfect match has infinite return loss and a reflection coefficient of zero. The poorest match equals 0 dB return loss and a reflection coefficient of 1. The notion of return loss is useful because some test equipment, particularly that used to determine antenna VSWR over a given frequency range, will produce results expressed in terms of return loss in decibel units. These examples are a special case of *network analysis,* which will be covered in a following section.

Effect of Mismatch

The perfect condition of no reflection occurs only when the load is purely resistive and equal to Z_0. Such a condition is called a **flat line** and indicates a VSWR of 1. It is highly desirable because all the generator power capability is getting to the load. Also, if reflection occurs, the voltage maximums along the line may exceed the cable's dielectric strength, causing a breakdown. In addition, the existence of a reflected wave means greater I^2R (power) losses, which can be a severe problem when the line is physically long. The energy contained in the reflected wave is absorbed by the generator (except for the I^2R losses) and not "lost" unless the generator is not perfectly matched to the line, in which case a re-reflection of the reflected wave occurs at the generator. In addition, a high VSWR also tends to accentuate noise problems and causes "ghosts" to be transmitted with video or data signals.

To summarize, then, the disadvantages of not having a perfectly matched (flat line) system are as follows:

1. The full generator power does not reach the load.
2. The cable dielectric may break down as a result of high-value standing waves of voltage (voltage nodes).
3. The existence of reflections (and re-reflections) increases the power loss in the form of I^2R heating, especially at the high-value standing waves of current (current nodes).
4. Noise problems are increased by mismatches.
5. "Ghost" signals can be created.

The VSWR can be determined quite easily, if the load is a known value of pure resistance, by the following equation:

$$\text{VSWR} = \frac{Z_0}{R_L} \quad \text{or} \quad \frac{R_L}{Z_0} \, (\text{whichever is larger}) \qquad \text{(12-30)}$$

where R_L is the load resistance. Whenever R_L is larger than Z_0, R_L is used in the numerator so that the VSWR will be greater than 1. It is not possible to have a VSWR of less than 1. For instance, on a 100-Ω line ($Z_0 = 100 \, \Omega$), a 200-Ω or 50-Ω R_L results in the same VSWR of 2. They both create the same degree of mismatch.

One other point should be made at this juncture. The VSWR is sometimes expressed in true ratio form, such as 2:1 for the example given in the previous paragraph. However, often the "to one" part (the part after the colon) is dropped because it is generally understood that VSWRs are related to the ideal condition of 1:1, representing a perfect match. For this reason, the VSWR is frequently expressed as a single number, in which case a VSWR expressed as 1 is equal to 1:1, a perfect match, and all other single-number expressions expressed similarly.

The higher the VSWR, the greater is the mismatch on the line. Thus, a low VSWR is the goal in a transmission line system except when the line is being used to simulate a capacitance, inductance, or tuned circuit. These effects are considered in the following sections.

EXAMPLE 12-7

A citizen's band transmitter operating at 27 MHz with 4-W output is connected via 10 m of RG-8A/U cable to an antenna that has an input resistance of 300 Ω. Determine

(a) The reflection coefficient.

(b) The electrical length of the cable in wavelengths (λ).

(c) The VSWR.

(d) The amount of the transmitter's 4-W output absorbed by the antenna.

(e) How to increase the power absorbed by the antenna.

SOLUTION

(a)
$$\Gamma = \frac{Z_L - Z_0}{Z_L + Z_0}$$

$$= \frac{300\Omega - 50\Omega}{300\Omega + 50\Omega} = \frac{5}{7} = 0.71 \qquad \text{(12-24)}$$

(b)
$$\lambda = \frac{v}{f}$$

$$= \frac{2.07 \times 10^8 \, \text{m/s}}{27 \times 10^6 \, \text{Hz}} = 7.67 \, \text{m} \qquad \text{(12-22)}$$

Because the cable is 10 m long, its electrical length is

$$\frac{10 \, \text{m}}{7.67 \, \text{m/wavelength}} = 1.3\lambda.$$

(c) Because the load is resistive,

$$\text{VSWR} = \frac{R_L}{Z_0}$$

$$= \frac{300\Omega}{50\Omega} = 6 \qquad \text{(12-30)}$$

An alternative solution, because Γ is known, is

$$\text{VSWR} = \frac{1 + \Gamma}{1 - \Gamma}$$

$$= \frac{1 + \dfrac{5}{7}}{1 - \dfrac{5}{7}} = \frac{\dfrac{12}{7}}{\dfrac{2}{7}} = 6. \qquad \textbf{(12-27)}$$

(d) The reflected voltage is Γ times the incident voltage. Because power is proportional to the square of voltage, the reflected power is $(5/7)^2 \times 4\ \text{W} = 2.04\ \text{W}$, and the power to the load is

$$P_{\text{load}} = 4\ \text{W} - P_{\text{refl}}$$

$$= 4\text{W} - 2.04\text{W} = 1.96\text{W}.$$

(e) You do not have enough theory yet to answer this question. Continue reading this chapter and read Chapter 14 on antennas.

Example 12-7 showed the effect of mismatch between line and antenna. The transmitted power of only 1.96 W instead of 4 W would seriously impair the effective range of the transmitter. It is little wonder that great pains are taken to get the VSWR as close to 1 as possible.

Quarter-Wavelength Transformer

One simple way to match a line to a resistive load is by use of a **quarter-wavelength matching transformer.** It is not physically a transformer but does offer the property of impedance transformation, as transformers do. To match a resistive load, R_L, to a line with characteristic impedance Z_0, a $\lambda/4$ section of line with characteristic impedance Z_0' should be placed between them. In equation form, the value of Z_0' is

$$Z_0' = \sqrt{Z_0 R_L}. \qquad \textbf{(12-31)}$$

The required $\lambda/4$ section to match the 50-Ω line to the 300-Ω resistive load of Example 12-7 would have $Z_0' = \sqrt{50\ \Omega \times 300\Omega} = 122\Omega$. This solution is shown pictorially in Figure 12-26. The input impedance looking into the matching section is 50 Ω. This is the termination on the 50-Ω cable, and thus the line is flat, as desired, from there on back to the generator. Remember that $\lambda/4$ matching sections are only effective working into a resistive load. The principle involved is that there will now be two reflected signals, equal but separated by $\lambda/4$. Because one of them travels $\lambda/2$ farther than the other, this 180° phase difference causes cancellation of the reflections.

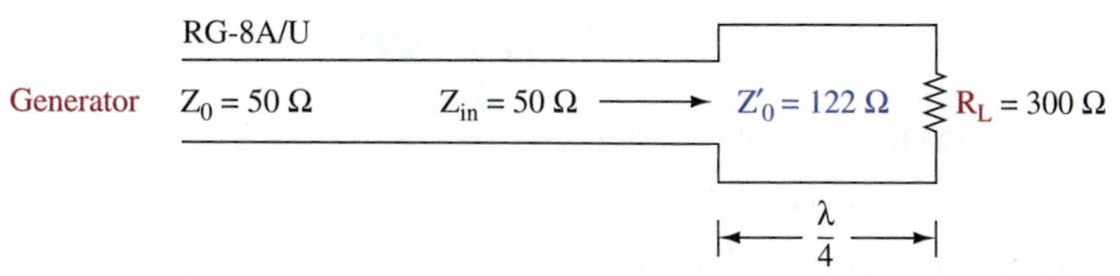

FIGURE 12-26 $\lambda/4$ **matching section for Example 12-7.**

Electrical Length

The **electrical length** of a transmission line is of importance to this overall discussion. Recall that when reflections occur, the voltage maximums occur at $\lambda/2$ intervals. If the transmission line is only $\lambda/16$ long, for example, the reflection still occurs, but the line is so short that almost no voltage variation along the line exists. This situation is illustrated in Figure 12-27.

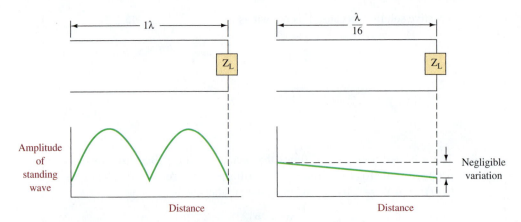

FIGURE 12-27   **Effect of line electrical length.**

Notice that the λ/16 length line has almost no "standing" voltage variations along the line because of its short electrical length, while the line that is one λ long has two complete variations over its length. Because of this, the transmission line effects we have been discussing are applicable only to electrically long lines—generally only those greater than λ/16 long.

Be careful that you understand the difference between electrical length and physical length. A line can be miles long physically and still be electrically short at low frequencies. For example, a telephone line carrying a 300-Hz signal has a λ of 621 mi. On the other hand, a 10-GHz signal has a λ of 3 cm, so even extremely small circuit interconnections behave as transmission lines and require application of transmission line theory.

12-8 THE SMITH CHART

Transmission Line Impedance

Transmission line calculations are very cumbersome from a purely mathematical standpoint. It is often necessary to know the input impedance of a line of a certain length with a given load. Referring back to Figure 12-21, we see that the impedance at any point on a line with standing waves is repetitive every half wavelength because the voltage and current waves are similarly repetitive. The impedance is, therefore, constantly changing along the line and is equal to the ratio of voltage to current at any given point. This impedance can be solved with the following equation for a lossless line:

$$Z_s = Z_0 \frac{Z_L + jZ_0\tan\beta s}{Z_0 + jZ_L\tan\beta s},$$
(12-32)

where Z_s = line impedance at a given point

Z_L = load impedance

Z_0 = line's characteristic impedance

βs = distance from the load to the point where it is desired to know the line impedance (electrical degrees).

Because the tangent function is repetitive every 180°, the result will similarly be repetitive, as we know it should be. If the line is $\frac{3}{4}\lambda$ long (270 electrical degrees), its impedance is the same as it would be at a point λ/4 from the load.

Another point can be made from the ideas discussed up to now. Section 12-7 established that with a resonant line (one not terminated in its characteristic impedance), the voltages and currents present on the line are out of phase and that the phase shift is continuous. Another way of looking at this result is to consider that the impedance along the line is *complex* in the mathematical sense of the term. That is to say, the impedance consists both of a *magnitude,* represented by the ratio of voltage to current at any point on the line, as well as a *phase shift* between the two quantities, which also varies along the line. The idea of complex impedance, consisting of both resistance and reactance, was reviewed in the discussion of the complex exponential in Section 8-5. The idea of complex impedance also pertains to the loads. We will see in Chapter 14 that an antenna can be modeled as a series circuit with resistive, inductive, and capacitive properties. In other words, the load can have the properties of complex impedance, and obtaining a complete picture of how complex loads interact with their sources requires us to have a tool that allows for the visual display of both the resistive (real) and reactive (imaginary) components of impedance. The Smith chart presents an elegant solution to the impedance-matching problems posed by complex sources and complex loads.

Smith Chart Introduction

Equation (12-32) can be solved, but the solution using the **Smith chart** is far more convenient and versatile. The Smith chart is essentially an impedance-matching tool for transmission lines. This impedance chart was developed by P. H. Smith in 1938 and is still widely used for line, antenna, and waveguide calculations in spite of the widespread availability of computers and programmable calculators.

Figure 12-28 illustrates the Smith chart. It contains two sets of lines. The lines representing constant resistance are circular and are all tangent to each other at the right-hand

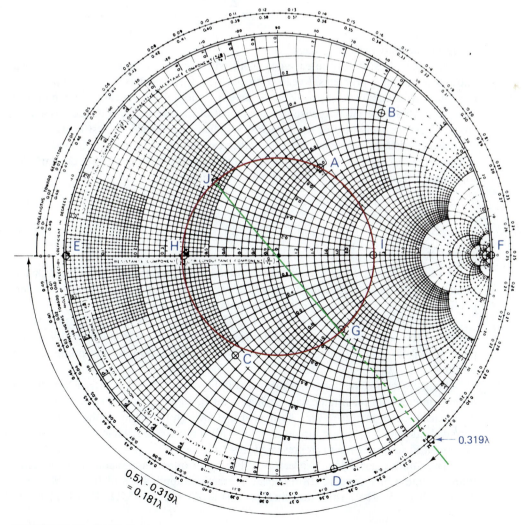

FIGURE 12-28   Smith chart.

end of the horizontal line through the center of the chart. The value of resistance along any one of these circles is constant, and its value is indicated just above the horizontal line through the center of the chart.

The second set of lines represents arcs of constant reactance. These arcs are also tangent to one another at the right-hand side of the chart. The values of reactance for each arc are labeled at the circumference of the chart and are positive on the top half and negative on the bottom half.

Several computer-aided design (CAD) programs are available that can accurately predict high-frequency circuit performance. It may therefore seem strange that the pencil-and-paper Smith chart is still used. The intuitive and graphical nature of Smith chart design is still desirable, however, because the effect of each design step is clearly apparent. The graphical Smith chart is used to represent network analyzer solutions, for example, as described in Section 12-10.

Using the Smith Chart

Impedance values in rectangular form (i.e., $Z = R \pm jX$; see Section 8-5) are usually **normalized** so that a single Smith chart can be used for a wide range of impedance and admittance values. Normalizing involves dividing all impedances by the characteristic impedance, Z_0, of the line. Therefore, if the impedance $100 + j50\Omega$ were to be plotted for a 50-Ω line, the normalized value would be calculated as $100/50 + j50/50 = 2 + j1$. The normalized impedance is then represented by a lowercase letter,

$$z = \frac{Z}{Z_0}, \tag{12-33}$$

where z is the normalized impedance.

In some cases, it will be easier to work with reciprocal quantities when solving impedance-matching problems. Recall from electrical fundamentals that the reciprocal of resistance, R, is conductance, G, which is an expression of the ease with which charge flows through a material. Likewise, the reciprocal of impedance, Z, is defined as admittance, Y, and the reciprocal of reactance, X, is called susceptance, B. Just as impedance consists of resistance and reactance, admittance can be shown to consist of conductance and susceptance. The Smith chart is ideally set up to work with any of these quantities as well as their reciprocals. When working with admittances,

$$y = \frac{Y}{Y_0}, \tag{12-34}$$

where y is the normalized admittance and Y_0 is the line's characteristic admittance. On the Smith chart, the location for admittance will always be one-half revolution away (diametrically opposite or 180°) from the plotted location of impedance.

The point A in Figure 12-28 is $z = 1 + j1$ or could be $y = 1 + j1$. It is the intersection of the one resistance circle and one reactive arc. Point B is $0.5 + j1.9$, while point C is $0.45 - j0.55$. Be sure that you now understand how to plot points on the Smith chart.

Point D (toward the bottom of the chart) in Figure 12-28 is $0 - j1.3$. All points on the circumference of the Smith chart, such as point D, represent pure reactance except for the points at either extreme of the horizontal line through the center of the chart. At the left-hand side, point E, a short circuit ($z = 0 + j0$) is represented, while at the right-hand side, point F, an open circuit is represented ($z = \infty$).

The Smith chart finds its greatest utility as an impedance-matching calculator. Points G and J in Figure 12-28 illustrate a practical application of the Smith chart. Assume that a 50-Ω transmission line has a load $Z_L = 65 - j55\Omega$. That load, when normalized, is

$$z_L = \frac{Z_L}{Z_0} = \frac{65}{50} - j\frac{55}{50} = 1.3 - j1.1.$$

Point G is plotted as $z_L = 1.3 - j1.1$. By drawing a circle through point G and using the chart center as the circle's center, the locations of all impedances along the transmission line are described by the circle. Recall that the impedance along a line varies but repeats every half wavelength. A full revolution around the circle corresponds to a half-wavelength (180°) movement on a line. A clockwise (CW) rotation on the chart means you are moving toward the generator, while a counterclockwise (CCW) rotation indicates movement toward the load. The scales on the outer circumference of the chart indicate the amount of movement in wavelengths. For example, moving from the load (point G) toward the generator to point H brings us to a point on the line where the impedance is purely resistive and equal to $z = 0.4$. As shown in Figure 12-28, that movement is equal to $0.5\lambda - 0.319\lambda = 0.181\lambda$. In other words, at a point 0.181λ from the load, the impedance on the line is purely resistive and has a normalized value of $z = 0.4$. In terms of actual impedance, the impedance is $Z = z \times Z_0 = 0.4 \times 50\Omega = 20\Omega$.

Moving another $\lambda/4$ or halfway around the circle from point H in Figure 12-28 brings you to another point of pure resistance, point I. At point I on the line, the impedance is $z = 2.6$. That also is the VSWR on the line. Wherever the circle drawn through z_L for a transmission line crosses the right-hand horizontal line through the chart center, that point is the VSWR that exists on the line. For this reason, the circle drawn through a line's load impedance is often called its VSWR circle.

The Smith chart allows for a very simple conversion of impedance to admittance, and vice versa. The value of admittance corresponding to any impedance is always diagonally opposite and the same distance from the center. For example, if an impedance is equal to $1.3 - j1.1$ (point G), then the admittance y is at point J and is equal to $0.45 + j0.38$ as read from the chart. Because $y = 1/z$, you can mathematically show that

$$\frac{1}{1.3 - j1.1} = 0.45 + j0.38$$

but the Smith chart solution is much less tedious. Recall that mathematical z to y and y to z transformations require first a rectangular to polar conversion, taking the inverse, and then a polar to rectangular conversion.

Other applications of the Smith chart are most easily explained by solution of actual examples.

EXAMPLE 12-8

Find the input impedance and VSWR of a transmission line 4.3λ long when $Z_0 = 100\ \Omega$ and $Z_L = 200 - j150\ \Omega$.

SOLUTION

First normalize the load impedance:

$$z_L = \frac{Z_L}{Z_0}$$
$$= \frac{200 - j150\Omega}{100\Omega} = 2 - j1.5. \tag{12-33}$$

That point is then plotted on the Smith chart in Figure 12-29, and its corresponding VSWR circle is also shown. That circle intersects the right-hand half of the horizontal line at 3.3 as shown, and therefore the VSWR = 3.3. Now, to find the line's input impedance, it is required to move from the load toward the generator (CW rotation) 4.3λ. Because each full revolution on the chart represents $\lambda/2$, it means that eight full rotations plus 0.3λ is necessary. Thus, just moving 0.3λ from the load in a CW direction will provide z_{in}. The radius extended through z_L

intersects the outer wavelength scale at 0.292λ. Moving from there to 0.5λ provides a movement of 0.208λ. Thus, additional movement of 0.092λ (0.3λ − 0.208λ) brings us to the generator and provides z_{in} for a 4.3λ line (or 0.3λ, 0.8λ, 1.3λ, 1.8λ, etc., length line). The line's z_{in} is read as

$$z_{in} = 0.4 + j0.57$$

from the chart, which in ohms is

$$Z_{in} = z_{in} \times Z_0 = (0.4 + j0.57) \times 100\ \Omega = 40\ \Omega + j57\ \Omega.$$

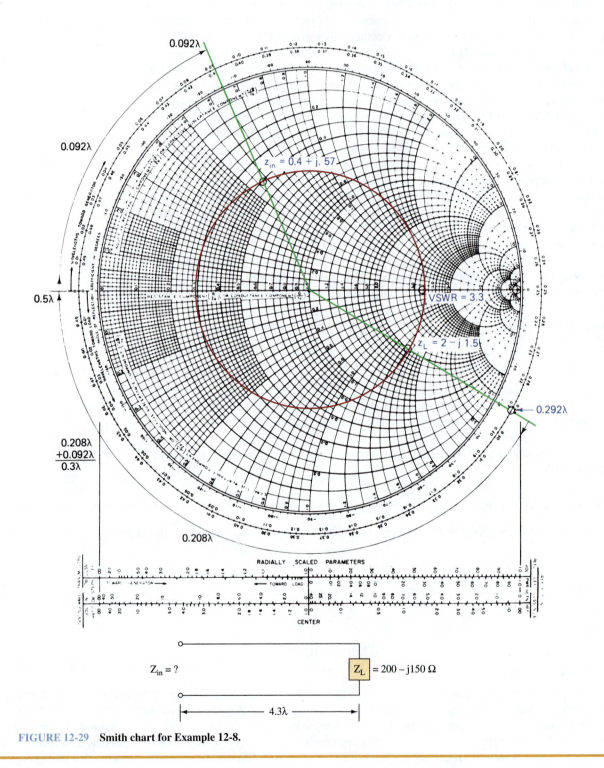

FIGURE 12-29 **Smith chart for Example 12-8.**

Corrections for Transmission Loss

Example 12-8 assumed the ideal condition of a lossless line. If line attenuation cannot be neglected, the incident wave gets weaker as it travels toward the load, and the reflected wave gets weaker as it travels back toward the generator. This causes the VSWR to get weaker as we approach the source end of the line. A true standing wave representation on the Smith chart would be a spiral. The correction for this condition is made using the "transm. loss, 1-dB steps" scale at the bottom left-hand side of the Smith chart. Other scales at the bottom of the chart can be used to provide the VSWR (which can be taken directly off the chart, as previously shown), the voltage and power reflection coefficient, and decibel loss information.

Matching Using the Smith Chart

Many of the calculations involved with transmission lines pertain to matching a load to the line and thus keeping the VSWR as low as possible. The following examples illustrate some of these situations.

EXAMPLE 12-9

A load of 75 Ω + $j50$ Ω is to be matched to a 50-Ω transmission line using a $\lambda/4$ matching section. Determine the proper location and characteristic impedance of the matching section.

SOLUTION

1. Normalize the load impedance:

$$z_L = \frac{Z_L}{Z_0} = \frac{75\ \Omega\ +\ j50\Omega}{50\Omega} = 1.5 + j1.$$

2. Plot z_L on the Smith chart and draw the VSWR circle. This is shown in Figure 12-30.

3. From z_L move toward the generator (CW) until a point is reached where the line is purely resistive. Recall that the $\lambda/4$ matching section works only between a pure resistance and the line.

4. Point A or B in Figure 12-30 could be used as the point where the matching section is inserted. We shall select point A because it is closest to the load. It is 0.058λ from the load. At that point the line should be cut and the $\lambda/4$ section inserted as shown in Figure 12-30.

5. The normalized impedance at point A is purely resistive and equal to $z = 2.4$. That also is the VSWR on the line with no matching. The actual resistance is 2.4 × 50 Ω = 120 Ω. Therefore, the characteristic impedance of the matching section is

$$Z_0' = \sqrt{Z_0 \times R_L}$$

$$= \sqrt{50\Omega \times 120\Omega} = 77.5\Omega. \qquad (12\text{-}29)$$

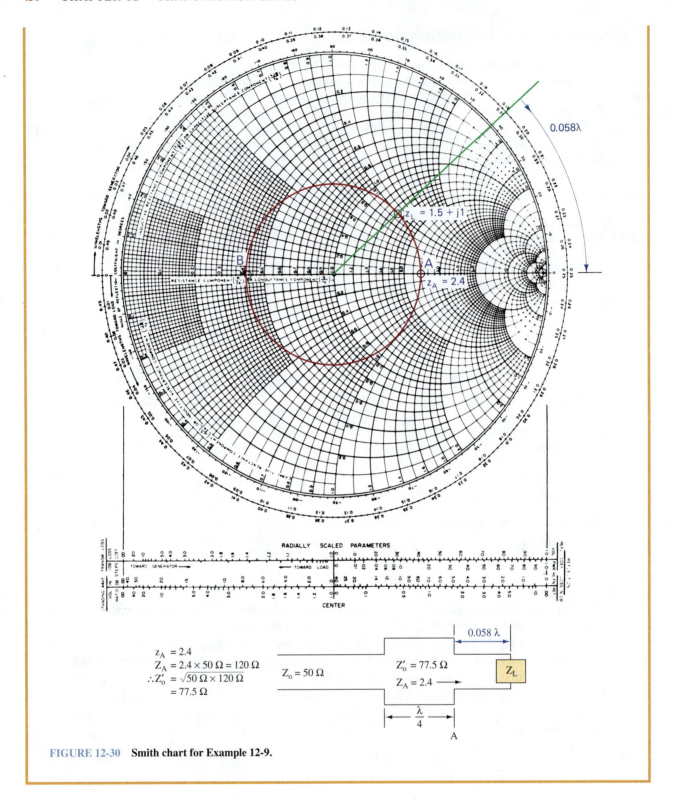

FIGURE 12-30 **Smith chart for Example 12-9.**

$z_A = 2.4$
$Z_A = 2.4 \times 50\ \Omega = 120\ \Omega$
$\therefore Z'_o = \sqrt{50\ \Omega \times 120\ \Omega}$
$\qquad = 77.5\ \Omega$

Stub Tuners

The use of short-circuited stubs is prevalent in matching problems. A **single-stub tuner** is illustrated in Figure 12-31(a). In it, the stub's distance from the load and location of its short circuit are adjustable. The **double-stub tuner** shown in Figure 12-31(b) has fixed stub locations, but the position of both short circuits is adjustable. The following example illustrates the procedure for matching with a single-stub tuner.

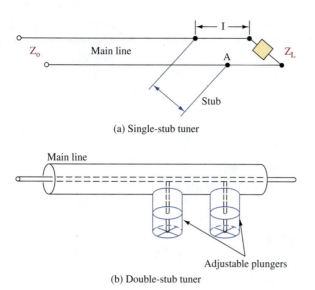

(a) Single-stub tuner

(b) Double-stub tuner

FIGURE 12-31   Stub tuners.

EXAMPLE 12-10

The antenna load on a 75-Ω line has an impedance of $50 - j100\Omega$. Determine the length and position of a short-circuited stub necessary to provide a match.

SOLUTION

1.
$$z_L = \frac{Z_L}{Z_o} = \frac{50 - j100\Omega}{75\Omega} = 0.67 - j1.33$$

2. Plot z_L on the Smith chart (Figure 12-29), and draw the VSWR circle.

   The following steps in this procedure require a bit of explanation. The operation of the single- or double-stub tuner is predicated on the idea that the removal (cancellation) of unwanted reactances can be achieved by placing an equivalent reactance of the opposite type in series with it. In other words, a series LC circuit with both capacitive and inductive reactances of, say, 100 Ω will have a net impedance of zero ohms because the reactances are 180° out of phase with each other and will, therefore, cancel. The stub operates on the principle that its presence in the system creates a reactance that is equal to but opposite in type to that of the reactance to be cancelled. However, for practical reasons, it is often difficult if not impossible to place reactances in series because doing so may require breaking into the installation to install the component. Also, physical components (inductors, capacitors) with the needed values may not be available because the values are very small, particularly at microwave frequencies. A more straightforward solution is to install *parallel* elements exhibiting known *susceptances* because these can often be included without modifying the underlying installation. The stub is effectively acting as a parallel-connected element exhibiting the susceptance needed to cancel out the undesired reactance. As will be shown in the next section, transmission lines can act as pure inductances or capacitances, depending on length and frequency. Since susceptance is the reciprocal of reactance, a parallel-connected transmission-line section exhibiting the correct susceptance will cancel out the undesired reactance in the same way that a series-connected reactance would. This Smith chart problem allows us to determine the two characteristics of the transmission line element needed to accomplish this goal: the length it needs to be in order to produce the desired susceptance, and the distance from the load it needs to be to cancel out the reflections produced by the original mismatch.

   To determine both quantities, we need to convert the series-circuit impedance or reactance elements to equivalent parallel-circuit admittances or susceptances. The Smith chart makes these determinations quite straightforward because the reciprocal quantities are always 180° removed from their counterparts. For this method to work, however, all quantities—including the impedance of a short circuit—must be converted to their reciprocal equivalents. The process is shown in the following steps.

3. Convert z_L to y_L by going to the side of the VSWR circle that is diagonally opposite to the location of z_L. Read $y_L = 0.27 + j0.59$.

4. Move from y_L to the point where the admittance is $1 \pm$ whatever j term results. That is point A in Figure 12-32. Read point A as $y_A = 1 + j1.75$ and note that point A is 0.093λ from the load. The short-circuited stub should be located 0.093λ from the load. The purpose of this step is to identify the pure reactance portion of the impedance mismatch. This is the quantity that will be "tuned out" by the stub in the next step.

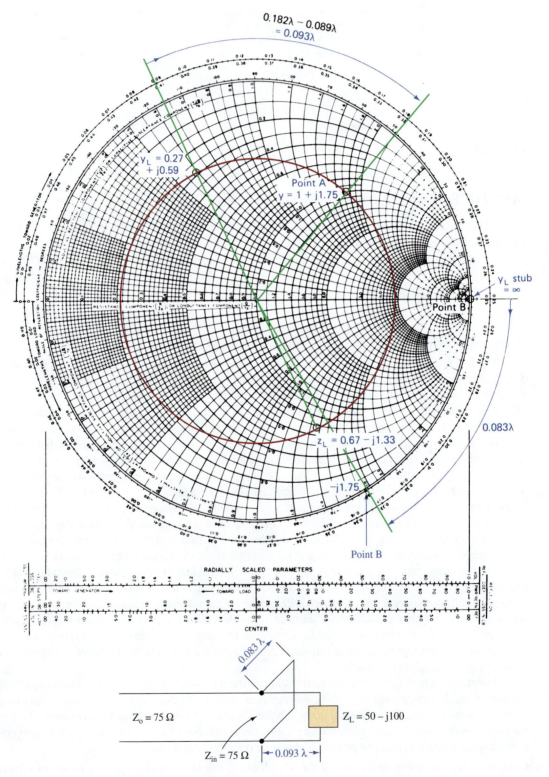

FIGURE 12-32   Smith chart for Example 12-10.

5. Now the $+ j1.75$ term at point A must be canceled by the stub admittance. If the stub admittance is $y_s = - j1.75$, the imaginary terms cancel because the total admittance at point A with the parallel stub is $(1 + j1.75 - j1.75) = 1$. Recall that parallel admittances are directly additive.

6. The load admittance of the short-circuited stub is infinite. That is plotted on the Smith chart as point B. From point B, move toward the generator until the stub admittance is $-j1.75$. That corresponds to a distance of 0.083λ and is the required length for the short-circuited stub.

7. A match is now accomplished because the total admittance at the stub location is 1. This means that

$$z = \frac{1}{y} = \frac{1}{1} = 1$$

or $Z = 1 \times 75\Omega = 75\Omega$, which matches the line to the point where the stub is connected back toward the generator.

The key to matching with the single-stub tuner of the previous example is moving back from the load until the admittance takes on a normalized real term of 1. Whatever j term (imaginary) results can then be canceled with a short-circuited stub made to look like the same j term with opposite polarity. While the Smith chart may at first be perplexing to work with, a bit of practice allows mastery in a short period of time.

12-9 TRANSMISSION LINE APPLICATIONS

Discrete Circuit Simulation

Transmission line sections can be used to simulate inductance, capacitance, and LC resonance. Particularly at microwave frequencies, the inductances and capacitances required are simply too small to be added to circuits as physical components. Rather, these quantities are created with transmission-line elements (sometimes fabricated on circuit boards) of more or less than a quarter wavelength at the frequency range of interest. Figure 12-33 tabulates the effects of transmission-line length for open and shorted sections with lengths equal to, less than, and greater than a quarter wavelength.

A shorted quarter-wavelength section looks like a parallel LC circuit, which behaves as an open circuit at its resonant frequency. A shorted section less than $\lambda/4$ looks like a pure inductance, and a section greater than $\lambda/4$ looks like a pure capacitance. These effects can be verified on a Smith chart by plotting z_L of 0 for a short-circuited line at the left center of the chart. Moving CW toward the generator less than $\lambda/4$ puts you on the top half of the chart, indicating a $+ j$ term and thus inductance. Moving exactly $\lambda/4$ puts you at the center-right point on the chart, indicating the infinite impedance of an ideal tank circuit, and moving past that point provides the $-j$ impedance of capacitance.

While open-circuit sections would seem to provide similar effects, they are seldom used. The open-circuited line tends to radiate a fair amount of energy off the end of the line so that total reflection does not take place. This causes the simulated circuit element to take on a resistive term, which greatly reduces its quality. These losses do not occur with shorted sections, and the simulated circuit has better quality than is possible using discrete inductors and/or capacitors. Short-circuited $\lambda/4$ sections offer Qs of about 10,000 as compared to a maximum of about 1000 using a very high-grade inductor and capacitor.

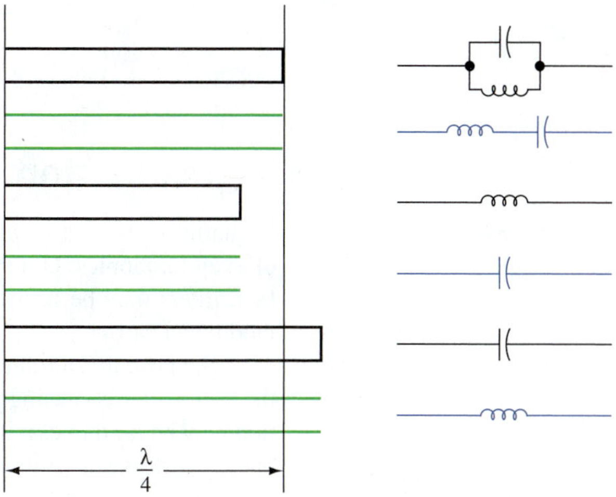

FIGURE 12-33   **Transmission line section equivalency.**

For this reason, it is normal practice to use transmission line sections to replace inductors and/or capacitors at frequencies above 500 MHz where the line section becomes short enough to be practically used. They are commonly found in the oscillator of UHF tuners for television; the tuners operate from about 500 to 800 MHz.

Baluns

As described in Section 12-2, parallel wire line generally carries two equal but 180° out-of-phase signals with respect to ground. Such a line is called a *balanced* line. In a coaxial line, the outer conductor is usually grounded so that the two conductors do not carry signals with the same relationship to ground. The coaxial line is therefore termed an *unbalanced* line.

It is sometimes necessary to change from an unbalanced to a balanced condition such as when a coaxial line feeds a balanced load like a dipole antenna. They cannot be connected directly together because the antenna lead connected to the cable's grounded conductor would allow the shield to become part of the antenna.

This situation is solved through use of an unbalanced-to-balanced transformer, which is usually called a *balun*. It can be a transformer as shown in Figure 12-34(a) and described in Section 12-2 or a special transmission line configuration as shown in Figure 12-34(b). The use of the standard transformer is limited because of excessive losses at high frequencies. The balun in Figure 12-34(b) does not suffer from that disadvantage. The inner conductor of the coaxial line is tapped at 180° ($\lambda/2$) from the end. The tap and the end of the inner conductor provide two equal but 180° out-of-phase signals, neither of which is grounded. Thus, they supply the required signals for the balanced line. These baluns are reversible because they function equally well in going from a balanced to an unbalanced condition.

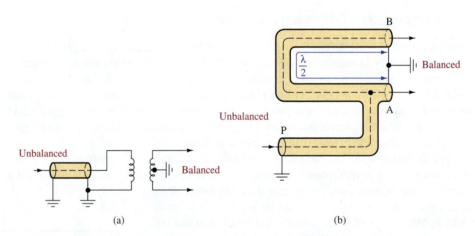

FIGURE 12-34 Baluns.

Transmission Lines as Filters

A quarter-wave section of transmission line can be used as an efficient filter or suppressor of even harmonics. Other types of filters may be used to filter out the odd harmonics. In fact, filters may be designed to eliminate the radiation of an entire single sideband of a modulated carrier.

Suppose that a transmitter is operating on a frequency of 5 MHz, and it is found that the transmitter is causing interference at 10 and 20 MHz. A shorted quarter-wave line section may be used to eliminate these undesirable harmonics. A quarter-wave line shorted at one end offers a high impedance at the unshorted end to the fundamental frequency. At a frequency twice the fundamental, such a line is a half-wave line, and at a frequency four times the fundamental, the line becomes a full-wave line. A half-wave or full-wave line offers zero impedance when its output is terminated in a short. Therefore, the radiation of even harmonics from the transmitter antenna can be eliminated almost completely by the circuit shown in Figure 12-35.

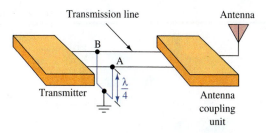

The resonant filter line, *AB*, is a quarter wave in length at 5 MHz and offers almost infinite impedance at this frequency. At the second harmonic, 10 MHz, the line *AB* is a half-wave line and offers zero impedance, thereby shorting this frequency to ground. The quarter-wave filter may be inserted anywhere along the nonresonant transmission line with a similar effect.

FIGURE 12-35 **Quarter-wave filters.**

Slotted Lines

One of the simplest and yet most useful measuring instruments at very high frequencies is the **slotted line.** As its name implies, it is a section of coaxial line with a lengthwise slot cut in the outer conductor. A pickup probe is inserted into the slot, and the magnitude of signal picked up is proportional to the voltage between the conductors at the point of insertion. The probe rides in a carriage along a calibrated scale so that data can be obtained to plot the standing wave pattern as a function of distance. From these data, the following information can be determined:

1. VSWR
2. Generator frequency
3. Unknown load impedance

Time-Domain Reflectometry

Time-domain reflectometry (TDR) is a system whereby a short-duration pulse is transmitted into a line. If monitored with an oscilloscope, the reflection of that pulse provides much information regarding the line. Of course, no reflection at all indicates an infinitely long line or (more likely) a line with a perfectly matched load and no discontinuities.

A very common problem with transmission line communications involves cable failure between communications terminals. These problems are usually a result of chemical erosion or a mechanical break. A time-domain reflectometer can be used to locate these problems. These instruments can pinpoint a fault within several feet at a distance of 10 mi by measuring the time a pulse takes to travel down a cable and return to the source and then calculating distance based on the known propagation velocity for the cable under test.

TDR is also often used in laboratory testing. In these cases a fast-rise-time, step signal is transmitted. The setup for this condition is shown in Figure 12-36(a). In Figure 12-36(b) the case for an open-circuited line is shown. Assuming a lossless line, the reflected wave is in phase and effectively results in a doubling of the incident step voltage. The time T shown is the time of incidence plus reflection and, therefore, should be divided by 2 to calculate the distance to the open circuit. Thus, the distance is propagation velocity (V_p) times $T/2$.

The case for a shorted line is shown in Figure 12-36(c). In this case the reflection is out of phase and results in cancellation. The situation of Z_L greater than Z_0 is shown in Figure 12-36(d), while Figure 12-36(e) shows the case for Z_L less than Z_0. The magnitudes shown in the scope display can be used to determine Z_L or the reflection coefficient Γ because

$$\Gamma = \frac{E_r}{E_i} \qquad (12\text{-}33)$$

and

$$\Gamma = \frac{Z_L - Z_0}{Z_L + Z_0}, \qquad (12\text{-}24)$$

and the final voltage, E_F, is

$$E_F = E_i + E_r$$
$$= E_i(1 + \Gamma)$$
$$= E_i\left(1 + \frac{Z_L - Z_0}{Z_L + Z_0}\right).$$

Figures 12-36(f) through (i) displays the TDR signal for complex impedances.

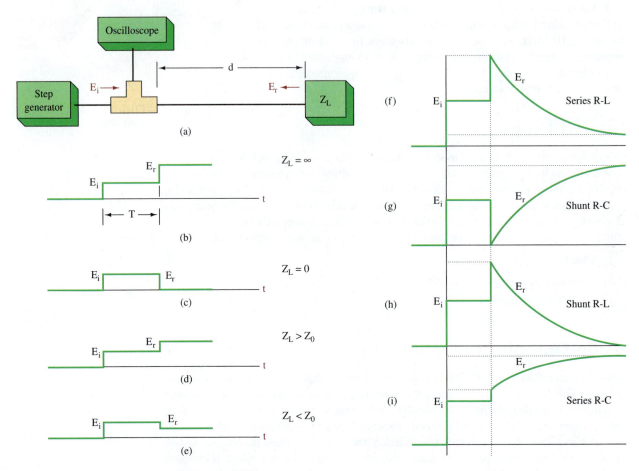

FIGURE 12-36   **Time-domain reflectometry (TDR).**

12-10 IMPEDANCE MATCHING AND NETWORK ANALYSIS

The transmission line theory described in this chapter highlights the critical role that impedance matching plays in communications systems. We have seen that maximum power transfer occurs when source and load impedances are matched and that any degree of impedance mismatch gives rise to signal reflections that may be undesired or that could lead to equipment damage in high-power systems. The Smith chart is a very useful tool for summarizing the complex impedance characteristics of transmission lines as well as for determining how to eliminate mismatched conditions. The effects analyzed in this chapter can also manifest themselves in devices or locations other than physical transmission lines. Particularly at microwave frequencies with their very short wavelengths (centimeters or millimeters), connectors, component leads, or circuit-board traces have physical dimensions that are at least a significant fraction of a wavelength, if not multiple wavelengths. Thus, the devices themselves take on the characteristics of transmission lines, and their behavior as such must be taken into account as part of the system design. Individual components or subsystems (such as amplifier modules designed to operate at microwave frequencies) will display complex impedances that often vary with frequency. Finally, designers of high-speed systems such as computers must concern themselves with signal-integrity issues that involve, among other considerations, the transmission-line-like behavior of interconnections between integrated circuits on motherboards as well as between computers and peripheral devices. With bus speeds now routinely running into the hundreds of megahertz, computers and other digital systems exhibit many of the same behaviors displayed by those operating at microwave frequencies. For all these reasons, it is often necessary to measure and analyze the complex impedance characteristics of components, circuits, and systems. These considerations fall within the purview of network analysis.

A *network analyzer* is a piece of test equipment used to characterize the linear impedance characteristics of devices under test (DUTs). DUTs can be circuits such as amplifiers or filters, loads such as antennas, or individual microwave components such as transistors or diodes. We saw earlier in this chapter that optimum performance is obtained when the impedances of all devices including loads are well matched to the impedances of all other devices in the system, including sources and transmission lines, and that empirical evidence of a good match is presented in the form of a high return loss, a low VSWR, and/or a reflection coefficient close to zero. Network analyzers allow users to characterize circuit behavior completely by measuring impedance-related characteristics at the input and output ports of the DUTs only. Essentially, the DUTs are treated as "black boxes" because, once the relevant parameters have been measured and calculations made, device or circuit behavior in any external environment can be predicted without regard to the internal configuration of the DUT.

A scalar network analyzer measures only the magnitude of DUT impedance, while the more sophisticated vector network analyzer (VNA) determines both the impedance magnitude and phase characteristics—that is, complex impedances—of DUTs. Both types are designed to work over wide frequency ranges. (The term "network" in this context is a holdover from the early days of electrical engineering, when circuits of all types were referred to as networks. This usage is distinct from the modern sense, where "networks" are perhaps most often thought of as interconnected systems of computers or communications equipment.) In very basic terms, the network analyzer is a sophisticated impedance ratio meter consisting of a swept radio-frequency source, a system of three receivers, a "test set" responsible for automated switching of signals and for separating the forward and reflected components of applied signals, a purpose-built computer for determining complex impedances and other parameters of interest, and a visual display for reporting results in graphical or Smith-chart form. Essentially, two receivers—the measurement receivers—produce voltages at their outputs that are directly related to the magnitudes of the signals applied to or reflected from the input and output DUT ports to which the receivers are connected. The third receiver serves as a reference by applying a known, calibrated signal to the two measurement receivers. The total voltages produced by the measurement receivers are therefore related to the ratios of measured to reference voltages at each DUT port, and these voltage ratios can be used to determine impedance as well as other characteristics such as return loss and reflection coefficient.

Microwave engineers often use so-called "scattering" or S-parameters to describe the linear transmission and reflection characteristics of signals applied to DUTs. If we first consider a network with one input port and one output port, we can develop a parameter set consisting of four variables, where two independent variables represent the excitation of the network from a source (the "stimulus"), and where two dependent variables represent the response of the network to the excitation. The behavior of such a network is analogous to the behavior of a flashlight beam shined through a window: though most of the light energy passes through the glass, a small fraction of the total incident light energy is reflected, visible evidence of which is the image of the illuminated flashlight in the window. The amount of light passing through the window can be quantified in terms of a "transmission coefficient"—the ratio of transmitted to reflected energy—whereas the much smaller reflection visible in the window is related to the reflection coefficient, Γ, familiar to us in the context of transmission line systems as the ratio of reflected to incident voltages. From the stimulus-response characteristics of electronic-circuit DUTs, which behave similarly to the window in the flashlight analogy, we can determine the transmission and reflection coefficients of our applied signals and can derive other important quantities such as VSWR and return loss from them.

Although a number of parameter sets—based on quantities such as admittance, conductance, and resistance—can be defined and shown to be related to each other, all except for S parameters suffer from a number of practical drawbacks. For example, the input and output to the device under test may need to be opened and then short circuited, which can be difficult to do, particularly at radio frequencies where lead inductances and capacitances make shorts and opens difficult to obtain and where unwanted oscillations are possible, thus invalidating measurements. S parameters, on the other hand, are determined with all ports resistively terminated. In a practical setup, the DUT is often placed in a transmission line between a 50-Ω source and a resistive load, thus making accurate measurements possible because oscillations are unlikely to occur. Also, because incident and reflected traveling waves do not vary in magnitude at points along a lossless transmission line, S parameters can be measured on DUTs located some distance from the measurement transducers (i.e., the source and reflection measurement points) as long as low loss transmission lines are used and as long as residual transmission line effects can somehow be "normalized out" or eliminated from subsequent calculations.

The parameter set created as a result of these transmitted and reflected signals can be arranged into a "scattering matrix," which describes all voltages incident to and reflected from all ports. In a manner similar to what happens when a full rack of billiard balls is hit with the cue ball and the balls in the rack travel in widely scattered directions, the scattering matrix models the behavior of signals applied to the DUT: signals incident to any network port similarly scatter and propagate through the network to be output from or reflected by any other port. In S parameter notation, incident and reflected signals are shown in subscript form, where the first subscript is the exit port and the second subscript is the incident port. Therefore, a network consisting of ports numbered 1 and 2 will have reflection coefficients denoted by S_{11} or S_{22} (that is, a signal incident to port 1 or 2, respectively, is also reflected back to the port from which it was incident); similarly, transmission coefficients are represented as S_{21} or S_{12}, depending on the incident and exit ports involved. In short, S parameters allow for practical measurements of quantities that would otherwise be difficult if not impossible to perform in systems where the DUT is physically quite large in comparison with the wavelengths of the signals applied to it and where measurements involving open- or short-circuit conditions are impractical.

Impedance analysis is not confined to microwave systems. Even nonelectronic DUTs exhibit many of the complex impedance properties of electronic systems. For example, in biomedical applications, viruses can be analyzed in blood samples through impedance analysis because chemical reactions take place when a known strain of a virus is added to a blood sample that already contains a virus. Different disease strains exhibit unique impedance signatures, and by characterizing impedance effects of blood samples across different frequencies, it is possible to detect specific strains of viruses. Likewise, so-called "body-composition" bathroom scales that purport to analyze relative levels of lean muscle mass to fat work on the principle that the complex impedance of the person being weighed (the DUT) is meaningfully related to the percentage of body fat. In industrial applications, electro-impedance spectroscopy employs network analysis principles to determine the corrosion of metals, such as aluminum and steel, which can damage industrial infrastructure as well as aircraft, ships, and cars. In the food industry, network analysis is used to analyze the relative moisture content of products such as cheeses to determine product quality.

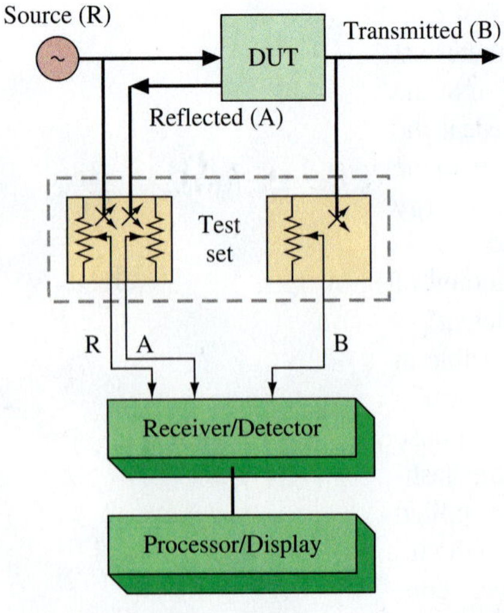

FIGURE 12-37 Conceptual block diagram of vector network analyzer.

Vector Network Analyzers and S Parameters

As described in the previous section, a vector network analyzer consists of a system of three receivers as well as a two-port "test set" that separates the forward and reflected components of applied signals through a system of directional couplers. The basic architecture of a vector network analyzer is shown in Figure 12-37. In addition to the signal separation function represented by the test set, the analyzer consists of a system of three receivers, designated R (for reference), A (for reflected), and B (for transmitted), as well as a processor and display system.

The test sets may be either true "S-parameter" sets, which allow for all four possible S parameters produced by a two-port device to be measured automatically (that is, without the need for the user to manually reverse the transmit and receive cables), or may be of the somewhat simpler "T/R" variety, which requires manual intervention to permit the measurement of all transmission and reflection parameters. With either type of test set, the analyzer works on the principle of impedance ratios shown in Figure 12-38. The reflected wave, measured by the A channel, and the transmitted wave, measured by the B channel, are converted to voltage ratios, each with respect to the known voltages created by the reference receiver. The amplitude and phase information of waves in both the A and B channels makes possible the determination of reflection and transmission characteristics of DUTs. The ratio approach allows for reflection and transmission measurements to be independent of absolute power and variations in source power versus frequency. Figure 12-38 also shows that familiar characteristics such as reflection coefficient and return loss can be derived from the ratio of reflected or transmitted signals to the reference.

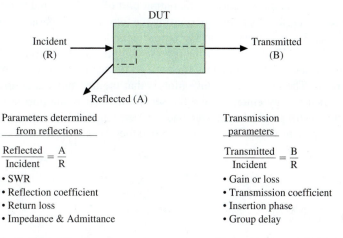

Parameters determined
from reflections

$$\frac{\text{Reflected}}{\text{Incident}} = \frac{A}{R}$$

- SWR
- Reflection coefficient
- Return loss
- Impedance & Admittance

Transmission
parameters

$$\frac{\text{Transmitted}}{\text{Incident}} = \frac{B}{R}$$

- Gain or loss
- Transmission coefficient
- Insertion phase
- Group delay

FIGURE 12-38   **Relationship of receiver measurements to (device under test) DUT characteristics.**

Figure 12-39 relates the transmitted and received signal ratios to their S parameter equivalents. Note, in particular, how the forward S parameters are determined by measuring the magnitude and phase of the incident, reflected, and transmitted signals when the unused port is terminated in a load whose impedance is equal to that of the test

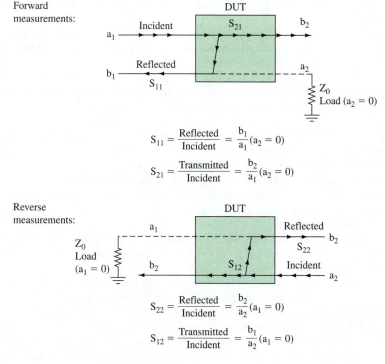

$$S_{11} = \frac{\text{Reflected}}{\text{Incident}} = \frac{b_1}{a_1}(a_2 = 0)$$

$$S_{21} = \frac{\text{Transmitted}}{\text{Incident}} = \frac{b_2}{a_1}(a_2 = 0)$$

$$S_{22} = \frac{\text{Reflected}}{\text{Incident}} = \frac{b_2}{a_2}(a_1 = 0)$$

$$S_{12} = \frac{\text{Transmitted}}{\text{Incident}} = \frac{b_1}{a_2}(a_1 = 0)$$

FIGURE 12-39   **Measurement of S parameters.**

system characteristic impedance. The use of ports terminated in the system impedance is one big advantage of S parameter measurements over other measurement-parameter schemes. So-called Z (resistance/impedance), Y (conductance/admittance), or H parameters (which are a mixture of conductances and admittances), require DUTs to be terminated either with open or short circuits. Making measurements of voltage or current at the input or output ports of DUTs can be difficult to do under open- or short-circuit conditions, because a circuit that appears open at low frequencies often acts as a reactive circuit at high frequencies. Such conditions can cause the DUT to oscillate and possibly self-destruct.

As can also be seen from Fig. 12-39, the S_{11} measurement is the input complex reflection coefficient or impedance of the DUT, and S_{21} is the forward complex transmission coefficient. By means of the automatic switching capability of the S parameter test set, the source can be switched to the output port of the DUT and the input port can be terminated in a perfect load, thus permitting the two reverse S parameters to be measured. Parameter S_{22} is equivalent to the output complex reflection coefficient (output impedance) of the DUT, while S_{12} is the reverse complex transmission coefficient. Note also that the number of S parameters for a given device is equal to the square of the number of ports. The numbering convention is that the first number following the S is the port at which energy emerges, and the second number is the port at which energy enters. With a vector network analyzer one can determine the effects on impedance of changes to component values or other circuit parameters as a function of frequency.

SUMMARY

The propagation of voltage and current along a transmission line is affected by line characteristics and by the impedances of the source and load. Transmission lines have a characteristic impedance determined by their physical construction. They exhibit properties of inductance and capacitance as well as resistance. These properties result in current flow within the line when a source voltage is applied. An infinite-length line, or one of finite length connected to a load whose resistance equals that of the line, is nonresonant, and all energy from the source will be fully absorbed by the load. Because of the time needed for the distributed line capacitances to acquire and release charge, energy can be seen to propagate down the line at a rate less than that of the speed of light. This rate is known as the velocity factor or propagation velocity, and it too is determined primarily by physical characteristics of the line.

An open- or short-circuited line, or one that is not terminated in its characteristic impedance, is known as a resonant line. Applied voltages and currents will propagate down the line, only to be reflected back toward the source when a discontinuity is encountered. The voltages of both the forward and reflected waves will add vectorially and will produce a standing wave voltage, so named because it appears stationary along the line. The greater the degree of mismatch, the larger this standing wave will be. The voltage standing-wave ratio is the ratio of maximum to minimum voltage of the standing wave and provides an easily measured indication of the severity of mismatch. Other ways to quantify the reflections are by means of the reflection coefficient, which is the ratio of reflected to forward voltage, and which may be a complex quantity (magnitude and phase angle), and the return loss, which is a ratio of forward to reflected power.

Voltages and currents on resonant lines are out of phase, and this phase angle varies along the line. This result implies also that there is an impedance variation along a resonant line. The voltage and current maxima and minima, as well as impedance and voltage standing wave ratio (VSWR), form repeating patterns every half wavelength. There is also a phase inversion between voltage and current every quarter wavelength. The property of phase inversion, where an open circuit appears as a short circuit one-quarter wavelength away and vice versa, can be used to produce quarter-wave stub filters and other devices.

Transmission-line behavior and impedance-matching characteristics can be modeled with the Smith chart, a special-purpose graphical analysis tool. The Smith chart displays complex impedance characteristics with resistance circles and reactance arcs. Because line characteristics repeat every half wavelength, a full revolution around the Smith chart represents a half wavelength movement along a transmission line. Once an impedance, in complex rectangular form, has been plotted, a number of line characteristics can be immediately determined. A circle drawn from the center of the chart through the plotted impedance point represents the VSWR of the line and shows the impedance at any point along a half-wavelength section. The magnitude and phase angle of the reflection coefficient as well as the magnitude of the voltage and power return losses can also be determined.

One of the most useful properties of the Smith chart is its ability to solve impedance-matching problems. Impedances can be matched in a number of ways. One is to use a quarter-wave section of line as a matching transformer. Another is to use a single- or double-stub tuner. The latter is particularly useful when the impedance to be matched has reactances that need to be tuned out. The stub tuner produces a match by acting as a parallel-connected susceptance sufficient to cancel out an unwanted reactance.

Through network analysis, the complex impedance characteristics of devices under test can be fully characterized without regard to their internal structures. A network analyzer is a piece of test equipment that can measure the impedance magnitude and, in some cases, its phase at all input and output ports over a range of frequencies. In this way, the impedance behavior of devices can be modeled and such familiar characteristics as reflection coefficient and return loss fully quantified to give a complete picture of device or subsystem behavior. Network analyzer measurements are made using scattering or S parameters, which are derived from ratios of transmitted or reflected amplitude to incident amplitude. The ratio approach allows for reflection and transmission measurements to be independent of absolute power and variations in source power versus frequency. S parameters also have the advantage of being the only parameter set in which measurements are made with all ports terminated with resistive loads. This characteristic allows for stable operation of the device under test, in contrast with other parameter sets that require tests to be conducted with ports open- or short-circuited. This latter condition can be difficult to achieve, particularly at microwave frequencies and can lead to erratic operation and possible destruction of the device. S parameters have the advantage of being stable and repeatable.

QUESTIONS AND PROBLEMS

SECTION 12-2

1. Define *transmission line*. If a simple wire connection can be a transmission line, why is an entire chapter of study devoted to it?

2. In general terms, discuss the various types of transmission lines. Include the advantages and disadvantages of each type.

*3. What would be the considerations in choosing a solid dielectric cable over a hollow pressurized cable for use as a transmission line?

*4. Why is an inert gas sometimes placed within concentric radio-frequency transmission cables?

5. Where are twisted-pair cables often used?

6. What is meant by the CAT6/5e designation?

7. Define *near-end crosstalk* and *attenuation* relative to testing twisted-pair cable.

8. Describe three additional test considerations for twisted-pair cable with the enhanced data capabilities.

9. Explain how an unbalanced line differs from a balanced line.

10. Explain how the pickup of unwanted signals on a balanced transmission line is attenuated when converted to an unbalanced signal.

11. A balanced transmission line picks up an undesired signal with a 5-mW level. After conversion to an unbalanced signal using a center-tapped transformer, the undesired signal is $0.011\mu W$. Calculate the common mode rejection ratio (CMRR). (56.6 dB)

SECTION 12-3

12. Draw an equivalent circuit for a transmission line, and explain the physical significance of each element.

13. Provide a physical explanation for the meaning of a line's characteristic impedance (Z_0).

14. Calculate Z_0 for a line that exhibits an inductance of 4 nH/m and 1.5 pF/m. (51.6 Ω)

---

*An asterisk preceding a number indicates a question that has been provided by the FCC as a study aid for licensing examinations.

15. Calculate the capacitance per meter of a 50-Ω cable that has an inductance of 55 nH/m. (22 pF/m)

16. In detail, explain how an impedance bridge can be used to determine Z_0 for a piece of transmission line.

*17. If the spacing of the conductors in a two-wire radio-frequency transmission line is doubled, what change takes place in the surge impedance (Z_0) of the line?

*18. If the conductors in a two-wire radio-frequency transmission are replaced by larger conductors, how is the surge impedance affected, assuming no change in the center-to-center spacing of the conductor?

*19. What determines the surge impedance of a two-wire radio-frequency transmission line?

20. Determine Z_0 for the following transmission lines:
   (a) Parallel wire, air dielectric with $D/d = 3$. (215 Ω)
   (b) Coaxial line, air dielectric with $D/d = 1.5$. (24.3 Ω)
   (c) Coaxial line, polyethylene dielectric with $D/d = 2.5$. (36.2 Ω)

21. List and explain the various types of transmission line losses.

*22. A long transmission line delivers 10 kW into an antenna; at the transmitter end, the line current is 5 A, and at the coupling house (load) it is 4.8 A. Assuming the line to be properly terminated and the losses in the coupling system to be negligible, what is the power lost in the line? (850 W)

23. Define *surge impedance*.

SECTION 12-4

24. Derive the equation for the time required for energy to propagate through a transmission line [Equation (12-19)].

*25. What is the velocity of propagation for radio-frequency waves in space?

26. A delay line using RG-8A/U cable is to exhibit a 5-ns delay. Calculate the required length of this cable. (3.39 ft)

27. Explain the significance of the velocity factor for a transmission line.

28. Determine the velocity of propagation of a 20-km line if the *LC* product is 7.5×10^{-12} s². (7.3×10^9 m/s)

29. What is the wavelength of the signal in Problem 28 if the signal's frequency is 500 GHz? (0.0146)

30. If the velocity of propagation of a 20-ft transmission line is 600 ft/s, how long will it take for the signal to get to the end of the line? (33.3 ms)

SECTION 12-5

31. Define a *nonresonant transmission line,* and explain what its traveling waves are and how they behave.

*32. An antenna is being fed by a properly terminated two-wire transmission line. The current in the line at the input end is 3 A. The surge impedance of the line is 500 Ω. How much power is being supplied to the line? (4.5 kW)

33. With the help of Figure 12-15, provide a charging analysis of a nonresonant line with an ac signal applied.

SECTION 12-6

34. Explain the properties of a resonant transmission line. What happens to the energy reaching the end of a resonant line? Are reflections a generally desired result?

35. A dc voltage from a 20-V battery with $R_s = 75$ Ω is applied to a 75-Ω transmission line at $t = 0$. It takes the battery's energy 10 μs to reach the load, which is an open circuit. Sketch current and voltage waveforms at the line's input and load.

36. Repeat Problem 35 for a short-circuited load.

37. What are *standing waves, standing wave ratio (SWR),* and *characteristic impedance,* in reference to transmission lines? How can standing waves be minimized?

38. With the help of Figure 12-19, explain how standing waves develop on a resonant line.

*39. If the period of one complete cycle of a radio wave is 0.000001 s, what is the wavelength? (300 m)

*40. If the two towers of a 950-kHz antenna are separated by 120 electrical degrees, what is the tower separation in feet? (345 ft)

SECTION 12-7

41. Define *reflection coefficient,* Γ, in terms of incident and reflected voltage and also in terms of a line's load and characteristic impedances.

42. Express standing wave ration (SWR) in terms of
(a) Voltage maximums and minimums.
(b) Current maximums and minimums.
(c) The reflection coefficient.
(d) The line's load resistance and Z_0.

43. Explain the disadvantages of a mismatched transmission line.

*44. What is the primary reason for terminating a transmission line in an impedance equal to the characteristic impedance of the line?

*45. What is the ratio between the currents at the opposite ends of a transmission line one-quarter wavelength long and terminated in an impedance equal to its surge impedance?

46. An single sideband (SSB) transmitter at 2.27 MHz and 200 W output is connected to an antenna $(R_{in} = 150\,\Omega)$ via 75 ft of RG-8A/U cable. Determine
(a) The reflection coefficient.
(b) The electrical cable length in wavelengths.
(c) The SWR.
(d) The amount of power absorbed by the antenna.

*47. What should be the approximate surge impedance of a quarter-wavelength matching line used to match a 600-Ω feeder to a 70-Ω (resistive) antenna?

SECTION 12-8

48. Calculate the impedance of a line 675 electrical degrees long. $Z_0 = 75\,\Omega$ and $Z_L = 50\,\Omega + j75\,\Omega$. Use the Smith chart *and* Equation (12-30) as separate solutions, and compare the results.

49. Convert an impedance, $62.5\,\Omega - j90\,\Omega$, to admittance mathematically *and* with the Smith chart. Compare the results.

50. Find the input impedance of a 100-Ω line, 5.35λ long, and with $Z_L = 200\,\Omega + j300\,\Omega$.

*51. Why is the impedance of a transmission line an important factor with respect to matching "out of a transmitter" into an antenna?

*52. What is *stub tuning?*

53. The antenna load on a 150-Ω transmission line is $225\,\Omega - j300\,\Omega$. Determine the length and position of a short-circuited stub necessary to provide a match.

54. Repeat Problem 53 for a 50-Ω line and an antenna of $25\,\Omega + j75\,\Omega$.

SECTION 12-9

55. Calculate the length of a short-circuited 50-Ω line necessary to simulate an inductance of 2 nH at 1 GHz.

56. Calculate the length of a short-circuited 50-Ω line necessary to simulate a capacitance of 50 pF at 500 MHz.

57. Describe two types of baluns, and explain their function.

*58. How may harmonic radiation of a transmitter be prevented?

*59. Describe three methods for reducing harmonic emission of a transmitter.

*60. Draw a simple schematic diagram showing a method of coupling the radio-frequency output of the final power amplifier stage of a transmitter to a two-wire transmission line, with a method of suppression of second and third harmonic energy.

61. Explain the construction of a slotted line and some of its uses.

62. Explain the principle of time-domain reflectometry (TDR) and some uses for this technique.

63. A pulse is sent down a transmission line that is not functioning properly. It has a propagation velocity of 2.1×10^8 m/s, and an inverted reflected pulse (equal in magnitude to the incident pulse) is returned in 0.731 ms. What is wrong with the line, and how far from the generator does the fault exist?

64. A fast-rise-time 10-V step voltage is applied to a 50-Ω line terminated with an 80-Ω resistive load. Determine Γ, E_F, and E_r. (0.231, 12.3 V, 2.3 V)

QUESTIONS FOR CRITICAL THINKING

65. With the help of Figure 12-12, provide a step-by-step explanation of how a dc voltage propagates through a transmission line.

66. An open-circuited line is 1.75λ. Sketch the incident, reflected, and resultant waveforms for both voltage and current at the instant the generator is at its peak negative value. Sketch and compare the waveforms for a short-circuited line.

67. You are asked to design a line "free of transmission line effects." You design one that is λ/16 long. How would you justify this design?

68. Match a load of 25 Ω + j75 Ω to a 50-Ω line using a quarter-wavelength matching section. Determine the proper location and characteristic impedance of the matching section. Repeat this problem for a $Z_L = 110\Omega - j50\Omega$ load. Provide *two* separate solutions.

CHAPTER 13

WAVE PROPAGATION

KEY TERMS

transducer
radio-frequency interference (RFI)
electromagnetic interference (EMI)
transverse
polarization
isotropic point source
wavefront
coefficient of reflection
diffraction
shadow zone
ground wave
surface wave
radio horizon
sky-wave propagation
skipping
isothermal region
critical frequency
critical angle
maximum usable frequency (MUF)
optimum working frequency
quiet zone
skip zone
fading
diversity reception
geostationary orbit
uplink
downlink
Earth station
transponder
attitude controls
subsatellite point
footprint
perigee
apogee
low Earth orbit (LEO)
satellites
look angle
differential GPS
code division multiple access (CDMA)
figure of merit (G/T)
satellite link budget
free-space path loss

13-1 ELECTRICAL TO ELECTROMAGNETIC CONVERSION

Early radios were often referred to as the "wireless." This new machine could speak without being "wired" to the source like the telegraph and telephone. The transmitter's output is coupled to its surrounding atmosphere and then intercepted by the receiver. We know that the atmosphere is *not* a conductor of electrons like a copper wire—air is, in fact, a very good insulator. Thus, the electrical energy fed into a transmitting antenna must be converted to another form of energy for transmission. In this chapter we shall study the effects of the transformed energy and its propagation.

The transmitting antenna converts its input electrical energy into electromagnetic energy. The antenna can thus be thought of as a **transducer**—a device that converts energy from one form to another. In that respect, a light bulb is very similar to an antenna. The light bulb also converts electrical energy into electromagnetic energy—light. The only difference between light and the radio waves we shall be concerned with is their frequency. Light is an electromagnetic wave at about 5×10^{14} Hz, while the usable radio waves extend from about 1.5×10^4 Hz up to 3×10^{11} Hz. The human eye is responsive to (able to perceive) the very narrow range of light frequencies, and consequently we are blind to the radio waves. Actually, that is an advantage because the great number of radio waves surrounding our Earth would otherwise paint a chaotic picture.

The receiving antenna intercepts the transmitted wave and converts it back into electrical energy. An analogous transducer is the photovoltaic cell, which also converts a wave (light) into electrical energy. Because a basic knowledge of waves is necessary to your understanding of antennas and radio communications, the following section is presented prior to your study of wave propagation.

13-2 ELECTROMAGNETIC WAVES

Electricity and electromagnetic waves are interrelated. An electromagnetic field consists of an electric field and a magnetic field. These fields exist with all electric circuits because any current-carrying conductor creates a magnetic field around the conductor, and any two points in the circuit with a potential difference (voltage) between them create an electric field. These two fields contain energy, but in circuits, this field energy is usually returned to the circuit when the field collapses. If the field does not fully return its energy to the circuit, it means the wave has been at least partially *radiated,* or set free, from the circuit. This radiated energy is undesired because it may cause interference with other electronic equipment in the vicinity. It is termed **radio-frequency interference (RFI)** if it is undesired radiation from a radio transmitter, and if from another source, it is termed **electromagnetic interference (EMI)** or, more simply, noise. In the case of a radio transmitter, it is hoped that the antenna efficiently causes the wave energy to be set free. The antenna is designed *not* to allow the electromagnetic wave energy to collapse back into the circuit.

In Chapter 1 we introduced the concept of an electromagnetic wave as a form of energy consisting of co-existing electric and magnetic fields. In what is surely one of the most profound insights in all of physics, the Scottish physicist James Clerk-Maxwell in the 1860s predicted the existence of electromagnetic waves and proposed that visible light was a form of electromagnetic energy. Maxwell integrated knowledge gained to that point by pioneers including Hans Christian Oersted, the Danish physicist who first noticed a relationship between electricity and magnetism, the French scientist André Ampère, who saw that magnetic effects exist around current-carrying conductors, and Michael Faraday, the English physicist who demonstrated that a changing current in one coil induced a current into an adjacent coil even if there was no direct electrical connection between them. Maxwell published an extensive mathematical analysis that included twenty equations describing the behavior of electric and magnetic fields. Oliver Heaviside of Great Britain and Heinrich Hertz of Germany continued Maxwell's original work, both experimentally and theoretically, after Maxwell's death. Ten years later, Hertz went on to demonstrate experimentally the existence of electromagnetic radiation (what we now call radio waves) at frequencies below those of visible light. Also in the two decades following Maxwell's death, Heaviside and Hertz combined and simplified Maxwell's original set of twenty equations. The result is the set of four *Maxwell's equations,* the differential form of which are shown here. The equations are the mathematical expression of four laws: Gauss's law for electric fields, Gauss's law for magnetic fields, Faraday's law, and the Ampère-Maxwell law:

Maxwell's Equations:

Gauss's Law: $\nabla \cdot \mathbf{E} = \dfrac{\rho}{\varepsilon_0}$ which states that an electric charge (ρ) produces an electric field ($\mathbf{E}$)

Gauss's Law for Magnetism: $\nabla \cdot \mathbf{B} = 0$ which states that there are no magnetic charges or magnetic monopoles, i.e., no isolated North and South poles.

Faraday's law: $\nabla \times \mathbf{E} = -\dfrac{\partial \mathbf{B}}{\partial t}$ which states that a changing magnetic field ($\mathbf{B}$) produces an electric field ($\mathbf{E}$)

Ampere's law: $\nabla \times \mathbf{B} = \mu_0 \mathbf{J} + \mu_0 \varepsilon_0 \dfrac{\partial \mathbf{E}}{\partial t}$ which states that an electric current ($\mathbf{J}$) or a changing electric field ($\mathbf{E}$) produces a magnetic field.

Maxwell's equations require knowledge of partial differential equations to appreciate fully. The mathematical solutions are beyond the scope of this text; however, the basic thrust can be understood intuitively and will help to articulate the underlying concept of electromagnetic radiation. Gauss's law for electric fields expresses the relationship between an electric field and an electric charge and is the formalization of the relationship between positive and negative charges introduced in your study of electrical fundamentals, where the electric field was shown to point away from positive charge toward negative charge. Gauss's law for magnetism states that there are no magnetic monopoles; although magnetic "charges" analogous to electric charges do exist in the form of poles, they do not do so in isolation. Instead, magnetic fields are the result of "magnetic dipoles," which are inseparably bound together to form closed loops of magnetic current.

The last two of Maxwell's equations are the field equations, and these form the basis of electromagnetic field theory. Faraday's law states that a time-varying (changing) magnetic field produces an electric field. This fact, known in Maxwell's era, is the basis for the operation of electric generators, in which the changing magnetic field produced by a rotating magnet creates an electric field in a surrounding conductor. What came next, though, was a truly profound insight. Maxwell essentially postulated that the converse to Faraday's law held true as well: that a changing electric field produces a magnetic field. Though he did not have the tools to demonstrate his predictions in the laboratory, Maxwell proposed a bold leap beyond the practical knowledge of the time by asserting, based on his theoretical predictions, a natural symmetry between changing electric and magnetic fields. The interaction of these changing fields is the basis for the self-replicating nature of radio waves. Stated concisely, a changing magnetic field will produce an electric field, and a changing electric field will produce a magnetic field. The fields thus become self-sustaining and propagate away from their source at the speed of light.

An electromagnetic wave is pictured in Figure 13-1. In it, $1\frac{1}{2}$ wavelengths of the electric field (E) and the magnetic field (H) are shown. The direction of propagation is shown to be perpendicular to both fields, which are also mutually perpendicular to each other. The wave is said to be **transverse** because the oscillations are perpendicular to the direction of propagation. The **polarization** of an electromagnetic wave is determined by the direction of its E field component. In Figure 13-1 the E field is vertical (y direction), and the wave is therefore said to be vertically polarized. As we will see, the antenna's orientation determines polarization. A vertical antenna results in a vertically polarized wave.

FIGURE 13-1
Electromagnetic wave.

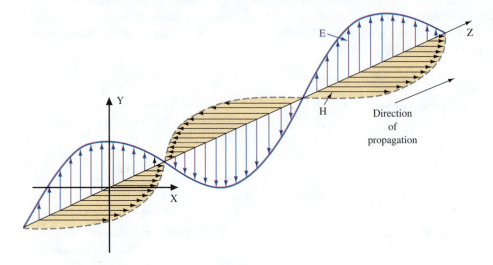

Wavefronts

If an electromagnetic wave were radiated equally in all directions from a point source in free space, a spherical wavefront would result. Such a source is termed an **isotropic point source**. A **wavefront** may be defined as a surface joining all points of equal phase. Two wavefronts are shown in Figure 13-2. An *isotropic* source radiates equally in all directions. The wave travels at the speed of light so that at some instant the energy will have reached the area indicated by wavefront 1 in Figure 13-2. The power density, $\mathscr{P}$ (in watts per square meter), at wavefront 1 is inversely proportional to the square of its distance, r (in meters), from its source, with respect to the originally transmitted power, P_t. Stated mathematically,

$$\mathscr{P} = \frac{P_t}{4\pi r^2}. \tag{13-1}$$

If wavefront 2 in Figure 13-2 is twice the distance of wavefront 1 from the source, then its power density in watts per unit area is just one-fourth that of wavefront 1. Any section of a wavefront is curved in shape. However, at appreciable distances from the source, small sections are nearly flat. These sections can then be considered as *plane wavefronts,* which simplifies the treatment of their optical properties provided in Section 13-3.

FIGURE 13-2 Antenna wavefronts.

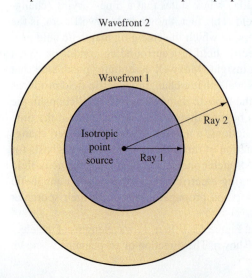

Characteristic Impedance of Free Space

The strength of the electric field, $\mathcal{E}$ (in volts per meter), at a distance r from a point source is given by

$$\mathcal{E} = \frac{\sqrt{30P_t}}{r}, \tag{13-2}$$

where P_t is the originally transmitted power in watts. This is one of Maxwell's equations.

Power density $\mathcal{P}$ and the electric field $\mathcal{E}$ are related to impedance in the same way that power and voltage relate in an electric circuit. Thus,

$$\mathcal{P} = \frac{\mathcal{E}^2}{\mathcal{Z}}, \tag{13-3}$$

where $\mathcal{Z}$ is the characteristic impedance of the medium conducting the wave. For free space, Equations (13-1) and (13-2) can be substituted into Equation (13-3) to give

$$\mathcal{Z} = \frac{\mathcal{E}^2}{\mathcal{P}} = \frac{30P_t}{r^2} \div \frac{P_t}{4\pi r^2} = 120\pi = 377\Omega. \tag{13-4}$$

Thus, you can see that free space has an intrinsic impedance similar to the characteristic impedance of a transmission line.

The characteristic impedance of any electromagnetic wave-conducting medium is provided by

$$\mathcal{Z} = \sqrt{\frac{\mu}{\varepsilon}}, \tag{13-5}$$

where μ is the medium's permeability and ε is the medium's permittivity.

For free space, $\mu = 1.26 \times 10^{-6}$ H/m and $\varepsilon = 8.85 \times 10^{-12}$ F/m. Substituting in Equation (13-5) yields

$$\mathcal{Z} = \sqrt{\frac{\mu}{\varepsilon}} = \sqrt{\frac{1.26 \times 10^{-6}}{8.85 \times 10^{-12}}} = 377\Omega,$$

which agrees with the result from Equation (13-4).

13-3 WAVES NOT IN FREE SPACE

Until now, we have discussed the behavior of waves in free space, which is a vacuum or complete void. We now consider the effects of our environment on wave propagation.

Reflection

Just as light waves are reflected by a mirror, radio waves are reflected by any medium such as metal surfaces or the Earth's surface. The angle of incidence is equal to the angle of reflection, as shown in Figure 13-3. Note that there is a change in phase of the incident and reflected waves, as seen by the difference in the direction of polarization. The incident and reflected waves are 180° out of phase.

Complete reflection occurs only for a theoretically perfect conductor and when the electric field is perpendicular to the reflecting element. For complete reflection, the **coefficient of reflection** ρ is 1 and is defined as the ratio of the reflected electric field intensity divided by the incident intensity. It is less than 1 in practical situations because of the absorption of energy by the nonperfect conductor and also because some of the energy actually propagates right through it.

The previous discussion is valid when the electric field is *not* normal to the reflecting surface. If it is fully parallel to the reflecting (conductive) surface, the electric field is

FIGURE 13-3 **Reflection of a wavefront.**

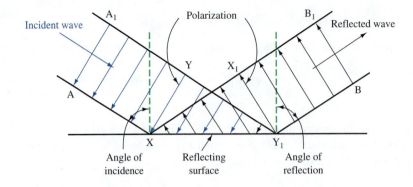

FIGURE 13-3 **Reflection of a wavefront.**

shorted out, and all of the electromagnetic energy is dissipated in the form of generated surface currents in the conductor. If the electric field is partially parallel to the surface, it will be partially shorted out.

If the reflecting surface is curved, as in a parabolic antenna, the wave may be analyzed using the appropriate optical laws with regard to focusing the energy, etc. This is especially true with respect to microwave frequencies, which are discussed in Chapter 14.

Refraction

Refraction of radio waves occurs in a manner akin to the refraction of light. Refraction occurs when waves pass from a medium of one density to another medium with a different density. A good example is the apparent bending of a spoon when it is immersed in water. The bending seems to take place at the water's surface, or exactly at the point where there is a change of density. Obviously, the spoon does not bend from the pressure of the water. The light forming the image of the spoon is bent as it passes from the water, a medium of high density, to the air, a medium of comparatively low density.

The bending (refraction) of an electromagnetic wave (light or radio wave) is shown in Figure 13-4. Also shown is the reflected wave. Obviously, the coefficient of reflection is less than 1 here because a fair amount of the incident wave's energy is propagated through the water—after refraction has occurred.

FIGURE 13-4 **Wave refraction and reflection.**

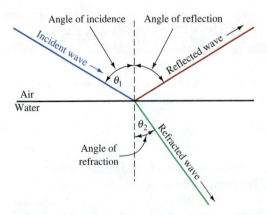

The angle of incidence, θ_1, and the angle of refraction, θ_2, are related by the following expression, which is Snell's law:

$$n_1 \sin \theta_1 = n_2 \sin \theta_2, \tag{13-6}$$

where n_1 is the refractive index of the incident medium and n_2 is the refractive index of the refractive medium. Recall that the refractive index for a vacuum is exactly 1 and it is approximately 1 for the atmosphere. For glass, it is about 1.5, and for water it is 1.33.

Diffraction

Diffraction is the phenomenon whereby waves traveling in straight paths bend around an obstacle. This effect is the result of Huygens's principle, advanced by the Dutch astronomer Christian Huygens in 1690. The principle states that each point on a spherical wavefront may be considered as the source of a secondary spherical wavefront. This concept is important to us because it explains radio reception behind a mountain or tall building. Figure 13-5 shows the diffraction process allowing reception beyond a mountain in all but a small area, which is called the **shadow zone**. The figure shows that electromagnetic waves are diffracted over the top and around the sides of an obstruction. The direct wavefronts that just get by the obstruction become new sources of wavefronts that start filling in the void, making the shadow zone a finite entity. The lower the frequency of the wave, the quicker is this process of diffraction (i.e., the shadow zone is smaller).

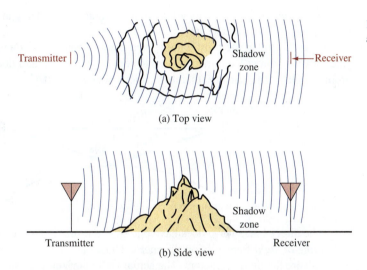

(a) Top view

(b) Side view

FIGURE 13-5 **Diffraction around an object.**

13-4 GROUND- AND SPACE-WAVE PROPAGATION

There are four basic modes of getting a radio wave from the transmitting to receiving antenna:

1. Ground wave
2. Space wave
3. Sky wave
4. Satellite communications

As we will see in the following discussions, the frequency of the radio wave is of primary importance in considering the performance of each type of propagation.

Ground-Wave Propagation

A **ground wave** is a radio wave that travels along the Earth's surface. It is sometimes referred to as a **surface wave**. The ground wave must be vertically polarized (electric field vertical) because the Earth would short out the electric field if horizontally polarized. Changes in terrain have a strong effect on ground waves. Attenuation of ground waves is directly related to the surface impedance of the Earth. This impedance is a function of conductivity and frequency. If the Earth's surface is highly conductive, the absorption of wave energy, and thus its attenuation, will be reduced. Ground-wave propagation is much better over water (especially salt water) than, say, a very dry (poor conductivity) desert terrain.

The ground losses increase rapidly with increasing frequency. For this reason ground waves are not very effective at frequencies above 2 MHz. Ground waves are, however, a very reliable communications link. Reception is not affected by daily or seasonal changes such as with sky-wave propagation.

Ground-wave propagation is the only way to communicate into the ocean with submarines. Extremely low-frequency (ELF) propagation is used. ELF waves encompass the range 30 to 300 Hz. At a typically used frequency of 100 Hz, the attenuation is about 0.3 dB/m. This attenuation increases steadily with frequency so that, at 1 GHz, a 1000-dB/m loss is sustained! Seawater has little attenuation to ELF signals, so these frequencies can be used to communicate with submerged submarines without their having to surface and be vulnerable to detection.

Space-Wave Propagation

The two types of space waves are shown in Figure 13-6. They are the direct wave and ground reflected wave. Do not confuse these with the ground wave just discussed. The direct wave is by far the most widely used mode of antenna communications. The propagated wave is direct from transmitting to receiving antenna and does not travel along the ground. The Earth's surface, therefore, does not attenuate it.

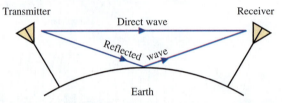

FIGURE 13-6   Direct and ground reflected space waves.

The direct space wave does have one severe limitation—it is basically limited to so-called *line-of-sight* transmission distances. Thus, the antenna height and the curvature of the Earth are the limiting factors. The actual **radio horizon** is about $\frac{4}{3}$ greater than the geometric line of sight because of diffraction effects and is empirically predicted by the following approximation:

$$d \cong \sqrt{2h_t} + \sqrt{2h_r}, \tag{13-7}$$

where d = radio horizon (mi)

h_t = transmitting antenna height (ft)

h_r = receiving antenna height (ft)

The diffraction effects cause the slight wave curvature, as shown in Figure 13-7. If the transmitting antenna is 1000 ft above ground level and the receiving antenna is 20 ft high, a radio horizon of about 50 mi results. This explains the coverage that typical broadcast FM and TV stations provide because they propagate directly by space wave propagation.

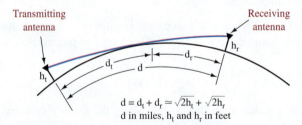

FIGURE 13-7   Radio horizon for direct space waves.

The reflected wave shown in Figure 13-6 is an example of *multipath* reception, first brought up in the context of wireless networks and mobile communications systems in Chapter 10. If the phase of these two received components is not the same, some degree of signal fading and/or distortion will occur. Phase shifts occur because the two (or more) signals arrive at the receiver at different times since the reflected signal has a father distance to travel. The reflected signal is weaker than the direct signal because of the inverse square-law relationship of signal strength to distance [Equation (13-1)] and because of losses incurred during reflection. Multipath effects can also result when both a direct and ground wave are received or when any two or more signal paths exist. Multipath is of particular concern in mobile systems because of the phenomenon of Rayleigh fading created as the result of many reflected signals causing constructive and destructive interference coupled with the relatively quick movement of the mobile subscriber unit and the short wavelengths involved at the 800-MHz or 1.9-GHz frequencies of operation.

13-5   SKY-WAVE PROPAGATION

One of the most frequently used methods of long-distance transmission is through the use of **sky-wave propagation**. Sky waves are those waves radiated from the transmitting antenna in a direction that produces a large angle with respect to the Earth. The sky wave has the ability to strike the ionosphere, be refracted from it to the ground, strike the ground, be reflected back toward the ionosphere, and so on. The refracting and reflecting action of the ionosphere and the ground is called **skipping**. An illustration of this skipping effect is shown in Figure 13-8. Because of the skip effect, communications over many thousands of miles—indeed, between continents—is possible without the need for infrastructure such as repeaters or satellites. Sky-wave communication is the reason for the wide coverage afforded by international short-wave broadcasters. In addition, amateur or "ham" radio operators rely on the skip effect to communicate with counterparts hundreds if not thousands of miles away on the high-frequency (HF) bands using transmitters with relatively modest power outputs.

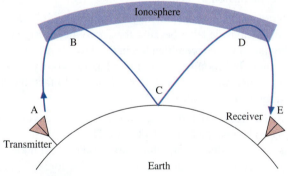

FIGURE 13-8   Sky-wave propagation.

The transmitted wave leaves the antenna at point *A,* is refracted from the ionosphere at point *B,* is reflected from the ground at point *C,* is again refracted from the ionosphere at point *D,* and arrives at the receiving antenna *E.* The critical nature of the sky waves and the requirements for refraction will be discussed thoroughly in this section.

To understand the process of refraction, the composition of the atmosphere and the factors that affect it must be considered. Insofar as electromagnetic radiation is concerned, there are only three layers of the atmosphere: the troposphere, the stratosphere, and the ionosphere. The troposphere extends from the surface of the Earth up to approximately 6.5 mi. The next layer, the stratosphere, extends from the upper limit of the troposphere to an approximate elevation of 23 mi. From the upper limit of the stratosphere to a distance of approximately 250 mi lies the region known as the ionosphere. Beyond the ionosphere is free space. The temperature in the stratosphere is considered to be a constant, unfluctuating value. Therefore, it is not subject to temperature inversions, nor can it cause significant refractions. The constant-temperature stratosphere is also called the **isothermal region**.

The ionosphere is appropriately titled because it is composed primarily of ionized particles. The density at the upper extremities of the ionosphere is very low and becomes progressively higher as it extends downward toward the Earth. The upper region of the ionosphere is subjected to severe radiation from the sun. Ultraviolet radiation from the sun causes ionization of the air into free electrons, positive ions, and negative ions. Even though the density of the air molecules in the upper ionosphere is small, the radiation particles from space are of such high energy at that point that they cause wide-scale ionization of the air molecules that are present. This ionization extends down through the ionosphere with diminishing intensity. Therefore, the highest degree of ionization occurs at the upper extremities of the ionosphere, while the lowest degree occurs in the lower portion of the ionosphere.

Ionospheric Layers

The ionosphere is composed of three layers designated, respectively, from lowest level to highest level as D, E, and F. The F layer is further divided into two layers designated F_1 (the lower layer) and F_2 (the higher layer). The presence or absence of these layers in the ionosphere and their height above the Earth vary with the position of the sun. At high noon, radiation from the sun in the ionosphere directly above a given point is greatest, while at night it is minimal. When the radiation is removed, many of the ions that were ionized recombine. The interval of time between these conditions finds the position and number of the ionized layers within the ionosphere changing. Because the position of the sun varies daily, monthly, and yearly with respect to a specified point on Earth, the exact characteristics of the layers are extremely difficult to predict. However, the following general statements can be made:

1. The D layer ranges from about 25 to 55 mi. Ionization in the D layer is low because it is the lowest region of the ionosphere (farthest from the sun). This layer has the ability to refract signals of low frequencies. High frequencies pass right through it but are partially attenuated in so doing. After sunset, the D layer disappears because of the rapid recombination of its ions.

2. The E layer limits are from approximately 55 to 90 mi high. This layer is also known as the Kennelly–Heaviside layer because these two men were the first to propose its existence. The rate of ionic recombination in this layer is rather rapid after sunset and is almost complete by midnight. This layer has the ability to refract signals of a higher frequency than were refracted by the D layer. In fact, the E layer can refract signals with frequencies as high as 20 MHz.

3. The F layer exists from about 90 to 250 mi. During the daylight hours, the F layer separates into two layers, the F_1 and F_2 layers. The ionization level in these layers is quite high and varies widely during the course of a day. At noon, this portion of the atmosphere is closest to the sun, and the degree of ionization is maximum. The atmosphere is rarefied at these heights, so the recombination of the ions occurs slowly after sunset. Therefore, a fairly constant ionized layer is present at all times. The F layers are responsible for high-frequency, long-distance transmission as the result of refraction for frequencies up to 30 MHz.

The relative distribution of the ionospheric layers is shown in Figure 13-9. With the disappearance of the D and E layers at night, signals normally refracted by these layers are refracted by the much higher layer, resulting in greater skip distances at night. The layers that form the ionosphere undergo considerable variations in altitude, density, and thickness, due primarily to varying degrees of solar activity. The F_2 layer undergoes the greatest variation because of solar disturbances (sunspot activity). There is a greater concentration of solar radiation in the Earth's atmosphere during peak sunspot activity, which recurs in 11-year cycles, as discussed in Chapter 1. During periods of maximum sunspot activity, the F layer is denser and occurs at a higher altitude. During periods of minimum sunspot activity, the lower altitude of the F layer returns the sky waves (dashed lines) to points relatively close to the transmitter compared with the higher altitude F layer occurring during maximum sunspot activity. Consequently, skip distance is affected by the degree of solar disturbance.

F layer
90–250 miles

Night Day

FIGURE 13-9   **Layers of the ionosphere.**

F_2 layer

F_1 layer

E layer

D layer

F layer

Earth

Radiation from sun

Night Day

D layer 25–55 miles
E layer 55–90 miles
F_1 layer 90–155 miles
F_2 layer < 250 miles

Effects of the Ionosphere on the Sky Wave

The ability of the ionosphere to return a radio wave to the Earth depends on the ion density, the frequency of the radio wave, and the angle of transmission. The refractive ability of the ionosphere increases with the degree of ionization. The degree of ionization is greater in summer than in winter and is also greater during the day than at night. As mentioned previously, abnormally high densities occur during times of peak sunspot activity.

CRITICAL FREQUENCY   If the frequency of a radio wave being transmitted vertically is gradually increased, a point is reached where the wave is not refracted sufficiently to curve its path back to Earth. Instead, these waves continue upward to the next layer, where refraction continues. If the frequency is sufficiently high, the wave penetrates all layers of the ionosphere and continues out into space. The highest frequency that is returned to Earth when transmitted vertically under given ionospheric conditions is called the **critical frequency**.

CRITICAL ANGLE   In general, the lower the frequency, the more easily the signal is refracted; conversely, the higher the frequency, the more difficult is the refracting or bending process. Figure 13-10 illustrates this point. The angle of radiation plays an important part in determining whether a particular frequency is returned to Earth by refraction from the ionosphere. Above a certain frequency, waves transmitted vertically continue into space. However, if the angle of propagation is lowered (from the vertical), a portion of the high-frequency waves below the critical frequency is returned to Earth. The highest angle at which a wave of a specific frequency can be propagated and still be returned (refracted) from the ionosphere is called the **critical angle** for that particular frequency. The critical angle is the angle that the wavefront path makes with a line extended to the center of the Earth. Refer to Figure 13-10, which shows the critical angle for 20 MHz. Any wave above 20 MHz (e.g., the 21-MHz wave shown) is not refracted back to Earth but goes through the ionosphere and into space.

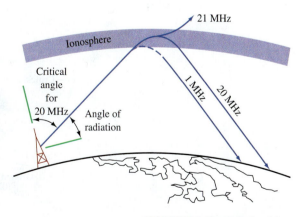

21 MHz

Ionosphere

Critical angle for 20 MHz

Angle of radiation

1 MHz

20 MHz

FIGURE 13-10   **Relationship of frequency to refraction by the ionosphere.**

MAXIMUM USABLE FREQUENCY   There is a best frequency for optimum communication between any two points at any specific condition of the ionosphere. As you can see in Figure 13-11, the distance between the transmitting antenna and the point at which the wave returns to Earth depends on the angle of propagation, which in turn is limited by the frequency. The highest frequency that is returned to Earth at a given distance is called the **maximum usable frequency (MUF)** and has an average monthly value for any given time of the year. The **optimum working frequency** is the one that provides the most consistent communication and is therefore the best one to use. For transmission using the F_2 layer, the optimum working frequency is about 85% of the MUF, while propagation via the E layer is consistent, in most cases, if a frequency near the MUF is used. Because ionospheric attenuation of radio waves is inversely proportional to frequency, using the MUF results in maximum signal strength.

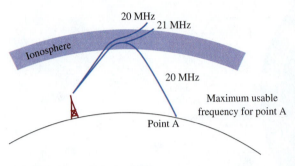

FIGURE 13-11   Relationship of frequency to critical angle.

Because of this variation in the critical frequency, nomograms and frequency tables are used to predict the maximum usable frequency for every hour of the day for every locality in which transmissions are made. This information is prepared from data obtained experimentally from stations scattered all over the world. All this information is pooled, and the results are tabulated in the form of long-range predictions that remove most of the guesswork from this type of radio communications.

The U.S. government transmits propagation data on a regular basis. The two stations are WWV, Fort Collins, Colorado, at 18 minutes past every hour on frequencies of 2.5, 5, 10, 15, and 20 MHz; and WWVH, Hawaii, on 5, 10 and 15 MHz, 45 minutes past every hour. These stations transmit the A and K indices and the solar flux, which can be used to predict MUF as well as other propagation characteristics. The K index, from 0 to 8, is a measure of the Earth's geomagnetic activity. A value above approximately 4 indicates a geomagnetic storm with severe effects on radio communications. The K index is updated every three hours and shows useful "trend" information. The A index is open-ended; that is, it has no maximum value, but readings above about 100 or so are rare. Values of perhaps 10 or lower indicate quiet conditions and good propagation. Based on the K index, the A index is updated every 24 hours at 1800 UT. Solar flux is a measure of sunspot activity. Like the A index, low values indicate good propagation.

SKIP ZONE   Between the point where the ground wave is completely dissipated and the point where the first sky wave returns, *no* signal will be heard. This area is called the **quiet zone** or **skip zone** and is shown in Figure 13-12. You can see that the skip zone occurs for a given frequency, when propagated at its critical angle. The skip zone is the

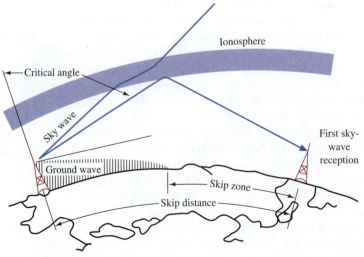

FIGURE 13-12   Skip zone.

distance from the end of ground-wave reception to the point of the first sky-wave reception. This occurs for the energy propagated at the critical angle. Similarly, the skip distance is the minimum distance from the transmitter to where the sky wave can be returned to Earth and also occurs for energy propagated at the critical angle.

FADING   **Fading** is a term used to describe variations in signal strength that occur at a receiver during the time a signal is being received. Fading may occur at any point where both the ground wave and the sky wave are received, as shown in Figure 13-13(a). The two waves may arrive out of phase, thus producing a cancellation of the usable signal. This type of fading is encountered in long-range communications over bodies of water where ground-wave propagation extends for a relatively long distance. In areas where sky-wave propagation is prevalent, fading may be caused by two sky waves traveling different distances, thereby arriving at the same point out of phase, as shown in Figure 13-13(b). Such a condition may be caused by a portion of the transmitted wave being refracted by the *E* layer while another portion of the wave is refracted by the *F* layer. A complete cancellation of the signal would occur if the two waves arrived 180° out of phase with equal amplitudes. Usually, one signal is weaker than the other, and therefore a usable signal may be obtained.

FIGURE 13-13   **Fading.**

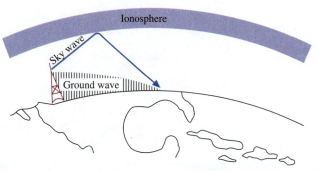

(a) Fading caused by arrival of ground wave and sky wave at the same point out of phase

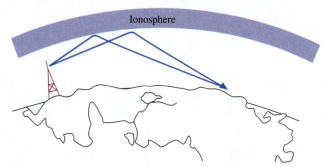

(b) Fading caused by arrival of two sky waves at the same point out of phase

Because the ionosphere causes somewhat different effects on different frequencies, a received signal may have phase distortion. As mentioned in Chapter 2, single sideband (SSB) is least susceptible to phase distortion problems. FM is so susceptible to these effects that it is rarely used below 30 MHz (where sky waves are possible). The greater the bandwidth, the greater the problem with phase distortion.

Frequency blackouts are closely related to certain types of fading, some of which are severe enough to blank out the transmission completely. The changing conditions in the ionosphere shortly before sunrise and shortly after sunset may cause complete blackouts at certain frequencies. The higher-frequency signals may then pass through the ionosphere, while the lower-frequency signals are absorbed.

Ionospheric storms (turbulent conditions in the ionosphere) often cause radio communications to become erratic. Some frequencies will be completely blacked out, while others may be reinforced. Sometimes these storms develop in a few minutes, and at other times they require as much as several hours to develop. A storm may last several days.

Tropospheric Scatter

Tropospheric scatter transmission can be considered as a special case of sky-wave propagation. Instead of aiming the signal toward the ionosphere, however, it is aimed at the troposphere. The troposphere ends just 6.5 mi above the Earth's surface. Frequencies from about 350 MHz to 10 GHz are commonly used with reliable communications paths of up to 400 mi.

The scattering process is illustrated in Figure 13-14. As shown, two directional antennas are pointed so that their beams intersect in the troposphere. The great majority of the transmitted energy travels straight up into space. However, by a little-understood process, a small amount of energy is *scattered* in the forward direction. As shown in Figure 13-14, some energy is also scattered in undesired directions. The best and most widely used frequencies are around 0.9, 2, and 5 GHz. Even then, however, the received signal is only one-millionth to one-billionth of the transmitted power. There is an obvious need for high-powered transmitters and extremely sensitive receivers. In addition, the scattering process is subject to two forms of fading. The first is a result of multipath transmissions within the scattering path, with the effect occurring as quickly as several times per minute. Atmospheric changes provide a second, but slower, change in the received signal strength.

FIGURE 13-14
Tropospheric scatter.

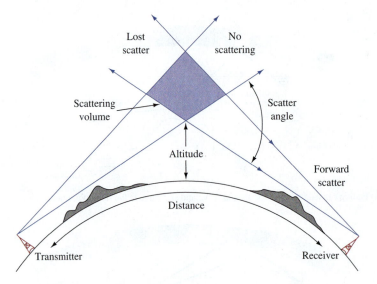

To accommodate these severe fading problems, some form of **diversity reception** is always used. This is the process of transmitting and/or receiving several signals and then either adding them all together at the receiver or selecting the best one at any given instant. The types of diversity reception utilized include one or combinations of the following:

Space diversity: comprising two or more receiving antennas separated by 50 wavelengths or more. The best-received signal at any instant is selected as input for the receiver.

Frequency diversity: transmission of the same information on slightly different frequencies. The different frequencies fade independently even when transmitted and received through the same antennas.

Angle diversity: transmission of information at two or more slightly different angles. This results in two or more paths based on illuminating different scattering volumes in the troposphere.

Polarization diversity: the capability of receiving horizontally and vertically polarized signals.

In spite of the high-power and diversity requirements and the more recent satellite communications, the use of tropospheric scatter continues since its first use in 1955. Tropospheric scatter provides reliable long-distance communication links in areas such as deserts and mountain regions and between islands, and it is used for voice and data links by the military and commercial users.

Our final category of wave propagation is that used in satellite communications (SATCOM). Communications via satellite is possible because of the placement of the satellites in **geostationary orbit** (sometimes called *geosynchronous* or *synchronous orbit*). This means that the satellite is located at a fixed point approximately 22,300 mi in altitude above the equator. At this altitude, the gravitational pulls of the Earth, the sun, and the moon work together, along with the centrifugal force caused by the satellite's rotation around the Earth to keep the satellite at a fixed location above the Earth. Of course, the satellite does drift (it moves in a figure-eight pattern) and must be periodically repositioned by on-board power thrusters to maintain the optimum location. But for us on Earth, the satellite appears to be stationary.

The satellite communication system consists of the following:

- **Uplink** (transmitter)
- Orbiting satellite
- **Downlink** (receiver)

The uplink and downlink are called an **Earth station** (ground base station), which will be typically transmitting and receiving data, video and/or audio, or it can be a receive-only site. Included in the Earth station are items such as the exciter, the high-power traveling wave tube amplifiers (TWTAs) (also called high-power amplifiers [HPAs]), a parabolic shaped reflector that is pointing at the satellites parked in geostationary orbit, and a receiver.

The satellites require a payload of antennas, transponders, and attitude controls for maintaining their location in geostationary orbit. The **transponder** is an electronic system performing reception, frequency translation, and retransmission. The **attitude controls** are used for orbital corrections (station keeping) on the satellite. These corrections are made approximately every 2 to 6 weeks. Geostationary satellites are parked over the equator (latitude = 0°) in orbit at a fixed longitude. The longitudinal position is referenced to Earth at the **subsatellite point**. The subsatellite point is the point on the Earth's surface where a line drawn from the satellite to the center of the Earth intersects the Earth's surface. The minimum spacing of the satellites is currently 2°.

An example of a geostationary satellite is the Boeing 601, first introduced in 1987. Boeing satellites are used for applications that include DirecTV, very small aperture (VSAT) business networks, as well as mobile satellite communications. The basic Boeing 601 configuration has up to 48 transponders and 4800 watts of power. The Boeing 601 HP (High Power version), first introduced in 1995, supports up to 60 transponders and provides up to 10,000 watts of power. A detailed picture of the satellite is provided in Figure 13-15.

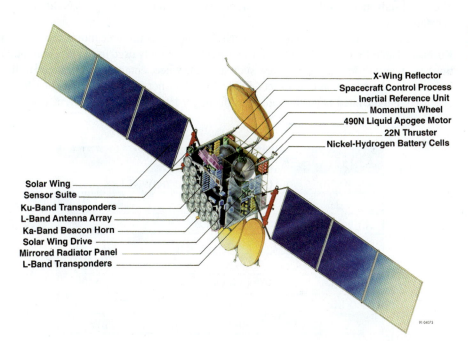

FIGURE 13-15 **A detailed view of the Boeing 601 satellite.** (Courtesy of Boeing.)

X-Wing Reflector
Spacecraft Control Process
Inertial Reference Unit
Momentum Wheel
490N Liquid Apogee Motor
22N Thruster
Nickel-Hydrogen Battery Cells

Solar Wing
Sensor Suite
Ku-Band Transponders
L-Band Antenna Array
Ka-Band Beacon Horn
Solar Wing Drive
Mirrored Radiator Panel
L-Band Transponders

Geostationary satellite communication has many advantages over terrestrial microwave communications and low-Earth orbit satellites. The geostationary satellite position has a fixed position with respect to the Earth; therefore, expensive tracking systems are not required. The path to and from the satellite is always available, except during certain weather conditions and solar disturbances. The satellite's transmission back to Earth covers a limited coverage area called a **footprint**, which is a geographical representation of a satellite's radiation pattern on the Earth. The footprint has contour lines showing areas of receiver power density expressed in dBW (dB watts). The radiation patterns are used to determine the expected satellite's signal strength when making link budget calculations (see Section 13-7). An example of a satellite footprint is provided in Figure 13-16.

FIGURE 13-16 **An example of a satellite footprint.**

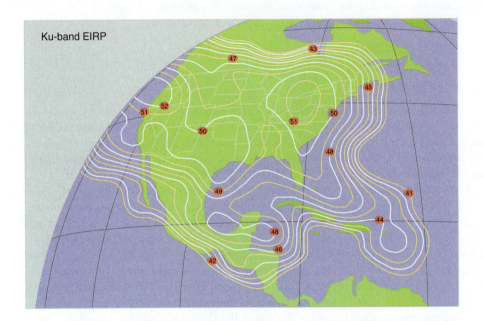

Some disadvantages to geostationary satellites are propagation delays because of the distance to and from the satellite (approximately 44,600 mi round trip). Satellite transmitters require more power because of the increased distance, resulting in additional transmitter costs. (It is difficult to fabricate high-power amplifiers that operate at high frequencies.) The last disadvantage is that there is a significant cost in maintaining a satellite parked in geostationary orbit.

The most common frequency bands used in satellite communications are C-band and Ku-band. Table 13-1 provides a list of frequencies commonly used in satellite communications for many different applications such as data, voice, entertainment, news, military, international broadcast, among others. Table 13-1 does not list all of the satellite frequencies used in the United States but provides the major frequency bands currently in use.

TABLE 13-1 • Satellite Frequency Bands		
BAND	**UPLINK (GHZ)**	**DOWNLINK (GHZ)**
L	1–2	Various
S	1.7–3	Various
C	5.9–6.4	3.7–4.2
X	7.9–8.4	7.25–7.75
Ku	14–14.5	11.7–12.2
Ka	27–30	17–20
	30–31	20–21

Orbital Patterns

The orbital patterns of satellites are elliptical. For each orbit there is a **perigee** (closest distance of the orbit to Earth) and an **apogee** (farthest distance of the orbit from Earth). This is shown in Figure 13-17.

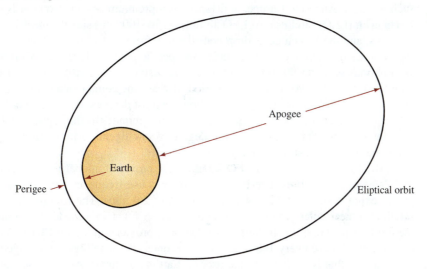

FIGURE 13-17   The perigee and apogee of a satellite's orbit.

Geostationary orbits use an equatorial orbit. Other possible satellite orbits are *polar* and *inclined orbits*. These orbits are shown in Figure 13-18. The geostationary satellites use an *equatorial orbit*.

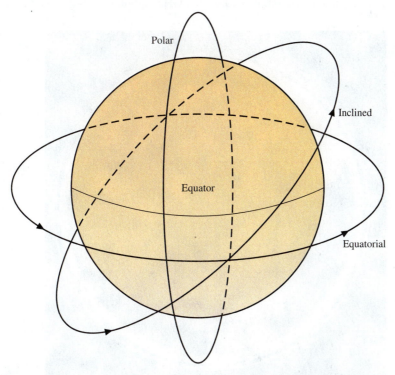

FIGURE 13-18   Orbital patterns for satellites.
(Courtesy of Iridium Satellite LLC.)

It is interesting to note that three *geostationary orbits,* parked 120° apart, can cover the entire Earth surface except for the polar regions above latitudes 76 N and 76 S. This assumes a minimum elevation of 5° for the receiving antennas. A satellite in a *polar orbit* can cover 100% of the Earth's surface. This is possible because the Earth is rotating as the satellite travels around the North Pole and the South Pole. Every location on Earth is visible to the satellite twice a day. *Inclined orbits* are used to reach the extreme northern and southern latitudes. The problem with inclined orbits is that the receiving station must track the satellite. An example of a highly inclined orbit is the Russian *Molniya,* which has a 63°

inclination angle and an orbital period of 12 hours. The apogee region is that where the satellite is above the Northern Hemisphere. At this point, the satellite is easiest to track. The satellite is visible to Earth stations from 4.5 to 10.5 hours each day.

Satellites in geosynchronous orbit have become numerous. International regulations limit their spacing to prevent interference. This puts the orbital slots over prime real estate, such as North America, Europe, and Japan, at a premium. Another option is to use **low Earth orbit (LEO) satellites**. At LEO altitudes (250–1000 mi) signal-time delay shrinks to 5–10 ms, and the launch costs drop considerably from that of the GEO satellites. These satellites are not stationary with respect to a specific point on Earth. They orbit the Earth with periods of about 90–100 minutes and are visible to an Earth station for only 5–20 minutes during each 90–100-minute period. If real-time communication is required, several LEO satellites are necessary. In addition, we must devise some method of handing off subscriber connections between satellites every few minutes as they appear and disappear over the horizon. This requires a high degree of intelligence within the network, much while the subscribers stay relatively still.

The Iridium system uses LEO satellites for providing global mobile satellite voice and data communication. The Iridium telephones use frequency-division/time-division multiplexing (FDMA/TDMA) with a data rate of 2.4 kbps. The Iridium system uses 66 satellites in near polar orbit (inclination angle of 86.4°) at an altitude of 485 miles above the Earth. A picture of the Iridium constellation is provided in Figure 13-19. The satellites orbit the Earth once every 100 minutes, 28 seconds. The satellites are arranged so that at least one satellite is available to the user on Earth at all times. At least four satellites are interlinked to each other so that communication with the Iridium ground station gateway is always available. The Iridium system uses L-band frequencies (1616–1626.5 MHz) for telephone and messaging services. The frequency link between satellites uses Ka-band (23.18–23.38 GHz) and uses Ka-band uplink and downlink frequencies to and from the ground station—downlink = 19.4–19.6 GHz; uplink = 29.1–29.3 GHz.

FIGURE 13-19 **A picture of the Iridium LEO satellite constellation.** (Courtesy of Iridium Satellite LLC.)

Azimuth and Elevation Calculations

The *azimuth* and *elevation* angles for the Earth station antenna must be calculated so that the correct satellite can be seen. This is called the **look angle**. The azimuth is the horizontal pointing angle of the Earth station antenna. The elevation is the angle at which we look up into the sky to see the satellite. To calculate the azimuth and elevation of a ground station antenna requires that the ground station latitude and longitude as well as the longitude of the satellite are known. The latitude and longitude of the Earth station can be obtained from U.S. Geological Survey maps or through the use of a Global Positioning System (GPS) receiver. Once the actual location is known, the elevation angle of the Earth station antenna can be calculated.

The equations for calculating the azimuth and elevation look angles are provided in Equations (13-8) and (13-9).

$$\tan(E) = \frac{\cos(G)\cos(L) - .1512}{\sqrt{1 - \cos^2(G)\cos^2(L)}}, \tag{13-8}$$

where E = elevation in degrees

S = satellite longitude in degrees

N = site longitude in degrees

$G = S - N$ in degrees

L = site latitude in degrees.

Next the azimuth can be calculated using Equation (13-9).

$$A = 180 + \arctan\left(\frac{\tan(G)}{\sin(L)}\right), \tag{13-9}$$

where A = azimuth of the antenna in degrees

S = satellite longitude in degrees

N = site longitude in degrees

L = site latitude in degrees

$G = S - N.$

Example 13-2 shows how to use Equations (13-8) and (13-9).

EXAMPLE 13-2

Calculate the azimuth and elevation angles for an Earth station (ground station) antenna given a satellite longitude of 83° west, a site longitude of 90° west, and a site latitude of 35° north.

Using Equation (13-9), the azimuth is equal to

$$A = 180 + \arctan\left(\frac{\tan(-7)}{\sin(35)}\right)$$

$$A = 180 + \arctan\left(\frac{-.128}{.5736}\right)$$

$$A = 168°.$$

The elevation angle is calculated using Equation (13-8)

$$\tan(E) = \frac{\cos(-7)\cos(35) - .1512}{\sqrt{1 - \cos^2(-7)\cos^2(35)}}$$

$$\tan(E) = \frac{.661846}{.582199} = 1.1368$$

$$\therefore E = \arctan(1.1368) = 48.663°.$$

Global Positioning System

The Global Positioning System (GPS) is another application made possible by satellite technology. It provides pinpoint geographic location information. GPS was originally used by the government and law enforcement agencies, but the availability of low-cost hand-held receivers has enabled personal use. You can now obtain your exact location when traveling by car or boat or when hiking. The GPS satellites transmit position data signals, and a GPS receiver processes and computes the time to receive each one. Doing this from four different satellites allows the receiver to determine your exact latitude and longitude.

GPS currently uses a constellation of twenty-eight satellites orbiting above the Earth at a distance of 10,900 mi. The GPS satellites complete an orbit about every 12 hours. The satellites transmit two signals, a coarse acquisition (C/A) signal transmitted on 1575.42 MHz, which is available for civilian use, and a precision code (P-code), transmitted on 1227.6 MHz and 1575.42 MHz, which are for military use only. GPS receivers measure the time it takes for the satellite signals to travel from the satellites to the receiver; from this information, the receiver can fix our position (i.e., locate where we are). It takes three satellites to fix our position in terms of latitude and longitude, whereas it takes four satellites to determine three-dimensional information: latitude, longitude, and elevation.

Civilian receivers can have a position accuracy of about 2 m, but this distance can vary. The position accuracy can be improved by using a technique called **differential GPS**. Receiver accuracy is improved by using a ground receiver at a known location to provide corrections to the satellite civilian signal error. With differential GPS, the accuracy of a GPS receiver can be improved to about 1 cm.

Multiplexing Techniques

A single satellite typically allows simultaneous communications among multiple users. Consider the situation shown in Figure 13-20. The satellite shown has a footprint (coverage area) as indicated. Some satellites use highly directional antennas so that the footprint may include two specific areas. For example, it may be desirable to use the same satellite to serve both Hawaii and the West Coast of the mainland United States. In that case, there is no sense in wasting downlink signal power over a large portion of the Pacific Ocean.

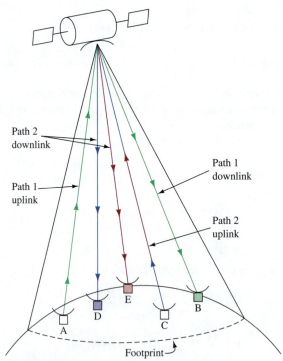

FIGURE 13-20   **Satellite footprint and multiple communications.**

In Figure 13-20, communication among five Earth stations is taking place simultaneously. Station A is transmitting to station B on path 1. Station C is transmitting to stations D and E on path 2. Control signals included with the original transmitted signals are used to allow reception at the appropriate receiver(s).

Two different multiplexing methods are commonly used to allow multiple transmissions with a single satellite. The early satellite systems all used **frequency-division multiple access (FDMA)**. In these systems, the satellite is a wideband receiver/transmitter that includes several frequency channels, much as how broadcast FM radio contains several channels. An Earth station that sends a signal indicating a desire to transmit is sent a control signal telling it on which available frequency to transmit. When the transmission is complete, the channel is released back to the "available" pool. In this fashion, a multiple access capability for the Earth stations is provided.

Most of the newer SATCOM systems use **time-division multiple access (TDMA)** as a means to allow a single satellite to service multiple Earth stations simultaneously. In TDMA, all stations use the same carrier frequency, but they transmit one or more traffic bursts in nonoverlapping time frames. This is illustrated in Figure 13-21, where three Earth stations are transmitting simultaneously but never at the same time. The traffic bursts are amplified by the satellite transponder and are retransmitted in a downlink beam that is received by the desired station(s). The computer control of these systems is rather elaborate, as you can well imagine.

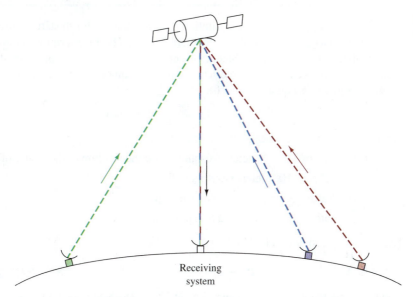

FIGURE 13-21   TDMA illustration.

Receiving system

TDMA offers the following advantages over FDMA systems:

- A single carrier. The traveling wave tube (TWTA) power amplifier in the satellite is much less subject to intermodulation problems and can operate at a higher power output when dealing with a smaller range of frequencies.

- Improved selectivity. In FDMA, the Earth station must transmit and receive on multiple frequencies and must provide a large number of frequency-selective up-conversion and down-conversion chains. In TDMA, the selectivity is accomplished in time rather than in frequency. This is much simpler and less expensive to accomplish.

- Suitability for digital communications. TDMA is ideally suited to digital communications because of the storage, rate conversions, and time-domain processing used in TDMA implementation. TDMA is also ideally suited to demand-assigned operation in which the traffic burst durations are adjusted to accommodate demand.

Earth Station Distance to and from the Satellites

In satellite communications, the distance from an Earth station to a satellite is used to estimate the time delay for a transmitted signal to travel from Earth to the satellite and back. The distance to the satellite varies from one Earth station site to another. It is common in satellite communications systems to have one channel on a satellite being used by many locations (*multiple access*). The technique used in a system such as this is TDMA. In this type of system, signals from the Earth stations must arrive at the satellite at fixed time intervals. If each Earth station is the same distance from the satellite, then the task of ensuring that the signals arrive at the satellite is simple. The reality is that Earth stations can be separated by thousands of miles and by several degrees in latitude and longitude. Therefore, the distance to the satellite and the time delay for the signal will vary. An explanation of how to calculate the distance from an Earth station to any satellite follows.

Table 13-2 provides necessary information for calculating the distance from a satellite to an Earth station. The information is provided in both kilometers and miles.

TABLE 13-2 • Earth Satellite Measurements		
MEASUREMENT	**KILOMETERS**	**MILES**
Mean equatorial Earth radius	6,378.155	3,963.2116
Distance from a satellite to a subsatellite point	35,786.045	22,236.4727
Distance to geostationary orbit from the center of the Earth	42,164.200	26,199.6843

Calculating the distance to a satellite requires that the Earth station latitude and longitude as well as the satellite's longitude are known. The JavaScript program available at http://web.nmsu.edu/~jbeasley/Satellite/ can be used for computing Earth station to and from distance as well as the round trip delay. The equation being used in the JavaScript program is given in Equation (13-10).

$$d = \sqrt{D^2 + R^2 - 2DR\cos\alpha\cos\beta}, \qquad \text{(13-10)}$$

where d = distance to the satellite (meters)

$D = 42.1642 \times 10^6$ meters (distance from the satellite to the center of the Earth)

$R = 6.378 \times 10^6$ meters (Earth's radius)

α = Earth station (site) latitude (in degrees)

β = satellite longitude–site longitude (in degrees).

Example 13-3 provides a look at how to apply Equation (13-10).

E X A M P L E   1 3 - 3

Calculate the distance from an uplink at 32°44′36″ N latitude and 106°16′37″ to a satellite parked in geostationary orbit at 99° W longitude.

Equation (13-10) requires that the Earth station latitude and longitude be express in degrees rather than in degrees, minutes, and seconds.

N LATITUDE 32°44′36″

$$\text{converted to degrees} = 32 + \frac{44}{60} + \frac{36}{3600} = 32.74333°$$

W LONGITUDE 106°16′37″

$$\text{converted to degrees} = 106 + \frac{16}{60} + \frac{37}{3600} = 106.2769448°.$$

Next, insert the data into Equation (13-10) and solve for the distance from the Earth station to the satellite.

$$d = \sqrt{D^2 + R^2 - 2DR\cos\alpha\cos\beta},$$

where d = distance to the satellite [meters]

$D = 42.1642 \times 10^6$ meters [distance from the satellite to the center of the Earth]

$R = 6.378 \times 10^6$ meters [Earth's radius]

$\alpha = 32.74333°$

$\beta = 99°\text{ W} - 106.27694 = -7.27694.$

Therefore

$$d = \sqrt{(42.1642 \times 10^6)^2 + (6.378 \times 10^6)^2 - (2)(42.1642 \times 10^6)(6.378 \times 10^6)}$$
$$\cos(32.74333)\cos(-7.27694)$$
$$d = 37,010 \times 10^6 \text{ meters.}$$

The time required for the signal to travel from the Earth station given in Example 13-3 to the satellite parked at 99° W can be calculated by dividing the distance (d) by the velocity of light (c) of 2.997925×10^5 km/s. The equation is written as follows:

$$\text{delay} = \frac{d}{c} = \frac{\text{distance}}{\text{velocity of light}} = \frac{d}{2.997925 \times 10^5}\,\text{km/s} \cong \frac{d}{3 \times 10^5}\,\text{km/s}$$

$$\text{The roundtrip delay} = \frac{2d}{c}. \tag{13-11}$$

Using Equation (13-11), the time delay for the signal to travel the distance to the satellite and the round trip delay will equal to

$$\text{delay} = \frac{d}{c} = \frac{37010.269\ \text{km}}{2.997925 \times 10^5\ \text{km/s}} = 0.123\ \text{seconds}$$

$$\text{The roundtrip delay} = \frac{2d}{c} = (2)(.123) = 0.2469\ \text{seconds}$$

It should be mentioned that a third multiplexing technique is being used. **Code division multiple access (CDMA)** also allows the use of just one carrier. In it, each station uses a different binary sequence to modulate the carrier. The control computer uses a "correlator" that can separate and "distribute" the various signals to the appropriate downlink station. (See Chapter 10 for discussions on CDMA.)

VSAT and MSAT Systems

Two other important areas of satellite communications are (1) very small aperture terminal (VSAT) fixed satellite communication systems and (2) ultrasmall aperture terminal mobile satellite (MSAT) systems. Technological advances and market demand have driven the development of these new markets. MSAT terminals, which can be called "VSATs on wheels," have several features in common with VSATs. An application of a MSAT system includes large, national trucking firms that use MSAT technology to maintain continuous communication with their trucks. Whereas VSATs take telecommunication services directly to fixed users, MSAT terminals take them to moving vehicles.

Conventional VSAT systems allow multiple inexpensive stations to be linked to a large, central installation. For example, Wal-Mart installed small aperture antenna systems (VSATs) at each of its stores and linked them with its central mainframe computer in Arkansas. This arrangement allows Wal-Mart to quickly convey data, such as what customers are buying and how much inventory is on hand. They can both supply each store with the items its customers are buying and speed up the checkout process. The VSAT dish antenna is typically 0.5–1.2 meters in diameter, and a transmitter power of just 2–3 W is sufficient.

VSAT-based systems are also available for home Internet services. WildBlue offers high-speed (1.5 Mbps-download, 256 kbps-upload) Internet access. WildBlue uses a Ka-band communication link that connects to the Telesat's Anik F2 satellite parked at 111.1° W longitude. The satellite uplink frequency is 29.5–30.0 GHz, and the downlink frequency is 19.7–20.2 GHz. The WildBlue minidish is an elliptical reflector measuring 25.6″ high and 29.1″ wide.

Figure 13-22 provides a pictorial representation of Chrysler's VSAT network. It connects the automaker's headquarters with more than six thousand dealerships and corporate facilities in North America. It is used to assist mechanics with repair and allows salespeople to order and confirm delivery dates for cars from a showroom computer. It also helps maintain proper inventories of automobiles and spare parts. Satellite systems have the advantage of simultaneous delivery of information to and from multiple sites through the use of TDMA techniques. The transmit power requirements are minimal, and all sites can share access.

Satellite Radio

The FCC allocated RF spectrum in the S-band (2.3 GHz) for the Digital Audio Radio Service (DARS) in 1992. In the U.S. the two satellite radio services, XM and Sirius, that were formerly competitors, have merged. XM began in 2001, and Sirius began in 2002. The

FIGURE 13-22 VSAT network.

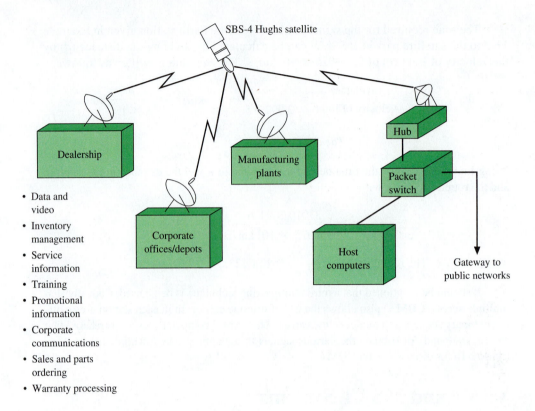

satellites used by these radio services orbit in geostationary or inclined orbital patterns. Both patterns were shown in Figure 13-18.

The XM Satellite Radio service uses two geostationary satellites. These satellites are parked approximately 22,300 mi above the Earth at a fixed location. The difference in the distance of the apogee and the perigee for geostationary satellites is minimal. The Sirius Satellite Radio service uses three satellites in an inclined orbit. Each satellite is above the continental United States at least 16 hours each day. The orbits of the satellites are arranged so that at least one satellite is always over the United Sates. The apogee for the Sirius satellites is 29,200 mi above the Earth, and, at this point, the satellites are over North America. The perigee for the Sirius satellites is 14,900 mi.

Reception of satellite radio services requires an antenna and custom chip sets to process the received signal. Reception of the Sirius Satellite Radio signal is made possible using diversity receivers (in addition to the special chip sets) and antennas. Receiver diversity is the reception of two signals from two satellites at any given time and the selection of the best one. This process is called *spatial diversity*. The Sirius satellites transmit the radio signal on three different frequencies in the 12.5 GHz band. Once again, the best-received signal is selected. The Sirius system also uses time diversity, which is provided by delaying the audio by about 4 seconds. This delay is accomplished by storing the satellite's digital data stream so that a momentary loss of signal does not interrupt the audio feed.

13-7 FIGURE OF MERIT AND SATELLITE LINK BUDGET ANALYSIS

Most satellite equipment providers, such as satellite television service providers, will provide the minimum equipment the end user needs, based on the user's geographic location and the satellite services required. In the case in which prepackaged equipment is not provided, there are two important equations that should always be examined when specifying a satellite Earth station system. These equations are as follows:

1. Figure of merit (G/T)
2. Satellite link budget (C/N)

This section presents the equations for calculating the **figure of merit (G/T)** and the **satellite link budget**. *Note:* An online satellite system calculator has been developed specifically for this textbook that uses the satellite calculations presented in the text. The URL for the online calculator is http://web.nmsu.edu/~jbeasley/Satellite/.

Not all satellite systems or configurations are equivalent. There may be a large antenna (reflector) but a poor amplifier on the front end. Another system could have a smaller antenna (reflector) but an excellent amplifier on the front end. Which is better? The figure of merit is a way to compare different Earth station receivers. The figure of merit takes into consideration the technical quality of each piece of the satellite Earth station equipment and enables the end user to obtain some measure of performance for the entire system. The final figure of merit can then be used as a comparison to other Earth stations.

With regard to the satellite link budget, satellite receivers will have a required carrier-to-noise (C/N) at the input. The satellite link budget is used to verify that the required C/N and signal level to the satellite receiver will be met to ensure the satellite receiver outputs a signal that meets specifications. A receive-signal level that does not meet required C/N specifications can result in excessive bit error rates (BER) for digital satellite receivers and an extremely noisy signal for analog receivers.

Figure of Merit

The figure of merit is used to provide a performance measure for different satellite Earth stations. Orbiting satellites also have a figure of merit (G/T), and this value is available from the satellite service provider. The larger the figure of merit (G/T), the better the Earth station system. The equation defining figure of merit is provided in Equation (13-12).

$$G/T = G - 10 \log(Ts), \tag{13-12}$$

where G/T = figure of merit (dB)

G = antenna gain (dBi)

Ts = sum of all T_{eq} (noise figure measurements).

Three critical components significantly contributing to noise figure (T_{eq}) should always be examined when selecting an Earth station. These are:

- The antenna
- LNA, LNB (LNC)

  LNA (low-noise amplifier)

  LNB (low-noise block-converter)

  LNC (low-noise converter)
- Receiver and passive components

The equivalent noise temperature for a satellite antenna can be obtained from the manufacturer. Typical noise temperatures for satellite antennas can be approximately 30K (Kelvin) or less. It is important to observe that the first amplifier stage in the Earth station, an LNA, an LNB, or an LNC dominates the sum of T_{eq}, noise figure measurements. LNAs, LNBs, and LNCs are sold according to their noise temperature specifications, and, typically, the lower the noise temperature, the more expensive the device.

An example of the need for a critical first stage amplifier is the selection of an LNA (low-noise amplifier) for a satellite receiver. The received voltage, obtained from a received satellite signal, is very small (μV) and a high-gain amplifier is required to use the signal. Therefore, for electronic systems amplifying small received signal voltage levels, the first stage needs to exhibit low-noise characteristics (small NF) and have a high gain (G). The third contribution to the sum of noise figures is from the receiver's passive components, but these noise figure contributions are typically very small and are demonstrated in Example 13-4.

Low-noise amplifiers are typically specified by their equivalent noise temperature rating T_e. The relationship for T_e and the noise figure, NF and noise factor, F are shown in Equations (13-13) and (13-14).

$$T_{eq} = T_o(F - 1), \tag{13-13}$$

where $T_o = 290\text{K (room temperature)}.$

$$NF(\text{dB}) = 10\log F = 10\log\left(\frac{T_{eq}}{T_o} + 1\right), \tag{13-14}$$

where $F = \dfrac{T_{eq}}{T_o} + 1.$

An example of calculating the figure of merit (G/T) for a satellite receiver is provided in Example 13-4, and calculating the free space path loss in Example 13-5.

EXAMPLE 13-4

Determine the figure of merit (G/T) for a satellite Earth station with the following parameters. Compare the figure of merit for this Earth station with another Earth station that has a 22.5-dB G/T rating.

Antenna gain = 45 dBi

Antenna noise temperature = 25K

LNB noise temperature = 70K

Noise temperature (receiver and passive components) = 2K

First calculate the sum of all of the noise temperature contributions.

$$T_s = (25 + 70 + 2)\text{K} = 97\text{K}$$

Use Equation (13-12) to calculate the figure of merit (G/T)

$$G/T = G - 10\log(T_s) \tag{13-12}$$
$$= 45 - 10\log(97)$$
$$= 45 - 10 \times 1.97$$
$$G/T = 25.13 \text{ dB}$$

Earth Station Comparison

The figure of merit for this Earth station (25.3 dB) is superior to the Earth station with the 22.5 dB G/T rating.

Satellite Link Budget Calculation

The next important equation that should always be examined when specifying a satellite Earth station calculates the satellite link budget. The satellite link budget is used to evaluate the quality of a satellite link signal in terms of the C/N and to make sure the satellite link will meet required C/N specifications. Satellite receivers will have some minimum C/N specification or a minimum input receive signal level. This minimum specification must be met for the satellite receiver to meet the maximum allowed BER or signal-to-noise specification. The total satellite link budget will be determined from both the *uplink budget* and the *downlink budget*.

An important parameter to evaluate for a satellite link is the **free-space path loss**. This is the attenuation of the RF signal as it propagates through space and the Earth's atmosphere to and from the satellite. There will be an uplink path loss (Earth station-to-satellite) and a downlink path loss (satellite-to-Earth station). The free-space path loss will be the biggest loss value (in dB) listed in a satellite link budget, with typical values ranging from 180 dB to 220 dB that depend on frequency and geographic location. The longer the distance the satellite signal has to travel, the greater the attenuation. It is also important to note that the free-space path loss is a function of the wavelength of the transmit frequency. This means that the smaller the wavelength (i.e., the higher the frequency), the greater the path loss. Free-space path loss can be calculated using Equation (13-15).

$$L_p(dB) = 20\log\left(\frac{4\pi d}{\lambda}\right), \tag{13-15}$$

where L_p = free space path loss (dB)

d = distance (meters)

λ = wavelength (meters).

EXAMPLE 13-5

Calculate the free-space path loss from an Earth station uplink to a satellite if the distance is 41.130383×10^6 m and the uplink frequency is 14.25 GHz.

(a)   Use Equation (13-15) to calculate the free-space path in dB.

The wavelength is calculated by

$$\lambda = \frac{c}{f} = \frac{2.997925 \times 10^8 \text{ m/s}}{14.25 \times 10^9 \text{ Hz}} = 0.0210381 \text{ m}.$$

The free-space path loss (L_p) expressed in dB is

$$L_p(dB) = 20\log\left(\frac{4\pi d}{\lambda}\right) = 20\log\left(\frac{4\pi 41.130383 \times 10^6}{0.0210381}\right) = 207.807 \text{ dB}.$$

The JavaScript program available at http://web.nmsu.edu/~jbeasley/Satellite/ can be used to compute the free-space path loss.

Now that all of the information is available regarding the gains and losses for the Earth station and the satellite link, a link budget can be prepared. The uplink and downlink budgets are determined by summing the gains and losses for the link. The satellite uplink budget will take into consideration the following:

UPLINK	
GAINS	**LOSSES**
Uplink power (EIRP)	Free-space path loss of the signal strength as it travels from the Earth station to the satellite
Satellite G/T	Atmospheric losses and possibly the pointing error of the Earth station
Boltzmann's constant adjustment	Bandwidth
Note: This value equals −228.6 dBW/K/Hz, which is obtained by taking $10\log(1.38 \times 10^{-23})$	

DOWNLINK	
GAINS	**LOSSES**
Satellite downlink power obtained from the satellite footprint	Free-space path loss of the signal strength as it travels from the satellite to the Earth station
Earth station G/T	Atmospheric losses and possibly the pointing error of the Earth station
Boltzmann's constant adjustment	Bandwidth
Note: This value equals −228.6 dBW/K/Hz, which is obtained by taking $10\log(1.38 \times 10^{-23})$	

Note: Commercially available satellite link budget calculators will include many more parameters in the link budget, but this example provides a good estimate of the expected C/N at the satellite and the Earth station.

The equation used in the JavaScript program for calculating the uplink C/N (dB), and downlink C/N (dB) are provided in Equations (13-16) and (13-17).

Uplink budget

$$C/N = 10\log A_t P_r - 20\log\left(\frac{4\pi d}{\lambda}\right) + 10\log\frac{G}{T_e} - 10\log L_a - 10\log K$$

$$- 10\log BW + 228.6 \text{ dBW/K/Hz}, \tag{13-16}$$

where A_t = Earth station transmit antenna gain (absolute)

   P_r = Earth station transmit power (watts)

   d = distance to the satellite from the Earth station (meters)

   λ = wavelength of the transmitted signal (meters/cycle)

G/T_e = satellite figure of merit

   L_a = atmospheric losses

   BW = Bandwidth.

Downlink budget

$$C/N = 10\log A_t P_r - 20\log\left(\frac{4\pi d}{\lambda}\right) - 10\log\frac{G}{T_e}$$

$$- 10\log L_a + 228.6 \text{ dBW/K/Hz}, \tag{13-17}$$

where A_t = satellite transmit antenna gain (absolute)

   P_r = satellite transmit power (watts)

   d = distance to the earth station from the satellite (meters)

   λ = wavelength of the transmitted signal (meters/cycle)

G/T_e = earth station figure of merit

   L_a = atmospheric losses.

It was previously mentioned that satellite receivers will have a required C/N and/or a minimum signal level at the receiver input to meet the system requirements. The link budget calculation is used to verify that the minimum requirements will be met for the uplink (Earth station to satellite) and for the downlink (satellite to Earth station). In the case in which the required C/N is not met, the equipment specified may need to modified. On the uplink side, this can require a larger antenna or an increased transmit power. On the downlink side, this can require a larger antenna or better LNA, LNB, or LNC. An example of using the JavaScript program to calculate the satellite link budget is provided in Example 13-6.

EXAMPLE 13-6

Use the online program at http://web.nmsu.edu/~jbeasley/Satellite/ to calculate the link budget for an Earth station located at 32°18′N latitude and 106°46′W longitude that will be linked to a satellite parked at 99°W longitude. A data rate of 10 Mbps using 8-PSK modulation is being used, which requires a bandwidth of 3.33 MHz–(65.22 dB). You are given the following information for the Earth station and satellite. The required C/N at the satellite is 6 dB, and the required C/N at the Earth station is 12 dB. Comment on the results obtained from the satellite link budget in regard to whether the received C/N is or is not acceptable.

Earth Station

Uplink frequency	14.274 GHz
Antenna Diameter	4.5 meters
Antenna Efficiency	0.6%
Earth Station G/T	30.6 dB/K
Transmit Power	3 W

Satellite

Downlink Frequency	11.974 GHz
Satellite EIRP	40.1 dBW
Satellite G/T	0.9 dB/K

Uplink—Solution

EIRP	+59.11 dBW
Path Loss	−205.89 dB
Satellite G/T	+0.9 dB/K
Bandwidth	−65.22 dB
Boltzmann's Constant	+228.6 dBW/K/Hz
Uplink C/N	16.59 dB

Downlink—Solution

EIRP	+40.1 dBW
Path Loss	−204.37 dB
Earth Station G/T	+30.6 dB/K
Bandwidth	−65.22 dB
Boltzmann's Constant	+228.6 dBW/K/Hz
Downlink C/N	27.7 dB

The calculated values for the uplink and the downlink C/N are both well within specified guidelines. There is sufficient margin for both the uplink and the downlink to allow for additional atmospheric losses and equipment degradation.

SUMMARY

Radio waves are a form of electromagnetic radiation, as is light. James Clerk-Maxwell predicted their existence, and Heinrich Hertz demonstrated electromagnetic waves at radio frequencies. All forms of electromagnetic radiation are capable of propagating in free space and consist of electric and magnetic field components at right angles to each other and transverse to the direction of propagation. Radio waves are self-sustaining and are the result of the interaction of changing electric and magnetic fields. Like light, radio waves are capable of being reflected, refracted, and diffracted.

Radio waves can take one or more routes of propagation from transmitter to receiver. Below 2 MHz, ground-wave propagation is significant. Ground waves follow the contours of the Earth and are not affected by mountain ranges or other obstructions to a significant degree. From 2 to 30 MHz, the high-frequency bands, sky-wave propagation is significant, and communication over hundreds or thousands of miles is possible because of the skip effect. The skip effect is the result of reflections from the ionosphere, a charged particle layer whose distance from the Earth's surface varies based on time of day and season. Also pertinent to high-frequency communications are the notions of critical frequency and critical angle. The critical frequency is the highest frequency returned to Earth when transmitted vertically, and the critical angle is the highest angle at which a radio wave at a specific frequency can be propagated and returned to Earth from the ionosphere. Related concepts are the maximum usable frequency, the highest frequency returned to Earth from the ionosphere between two specific transmit/receive points, and the optimum working frequency, which provides for the most consistent communication path.

Another important form of radio-wave propagation is that embodied by satellite systems. Some satellites use geosynchronous orbits to maintain a constant relative position to Earth, while others operate in low-Earth orbit. They use a variety of multiplexing techniques to increase capacity; digital satellite systems are becoming increasingly reliant on time-division multiple-access (TDMA) and code-division multiple-access (CDMA) multiplexing because of the advantages over frequency-division multiple access (FDMA). Also becoming increasingly widespread are very small aperture terminal (VSAT) for fixed installations and ultrasmall aperture terminal (MSAT) for mobile installations. Finally, a successful satellite installation is contingent on maintaining an adequate carrier-to-noise (C/N) ratio, and this is in turn dependent on figure-of-merit and link budget calculations.

QUESTIONS AND PROBLEMS

SECTION 13-1

1. Explain why an antenna can be thought of as a transducer.

2. List the similarities and dissimilarities between light waves and radio waves.

SECTION 13-2

3. What are the two components of an electromagnetic wave? How are they created? Explain the two possible things that can happen to the energy in an electromagnetic wave near a conductor.

*4. What is *horizontal and vertical polarization* of a radio wave?

*5. What kinds of fields emanate from a transmitting antenna, and what relationships do they have to each other?

6. Define *wavefront.*

7. Calculate the power density in watts per square meter (on Earth) from a 10-W satellite source that is 22,000 mi from Earth. (6.35×10^{-16} W/m^2)

8. Calculate the power received from a 20-W transmitter, 22,000 mi from Earth, if the receiving antenna has an effective area of 1600 mi^2. (2.03×10^{-12} W)

9. Calculate the electric field intensity, in volts per meter, 20 km from a 1-kW source. How many decibels down will that field intensity be if the distance is an additional 30 km from the source? (8.66 mV/m, 7.96 dB)

10. Calculate the characteristic impedance of free space using two different methods.

*11. How does the field strength of a standard broadcast station vary with distance from the antenna?

12. Define *permeability.*

SECTION 13-3

13. In detail, explain the process of reflection for an electromagnetic wave.

14. With the aid of Snell's law, fully explain the process of refraction for an electromagnetic wave.

15. What is *diffraction* of electromagnetic waves? Explain the significance of the shadow zone and how it is created.

16. Write the equation for the coefficient of reflection. ($\rho = \varepsilon_r / \varepsilon_i$)

17. Define *refraction.*

18. Define *shadow zone.*

SECTION 13-4

19. List the three basic modes whereby an electromagnetic wave propagates from a transmitting to a receiving antenna.

20. Describe ground-wave propagation in detail.

21. Explain why ground-wave propagation is more effective over seawater than desert terrain.

*22. What is the relationship between operating frequency and ground-wave coverage?

*23. What are the lowest frequencies useful in radio communications?

24. Fully explain space-wave propagation. Explain the difference between a direct and reflected wave.

SECTION 13-5

25. List the course of events in the process of sky-wave propagation.

26. Provide a detailed discussion of the ionosphere—its makeup, its layers, its variations, and its effect on radio waves.

*27. What effects do sunspots and the aurora borealis have on radio communications?

---

*An asterisk preceding a number indicates a question that has been provided by the FCC as a study aid for licensing examinations.

28. Define and describe *critical frequency, critical angle,* and *maximum usable frequency* (MUF). Explain their importance to sky-wave communications.

29. What is the optimum working frequency, and what is its relationship to the MUF?

30. What frequencies have substantially straight-line propagation characteristics analogous to those of light waves and are unaffected by the ionosphere?

31. What radio frequencies are useful for long-distance communications requiring continuous operation?

32. In radio transmissions, what bearings do the angle of radiation, density of the ionosphere, and frequency of emission have on the length of the skip zone?

33. Why is it possible for a sky wave to "meet" a ground wave 180° out of phase?

34. What is the process of tropospheric scatter? Explain under what conditions it might be used.

*35. What is the purpose of a diversity antenna receiving system?

36. List and explain three types of diversity reception schemes.

37. What is skipping?

38. Define *fading.*

39. What happens when a signal is above the critical frequency?

SECTION 13-6

40. What is *satellite communications?* List reasons for their increasing popularity.

41. Explain the differences between GEO and LEO satellite systems. Describe the advantages and disadvantages of each system.

42. Describe a typical VSAT installation. How does it differ from an MSAT system?

43. Explain the methods of multiplexing in SATCOM systems, and provide the advantages of TDMA over FDMA.

44. Use the satellite footprint of the Telstar 5 provided in Figure 13-16 to determine the expected EIRP for the signal in your area.

45. An Earth station is located at 98° W longitude and 35.1° N latitude. Determine the azimuth and elevation angles for the Earth station if the antenna is to be pointed at a satellite parked at 92° W longitude.

46. What two signals does a GPS satellite transmit? How are they used, and what frequencies are being used?

47. What is the distance to a satellite parked at 69° W longitude from an Earth station located at 29° N latitude and 110°36'20" W longitude? Calculate the round-trip time delay for a signal traveling from the Earth station to the satellite and back.

48. Define *apogee* and *perigee.*

49. What is the altitude and orbital period of the Iridium LEO satellites? How many satellites are there, and what type of orbital pattern is used?

SECTION 13-7

50. Calculate the noise factor (NF) in dB for a 100° LNA.

51. Determine the figure of merit for a satellite Earth station with the following specifications:

    Antenna Gain—48 dBi
    Reflector noise temperature—28 K
    LNA noise temp—55 K
    Noise temp. (various components)—3 K

52. Calculate the free-space path loss for a link between a satellite parked at 89° W longitude and an Earth station at 29° N latitude and 11036'20" W longitude. The downlink frequency is 11.974 GHz.

53. Prepare a satellite link budget for an Earth station located at 35° 10' N latitude and 99°15' W longitude that will be linked to a satellite parked at 91° W longitude. A data rate of 6 Mbps using 8-PSK modulation is being used, which requires a bandwidth of 2 MHz (63 dB). The required C/N at the satellite is 8 dB, and the required C/N at the Earth station is 15 dB. Comment on the results obtained from the satellite link budget in regard to whether the received C/N is or is not acceptable.

Earth Station

Uplink frequency	14.135 GHz
Antenna Diameter	5.0 meters
Antenna Efficiency	0.65
Earth Station G/T	31.2 dBK
Transmit EIRP	62 dBW (4.5 W)

Satellite

Downlink Frequency	11.752 GHz
Satellite EIRP	38.2 dBW
Satellite G/T	0.9 dBK

QUESTIONS FOR CRITICAL THINKING

54. A user complains about "interference." How can you determine whether this is electromagnetic interference (EMI) or radio-frequency interference (RFI)?

55. Calculate the radio horizon for a 500-ft transmitting antenna and a receiving antenna of 20 ft. Calculate the required height increase for the receiving antenna if a 10 percent increase in radio horizon were required. (37.9 mi, 31.2 ft)

56. In the strictest sense, define *skip distance* and *skip zone*.

57. You will be receiving sky waves. In what ways can you anticipate fading to occur?

CHAPTER 14

ANTENNAS

CHAPTER OUTLINE

KEY TERMS

reciprocity
polarization
half-wave
dipole
radiation field
induction field
near field
far field
radiation pattern
omnidirectional
directional
beamwidth
antenna gain
dBi
dBd
radiation resistance
corona discharge
feed line
delta match
monopole antenna
image antenna
counterpoise
loading coil
antenna array

parasitic array
driven array
reflector
director
lobes
front-to-back ratio (F/B ratio)
collinear array
phased array
null
twin lead
millimeter (mm) waves
circular horn
pyramidal horn
sectoral horn
microwave dish
prime focus feed
Cassegrain feed
offset feed
polar pattern
effective aperture
radome
zoning
patch antenna

In this chapter we introduce the fundamentals of antennas and describe the most commonly encountered types. Chapter 13 established the idea that an antenna is a *transducer,* a device that converts one form of energy to another. A transmit antenna is a conductor or system of conductors that provides a transition from a guided wave of electrical energy to an electromagnetic wave propagating in free space. Likewise, a receive antenna, also consisting of a single conductor or an array of conductors, converts an electromagnetic wave cutting across it back to an electrical signal in the form of alternating voltages and currents. The electromagnetic wave was also shown to be composed of both electric and magnetic fields that exist at right angles to each other and at right angles to the direction of propagation. Electromagnetic energy propagates outward equally in all directions from an isotropic point source. Such a source is a theoretical construct, however, not a physically realizable component. This chapter extends the idea of the point-source radiator to actual antennas.

We will demonstrate shortly that systems of wires and reflectors can be used to focus electromagnetic energy in a preferred direction, but in its most basic form an antenna is a passive element consisting of nothing more than a single conductor. Electromagnetic energy is produced as the result of accelerating and decelerating electric charges moving within the conductor. A sinusoidal voltage and current applied to the conductor causes the charges residing within it to move. The accelerating charges create voltage potentials and, hence, electric fields. The presence of a potential difference is sufficient to create current flows, which give rise to a moving magnetic field at right angles to the electric field. The moving magnetic field so created begets another electric field some distance away from it, and the process continues as the two fields reproduce in tandem while propagating away from the conductor that created them. The wave created from moving electric and magnetic fields thereby propagates from its point of origin through space to its destination.

The process of converting electrical to electromagnetic energy causes an antenna to appear as a load at the end of the transmission line to which it is attached. Just as a resistor acts as a load by converting electrical to thermal energy, the antenna acts as a load by converting electrical to electromagnetic energy. Further, the antenna has the impedance characteristics of a series circuit with inductive or capacitive reactance as well as resistance. At its resonant frequency the antenna will act as a purely resistive load, but at frequencies above or below resonance it will exhibit reactive characteristics as well. Any mismatch, resistive or reactive, manifests itself in the form of reflections on the transmission line. This less-than-perfect match does not mean that the antenna will not work. Rather, it simply means that power transfer is less than maximum. It also implies that the antenna will be effective over a span of frequencies, and, indeed, the antenna can be said to have an effective bandwidth. To have adequate signal strength at the receiver, either the power transmitted must be extremely high or the efficiency of the transmitting and receiving antennas must be high because of the high losses in wave travel between the transmitter and the receiver.

Any receiving antenna transfers energy from the atmosphere to its terminals with the same efficiency with which it transfers energy from the transmitter into the atmosphere. This property of interchangeability for transmitting and receiving operations is known as antenna **reciprocity**. Antenna reciprocity occurs because antenna characteristics are essentially the same regardless of whether an antenna is sending or receiving electromagnetic energy. Because of reciprocity, we will generally treat antennas from the viewpoint of the transmitting antenna, with the understanding that the same principles apply equally well when the antenna is used for receiving electromagnetic energy.

Effective antenna operation requires that the transmitting and receiving antennas have the same polarization. **Polarization** is defined in terms of the orientation of the electric field, which is the same as the antenna's physical configuration. Thus, a vertical antenna will transmit a vertically polarized wave. The received signal is theoretically zero if a vertical E field cuts through a horizontal receiving antenna.

The received signal strength of an antenna is usually described in terms of the electric field strength. If a received signal induces a 10-μV signal in an antenna 2 m long, the field strength is 10 μV/2 m, or 5 μV/m. Recall from Chapter 13 that the received field strength is inversely proportional to the distance from the transmitter [Equation (13-2)].

14-2 HALF-WAVE DIPOLE ANTENNA

Any antenna having an electrical length of one-half wavelength at the applied frequency is called a half-wave dipole antenna. Half-wave dipole antennas are predominantly used with frequencies above 2 MHz. It is unlikely that a half-wave dipole antenna will be found in applications below 2 MHz because at these low frequencies this antenna is physically too large. Consider a half-wave dipole antenna for a 60-Hz signal.

$$\lambda = \frac{c}{f} = \frac{186,000 \text{ mi/s}}{60} = 3100 \text{ mi}$$

A A$\frac{1}{2}\lambda$ antenna for 60 Hz is therefore 3100 mi/2, or 1550 mi!

Development of the Half-Wave Dipole Antenna

The open-circuited, two-wire transmission line introduced in Chapter 12 and shown in Figure 14-1 suffers from excessive radiation at high frequencies. Radiation from a transmission line is undesirable since a perfect transmission line would have no losses. However, the two-wire transmission line can be made into an effective antenna by spreading the conductors such that they are no longer parallel but rather 180° apart from each other. For this reason, an analysis of the open-ended, quarter-wave transmission line will furnish an excellent introduction for understanding basic antenna theory.

Voltage at the open end of the line is maximum, while current is zero, regardless of the wavelength of the line. The opposite relationship between voltage and current holds true for a shorted line. On either the open or shorted line, standing waves will be produced. Because the voltage applied to the line is sinusoidal, the line will constantly be charging and discharging. Current will be flowing in the line continuously. Because the current at the end of an open-circuited line is minimum, a quarter-wave back (facing toward the source), the current must be maximum. The impedance at the sending end is low, and the impedance at the open circuit is high. At the open end, E is high and I is very low. This causes the impedance, Z, which is equal to E/I, to be very high. The opposite situation exists at the sending end. The standing waves of current and voltage are shown on the quarter-wave section in Figure 14-1.

It is desirable to have maximum radiation from an antenna. Under such conditions all energy applied to the antenna would be converted to electromagnetic waves and radiated. This maximum radiation is not possible with the two-wire transmission line because the magnetic field surrounding each conductor of the line is in a direction that opposes the lines of force about the other conductor. Under these conditions, the quarter-wave transmission line proves to be an unsatisfactory antenna because the opposing forces cancel; however, with only a slight physical modification, this section of transmission line can be transformed into a relatively efficient antenna. This transformation is accomplished by bending each line outward 90° to form a **half-wave**, or a λ/2 **dipole**, as shown in Figure 14-2.

The antenna shown in Figure 14-2 is composed of two quarter-wave sections. The electrical distance from the end of one to the end of the other is a half-wavelength. If voltage is applied to the line, the current is maximum at the input and minimum at the ends. The voltage is maximum between the ends, and minimum between the input terminals.

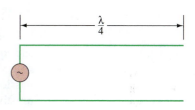

FIGURE 14-1 Quarter-wave transmission line segment (open-ended).

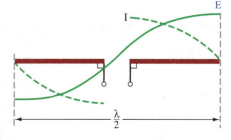

FIGURE 14-2 Basic half-wave dipole antenna.

Half-Wave Dipole Antenna Impedance

An impedance value may be specified for a half-wave antenna thus constructed. The impedance is the ratio of voltage applied to the antenna to the current flowing in it at any point. Generally, the impedance at the ends is maximum, while that at the input is minimum. Consequently, the impedance value varies from a minimum value at the generator to a maximum value at the open ends. An impedance curve for the half-wave antenna is

FIGURE 14-3 Impedance along a half-wave antenna.

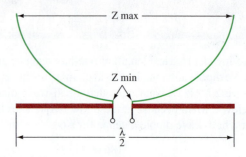

shown in Figure 14-3. Notice that the line has different impedance values for different points along its length. The impedance values for half-wave antennas vary from about 2500 Ω at the open ends to 73 Ω at the source ends.

Radiation and Induction Field

Feeding the Hertz antenna at the center results in an input impedance that is purely resistive and equal to 73 Ω. Recall that with an open-circuited λ/4 transmission line, the input impedance was 0 Ω, and, therefore, it could not absorb power. By spreading the open λ/4 transmission line out into a half-wave dipole antenna, its input impedance has taken on a finite resistive value. It can now absorb power, but the question is, how? The answer is that it can now efficiently accept electrical energy and radiate it into space as electromagnetic waves.

Figure 14-4 is an illustration of the process by which an electric field detaches itself from an antenna and launches into space. The magnetic field, though present, is not shown. In Figure 14-4(a), the electric-field lines of force created as the result of opposite-type charges in the antenna conductor extend from the point of maximum positive potential to the point of maximum negative potential. The direction of the arrowheads indicates the polarity, and because they have the same polarity, the lines exert a repulsive force, as indicated by the outer lines being stretched away from the inner one. Figure 14-4(b) illustrates that the electric force lines start to form closed loops as the applied voltage drops and the charges, formerly separated, come together. Notice that the arrowheads are still facing in the same direction, implying that the polarity is the same and that the lines of force repel each other. Thus, the centers of the lines still face an outward force.

FIGURE 14-4 Electric field detaching itself from antenna.

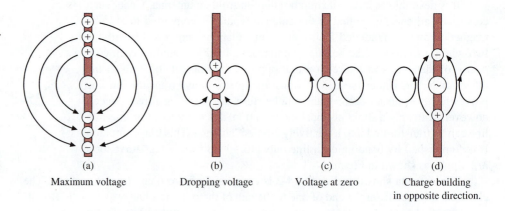

| (a) | (b) | (c) | (d) |
| Maximum voltage | Dropping voltage | Voltage at zero | Charge building in opposite direction. |

Figures 14-4(a) and (b) illustrate what takes place during the first half-cycle alternation of applied voltage. As the applied voltage approaches zero at the end of the half-cycle, some force lines collapse back into the dipole, but other lines form complete loops, as shown in Figure 14-4(c). Notice the direction of the arrowheads on the loops; they are both facing up. During the next half cycle, the applied voltage increases, but its polarity has reversed. The force lines build up, and again they extend from the positive charge to the negative charge. However, during the second half cycle, the charge lines developing around the antenna also have the polarity indicated by their upward-facing arrowheads. Because these newly developing charge lines have the same polarity as that of the closed loops next to the antenna and they are expanding as the amplitude of the applied energy is increasing, the expanding charge lines repel the closed loops and force them out into space, creating a radiated field. This radiated field is appropriately termed the **radiation field**. Antennas also have an **induction field** associated with them. It is the portion of field energy that *does* collapse back into the antenna and is therefore limited to the zone immediately surrounding the antenna. Its effect becomes negligible at a distance more than about one-half wavelength from the antenna.

Other designators for antenna fields are the **near field** and the **far field**. The far-field region begins when the distance

(a)
$$R_{ff} = 1.6\lambda: \quad \frac{D}{\lambda} < 0.32,$$
(14-1a)

(b)
$$R_{ff} = 5D: \quad 0.32 < \frac{D}{\lambda} < 2.5,$$
(14-1b)

(c)
$$R_{ff} = \frac{2D^2}{\lambda}: \quad \geq 2.5\lambda,$$
(14-1c)

where R_{ff} = far field distance from the antenna [meters]

   D = dimension of the antenna [meters]

   λ = wavelength of the transmitted signal [meters/cycle].

The near-field region is any distance less than R. The effects of the induction field are negligible in the far field.

EXAMPLE 14-1

Determine the distance from a $\lambda/2$ dipole to the boundary of the far field region if the $\lambda/2$ dipole is used in a 150-MHz communications system.

SOLUTION

The wavelength (λ) for a $\lambda/2$ dipole at 150 MHz is approximately

$$\lambda = \frac{3 \times 10^8}{150 \times 10^6} = 2 \frac{m}{cycle}.$$

Therefore $\lambda/2 = 1$ m, which is the antenna's dimension (D).

$$\frac{D}{\lambda} = \frac{1}{2} = 0.5.$$

Therefore, select Equation (14-1b).

$$R_{ff} = 5D = 5(1) = 5 \text{ m}.$$

Therefore, the boundary for the far-field region is any distance greater than 1 m from the antenna. In this case, the far-field distance is equal to the diameter (D) of the $\lambda/2$ dipole.

EXAMPLE 14-2

Determine the distance from a parabolic reflector with diameter (D) = 4.5 m to the boundary of the far-field region if the parabolic reflector is used for Ku-band transmission of a 12-GHz signal.

SOLUTION

The wavelength (λ) for a 12-GHz signal is approximately

$$\lambda = \frac{3 \times 10^8}{12 \times 10^9} = 0.025 \frac{m}{cycle}$$

$D = 4.5$ meter.

$$\frac{D}{\lambda} = \frac{4.5}{.025} = 180.$$

Therefore, select Equation (14-1c).

$$R > \frac{2(4.5)^2}{0.025} = 1620 \text{ m}.$$

Therefore, the boundary for the far field region for this parabolic reflector is a distance greater than 1620 m from the antenna. The far-field boundary for high-gain antennas (e.g., a parabolic reflector) will always be greater than for low-gain antennas (e.g., a dipole antenna).

Resonance

As already established, a dipole that is electrically one-half wavelength long exhibits a purely resistive impedance. The resistive condition will hold true at any point along the antenna, though the impedance magnitude will vary. Also, and as shown in Figure 14-3, the impedance is minimum if the antenna is center-fed. An antenna with a purely resistive impedance is said to exhibit the condition of *resonance,* and the resonant condition holds true when the antenna is electrically one-half wavelength, or a multiple of a half-wavelength, at the frequency applied to it. At frequencies where the antenna is no longer a half wavelength, its impedance becomes *complex,* having both resistive and reactive properties. If the antenna and feed-line impedances do not match, either because their resistances are different or because of the presence of reactive components, then the reflections introduced as a result will give rise to standing waves on the transmission line, and the reflections will propagate back toward the source. For antennas designed to operate at one frequency only, such as transmit antennas used in broadcasting applications, considerable effort will be expended to effect an ideal resistive match and, hence, a voltage standing-wave ratio (VSWR) close to 1:1. However, transmit or receive antennas intended to operate over a range of frequencies will, of course, still only be resonant at one frequency, which means there will be some degree of mismatch (hence, standing waves and a VSWR greater than 1:1) apparent over the operating frequency range. The presence of some mismatch is not necessarily a problem, and an antenna can still perform effectively when operating at other than its resonant frequency. This is frequently a point of confusion: The antenna will work as long as there is current flow and it has the ability to radiate; however, maximum power transfer will only take place when source, feed-line, and antenna impedances are equal.

Looking at this issue from another perspective, the antenna can be said to have an operating *bandwidth,* or range of frequencies over which the VSWR is acceptably low. How low is low enough? This is largely a design choice. A VSWR of 1.5:1 causes about 4%, and at 2:1 VSWR about 10%, of the incident power to be reflected. A 2:1 VSWR may be perfectly acceptable for receive applications because the power levels are so low. However, a 1.5:1 VSWR present on the transmission line of a 50-kW AM broadcast station will result in 2 kW being reflected back to the final-amplifier output stage of the station transmitter. Clearly in the latter case, then, the antenna will be carefully matched to the impedance of the transmitter and line to achieve a VSWR as close to 1:1 as possible.

Radiation Patterns

The radiation pattern for the $\lambda/2$ dipole antenna is shown in Figure 14-5(a). A **radiation pattern** is an indication of radiated field strength around the antenna. The pattern shown in Figure 14-5(a) shows that maximum field strength for the $\lambda/2$ dipole occurs at right angles to the antenna, while virtually zero energy is launched "off the ends." So if you wish to communicate with someone, the best results would be obtained when he or she is in the

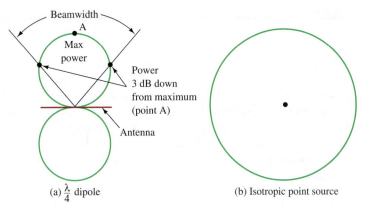

FIGURE 14-5 **Radiation patterns.**

(a) $\frac{\lambda}{4}$ dipole   (b) Isotropic point source

direction of A or 180° opposite; the person should not be located off the ends of the antenna. Recall from Chapter 13 that we considered an isotropic source of waves. Its radiation pattern is spherical, or as shown in two dimensions [Figure 14-5(b)], it is circular or **omnidirectional**. The half-wave dipole antenna is termed **directional** because it concentrates energy in certain directions at the expense of lower energy in other directions.

Another important concept is an antenna's **beamwidth**. It is the angular separation between the half-power points on its radiation pattern. It is shown for the $\lambda/2$ dipole in Figure 14-5(a). A three-dimensional radiation pattern cross section for a vertically polarized $\lambda/2$ dipole is shown in Figure 14-6. You can see that it is a doughnut-shaped pattern. If the antenna were mounted close to ground, the pattern would be altered by the effects of ground-reflected waves.

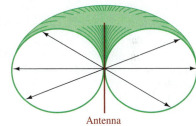

FIGURE 14-6 **Three-dimensional radiation pattern for a half-wave dipole.**

Antenna Gain

The half-wave dipole antenna has *gain* with respect to the theoretical isotropic radiator. **Antenna gain** is different from amplifier gain because feeding 50 W into a dipole does not result in more than 50 W of radiated field energy. Instead, antenna gain is achieved by focusing its radiated energy in a preferred direction at the expense of other directions. The principle is the same as what happens when a reflector is placed behind a light: The light appears brighter in front of the bulb and reflector because reflected light energy adds to the energy from the bulb. However, no energy is radiated behind the reflector. The total amount of energy has not increased; it is merely focused in some directions at the expense of others.

Antenna gain is based on the same concept. Antennas are passive devices, so gain for an antenna is not created by applying more energy from a power source, as would be the case with an amplifier. Instead, antenna gain is represented with respect to a reference antenna, which is often the point-source radiator with its theoretical spherical radiation pattern. The dipole, therefore, has a gain relative to the isotropic radiator because dipole energy is radiated in the doughnut-shaped pattern of Figure 14-6 rather than equally in all directions. Antenna gain is always expressed in terms of the direction of maximum radiation. The half-wave dipole antenna has a 2.15-dB gain (at right angles to the antenna) as compared to an isotropic radiator. However, because a perfect isotropic radiator cannot be practically realized, the $\lambda/2$ dipole antenna is sometimes taken as the standard reference to which all other antennas are compared with respect to their gain.

For an antenna whose gain is provided with respect to an isotropic radiator, the gain is expressed in decibel units with the letter *i* appended to the designation, or **dBi**. In other words, the half-wave dipole antenna's gain can be expressed as 2.15 dBi. If an antenna's gain is given in decibels with respect to a dipole, it is expressed as **dBd**. This occurs somewhat less often than dBi in antenna literature. The gain of an antenna in dBi is 2.15 dB more than when expressed in dBd. Thus, an antenna with a gain of 3 dBd has a gain of 5.15 dBi (3 dB + 2.15 dB).

EFFECTIVE RADIATED POWER   When the gain of an antenna is multiplied by its power input, the result is termed its *effective radiated power* (ERP). For instance, an antenna with a gain of 7 and fed with 1 kW has an ERP of 7 kW. In system-design applications, ERP is used to predict coverage and takes into account all system gains and losses. Effective radiated power also pertains to the power output of FM and TV broadcast stations. For example, an FM radio station licensed for 100 kW most likely achieves that power in part by using a very high-gain antenna that essentially focuses all the transmitted energy along the horizon and parallel to the Earth's surface, where the listeners are. This additional energy comes from refocusing the energy that would have gone upward toward space and downward into the region directly below the antenna's location on its tower.

Determining ERP is straightforward if all power levels are converted to decibel form. Then, the problem becomes one of simple addition and subtraction, as the following example shows.

EXAMPLE 14-3

A 50-W transmitter is connected to an antenna with a gain specification of 9 dBd. Transmission-line losses and filter/connector losses together add up to 3 dB. What is the ERP?

SOLUTION

Conversion of all power levels to dB units reduces this problem to one of addition and subtraction. Converting the transmitter output power to dBm, we obtain

$$dBm = 10 \log\left(\frac{50}{1 \times 10^{-3}}\right)$$
$$= +47 \text{ dBm.}$$

Therefore, the ERP is 47 dBm + 9 dBd − 3 dB = 53 dBm. Because 53 dBm represents a 6-dB increase from the transmitter output power, the ERP is four times that of the transmitter output (remember that each 3-dB power step represents a doubling of power, so 6 dB can be thought of as two 3-dB steps or doubling the power twice) or 200 W.

The above example used an antenna whose gain was referenced with respect to a dipole, as evidenced by the dBd designation. This is common practice in designs for systems in the VHF and UHF ranges. However, and particularly in microwave applications, antenna gains are usually referenced with respect to an isotropic radiator because the short wavelengths involved more closely represent the behavior of a point-source antenna. Effective-power calculations involving antennas with gains expressed in dBi produce results termed *effective isotropic radiated power* (EIRP). The EIRP will always be 2.15-dB higher than the ERP because the gain introduced by the dipole reference has not been taken into account. Put another way, the 9-dBd-gain antenna used for Example 14-3 would have a gain of 11.15 dBi. In a manner of speaking, the higher number is somewhat illusory because it is not possible to build an isotropic radiator, only a dipole.

RECEIVED POWER   The amount of power received by an antenna through free space can be predicted by the following:

$$P_r = \frac{P_t G_t G_r \lambda^2}{16\pi^2 d^2},$$ (14-2)

where P_r = power received (watts)

P_t = power transmitted (watts)

G_t = transmitting antenna gain (ratio, not dB) compared to isotropic radiator

G_r = receiving antenna gain (ratio, not dB) compared to isotropic radiator

λ = wavelength (meters)

d = distance between antennas (meters).

EXAMPLE 14-4

Two $\lambda/2$ dipoles are separated by 50 km. They are "aligned" for optimum reception. The transmitter feeds its antenna with 10 W at 144 MHz. Calculate the power received.

SOLUTION

The two dipoles have a gain of 2.15 dB. That translates into a gain ratio of $\log^{-1} 2.15$ dB = 1.64.

$$P_r = \frac{P_t G_t G_r \lambda^2}{16\pi^2 d^2} \tag{14-2}$$

$$= \frac{10\ \text{W} \times 1.64 \times 1.64 \times \left(\dfrac{3 \times 10^8\ \text{m/s}}{144 \times 10^6}\right)^2}{16\pi^2 \times (50 \times 10^3\ \text{m})^2}$$

$$= 2.96 \times 10^{-10}\ \text{W}.$$

The received signal in Example 14-4 would provide a voltage of 147 μV into a matched 73-Ω receiver system [$(P = V^2/R)$, $V = \sqrt{(2.96 \times 10^{-10}\ \text{W} \times 73\ \Omega)} = 147\ \mu$V]. This is a relatively strong signal because receivers can often provide a usable output with less than a 1-μV signal.

Polar Plots

A more complete picture of antenna radiation characteristics, including gain and bandwidth, is given with a *polar radiation plot*. Actually, two plots are required to give a full indication of propagation in three dimensions. The polar plots represent a cross-sectional view and attempt to show the three-dimensional radiation pattern in two dimensions. Antenna manufacturers publish polar plots in their specification literature. Figure 14-7 shows a blank polar diagram for use in plotting radiation patterns over a full 360° field of view.

The strength of the radiation is represented by the distance from the center of the plot and is expressed as a decibel reduction with respect to maximum radiation; in other words, the outer radius labeled "0 dB" represents the maximum radiation emitted from the antenna, which is in the direction of maximum antenna gain, and the circles inside the outermost circle represent reductions from maximum in 1-dB increments. Most manufacturers adhere to the convention of plotting the direction of maximum gain at 0°, which is facing north in the figure, but the 0° point may face in a different direction. In Figure 14-8(a), the cross-sectional view of the doughnut shaped radiation pattern for a half-wave dipole has been redrawn on a polar radiation plot with the axes of maximum radiation facing east-west, as would be the case for a vertically oriented antenna. The view in Figure 14-8 (a) is an *elevation* view, which is the view you would see if you sliced the doughnut in half vertically (imagine the radiation doughnut extending into and out of the page) and stood in front of the half extending behind the page. Figure 14-8 (b) shows the *azimuth* view, which is the view obtained if you were to look down on the doughnut from above and imagine it sliced lengthwise as though you were slicing a bagel. The azimuth view shows that the

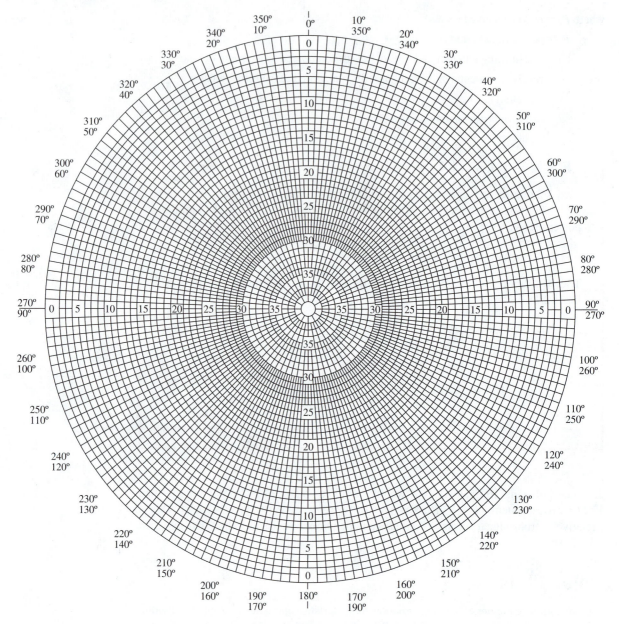

FIGURE 14-7   **Polar diagram for plotting antenna radiation patterns.**

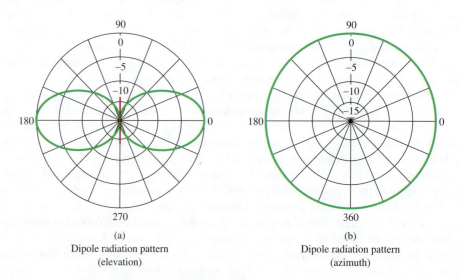

(a)
Dipole radiation pattern
(elevation)

(b)
Dipole radiation pattern
(azimuth)

FIGURE 14-8   **Polar diagrams for half-wave dipole.**

radiation pattern for a half-wave dipole is *omnidirectional,* or equal in all directions. If the dipole is oriented vertically, the view of Figure 14-8(b) would represent the radiation pattern extending parallel to the Earth's surface and toward the horizon over the full 360°. As we shall see shortly, other antenna configurations exhibit directional patterns in either the elevation or azimuth planes, or both.

One final point: the polar plots shown do not provide enough information to determine the absolute gain (in dB) of the antenna because the reference was not given. In other words, do not confuse the "0 dB" on a plot of this nature with the radiation from a reference isotropic radiator or dipole. Some manufacturers may elect to draw plots with the 0 dB point shown somewhere in the interior of the circle and a notation indicating the appropriate reference. Plots drawn from this perspective would then show positive decibel values extending toward the outer radius, which represent gain over that of the reference, and negative decibel values toward the center, representing loss. One form is not necessarily superior to another, but when comparing antenna specifications from various manufacturers, be careful not to "read in" more information than has explicitly been stated.

14-3 RADIATION RESISTANCE

The portion of an antenna's input impedance that is the result of power radiated into space is called the **radiation resistance**, R_r. Note that R_r is not the resistance of the conductors that form the antenna. It is simply an effective resistance that is related to the power radiated by the antenna. Since a relationship exists between the power radiated by the antenna and the antenna current, radiation resistance can be mathematically defined as the ratio of total power radiated to the square of the effective value of antenna current, or

$$R_r = \frac{P}{I^2}, \qquad \text{(14-3)}$$

where R_r = radiation resistance (Ω)

I = effective rms value of antenna current at the feed point (A)

P = total power radiated from the antenna.

It should be mentioned at this point that not all of the energy absorbed by the antenna is radiated. Power may be dissipated in the actual antenna conductor by high-powered transmitters, by losses in imperfect dielectrics near the antenna, by eddy currents induced in metallic objects within the antenna's induction field, and by arcing effects in high-powered transmitters. These arcing effects are termed **corona discharge**. If these losses are represented by one lumped value of resistance, R_d, and the sum of R_d and R_r is called the antenna's total resistance, R_T, the antenna's efficiency can be expressed as

$$\eta = \frac{P_{\text{transmitted}}}{P_{\text{input}}} = \frac{R_r}{R_r + R_d} = \frac{R_r}{R_T}. \qquad \text{(14-4)}$$

Effects of Antenna Length

The radiation resistance varies with antenna length, as shown in Figure 14-9. For a half-wave antenna, the radiation resistance measured at the current maximum (center of the antenna) is approximately 73 Ω. For a quarter-wave antenna, the radiation resistance measured at its current maximum is approximately 36.6 Ω. These are free-space values, that is, the values of radiation resistance that would exist if the antenna were completely isolated so that its radiation pattern would not be affected by ground or other reflections.

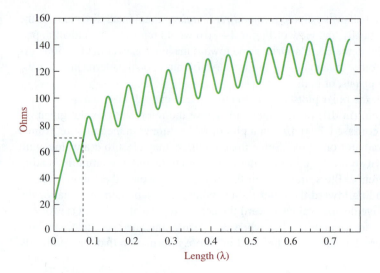

Ground Effects

For practical antenna installations, the height of the antenna above ground affects radiation resistance. Changes in radiation resistance occur because of ground reflections that intercept the antenna and alter the amount of antenna current flowing. Depending on their phase, the reflected waves may increase antenna current or decrease it. The phase of the reflected waves arriving at the antenna, in turn, is a function of antenna height and orientation.

At some antenna heights, it is possible for a reflected wave to induce antenna currents in phase with transmitter current so that total antenna current increases. At other antenna heights, the two currents may be 180° out of phase so that total antenna current is less than if no ground reflection occurred.

With a given input power, if antenna current increases, the effect is as if radiation resistance decreases. Similarly, if the antenna height is such that the total antenna current decreases, the radiation resistance is increased. The actual change in radiation resistance of a half-wave antenna at various heights above ground is shown in Figure 14-10. The radiation resistance of the horizontal antenna rises steadily to a maximum value of 90 Ω at a height of about three-eighths wavelength. The resistance then continues to rise and fall around an average value of 73 Ω, which is the free-space value. As the height is increased, the amount of variation keeps decreasing.

FIGURE 14-10 **Radiation resistance of half-wavelength antennas at various heights.**

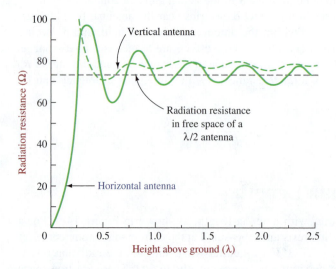

The variation in radiation resistance of a vertical antenna is much less than that of the horizontal antenna. The radiation resistance (dashed line in Figure 14-10) is a maximum value of 100 Ω when the center of the antenna is a quarter-wavelength above ground. The value falls steadily to a minimum value of 70 Ω at a height of a half-wavelength above ground. The value then rises and falls by several ohms about an average value slightly above the free-space value of a horizontal half-wave antenna.

Since antenna current is affected by antenna height, the field intensity produced by a given antenna also changes. In general, as the radiation resistance is reduced, the field intensity increases, whereas an increase in radiation resistance produces a drop in radiated field intensity.

Electrical versus Physical Length

If an antenna is constructed of very thin wire and is isolated in space, its electrical length corresponds closely to its physical length. In practice, however, an antenna is never isolated completely from surrounding objects. For example, the antenna will be supported by insulators with a dielectric constant greater than 1. The dielectric constant of air is arbitrarily assigned a numerical value equal to 1. Therefore, the velocity of a wave along a conductor is always slightly less than the velocity of the same wave in free space, and the physical length of the antenna is less (by about 5%) than the corresponding wavelength in space. The physical length can be approximated as about 95% of the calculated electrical length.

EXAMPLE 14-5

We want to build a λ/2 dipole to receive a 100-MHz broadcast. Determine the optimum length of the dipole.

SOLUTION

At 100 MHz,

$$\lambda = \frac{c}{f} = \frac{3 \times 10^8 \text{ m/s}}{100 \times 10^6 \text{ Hz}} = 3 \text{ m.}$$

Therefore, its electrical length is λ/2, or 1.5 m. Applying the 95% correction factor, the actual optimum physical length of the antenna is

$$0.95 \times 1.5 \text{ m} = 1.43 \text{ m.}$$

The result of the preceding example is also obtained by using the following formula:

$$L = \frac{468}{f(MHz)}, \tag{14-5}$$

where L is dipole length in feet. For Example 14-4, it would give $L = \frac{468}{100} = 4.68$ ft, which is equal to 1.43 m.

Effects of Nonideal Length

The 95% correction factor is an approximation. If ideal results are desired, a trial-and-error procedure is used to find the exact length for optimum antenna performance. If the antenna length is not the optimum value, its input impedance looks like a capacitive circuit or an inductive circuit depending on whether the antenna is shorter or longer than the specified wavelength. A half-wave dipole antenna slightly longer than a half-wavelength acts like an inductive circuit, and an antenna slightly shorter than a half-wavelength appears to the source as a capacitive circuit. Compensation for additional length can be made by cutting the antenna down to proper length or by tuning out the inductive reactance by adding a capacitance in series. This added X_c completely cancels the inductive reactance, and the source then sees a pure resistance, provided the proper size capacitor is used. If an antenna is shorter than the required length, the source end of the line appears capacitive. This condition may be corrected by adding inductance in series with the antenna input.

Antenna descriptions can be based in part on the point at which the transmission line connects to the antenna. Connect the line to the end of the antenna and we have an end-fed antenna; connect it to the center and it is called center fed. If the transmission line joins the antenna at a high-voltage point, the antenna is said to be voltage fed. Conversely, connect to a high-current point and we have a current-fed antenna. All of these types are shown in Figure 14-11.

FIGURE 14-11 **(a) Current feed and (b) voltage feed.**

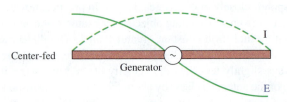

(a) Generator at current maximum means current feed

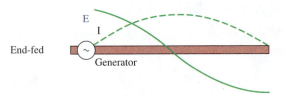

(b) Generator at voltage maximum means voltage feed

It is seldom possible to connect a generator directly to an antenna. It is usually necessary to transfer energy from the generator (transmitter) to the antenna by use of a transmission line (also called an antenna **feed line**). Such lines may be resonant, nonresonant, or a combination of both types.

Resonant Feed Line

The resonant transmission line is not widely used as an antenna feed method because it tends to be inefficient and is very critical with respect to its length for a particular operating frequency. In certain high-frequency applications, however, resonant feeders sometimes prove convenient.

In the current-fed antenna with a resonant line, shown in Figure 14-12, the transmission line is connected to the center of the antenna. This antenna has a low impedance at the center and, like the voltage-feeding transmission line, has standing waves on it. Constructing it to be exactly a half-wavelength causes the impedance at the sending end to be low. A series resonant circuit is used to develop the high currents needed to excite the line. Adjusting the capacitors at the input compensates for slight irregularities in line and antenna length.

Although this example of an antenna feed system is a simple one, the principles described apply to antennas and to lines of any length provided both are resonant. The line connected to the antenna may be either a two-wire or coaxial line. In high-frequency applications, the coaxial line is preferred because it has lower radiation loss.

One advantage of connecting a resonant transmission line to an antenna is that it makes impedance matching unnecessary. In addition, providing the appropriate resonant circuit at the input can compensate for any irregularities in either the line or the antenna. Its disadvantages are increased power losses in the line resulting from high standing waves of current, increased probability of arc-over because of high

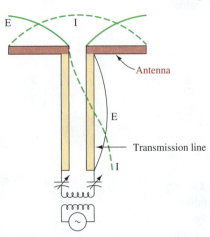

FIGURE 14-12 **Current feed with resonant line.**

standing waves of voltage, very critical length, and production of radiation fields by the line because of the standing waves on it.

Nonresonant Feed Line

The nonresonant feed line is the more widely used technique. The open-wire line, the shielded pair, the coaxial line, and the twisted pair may be used as nonresonant lines. This type of line has negligible standing waves if it is properly terminated in its characteristic impedance at the antenna end. It has a great advantage over the resonant line because its operation is practically independent of its length.

The illustrations in Figure 14-13 show the excitation of a half-wave antenna by non-resonant lines. If the input to the center of the antenna in Figure 14-13(a) is 73 Ω and if the coaxial line has a characteristic impedance of 73 Ω, a common method of feeding this antenna is accomplished by connecting directly to the center of the antenna. This method of connection produces no standing waves on the line when the line is matched to a generator. Coupling to a generator is often made through a simple untuned transformer secondary.

Another method of transferring energy to the antenna is through the use of a twisted-pair line, as shown in Figure 14-13(b). It is used as an untuned line for low frequencies. The twisted pair is not used at higher frequencies because of excessive losses occurring in the insulation. The characteristic impedance of such lines is about 70 Ω.

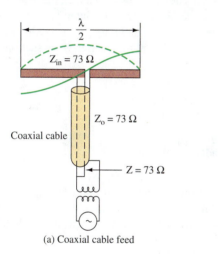

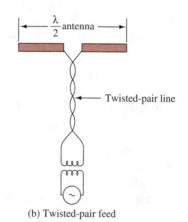

(a) Coaxial cable feed

(b) Twisted-pair feed

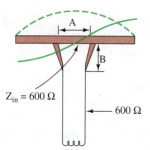

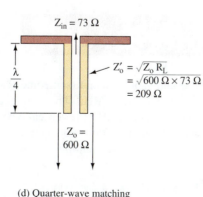

(c) Delta match

(d) Quarter-wave matching transformer

FIGURE 14-13 Feeding antennas with nonresonant lines.

Delta Match

When a line does not match the impedance of the antenna, it is necessary to use special impedance matching techniques such as those discussed with Smith chart applications in Chapter 12. An example of an additional type of impedance matching device is the

delta match, shown in Figure 14-13(c). The open, two-wire transmission line inherently does not have a characteristic impedance (Z_0) low enough to match a center-fed dipole with $Z_{in} = 73 \ \Omega$. Practical values of Z_0 for such lines lie in the range 300 to 700 Ω. To provide the required impedance match, a delta section (shown in Figure 14-13(c)) is used. This match is obtained by spreading the transmission line as it approaches the antenna. In the example given, the characteristic impedance of the line is 600 Ω, and the center impedance of the antenna is 73 Ω. As the end of the transmission line is spread, its characteristic impedance increases. Proceeding from the center of the antenna to either end, a point will be reached where the antenna impedance equals the impedance at the output terminals of the delta section. Recall that the antenna impedance increases as you move from its center to the ends. The delta section is then connected at this distance to either side of the antenna center.

The delta section becomes part of the antenna and, consequently, introduces radiation loss (one of its disadvantages). Another disadvantage is that trial-and-error methods are usually required to determine the dimensions of the A and B sections for optimum performance. Both the distance between the delta output terminals (its width) and the length of the delta section are variable, so adjustment of the delta match is difficult.

Quarter-Wave Matching

Still another impedance-matching device is the quarter-wave transformer, or matching transformer, as shown in Figure 14-13(d). This device is used to match the low impedance of the antenna to the line of higher impedance. Recall from Chapter 12 that the quarter-wave matching section is effective only between a line and purely resistive loads.

To determine the characteristic impedance (Z_0') of the quarter-wave section, the following formula from Chapter 12 is used.

$$Z_0' = \sqrt{Z_0 R_L},$$

(12-31)

where $Z_0' = $ characteristic impedance of the matching line

$Z_0 = $ impedance of the feed line

$R_L = $ resistive impedance of the radiating element.

For the example shown, Z_0' has a value slightly over 209 Ω. With this matching device, standing waves will exist on the $\lambda/4$ section but not on the 600-Ω line. Recall from Chapter 12 the use of stub-matching techniques as another alternative.

This matching technique is useful for narrowband operation, while the delta section is more broadband in operation.

14-5 MONOPOLE ANTENNA

The **monopole antenna** (sometimes called a vertical antenna) is used primarily with frequencies below 2 MHz. The difference between the vertical antenna and the half-wave dipole antenna is that the vertical type requires a conducting path to ground, and the half-wave dipole type does not. The monopole antenna is usually a quarter-wave grounded antenna or any odd multiple of a quarter-wavelength.

Effects of Ground Reflection

A monopole antenna used as a transmitting element is shown in Figure 14-14. The transmitter is connected between the antenna and ground. The actual length of the antenna is one quarter-wavelength. However, this type of antenna, by virtue of its connection to ground, uses the ground as the other quarter-wavelength, making the antenna electrically a half-wavelength. This is so because the Earth is considered to be a good conductor. In fact, there is a reflection from the Earth that is equivalent to the radiation that would be realized if another quarter-wave section were used. The reflection from the ground looks as if it is coming from a λ/4 section beneath the ground. This is known as the **image antenna** and is shown in Figure 14-14. With the monopole antenna, which is a quarter-wave in actual physical length, half-wave operation may be obtained. All the voltage, current, and impedance relationships characteristic of a half-wave antenna also exist in this antenna. The only exception is the input impedance, which is approximately 36.6 Ω at the base. The effective current in the monopole grounded antenna is maximum at the base and minimum at the top, while voltage is minimum at the bottom and maximum at the top.

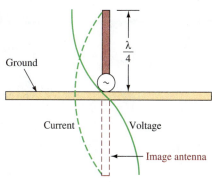

FIGURE 14-14 Grounded monopole antenna.

When the conductivity of the soil in which the monopole antenna is supported is very low, the reflected wave from the ground may be greatly attenuated, which is undesirable. To overcome this disadvantage, the site can be moved to a location with high-conductivity soil, such as damp areas. If moving the site is impractical, provisions must be made to improve the reflecting characteristics of the ground by installing a buried ground screen.

The Counterpoise

When an actual ground connection cannot be used because of the high resistance of the soil or a large buried ground screen is impractical, a **counterpoise** may replace the usual direct ground connection. This is required for monopole antennas mounted on the top of tall buildings. The counterpoise consists of a structure made of wire erected a short distance above the ground and *insulated from the ground*. The size of the counterpoise should be at least equal to, and preferably larger than, the size of the antenna.

The counterpoise and the surface of the ground form a large capacitor. This capacitance causes antenna current to be collected in the form of charge and discharge currents. The end of the antenna normally connected to ground is connected through the large capacitance formed by the counterpoise. If the counterpoise is not well insulated from ground, the effect is much the same as that of a leaky capacitor, with a resultant loss greater than if no counterpoise were used.

Although the shape and size of the counterpoise are not particularly critical, it should extend for equal distances in all directions. When the antenna is mounted vertically, the counterpoise may have any simple geometric pattern, like those shown in Figure 14-15. The counterpoise is constructed so that it is nonresonant at the operating frequency. The operation realized by use of either the well-grounded monopole antenna or the monopole antenna using a counterpoise is the same as that of the half-wave antenna of the same polarization.

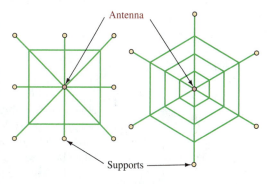

FIGURE 14-15 Counterpoise (top view).

Radiation Pattern

The radiation pattern for a monopole antenna is shown in Figure 14-16(a). It is omnidirectional in the ground plane but falls to zero off the antenna's top. Thus, a large amount of energy is launched as a ground wave, but appreciable sky-wave energy also exists. By increasing the vertical height to λ/2, the ground-wave strength is increased, as shown in Figure 14-16(b). The maximum ground-wave strength is obtained by using a length slightly less than 5/8λ. Any greater length produces high-angle radiation of increasing strength, and horizontal radiation is reduced. At a height of 1λ, there is no ground wave.

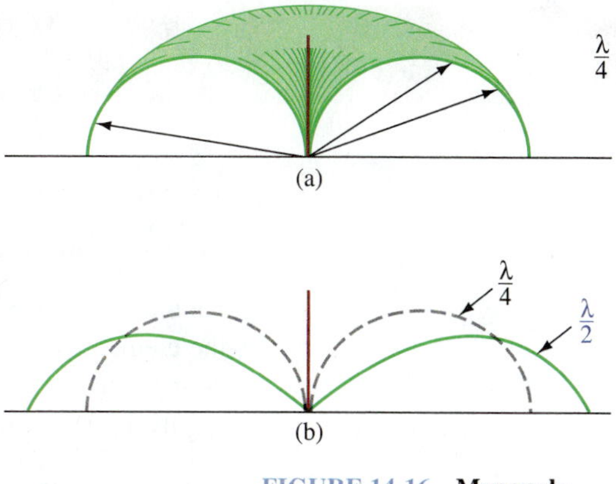

Loaded Antennas

In many low-frequency applications, it is not practical to use an antenna that is a full quarter-wavelength. This is especially true for mobile transceiver applications. Monopole antennas less than a quarter-wavelength have an input impedance that is highly capacitive,

FIGURE 14-16 Monopole antenna radiation patterns.

and they become inefficient radiators. The reason for this is that a highly reactive load cannot accept energy from the transmitter. It is reflected and sets up high standing waves on the feeder transmission line. An example of this is a $\lambda/8$ vertical antenna, which exhibits an input impedance of about $8\ \Omega - j500\ \Omega$ at its base.

To remedy this situation, the *effective* height of the antenna should be $\lambda/4$, and this can be accomplished with several different techniques. Figure 14-17 shows a series inductance that is termed a **loading coil**. It is used to tune out the capacitive appearance of the antenna. The coil–antenna combination can thus be made to appear resonant (resistive) so that it can absorb the full transmitter power. The inductor can be variable to allow adjustment for optimum operation over a range of transmitter frequencies. Notice the standing wave of current shown in Figure 14-17. It has maximum amplitude at the loading coil and thus does not add to the radiated power. This results in heavy I^2R losses in the coil instead of this energy being radiated. However, the transmission line feeding the loading coil/antenna is free of standing waves when the loading coil is properly tuned.

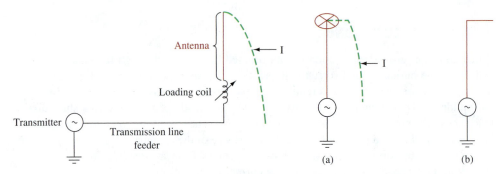

FIGURE 14-17   **Monopole antenna with loading coil.**

FIGURE 14-18   **Top-loaded monopole antennas.**

A more efficient solution is the use of top loading, as shown in Figure 14-18(a). Notice that the high-current standing wave now exists at the base of the antenna so that maximum possible radiation now occurs. The metallic *spoked wheel* at the top adds shunt capacitance to ground. This additional capacitance reduces the antenna's capacitive reactance because C and X_c are inversely related. The antenna can, therefore, be made nearly resonant with the proper amount of top loading. This does not allow for convenient variable frequency operation as with the loading coil, but it is a more efficient radiator. The *inverted L* antenna in Figure 14-18(b) accomplishes the same goal as the top-loaded antenna but is usually less convenient to construct physically.

14-6 ANTENNA ARRAYS

Half-Wave Dipole Antenna with Parasitic Element

An **antenna array** is one that has more than one element or component. If one or more of the elements is not electrically connected, it is called a **parasitic array**. If all elements are connected, the array is a **driven array**. The most elementary antenna array is shown in Figure 14-19(a). It consists of a simple half-wave dipole and a nondriven (not electrically connected) half-wave element located a quarter-wavelength behind the dipole.

The dipole radiates electromagnetic waves with the usual bidirectional pattern. However, the energy traveling toward the parasitic element, upon reaching it, induces voltages and currents but incurs a 180° phase shift in the process. These voltages and currents cause the parasitic element also to radiate a bidirectional wave pattern. However, because of the 180° phase shift within the parasitic element, the energy traveling away from the driven

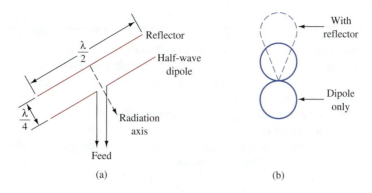

FIGURE 14-19 **Elementary parasitic array.**

element cancels that from the driven element. The energy from the parasitic element traveling toward the driven element reaches it in phase and causes a doubling of energy propagated in that direction. This effect is shown by the radiation pattern in Figure 14-19(b). The parasitic element is also termed a **reflector** because it effectively "reflects" energy from the driven element. Notice that this simple array has resulted in a more directive antenna and thus exhibits gain with respect to a standard half-wave dipole antenna.

Let us consider *why* the energy from the reflector gets back to the driven element in phase and thus reinforces propagation in that direction. Recall that the initial energy from the driven element travels a quarter-wavelength before reaching the reflector. This is equivalent to 90 electrical degrees of phase shift. An additional 180° of phase shift occurs from the induction of voltage and current into the reflector. The reflector's radiated energy back toward the driven element experiences another 90° of phase shift before reaching the driven element. Thus, a total phase shift of 360° (90° + 180° + 90°) results so that the reflector's energy reaches the driven element in phase.

Yagi–Uda Antenna

The Yagi–Uda antenna consists of a driven element and two or more parasitic elements. It is named for Shintaro Uda and Hidetsugu Yagi, the two Japanese scientists who were instrumental in its development. The version shown in Figure 14-20(a) has two parasitic elements: a reflector and a director. A **director** is a parasitic element that serves to "direct" electromagnetic energy because it is in the direction of the propagated energy with respect to the driven element. The radiation pattern is shown in Figure 14-20(b). Notice the two side **lobes** of radiated energy that result. They are generally undesired, as is the small amount of reverse propagation. The difference in gain from the forward to the reverse direction is defined as the **front-to-back ratio** (*F/B* ratio). For example, the pattern in Figure 14-20(b) has a forward gain of 12 dB and a −3 dB gain (actually, loss, because it is a negative gain) in the reverse direction. Its *F/B* ratio is therefore [12 dB − (−3 dB)], or 15 dB.

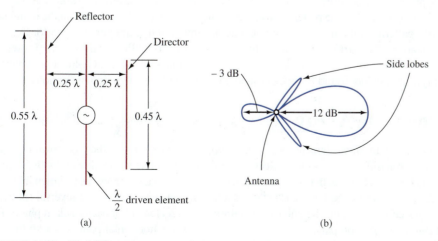

FIGURE 14-20 Yagi-Uda antenna.

This Yagi–Uda antenna provides about 10 dB of power gain with respect to a half-wavelength dipole reference. This is somewhat better than the approximate 3-dB gain of the simple array shown in Figure 14-19. In practice, the Yagi–Uda antenna often consists of one reflector and two or more directors to provide even better gain characteristics. They are often used as HF transmitting antennas and receiving antennas for single VHF or UHF television channels.

The analysis of how the radiation patterns of these antennas result is rather complex and cannot be simply accomplished, as was done for the simple array shown in Figure 14-19. More often than not, the lengths and spacings of the parasitic elements are the result of experiments rather than theoretical calculations.

Driven Collinear Array

A **driven array** is a multielement antenna in which all the elements are excited through a transmission line. A four-element collinear array is shown in Figure 14-21(a). A **collinear array** is any combination of half-wave elements in which all the elements are placed end to end to form a straight line. All elements are excited so that their fields are in phase (additive) for points perpendicular to the array. This is accomplished by the $\lambda/2$ length of transmission line (a $\lambda/4$ twisted pair) between the elements on both sides of the feed point. They are twisted so that the fields created by the line cancel each other to minimize losses.

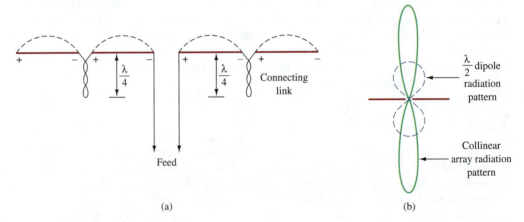

FIGURE 14-21   **Four-element collinear array.**

The radiation pattern for this antenna is provided in Figure 14-21(b). Energy off the ends is canceled from the $\lambda/2$ spacing (cancellation) of elements, but reinforcement takes place perpendicular to the antenna. The resulting radiation pattern thus has gain with respect to the standard half-wave dipole antenna radiation pattern shown with dashed lines in Figure 14-21(b). It has gain at the expense of energy propagated away from the antenna's perpendicular direction. The full three-dimensional pattern for both antennas is obtained by revolving the pattern shown about the antenna axis. This results in the doughnut-shaped pattern for the half-wave dipole antenna and flattened doughnut shape for this collinear array. The array is a more directive antenna (smaller beamwidth). Increased directivity and gain are obtained by adding more collinear elements. Collinear antennas of this type are widely found in repeater installations for two-way radio applications.

Broadside Array

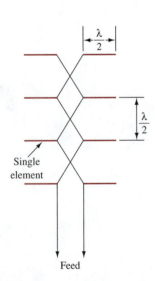

FIGURE 14-22   **Eight-element broadside array.**

If a group of half-wave elements is mounted vertically, one over the other, as shown in Figure 14-22, a broadside array is formed. Such an array provides greater directivity in both the vertical and horizontal planes than does the collinear array. With the arrangement shown in Figure 14-22, the separation between each stack is a half-wavelength. The signal reversal shown in the connecting wires puts the voltage and current in each element of each stack in phase. The net resulting radiation pattern is a directive pattern in the horizontal plane (as with the collinear array) but also a directive pattern in the vertical plane (in contrast to the collinear array).

Vertical Array

You have probably noticed that some standard broadcast AM stations use three or more vertical antennas lined up in a row with equal spacing between them. The radiation pattern of a single vertical antenna is omnidirectional in the horizontal plane, which may be undesirable because of the possibility of interference to adjacent-channel stations or because the listeners are concentrated within a small region. For instance, it doesn't make sense for a New York City station to beam half of its energy to the Atlantic Ocean. By properly controlling the phase and power level into each of the towers, almost any radiation pattern desired can be obtained. Thus, the energy that would have been wasted over the Atlantic Ocean can be redirected to the areas of maximum population density.

This arrangement is called a **phased array** because controlling the phase (and power) to each element results in a wide variety of possible radiation patterns. A station may easily change its pattern at sunrise and sunset because increased sky-wave coverage at night might interfere with a distant station operating at about the same frequency. To give an idea of the countless radiation patterns possible with a phased array, refer to Figure 14-23. It shows the radiation patterns obtainable with just two $\lambda/4$ vertical antennas with variable spacing and input voltage phase. The patterns are simply the vector sum of the instantaneous field strengths from each individual antenna.

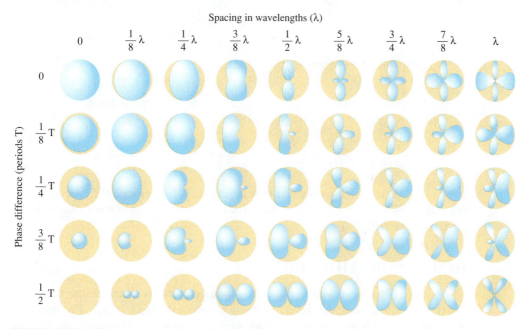

FIGURE 14-23   **Phase-array antenna patterns.**

14-7 SPECIAL-PURPOSE ANTENNAS

Log-Periodic Antenna

The log-periodic antenna is a special case of a driven array. It was first developed in 1957 and has proven so desirable that its many variations now make up an entire class of antennas. It provides reasonably good gain over an extremely wide range of frequencies. Therefore, it is useful for multiband transceiver operation and as a TV receiving unit to cover the entire VHF and UHF bands. It can be termed a wide-bandwidth or broadband antenna. Bandwidth is not to be confused with beamwidth in this situation.

Antenna *bandwidth* is defined with respect to its design frequency, often termed its *center frequency.* If a 100-MHz (center frequency) log-periodic antenna's transmitted or received power is 3 dB down at 50 MHz and 200 MHz, its bandwidth is 200 MHz − 50 MHz, or 150 MHz. This measurement is made in the direction of highest antenna directivity.

The most elementary form of log-periodic antenna is shown in Figure 14-24(a). It is termed a log-periodic dipole array and derives its name from the fact that its important characteristics are periodic with respect to the logarithm of frequency. This is true of its impedance, its standing-wave ratio (SWR) with a given feed line, and the strength of its radiation pattern. For instance, its input impedance is shown to be nearly constant (but periodic) as a function of the log of frequency in Figure 14-24(b).

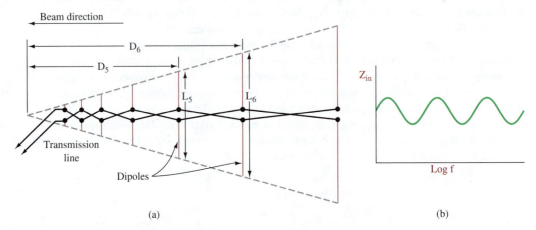

(a)   (b)

FIGURE 14-24   **Log-periodic dipole array.**

The log-periodic array in Figure 14-24(a) consists of several dipoles of different lengths and spacings. The dipole lengths and spacings are related by

$$\frac{D_1}{D_2} = \frac{D_2}{D_3} = \frac{D_3}{D_4} = \frac{D_4}{D_5} \cdots = \tau = \frac{L_1}{L_2} = \frac{L_2}{L_3} = \frac{L_3}{L_4} = \frac{L_4}{L_5} \cdots \qquad (14\text{-}6)$$

where τ is called the *design ratio* with a typical value of 0.7. The range of frequencies over which it is useful is determined by the frequencies at which the longest and shortest dipoles are a half-wavelength.

Small-Loop Antenna

A loop antenna is a turn of wire whose dimensions are normally much smaller than 0.1λ. When this condition exists, the current in it may all be considered in phase. This results in a magnetic field that is everywhere perpendicular to the loop. The resulting radiation pattern is sharply bidirectional, as indicated in Figure 14-25, and is effective over an extremely wide range of frequencies—those for which its diameter is about $\lambda/16$ or less. The antenna is usually circular, but any shape is effective.

FIGURE 14-25   **Loop antenna.**

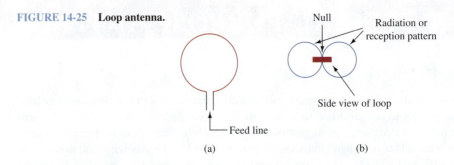

(a)   (b)

Because of its sharp **null** (see Figure 14-25), small size, and broadband characteristics, the loop antenna's major application is in direction-finding (DF) applications. The goal is to determine the direction of some particular radiation. Generally, readings from two different locations are required because of the antenna's bidirectional pattern. If the two locations are far enough apart, the distance and direction of the radiation source can be calculated using trigonometry. Since the signal falls to zero much more sharply than it peaks, the nulls are used in the DF applications. While other antennas with directional characteristics can be used in DF, the loop's small size seems to outweigh the gain advantages of larger directive antennas.

Ferrite Loop Antenna

The familiar ferrite loop antenna found in most broadcast AM receivers is an extension of the basic loop antenna just discussed. The effect of using a large number of loops wound about a highly magnetic core (usually ferrite) serves to increase greatly the effective diameter of the loops. This forms a highly efficient receiving antenna, considering its small physical size compared to the hundreds of feet required to obtain a quarter-wavelength for the broadcast AM band. The directional characteristics of this antenna are verified by the fact that a portable AM receiver can usually be oriented to *null* out reception of a station. You should now be able to determine a line through which that broadcasting station exists when the null is detected.

Folded Dipole Antenna

Recall that the standard half-wavelength dipole antenna has an input impedance of 73 Ω. Recall also that it becomes very inefficient whenever it is not used at the frequency for which its length equals $\lambda/2$ (i.e., it has a narrow bandwidth). The folded dipole antenna shown in Figure 14-26(a) offers the same radiation pattern as the standard half-wave dipole antenna but has an input impedance of 288 Ω (approximately 4 × 73 Ω) and offers relatively broadband operation.

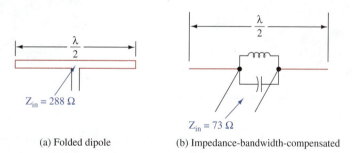

FIGURE 14-26 **Dipoles.**

(a) Folded dipole   (b) Impedance-bandwidth-compensated

   A standard half-wave dipole antenna can provide the same broadband characteristics as the folded dipole by incorporating a parallel tank circuit, as shown in Figure 14-26(b). With the tank circuit resonant at the frequency corresponding to the antenna's $\lambda/2$ length, the tank presents a very high resistance in parallel with the antenna's 73 Ω and has no effect. However, as the frequency goes down, the antenna becomes capacitive, while the tank circuit becomes inductive. The net result is a resistive overall input impedance over a relatively wide frequency range.

   The folded dipole is a useful receiving antenna for broadcast FM and for VHF TV. Its input impedance matches well with the 300-Ω input impedance terminals common to these receivers. It can be inexpensively fabricated by using a piece of standard 300-Ω parallel wire transmission line, commonly called **twin lead**, cut to $\lambda/2$ at midband and shorting together the two conductors at each end. Folded dipoles are also invariably used as the driven element in Yagi–Uda antennas. This helps to maintain a reasonably high input impedance because the addition of each director lowers this array's input impedance. It also gives the antenna a broader band of operation.

   In applications where a folded dipole with other than a 288-Ω impedance is desired, a larger-diameter wire for one length of the antenna is used. Impedances up to about 600 Ω are possible in this manner.

Slot Antenna

Coupling RF energy into a slot in a large metallic plane can result in radiated energy with a pattern similar to a dipole antenna mounted over a reflecting surface. The length of the slot is typically one half-wavelength. These antennas function at UHF and microwave frequencies with energy coupled into the slot by waveguides or coaxial line feed connected directly across the short dimension of a rectangular slot. These antennas are commonly used in modern aircraft in an array module as shown in Figure 14-27(a). This 32-element (slot) array shows half the slots filled with dielectric material (to provide the required smooth airplane surface) and the others open to show the phase-shifting circuitry used to drive the slots. The rear view in Figure 14-27(b) shows the coaxial feed connectors used for this antenna array.

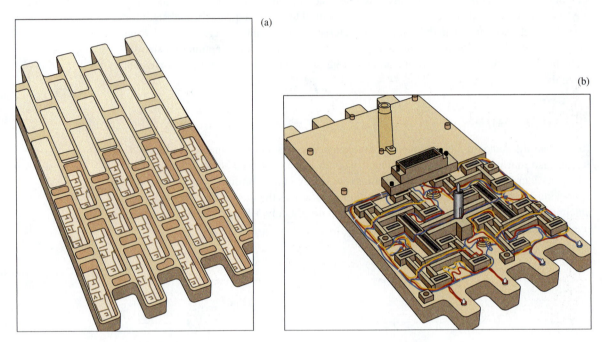

(a)

(b)

FIGURE 14-27   **Slot antenna array.**

Phase-shifting networks control the individual drive to each slot. Proper phasing allows production of a directive radiation pattern that can be swept through a wide angle without physically moving the antenna. This allows a convenient mobile scanning radar system without mechanical complexities. These *phased array* antennas are typically built right into the wings of aircraft, with the dielectric window filling eliminating aerodynamic drag.

14-8 MICROWAVE ANTENNAS

The antennas studied up to this point bear little resemblance to those used for microwave frequencies (>1 GHz). Microwave antennas actually use optical theory more than standard antenna theory. These antennas tend to be highly directive and, therefore, provide high gain compared to the reference half-wavelength dipole. The reasons for this include the following:

1. Because of the short wavelengths involved, the physical sizes required are small enough to allow "peculiar" arrangements not practical at lower frequencies.

2. There is little need for omnidirectional patterns because no broadcasting takes place at these frequencies. Microwave communications are generally of a point-to-point nature. The exception is telemetry applications.

3. Because of increased device noise at microwave frequencies, receivers require the highest possible input signal. Highly directional antennas (and thus high gain) make this possible.

4. Microwave transmitters are limited in their output power either for cost reasons or because of the unavailability of devices that produce high powers at microwave frequencies. Low output power is compensated for by a highly directional antenna system.

Microwaves are divided into bands as shown in Table 14-1. The frequencies above 40 GHz are called **millimeter (mm) waves** because their wavelength is described in millimeters.

TABLE 14-1 • Microwave Frequency Designations	
BAND	**FREQUENCY (GHz)**
L	1–2
S	2–4
C	4–8
X	8–12
Ku	12–18
K	18–27
Ka	27–40

Horn Antenna

Open-ended sections of waveguides can be used as radiators of electromagnetic energy. The three basic forms of horn antennas are shown in Figure 14-28. They all provide a gradual flare to the waveguide to allow maximum radiation and thus minimum reflection back into the guide.

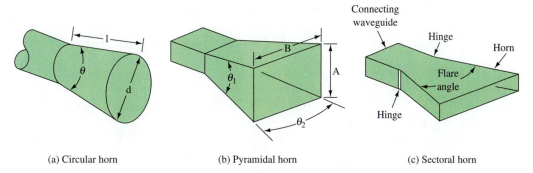

(a) Circular horn   (b) Pyramidal horn   (c) Sectoral horn

FIGURE 14-28   Horn antennas.

An open-circuit waveguide theoretically reflects 100% of the incident energy. In practice, however, the open-circuit guide "launches" a fair amount of energy, while the short-circuit guide does provide the theoretical 100% reflection. By gradually flaring out the open circuit, the goal of total radiation is nearly attained. The flared end of the horn antenna acts as an impedance transformer between the waveguide and free space. For a proper transformation ratio, the linear dimension of each side must be at least a half-wavelength.

The **circular horn** in Figure 14-28(a) provides efficient radiation from a circular waveguide. The flare angle θ and length l are important to the amount of gain it can provide. Generally, the greater the diameter d, the greater the gain.

For the **pyramidal horn** in Figure 14-28(b), the radiation pattern depends on the area of the aperture. The effect of horn length is similar to that with the circular horn. Wider horizontal patterns are obtained by increasing θ_2, while wider vertical patterns are possible by increasing θ_1. In Figure 14-28(b), when the ratio $B/A = 1.35$, a symmetrical radiation pattern is realized.

The **sectoral horn** in Figure 14-28(c) has the top and bottom walls at a 0° flare angle. The side walls are sometimes hinged (as shown) to provide adjustable flare angles. Maximum radiation occurs for angles between 40° and 60°.

The horns just described provide a maximum gain on the order of 20 dB compared to the half-wavelength dipole reference. While they do not provide the amounts of gain of subsequently described microwave antennas, their simplicity and low cost make them popular for noncritical applications.

The Parabolic Reflector Antenna

The *parabolic reflector antenna* is one of the most common microwave frequency antennas in use in satellite and terrestrial communication systems. The name of the antenna comes from the geometric shape of the antenna, which is a paraboloid. The key reasons for the popularity of a parabolic reflector are its high gain and directivity. The ability of a paraboloid to focus light rays or sound waves at a point is common knowledge. Some common applications include dentists' lights, flashlights, and automobile headlamps. The same ability is applicable to electromagnetic waves of lower frequency than light as long as the paraboloid's mouth diameter is at least ten wavelengths. This precludes their use at low frequencies but allows use at microwave frequencies.

A parabola has a focal point, and any wave traveling perpendicular to the directrix (see Figure 14-29) that strikes the reflector will be reflected to the focal point. The opposite is also true when a wave leaves the focal point and strikes the reflector and is reflected perpendicular to the directrix. The focal point for a parabolic reflector can be calculated if the reflector diameter (*D*) and depth (h) are known. Equation (14-7) can be used to determine the focal length (i.e., the distance the antenna element must be placed out from the center of the reflector for maximum performance). An example of calculating the focal length is provided in Example 14-6.

FIGURE 14-29 **Location of the focal point for a parabolic reflector.**

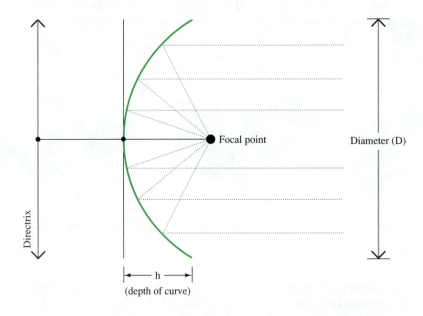

Directrix

● Focal point

Diameter (D)

|← h →|
(depth of curve)

EXAMPLE 14-6

Determine the focal length of a parabolic reflector with a diameter of 3 m and a curve depth of 0.3 m.

SOLUTION

$$\text{Focal length } (f) = \frac{D}{16\,h} = \frac{3}{16(0.3)} = 0.625 \text{ m} \qquad (14\text{-}7)$$

The focal length is 0.625 m out from the center of the parabolic reflector.

There are various methods of feeding the **microwave dish**, as the paraboloid antenna is commonly called. Figure 14-30(a) shows the dish being fed with a simple dipole–reflector combination at the paraboloid's focus. A horn-fed version is shown in Figure 14-30(b). A common type of satellite antenna is the **prime focus feed**, shown in Figure 14-30(c). In this system, the antenna and amplifier are placed at the focal point of the reflector. The antenna and amplifier are supported with struts attached to the edge of the reflector. The **Cassegrain feed** in Figure 14-30(d) is used to shorten the length of the feed mechanism in highly critical satellite communication applications. It uses a hyperboloid secondary reflector, whose focus coincides with that of the paraboloid. Those transmitted rays obstructed by the hyperboloid are generally such a small percentage as to be negligible. Another antenna type is the **offset feed**, shown in Figure 14-30(e). Offset feed antennas are often used in home satellite reception and are desirable because the receive signal is not affected by blockage of the cabling and support hardware.

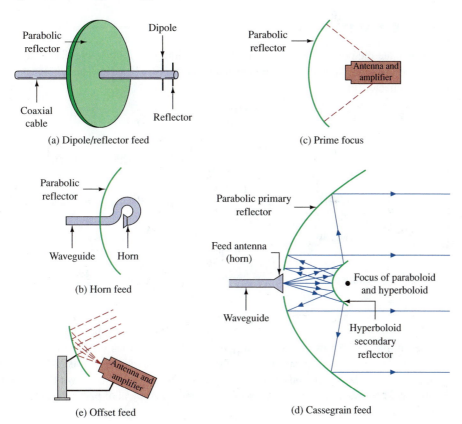

FIGURE 14-30   Microwave dish antennas.

(a) Dipole/reflector feed

(c) Prime focus

(b) Horn feed

(e) Offset feed

(d) Cassegrain feed

The approximate gain of a parabolic reflector antenna can be calculated using Equation (14-8). Equation (14-9) provides the antenna gain calculation expressed in terms of dBi, where dBi is the gain expressed in dB relative to an isotropic radiator. This equation shows us that the gain of the antenna (A_p) increases as wavelength (λ) of the radio wave decreases. In fact, the gain of the antenna increases proportionally to the square of the antenna diameter (D).

$$A_p = k\frac{(\pi D)^2}{\lambda^2}. \tag{14-8}$$

$$A_p(\text{dBi}) = 10 \log k\frac{(\pi D)^2}{\lambda^2}. \tag{14-9}$$

A_p = power gain with respect to an isotropic radiator

D = antenna diameter (meters)

λ = free-space wavelength of the carrier frequency

k = reflection efficiency (typical value 0.4 to 0.7)

The signal radiating from a parabolic reflector tends to spread as the signal leaves the antenna and propagates through space. This spreading is very much like that of the beam of a flashlight; that is, the farther away the light, the wider the beam. The 3-dB bandwidth (half-power point) for a parabolic dish antenna is approximated by Equation (14-10). This equation shows us that as the frequency increases (wavelength decreases), the beamwidth gets narrower. Both equations also show us that the beamwidth narrows, increasing the diameter of the parabolic reflector. Examples of how to calculate the power gain and the beamwidth of a parabolic antenna system are provided in Example 14-7.

$$\text{beamwidth [degrees]} = \frac{21 \times 10^9}{fD} \approx \frac{70\lambda}{D}, \qquad (14\text{-}10)$$

f = frequency (Hz)

D = antenna diameter (meters)

λ = free-space carrier wavelength

EXAMPLE 14-7

Calculate the power gain (dBi) and the beamwidth of a microwave dish antenna with a 3-m mouth diameter when the antenna is used at 10 GHz. The efficiency of the reflector (k) = 0.6.

$$A_p(\text{dBi}) = 10 \log k \frac{(\pi D)^2}{\lambda^2}$$

$$\lambda = \frac{c}{f} = \frac{2.997925 \times 10^8}{10 \times 10^9}\text{m/s} = 0.0299 \approx 0.03 \text{ m}$$

$$A_p(\text{dBi}) = 10 \log 0.6 \frac{[(\pi)(3)]^2}{0.03^2} = 49.94 \text{ dBi}$$

$$\text{beamwidth} = \frac{70\lambda}{D}$$

$$\text{beamwidth} = \frac{(70)(0.03)}{(3)} = 0.7°.$$

Example 14-7 shows the extremely high gain capabilities of these antennas. This antenna has a 49.94 dBi gain at 10 GHz, which equates to a gain of ~100,000. This power gain is effective, however, only if the receiver is within the 0.7° beamwidth of the dish. Figure 14-31 shows a **polar pattern** for this antenna. It is typical of parabolic antennas and

FIGURE 14-31 **Polar pattern for parabolic antenna in Example 14-6.**

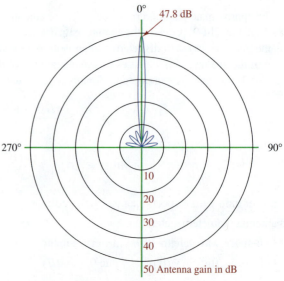

shows the 47.8-dB gain at the 0° reference. Notice the three side lobes on each side of the main one. As you might expect from the antenna's physical construction, there can be no radiated energy from 90° to 270°.

Parabolic reflectors also have a rating for the **effective aperture**. This relationship enables us to measure the effective signal capture area that a parabolic reflector provides for a given diameter and efficiency. This relationship is written as:

$$A_e = k\pi \left(\frac{D}{2}\right)^2,$$ (14-11)

where k = reflection efficiency (provided by the manufacturer)

D = reflector diameter (meters).

An example using Equation (14-11) is provided in Example 16-3.

EXAMPLE 14-8

Calculate the aperture efficiency (A_e) of a parabolic reflector, which has a diameter (D) of 4.5 meters and an efficiency factor (k) of 62%. Compare this measurement with the ideal capture area.

$$A_e = k\pi \left(\frac{D}{2}\right)^2 \text{ [m}^2\text{]}$$

$$A_e = (.62)(\pi)\left(\frac{4.5}{2}\right)^2 \text{ [m}^2\text{]}$$

$$= 9.86 \text{ m}^2$$

The ideal capture area for a 4.5-m parabolic antenna is $(\pi)\left(\frac{4.5}{2}\right)^2 = 15.9 \text{ m}^2$.

The effective capture area for a 4.5 m parabolic reflector is much less than the ideal value. This loss in capture area is attributed to any obstructions (cabling and mechanical hardware) and the nonideal shape of the manufactured parabolic antenna.

Microwave dish antennas are widely used in satellite communications because of their high gain; they are also used for satellite tracking and radio astronomy. They are sometimes used in point-to-point line-of-sight radio links. Often these antennas have a "cover" over the dish. This is a low-loss dielectric material known as a **radome**. Its purpose may be maintenance of internal pressure or, more simply, environmental protection. The construction of a bird's nest within the dish is undesirable for the bird as well as the antenna user.

Lens Antenna

We have all witnessed the effect of focusing the sun's rays into a point using a simple magnifying glass. The effect can also be accomplished with microwave energy, but because of the much higher wavelength, the lens must be large and bulky to be effective. The same effect can be obtained with much less bulk using the principle of **zoning**, as shown in Figure 14-32.

If an antenna launches energy from the focus as shown in Figure 14-31, its spherical wavefront is converted into a plane (and thus highly directive) wave. The inside section of the lens is made thick at the center and thinner toward the edge to permit the lagging portions of the spherical wave to catch up with the faster portions at the center of the wavefront. The other lens sections have the same effect, but they work on the principle that for all sections of the wavefront to be in phase, it is not necessary for all paths to be the same. A 360° phase difference (or multiple) will provide correct phasing. Thus, the plane wavefronts are made up of parts of two, three, or more of the spherical wavefronts.

FIGURE 14-32 **Zoned lens**
antenna.

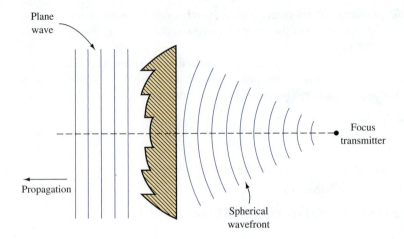

Obviously the thickness of the steps is critical because of its relationship to wavelength. This is, therefore, not a broadband antenna, as is a simple magnifying glass antenna. However, the savings in bulk and expense justify the use of the zoned lens. Keep in mind that these antennas need not be glass because microwaves pass through any dielectric material, though at a reduced velocity compared to free space.

Patch Antenna

The **patch antenna** is simply a square or round "island" of conductor on a dielectric substrate backed by a conducting ground plane. A square patch antenna is shown in Figure 14-33. The square's side is made equal to one half-wavelength and the antenna has a bandwidth less than 10% of its resonant frequency. The circular patch antenna is constructed with a diameter equal to about 0.6 wavelength and has about half the bandwidth or less than 5% of its resonant frequency.

FIGURE 14-33 **Patch antenna.**

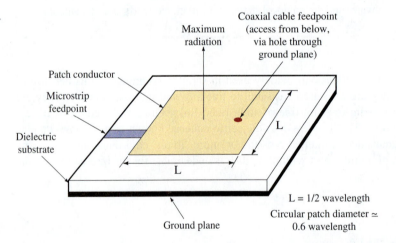

Notice the antenna feedpoint in Figure 14-33. The process of matching impedances between a coaxial cable and the antenna can be precisely achieved with proper positioning. The same can be said if the feed is to be with a microstrip line as shown in blue in Figure 14-33. Keep in mind that the antenna will be fed by one or the other method (coaxial or microstrip) but not both. The radiation pattern is circular and transverse (at right angles) to the antenna away from the ground plane.

The patch antenna is extremely cheap to fabricate using printed circuit boards (PCBs) as the dielectric substrate and using standard microstrip fabrication techniques. In fact, a large number of patch antennas can be fabricated easily on a single PCB so that a phased array antenna can be manufactured inexpensively. As described earlier, a phased array consists of multiple antennas where each antenna signal can be controlled for power and phase. This allows the transmitted (or received) signal to be electronically "steered."

At the end of Chapter 13, a path-loss and link budget analysis was presented for a satellite-based communications link. We present now a similar analysis, albeit somewhat simplified, for a terrestrial (i.e., ground-based, as opposed to satellite) "point-to-point" microwave-based system. This example will tie together many concepts presented in previous chapters and will allow us to illustrate some questions that must be answered in preparing radio system designs. Though this discussion is somewhat simplified for purposes of clarity, the example presented here is illustrative of the kinds of questions to be answered and the assumptions that must be made. It also illustrates the systems approach to communications work. The bottom-line question to be answered is quite straightforward: Will my system work with the degree of reliability my application requires?

Many terrestrial communications systems, particularly those operating at microwave frequencies, involve point-to-point links, where the path of interest is directly between transmitter and receiver, both of which are installed at fixed locations. Such systems can be contrasted with broadcast stations and other "point-to-multipoint" configurations such as cellular telephone systems, in which the region of interest is a large area around the transmitter. Point-to-point links are usually established for a specific purpose, such as linking a cellular phone base station with the carrier's central-office switching equipment or transferring data from a main to an auxiliary, remotely located computer mainframe, in which the maintenance of a highly reliable communications path is of paramount concern.

Terrestrial microwave systems, in particular, generally operate with relatively low transmitter powers, anywhere from 0.5 W or so (+27 dBm) up to a maximum of a few watts, perhaps 5 or 10 (+37 to 40 dBm). These low powers are offset in large part by the extremely high gains of the transmit and receive antennas, which are achieved by focusing the transmitted energy over an extremely narrow beamwidth, often on the order of 1 to 3°, in contrast with the 30 to 60° bandwidths for antennas at VHF and UHF. Such systems may operate on frequency bands for which an FCC license is required or, increasingly, may use the direct-sequence spread-spectrum techniques covered in Chapter 8 to operate on the unlicensed industrial, scientific, and medical (ISM) bands at 2.4 and 5.8 GHz. The advantage to the latter approach, of course, is that an FCC station license is not required. This consideration is significant. Frequency congestion, particularly in populated areas, means that, often, no new frequency allocations are available for licensure; new entrants in such situations are faced with the prospect of negotiating for and buying allocations from incumbent users, often at significant expense. The downside to the ISM bands is that, as they become more heavily used, the background noise of other spread-spectrum systems operating in the same frequency bands will limit operational range. However, even with transmitter powers as low as +17 to +23 dBm (50 mW to 200 mW), the legal maximum powers for unlicensed ISM use depending on band, reliable operation over distances of 10 to 20 mi or more is possible provided that a clear line of sight can be obtained.

Radio transceivers are available from several manufacturers that provide very high data rates—often up to DS3 levels—on unlicensed bands. Thus, such radios could be used to link a remote cellular base station site to the central switching office, or could be used to provide a high bandwidth, wireless Ethernet, wide-area network between several locations on a college or corporate campus, for example. These types of installations often imply the need for a very high degree of reliability, and a critical question in link design is to identify possible points of failure to minimize the possibility of extended outages. With proper engineering of the radio path, radios designed for this type of application are capable of achieving at least 99.999% communications reliability, which is equivalent to a down time of less than 5 minutes, 15 seconds per year. This so-called "five nines" reliability specification is sometimes referred to as "carrier class" because it is the overall availability target for which the landline telephone network is designed.

A crucial specification for achieving the desired reliability specification is the *threshold sensitivity* of the receiver, which is the minimum signal level that must be present at the receiver antenna input to produce an acceptable bit-error rate (BER). Recall from Chapter 8 that BER is expressed in terms of the number of bits received in error over a one-second period and is expressed as a number with a negative exponent. A BER of 10^{-6} may be acceptable for voice-grade communications, while lower BERs may be required for other applications. A typical receiver threshold sensitivity would be on the order of -80 dBm for a BER of 10^{-6}; the threshold is dependent on data rate—DS3 levels will require more robust signal levels than those at fractional T1 data rates, for example. Because of forward-error-correction capability built into the receiver, it will be able to recover data from errors introduced very close to this threshold; below the threshold, where the error correction function cannot keep up, the received data become unusable very quickly—often within 2 dB or so. This "cliff effect" is typical for digital systems of all types employing error correction algorithms.

Because microwave communications are line of sight, a *path study* must be made to ensure that no obstructions lie anywhere in the line between transmitter and receiver. Obstructions can take the form of buildings, mountains, foliage, altitude changes, or the curvature of the Earth. Historically, performing a path study has involved, at a minimum, determining elevations at regular intervals between the transmitter and receiver with topographic maps. Often, surveyors would be employed to visit the proposed locations and would then plot any possible obstructions between transmitter and receiver on elevation/distance maps. There are now online resources and digitized terrain maps that greatly simplify this process if the exact latitudes and longitudes of both the transmit and receive antennas are known. At the very least the curvature of the Earth must be taken into account. The maximum distance is governed by the radio horizon, which can be approximated with Equation (13-7).

In addition to ensuring that nothing obstructs the communications path, we must also ensure that no objects are close enough to produce diffraction effects. These effects can cause the signal to be partially regenerated and, because of differences in distance between the direct and diffracted paths, the direct and diffracted signals can be received out of phase with each other and can partially or completely cancel. The area in which important diffraction-related effects can occur is referred to as the *first Fresnel zone*. Figure 14-34 shows that Fresnel zones are football-shaped regions in the path where diffraction can be significant. A rule of thumb is to ensure an unobstructed path within a distance of 60% of the first Fresnel zone. This 60% distance can be determined with the following formula:

$$R = 10.4\sqrt{\frac{d_1 d_2}{f(d_1 + d_2)}},$$

where

$R =$ required clearance from the obstacle (meters)

$f =$ frequency (gigahertz)

$d_1 =$ distance to the antenna nearer the obstacle (kilometers)

$d_2 =$ distance to the antenna farther from the obstacle (kilometers).

FIGURE 14-34 Illustration of Fresnel zones.

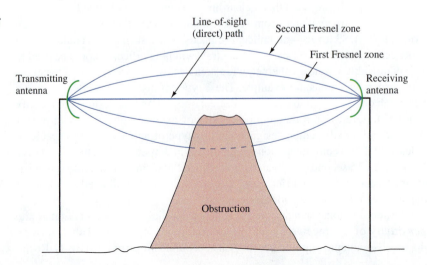

Once the path study is complete and we have satisfied ourselves that no obstructions or Fresnel zone issues will affect us, we can determine the free-space loss resulting from signal propagation. Recall from Chapter 13 that electromagnetic energy radiates outward in a spherical fashion from a point-source radiator; therefore, the total power density resident at the point source (or a practical antenna) spreads out over a larger and larger area as the sphere propagates away from the source. Any given wavefront (i.e., portion of the expanding sphere), therefore, has an ever-smaller amount of the total energy contained within it. Put another way, the power density within a wavefront falls off in accordance with the inverse-square relationship first shown as Equation (13-1). This inverse-square relationship helps explain why there is so much attenuation of signal strength (often 120 dB or more) between transmitter and receiver and why transmitter powers in the tens of thousands of watts produce receiver voltage levels in the millivolt or microvolt range some distance away.

Much of our signal attenuation is caused by propagation in free space. This attenuation can be calculated as

$$\text{Loss (dB)} = 32.44 + 20 \log d + 20 \log f,$$

where d is distance *in kilometers* and f is frequency *in megahertz* (thus, 5.8 GHz would be converted to 5800 MHz for use in the above formula).

Partially offsetting the free-space losses are the gains produced by the transmit and receive antennas. However, losses from the transmission lines, connectors, filters, and splitters must also be considered. As described in the previous section, microwave antennas are generally parabolic dishes with high gain and narrow beamwidth. Typical gain specifications at 5.8 GHz are 30 dBi for 2-ft-diam dishes and 38 dBi for 8-ft-diam dishes. Although larger-diameter antennas have higher gains, in a practical installation, increases in gain must be balanced against other factors: higher cost, increased weight, higher wind loading factors, as well as "intangibles," such as aesthetics and the ease or difficulty of obtaining permits or zoning approval for larger and more conspicuous antennas rather than smaller and more easily disguised antennas. (Never underestimate the "neighbor acceptance factor.")

All radio systems must be designed to be resistant to *fading,* that is, reductions in signal strength caused by a host of factors: weather, multipath fading, temporary obstructions (such as vehicles parked near the transmit or receive antennas), refraction/diffraction effects, and the like. Weather, particularly moisture, is the biggest enemy of microwave links, and weather effects become more pronounced at high frequencies. The *fade margin* is the difference between the received signal level (RSL, generally expressed in dBm) and the receiver's specified minimum acceptable, or threshold, signal level. For example, if a receiver had a receive threshold of −80 dBm, and the link were designed for an average RSL of −40 dBm, the fade margin would be 40 dB. Fade margins of 40 dB are common, particularly in humid or foggy areas, or where part of the path is over water. An absolute minimum fade margin would be about 15 dB; such a low margin would only be appropriate in very dry climates or where high reliability is not of concern.

Other factors to consider are system losses, of which the most important are transmission-line, filter network, and connector losses. Transmission-line losses can be significant at microwave frequencies, particularly if long runs are involved. For example, at 5.8 GHz, 100 ft of high-quality, half-inch coaxial cable will have between 3 and 6 dB of loss, and several hundred feet of cable may be required if the transmitter is located in a communications shelter and the antenna is located at the top of a tall tower. Remember also that this length of cable may be required at both ends as well. For this reason, waveguide (to be discussed in the next chapter) may be called for to minimize cable losses, even though waveguide is expensive and significantly more difficult to install than coaxial cable.

Now that we have identified the most important characteristics, we can begin to design our system. Remember that the question to be addressed is how well this system will perform. Answering this question requires us to determine the received signal level (RSL) and the fade margin.

For purposes of illustration, let us assume some system specifications. Again, some factors, such as ground permittivity and effects of buildings or foliage, have been deliberately left out for the sake of clarity. An in-depth study would need to take factors such as these into consideration, but the underlying methodology is the same. If the predicted RSL is sufficiently above the threshold sensitivity to afford an adequate fade margin, then the system will work to the desired reliability specification.

Our system will operate in the 5.8-GHz ISM band. The transmitter output power is +23 dBm, the maximum allowable for unlicensed operation, and 4-ft dishes, each with a specified gain of +33 dBi, will be used at each end. Assume further that coaxial-cable transmission lines are used, and that the lines, their associated connectors, and any filtering networks installed at either end contribute 4 dB of loss at each end (this is a conservative estimate of loss—it is likely to be larger). Further, we can assume the two ends of the link are 10 km (6.3 mi) apart.

The calculations are shown in Figure 14-35. At one end, the transmitter output power of +23 dBm is reduced by 4 dB to +19 dBm to account for system losses. At the antenna, the +19 dBm input signal is increased by the gain of the antenna, so 19 dBm + 33 dBi = 52 dBm, for an EIRP of slightly less than 200 W. This output power is attenuated by the path loss:

$$\text{Loss (dB)} = 32.44 + 20 \log (10) + 20 \log (5800)$$
$$= 32.44 + 20 + 75.27$$
$$= 127.71 \text{ dB.}$$

Because this result represents a loss, it must be subtracted from the EIRP determined earlier to determine the signal strength at the receive antenna. Thus, +52 dBm − 127.71 dB = −75.71 dBm. This last result is the signal strength at the receive antenna. Assuming the receive antenna has the same gain characteristics as the transmit antenna, then the RSL is −75.71 dBm + 33 dBi − 4 dB = −46.7 dBm. For a receiver threshold sensitivity of −80 dBm, the fade margin is −46.7 − (−80) = 33.3 dB.

FIGURE 14-35 Example path-loss calculations.

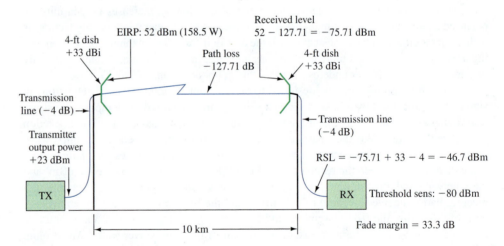

Is this fade margin adequate? Ultimately, that is a judgment call. For high-reliability operation in a humid environment or where part of the path is over water, a 40-dB or even greater fade margin may be required. However, as the above example shows, some changes must be made, either an increase in antenna size or, perhaps, a reduction in transmission line losses accomplished either by using waveguide or by placing the transmitter "head" (containing the RF components, including final amplifier) next to the antenna itself, rather than housing the radio in a communications shelter. One option that is generally not available is to increase the transmitter power output, because the output power originally assumed was the maximum permitted by law.

One final point should be made. Several computer models are available to assist system designers with coverage predictions over a given geographic area. One of the most widely used is the *Longley-Rice* methodology, developed in 1968 by Anita Longley and Phil Rice, scientists at what is now the National Institute of Standards and Technology. This model takes into account such factors as terrain conditions, foliage, the degree to which an area has been "urbanized" (to quantify multipath effects from buildings), and other factors such as frequency bands to be deployed. Other, more special-purpose models are available. Broadcasters and cellular system operators use Longley-Rice models to predict coverage areas of existing and proposed systems.

SUMMARY

An antenna is a transducer that converts energy from electrical form to electromagnetic form and vice versa. Operation of the most basic antenna, the half-wave dipole, can be viewed as an extension of the quarter-wave open-ended transmission-line section examined in Chapter 12. Energy normally contained within the parallel-conductor transmission line is allowed to radiate into free space when the quarter-wavelength sections are spread 180° apart. Electromagnetic radiation is thus created, and this radiation is represented in the form of waves that propagate outward from the antenna into free space.

A half-wavelength section of conductor is electrically resonant at its frequency of operation and presents a purely resistive load. At frequencies above and below resonance, the conductor exhibits reactive characteristics that contribute to impedance mismatches and a voltage standing-wave ratio (VSWR) higher than 1:1. Also, conditions that cause currents to be reflected back into the antenna, such as metallic or other conducting objects in the vicinity, will create mismatches and cause the antenna radiation resistance to change. The radiation resistance represents the ratio of applied source voltage to antenna current or, alternatively, the radiated power divided by the square of the rms current at the feed point. Further, the radiation resistance is an expression of the power that is converted to electromagnetic form and radiated into space.

Antennas are inherently passive devices, but practical, physically realizable antennas exhibit gain in comparison with an isotropic radiator (a theoretical point source antenna) by focusing their energy in a preferred direction. Dipoles have 2.15 dB of gain over isotropic radiators, and other types of antennas have gain with respect to the dipole. This gain can be achieved in a number of ways, but all involve concentrating radiated energy in some directions at the expense of others.

Vertical antennas are generally one-quarter wavelength long, with a suitably situated ground plane forming the other quarter wavelength needed to create a resonant, half-wavelength antenna. Antennas can use loading coils or capacitive elements, as well as matching networks, to tune out reactive effects.

Antenna arrays consist of multiple elements and fall into either of two categories. Parasitic arrays have one driven element and one or more parasitic elements. A Yagi-Uda antenna is an example of a parasitic array that has a reflective element and several directive elements to produce a great deal of gain, over 10 dB in many cases, in the forward direction of radiation. Driven arrays apply energy to all elements. Examples are collinear and broadside arrays as well as the arrays of vertically polarized transmission towers used in AM broadcasting applications.

There are a number of special-purpose antennas used as well. Among these are log-periodic antennas, which feature very wide bandwidth and reasonably good gain over a wide range of frequencies. Other antennas include the loop antenna, useful for direction-finding purposes because of its sharp, bidirectional radiation pattern, and the ferrite loop and folded dipole antennas, both used in broadcast radio receivers.

Antennas used at microwave frequencies constitute a distinct class unto themselves because of the very short wavelengths involved. The most basic type is the horn antenna, which because of its simple construction is able to provide on the order of 20 dB of gain and is used in noncritical applications. The most widely used microwave antennas are dish antennas, which work on the principle of the parabolic reflector to focus energy onto a dipole receiving element. Dish antennas are capable of very high signal gains by focusing energy onto a very narrow beamwidth. Patch antennas can be easily fabricated on printed-circuit-board (PCB) material and are ideally suited for portable, mass-market applications. Arrays of patch antennas can be fabricated to permit the economical steering of energy for various purposes.

A link-budget analysis for a microwave point-to-point system brings together many of the concepts presented up to this point and represents a true systems approach to communications. Designers must consider questions of reliability, spectrum availability, permissible power levels, and antenna size, among others. Point-to-point systems make up for their relatively low powers by using very-high-gain dish antennas. Most power losses, amounting to 120 dB or more, occur in the path from transmitter to receiver. System designers must also consider obstructions within the first Fresnel zone and provide for an adequate fade margin to maintain reliable connections in the face of weather and other environmental changes.

QUESTIONS AND PROBLEMS

SECTION 14-1

*1. How should a transmitting antenna be designed if a vertically polarized wave is to be radiated, and how should the receiving antenna be designed for best performance in receiving the ground wave from this transmitting antenna?

*2. If a field intensity of 25 mV/m develops 2.7 V in a certain antenna, what is its effective height? (108 m)

*3. If the power of a 500-kHz transmitter is increased from 150 W to 300 W, what would be the percentage change in field intensity at a given distance from the transmitter? What would be the decibel change in field intensity? (141%, 3 dB)

*4. If a 500-kHz transmitter of constant power produces a field strength of 100 μV/m at a distance of 100 mi from the transmitter, what would be the theoretical field strength at a distance of 200 mi from the transmitter? (50 μV/m)

*5. If the antenna current at a 500-kHz transmitter is reduced 50%, what would be the percentage change in the field intensity at the receiving point? (50%)

*6. Define *field intensity*. Explain how it is measured.

*7. Define *polarization* as it refers to broadcast antennas.

8. Explain how antenna reciprocity occurs.

SECTION 14-2

9. Explain the development of a half-wave dipole antenna from a quarter-wavelength, open-circuited transmission line.

*10. Draw a diagram showing how current varies along a half-wavelength dipole antenna.

*11. Explain the voltage and current relationships in a one-wavelength antenna, one half-wavelength (dipole) antenna, and one quarter-wavelength *grounded* antenna.

*12. What effect does the magnitude of the voltage and current at a point on a half-wave antenna in *free space* (a dipole) have on the impedance at that point?

*13. Can either of the two fields that emanate from an antenna produce an electromagnetic field (EMF) in a receiving antenna? If so, how?

14. Draw the three-dimensional radiation pattern for the half-wave dipole antenna, and explain how it is developed.

15. Define antenna *beamwidth*.

*16. What is the effective radiated power of a television broadcast station if the output of the transmitter is 1000 W, antenna transmission line loss is 50 W, and the antenna power gain is 3? (2850 W)

17. A $\lambda/2$ dipole is driven with a 5-W signal at 225 MHz. A receiving dipole 100 km away is aligned so that its gain is cut in half. Calculate the received power and voltage into a 73-Ω receiver. (7.57 pW, 23.5 μV)

18. An antenna with a gain of 4.7 dBi is being compared with one having a gain of 2.6 dBd. Which has the greater gain?

19. Explain why a monopole antenna is used below 2 MHz.

20. Explain what is meant by half-wave dipole. Calculate the length of a 100-MHz $\frac{2}{3}\lambda$ antenna.

21. Determine the distance from a $\lambda/2$ dipole to the boundary of the far-field region if the $\lambda/2$ dipole is being used in the transmission of a 90.7-MHz FM broadcast band signal. ($R = 1.653$ m)

22. Determine the distance from a parabolic reflector of diameter $D = 10$ m to the boundary of the far-field region. The antenna is being used to transmit a 4.1-GHz signal. ($R = 2733.3$ m)

23. Define *near* and *far fields*.

SECTION 14-3

24. Define *radiation resistance* and explain its significance.

*25. The ammeter connected at the base of a vertical antenna has a certain reading. If this reading is increased 2.77 times, what is the increase in output power? (7.67)

26. How is the operating power of an AM transmitter determined using antenna resistance and antenna current?

27. Explain what happens to an antenna's radiation resistance as its length is continuously increased.

28. Explain the effect that ground has on an antenna.

29. Calculate the efficiency of an antenna that has a radiation resistance of 73 Ω and an effective dissipation resistance of 5 Ω. What factors could enter into the dissipation resistance? (93.6%)

*30. Explain the following terms with respect to antennas (transmission or reception):
  (a) Physical length.
  (b) Electrical length.
  (c) Polarization.
  (d) Diversity reception.
  (e) Corona discharge.

31. What is the relationship between the electrical and physical length of a half-wave dipole antenna?

*32. What factors determine the resonant frequency of any particular antenna?

*33. If a vertical antenna is 405 ft high and is operated at 1250 kHz, what is its physical height expressed in wavelengths? (0.54λ)

*34. What must be the height of a vertical radiator one half-wavelength high if the operating frequency is 1100 kHz? (136 m)

SECTION 14-4

35. What is an antenna feed line? Explain the use of resonant antenna feed lines, including advantages and disadvantages.

36. What is a nonresonant antenna feed line? Explain its advantages and disadvantages.

37. Explain the operation of a delta match. Under what conditions is it a convenient matching system?

*38. Draw a simple schematic diagram of a push-pull, neutralized radio-frequency amplifier stage, coupled to a vertical antenna system.

*39. Show by a diagram how a two-wire radio-frequency transmission line may be connected to feed a half-wave dipole antenna.

*40. Calculate the characteristic impedance of a quarter-wavelength section used to connect a 300-Ω antenna to a 75-Ω line. (150 Ω)

41. Explain how delta matching is accomplished.

SECTION 14-5

*42. Which type of antenna has a minimum of directional characteristics in the horizontal plane?

*43. If the resistance and the current at the base of a monopole antenna are known, what formula can be used to determine the power in the antenna?

*44. What is the difference between a half-wave dipole and a monopole antenna?

*45. Draw a sketch and discuss the horizontal and vertical radiation patterns of a quarter-wave monopole antenna. Would this also apply to a similar type of receiving antenna?

46. What is an image antenna? Explain its relationship to the monopole antenna.

*47. What would constitute the ground plane if a quarter-wave grounded (whip) antenna, 1 m in length, were mounted on the metal roof of an automobile? Mounted near the rear bumper of an automobile?

*48. What is the importance of the ground radials associated with standard broadcast antennas? What is likely to be the result of a large number of such radials becoming broken or seriously corroded?

*49. What is the effect on the resonant frequency of connecting an inductor in series with an antenna?

*50. What is the effect on the resonant frequency of adding a capacitor in series with an antenna?

*51. If you desire to operate on a frequency lower than the resonant frequency of an available monopole antenna, how may this be accomplished?

*52. What is the effect on the resonant frequency if the physical length of a λ/2 dipole antenna is reduced?

*53. Why do some standard broadcast stations use top-loaded antennas?

*54. Explain why a *loading coil* is sometimes associated with an antenna. Under this condition, would absence of the coil mean a capacitive antenna impedance?

SECTION 14-6

55. Explain how the directional capabilities of the elementary antenna array shown in Figure 14-19 are developed.

56. Define the following terms:
   (a) Driven elements.
   (b) Parasitic elements.
   (c) Reflector.
   (d) Director.

57. Calculate the ERP from a Yagi–Uda antenna (illustrated in Figure 14-20) driven with 500 W. (2500 W)

58. Calculate the *F/B* ratio for an antenna with
   (a) Forward gain of 7 dB and reverse gain of −3 dB.
   (b) Forward gain of 18 dB and reverse gain of 5 dB.

59. Sketch a Yagi–Uda configuration.

60. Describe the physical configuration of a collinear array. What is the effect of adding more elements to this antenna?

61. Describe the physical configuration of a broadside array. Explain the major advantage they have compared to collinear arrays.

*62. What is the direction of maximum radiation from two vertical antennas spaced $\lambda/2$ and having equal currents in phase?

*63. How does a directional antenna array at an AM broadcast station reduce radiation in some directions and increase it in other directions?

*64. What factors can cause the directional pattern of an AM station to change?

65. Define *phased array*.

66. Explain how a parasitic array can be developed.

SECTION 14-7

67. Describe the major characteristics of a log-periodic antenna. What explains its widespread use? Explain the significance of its shortest and longest elements.

*68. Describe the directional characteristics of the following types of antennas:
   (a) Horizontal half-wave dipole antenna.
   (b) Vertical half-wave dipole antenna.
   (c) Vertical loop antenna.
   (d) Horizontal loop antenna.
   (e) Monopole antenna.

*69. What is the directional reception pattern of a loop antenna?

70. What is a ferrite loop antenna? Explain its application and advantages.

71. What is the radiation resistance of a standard folded dipole? What are its advantages over a standard dipole? Why is it usually used as the driven element for Yagi–Uda antennas instead of the half-wave dipole antenna?

72. Describe the operation of a slot antenna and its application with aircraft in a driven array format.

73. An antenna has a maximum forward gain of 14 dB at its 108-MHz center frequency. Its reverse gain is −8 dB. Its beamwidth is 36° and the bandwidth extends from 55 to 185 MHz. Calculate:
   (a) Gain at 18° from maximum forward gain. (11 dB)
   (b) Bandwidth. (130 MHz)
   (c) *F/B* ratio. (22 dB)
   (d) Maximum gain at 185 MHz. (11 dB)

74. Explain the difference between antenna beamwidth and bandwidth.

SECTION 14-8

75. Microwave antennas tend to be highly directive and provide high gain. Discuss the reasons for this.

76. What is a horn antenna? Provide sketches of three basic types, and explain their important characteristics.

*77. Describe how a radar beam is formed by a paraboloidal reflector.

78. With sketches, explain three different methods of feeding parabolic antennas.

79. A160-ft-diam parabolic antenna is driven by a 10-W transmitter at 4.3 GHz. Calculate its effective radiated power (ERP) and its beamwidth. (29.3 MW, 0.10°)

80. A parabolic antenna has a 0.5° beamwidth at 18 GHz. Calculate its gain in dB. (50.7)

81. What is a radome? Explain why its use is often desirable in conjunction with parabolic antennas.

82. Explain the principles of a zoned lens antenna, including the transformation of a spherical wave into a plane wave.

83. The antenna in Figure 14-32 has a resonant frequency of 1.3 GHz. Calculate the bandwidth of this patch antenna. ($\approx$130 MHz)

84. The beamwidth of a 4.0 GHz signal is 1.1°. Determine the dB power gain. (43.8 dB)

85. Determine the focal length of a 5-m parabolic reflector with a curve depth of 1.23 m.

86. Determine the approximate gain of a 10-m antenna operating at 14 GHz. Express your answer in dB.

87. Determine the approximate beamwidth of a 10-m parabolic antenna operating at (a) 14 GHz and (b) 4 GHz. Compare the two results.

88. Find the effective aperture of a 3-m parabolic antenna, given a reflection efficiency of 0.6.

QUESTIONS FOR CRITICAL THINKING

*89. A ship radio-telephone transmitter operates on 2738 kHz. At a distant point from the transmitter, the 2738-kHz signal has a measured field of 147 mV/m. The second harmonic field at the same point is measured as 405 μV/m. How much has the harmonic emission been attenuated below the 2738-kHz fundamental? (51.2 dB)

*90. You are asked to calculate effective radiated power. What data do you need to collect and how do you perform the calculation?

91. Design a log-periodic antenna to cover the complete VHF TV band from 54 MHz to 216 MHz. Use a design factor (τ) of 0.7, and provide a scaled sketch of the antenna with all dimensions indicated.

92. A loop antenna used for DF purposes detects a null from a signal with the loop rotated 35° counterclockwise (CCW) from a line of latitude. When the antenna is moved 3 mi west along the same line of latitude, it detects a null from the same signal source when rotated 45° clockwise (CW) from the line of latitude. You have been asked to identify the exact location of the signal source with respect to the two points when readings were taken. Provide this information. (You may use a sketch.)

CHAPTER 15

WAVEGUIDES AND RADAR

KEY TERMS

waveguide
characteristic wave impedance
vane
radar
echo signal
pulse repetition frequency (PRF)
pulse repetition rate (PRR)
pulse repetition time (PRT)
rest time

receiver time
radar mile
second return echoes
maximum usable range
double range echoes
peak power
duty cycle
Doppler effect
stripline
microstrip
dielectric waveguide

15-1 COMPARISON OF TRANSMISSION SYSTEMS

The mode of energy transmission chosen for a given application normally depends on the following factors: (1) initial cost and long-term maintenance, (2) frequency band to be used and its information-carrying capacity, (3) selectivity or privacy offered, (4) reliability and noise characteristics, and (5) power level and efficiency. Naturally, any one mode of energy transmission has only some of the desirable features. Choosing the mode of energy transmission best suited for a particular application, therefore, becomes a matter of sound technical judgment.

Transmission lines, antennas, and fiber optics are the more commonly known means of high-frequency transmission, but waveguides also play an important role. The following examples show that each method of transmission has its proper place. Fiber optics has been left out of this comparison, but this mode of transmission is discussed fully in Chapter 16.

It is desired to transmit a 1-GHz signal between two points 30 mi apart. If the received energy in each case were chosen to be 1 nW (10^{-9} W), then for comparison it would be found that for reasonably typical installations, the required *transmitted* power would be on the order of

1. *Transmission lines:* 10^{1500} nW (15,000-dB loss)
2. *Waveguides:* 10^{150} nW (1500-dB loss)
3. *Antennas:* 100 mW (80-dB loss)

Clearly, the transmission of energy without any electrical conductors (antennas) exceeds the efficiency of waveguides and transmission lines by many orders of magnitude.

If the transmission path length of the preceding example were shortened by a factor of 100:1, to a distance of 1500 ft, the comparison becomes

1. *Transmission lines:* 1 MW (150-dB loss)
2. *Waveguides:* 30 nW (15-dB loss)
3. *Antennas:* 10 μW (40-dB loss)

Quite clearly the waveguide now surpasses either the transmission line or antenna for efficiency of energy transfer.

A comparison of the energy input required versus distance to obtain a received power of 1 nW for these three modes of energy transmission is shown in Figure 15-1. The frequency is 1 GHz, and the results are expressed on a decibel scale with a 0-dB reference at the required receiver power level of 1 nW. The dashed section of the antenna curve, somewhat beyond 30 mi, indicates that the attenuation becomes severe beyond the line-of-sight distance, which is typically 50 mi.

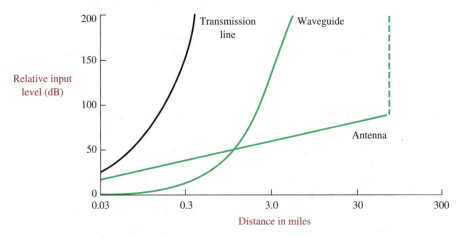

FIGURE 15-1 **Input power required versus distance for fixed receiver power.**

One final comparison, and then a specific look at waveguides. Transmission of energy down to zero frequency is practical with transmission lines, but waveguides, antennas, and fiber optics inherently have a practical low-frequency limit. In the case of antennas, this limit is about 100 kHz, and for waveguides, it is about 300 MHz. Fiber-optic transmissions take place at the frequency of light, or greater than 10^{14} Hz! Theoretically the antennas and waveguides could be made to work at arbitrarily low frequencies, but the physical sizes required would become excessively large. However, with the low gravity and lack of atmosphere on the moon, it may be feasible to have an antenna 10 mi high and 100 mi long for frequencies as low as a few hundred cycles per second. As an indication of the sizes involved, it may be noted that for either waveguides or antennas, the important dimension is normally a half-wavelength. Thus, a waveguide for a 300-MHz signal would be about the size of a roadway drainage culvert, and an antenna for 300 MHz would be about $1\frac{1}{2}$ ft long.

15-2 TYPES OF WAVEGUIDES

Any surface separating two media of distinctly different conductivities or permittivities has a guiding effect on electromagnetic waves. For example, a rod of dielectric material, such as polystyrene, can carry a high-frequency wave, somewhat as a glass fiber conducts a beam of light. These phenomena will be further explored in Section 15-10 and Chapter 16. The best guiding surface, however, is that between a good dielectric and a good conductor.

In a broad sense, all kinds of transmission lines, including coaxial cables and parallel wires, are waveguides. In practice, however, the term **waveguide** has come to signify a hollow metal tube or pipe used to conduct electromagnetic waves through its interior. They were first used extensively in radar sets during World War II, operating at wavelengths of between 10 and 3 cm. They are commonly called *plumbing* in the trade.

A waveguide can be almost any shape. The most popular shape is rectangular, but circular and even more exotic shapes are used. We shall mainly study the rectangular waveguide operating in the TE_{10} mode. We shall learn more about this terminology shortly.

Like coaxial lines, waveguides are perfectly shielded—hence, no radiation loss. The attenuation of a hollow pipe is less, and the power capacity is greater, than that of a coaxial line of the same size at the same frequency. Most of the copper loss of a coaxial line occurs in the thin inner conductor; hence, its elimination in a waveguide reduces attenuation and increases the power capacity. It also simplifies the construction and makes the line more rugged.

Waveguide Operation

A rigorous mathematical demonstration of waveguide operation is beyond our intentions. A practical explanation is possible by starting out with a normal two-wire transmission line. Recall from Chapter 12 that a quarter-wavelength shorted stub looks like an open circuit at its input. For this reason, the shorted stub is often used as an insulating support for transmission lines. If an infinite number of these supports were added both above and below the two-wire transmission line, as shown in Figure 15-2, you can visualize it turning into a rectangular waveguide. If the shorted stubs were less than a quarter-wavelength, operation would be drastically impaired. The same is true of a rectangular waveguide. The *a* dimension of a waveguide, shown in Figure 15-3, must be at least a half-wavelength at the operating frequency, and the *b* dimension is normally about one-half the *a* dimension.

The wave that is propagated by a waveguide is electromagnetic and, therefore, has electric (*E* field) and magnetic (*H* field) components. In other words, energy propagates down a waveguide in the form of a radio signal. The configuration of these two fields determines the mode of operation. If no component of the *E* field is in the direction of propagation, we say that it is in the TE, or *transverse electric* mode. *Transverse* means "at right angles." TM, or *transverse magnetic,* is the mode of waveguide operation whereby the magnetic field has no component in the direction of propagation. Two-number subscripts normally follow the TE or TM designations, and they can be interpreted as follows:

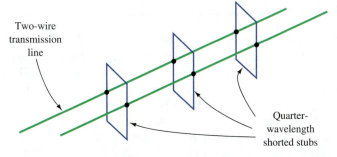

Two-wire
transmission
line

Quarter-
wavelength
shorted stubs

FIGURE 15-2   **Transforming a transmission line into a waveguide.**

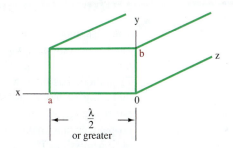

$\dfrac{\lambda}{2}$
or greater

FIGURE 15-3   **Waveguide dimension designation.**

For TE modes, the first subscript is the number of one half-wavelength E-field patterns along the a (longest) dimension, and the second subscript is the number of one half-wavelength E-field patterns along the b dimension. For TM modes, the number of one half-wavelength H fields along the a and b dimensions determines the subscripts. Refer to Figure 15-4 for further illustration of this process.

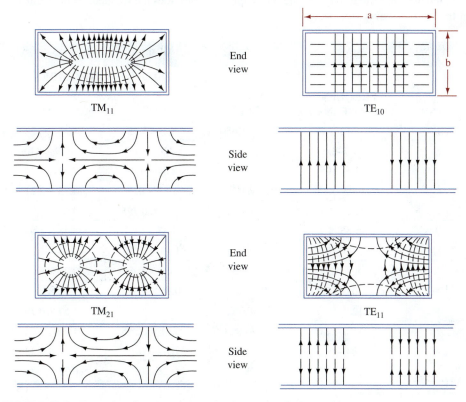

End view

Side view

TM$_{11}$

TE$_{10}$

End view

Side view

TM$_{21}$

TE$_{11}$

FIGURE 15-4   **Examples of modes of operation in rectangular waveguides.**

The electric field is shown with solid lines, and the magnetic field is shown with dashed lines. Notice the end view for the TE$_{10}$ mode in Figure 15-4. The electric field goes from a minimum at the ends along the a dimension to a maximum at the center. This is equivalent to one half-wavelength of E field along the a dimension, while no component exists along the b dimension. This is, therefore, called the TE$_{10}$ mode of operation. In the TM$_{21}$ mode of operation, note that along the a dimension the H field (dashed lines) goes from zero to maximum to zero to maximum to zero. That is two half-wavelengths. Along the b dimension, one half-wavelength of the H field occurs—thus the TM$_{21}$ designation. Note that in the side views, the H fields (dashed lines) are not shown for the sake of simplicity. In these side views, the TE modes have no E field in the direction of propagation (right to left or left to right), while in the TM modes the E field (solid lines) does exist in the propagation direction.

Dominant Mode of Operation

The TE$_{10}$ mode of operation is called the dominant mode because it is the most "natural" one for operation. A waveguide is often thought of as a high-pass filter because only very high frequencies, those above the waveguide cutoff frequency, can be propagated. The TE$_{10}$ mode has the lowest cutoff frequency of any of the possible modes of propagation, including both TM and TE types. It is of special interest because there will exist a frequency range between its cutoff frequency (f_c, the lowest frequency that a given waveguide propagates) and that of the next higher-order mode in which this is the only possible mode of transmission. Thus, if a waveguide is excited within this frequency range, energy propagation must take place in the dominant mode, regardless of the way in which the guide is excited. Control of the mode of operation is important in any practical transmission system, and thus the TE$_{10}$ mode has a distinct advantage over the other possible modes in a rectangular waveguide. Even more important, however, is the fact that TE$_{10}$ operation allows use of the physically smallest waveguide for a given frequency of operation.

The dimensions for an RG-52/U waveguide are 0.9 × 0.4 in. This is one of the standard sizes used in the X-band frequency range and is usually called the X-band waveguide. The recommended frequency range for this waveguide size is 8.2 to 12.4 GHz. As a result of this limited range of usefulness, standard sizes of waveguides have been established, each having a specified frequency range. Table 15-1 provides the frequency range and size of the various waveguide bands.

TABLE 15-1 • Waveguide Bands/Sizes			SIZE OF WAVEGUIDE	
BAND	FREQUENCY RANGE IN GHz	TYPE	IN.	CM
L	1.12–1.7	WR650	6.5 × 3.25	16.5 × 8.26
S	1.7–2.6	WR430	4.3 × 2.15	10.9 × 8.6
S	2.6–3.95	WR284	2.84 × 1.34	7.21 × 3.40
G	3.95–5.85	WR187	1.87 × 0.87	4.75 × 2.21
C	4.9–7.05	WR159	1.59 × 0.795	4.04 × 2.02
J	5.85–8.2	WR137	1.37 × 0.62	3.48 × 1.57
H	7.05–10.0	WR112	1.12 × 0.497	2.84 × 1.26
X	8.2–12.4	WR90	0.9 × 0.4	2.29 × 1.02
M	10.0–15.0	WR75	0.75 × 0.375	1.91 × 0.95
P	12.4–18.0	WR62	0.62 × 0.31	1.57 × 0.79
N	15.0–22.0	WR51	0.51 × 0.255	1.30 × 0.65
K	18.0–26.5	WR42	0.42 × 0.17	1.07 × 0.43
R	26.5–40.0	WR28	0.28 × 0.14	0.71 × 0.36

Note: WR = waveguide rectangular

The formula for cutoff wavelength is

$$\lambda_{co} = 2a \tag{15-1}$$

for the TE$_{10}$ mode, where a is the long dimension of the waveguide rectangle. Thus, for an RG-52/U waveguide, λ_{co} is 2(0.9) or 1.8 in., or 4.56 cm. Therefore,

$$f_{co} = \frac{c}{\lambda_{co}} = \frac{3 \times 10^{10} \text{ cm/s}}{4.56 \text{ cm}} = 6.56 \text{ GHz}. \tag{15-2}$$

The lowest frequency of propagation (without considerable attenuation) is 6.56 GHz, but the recommended range is 8.2 to 12.4 GHz. The next higher-order mode is the TE$_{20}$, which has a cutoff frequency of 13.1 GHz. Thus, within the frequency range from 6.56 to 13.1 GHz, only the TE$_{10}$ mode can propagate within the X-band waveguide, in the ordinary sense of the word.

15-3 PHYSICAL PICTURE OF WAVEGUIDE PROPAGATION

For a wave to exist in a waveguide, it must satisfy Maxwell's equations throughout the waveguide. These mathematically complex equations are beyond the scope of this book, but one boundary condition of these equations can be put into plain language: There can be no tangential component of electric field at the walls of the waveguide. This makes sense because the conductor would then *short* out the *E* field. An exact solution for the field existing within a waveguide is a relatively complicated mathematical expression. It is possible, however, to obtain an understanding of many of the properties of waveguide propagation from a simple physical picture of the mechanisms involved. The fields in a typical TE_{10} waveguide can be considered as the resultant fields produced by an ordinary plane electromagnetic wave that travels back and forth between the sides of the guide, as illustrated in Figure 15-5.

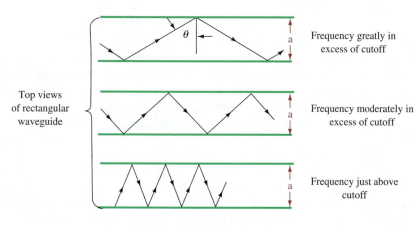

Frequency greatly in excess of cutoff

Top views of rectangular waveguide

Frequency moderately in excess of cutoff

Frequency just above cutoff

FIGURE 15-5 **Paths followed by waves traveling back and forth between the walls of a waveguide.**

The electric and magnetic component fields of this plane wave are in time phase but are geometrically at right angles to each other and to the direction of propagation. Such a wave travels with the velocity of light and, upon encountering the conducting walls of the guide, is reflected with a phase reversal of the electric field and with an angle of reflection equal to the angle of incidence. A picture of the wavefronts involved with such propagation for a rectangular waveguide is shown in Figure 15-6.

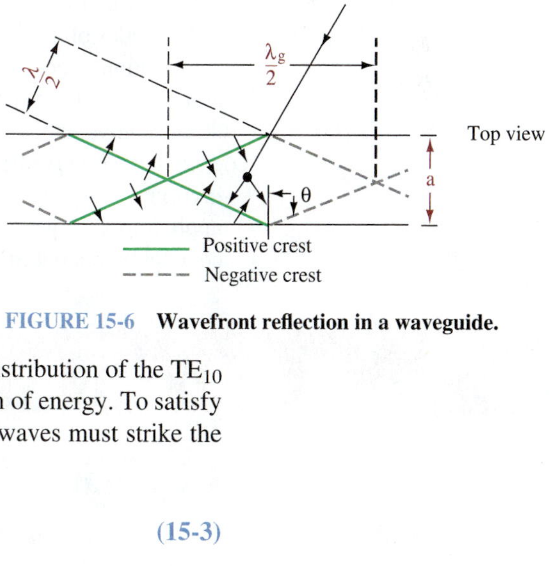

When the angle θ (see Figure 15-5) is such that the successive positive and negative crests traveling in the same direction just fail to overlap inside the guide, it can be shown that the summation of the various waves and their reflections leads to the field distribution of the TE_{10} mode, which travels down the waveguide and represents propagation of energy. To satisfy the necessary boundary conditions for propagation, the component waves must strike the sides of the rectangular waveguide at an angle equal to

Top view

—— Positive crest
- - - - Negative crest

FIGURE 15-6 **Wavefront reflection in a waveguide.**

$$\cos \theta = \frac{\lambda}{2a}, \tag{15-3}$$

where a is the width of the waveguide and λ is the wavelength of the wave on the basis of the velocity of light.

Because the component waves that can be considered as building up the actual field in the waveguide all travel at an angle with respect to the axis of the guide, the rate at which energy propagates down the guide is less than the velocity of light. This velocity with which energy propagates is termed group velocity (V_g) and in the case of Figure 15-6 is given by the relation

$$\frac{V_g}{C} = \sin \theta = \sqrt{1 - \left(\frac{\lambda}{2a}\right)^2}. \tag{15-4}$$

The guide wavelength (λ_g) is greater than the free-space wavelength (λ). A study of the $\lambda_g/2$ and $\lambda/2$ shown in Figure 15-6 should help in visualizing this situation. Thus,

$$\frac{\text{wavelength in guide}}{\text{wavelength in free space}} = \frac{\lambda_g}{\lambda} = \frac{1}{\sin\theta} \tag{15-5}$$

and therefore

$$\frac{c}{V_g} = \frac{\lambda_g}{\lambda} = \frac{1}{\sqrt{1 - (\lambda/2a)^2}}. \tag{15-6}$$

In Smith chart solutions of waveguide problems, λ_g should be used for making moves, not the free-space wavelength λ. The velocity with which the wave appears to move past the guide's side wall is termed the phase velocity, V_p. It has a value greater than the speed of light. It is only an "apparent" velocity; however, it is the velocity with which the wave is changing phase at the side wall. The phase and group velocities, V_p and V_g, respectively, are related by the fact that

$$\sqrt{V_p V_g} = \text{velocity of light.} \tag{15-7}$$

As the wavelength is increased, the component waves must travel more nearly at right angles to the axis of the waveguide, as shown in the bottom portion of Figure 15-5. This causes the group velocity to be lowered and the phase velocity to be still greater than the velocity of light, until finally one has $\theta = 0°$. The component waves then bounce back and forth across the waveguide at right angles to its axis and do not travel down the guide at all. Under these conditions, the group velocity is zero, the phase velocity becomes infinite, and propagation of energy ceases. This occurs at the frequency that was previously defined as the cutoff frequency, f_{co}. The cutoff frequency for the TE_{10} mode of operation can be determined from the relationship given in Equation (15-1). Note that the waveguide acts as a high-pass filter, with the cutoff frequency determined by the waveguide dimensions. To obtain propagation, the waveguide must have dimensions comparable to a half-wavelength, and that limits its practical use to frequencies above 300 MHz.

At frequencies very much greater than cutoff frequency, it is possible for the higher-order modes of transmission to exist in a waveguide. Thus, if the frequency is high enough, propagation of energy can take place down the guide when the system of component waves that are reflected back and forth has the form of the TE_{20} mode. It has a field distribution that is equivalent to two distributions of the dominant TE_{10} mode placed side by side, but each with reversed polarity. This conceptual presentation of waveguide propagation, involving a wave suffering successive reflections between the sides of the guide, can be applied to all types of waves and to other than rectangular guides. The way in which the concept works out in these other cases is not so simple, however, as for the TE_{10} mode.

15-4 OTHER TYPES OF WAVEGUIDES

Circular

The dominant mode (TE_{10}) for rectangular waveguides is by far the most widely used. The use of other modes or of other shapes is extremely limited. However, the use of a circular waveguide is found in radar applications where it is necessary to have a continuously rotating section like that in Figure 15-7. Modes in circular waveguides can be rotationally symmetrical, which means that a radar antenna can physically rotate with no electrical disturbance. While a circular waveguide is actually simpler to manufacture than a rectangular one, for a given frequency of operation, its cross-sectional area must be more than double that of a rectangular guide. It is, therefore, more expensive and takes up more space than a rectangular guide. Typical radar systems, therefore, consist of a main run with a rectangular waveguide and a circular rotating joint. The transition between rectangular and circular waveguides is accomplished with the circular-to-rectangular taper shown in Figure 15-8. The transition is accomplished as gradually as possible to minimize reflections.

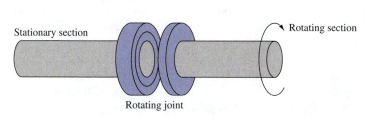

Stationary section   Rotating section

Rotating joint

FIGURE 15-7   **Circular waveguide rotating joint.**

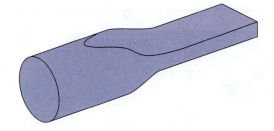

FIGURE 15-8   **Circular-to-rectangular taper.**

One of the limitations of circular waveguides relates to the range of frequencies it can propagate. From Table 15-1, we can see that rectangular waveguide has a useful bandwidth of about 50% of the frequency range that can be propagated. For example, at L-band frequencies, the range is 1.12 to 1.7 GHz, or 0.58 GHz, which is slightly less than half of the center frequency (1.41 GHz) of the waveguide frequency range.

$$\text{center frequency} = 1.12 \text{ GHz} + \frac{(1.7 - 1.12) \text{ GHz}}{2} = 1.41 \text{ GHz}$$

In this case, 0.58 GHz is 41% of the center frequency:

$$\frac{0.58 \text{ GHz}}{1.41 \text{ GHz}} \times 100\% = 41\%.$$

A study of the modes in a circular waveguide shows the bandwidth to be about 15%, a much-reduced amount compared to a rectangular waveguide.

Ridged

Two types of ridged waveguides are shown in Figure 15-9. Although it is more expensive to manufacture than a standard rectangular waveguide, it does provide one unique advantage. It allows operation at lower frequencies for a given set of outside dimensions, which means that smaller overall external dimensions are made possible. This property is advantageous in applications where space is at a premium, such as space probes and the like. A ridged waveguide has greater attenuation, and this, combined with its higher cost, limits it to special applications. This means that a ridged waveguide has the greater bandwidth as a percentage of center frequency.

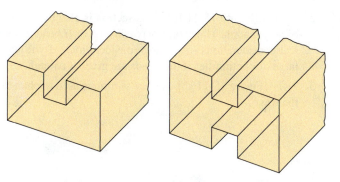

FIGURE 15-9   **Ridged waveguides.**

Flexible

It is sometimes desirable to have a section of waveguide that is flexible, as shown in Figure 15-10. This configuration is often useful in the laboratory or in applications where continuous flexing occurs. Flexible waveguides consist of spiral-wound ribbons of brass or copper. The outside section is covered with a soft dielectric such as rubber to maintain air- and watertight conditions and prevent dust contamination, which can

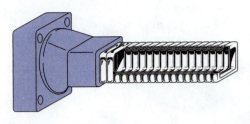

FIGURE 15-10 **Flexible waveguide.**

encourage arcs to form from one side to the other in high-power situations such as radar. A waveguide is often pressurized with nitrogen to prevent such contamination. In the case of a leak, gas rushes out, preventing anything from entering.

Corrosion would cause an increase in attenuation through surface current losses and increased reflections. In critical applications, waveguides are filled with inert gas and/or their inside walls are coated with noncorrosive (but expensive) metals such as gold or silver.

15-5 OTHER WAVEGUIDE CONSIDERATIONS

Waveguide Attenuation

Waveguides are capable of propagating huge amounts of power. For example, typical X-band (0.9 × 0.4 in.) waveguides can handle 1 MW if operated at 1.5 times f_{co} and an air dielectric strength of 3×10^6 V/m is assumed. At frequencies below cutoff, the attenuation in any waveguide is very large, as previously explained. At frequencies above cutoff, the guide supports traveling waves, and they are slightly attenuated because of losses in the conducting walls and in the dielectric that fills the guide. For air-filled guides, the dielectric loss is normally negligible, but if dielectric other than air is used, these losses are often greater than the conductor losses. As the frequency increases, attenuation drops to a broad minimum and then increases slowly with increasing frequency.

The conductor losses are governed in part by the skin effect described in Chapter 12. (At high frequencies the current tends to flow only at the surface of a conductor.) Current that flows in the guide walls is concentrated near the inner surface.

Bends and Twists

It is often necessary to change the physical direction of propagation or the wave's polarization in waveguides.

1. *H bend* [Figure 15-11(a)]: It is used to change the physical direction of propagation. It derives its name from the fact that the *H* lines are bent in this transition, while the *E* lines remain vertical for the dominant mode.

2. *E bend* [Figure 15-11(b)]: This section is also used to change the physical direction of propagation. The choice between an *E* or *H* bend is normally governed by mechanical considerations (plumbing considerations, if you will) because neither produces large discontinuities if the bends are gradual enough.

3. *Twist* [Figure 15-11(c)]: A twist section is used to change the plane of polarization of the wave.

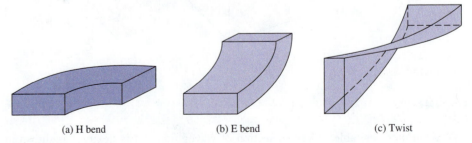

(a) H bend   (b) E bend   (c) Twist

FIGURE 15-11 **Waveguide bends and twists.**

You can see that any desired angular orientation of the wave may be obtained with an appropriate combination of the three types of sections just discussed.

Tees

1. *Shunt tee* [Figure 15-12(a)]: A shunt tee is so named because of the side arm shunting of the *E* field for TE modes, which is analogous to voltage in a transmission line. You can see that if two input waves at arms *A* and *B* are in phase, the portions transmitted into arm *C* will be in phase and thus will be additive. On the other hand, an input at *C* results in two equal, in-phase outputs at *A* and *B*. Of course, the *A* and *B* outputs have half the power (neglecting losses) of the *C* input.

2. *Series tee* [Figure 15-12(b)]: If you consider the *E* field of an input at *D*, you should be able to visualize that the outputs at *A* and *B* are equal and 180° out of phase, as shown. Once again, the two outputs are equal but are now 180° out of phase. The series tee is often used for impedance matching just as the single-stub tuner is used for transmission lines. In that case, arm *D* contains a sliding piston to provide a short circuit at any desired point.

3. *Hybrid or magic tee* [Figure 15-12(c)]: This is a combination of the first two tees mentioned and exhibits properties of each. From previous consideration of the shunt and series tees, you can see that if two equal signals are fed into arms *A* and *B* in phase, there is cancellation in arm *D* and reinforcement in arm *C*. Thus, all the energy will be transmitted to *C* and none to *D*. Similarly, if energy is fed into *C*, it will divide evenly between *A* and *B*, and none will be transmitted to *D*. The hybrid tee has many interesting applications.

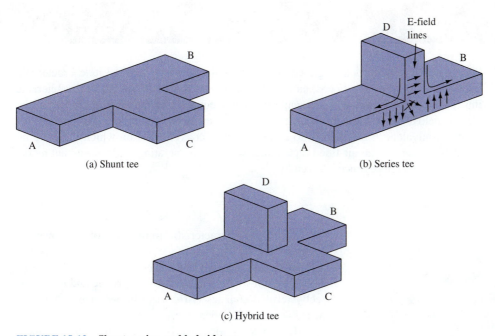

(a) Shunt tee

(b) Series tee

(c) Hybrid tee

FIGURE 15-12   **Shunt, series, and hybrid tees.**

A typical hybrid tee application is illustrated in Figure 15-13. It is functioning as a transmit/receive (TR) switch, which allows a single antenna to be used for both transmission and reception. The transmitter's output is fed into arm *C*, where it splits between the *A* and *B* outputs, with almost no power going to the sensitive receiver at *D*. When the antenna receives a signal at *B*, energy is sent to the receiver at *D* as well as to arms *A* and *C*. The low received power does no damage to the powerful transmitter output. The matched load at *A* is necessary to prevent reflections. Problem 60 at the end of the chapter introduces you to another hybrid tee application.

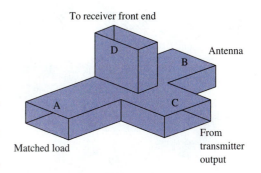

FIGURE 15-13   **Hybrid-tee TR switch.**

Tuners

A metallic post inserted in the broad wall of a waveguide provides a lumped reactance at that point. The action is similar to the addition of a shorted stub along a transmission line. When the post extends less than a quarter-wavelength, it appears capacitive, while exceeding a quarter-wavelength makes it appear inductive. Quarter-wavelength insertion causes a series resonance effect whose sharpness (Q) is inversely proportional to the diameter of the post.

The primary usage of posts is in matching a load to a guide to minimize the voltage standing-wave ratio (VSWR). The most often used configurations are shown in Figure 15-14.

1. *Slide-screw tuner* [Figure 15-14(a)]: The slide-screw tuner consists of a screw or metallic object of some sort protruding vertically into the guide and adjustable both longitudinally and in depth. The effect of the protruding object is to produce shunting reactance across the guide—thus, it is analogous to a single-stub tuner in transmission line theory.

2. *Double-slug tuner* [Figure 15-14(b)]: This type of tuner involves placing two metallic objects, called slugs, in the waveguide. The necessary two degrees of freedom to effect a match are obtained by making adjustable both the longitudinal position of the slugs and the spacing between them. Thus, it is somewhat analogous to the transmission line double-stub tuner but differs because the position of the slugs and not the effective shunting reactance is variable.

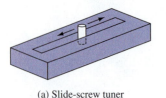

(a) Slide-screw tuner

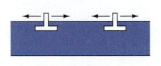

(b) Double-slug tuner

FIGURE 15-14   Tuners.

15-6 TERMINATION AND ATTENUATION

Because a waveguide is a single conductor, it is not as easy to define its characteristic impedance (Z_0) as it is for a coaxial line. Nevertheless, you can think of the characteristic impedance of a waveguide as being approximately equal to the ratio of the strength of the electric field to the strength of the magnetic field for energy traveling in one direction. This ratio is equivalent to the voltage-to-current ratio in coaxial lines on which there are no standing waves. For an air-filled rectangular waveguide operating in the dominant mode, its characteristic impedance is given by

$$Z_0 = \frac{\mathcal{L}}{\sqrt{1 - (\lambda/2a)^2}}, \qquad (15\text{-}8)$$

where $\mathcal{L}$ is the characteristic impedance of free space = $120\pi = 377\ \Omega$. The guide's characteristic impedance is affected by the frequency of the energy in it because $\lambda = c/f$. Therefore, the guide's impedance is variable and more correctly termed **characteristic wave impedance** rather than just characteristic impedance.

On a waveguide there is no place to connect a fixed resistor to terminate it in its characteristic (wave) impedance as there is on a coaxial cable. But several special arrangements accomplish the same result. One consists of filling the end of the waveguide with graphited sand, as shown in Figure 15-15(a). As the fields enter the sand, currents flow in it. These currents create heat, which dissipates the energy. None of the energy dissipated as heat is reflected back into the guide. Another arrangement [Figure 15-15(b)] uses a high-resistance rod, which is placed at the center of the E field. The E field (voltage) causes current to flow through the rod. The high resistance of the rod dissipates the energy as an I^2R loss.

Still another method for terminating a waveguide is to use a wedge of resistive material [Figure 15-15(c)]. The plane of the wedge is placed perpendicular to the magnetic

Energy

Waveguide

(a)   Graphited sand

Energy

Waveguide

(b)   Resistive rod

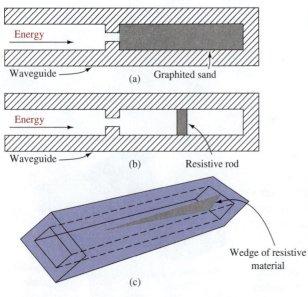

Wedge of resistive material

(c)

FIGURE 15-15   Termination for minimum reflections.

lines of force. When the *H* lines cut the wedge, a voltage is induced in it. The current produced by the induced voltage flowing through the high resistance of the wedge creates an I^2R loss. This loss is dissipated in the form of heat. This permits very little energy reaching the closed end to be reflected.

Each of the preceding terminations is designed to match the impedance of the guide to ensure a minimum of reflection. On the other hand, there are many instances where it is desirable for all the energy to be reflected from the end of the waveguide. The best way to accomplish this is simply to attach or weld a metal plate at the end of the waveguide.

Variable Attenuators

Variable attenuators find many uses at microwave frequencies. They are used to (1) isolate a source from reflections at its load to preclude frequency pulling; (2) adjust the signal level, as in one arm of a microwave bridge circuit; and (3) measure signal levels, as with a calibrated attenuator.

There are two versions of variable attenuators:

1. *Flap attenuator* [Figure 15-16(a)]*:* Attenuation is accomplished by insertion of a thin card of resistive material (often referred to as a **vane**) through a slot in the top of the guide. The amount of insertion is variable, and the attenuation can be made approximately linear with insertion by proper shaping of the resistance card. Notice the tapered edges, which minimize unwanted reflections.

2. *Vane attenuator* [Figure 15-16(b)]*:* In this type of attenuator the resistance card or vanes move in from the sides, as shown in the figure. You can see that the losses (and thus attenuation) are at a minimum when the vanes are close to the side walls where *E* is small, and maximum when the vanes are in the center.

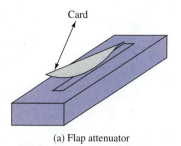

(a) Flap attenuator

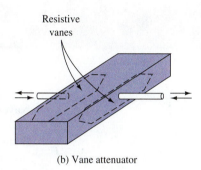

(b) Vane attenuator

FIGURE 15-16   **Attenuators.**

15-7 DIRECTIONAL COUPLER

The two-hole directional coupler consists of two pieces of waveguide with one side common to both guides and two holes in this common side. Its function is analogous to directional couplers used for transmission lines. The sections may be arranged physically either side by side or one over the other. The directional properties of such a device can be seen by looking at the wave paths labeled *A*, *B*, *C*, and *D* in Figure 15-17.

Waves *A* and *B* follow equal-length paths and thus combine in phase in the secondary guide. If the spacing between holes is $\lambda_g/4$, waves *C* and *D* (which are of equal strength) are $\lambda_g/2$ or 180° out of phase and thus cancel. Therefore, if the field within the main guide consists of a superposition of incident and reflected waves, a certain fraction of the wave moving left to right will be coupled out through the secondary guide, and the

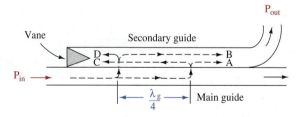

FIGURE 15-17   **Two-hole directional coupler.**

same fraction of the right-to-left wave will be dissipated in the vane. The wave traveling from right to left causes this energy to be cancelled in the secondary guide's output because of the 180° phase shift caused by the $\lambda_g/2$ path difference of its two equal components induced through the two coupling holes. This type of coupler is frequency-sensitive because the spacing between holes must be $\lambda_g/4$ or an odd multiple thereof. The addition of more holes properly spaced can improve both the operable frequency range and directivity. This is known as a multihole coupler.

Thus, we see that a directional coupler transfers energy from a primary to an adjacent—otherwise independent—secondary waveguide for energy traveling in the main guide in one direction only. The energy that flows toward the left in the secondary guide is absorbed by the vane in Figure 15-17, which is a matched load to prevent reflections.

The ratio of P_{out} and the incident power, P_{in}, is known as the *coupling:*

$$\text{coupling (dB)} = 10 \log \frac{P_{in}}{P_{out}}. \qquad \textbf{(15-9)}$$

We now can understand that a directional coupler can distinguish between the waves traveling in opposite directions. It can be arranged to respond to either incident or reflected waves. By connecting a microwave power meter to the output of the secondary guide, a measure of power flow can be made. The coupling is normally less than 1% so that the power meter has negligible loading effect on the operation in the main guide. By physically reversing the directional coupler, a power flow in the opposite direction is determined, and the level of reflections and standing-wave ratio (SWR) can be determined.

15-8 COUPLING WAVEGUIDE ENERGY AND CAVITY RESONATORS

We have described waveguide operation in terms of E and M fields, but how do we form these fields within the guide? In other words, how do we get energy into and out of a waveguide? Fundamentally, there are three methods of coupling energy into or out of a waveguide: *probe, loop,* and *aperture.* Probe, or capacitive, coupling is illustrated in Figure 15-18. Its action is the same as that of a quarter-wave monopole antenna. When the probe is excited by a radio-frequency (RF) signal, an electric field is set up [Figure 15-18(a)]. The probe should be located in the center of the *a* dimension and a quarter-wavelength, or odd multiple of a quarter-wavelength, from the short-circuited end, as illustrated in Figure 15-18(b). This is a point of maximum E field and, therefore, is a point of maximum coupling between the probe and the field. Usually, the probe is fed with a short length of coaxial cable. The outer conductor is connected to the waveguide wall, and the probe extends into the guide but is insulated from it, as shown in Figure 15-18(c). The degree of coupling may be varied by varying the length of the probe, removing it from the center of the E field, or shielding it.

FIGURE 15-18 **Probe, or capacitive, coupling.**

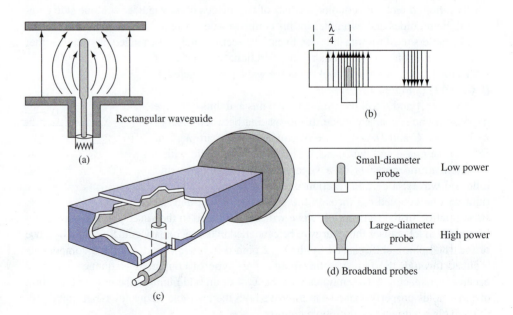

In a pulse-modulated radar system, there are wide sidebands on either side of the carrier frequency. For a probe not to discriminate too sharply among frequencies that differ from the carrier frequency, a wideband probe may be used. This type of probe is illustrated in Figure 15-18(d) for both low- and high-power usage.

Figure 15-19 illustrates loop, or inductive, coupling. The loop is placed at a point of maximum H field in the guide. As shown in Figure 15-19(a), the outer conductor is connected to the guide, and the inner conductor forms a loop inside the guide. The current flow in the loop sets up a magnetic field in the guide. This action is illustrated in Figure 15-19(b). As shown in Figure 15-19(c), the loop may be placed in several locations. The degree of loop coupling may be varied by rotation of the loop.

The third method of coupling is aperture, or slot, coupling. This type of coupling is shown in Figure 15-20. Slot A is at an area of maximum E field and is a form of electric field coupling. Slot B is at an area of maximum H field and is a form of magnetic field coupling. Slot C is at an area of maximum E and H field and is a form of electromagnetic coupling.

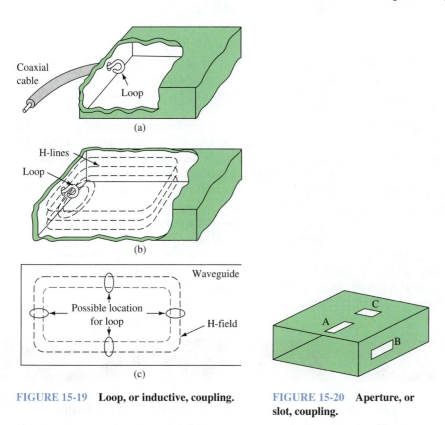

FIGURE 15-19   Loop, or inductive, coupling.

FIGURE 15-20   Aperture, or slot, coupling.

Cavity Resonators

Circuits composed of lumped inductance and capacitance elements may be made to resonate at any frequency from less than 1 Hz to many thousand megahertz. At extremely high frequencies, however, the physical sizes of the inductors and capacitors become extremely small. Also, losses in the circuit become extremely great. Resonant devices of different construction are therefore preferred at extremely high frequencies. In the UHF range, sections of parallel wire or coaxial transmission line are commonly employed in place of lumped constant resonant circuits, as was shown in Chapter 12. In the microwave region, cavity resonators are used. Cavity resonators are metal-walled chambers fitted with devices for admitting and extracting electromagnetic energy. The Q of these devices may be much greater than that of conventional LC tank circuits.

Although cavity resonators, built for different frequency ranges and applications, have various physical forms, the basic principles of operation are essentially the same for all. Resonant cavity walls are made of highly conductive material and enclose a good dielectric, usually air. One example of a cavity resonator is the rectangular box shown in Figure 15-21. It may be thought of as a section of rectangular waveguide closed at both ends by conducting plates. Because the end plates are short circuits for waves traveling in the Z direction, the cavity is analogous to a transmission line section with

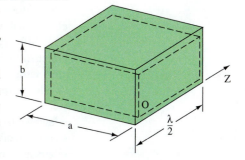

FIGURE 15-21   Rectangular waveguide resonator.

short circuits at both ends. Resonant modes occur at frequencies for which the distance between end plates is a half-wavelength or multiple of the half-wavelength.

Cavity modes are designated by the same numbering system that is used with waveguides, except that a third subscript is used to indicate the number of half-wave patterns of the transverse field along the axis of the cavity (perpendicular to the transverse field). The rectangular cavity is only one of many cavity devices useful as high-frequency resonators. By appropriate choice of cavity shape, advantages such as compactness, ease of tuning, simple mode spectrum, and high Q may be secured as required for special applications. Coupling energy to and from the cavity is accomplished just as for the standard waveguide, as shown in Figure 15-19.

Cavity Tuning

The resonant frequency of a cavity may be varied by changing any of three parameters: *cavity volume, cavity inductance,* or *cavity capacitance.* Although the mechanical methods for tuning cavity resonators may vary, they all make use of the electrical principles explained below.

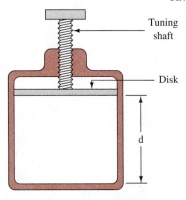

Figure 15-22 illustrates a method of tuning a cylindrical-type cavity by varying its volume. Varying the distance d results in a new resonant frequency. Increasing distance d lowers the resonant frequency, while decreasing d causes an increase in resonant frequency. The movement of the disk may be calibrated in terms of frequency. A micrometer scale is usually used to indicate the position of the disk, and a calibration chart is used to determine frequency.

A second method for tuning a cavity resonator is to insert a nonferrous metallic screw (such as brass) at a point of maximum H field. This decreases the permeability of the cavity and decreases its effective inductance, which raises its resonant frequency. The farther the screw penetrates into the cavity, the higher is the resonant frequency. A paddle can be used in place of the screw. Turning the paddle to a position more nearly perpendicular to the H field increases resonant frequency.

Labels in figure: Tuning shaft, Disk, d

FIGURE 15-22 **Cavity tuning by volume.**

15-9 RADAR

The first practical use of waveguides occurred with the development of radar during World War II. The high powers and high frequencies involved in these systems are much more efficiently carried by waveguides than by transmission lines. The word **radar** is an acronym formed from the words *ra*dio *d*etection *a*nd *r*anging. Radar is a means of employing radio waves to detect and locate objects such as aircraft, ships, and land masses. Location of an object is accomplished by determining the distance and direction from the radar equipment to the object. The process of locating objects requires, in general, the measurement of three coordinates: range, angle of azimuth (horizontal direction), and angle of elevation.

A radar set consists fundamentally of a transmitter and a receiver. When the transmitted signal strikes an object (target), some of the energy is sent back as a reflected signal. The small-beamwidth transmit/receive antenna collects a portion of the returning energy (called the **echo signal**) and sends it to the receiver. The receiver detects and amplifies the echo signal, which is then used to determine object location.

Military use of radar includes surveillance and tracking of air, sea, land, and space targets from air, sea, land, and space platforms. It is also used for navigation, including aircraft terrain avoidance and terrain following. Many techniques and applications of radar developed for the military are now found in civilian equipment. These applications include weather observation, geological search techniques, and air traffic control units, to name just a few. All large ships at sea carry one or more radars for collision avoidance and navigation. Certain frequencies see better through rain; others resolve closely spaced targets better; still others are suited for long-range operation. Generally, the larger a radar antenna, the better the system's resolution. In space, radars are used for spacecraft rendezvous, docking, and landing, as well as for remote sensing of the Earth's environment and planetary exploration.

Radar Waveform and Range Determination

A representative radar pulse (waveform) is shown in Figure 15-23. The number of these pulses transmitted per second is called the **pulse repetition frequency (PRF)** or **pulse repetition rate (PRR)**. The time from the beginning of one pulse to the beginning of the next pulse is called the **pulse repetition time (PRT)**. The PRT is the reciprocal of the PRF (PRT = 1/PRF). The duration of the pulse (the time the transmitter is radiating energy) is called the *pulse width* (PW). The time between pulses is called **rest time** or **receiver time**. The pulse width plus the rest time equals the PRT (PW + rest time = PRT). For radar to provide an accurate directional picture, a highly directive antenna is necessary. The desired directivity can be provided only by microwave antennas (see Chapter 14), and thus the RF energy shown in Figure 15-23 is usually in the GHz (microwave) range.

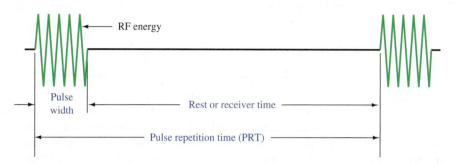

FIGURE 15-23 **Radar pulses.**

The distance to the target (range) is determined by the time required for the pulse to travel to the target and return. The velocity of electromagnetic energy is 186,000 statute mi/s, or 162,000 nautical mi/s. (A nautical mile is the accepted unit of distance in radar and is equal to 6076 ft.) In many instances, however, measurement accuracy is secondary to convenience, and as a result a unit known as the **radar mile** is commonly used. A radar mile is equal to 2000 yd, or 6000 ft. The small difference between a radar mile and a nautical mile introduces an error of about 1% in range determination.

For purposes of calculating range, the two-way travel of the signal must be taken into account. It can be found that it takes approximately 6.18 μs for electromagnetic energy to travel 1 radar mile. Therefore, the time required for a pulse of energy to travel to a target and return is 12.36 μs/radar mile. The range, in miles, to a target may be calculated by the formula

$$\text{range} = \frac{\Delta t}{12.36},$$ (15-10)

where Δt is the time between transmission and reception of the signal in microseconds. For shorter ranges and greater accuracy, however, range is measured in meters.

$$\text{range (meters)} = \frac{c\Delta t}{2},$$ (15-11)

where c is the speed of light and Δt is in seconds.

Radar System Parameters

Once the radar emits the pulse of electromagnetic energy, a sufficient length of time must elapse to allow any echo signals to return and be detected before the next pulse is transmitted. Therefore, the longest range at which targets are expected determines the PRT of the radar. If the PRT were too short (PRF too high), signals from some targets might arrive after the transmission of the next pulse. This could result in ambiguities in measuring range. Echoes that arrive after the transmission of the next pulse are called **second return echoes** (also *second time around* or *multiple time around echoes*). Such an echo would appear to be at a much shorter target range than actually exists and could be misleading if not identified as a second return echo. The range beyond which targets appear as second return echoes is called the *maximum unambiguous range*. Maximum unambiguous range may be calculated by the formula

$$\text{maximum unambiguous range} = \frac{\text{PRT}}{12.2},\qquad(15\text{-}12)$$

where range is in miles and the PRT is in microseconds. Figure 15-24 illustrates the principles of the second return echo.

FIGURE 15-24 Second return echo.

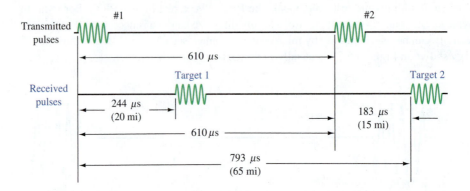

Figure 15-24 shows a signal with a PRT of 610 μs, which results in a maximum unambiguous range of 50 mi. Target number 1 is at a range of 20 mi. Its echo signal takes 244 μs to return. Target number 2 is actually 65 mi away, and its echo signal takes 793 μs to return. However, this is 183 μs after the next pulse was transmitted; therefore, target number 2 appears to be a weak target 15 mi away. Thus, the maximum unambiguous range is the **maximum usable range** and will be referred to from now on as simply *maximum range*. (It is assumed here that the radar has sufficient power and sensitivity to achieve this range.)

If a target is so close to the transmitter that its echo is returned to the receiver before the transmitter is turned off, the transmitted pulse will mask the reception of the echo. In addition, almost all radars utilize an electronic device to block the receiver for the duration of the transmitted pulse. However, **double range echoes** are frequently detected when there is a large target close by. Such echoes are produced when the reflected beam is strong enough to make a second trip, as shown in Figure 15-25. Double range echoes are weaker than the main echo and appear at twice the range.

FIGURE 15-25 Double range echo.

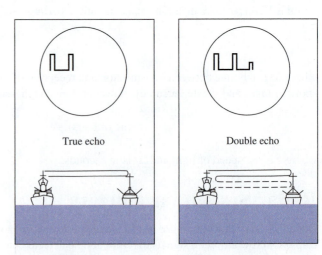

Minimum range is measured in meters and may be calculated by the formula

$$\text{minimum range} = 150\,\text{PW},\qquad(15\text{-}13)$$

where range is in meters and pulse width (PW) is in microseconds. Typical pulse widths range from fractions of a microsecond for short-range radars to several microseconds for high-power long-range radars.

A radar transmitter generates RF energy in the form of extremely short pulses with comparatively long intervals of rest time. The useful power of the transmitter is that contained in the radiated pulses and is termed the **peak power** of the system. Because the radar transmitter is resting for a time that is long with respect to the pulse time, the average power delivered during one cycle of operation is relatively low compared with the peak power available during the pulse time.

The **duty cycle** of radar is

$$\text{duty cycle} = \frac{\text{pulse width}}{\text{pulse repetition time}}. \tag{15-14}$$

For example, the duty cycle of a radar having a pulse width of 2 μs and a pulse repetition time of 2 ms is

$$\frac{2 \times 10^{-6}}{2 \times 10^{-3}} \text{ or } 0.001.$$

Similarly, the ratio between the average power and peak power may be expressed in terms of the duty cycle. In a system with peak power of 200 kW, a PW of 2 μs, and a PRT of 2 ms, a peak power of 200 kW is supplied to the antenna for 2 μs, while for the remaining 1998 μs the transmitter output is zero. Because average power equals peak power times duty cycle, the average power equals $(2 \times 10^5) \times (1 \times 10^{-3})$, or 200 W.

High peak power is desirable for producing a strong echo over the maximum range of the equipment. Conversely, low average power enables the transmitter output circuit components to be made smaller and more compact. Thus, it is advantageous to have a low duty cycle. A short pulse width is also advantageous with respect to being able to "see" (resolve) closely spaced objects.

Basic Radar Block Diagram

A block diagram of a basic radar system is shown in Figure 15-26. The *timer* (also called *trigger generator* or *synchronizer*) in the modulator block controls the PRF. The pulse-forming circuits in the modulator are triggered by the timer and generate high-voltage

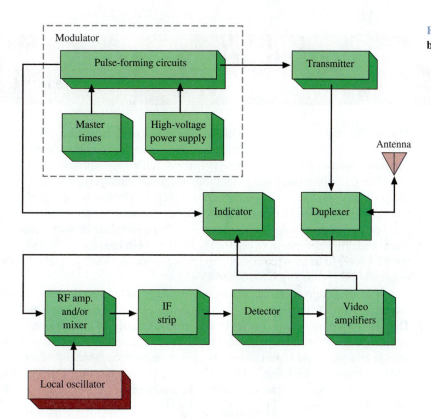

FIGURE 15-26 **Radar system block diagram.**

pulses of rectangular shape and short duration. These pulses are used as the supply voltage for the transmitter and, in effect, turn it on and off. The modulator, therefore, determines the pulse width of the system. The transmitter generates the high-frequency, high-power RF carrier and determines the carrier frequency. The duplexer is an electronic switch that allows the use of a common antenna for both transmitting and receiving. It prevents the strong transmitted signal from being received by the sensitive receiver. The receiver section is basically a conventional superheterodyne receiver. In older radars, no RF amplifier is found because of noise problems with the RF amplifiers of the World War II era.

Doppler Effect

The **Doppler effect** is the phenomenon whereby the frequency of a reflected signal is shifted if there is relative motion between the source and reflecting object. This is the same effect whereby the pitch of a train's whistle is shifted as the train moves toward and then away from the listener. Doppler radar, or CW radar, is always on. It is not turned off and on as pulsed radar is, hence the name, continuous wave. Only moving targets are "seen" by CW radar because only moving targets cause a Doppler shift. CW radars use two antennas, one each for transmitting and receiving.

The relative velocity between transmitter and target determines the amount of frequency shift encountered. It is predicted by

$$f_d = \frac{2v \cos \theta}{\lambda},$$

(15-15)

where f_d = frequency change between transmitted and reflected signal

v = relative velocity between radar and target

λ = wavelength of transmitted wave

θ = angle between target direction and radar system.

If you have ever received a speeding ticket, in a radar trap, you now have a better understanding of your downfall.

15-10 MICROINTEGRATED CIRCUIT WAVEGUIDING

The field of communications now makes heavy use of the microwave frequencies from 1 up to 300 GHz. At microwave frequencies, even the shortest circuit connections must be carefully considered because of the extremely small wavelengths involved.

The thin-film hybrid and monolithic integrated circuits used at microwave frequencies are called microwave integrated circuits (MICs). Obviously the use of short chunks of coaxial transmission line or waveguides is not practical for the required connections of mass-produced miniature circuits. Instead, either a **stripline** or **microstrip** connection is often used. They are shown in Figure 15-27. They both lend themselves to mass-produced circuitry and can be thought of as a cross between waveguides and transmission lines with respect to their propagation characteristics.

The stripline consists of two ground planes (conductors) that "sandwich" a smaller conducting strip with constant separation by a dielectric material (printed circuit board). The two types of microstrip shown in Figure 15-27 consist of either one or two conducting strips separated from a single ground plane by a dielectric. One conducting strip is analogous to an unbalanced transmission line. The two-conducting-strip version is analogous to a balanced transmission line. Although stripline offers somewhat better performance due to lower radiation losses, the simpler and thus more economical microstrip is the prevalent construction technique.

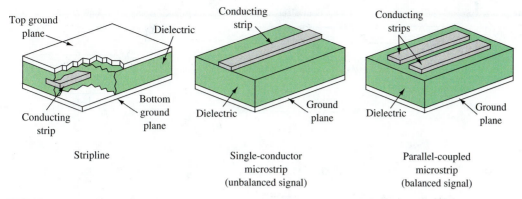

FIGURE 15-27   **Stripline and microstrip.**

In either case, the losses exceed those of either waveguides or coaxial transmission lines, but the miniaturization and cost savings far outweigh the loss considerations. This is especially true when the very short connection paths are considered.

As with waveguides and transmission lines, the characteristic impedances of stripline and microstrip are determined by physical dimensions and the type of dielectric. The most often used dielectric is alumina, with a relative dielectric constant of 9.6. Proper impedance matching, to minimize standing waves, is still an important consideration.

Figure 15-28 provides end views of the three lines. The formulas for calculating Z_0 for them are provided in the figure. In the formulas, ln is the natural logarithm and ε is the dielectric constant of the board.

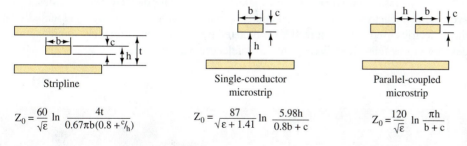

$$Z_0 = \frac{60}{\sqrt{\varepsilon}} \ln \frac{4t}{0.67\pi b(0.8 + c/h)}$$

$$Z_0 = \frac{87}{\sqrt{\varepsilon + 1.41}} \ln \frac{5.98h}{0.8b + c}$$

$$Z_0 = \frac{120}{\sqrt{\varepsilon}} \ln \frac{\pi h}{b + c}$$

FIGURE 15-28   **Characteristic impedance.**

Microstrip Circuit Equivalents

Microstrip can be used to simulate circuit elements just as previously discussed for transmission lines and waveguides. The physical layout for some single-conductor microstrip simulations is shown in Figure 15-29. The series capacitance in Figure 15-29(c) shows that an actual break in the conductor is used. This concept can be extended to allow coupling between two microstrips by running the two conductors close together. The amount of coupling can be accurately controlled by the length and spacing of the parallel segments, as shown in Figure 15-29(f). The configuration in Figure 15-29(e) simulates a series *LC* circuit connected to ground. Almost any type of *LC* circuit can be fabricated with microstrip.

Dielectric Waveguide

A more recent contender for "wiring" of miniature millimeter wavelength circuits is the **dielectric waveguide**. Its operation depends on the principle that two dissimilar dielectrics have a guiding effect on electromagnetic waves.

FIGURE 15-29 **Microstrip circuit equivalents.**

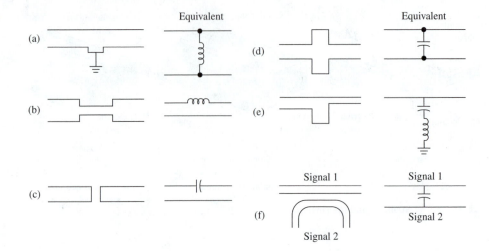

Dielectric waveguide

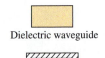

Dielectric-filled waveguide

FIGURE 15-30 **Dielectric waveguide and dielectric-filled waveguide.**

The dielectric waveguide should not be confused with the dielectric-filled waveguide. Both are shown in Figure 15-30. A regular metallic waveguide is sometimes filled with dielectric because it decreases the size necessary to allow propagation of a given frequency.

The dielectric waveguide is obviously easy to mass-produce within integrated circuits and offers an advantage over microstrip. At frequencies above 20 to 30 GHz, the losses with microstrip become excessive for many system applications compared to the dielectric waveguide. For example, at 60 GHz microstrip typically attenuates 0.15 dB/cm, while the dielectric waveguide attenuates only 0.06 dB/cm. The figure for a standard rectangular waveguide at 60 GHz is about 0.02 dB/cm but would be used only in systems where cost is not a factor.

Alumina is commonly used as the dielectric material for dielectric waveguides. However, semiconductors such as silicon and gallium arsenide (GaAs) will undoubtedly be the dielectrics used in the future. This is dictated by the fact that ultimately semiconductor devices will be fabricated directly into the dielectric waveguide.

15-11 TROUBLESHOOTING

After completing this section, you should be able to troubleshoot waveguide systems. Waveguide problems are very similar to ordinary transmission line problems. The test equipment may look different, but it is doing the same things.

A word of caution: Waveguide is commonly used to carry large amounts of microwave power. Microwaves are capable of burning skin and damaging eyesight. Never work on waveguide runs or antennas connected to a transmitter or radar until you are sure the system is off and cannot be turned on by another person.

Some Common Problems

1. The joints or flanges between two waveguide sections are the most likely source of a problem. Waveguides are sometimes pressurized to increase the power rating of the guide and keep water out. Improperly fitted joints can let water in and gas out. Water raises the VSWR, which can damage most microwave tubes.

   There are two types of flanges: choke and cover. Choke flanges have a groove cut into the face of the flange to keep microwave energy from escaping. There is a second groove for a gasket. Cover flanges are simply smooth. Both must be clean and flat. The screws must be the correct size because they help align the two pieces and keep them tightly sealed together.

2. Arcs can occur at improperly fitted joints and actually burn holes in the guide. Arcs occur in the waveguide under high power if some component such as the antenna has failed. You might see evidence of an arc on the broad wall right in the center of the guide.

3. Worn-out components must be checked and replaced. Radar antennas generally have one or more rotary joints. Rotary joints have bearings and sometimes moving contacts. Rotary joints sometimes fail only after the transmitter has had time to heat them sufficiently. By the time the technician has opened the guide and installed test equipment, the joint will cool and test well. It is best to have a directional coupler in line while running the transmitter and watch for an increase in reflected power.

Rotary joints can also be tested on the bench by connecting the joint to a dummy load and measuring VSWR while turning the joint. Still, there is no substitute for the operational test.

Flexible waveguide is subject to cracking and corrosion. Normally, the loss of a 2-ft-long piece of rigid waveguide is so low that the loss of the guide is extremely difficult to measure. Connecting a dummy load to the guide and measuring VSWR while flexing it can usually detect a bad piece of flexible guide.

Test Equipment

Figure 15-31 shows how to connect the test equipment for a VSWR test. First connect the power meter to the forward (incident) power coupler and note the reading. Then connect the power meter to the reflected coupler and note the reading.

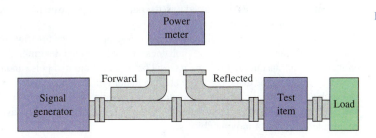

FIGURE 15-31 **VSWR test.**

Reflected power should be very low, and the VSWR should be nearly 1. VSWR is given by the following equation:

$$\text{VSWR} = \frac{1 + \sqrt{\dfrac{P_r}{P_i}}}{1 - \sqrt{\dfrac{P_r}{P_i}}}, \qquad (15\text{-}16)$$

where P_i = incident power

P_r = reflected power.

The same test equipment is used to measure loss by reversing the reflected coupler and putting the test item between the couplers, as shown in Figure 15-32. No two couplers are exactly alike, so you must first connect them together without the test item and determine the difference. You should be able to make the loss measurement to within 0.2 dB in this manner.

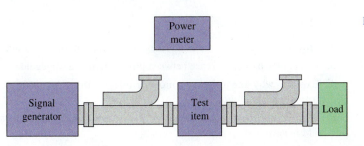

FIGURE 15-32 **Loss test.**

SUMMARY

Waveguides are hollow pipes or tubes used for the transfer of electromagnetic energy at microwave frequencies from a transmitter to an antenna. Waveguides exhibit extremely low loss and are capable of handling very high powers. The mode of waveguide operation is distinctly different from that of two-wire transmission lines. In two-wire lines or free space, the mode is transverse electric-magnetic (TEM), which means that both the electric and magnetic components of the wave are significant in their propagation. In contrast, waveguide propagation is either transverse electric (TE) or transverse magnetic (TM), in which either the electric- or magnetic-field components are significant for energy transfer, but not both.

Waveguide dimensions are critical and determine the cutoff frequency, which is the lowest frequency at which the guide will couple energy without excessive loss. For the most widely used waveguide shape, rectangular, the longer dimension must be a minimum of one-half wavelength to satisfy the boundary condition that the electric field be zero at the waveguide edges. The guide effectively acts as a high-pass filter, allowing frequencies above the cutoff frequency to pass. Waveguides are capable of multiple modes of operation. Modes of operation are denoted by two-number subscripts appended to the TE or TM direction, where the first number represents the number of half-wavelength modes along the longer dimension (representing the number of points where the concentrations of electric or magnetic field energy are highest), and the second subscript representing the number of half-wavelength modes in the shorter dimension. The simplest, and generally the most-efficient mode, is the TE_{10} mode. Higher-order modes allow higher frequency operation in a given-size waveguide, but at the expense of efficiency. Probes, loops, and couplers are used to couple energy in and out of waveguide and other microwave components, such as resonant cavities.

Calculation of characteristic impedances for waveguide is less straightforward than that for coaxial transmission line, but it is possible. Also, because of the microwave frequencies involved and the rigid construction of most types of line, various types of connectors, transitions, and loads are used. Among these are twists, tees, and tuners. Resonant cavities are used to provide tuned elements, and discontinuities introduced into the path of energy create impedance disturbances that are used for impedance-matching purposes.

One of the original and still most widely used applications for waveguide is in radar installations. Objects can be detected, and their location and distance determined with high-power pulses emitted from a transmitter. The time interval between the emitted pulse and the return or echo pulse is used to calculate the distance. There are two major types of radar: pulse and Doppler. Characteristics such as pulse repetition rate, frequency, and time are used to determine the resolution, distance, and accuracy of pulsed-radar systems. Doppler radar is a continuous-wave form of radar used to determine the speed of moving objects.

Waveguides and transmission lines on a very small scale fall into the categories of stripline and microstrip. Stripline is a two-conductor form of transmission line, best thought of as compressed coaxial cable, whereas microstrip is a single conductor on an insulating substrate, such as printed-circuit board material, with an underlying ground plane on the opposite side of the substrate. Both forms lend themselves well to mass fabrication on printed-circuit boards and miniaturization. In addition, impedance-matching elements can be fabricated by carefully fabricating the conductors such that their distances from each other, as well as their dimensions, provide the desired impedances.

QUESTIONS AND PROBLEMS

SECTION 15-1

1. Discuss the relative merits and drawbacks of using antennas, waveguides, and transmission lines as the media for a communications link.

SECTION 15-2

2. Provide a broad definition of a waveguide. What is normally meant by the term *waveguide*?

3. Explain the basic difference between propagation in a waveguide versus a transmission line.

4. Explain why the different mode configurations are termed *transverse electric* or *transverse magnetic*.

5. What are the *modes* of operation for a waveguide? Explain the subscript notation for TE and TM modes.

6. What is the *dominant mode* in rectangular waveguides? What property does it have that makes it dominant? Show a sketch of the electric field at the mouth of a rectangular waveguide carrying this mode.

7. Describe the significance of the cutoff wavelength.

8. A rectangular waveguide is 1 cm by 2 cm. Calculate its cutoff frequency, f_{co}. (7.5 GHz)

9. How does energy propagate down a waveguide? Explain what determines the angle this energy makes with respect to the guide's sidewalls.

10. For TE_{10}, $a = \lambda/2$. What is a for TE_{20}? (Assume a rectangular waveguide.)

SECTION 15-3

11. Why is the velocity of energy propagation usually significantly less in a waveguide than in free space? Calculate this velocity (V_g) for an X-band waveguide for a 10-GHz signal. Calculate guide wavelength (λ_g) and phase velocity (V_p) for these conditions. (2.26×10^8 m/s, 3.98 cm, 3.98×10^8 m/s)

12. Why are free-space wavelength (λ) and guide wavelength (λ_g) different? Explain the significance of this difference with respect to Smith chart calculations.

*13. Why are rectangular cross-sectional waveguides generally used in preference to circular cross-sectional waveguides?

SECTION 15-4

14. Why are circular waveguides used much less than rectangular ones? Explain the application of a circular rotating joint.

15. Describe the advantages and disadvantages of a ridged waveguide.

16. Describe the physical construction of a flexible waveguide, and list some of its applications.

17. Describe some advantages a ridged waveguide has over a rectangular waveguide.

SECTION 15-5

18. List some of the causes of waveguide attenuation. Explain their much greater power handling capability as compared to coaxial cable of similar size.

*19. Describe briefly the construction and purpose of a waveguide. What precautions should be taken in the installation and maintenance of a waveguide to ensure proper operation?

20. Why are waveguide bend and twist sections constructed to alter the direction of propagated energy gradually?

21. Describe the characteristics of shunt and series tee sections. Explain the operation of a hybrid tee when it is used as a TR switch.

22. Discuss several types of waveguide tuners in terms of function and application.

SECTION 15-6

23. Verify the characteristic wave impedance of 405 Ω for the data given in Problem 59.

24. Calculate the characteristic wave impedance for an X-band waveguide operating at 8, 10, and 12 GHz. (663 Ω, 501 Ω, 450 Ω)

25. Explain various ways of terminating a waveguide to minimize and maximize reflections.

26. Describe the action of flap and vane attenuators.

SECTION 15-7

27. Describe in detail the operation of a directional coupler. Include a sketch with your description. What are some applications for a directional coupler? Define the *coupling* of a directional coupler.

28. Calculate the coupling of a directional coupler that has 70 mW into the main guide and 0.35 mW out the secondary guide. (23 dB)

---

*An asterisk preceding a number indicates a question that has been provided by the FCC as a study aid for licensing examinations.

SECTION 15-8

29. Explain the basics of capacitively coupling energy into a waveguide.

30. Explain the basics of inductively coupling energy into a waveguide.

31. What is slot coupling? Describe the effect of varying the position of the slot.

*32. Discuss the following with respect to waveguides:
    (a) Relationship between frequency and size.
    (b) Modes of operation.
    (c) Coupling of energy into the waveguide.
    (d) General principles of operation.

33. What is a cavity resonator? In what ways is it similar to an *LC* tank circuit? How is it dissimilar?

*34. Explain the operating principles of a cavity resonator.

*35. What are waveguides?

36. Describe a means whereby a cavity resonator can be used as a waveguide frequency meter.

37. Explain three methods of tuning a cavity resonator.

SECTION 15-9

*38. Explain briefly the principle of operation for a radar system.

*39. Why are waveguides used in preference to coaxial lines for the transmission of microwave energy in radar installations?

40. With respect to a radar system, explain the following terms:
    (a) Target.
    (b) Echo.
    (c) Pulse repetition rate.
    (d) Pulse repetition time.
    (e) Pulse width.
    (f) Rest time.
    (g) Range.

41. Calculate the range in miles and meters for a target when Δt is found to be 167 μs. (13.5 mi, 25,050 m)

*42. What is the distance in nautical miles to a target if it takes 123 μs for a radar pulse to travel from the radar antenna to the target and back to the antenna, and be displayed on the pulse-position-indicator (PPI) scope? (10 mi)

43. What are double range echoes?

44. Why does a radar system have a minimum range? Calculate the minimum range for a system with a pulse width of 0.5 μs.

45. In detail, discuss the various implications of duty cycle for a radar system.

*46. What is the peak power of a radar pulse if the pulse width is 1 μs, the pulse repetition rate is 900, and the average power is 18 W? What is the duty cycle? (20 kW, 0.09%)

47. For the radar block diagram in Figure 15-26, explain the function of each section.

48. A police radar speed trap functions at a frequency of 1.024 GHz in direct line with your car. The reflected energy from your car is shifted 275 Hz in frequency. Calculate your speed in miles per hour. Are you going to get a ticket? (90 mph, *yes!*)

49. What is the Doppler effect? What are some other possible uses for it other than police speed traps?

50. Why is a Doppler radar often called a CW radar?

SECTION 15-10

51. Using sketches, explain the physical construction of stripline, single-conductor microstrip, and parallel-coupled microstrip. Discuss their relative merits, and also compare them to transmission lines and waveguides.

52. What is a dielectric waveguide? Discuss its advantages and disadvantages with respect to regular waveguides.

53. Calculate Z_0 for stripline constructed using a circuit board with a dielectric constant of 2.1, $b = 0.1$ in., $c = 0.006$ in., and $h = 0.08$ in. The conductor is spaced equally from the top and bottom ground planes. (50 Ω)

54. Make a sketch of a single-conductor microstrip that simulates an inductor to ground followed by a series capacitance.

SECTION 15-11

55. Describe the proper procedure for troubleshooting waveguides.

56. Explain how to prevent arcs.

57. Explain where problems are most likely to occur with waveguides and describe a process to prevent these problems.

58. Describe how to test a waveguide.

QUESTIONS FOR CRITICAL THINKING

59. A 9-GHz signal is operating in the dominant mode in a rectangular waveguide 3 by 4.5 cm. The characteristic (wave) impedance is 405 Ω. Provide a report that includes the λ_g, λ, V_g, and V_p for this system; the SWR that would be caused by a horn antenna load of 350 Ω + $j100$ Ω; and the impedance in the guide 4 cm from the antenna load. Include a Smith chart analysis with the report.

60. Can a hybrid tee be used to feed the first stage of a microwave receiver (the mixer—no RF stage) with the antenna signal and local oscillator signal without any local oscillator radiation off the receiving antenna? Provide a sketch to illustrate.

61. Analyze the relationship between multiple targets and *maximum range*. Use the calculated maximum unambiguous range for a radar system with PRT equal to 400 μs to illustrate.

62. You must use a directional coupler to measure VSWR and to determine the loss introduced by a device in a waveguide system. Describe the tests you will use and how they differ from each other.

CHAPTER 16

FIBER OPTICS

CHAPTER OUTLINE

KEY TERMS

refractive index
infrared light
optical spectrum
O-, E-, S-, C-, L-, and U-bands
core
cladding
numerical aperture (NA)

multimode fibers
step-index
pulse dispersion
graded-index fiber
single-mode
long-haul
mode field diameter
zero-dispersion wavelength

attenuation
scattering
absorption
macrobending
microbending
dispersion
modal dispersion
chromatic dispersion
polarization mode
dispersion compensating fiber
fiber Bragg grating
coherent
distributed feedback (DFB) laser
dense wavelength division multiplex (DWDM)
vertical cavity surface emitting lasers (VCSELs)
tunable lasers
light pipe
glass
isolator

received signal level (RSL)
dark current
fusion splicing
mechanical splices
index-matching gel
SC
ST
small-form factor
long-haul system
backhoe fading
optical time-domain reflectometer (OTDR)
event
synchronous optical network (SONET)
optical carrier (OC)
synchronous transport signals (STS)
fiber to the curb (FTTC)
fiber to the home (FTTH)
air fiber

Significant advances in the development and manufacture of fiber-optic systems have made them the latest frontier in the field of telecommunications. They are being used extensively for both military and commercial data links and have replaced a lot of copper wire. They have also taken over almost all the point-to-point long-distance communications traffic previously handled by microwave and satellite links, particularly transoceanic.

A fiber-optic communications system is surprisingly simple, as shown in Figure 16-1. It is composed of the following elements:

1. A fiber-optic transmission strand can carry the signal (in the form of a light beam modulated by an analog waveform or by digital pulses) a few feet or even hundreds or thousands of miles. A cable may contain three or four hairlike fibers or a bundle of hundreds of such fibers.

2. A source of invisible infrared radiation—usually a light-emitting diode (LED) or a solid-state laser—that can be modulated to impress digital data or an analog signal on the light beam.

3. A photosensitive detector to convert the optical signal back into an electrical signal at the receiver. The most often used detectors are *p-i-n* or avalanche photodiodes.

4. Efficient optical connectors at the light source–cable interface and at the cable–photodetector interface. These connectors are also critical when splicing the optical cable because of excessive losses that can occur at connections.

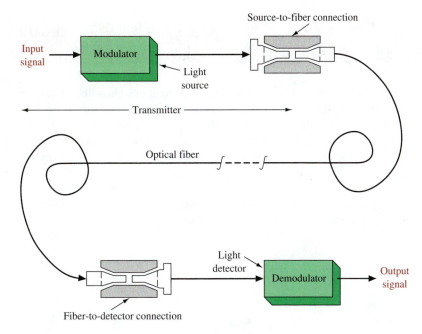

FIGURE 16-1 Fiber-optic communication system.

The advantages of optical communications links compared to waveguides or copper conductors are enormous and include the following:

1. **Extremely wide system bandwidth:** The intelligence is impressed on the light by varying the light's amplitude. Since the best LEDs have a 5-ns response time, they provide a maximum bandwidth of about 100 MHz. Using laser light sources, however, bandwidths of up to 10 Gbps are possible on a single glass fiber, and several lasers can be combined on one fiber. The amount of information multiplexed on such a system, in the tens of Gbps, is indeed staggering.

2. **Immunity to electrostatic interference:** External electrical noise and lightning do not affect energy in a fiber-optic strand. This is true only for the optical strands, however, not the metallic cable components or connecting electronics.

3. **Elimination of crosstalk:** The light in one glass fiber does not interfere with, nor is it susceptible to, the light in an adjacent fiber. Recall that crosstalk results from the electromagnetic coupling between two adjacent copper wires (see Chapter 12).

4. **Lower signal attenuation than other propagation systems:** Typical attenuation of a fiber-optic strand varies from 0.1 to 0.008 dB per 100 ft, depending on the wavelength of operation. By way of contrast, the loss of RG-6 and RG-59 75-ohm coaxial cable at 1 GHz is approximately 11.5 dB per 100 feet. The loss of $\frac{1}{2}''$ coaxial cable is approximately 4.2 dB per 100 ft.

5. **Substantially lighter weight and smaller size:** The U.S. Navy replaced conventional wiring on the A-7 airplane with fiber that carries data between a central computer and all its remote sensors and peripheral avionics. In this case, 224 ft of fiber optics weighing 1.52 lb replaced 1900 ft of copper wire weighing 30 lb.

6. **Lower costs:** Optical-fiber costs are continuing to decline. The costs of many systems are declining with the use of fiber, and that trend is accelerating.

7. **Safety:** In many copper wired systems, the potential hazard of short circuits requires precautionary designs, whereas the dielectric nature of optic fibers eliminates the spark hazard.

8. **Corrosion:** Glass is basically inert, so the corrosive effects of certain environments are not a problem.

9. **Security:** Because of its immunity to and from electromagnetic coupling and radiation, optical fiber can be used in most secure environments. Although it can be intercepted or tapped, it is difficult to do so.

16-2   THE NATURE OF LIGHT

Before one can understand the propagation of light in a glass fiber, it is necessary to review some basics of light refraction and reflection. The speed of light in free space is 3×10^8 m/s but is reduced in other media. The reduction as light passes into denser material results in refraction of the light. Refraction causes the light wave to be bent, as shown in Figure 16-2(a). The speed reduction and subsequent refraction is different for each wavelength, as shown in Figure 16-2(b). The visible light striking the prism causes refraction at both air/glass interfaces and separates the light into its various frequencies (colors) as shown. This same effect produces a rainbow, with water droplets acting as prisms to split the sunlight into the visible spectrum of colors (the various frequencies).

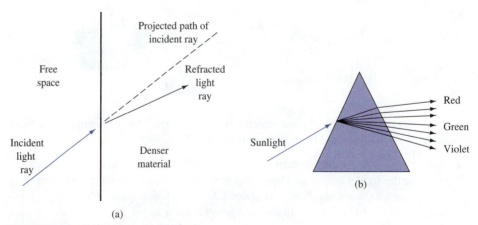

FIGURE 16-2   Refraction of light.

The amount of bend provided by refraction depends on the **refractive index** of the two materials involved. The refractive index, n, is the ratio of the speed of light in free space to the speed in a given material. It is slightly variable for different frequencies of light, but for most purposes a single value is accurate enough. The refractive index for free space (a vacuum) is 1.0, while air is 1.0003, and water is 1.33; for the various glasses used in fiber optics, it varies between 1.42 and 1.50.

Snell's law [Equation (13-6) from Chapter 13] predicts the refraction that takes place when light is transmitted between two different materials:

$$n_1 \sin \theta_1 = n_2 \sin \theta_2. \tag{13-6}$$

This effect was shown in Figure 13-4. Figure 16-3 shows the case where an incident ray is at an angle so that the refracted ray goes along the interface and so θ_2 is 90°. When θ_2 is 90°, the angle θ_1 is at the critical angle (θ_c) and defines the angle at which the incident rays no longer pass through the interface. When θ_1 is equal to or greater than θ_c, all the incident light is reflected and the angle of the incidence equals the angle of reflection, as we saw in Figure 13-4.

The frequency of visible light ranges from about 4.4×10^{14} Hz for red up to 7×10^{14} Hz for violet.

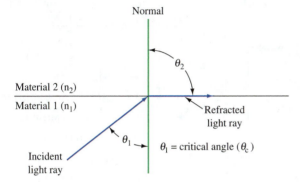

FIGURE 16-3   Critical angle.

EXAMPLE 16-1

Calculate the wavelengths of red and violet light.

SOLUTION

For red,

$$\lambda = \frac{c}{f}$$

$$= \frac{3 \times 10^8 \text{ m/s}}{4.4 \times 10^{14} \text{ Hz}} = 6.8 \times 10^{-7} \text{ m}$$

$$= 0.68 \ \mu\text{m or } 0.68 \text{ micron or } 680 \text{ nm.}$$

For violet,

$$\lambda = \frac{3 \times 10^8 \text{ m/s}}{7 \times 10^{14} \text{ Hz}} = 0.43 \text{ micron or } 430 \text{ nm.}$$

In the fiber-optics industry, spectrum notation is stated in nanometers (nm) rather than in frequency (Hz) simply because it is easier to use, particularly in spectral-width calculations. A convenient point of commonality is that 3×10^{14} Hz, or 300 THz, is equivalent to 1 μm, or 1000 nm. This relationship is shown in Figure 16-4. The one exception to this naming convention comes up when discussing dense wavelength division multiplexing (DWDM), which is the transmission of several optical channels, or wavelengths, in the 1550-nm range, all on the same fiber. For DWDM systems, notations, and particularly channel separations, are stated in terahertz (THz). Wave division multiplexing (WDM) systems are discussed in Section 16-9.

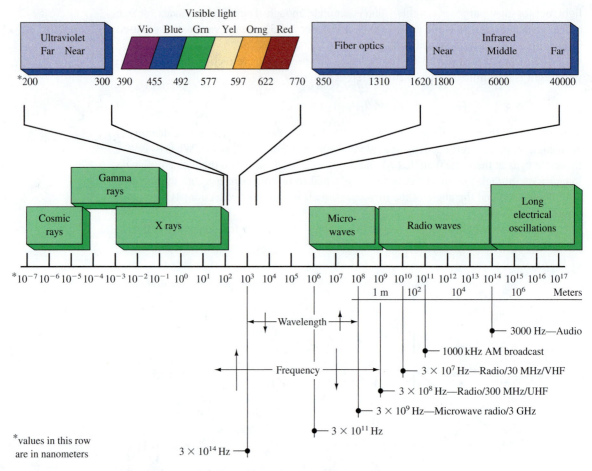

FIGURE 16-4   **The electromagnetic wavelength spectrum.**

An electromagnetic wavelength spectrum chart is provided in Figure 16-4. The electromagnetic light waves just below the frequencies in the visible spectrum are called **infrared light** waves. Whereas visible light has a wavelength from approximately 390 nm up to 770 nm, infrared light extends from 680 nm up to the wavelengths of the microwaves. For the frequencies above visible light, the electromagnetic spectrum includes the ultraviolet (UV) rays and X rays. The frequencies from the infrared on up are termed the **optical spectrum**.

The most commonly used wavelengths in today's fiber-optic systems are 750 to 850 nm, 1310 nm, and 1530 to 1560 nm. However, industry has categorized the entire spectrum in terms of **O-, E-, S-, C-, L-, and U-bands**. Fixed fiber-optic wavelength specifications are simply stated in terms of fixed wavelengths as 850, 1310, or 1550 nm.

Construction of the Fiber Strand

Typical construction of an optical fiber is shown in Figure 16-5. The **core** is the portion of the fiber strand that carries the transmitted light. Its chemical composition is simply a very pure glass: silicon dioxide, doped with small amounts of germanium, boron, and phosphorous. Plastic fiber is used only in short lengths in industrial applications because of its high attenuation. (See Section 16-3 for more information on plastic fiber.) The **cladding** is the material surrounding the core. It is almost always glass, although plastic cladding of a glass fiber is available but rarely used. In any event, the refractive index for the core and the cladding are different. The cladding must have a lower index of refraction to keep the light in the core. A plastic coating surrounds the cladding to provide protection.

As shown in Figure 16-6(a), propagation results from the continuous reflection at the core/clad interface so that the ray "bounces" down the fiber length by the process of total

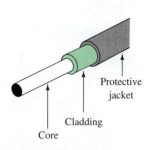

FIGURE 16-5   **Single-fiber construction.**

TABLE 16-1 • The Optical Bands	
BAND	WAVELENGTH RANGE (NM)
O	1260 to 1360
E	1360 to 1460
S	1460 to 1530
C	1530 to 1565
L	1565 to 1625
U	1625 to 1675

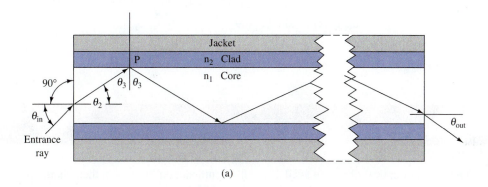

(a)

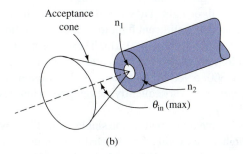

(b)

FIGURE 16-6   (a) Development of numerical aperture; (b) acceptance cone.

internal reflection (TIR). If we consider point P in Figure 16-6(a), the critical angle value for θ_3 is, from Snell's law,

$$\theta_c = \theta_3(\text{min}) = \sin^{-1}\frac{n_2}{n_1}.$$

Because θ_2 is the complement of θ_3,

$$\theta_2(\text{max}) = \sin^{-1}\frac{(n_1^2 - n_2^2)^{1/2}}{n_1}.$$

Now applying Snell's law at the entrance surface and because $n_{\text{air}} \simeq 1$, we obtain

$$\sin\theta_{\text{in}}(\text{max}) = n_1 \sin\theta_2(\text{max}).$$

Combining the two preceding equations yields

$$\sin\theta_{\text{in}}(\text{max}) = \sqrt{n_1^2 - n_2^2} \qquad (16\text{-}1)$$

Therefore, θ_{in} (max) is the largest angle with the core axis that allows propagation via total internal reflection. Light entering the cable at larger angles is refracted through the core/clad interface and lost. The value $\sin\theta_{\text{in}}$ (max) is called the **numerical aperture (NA)** and defines the half-angle of the cone of acceptance for propagated light in the fiber. This is shown in Figure 16-6(b). The preceding analysis might lead you to think that crossing over θ_{in} (max) causes an abrupt end of light propagation. In practice, however, this is not true; thus, fiber manufacturers usually specify NA as the acceptance angle where the output light is no greater than 10 dB down from the peak value. The NA is a basic specification of a fiber provided by the manufacturer that indicates its ability to accept light and shows how much light can be off-axis and still be propagated.

EXAMPLE 16-2

An optical fiber and its cladding have refractive indexes of 1.535 and 1.490, respectively. Calculate NA and $\theta_{in}(max)$.

SOLUTION

$$
\begin{aligned}
NA = \sin \theta_{in}(max) &= \sqrt{n_1^2 - n_2^2} \\
&= \sqrt{(1.535)^2 - (1.49)^2} = 0.369
\end{aligned}
\qquad (16\text{-}1)
$$
$$
\begin{aligned}
\theta_{in}(max) &= \sin^{-1} 0.369 \\
&= 21.7°
\end{aligned}
$$

16-3 OPTICAL FIBERS

Three types of optical fibers are available, with significant differences in their characteristics. The first communication-grade fibers (early 1970s) had light-carrying core diameters about equal to the wavelength of light. They could carry light in just a single waveguide mode. The difficulty of coupling significant light into such a small fiber led to the development of fibers with cores of about 50 to 100 μm. These fibers support many waveguide modes and are called **multimode fibers**. The first commercial fiber-optic systems used multimode fibers with light at 800 to 900 nm wavelengths. A variation of the multimode fiber, termed *graded-index fiber,* was subsequently developed. This development afforded greater bandwidth capability.

As the technology became more mature, the single-mode fibers were found to provide lower losses and even higher bandwidth. This has led to their use at 1300 and 1500 nm in many telecommunications applications. The new developments have not made old types of fiber obsolete. The application now determines the type used. The following major criteria affect the choice of fiber type:

1. Signal losses, with respect to distance
2. Ease of light coupling and interconnection
3. Bandwidth

Multimode Step-Index Fiber

A fiber showing three different modes (i.e., multimode) of propagation is presented in Figure 16-7. The lowest-order mode is seen traveling along the axis of the fiber, and the middle-order mode is reflected twice at the interface. The highest-order mode is reflected many times and makes many trips across the fiber. This type of fiber is called **step-index** because of the abrupt change in the refractive index at the core-cladding boundary. As a result of the variable path lengths, the light entering the fiber takes a variable length of time to reach the

FIGURE 16-7 Modes of propagation for step-index fiber.

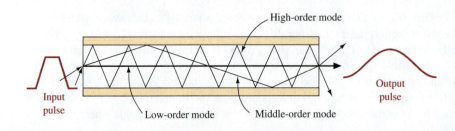

detector. This results in a pulse-broadening or dispersion characteristic, as shown in Figure 16-7. This effect is termed **pulse dispersion** and limits the maximum distance and rate at which data (pulses of light) can be practically transmitted. Also note that the output pulse has reduced amplitude as well as increased width. The greater the fiber length, the worse this effect. As a result, manufacturers rate their fiber in bandwidth per length, such as 400 MHz/km. That fiber can successfully transmit pulses at the rate of 400 MHz for 1 km, 200 MHz for 2 km, and so on. In fact, current networking standards limit multimode fiber distances to 2 km. Longer transmission paths can be attained by locating regenerators at appropriate locations. Step-index multimode fibers are rarely used in telecommunications because of their very high amounts of pulse dispersion and minimal bandwidth capability.

Multimode Graded-Index Fiber

In an effort to overcome the pulse dispersion problem, the **graded-index fiber** was developed. In the manufacturing process for this fiber, the index of refraction is tailored to follow the parabolic profile shown in Figure 16-9(c). This tailoring results in low-order modes traveling through the center (Figure 16-8). High-order modes see lower index of refraction material farther from the core axis, and thus the velocity of propagation increases away from the center. Therefore, all modes, even though they take various paths and travel different distances, tend to traverse the fiber length collectively in less time than in step index fiber. These fibers can therefore handle higher bandwidths and/or provide longer lengths of transmission before pulse dispersion effects destroy intelligibility and introduce bit errors.

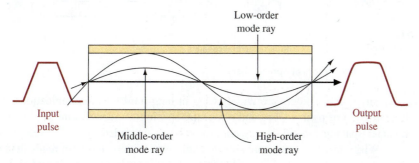

FIGURE 16-8 **Modes of propagation for graded-index fiber.**

In the telecommunications industry, two core sizes for graded-index fiber are commonly used: 50 and 62.5 μm. Both have 125-μm cladding. The large core diameter and the high NA of these fibers simplify input cabling and allow the use of relatively inexpensive connectors. Fibers are specified by the diameters of their core and cladding. For example, the fibers just described would be called 50/125 fiber and 62/125 fiber.

The 62.5 μm fiber has largely been standardized for data networks. Typical bandwidths at 850 nm are up to 180 MHz/km, and at 1300 nm they are up to 600 MHz/km. The 50 μm fiber has more recently become standardized because of the advent of Gbit and 10 Gbit networks and systems. The smaller core allows greater bandwidth: up to 600 MHz/km at 850 nm and 1000 MHz/km at 1300 nm.

Single-Mode Fibers

A technique used to minimize pulse dispersion effects is to make the core extremely small—on the order of a few micrometers. This type accepts only a low-order mode, thereby allowing operation in high-data-rate, long-distance systems. This fiber is typically used with high-power, highly directional modulated light sources such as a laser. Fibers of this variety are called **single-mode**, or monomode, fibers. Core diameters of only 7 to 10 μm are typical.

This type of fiber is also termed a **step-index** fiber. Step index refers to the abrupt change in refractive index from core to clad, as shown in Figure 16-9. The single-mode fiber, by definition, carries light in a single waveguide mode. A single-mode fiber transmits a single mode for all wavelengths longer than the cutoff wavelength (λ_c). A typical cutoff wavelength is 1260 nm. At wavelengths shorter than the cutoff, the fiber supports two or more modes and becomes multimode in operation.

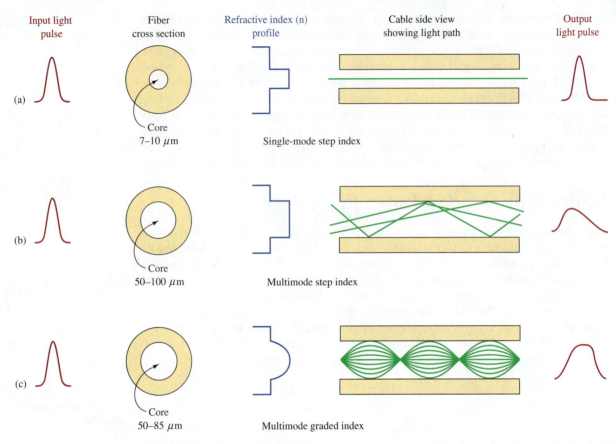

| Input light pulse | Fiber cross section | Refractive index (n) profile | Cable side view showing light path | Output light pulse |

(a) — Core 7–10 μm — Single-mode step index

(b) — Core 50–100 μm — Multimode step index

(c) — Core 50–85 μm — Multimode graded index

FIGURE 16-9 **Types of optical fiber.**

Single-mode fibers are widely used in **long-haul** telecommunications. They permit transmission of over 1 Gbps and a repeater spacing of over 80 km. These bandwidth and repeater-spacing capabilities are constantly being upgraded by new developments.

When describing the core size of single-mode fibers, the term **mode field diameter** is the more commonly used term. Mode field diameter is the actual guided optical power distribution diameter. In a typical single-mode fiber—the mode field diameter is 1 μm or so larger than the core diameter—the actual value depends on the wavelength being transmitted. In fiber specification sheets, the core diameter is stated for multimode fibers, but the mode field diameter is typically stated for single-mode fibers.

Figure 16-9 provides a summary of the three types of fiber discussed, including typical core/clad relationships, refractive index profiles, and pulse-dispersion effects.

Fiber Classification

The various types of fiber strands, both multimode and single mode, are categorized by the Telecommunications Industry Association according to the lists provided in Tables 16-2

TABLE 16-2 • Multimode Classifications (by the Refractive Index Profile and Composition)			
CLASS	**INDEX**	**CORE**	**CLADDING**
Ia	Graded	Glass	Glass
Ib	Quasi-graded	Glass	Glass
Ic	Step	Glass	Glass
IIa	Step	Glass	Plastic cladding retained for connectorization
IIb	Step	Glass	Plastic cladding removed for connectorization
III	Step/graded	Plastic	Plastic

TABLE 16-3 • Single-Mode Classifications

CLASS	DISPERSION CHARACTERISTIC	REGION OF ZERO-DISPERSION WAVELENGTH
IVa	Unshifted	1310 nm
IVb	Shifted	1550 nm
IVc	Flattened	Low values in both 1310 and 1550 nm ranges
IVd	Near zero	Adjacent to but outside the 1530–1560-nm operational range

and 16-3. The International Electrotechnical Commission also classifies multimode fiber in accordance with performance capability. It is listed as OM-1, OM-2, and OM-3 and is used in conjunction with the type of data transceiver, wavelength, data protocol, and span distance. OM-1 is the standard grade, OM-2 is better, whereas OM-3 is an enhanced high-performance grade for 10-Gbit networks. Single-mode fiber is classified by dispersion characteristic and the **zero-dispersion wavelength**, which is the point where material and waveguide dispersion cancel one another. A general comparison of single-mode and multimode fiber is provided in Table 16-4.

TABLE 16-4 • Generalized Comparisons of Single-Mode and Multimode Fiber

FEATURE	SINGLE-MODE	MULTIMODE
Core size	Smaller (7.5 to 10 μm)	Larger (50 to 100 μm)
Numerical aperture	Smaller (0.1 to 0.12)	Larger (0.2 to 0.3)
Index of refraction profile	Step	Graded
Attenuation (dB/km) (a function of wavelength)	Smaller (0.25 to 0.5 dB/km)	Larger (0.5 to 4.0 dB/km)
Information-carrying capacity (a function of distance)	Very large	Small to medium
Usage	Long-haul carriers and CATV, CCTV	Short-haul LAN
Capacity/distance characterization	Expressed in bits/second (bps)	BW in MHz/km
Which to use (this is a judgment call)	Over 2 km	Under 2 km

Plastic Optical Fiber

Plastic fiber is used in short-range markets such as sensors, robotics, displays, automotive applications, and, to a limited extent, in data links under 100 m. It has the same advantages as glass fiber versus copper except for two primary exceptions: high loss and low bandwidth.

Features

Materials—Polymers such as polymethyl acrylate

Core size—Up to 1000 μm

Numerical aperture—0.3 to 0.8

Bandwidth—Up to 3 Gb/s at 100 m but more realistically a few hundred megabits at a few hundred meters

Attenuation—120 to 180 dB/km but optimized at a 650-nm wavelength

Plastic fibers are well supported by connectorization and splicing components, which are not as critical as glass. This results in a less expensive installation.

16-4 FIBER ATTENUATION AND DISPERSION

There are two key distance-limiting parameters in fiber-optic transmissions: attenuation and dispersion.

Attenuation

Attenuation is the loss of power introduced by the fiber. This loss accumulates as the light is propagated through the fiber strand. The loss is expressed in dB/km (decibels per kilometer) of length. The loss, or attenuation, of the signal results from the combination of four factors: scattering, absorption, macrobending, and microbending.

Scattering: This is the primary loss factor over the three wavelength ranges used in telecommunications systems. It accounts for 85% of the loss and is the basis for the attenuation curves and values, such as that shown in Figure 16-10 and industry data sheets. The scattering is known as Rayleigh scattering and is caused by refractive index fluctuations. Rayleigh scattering decreases as wavelength increases, as shown in Figure 16-10.

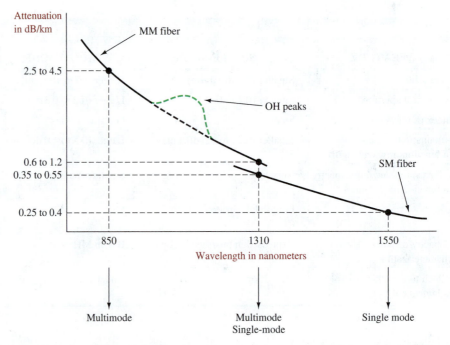

FIGURE 16-10 **Typical attenuation of cabled fiber strand.**

Absorption: The second loss factor is a composite of light interaction with the atomic structure of the glass. It involves the conversion of optical power to heat. One portion of the absorption loss is caused by the presence of OH hydroxol ions dissolved in the glass during manufacture. These ions cause the water attenuation or OH peaks shown in Figure 16-10 and other attenuation curves in older manufactured fiber. A recent and significant development in the manufacture of optical fiber has been the removal of these hydroxol ions, particularly in the 1380-nm region. By eliminating the attenuation peak, the fiber can effectively be used continuously from 1260 to 1675 nm, a significant increase in the bandwidth capability of the newer fiber (see Table 16-1).

Macrobending: The loss caused by the light mode breaking up and escaping into the cladding when the fiber bend becomes too tight. As the wavelength increases, the loss in a bend increases. Although losses are in fractions of decibels, the bend radius in small splicing trays and patching enclosures should be kept as large as possible.

Microbending: A type of loss caused by mechanical stress placed on the fiber strand, usually in terms of deformation resulting from too much pressure being applied to the cable. For example, excessively tight tie wrap or clamps contribute to this loss. This loss is noted in fractions of a decibel.

Dispersion

Dispersion, or pulse broadening, is the second of the two key distance-limiting parameters in a fiber-optic transmission system. It is a phenomenon in which the light pulse spreads out in time as it propagates along the fiber strand. This results in a broadening of the pulse. If the pulse broadens excessively, it can blend into the adjacent digital time slots and cause bit errors. Pulse dispersion is measured in terms of picoseconds (ps) of pulse broadening per spectral width expressed in nanometers (nm) of the pulse times fiber length (km). The total dispersion is then obtained by multiplying the pulse dispersion value times the length of the fiber (L). Manufacturers' specification sheets are available that provide an estimate of the dispersion for a given wavelength. A summary of values of dispersion for the primary wavelength used in fiber-optic transmission is provided in Table 16-5. The effects of dispersion on a light pulse are shown in Figure 16-11.

TABLE 16-5 • Dispersion Values for Common Optical Wavelengths for Class IVa Fiber	
WAVELENGTH (NM)	**PULSE DISPERSION [PS/(NM · KM)]**
850	80 to 100
1310	±2.5 to 3.5
1550	+17

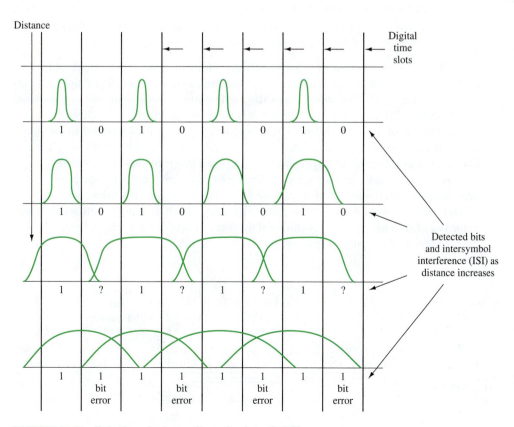

FIGURE 16-11   Pulse broadening or dispersion in optical fibers.

The equation for calculating the total dispersion is

$$\text{pulse dispersion} = \text{ps/(nm} \cdot \text{km)} \times \Delta\lambda. \tag{16-2}$$

The pulse dispersion value is obtained from Table 16-5:

$$\Delta\lambda = \text{spectral width of the light source}$$

$$\text{total dispersion} = \text{pulse dispersion} \times \text{length (km)}. \tag{16-3}$$

EXAMPLE 16-3

Determine the amount of pulse spreading of an 850-nm LED that has a spectral width of 22 nm when run through a fiber 2 km in length. Use a dispersion value of 95 ps/(nm · km).

SOLUTION

The dispersion value is $1 = 95$ ps/(nm × km). Use Equation (16-2), $L = 2$ km, and $\Delta\lambda = 22$ nm.

$$\text{pulse dispersion} = \text{ps/(nm} \cdot \text{km)} \times \Delta\lambda = (95)(22) = 2090 \text{ ps/km.} \tag{16-2}$$

$$\text{total pulse dispersion} = \text{pulse dispersion} \times \text{length}(L)$$
$$= (2090)(2) = 4.18 \text{ ns/km.} \tag{16-3}$$

There are three types of dispersion: modal, chromatic, and polarization.

Modal dispersion: The broadening of a pulse due to different path lengths taken through the fiber by different modes.

Chromatic dispersion: The broadening of a pulse due to different propagation velocities of the spectral components of the light pulse.

Polarization mode dispersion: The broadening of a pulse due to the different propagation velocities of the X and Y polarization components of the light pulse.

Modal dispersion occurs predominantly in multimode fiber. From a light source, the light modes can take many paths as they propagate along the fiber. Some light rays do travel in a straight line, but most take variable-length routes. As a result, the rays arrive at the detector at different times, and the result is pulse broadening. This is shown in Figure 16-11. The use of graded-index fiber greatly reduces the effects of modal dispersion and therefore increases the bandwidth to about 1 GHz/km. On the other hand, single-mode fiber does not exhibit modal dispersion because only a single mode is transmitted.

A second, equally important type of dispersion is chromatic. Chromatic dispersion is present in both single-mode and multimode fibers. Basically, the light from both lasers and LEDs produces several different-wavelength light rays. Each light ray travels at a different velocity and, as a result, these rays arrive at the receiver detector at different times, causing the broadening of the pulse (see Figures 16-11 and 16-12).

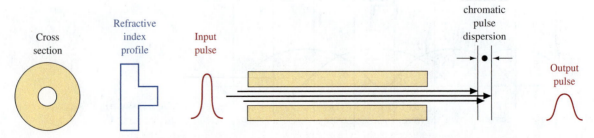

FIGURE 16-12 **Spectral component propagation: single-mode, step index.**

There is a point where dispersion is actually at zero, this being determined by the refractive index profile. This minimum-dispersion point happens near 1310 nm and is called the zero-dispersion wavelength. Altering the refractive index profile shifts this zero-dispersion wavelength to the 1550-nm region. Such fibers are called dispersion-shifted. This is significant because the 1550-nm region exhibits a lower attenuation than at 1310 nm. This reduction in attenuation becomes an operational advantage, particularly to long-haul carriers, because repeater and regenerator spacing can be maximized with minimum attenuation and minimum dispersion in the same wavelength region.

To illustrate chromatic dispersion further, Figure 16-12 shows a step-index single-mode fiber with different spectral components propagating directly along the core. Because they travel at different velocities, they arrive at the receiver detector at different times, causing a wider pulse to be detected than was transmitted. Again, this broadening is measured in picoseconds per kilometer of length times the spectral width in nanometers, as presented in Equation (16-3).

Polarization mode is the type of dispersion found in single-mode systems and becomes of particular concern in long-haul, high-data-rate digital and high-bandwidth analog video systems. In a single-mode fiber, the single propagating mode has two polarizations, horizontal and vertical, or X axis and Y axis. The index of refraction can be different for the two components, and this affects their relative velocity. This is shown in Figure 16-13.

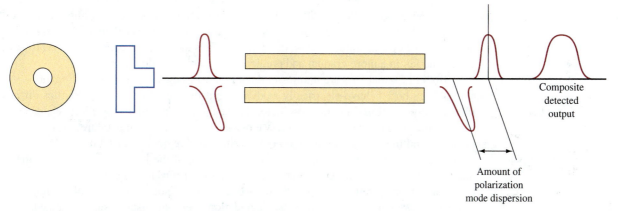

FIGURE 16-13   **Polarization mode dispersion in single-mode fiber.**

Dispersion Compensation

A considerable amount of fiber in use today is of the class IVa variety, installed in the 1980s and early 1990s. These fibers were optimized to operate at the 1310-nm region, which means that their zero-dispersion point was in the 1310-nm wavelength range. As a result of the considerable and continuous network expansion needs in recent years, it is often desired to add transmission capacity to the older fiber cables by using the 1550-nm region, particularly because the attenuation at 1550 nm is less than at 1310 nm. One major problem arises at this point. The dispersion value of the class IVa fiber in the 1550-nm region is approximately +17 ps/nm · km, which severely limits its distance capability.

To overcome this problem, a fiber was developed to provide approximately −17 ps of dispersion in the 1550-nm range. Called **dispersion compensating fiber**, it acts like an equalizer, negative dispersion canceling positive dispersion. The result is close to zero dispersion in the 1550-nm region. This fiber consists of a small coil normally placed in the equipment rack just prior to the optical receiver input. This does introduce some insertion loss (3 to 10 dB) and may require the addition of an optical-line amplifier.

Also on the market is a **fiber Bragg grating**. This technology involves etching irregularities onto a short strand of fiber, which changes the index of refraction and, in turn, accelerates slower wavelengths toward the output. This results in a less dispersed, or narrower, light pulse, minimizing intersymbol interference (ISI).

16-5 OPTICAL COMPONENTS

Two kinds of light sources are used in fiber-optic communication systems: the diode laser (DL) and the high-radiance LED. In designing the optimum system, the special qualities of each light source should be considered. DLs and LEDs bring to systems different characteristics:

1. Power levels
2. Temperature sensitivities
3. Response times
4. Lifetimes
5. Characteristics of failure

The diode laser is a preferred source for moderate-band to wideband systems. It offers a fast response time (typically less than 1 ns) and can couple high levels of useful optical power (usually several mW) into an optical fiber with a small core and a small numerical aperture. Recent advances in DL fabrication have resulted in predicted lifetimes of 10^5 to 10^6 hours at room temperature. Earlier DLs were of such limited life that it restricted their use. The DL is usually used as the source for single-mode fiber because LEDs have a low input coupling efficiency.

Some systems operate at a slower bit rate and require more modest levels of fiber-coupled optical power (50 to 250 μW). These applications allow the use of high-radiance LEDs. The LED is cheaper, requires less-complex driving circuitry than a DL, and needs no thermal or optical stabilizations. In addition, LEDs have longer operating lives (10^6 to 10^7 h) and fail in a more gradual and predictable fashion than do DLs.

Both LEDs and DLs are multilayer devices most frequently fabricated of AlGaAs on GaAs. They both behave electrically as diodes, but their light-emission properties differ substantially. A DL is an optical oscillator; hence it has many typical oscillator characteristics: a threshold of oscillation, a narrow emission bandwidth, a temperature coefficient of threshold and frequency, modulation nonlinearities, and regions of instability.

The light output wavelength spread, or spectrum, of the DL is much narrower than that of LEDs: about 1 nm compared with about 40 nm for an LED. Narrow spectra are advantageous in systems with high bit rates because the dispersion effects of the fiber on pulse width are reduced, and thus pulse degradation over long distances is minimized.

Light is emitted from an LED as a result of the recombining of electrons and holes. Electrically, an LED is a *pn* junction. Under forward bias, minority carriers are injected across the junction. Once across, they recombine with majority carriers and give up their energy. The energy given up is about equal to the material's energy gap. This process is radiative for some materials (such as GaAs) but not so for others, such as silicon. LEDs have a distribution of nonradiative sites—usually crystal lattice defects, impurities, and so on. These sites develop over time and explain the finite life/gradual deterioration of light output.

Figure 16-14 shows the construction of a typical semiconductor laser used in fiber-optic systems. The semiconductor laser uses the properties of the junction between heavily doped

FIGURE 16-14
Semiconductor laser.

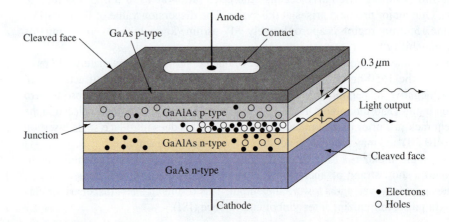

layers of *p*- and *n*-type materials. When a large forward bias is applied, many free holes and electrons are created in the immediate vicinity of the junction. When a hole and electron pair collide and recombine, they produce a photon of light. The *pn* junction in Figure 16-14 is sandwiched between layers of material with different optical and dielectric properties.

The material that shields the junction is typically aluminum gallium arsenide, which has a lower index of refraction than gallium arsenide. This difference "traps" the holes and electrons in the junction region and thereby improves light output. When a certain level of current is reached, the population of minority carriers on either side of the junction increases, and photon density becomes so high that they begin to collide with already excited minority carriers. This causes a slight increase in the ionization energy level, which makes the carrier unstable. It thus recombines with a carrier of the opposite type at a slightly higher level than if no collision had occurred. When it does, two equal-energy photons are released.

The carriers that are "stimulated" (indeed, *laser* is an acronym for *l*ight *a*mplification by *s*timulated *e*mission of *r*adiation) as described in the preceding paragraph may reach a density level so that each released photon may trigger several more. This creates an avalanche effect that increases the emission efficiency exponentially with current above the initial emission threshold value. Placing mirrored surfaces at each end of the junction zone usually enhances this behavior. These mirrors are parallel, so generated light bounces back and forth several times before escaping. The mirrored surface where light emits is partially transmissive (i.e., partially reflective).

The laser diode functions as an LED until its threshold current is reached. At that point, the light output becomes **coherent** (spectrally pure or only one frequency), and the output power starts increasing rapidly with increases in forward current. This effect is shown in Figure 16-15.

The typical spectral purity of these lasers yields a line width of about 1 nm versus about 40 nm for LED sources. Recall that this is critical for minimizing pulse dispersion. The wavelength of light generated is determined by the materials used. The "short-wavelength" lasers at 780 to 900 nm use gallium arsenide (GaAs) and aluminum gallium arsenide (AlGaAs). "Long-wavelength" (infrared) devices at 1300 to 1600 nm are made of layers of indium gallium arsenide phosphide (InGaAsP) and indium phosphide (InP).

A new device, called a **distributed feedback (DFB) laser** uses techniques that provide optical feedback in the laser cavity. This enhances output stability, which produces a narrow and more stable spectral width. Widths are in the range of 0.01 to 0.1 nm. This allows the use of more channels in **dense wavelength division multiplex (DWDM)** systems.

Another recent development is an entirely new class of laser semiconductors called **vertical cavity surface emitting lasers (VCSELs)**. These lasers can support a much faster signal rate than can LEDs, including gigabit networks. They do not have some of the operational and stability problems of conventional lasers; however, VCSELs have the simplicity of LEDs with the performance of lasers. Their primary wavelength of operation is in the 750- to 850-nm region, although development work is underway in the 1310-nm region. Reliabilities approaching 10^7 hours are projected. Table 16-6 provides a comparison of laser and LED optical transmitters.

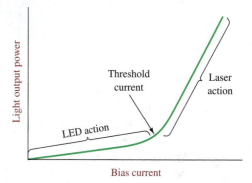

FIGURE 16-15   Light output versus bias current for a laser diode.

Modulating the Light Source

Most fiber-optic communication occurs using digital pulse (on–off) systems. Pulse-code modulation is most often used with return-to-zero or Manchester coding (Chapter 9). The transmission of analog signals can be accomplished by varying the amplitude of the light output. This is used largely by CATV systems and can be described as an amplitude modulation (AM) system. A very simple AM system using an LED light source is shown in Figure 16-16. The use of frequency modulation is not possible because the frequency of the light output from a laser or LED cannot be varied in a modulation sense. However, there is a class of lasers called **tunable lasers** in which the fundamental wavelength can be

TABLE 16-6 • A Comparison of Laser and LED Optical Transmitters

	LASER	LED
Usage	High bit rate, long haul	Low bit rate, short haul, LAN
Modulation rates	<40 Mbps to gigabits	<400 Mbps
Wavelength	Single-mode at 1310 and 1550 nm	Single-mode and multimode at 850/1310 nm
Rise time	<1 ns	10 to 100 ns
Spectral width	<1 nm up to 4 nm	40 to 100 nm
Spectral content	Discrete lines	Broad spectrum/continuous
Power output	0.3 to 1 mW (−5 to 0 dBm)	10 to 150 μW (−20 to −8 dBm)
Reliability	Lower	Higher
Linearity	40 dB (good)	20 dB (moderate)
Emission angle	Narrow	Wide
Coupling efficiency	Good	Poor
Temperature/humidity	Sensitive	Not sensitive
Durability/life	Medium (10^5 hours)	High (>10^6 hours)
Circuit complexity	High	Low
Cost	High	Low

Note: The values shown depend, to some extent, on the associated electronic circuitry.

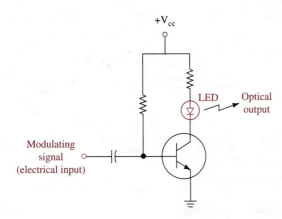

FIGURE 16-16
LED modulator.

shifted a few nanometers, but not from a modulation point of view. The primary market for these devices is in a network operations environment where DWDM is involved. Traffic routing is often made by wavelength, and, as such, wavelengths or transmitters must be assigned and reassigned to accommodate dynamic routing or networking, bandwidth on demand, seamless restoration (serviceability), optical packet switching, and so on. Tunable lasers are used along with either passive or tunable WDM filters.

Intermediate Components

The typical fiber-optic telecommunications link is—as shown in Figure 16-1—a light source or transmitter and light detector or receiver, interconnected by a strand of optical **fiber**, or **light pipe**, or **glass**. An increasing number of specialized networks and system applications have various intermediate components along the span between the transmitter and the receiver. A brief review of these devices and their uses is provided.

ISOLATORS An **isolator** is an in-line passive device that allows optical power to flow in one direction only. Typical forward-direction insertion losses are less than 0.5 dB, with reverse-direction insertion losses of at least 40 to 50 dB. They are polarization-independent and available for all wavelengths. One popular use is preventing reflections caused by optical span irregularities getting back into the laser transmitter. Distributed feedback lasers are particularly sensitive to reflections, which can result in power instability, phase noise, line-width variations, etc.

ATTENUATORS Attenuators are used to reduce the **received signal level (RSL)**. They are available in fixed and variable configurations. The fixed attenuators are for permanent use in point-to-point systems to reduce the RSL to a value within the receiver's dynamic range. Typical values of fixed attenuators are 3 dB, 5 dB, 10 dB, 15 dB, and 20 dB. Variable attenuators are typically for temporary use in calibration, testing, and laboratory work but more recently are being used in optical networks, where changes are frequent and programmable.

BRANCHING DEVICES   Branching devices are used in simplex systems where a single optical signal is divided and sent to several receivers, such as point-to-multipoint data or a cable TV distribution system. They can also be used in duplex systems to combine or divide several inputs. The units are available in single-mode and multimode units. The primary optical parameters are insertion loss and return loss, but the values for each leg may vary slightly because of differences in the device-mixing region.

SPLITTERS   Splitters are used to split, or divide, the optical signal for distribution to any number of places. The units are typically simplex and come in various configurations, such as 1×4, 1×8, . . ., 1×64.

COUPLERS   Couplers are available in various simplex or duplex configurations, such as 1×2, 2×2, 1×4, and various combinations up to 144×144. There are both passive and active couplers, the latter most often associated with data networks. Couplers can be wavelength dependent or independent.

WAVELENGTH DIVISION MULTIPLEXERS   Wavelength division multiplexers combine or divide two or more optical signals, each having a different wavelength. They are sometimes called *optical beamsplitters*. They use dichroic filtering, which passes light selectively by wavelength, or diffraction grating, which refracts light beams at an angle, selectively by wavelength. An additional optical parameter of importance is port-to-port crosstalk coupling, where wavelength number 1 leaks out of or into the port of wavelength number 2. A port is the input or output of the device.

OPTICAL-LINE AMPLIFIERS   Optical-line amplifiers are not digital regenerators, but analog amplifiers. Placement can be at the optical transmitter output, midspan, or near the optical receiver. They are currently used by high-density long-haul carriers, transoceanic links, and, to some extent, the cable TV industry.

Detectors

The devices used to convert the transmitted light back into an electrical signal are a vital link in a fiber-optic system but are often overlooked in favor of the light source and fibers. However, simply changing from one photodetector to another can increase the capacity of a system by an order of magnitude. Because of this, current research is accelerating to allow production of improved detectors. For most applications, the detector used is a *p-i-n* diode. The avalanche photodiode is also used in photodetector applications.

Just as a *pn* junction can be used to generate light, it can also be used to detect light. When a *pn* junction is reversed-biased and under dark conditions, very little current flows through it. The current that does flow is termed the **dark current**. However, when light shines on the device, photon energy is absorbed and hole–electron pairs are created. If the carriers are created in or near the junction depletion region, they are swept across the junction by the electric field. This movement of charge carriers across the junction causes a current flow in the circuitry external to the diode and is proportional to the light power absorbed by the diode.

The important characteristics of light detectors are:

1. **Responsivity:** This is a measure of output current for a given light power launched into the diode. It is given in amperes per watt at a particular wavelength of light.
2. **Dark current:** This is the thermally generated reverse leakage current (under dark conditions) in the diode. In conjunction with the response current as predicted by device responsivity and incident power, it provides an indication of on–off detector output range.
3. **Response speed:** This determines the maximum data-rate capability of the detector.

4. **Spectral response:** This determines the responsivity that is achieved relative to the wavelength at which responsivity is specified. Figure 16-17 provides a spectral response versus light wavelength for a typical *p-i-n* photodiode. The curve shows that its relative response at 900 nm (0.9 μm) is about 80 percent of its peak response at 800 nm.

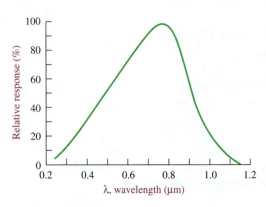

FIGURE 16-17 Spectral response of a *p-i-n* diode.

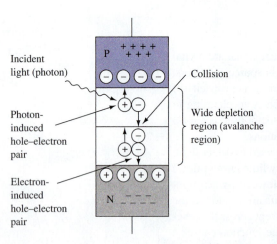

FIGURE 16-18 *p-i-n* diode.

Figure 16-18 shows the construction of a *p-i-n* diode used as a photodetector. As mentioned previously, light falling on a reverse-biased *pn* junction produces hole–electron pairs. The ability of a generated hole–electron pair to contribute to current flow depends on the hole and electron being rapidly separated from each other before they collide and cancel each other out. The reverse-biased diode creates a depletion region at the *pn* junction. The reverse-biased junction can be thought of as a capacitor, with the depletion region acting as the dielectric. The hole and electron created in the depletion region are rapidly pulled apart by the *p* and *n* materials that act as the capacitor's plates. Widening the depletion region gives more opportunity for hole–electron pairs to form and thus enhances the photodetector operation. The intrinsic (*i*) layer of the *p-i-n* diode in Figure 16-18 performs that function. The intrinsic layer is a very lightly doped semiconductor material.

The operation of the avalanche photodiode is illustrated in Figure 16-19. The diode is operated at a reverse voltage near the breakdown of the junction. At that potential, the electrons can be pulled from the atomic structure. With a small amount of additional energy, electrons are dislodged from their orbits, producing free electrons and resulting holes. As shown in Figure 16-19, a photon of light incident on the junction generates a hole–electron pair in the depletion region. Because of the large electric field, the electron movement is accelerated and the electrons collide with other bound electrons. These collisions create additional hole–electron pairs that are also accelerated, which produces still more hole–electron pairs, and an avalanche multiplication process (gain) occurs. One electron may produce up to 100 electrons in the avalanche photodiode. The avalanche photodiode is 5 to 7 dB more sensitive than the *p-i-n* diode. This advantage is maintained except when extremely high data rates exceeding 4 Gbps are experienced. In these cases, the better frequency response of the *p-i-n* diode favors its use.

FIGURE 16-19
Avalanche photodiode.

It should be noted that a second role for light detectors in fiber-optic systems exists. Detectors are used to monitor the output of laser diode sources. A detector is placed in proximity to the laser's light output. The generated photocurrent is used in a circuit to keep the light output of the laser constant under varying temperature and bias conditions. This is necessary to keep the laser just above its threshold forward bias current and to enhance its lifetime by not allowing the output to increase to higher levels. Additionally, the receiver does not want to see the varying light levels of a noncompensated laser.

The output current of photodiodes is at a very low level—on the order of 10 nA up to 10 μA. As a result, the noise benefits of fiber optics can be lost at the receiver connection between diode and amplifier. Proper design and shielding can minimize that problem, but an alternative solution is to integrate the first stage of amplification into the same circuit as the photodiode. These integrations are termed *integrated detector preamplifiers* (IDPs) and provide outputs that can drive TTL logic circuits directly. A comparison of *p-i-n* and APD detectors is provided in Table 16-7.

TABLE 16-7 • A Comparison of Detectors		
PARAMETER	***P-I-N***	**APD**
Bandwidth	Low bit rate <200 Mbps	High bit rate >200 Mbps to Gbps
Wavelength	850 and 1310 nm	1310 and 1550 nm
Sensitivity	Low, −35 dBm to −40 dBm	High, −45 dBm
Dynamic range	Low	High
Dark current	High	Low, less noise
Circuit complexity	Low	Medium
Temperature sensitivity	Low	High
Cost	Low	High
Life	10^9 hours	10^6 hours
Photon and electron conversion gain	1	3 to 5
Operating voltages	Low	High

Note: The values depend, to some extent, on the associated electronic circuitry.

16-6 FIBER CONNECTIONS AND SPLICES

Optical fiber is made of ultrapure glass. Optical fiber makes window glass seem opaque by comparison. It is therefore not surprising that the process of making connections from light source to fiber, fiber to fiber, and fiber to detector becomes critical in a system. The low-loss capability of the glass fiber can be severely compromised if these connections are not accomplished in exacting fashion.

Optical fibers are joined either in a permanent fusion splice or with a connector. The connector allows repeated matings and unmatings. Above all, these connections must lose as little light as possible. Low loss depends on correct alignment of the core of one fiber to another, or to a source or detector. Loss occurs when two fibers are not perfectly aligned within a connector. Axial misalignment typically causes the greatest loss—about 0.5 dB for a 10% displacement. This condition and other loss sources are illustrated in Figure 16-20. Most connectors leave an air gap, as shown in Figure 16-20(c). The amount of gap affects loss because light leaving the transmitting fiber spreads conically. Angular misalignment [Figure 16-20(b)] can usually be well controlled in a connector.

The losses created by rough end surfaces shown in Figure 16-20(d) are often caused by a poor cut, or "cleave," but can be minimized by polishing. Polishing typically takes place after a fiber has been placed in a connector.

The source of connection losses shown in Figures 16-20(a) to (d) can, for the most part, be controlled by a skillful cable splicer. There are four other situations that can cause additional connector or splice loss. These are shown in Figures 16-20(e), (f), (g), and (h). These are related to the nature of the fiber strand at the point of connection and are usually beyond the control of the cable splicer. The effect of

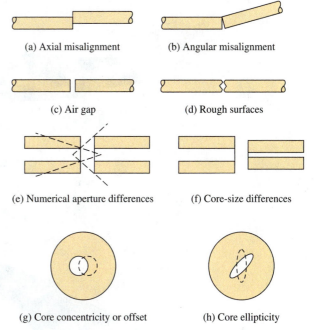

(a) Axial misalignment   (b) Angular misalignment

(c) Air gap   (d) Rough surfaces

(e) Numerical aperture differences   (f) Core-size differences

(g) Core concentricity or offset   (h) Core ellipticity

FIGURE 16-20 **Sources of connection loss.**

these losses can be minimized somewhat by the use of a rotary mechanical splice, which by the joint rotation will get a better core alignment.

There are two techniques to consider for splicing. **Fusion splicing** is a long-term method in which two fibers are fused or welded together. The two ends are stripped of their coating, cut or cleaved, and inserted into the splicer. The ends of the fiber are aligned and an electric arc is fired across the ends, melting the glass and fusing the two ends together. There are both manual and automatic splicers; the choice usually depends on the number of splices to be done on a given job, technician skill levels available, and, of course, the budget. Typical insertion losses of less than 0.1 dB—frequently in the 0.05-dB range—can be achieved consistently.

Mechanical splices can be permanent and an economical choice for certain fiber-splicing applications. Mechanical splices also join two fibers together, but they differ from fusion splices because an air gap exists between the two fibers. This results in a glass–air–glass interface, causing a severe double change in the index of refraction. This change results in an increase in insertion loss and reflected power. The condition can be minimized by applying an **index-matching gel** to the joint. The gel is a jellylike substance that has an index of refraction much closer to the glass than air. Therefore, the index change is much less severe.

Mechanical splices are universally popular for repair and for temporary or laboratory work. They are quick, cheap, easy, and quite appropriate for small jobs. The best method for splicing depends on the application, including the expected future bandwidth (i.e., gigabit), traffic, the job size, and economics.

Fiber Connectorization

For fiber connectorization, there are several choices on the market. Currently popular are the **SC** and **ST**. Smaller connectors called **small-form factor** connectors are also on the market. The connectors are about one-half the size of conventional SC and ST units and are intended for use in local area networks in the home and office. Three designs, the types LC, MT-RJ, and VF-45, are also recognized by the Telecommunications Industry Association. Examples of SC, ST, and MT-RJ connectors are provided in Figures 16-21(a), (b), and (c). Some general requirements for fiber connectors are provided in Table 16-8.

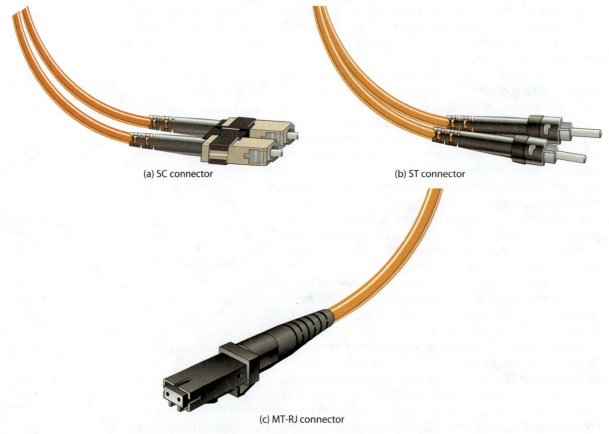

(a) SC connector

(b) ST connector

(c) MT-RJ connector

FIGURE 16-21   Fiber connectors.

TABLE 16-8 • General Fiber Connector Requirements
Easy and quick to install.
Low insertion loss. A properly installed connector has as little as 0.25 dB insertion loss.
High return loss greater than 50 dB. This is increasingly important in gigabit networks, DWDM systems, high-bandwidth video, and so on.
Repeatability.
Economical.

In preparing the fiber for splicing or connectorization, only the coating is removed from the fiber strand. The core and the clad are not separable. The 125-μm clad diameter is the portion that fits into the splice or connector, and therefore most devices can handle both single and multimode fiber.

Sometimes the issue of splicing together fibers of different core sizes arises. The one absolute rule is this: Do not splice single and multimode fiber together! Similarly, good professional work does not allow different sizes of multimode fiber to be spliced together. However, in an emergency, different sizes of multimode fiber can be spliced together if the following limitations are recognized:

> When transmitting from a small- to a larger-core diameter, there will be minimal, if any, increase in insertion loss. However, when the transmission is from a larger to a smaller core size, there will be added insertion loss, and a considerable increase in reflected power should be expected.

Industrial practice has confirmed the acceptability of different-core size interchangeability for emergency repairs in the field, mainly as the result of lab tests with 50- and 62.5-μm multimode fiber for a local area net environment.

16-7 SYSTEM DESIGN AND OPERATIONAL ISSUES

When designing a fiber-optic transmission link, the primary performance issue is the bit error rate (BER) for digital systems and the carrier-to-noise (*C/N*) ratio for analog systems. In either case, performance degrades as the link length increases. As stated in Section 16-4, attenuation and dispersion are the two distance-limiting factors in optical transmissions. The distance limit is the span length at which the BER or *C/N* degrades below some specified point. From an engineering point of view, there are two different types of environments for fiber links: long-haul and local-area networks (LANs).

A **long-haul system** is the intercity or interoffice class of system used by telephone companies and long-distance carriers. These systems typically have high channel density and high bit rate, are highly reliable, incorporate redundant equipment, and involve extensive engineering studies.

LANs take a less strict position on the issues stated under long-haul applications. They typically have lower channel capacity and minimal redundancy and are restricted to building-to-building or campus environments. Some LANs are becoming very large, including metropolitan-area networks (MANs) and wide-area networks (WANs); as such, they usually rely on long-haul carriers for their connectivity.

From a design standpoint, those involved in long-haul work perform the studies on a per-link basis. LANs typically are prespecified and preengineered as to length, bit-rate capability, performance, and so on. The following is an example of a system designed for an installation typical of a long distance communication link.

In this example, each of the many factors that make up the link calculation, power budget, or light budget is discussed along with its typical contribution. A minimal received signal level (RSL) must be obtained to ensure that the required BER is satisfied. For example, if the minimum RSL is −40 dBm for a BER of 10^{-9}, then this value is the required received optical power. If, after the initial calculations are completed, the projected performance is not as expected, then go back and adjust any of the parameters, recognizing that there are trade-off issues.

Refer to the system design shown in Figures 16-22 and 16-23:

FIGURE 16-22 System design.

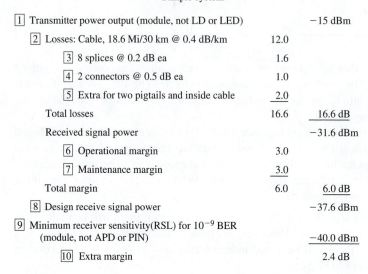

ATTENUATION OR LINK LOSS
Sample system

☐1 Transmitter power output (module, not LD or LED)		−15 dBm
☐2 Losses: Cable, 18.6 Mi/30 km @ 0.4 dB/km	12.0	
☐3 8 splices @ 0.2 dB ea	1.6	
☐4 2 connectors @ 0.5 dB ea	1.0	
☐5 Extra for two pigtails and inside cable	2.0	
Total losses	16.6	16.6 dB
Received signal power		−31.6 dBm
☐6 Operational margin	3.0	
☐7 Maintenance margin	3.0	
Total margin	6.0	6.0 dB
☐8 Design receive signal power		−37.6 dBm
☐9 Minimum receiver sensitivity(RSL) for 10^{-9} BER (module, not APD or PIN)		−40.0 dBm
☐10 Extra margin		2.4 dB

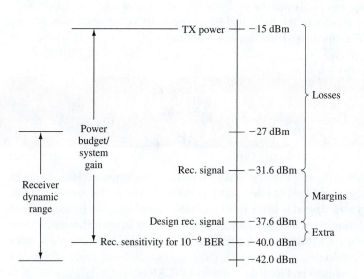

1. **Transmitter power output:** A value usually obtained from the manufacturer's specification or marketing sheet. *Caution:* Be sure the value is taken from the output port of the transmit module or rack. This is point 1 in Figures 16-22 and 16-23. This is the point where the user can access the module for measurement and testing. Otherwise the levels can be off as much as 1 dB because of pigtail or coupling losses between the laser or LED and the actual module output.

2. **Cable losses:** The loss in dB/km obtained from the cable manufacturer's sheet. This value is multiplied by the length of the cable run to obtain the total loss. An example of this calculation is provided in Example 16-4.

   Note that the actual fiber length can exceed the cable run length by 0.5% to 3% because of the construction of the fiber cable (in plastic buffer tubes). Fiber cables are loosely enclosed in buffer tubes to isolate the fiber from construction stress when the cable is pulled.

3. **Splice losses:** Values depend on the method used for splicing as well as the quality of splicing provided by the technician. Losses can vary from 0.2 dB to 0.5 dB per splice.

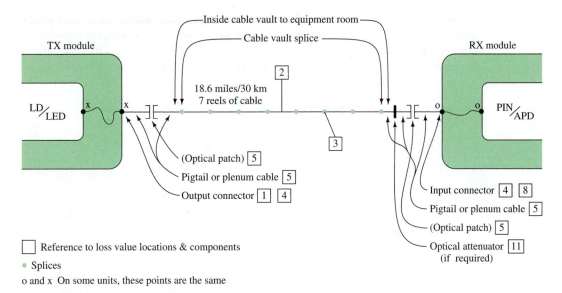

FIGURE 16-23   A graphical view of the system design problem shown in Figure 16-22.

4. **Connector losses:** A value depending on the type and quality of the connector used as well as the skill level of the installer. Losses can vary from 0.25 dB to 0.5 dB.

5. **Extra losses:** A category used for miscellaneous losses in passive devices such as splitters, couplers, WDM devices, optical patch panels, and so on.

6. **Operational margin:** Accounts for system degradation due to equipment aging, temperature extremes, power supply noise and instability, timing errors in regenerators, etc.

7. **Maintenance margin:** Accounts for system degradation from the addition of link splices, added losses because of wear and misalignment of patch cords and connectors, etc. This includes the loss generated from repairing cables that have been dug up by a backhoe.

    (The term **backhoe fading** is used to indicate that the system has had total loss of data flow because the fiber cable has been dug up by a backhoe.)

8. **Design receive signal power:** A value obtained from summing the gains and losses in items 1–7. This value should exceed the specified RSL, as specified in item 9.

9. **Receiver sensitivity for a 10^{-9} BER:** The minimum RSL for the receiver to perform at the specified BER. If the design receive signal power (item 8) does not meet this requirement, then adjustments must be made. For example, transmit power can be increased, splice loss estimates can be reduced, a more optimistic maintenance margin can occur, and so on. Also, the receiver may have a maximum RSL, and a system may require attenuators to decrease the RSL so that it falls within a specified operating range (receiver dynamic range).

10. **Extra margin:** The difference between the design receive signal power (item 8) and the receiver sensitivity (item 9). Item 8 should be greater than item 9; for example, −37.6 dBm is larger than −40 dBm. One or two dB is a good figure.

11. **Optional optical attenuator:** A place where an optional attenuator can be installed and later removed as aging losses begin to increase.

EXAMPLE 16-4

Determine the loss in dB for a 30-km fiber cable run that has a loss of 0.4 dB/km.

SOLUTION

$$\text{total cable loss} = 30 \text{ km} \times 0.4 \text{ dB/km} = 12 \text{ dB}.$$

A graphical view of the previous system design problem is shown in Figure 16-23. Figure 16-24 provides another way to describe the system design problem graphically. Notice that the values are placed along the distance covered by the fiber. This provides the maintenance staff and the designer with a clear picture of how the system was put in place.

FIGURE 16-24 An alternative view of the system design problem.

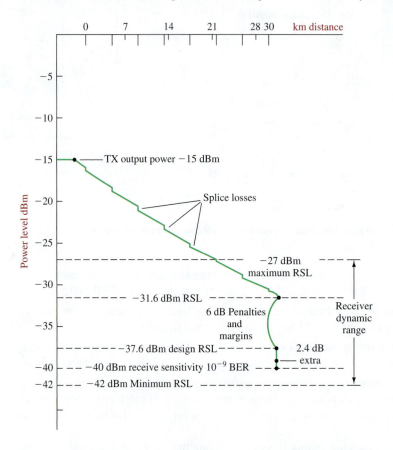

Another system design consideration is dispersion, the second distance-limiting factor in a fiber-optic system. The concept of dispersion was first examined in Section 16-4. The practical significance of dispersion is that it is desirable to have the operating wavelength of a fiber-optic system in the zero dispersion wavelength region (see Table 16-3). The formula for calculating the fiber span length when taking dispersion into account is provided in Equation (16-4).

$$L = \frac{440,000}{BR \times D \times SW},$$

(16-4)

where L = span length (kilometers)

BR = line bit rate (megabits/second)

D = cable dispersion (picoseconds/nanometer/kilometer)

SW = spectral width of the transmitter (nanometers)

440,000 = an assumed Gaussian constant based on a 3 dB optical bandwidth using a full-width-half-maximum (FWHM) pulse shape.

Example 16-5 demonstrates how to use Equation (16-4).

EXAMPLE 16-5

Determine the fiber span length given the following two sets of manufacturers' information. Compare the results for the two span length calculations.

(a) Line bit rate = 565 Mbps

Cable dispersion = 3.5 ps/nm/km

Transmitter spectral width = 4 nm

(b) Line bit rate = 1130 Mbps

   Cable dispersion = 3.5 ps/nm/km

   Transmitter spectral width = 2 nm

SOLUTION

(a) $L = \dfrac{440,000}{565 \times 3.5 \times 4} = 55.6 \text{ km}$

(b) $L = \dfrac{440,000}{1130 \times 3.5 \times 2} = 55.6 \text{ km}$

By reducing the transmitter's spectral width from 4 nanometers to 2 nanometers, the bit rate can be doubled without having to upgrade the cable facility.

Dispersion is typically a single-mode, long-haul, high-bit-rate consideration. The person planning the system should seek advice from cable and optoelectronic equipment manufacturers and experienced system designers.

16-8 CABLING AND CONSTRUCTION

This section provides a brief outline of the issues associated with the exterior or interior installation of fiber cable. Even though the installation techniques are well established, new products and tools are being brought to the market to improve the installation. Trade journals and Internet sites provide an excellent and reliable way to keep informed about the changes.

Exterior (Outdoor) Installations

Fiber can be installed on poles or underground in ducts, in utility tunnels, or by direct burial. You must be aware of the exposure factors for each installation. These factors can include temperature, humidity, chemicals, rodents, abrasion, water, ice, wind, mechanical vibration, and lightning, to name a few. Some of the ways to protect an exterior cable are provided by armored cable, a water-resistant sheath, close adherence to pulling tensile and bend radius specifications, and frequent grounding of metallic cable components.

Interior (Indoor) Installations

The environment for interior installations is usually well controlled. However, exposure factors include mechanical vibrations, heat, and possibly fire. In all cases, a cable with a fire-retardant sheath and sheaths that generate low or minimal smoke and toxic fumes are required. Plenum cable is available for installation in air ducts, air-handling spaces, and raised floors. Manufacturers' data sheets and local code requirements should be referenced to find the proper cable for an installation.

Testing the Fiber Installation

Figures 16-25(a) and (b) shows traces obtained from an **optical time-domain reflectometer (OTDR)** for two different sets of multimode fibers. In field terms, this is called "shooting" the fiber. The OTDR sends a light pulse down the fiber and measures the reflected light. The OTDR enables the installer or maintenance crew to verify the quality of each fiber span and obtain some measure of performance. The X axis on the traces indicates the distance, whereas the Y axis indicates the measured optical power value in dB. Both OTDR traces are for 850-nm multimode fiber.

FIGURE 16-25 **An OTDR trace of an 850-nm fiber.**

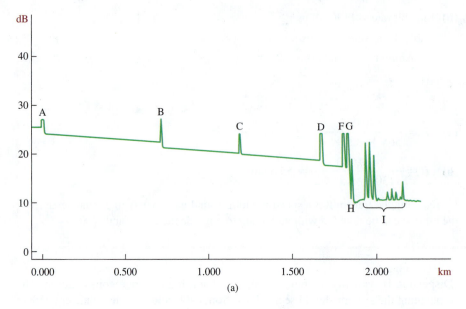

(a)

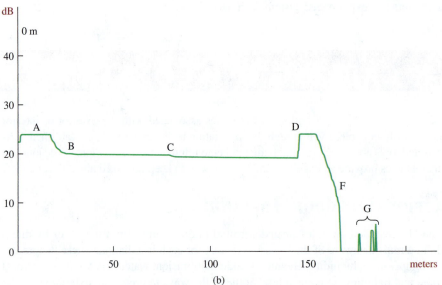

(b)

In Figure 16-25(a), Point A is a "dead" zone or a point too close to the OTDR for a measurement to be made. The measured value begins at about 25 dBm and decreases in value as the distance traveled increases. An **event**, or a disturbance in the light propagating down the fiber, occurs at point B. This is an example of what a poor-quality splice looks like (with regard to reflection as well as insertion loss). Most likely, this is a mechanical splice. The same type of event occurs at points C and D. These are also most likely mechanical splices. Points F and G are most likely the jumpers and patch-panel connections at the fiber end. The steep drop at point H is actually the end of the fiber. Point I is typical noise that occurs at the end of an "unterminated" fiber. Notice at point G that the overall value of the trace has dropped to about 17 dBm. There has been about 8 dB of optical power loss in the cable in a 1.7 km run.

An OTDR trace for another multimode fiber is shown in Figure 16-25(b). The hump at point A is basically a "dead" zone. The OTDR cannot typically return accurate measurement values in this region. This is common for most OTDRs, and the dead zone varies for each OTDR. The useful trace information begins at point B with a measured value of 20 dBm. Point C shows a different type of event. This type of event is typical of coiled fiber, or fiber that has been tightly bound, possibly with a tie-wrap, or that has had some other disturbance affecting the integrity of the fiber. Points D and F are actually the end of the fiber. At point D the trace level is about 19 dBm for a loss of about 1 dB over the 150-m run. Point G is just the noise that occurs at the end of a "terminated" fiber.

The need for increased bandwidth is pushing the fiber-optic community into optical networking solutions that are almost beyond the imagination of even the most advanced networking person. Optical solutions for long-haul, metropolitan-area, and local-area networks are available. Cable companies are already using the high-bandwidth capability of fiber to distribute television programming as well as internet data throughout their service areas.

The capital cost differences between a fiber system and a coaxial cable system are diminishing, and the choice of networking technology for new networks is no longer just budgetary. Fiber has the capacity to carry more bandwidth; as the fiber infrastructure cost decreases, fiber will be chosen to carry the data. Of course, the copper infrastructure is already in place, and new developments are providing tremendous increases in data speed over copper. However, optical fiber is smaller and eases the installation in already crowded ducts and conduits. And security is enhanced because it is difficult to tap optical fiber without detection.

Defining Optical Networking

Optical networks are becoming a major part of data delivery in homes, in businesses, and for long-haul carriers. The telecommunications industry has been using fiber for carrying long-haul traffic for many years. Some major carriers are merging with cable companies so that they are poised to provide high-bandwidth capabilities to the home. Developments in optical technologies are reshaping the way we will use fiber in future optical networks.

Yes, fiber provides additional bandwidth, but do we keep using the same approaches to solve networking problems? The answer is no; we need a new set of rules to define optical networking. Sprint Corporation has defined a new foundation for optical networking ["Changing the rules for developing optical solutions," *Lightwave* (October 1999)]. Five of the rules for optical networking are summarized as follows:

1. The next generation of optical networks must be able to carry multiple protocols. For example, optical networks should be able to carry IP Internet traffic and asynchronous transfer mode (ATM).

2. The architecture for the next generation of optical networks must be flexible.

3. The network must be manageable, including diagnostic capabilities for signal quality and faults.

4. The data transport must provide high speed and be invisible to the user. For example, the user should not have to be concerned with how the data are being transported or what protocol is being used.

5. The implementation of optical networks must provide for compatible interfacing with today's data-transport methodologies while providing the flexibility to incorporate future developments.

In addition to these five rules, the issues of chromatic and polarization mode dispersion become an increasing problem because of the need for greater transmission capability coupled with the restricted economic capability to install more fiber.

Basically, these rules for optical networking maintain a level of reliability and flexibility in the transport of data. But there is a new slant with optical networks. DWDM and tunable lasers have changed the way optical networks can be implemented. It is now possible to transport many wavelengths over a single fiber. Lab tests at AT&T have successfully demonstrated the transmission of 1,022 wavelengths over a single fiber; however, conventional systems are limited to approximately 32 wavelengths.

The transport of multiple wavelengths over a single fiber opens up the possibilities to routing or switching many different data protocols over the same fiber but on different wavelengths. The development of cross connects that allow data to arrive on one wavelength and leave on another opens other possibilities.

Synchronous optical network (SONET) is currently the standard for the long-haul optical transport of telecommunications data. SONET defines a standard for:

- Increase in network reliability
- Network management
- Defining methods for the synchronous multiplexing of digital signals such as DS-1 (1.544 Mbps) and DS-3 (44.736 Mbps)
- Defining a set of generic operating/equipment standards
- Flexible architecture

SONET specifies the various **optical carrier (OC)** levels and the equivalent electrical **synchronous transport signals (STS)** used for transporting data in a fiber-optic transmission system. Optical network data rates are typically specified in terms of the SONET hierarchy. Table 16-9 lists the more common data rates.

TABLE 16-9 • SONET Hierarchy Data Rates

SIGNAL	BIT RATE	CAPACITY
OC-1 (STS-1)	51.840 Mbps	28 DS-1s or 1 DS-3
OC-3 (STS-3)	155.52 Mbps	84 DS-1s or 3 DS-3s
OC-12 (STS-12)	622.080 Mbps	336 DS-1s or 12 DS-3s
OC-48 (STS-48)	2.48832 Gbps	1344 DS-1s or 48 DS-3s
OC-192 (STS-192)	9.95328 Gbps	5376 DS-1s or 192 DS-3s
OC-768 (STS-768)	39.81312 Gbps	768 DS-3s

OC: Optical carrier   DS-1: 1.544 Mbps
STS: Synchronous transport signals   DS-3: 44.736 Mbps

The architectures of fiber networks for the home include providing **fiber to the curb (FTTC)** and **fiber to the home (FTTH)**. Both are being deployed today. These developments provide high bandwidth to a location with proximity to the home and provides a high-speed data link, via copper (twisted pair), using very high-data digital subscriber line (VDSL). This is a cost-effective way to provide large-bandwidth capabilities to a home. FTTH will provide unlimited bandwidth to the home; however, the key to its success is the development of a low-cost optical-to-electronic converter in the home and laser transmitters that are tunable to any desired channel.

Conventional high-speed Ethernet local area networks operating over fiber use the numerics listed in Table 16-10 for describing the network configuration.

Fiber helps to eliminate the 100 m distance limit associated with unshielded twisted-pair (UTP) copper cable. This is possible because fiber has a lower attenuation loss. In a star network, the computer and the hub (or switch) are directly connected. If the fiber is

TABLE 16-10 • Ethernet/Fiber Numerics

NUMERIC	DESCRIPTION
10Base F	10-Mbps Ethernet over fiber—generic specification for fiber.
10BaseFB	10-Mbps Ethernet over fiber—part of the IEEE 10BaseF specification. Segments can be up to 2 km.
10BaseFL	10-Mbps Ethernet over fiber—segments can be up to 2 km in length. It replaces the FOIRL specification.
10BaseFP	A passive fiber star network. Segments can be up to 500 m in length.
100BaseFX	A 100-Mbps fast Ethernet standard that uses two fiber strands.
1000BaseLX	Gigabit Ethernet standard that uses two fiber strands.

Note: multimode fiber—2 km length; single-mode fiber—10 km.

used in a star network, a media converter may be required. The media converter converts the electronic signal to an optical signal, and vice versa. A media converter is required at both ends, as shown in Figure 16-26.

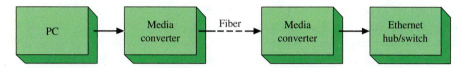

FIGURE 16-26   **An example of connecting a PC to an Ethernet hub or switch via fiber.**

Another example of how fiber is currently used in Ethernet LANs is for the high-speed transport of data, point-to-point, over longer distances. For example, the output of an Ethernet switch might be sent via fiber to a local router, as shown in Figure 16-27(a). In this example, the inputs to the Ethernet switch are 10BaseT (10-Mbps twisted-pair) lines coming from computers on the LAN. The output of the Ethernet switch leaves via fiber at a 100-Mbps data rate (100BaseFX). The fiber makes it easy to increase the data rate over increased distances. As shown in Figure 16-27(b), multiple-switch outputs, connected through a router, might be connected to a central router via fiber at a gigabit data rate.

The fiber provides substantially increased bandwidth for the combined traffic of the Ethernet switches and PCs [Figures 16-27(a) and (b)]. Fiber has greater capacity, which enables greater bits per second (bps) transfer rates, minimizes congestion problems, and provides tremendous growth potential for each of the fiber runs.

Conventional high-speed Ethernet local area networks operating over fiber use the numerics listed in Table 16-10 for describing the network configuration.

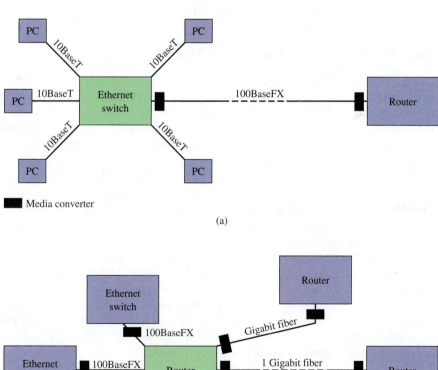

FIGURE 16-27   **Examples of point-to-point connections using fiber in local area networks.**

Air Fiber

Another form of optical networking involves the propagation of laser energy through the atmosphere, a line-of-sight technique similar to microwave radio. This application (usually called **air fiber**, free space optics, or a similar expression) uses a parabolic lens to focus the laser energy in a narrow beam. The beam is then aimed through the air to a receive parabolic lens a short distance away. This short distance is conservatively about 3 km, or a little farther depending on laser power and detector sensitivity used as well as the reliability/bit error rate desired.

The normal long-term 99.999% reliability is achievable, but it is difficult. The expected degrading effects of simple rain and fog are often not particularly troublesome, but when a high-moisture-density cell crosses the propagation path, high signal attenuation can be noted. A good weather pattern study is advisable when planning an optical path.

The optical transmission equipment needs a stable mounting platform because of the degrading effects of building movement and vibration. Most optical platforms have auto-tracking options available to optimize antenna (lens) alignment constantly and minimize bit errors and outages.

Wavelengths available are from 800 to 1500 nm, all with their own pros and cons. System planners should be aware of laser safety when planning open-air optical spans.

The use of this media for networking is ideal between tall buildings in urban areas, in short metro spans, and on industrial and college campuses. It is particularly well suited for temporary service and disaster recovery operations. An additional advantage is that FCC licensing is not required. These applications are enhanced because of the potential for monetary savings in construction cost, physical infrastructure disruptions, and additional fill of cable duct facilities. This type of optical networking equipment is capable of handling a wide variety of data protocols, such as FDDI, DS–3, ATM, and gigabit formats.

Fiber Distributed Data Interface

The American National Standards Institute (ANSI) developed the Fiber Distributed Data Interface (FDDI) standard that is now in widespread use. FDDI utilizes two 100-Mbps token-passing rings. The two independent counter-rotating rings are connected to a certain number of nodes (stations) in the network. The primary ring connects only *class A* stations—those offering a high level of protection because of their ability to transfer operation to the secondary ring should the primary ring fail. The secondary ring reaches all stations and carries data in the opposite direction of the primary ring. The secondary ring can also be used with the primary ring operating to allow increased data throughput. The switchover to the secondary path in the event of failure is accomplished by a pivoting spherical mirror within a *dual-bypass* switch. The changeover takes 5 to 10 μs, and a loss of about 1 dB results from the presence of the dual-bypass switch.

Stations on the FDDI ring can be separated by up to 2 km as long as the average distance between nodes is less than 200 m. These limits are imposed to minimize the time it takes a signal to move around the ring. A total of 1000 physical connections and a total fiber path of 200 km are allowed. This allows 500 stations because each represents two physical connections. The type of fiber used is not dictated and is chosen by the user based on the performance required. The fibers used most often are 62.5/125 or 100/140 multimode fiber. LEDs are specified as light sources at 1300 nm.

16-10 SAFETY

Any discussion of fiber optics is not complete unless it addresses safety issues, even if only briefly. As the light propagates through a fiber, two factors will further attenuate the light if there is an open or break.

1. A light beam will disperse or fan out from an open connector.

2. If a damaged fiber is exposed on a broken cable, the end will likely be shattered, which will considerably disperse the light. In addition, there will be a small amount of attenuation from the strand within the cable, plus any connections or splices along the way.

However, two factors can increase the optical power at an exposed fiber end.

1. There could be a lens in a pigtail that could focus more optical rays down the cable.

2. In the newer DWDM systems, several optical signals are in the same fiber; although separate, they are relatively close together in wavelength. The optical power incident upon the eye is then multiplied.

Be aware of two factors:

1. The eye can't see fiber-optic communications wavelengths, so there is no pain or awareness of exposure. However, the retina can still be exposed and damaged. (Refer to Figure 16-4, the electromagnetic spectrum.)

2. Eye damage is a function of the optical power, wavelength, source or spot diameter, and the duration of exposure.

For those working on fiber-optic equipment:

1. *NEVER* look into the output connector of energized test equipment. Such equipment can have higher powers than the communications equipment itself, particularly OTDRs.

2. If you need to view the end of a fiber, *ALWAYS turn off the transmitter,* particularly if you don't know whether the transmitter is a laser or LED, because lasers are higher power sources. If you are using a microscope to inspect a fiber, the optical power will be multiplied.

From a mechanical point of view:

1. Good work practices are detailed in safety, training, and installation manuals. *READ AND HEED*.

2. Be careful with machinery, cutters, chemical solvents, and epoxies.

3. Fiber ends are brittle and break off easily, including the ends cut off from splicing and connectorization. These ends are extremely difficult to see and can become "lost" and/or easily embedded in your finger. You won't know until your finger becomes infected. Always account for all scraps.

4. Use safety glasses specifically designed to protect the eye when working with fiber-optic systems.

5. Obtain and *USE* an optical safety kit.

6. Keep a *CLEAN* and orderly work area.

In all cases, be sure the craft personnel have the proper training for the job!

16-11 TROUBLESHOOTING

Today, optical fiber is the infrastructure of many communications hubs. Fiber carries billions of telephone calls a day. Optical fiber makes up the backbone structure of many local area networks currently in use. In this section we will look at planning an optical-fiber installation and maintaining it once it is in place.

After completing this section you should be able to

- Draw a fiber link showing all components
- Explain the use of the optical power meter
- Describe rise-time measurement
- Troubleshoot fiber-optic data links

System Testing

Once a system is installed, it should be tested thoroughly to ensure compliance with the contract specifications and performance in accordance with the manufacturer's manual. The following lists of tests provide a good measure of performance. Also, they are the start of a maintenance database for future reference. Not all these tests apply to all optical networks. In the realm of testing and evaluation, you will find that testing is expensive; from an experience perspective, you will find that not testing is more expensive.

General Guidelines

Tests on the cable plant itself should include:

- Measuring fiber insertion loss, which should be compared to the engineering system design. An optical test set is preferred, but an OTDR can be used to perform simple tests.
- Gathering OTDR traces, noting the loss slope plus return loss.
- Testing of all wavelengths planned or projected for use.

Overall system tests should include:

- Bit error rate (BER) tests
- Central wavelength
- Spectral width tests
- Transmitter output power (average, not peak)
- Receiver sensitivity (to some BER)
- Input voltage tolerance
- Protection and alarms
- System restoration

Note: Specifics on the requirements and guidelines for these tests are available from the manufacturers.

Losses in an Optical-Fiber System

The optical-fiber system in Figure 16-28 has an emitter, two connectors, the fiber, and the detector. The proper performance of this fiber link depends on the total power losses of the light signal through the link being less than the specified maximum allowable loss. Power is lost in all of the components that make up the system. A connector may have a power loss of 1.5 dB and a splice with 0.5 dB, and the fiber cable itself will also attenuate the light signal. As an example, if a fiber system's maximum allowable losses are 20 dB, and total power losses add up to 17 dB, the system still has a 3-dB working margin. Of course, this is a small working margin and does not take into account emitters and detectors weakening over time.

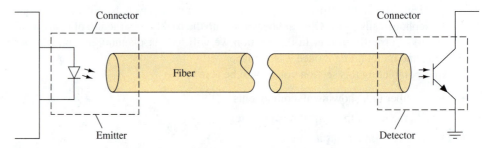

FIGURE 16-28   **A fiber link showing emitter, detector, connectors, and fiber cable.**

Calculating Power Requirements

A power budget should be prepared when a fiber-based system is installed. The power budget specifies the maximum losses that can be tolerated in the fiber system. This helps ensure that losses stay within the budgeted power allocation. Once the optical-fiber system is in place, the optical power meter is used to determine the actual power being lost in the system. A calibrated light source injects a known amount of light into the fiber, and the optical power meter connected to the other end of the fiber measures the light power reaching it. Periodic checks should be scheduled as preventive maintenance to keep the fiber system in peak performance. Weakening emitters should be replaced before they degrade system performance.

Connector and Cable Problems

Some of the problems associated with fiber-optic links are caused by contact of a foreign substance with the fiber (even the oil from your skin can cause serious trouble). Connectors and splices are potential trouble spots. A back-biased photodiode and an op-amp can be used as a relative signal strength indicator. Looking for the signal while gently flexing cables and connectors can help pinpoint problem areas.

Characteristics of Light-emitting Diodes and Diode Lasers

These special-purpose diodes are nonetheless diodes and should exhibit the familiar exponential I versus V curve. These diodes do not draw current when forward biased until the voltage reaches about 1.4 V. Some ohmmeters do not put out sufficient voltage to turn an LED on; you may have to use a power supply, a current-limiting resistor, and a voltmeter to test the diode.

Reverse voltage ratings are very low compared to ordinary silicon rectifier diodes—as little as 6 V. More voltage may destroy the diode.

A Simple Test Tool

Some systems use visible wavelengths; most use invisible infrared. Another diode of the same type or of similar emission wavelength can be used as a detector to check for output. Use a meter in current mode, not voltage, and compare a good system to the troublesome one.

To increase sensitivity, a simple current-to-voltage converter circuit, made with an op-amp, a feedback resistor, and the detector diode pumping current to the op-amp input, converts the current from the detector diode to a voltage out of the op-amp. The circuit for this is shown in Figure 16-29. If signal levels are high, just a resistor across the detector diode is appropriate. Remember to keep the bias voltage small enough so that the voltage developed is well below the maximum reverse voltage allowed for the diode.

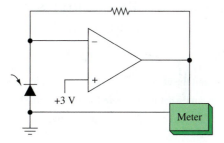

FIGURE 16-29   Light probe.

If you wish to see the signal modulation, try using an oscilloscope in place of a simple multimeter. A less quantitative check for emitted output can be made using a test card of the type used in TV-repair shops to check for output from infrared remote controls. These cards are coated with a special chemical that, in the simultaneous presence of visible and infrared illumination, emits an orange glow.

SUMMARY

In Chapter 16, we introduced the field of fiber optics. We learned that many applications exist in electronic communications for these optical devices. The major topics you should now understand include:

- the advantages offered by fiber-optic communication
- the analysis and properties of light waves
- the physical and optical characteristics of optical fibers, including multimode, graded index, and single-mode fibers
- the attenuation and dispersion effects in fiber
- the description and operation of the diode laser (DL) and high-radiance light-emitting diode (LED) light sources
- the application of *p-i-n* diodes as light detectors
- the description of common techniques used to connect fibers
- the general applications of fiber-optic systems
- the power considerations and calculations in fiber-optic systems
- the usage of fiber optics in local-area networks (LANs)
- the description of LAN components, including wavelength-dependent and independent couplers and optical switches

QUESTIONS AND PROBLEMS

SECTION 16-1

1. List the basic elements of a fiber-optic communications system. Explain its possible advantages compared to a more standard communications system.
2. List five advantages of an optical communications link.

SECTION 16-2

3. Define *refractive index*. Explain how it is determined for a material.
4. A green light-emitting diode (LED) light source functions at a frequency between red and violet. Calculate its frequency and wavelength. (5.7×10^{14} Hz, 526 nm)
5. A fiber cable has the following index of refraction: core, 1.52, and cladding, 1.31. Calculate the numerical aperture for this cable. (0.77)
6. Determine the critical angle beyond which an underwater light source will not shine into the air. (48.7°)
7. Define *infrared light* and the *optical spectrum*.
8. What are the six fixed wavelengths commonly used today?
9. What are the wavelength ranges for the optical bands O, E, S, C, L, and U?
10. Draw a picture of the construction of a single fiber.

SECTION 16-3

11. Define *pulse dispersion* and the effect it has on the transmission of data.
12. What is multimode fiber, and what is the range for the core size?
13. Why was graded index fiber developed? What are the two typical core sizes and the cladding size for graded index fiber?
14. What are the applications for single-mode fibers?

15. What are the core/cladding sizes for single-mode fibers?

16. Define *mode field diameter* for fiber-optic cable.

17. Define *zero-dispersion wavelength* for fiber-optic cable.

18. Describe two applications that would be suitable for using plastic optical fiber.

SECTION 16-4

19. What are the two key distance-limiting parameters in fiber-optic transmission?

20. What are the four factors that contribute to attenuation?

21. Define *dispersion*. What are typical dispersion values for 850-, 1310-, and 1550-nm-wavelength fibers?

22. Determine the amount of pulse spreading of an 850-nm LED that has a spectral width of 18 nm when run through a 1.5-km filter. Use a pulse-dispersion value of 95 ps/(nm · km). (2.565 ns/km)

23. What are the three types of dispersion?

24. What is dispersion-compensating fiber?

SECTION 16-5

25. Compare the diode laser (DL) and the LED for use as light sources in optical communication systems.

26. Explain the process of lasing for a semiconductor DL. What is varied to produce light at different wavelengths?

27. Define *dense wavelength division multiplexing (DWDM)*.

28. What are tunable lasers and what is the primary market for them?

29. What are isolators?

30. Attenuators are used to do what?

31. List five intermediate components for fiber-optic systems.

32. What is an optical detector?

33. What are the benefits of a distributed feedback laser?

34. List the advantages of vertical cavity surface emitting lasers (VCSEL).

SECTION 16-6

35. List the eight sources of connection loss in fiber.

36. Compare the advantages and disadvantages of fusion and mechanical splicing. Which would you select if you were splicing many fiber strands? Explain your choice.

37. List the three most popular fiber-end connectors.

38. Describe the procedure for preparing the fiber for splicing or connectorization.

39. What are the general rules for splicing single-mode and multimode fiber together?

SECTION 16-7

40. What are the primary performance issues when designing a fiber-optic transmission link?

41. Define a *long-haul* and a *local-area network*.

42. Define *received signal level (RSL)*.

43. Define *maintenance margin*.

44. When testing a fiber with an optical time-domain reflectometer (OTDR), it was determined that the actual length of fiber used for a 20-km span was 20.34 km. Is this actually possible and why?

45. Determine the cable loss in dB for a 10-km fiber run. The fiber has a loss of 0.4 dB/km. (4 dB)

46. For power budgeting of fiber-optic transmission systems, what are four components that contribute to power loss between a transmitter and a receiver?

SECTION 16-8

47. List four tips for installing fiber-optic cable.

48. What is an OTDR and how is it used?

49. Examine the OTDR trace provided in Figure 16-32. Explain the trace behavior at points A, B, C, D, and E.

SECTION 16-9

50. What are the changes in optical solutions that may greatly affect the design of optical networks?

51. Define *fiber to the curb (FTTC)* and *fiber to the home (FTTH)*.

52. What is OC-192?

53. Describe how fiber can be used in a local-area network (LAN) to increase the data capacity and potentially minimize congestion problems.

54. What is Fiber Distributed Data Interface (FDDI)? Provide a basic description of an FDDI system.

55. A seven-station FDDI system has spacings of 150 m, 17 m, 270 m, 235 m, and 320 m for six of the stations. Determine the maximum spacing for the seventh station. (208 m)

QUESTIONS FOR CRITICAL THINKING

56. Analyze the numerical aperture (NA) and cutoff wavelengths for single-mode fiber with a core of 2.5 μm and refraction indexes of 1.515 and 1.490 for core and cladding, respectively. (0.274, 1.73 μm)

57. A system operating at 1550 nm exhibits a loss of 0.35 dB/km. If 225 μW of light power is fed into the fiber, analyze the received power through a 20-km section. (44.9 μW)

58. A fiber-optic system uses a cable with an attenuation of 3.2 dB/km. It is 1.8 km long and has one splice with an 0.8-dB loss. Due to the source/receiver connection, it has a 2-dB loss at both transmitter and receiver. It requires 3 μW of received optical power at the detector. Report on the level of optical power required from the light source. (34.1 μW)

59. Provide a complete power budget analysis for a system with the following losses and specifications:

Losses:

Pigtail losses:	6.5 dB
Two connections:	1.0 dB each
Three splices:	0.5 dB each
20 km of fiber:	0.35 dB/km

Specifications:

Laser power output:	−2 dBm
Minimum RSL:	−33 dBm
Maximum RSL:	−22 dBm
Maintenance margin:	3 dB
Power margin:	1 dB
Operational margin:	3 dB

FCC General Radiotelephone Operator License (GROL) Requirements

After completing a course in electronics communications you will be ready to apply for and take the FCC General Radiotelephone Operator License (GROL) examination. The GROL is an internationally recognized industry benchmark that can be used to demonstrate your knowledge and skill in the field of electronics communications. The GROL is required to adjust, maintain, or internally repair FCC licensed radiotelephone transmitters in the aviation, maritime, and international fixed public radio services. It conveys all the operating authority of the Marine Radio Operators Permit (MROP).

The GROL is required to operate the following:

- Any maritime land radio station or compulsorily equipped ship radiotelephone station operating with more than 1500 watts of peak envelope power
- Voluntarily equipped ship and aeronautical (including aircraft) stations with more than 1000 watts of peak envelope power.

Even if you do not plan to work on equipment in an environment where a GROL is required by law, possession of the license often provides a competitive advantage when job-hunting. Many employers use the GROL as a means of establishing a level of competency when screening applicants for positions within the communications field.

To be eligible for a GROL you must

1. be a legal resident of (or otherwise eligible for employment in) the United States;
2. be able to receive and transmit spoken messages in English; and
3. pass the *Written Examination for Element 1* (or provide proof of a current Marine Radio Operator Permit) and *Written Examination for Element 3*.

Written Element 1

This examination covers basic radio law and operating procedures with which maritime radio operators must be familiar. Questions on this element concern provisions of laws, treaties, regulations, and operating procedures and practices generally followed or required in communicating by means of radiotelephone stations. There are 144 questions in the practice pool with 24 selected for the examination. The minimum passing score is 75% (18 questions answered correctly).

Written Element 3

This examination is composed of questions concerning electronic fundamentals and techniques required to adjust, repair, and maintain radio transmitters and receivers at stations licensed by the FCC in the aviation, maritime, and international fixed public radio services. There are 600 questions in the practice pool with 100 selected for the examination. The minimum passing score is 75% (75 questions answered correctly).

The following list shows the subject-matter breakdown for Element 3 questions. The first number in parentheses represents the total number of questions on this subject in the practice pool, while the second number represents the actual number of questions on this subject in the Element 3 examination. Where this subject is typically covered is also shown. Note that almost 50% of all Element 3 questions are covered in this text.

- Principles (48, 8) Covered in DC/AC and Semiconductor texts
- Electrical Math (60, 10) Covered in DC/AC texts
- Components (60, 10) Covered in AC and Semiconductor texts
- Circuits (24, 4) Covered in AC, Semiconductor texts and Chapter 4 of this text
- Digital Logic (48, 8) Covered in Digital Logic texts
- Receivers (60, 10) Covered in Section 1.5, 4.3 and Chapter 6 of this text
- Transmitters (36, 6) Covered in Chapter 5 of this text
- Modulation (18, 3) Covered in Chapters 2 and 3 of this text
- Power Sources (18, 3) Covered in DC/AC texts
- Antennas (30, 5) Covered in Chapters 13 and 14 of this text
- Aircraft (36, 6) Covered in Chapter 13 of this text and Aircraft texts
- Installation, Maintenance & Repair (48, 8) Covered in DC/AC texts
- Communications Technology (18, 3) Covered in Section 10.3 of this text
- Marine (30, 5) Covered in Marine texts, and Section 5.3, 6.3, and 7.6 of this text
- Radar (30, 5) Covered in Microwave texts and Chapter 15 of this text
- Satellite (24, 4) Covered in Microwave texts
- Safety (12, 2) Covered in Electronic lab manuals and Section 16.10 of this text

Written Element 8 (Optional)

The Ship Radar Endorsement is an optional endorsement placed only on the GROL or on First or Second Class Radiotelegraph Operator's certificates. Only persons whose commercial radio operator license bears this endorsement may repair, maintain, or internally adjust ship radar equipment.

To be eligible for this endorsement, you must

1. hold (or qualify for) a First or Second Class Radiotelegraph Operator's Certificate or a GROL, and
2. pass the Written Element 8 examination.

The Element 8 examination consists of 50 questions from a practice pool of 300 questions covering specialized theory and practice applicable to the proper installation, servicing and maintenance of ship radar equipment in general use for marine navigational purposes. The minimum passing score is 75% (38 questions answered correctly).

Before the FCC Exam

1. **Preparation.**
   Obtain current study materials.
   Please note that private contractors administer all GROL exams using questions and test procedures established by the FCC. Several organizations act as test administrators; an up-to-date list and schedule of fees can be found at http://wireless.fcc.gov/commoperators/index.htm?job=cole. One is the National Association of Radio and Telecommunications Engineers (NARTE). Current information can be found at www.narte.org. You can also check out Prep Courses on the iNARTE Test Center Listing. Your local school or college may offer one.

2. **Choose your testing time and location.**
   If using the NARTE GROL service, find your location in the iNARTE Test Center listing or the on-line Registration Form.

3. **Register with iNARTE.**

   Registering with iNARTE costs US $65 per sitting, for up to three exam Elements per sitting. The sitting time limit is four hours. Example: For $65, you can register to take Elements 1, 3 and 8. You will have a maximum of 4 hours to complete all three exams. You may use the online registration form, or mail your registration form and $65 (non-refundable) testing fee to iNARTE to 840 Queen Street, New Bern, NC 28560.

4. **iNARTE will coordinate your examination appointment.**

   Written confirmation of your exam appointment will be sent to you by mail.

5. **Plan to make your FCC exam appointment.**

   Examinations are usually held on weekdays. iNARTE's exam fee is non-refundable. Examinees scheduling their examination date directly with the Test Center must schedule the examination within 30 days from receipt of their authorization letter from iNARTE. Otherwise, rescheduling of an examination will require a new registration and payment of the $65 registration fee. A scheduled examination may be rescheduled ONE TIME without an additional charge by notification to iNARTE at least 7 business days prior to the examination. Rescheduling with less than 7 days notification requires that you pay an additional fee unless you can show undue hardship or extenuating circumstances beyond your control (subject to approval by iNARTE). Rescheduling of the exam must be done within 30 days.

6. **The GROL is issued for the lifetime of the holder.**

ACRONYMS AND ABBREVIATIONS

A

AAL	ATM adaptation layer
ac	alternating current
ACA	adaptive channel allocation
ACIL	trade association (formerly the American Council of Independent Laboratories)
ACK	acknowledgment
ACL	advanced CMOS logic
ACM	address complete message
ACR	attenuation-to-crosstalk ratio
ADC	analog-to-digital; analog-to-digital converter
ADCCP	advanced digital communications control protocol
ADSL	asymmetric digital subscriber line
AF	audio frequency
AFC	automatic frequency control
AFSK	audio-frequency shift keying
AGC	automatic gain control
AGCH	access grant channel
AIAA	American Institute of Aeronautics and Astronautics
AlGaAs	aluminum gallium arsenide
ALC	automatic level control
ALU	arithmetic logic unit
AM	amplitude modulation
AMI	alternate-mark inversion
AML	automatic-modulation-limiting
AMPS	Advanced Mobile Phone Service
ANL	automatic noise limiter
ANM	answer message
ANSI	American National Standards Institute
APC	angle-polished connectors
APCO	Association of Public-Safety Communications Officials
APD	avalanche photodiode
AP-S	Antennas and Propagation Society
ARPA	Advanced Research Projects Agency (now DARPA)
ARQ	automatic repeat request
ARRL	American Radio Relay League
ASCII	American Standard Code for Information Interchange
ASIC	application-specific integrated circuit
ASK	amplitude-shift keying

ASSP	application-specific standard products
ATC	adaptive transform coding
ATE	automatic test equipment
ATG	automatic test generation
ATM	asynchronous transfer mode
ATSC	Advanced Television Systems Committee
ATV	advanced television
AWGN	additive white Gaussian noise

B

b	bit
B	byte
B8ZS	bipolar 8 zero substitution
BAW	bulk acoustic wave
BBNS	broadband network services
BCC	block check character
BCCH	broadcast control channel
BCH	Bose–Chaudhuri–Hocquenghem (BCH) codes
BCD	binary-coded decimal
B-CDMA	broadband CDMA
BCI	broadcast interference
BeCu	beryllium copper
B8ZS	bipolar 8 zero substitution
BER	bit error rate
BERT	bit-error-rate tester
BFO	beat-frequency oscillator
BiCMOS	bipolar-CMOS
BIOS	basic input/output system
BIS	buffer information specification
B-ISDN	broadband integrated-services digital network (an ATM protocol model)
BJT	bipolar junction transistor
BOP	bit-oriented protocol
bps	bits per second
BPSK	binary phase-shift keying
BPV	bipolar variation
BRI	basic-rate interface
BS	base station
BSC	base-station controller
BSS	Broadcasting Satellite Service
BW	bandwidth
BWA	broadband wireless access
BWO	backward-wave oscillator

C

CAD	computer-aided design
CAE	computer-aided engineering
CAM	computer-aided manufacturing
CAT	computer-aided test
CAT5	category 5
CAT6/5e	category 6 and category 5e
CATV	community-access (cable) television
CBCH	cell broadcast channel
CBIR	committed burst information rate
CCA	clear-channel assortment
CCD	charge-coupled device
CCITT	Consultative Committee on International Telephone & Telegraph
CCK	complementary code keying
CCW	counterclockwise
CD	carrier detect *or* compact disc
CDM	code-division multiplex
CDMA	code-division multiple access
CDMA2000	a 3G wireless development popular in the United States
CDPD	cellular digital packet data
CE	compliance engineering
CELP	code-excited linear prediction (coding)
CHBT	complementary heterojunction bipolar transistor
C/I	carrier/interference ratio
CIC	circuit identification code
CMOS	complementary metal-oxide semiconductor
CMRR	common mode rejection ratio
C/N	carrier-to-noise
CODEC	coder/decoder
COFDM	OFDM with channel coding
COP	character-oriented protocol
CPE	customer premise equipment
CPU	central processing unit
CRC	cyclic redundancy check
CSMA	carrier sense multiple access
CSMA/CA	carrier sense multiple access collision avoidance
CSMA/CD	CSMA with collision detection
CSU/DSU	channel service unit/data service unit
CTCSS	Continuous Tone Coded Squelch System
CTI	computer telephone integration
CTIA	Cellular Telecommunications Industry Association
CTS	clear-to-send
CT2	second-generation cordless telephone
CVBS	composite video blanking and synchronization
CVD	chemical-vapor deposition
CW	continuous wave *or* clockwise

D

DAC	digital-to-analog; digital-to-analog converter
DARPA	Defense Advanced Research Projects Agency
DARS	Digital Audio Radio Service
DAS	data-acquisition system
dB	decibel
dBc	decibels with respect to carrier
dBi	antenna gain in decibels, with respect to isotropic antenna
dBm	decibels with respect to 1 mW
DBPSK	differential binary phase-shift keying
DBR	distributed Bragg reflector
DBS	direct-broadcast satellite
dc	direct current
DCCH	digital control channel
DCD	digital carrier detect
DCE	data communications equipment
DCR	direct current receiver
DCS	Digital Coded Squelch
DDC	direct digital control
DDCMP	digital data communications message protocol
DDS	direct digital synthesizer (or synthesis) *or* digital-data systems
DECT	Digital European Cordless Telecommunications
DELTIC	delay-line time compression
DES	Data Encryption Standard
DF	direction finding
DFB	distributed feedback
DFD	digital frequency discriminators
DI	dielectric isolation
diam	diameter
DIL	dual in-line
DIP	dual in-line package
DL	diode laser
DLVA	detector log video amplifier
DMA	direct memory access
DMM	digital multimeter
DMT	data-modulation technique *or* discrete multitone
DMUX	demultiplexer
DNL	differential nonlinearity
DOCR	digital on-channel repeater
DOD	direct outward dialing
DPC	destination point code
DPDT	double-pole, double-throw
DPSK	differential phase-shift keying
DPST	double pole, single throw
DQPSK	differential quadrature phase-shift keying
DRAM	dynamic random-access memory
DRO	dielectric resonator oscillator
DSL	digital subscriber line
DSO	digital storage oscilloscope
DSP	digital signal processing
DSR	data set ready
DSSS	direct-sequence spread spectrum
DTCXO	digital temperature-compensated crystal oscillator
DTE	data terminal equipment
DTH	digital to home

DTMF	dual-tone multifrequency
DTR	data terminal ready
DTV	digital television
DUT	device under test
DVB	digital video broadcast
DVM	digital voltmeter
DWDM	dense wavelength-division multiplexer

E

E_b	energy per bit *or* bit energy
EBCDIC	Extended Binary Coded Decimal Interchange Code
ECC	error-correction coding
ECL	emitter-coupled logic
EDA	electronic design automation
EDC	error detection and correction
EDFA	erbium-doped fiber amplifier
EDR	Enhanced Data Rate
EEPROM	electrically erasable programmable read-only memory
EHF	extremely high frequency
EIA	Electronic Industries Association
EIRP	effective isotropic radiated power
EISA	extended industry standard architecture
EKG	electrocardiogram
ELF	extremely low frequency
EM	electromagnetic
EMC	electromagnetic compatibility
EMF	electromotive force
EMI	electromagnetic interference
ENOB	effective number of bits
ENR	excess noise ratio
EOM	end-of-message
EPROM	erasable programmable read-only memory
ERP	effective radiated power
ESD	electrostatic discharge
ESF	extended superframe
ESI	equivalent series inductance
ESMR	enhanced specialized mobile radio
ESR	equivalent series resistance
ETACS	extended total access communications systems
ETDMA	enhanced time-division multiple access
E-3	industry standard for ATM (34.736 Mb/s)
ETSI	European Telecommunications Standards Institute
eV	electron volts
EVM	error vector magnitude

F

FACCH	fast associated control channel
F/B	front-to-back
FCC	Federal Communications Commission
FCCH	frequency control channel
FDD	frequency division duplex
FDDI	Fiber-Distributed Data Interface
FDM	frequency division multiplex

FDMA	frequency-division multiple access
FEC	forward error-correction (*or* control)
FEM	finite-element modeling
FER	frame error rate
FET	field-effect transistor
FFSK	fast frequency-shift keying
FFT	fast Fourier transform
FHMA	frequency-hopping multiple access
FHSS	frequency-hopping spread spectrum
FIFO	first-in, first-out
FIR	finite-impulse response
FISU	fill-in signaling unit
FITS	failures in 10^9 hours
FLOPs	floating-point operations
FM	frequency modulation or frequency-modulated
4FSK	four-level frequency-shift keying
FPGA	field-programmable gate array
FQPSK	filtered quadrature phase-shift keying
FSF	frequency scaling factor
FSK	frequency-shift keying
FSR	full-scale range
FTTC	fiber to the curb
FTTH	fiber to the home
FT1	fractional T1

G

GaAs	gallium arsenide
GEO	geostationary earth orbit
GFSK	Gaussian frequency-shift keying
GHz	gigahertz
GMSK	Gaussian minimum-shift keying
GPS	Global Positioning System
GROL	General Radiotelephone Operator License
GSGSG	ground-signal ground-signal ground
GSM	Global System for Mobile Communications
GSSG	ground-signal signal-ground
G/T	figure of merit

H

HBT	heterojunction bipolar transistor
HDLC	high-level data link control
HDSL	high-bit-rate digital subscriber line
HDTV	high-definition television
HEMT	high-electron mobility transistor
HF	high frequency
HFC	hybrid fiber coaxial
HPA	high-power amplifier
HSUPA	high-speed download packet access
HTS	high-temperature superconductor
HVAC	heating, ventilating, and air conditioning
Hz	Hertz, originally cycles per second

I

IAGC	instantaneous automatic gain control
IAM	initial address message
IANA	Internet Assigned Numbers Authority

IBOC	in-band on-channel
IBIS	input/output buffer information specification
IC	integrated circuit
ICW	interrupted continuous wave
IDP	integrated detector preamplifer
IDSL	ISDN digital subscriber line
IEEE	Institute of Electrical and Electronics Engineers
IESS	Intelsat Earth Station Standards
IF	intermediate frequency
IFFT	inverse FFT
IFM	instantaneous frequency measurement
IHFM	Institute of High Fidelity Manufacturers
IIR	infinite impulse response
IM	intermodulation
IMD	intermodulation distortion
IMPATT	impact ionization avalanche transit time diode
IMTS	improved mobile telephone service
IMT-2000	international mobile telecommunications
InGaAs	indium gallium arsenide
INL	integral nonlinearity
InP	indium phosphide
INTELSAT	International Telecommunications Satellite Organization
I/O	input/output
IOC	integrated optical circuit
IP	Internet protocol
I/Q	in-phase/quadrature
IQST	INTELSAT qualified satellite terminals
IR	infrared
IrDA	Infrared Data Association
IS	international standards
ISA	industry-standard architecture
ISDN	integrated services digital network
IS-54	Interim Standard 54 (dual-mode TDMA/AMPS)
ISHM	International Society for Hybrid Microelectronics
ISI	intersymbol interference
ISL	intersatellite link
ISM	industrial, scientific, and medical
IS-95	Interim Standard 95 (dual-mode CDMA/AMPS)
ISP	Internet service provider
ISUP	ISDN user part
ITFS	instructional television fixed services
ITS	intelligent transportation systems
ITU	International Telecommunication Union
ITV	instructional television

J

JDC	Japanese digital cellular
JFET	junction field-effect transistor

K

kHz	kilohertz

L

LAN	local-area network
LC	inductive-capacitive *or* liquid crystal
LCC	leadless ceramic chip carrier
LCD	liquid-crystal display
LDCC	leaded ceramic chip carrier
LDMOS	laterally diffused metal oxide silicon
LED	light-emitting diode
LEO	low Earth orbit satellite
LF	low frequency
LHCP	left-hand circular polarization
LiIon	lithium ion
LMDS	local multichannel distribution system
LMR	land mobile radio
LNA	low-noise amplifier
LNB	low-noise block down-converter
LNBF	low-noise block feedhorn
LO	local oscillator
LOS	line of sight
LPF	low-pass filter
LPTV	low-power television
LRC	longitudinal redundancy check
lsb	least-significant bit
LSI	large-scale integration
LSSU	link status signaling unit
LTE	long-term evolution
LTP	long-term prediction
LVDS	low-voltage differential signaling

M

MAC	media access control
MAP	mobile application part
MBE	molecular beam epitaxy
MCA	multichannel amplifier
MCW	modulated continuous wave
MDAC	multiplying digital-to-analog converter
MDS	multipoint distribution systems
MDSL	medium-speed digital subscriber line
MER	modulation error ratio
MESFET	metal semiconductor field-effect transistor
MFLOPS	million floating-point operations per second
MIC	microwave integrated circuit
MIL	military specification
MIMO	multiple-input multiple-output
MIPS	million instructions per second
MMDS	multichannel, multipoint distribution systems
MMIC	monolithic microwave integrated circuit
MOCVD	metal-organic chemical-vapor deposition
modem	modulator/demodulator
MOS	metal-oxide semiconductor
MOSFET	metal-oxide semiconductor field-effect transistor
MPSD	masked-programmable system devices
MPSK	minimum phase-shift keying
MROP	Marine Radio Operators Permit
MSA	metropolitan statistical area
MSAT	mobile satellite

msb	most significant bit
MSK	minimum-shift keying
MSPS	million samples per second
MSAT	mobile satellite
MSS	mobile satellite service
MTA	major trading area
MTBF	mean time between failures
MTP	message transfer part
MTSO	mobile telephone switching office
MTTF	mean time to failure
MUF	maximum usable frequency
MUX	multiplexer
MVDS	microwave video-distribution system

N

NA	numerical aperture
NAB	National Association of Broadcasters
NADC	North American Digital Cellular
NAK	negative acknowledgment
NAMPS	narrowband Advanced Mobile Phone Service
NARTE	National Association of Radio and Telecommunications Engineers
NASA	National Aeronautics and Space Administration
NBX	network branch exchange
NCO	numerically controlled oscillator
NEMA	National Electrical Manufacturers Association
NEMI	National Electronics Manufacturing Initiative, Inc.
NEXT	near-end crosstalk
NF	noise figure
NIC	network interface card
NiCd	nickel cadmium
NiMH	nickel metal hydride
NIST	National Institute of Standards & Technology (formerly NBS)
NLSP	network-link services protocol
NLOS	nonline-of-sight
NMT-900	Nordic Mobile Telephone
NNI	network-node interface
NRZ	nonreturn-to-zero code
NRZI	nonreturn-to-zero inverted code
NRZ-L	nonreturn-to-zero low
NTSC	National Television Systems Committee (US television broadcast standard)

O

OC	optical carrier
OC-48	2.4-Gb/s optical-carrier industry standard
OC-192	Optical Carrier 192
OCR	optical character recognition
OC-3	155-Mb/s optical-carrier industry standard
OC-12	622-Mb/s optical-carrier industry standard
OCXO	oven-controlled crystal oscillator
OEM	original-equipment manufacturer
OFDM	orthogonal frequency-division multiplexing

OOK	on-off keying (modulation)
OMAP	operations, maintenance, and administration part
op-amp	operational amplifier
OPC	origination point code
OPDAR	optical radar
OQPSK	offset quadrature phase-shift keying *or* orthogonal quadrature phase-shift keying
OSI	open system interconnection
OTA	over the air
OTDR	optical time-domain reflectometer

P

PABX	private automatic branch exchange
PACS	personal advanced communications systems
pACT	personal Air Communications Technology
PAE	power-added efficiency
PAL	phase-alternation-line (a 625-line 50-field color television system)
PAM	pulse-amplitude modulation
PAN	personal-area network
PBX	private branch exchange
PC	personal computer
PC	convex-polished
PCB	printed circuit board
PCH	paging channel
PCI	peripheral component interconnect
PCIA	Personal Communications Industry Association
PCM	pulse-code modulation
PCMCIA	Portable Computer Memory Card International Association
PCN	personal communications network
PCS	personal communications services *or* plastic-clad silica (fiber)
PCU	programmer control unit
PDA	personal digital assistant
PDBM	pulse-delay binary modulation
PDC	personal digital cordless
PDF	probability density function
PDH	piesiochronous digital hierarchy
PDM	pulse-duration modulation
PECL	positive emitter-coupled logic
PEP	peak envelope power
PFM	pulse-frequency modulation
PGBM	pulse-gated binary modulation
PHEMT	pseudomorphic high-electron-mobility transistor
PHP	Personal HandyPhone
PHS	Personal HandyPhone System
PICD	personal information and communication device
PIM	passive intermodulation
PIN	personal identification number *or* positive-intrinsic-negative
pixel	picture element
PLCC	plastic leaded-chip carrier
PLD	programmed logic device

PLL	phase-locked loop
PLM	pulse-length modulation
PLMR	public land mobile radio
PLO	phase-locked oscillator
PM	phase modulation
PMR	professional mobile radio
PN	pseudonoise
PolSK	polarization-shift keying
POTS	plain old telephone service
p-p	peak-to-peak
PPB	parts per billion
PPBM	pulse-polarization binary modulation
PPI	pulse-position-indicator
PPM	parts per million *or* periodic permanent magnet *or* pulse-position modulation
PPP	point-to-point protocol
PQFP	plastic quad-leaded flat pack
PRBS	pseudorandom-bit sequence
PRF	pulse repetition frequency
PRI	pulse repetition interval
PRK	phase-reversal keying
PRL	preferred roaming list
PRML	partial-response maximum likelihood
PRR	pulse repetition rate
PRT	pulse repetition time
PSIP	Program and System Information Protocol
PSK	phase-shift keying
PSNEXT	power sum NEXT test
PSTN	public switched telephone network
PTFE	polytetrafluoroethylene
PTM	pulse-time modulation
PVC	permanent virtual connection
PWM	pulse-width modulation

Q

Q	quality factor
QAM	quadrature amplitude modulation
QFP	quad flat pack
QoS	quality of service
QPSK	quadrature phase-shift keying
QSOP	quarter-sized outline package
QUIL	quad in-line

R

RAC	reflective array compressor
RACH	random access channel
RADAR	radio detecting and ranging
RAM	random-access memory
RC	resistance-capacitance
RD	receive data
REL	release message
RF	radio frequency
RFC	radio-frequency choke
RFI	radio-frequency interference
RFID	radio-frequency identification
RHCP	right-hand circular polarization
RI	ring indicator
RIC	remote intelligent communications

RIS	random interleaved sampling
RISC	reduced-instruction set computer
RLC	release complete message
rms	root mean square
ROM	read-only memory
RPE-LPC	regular pulse excitation-linear prediction coding
RS	Reed-Solomon
RSA	rural statistical area
RS-422, RS-485	balanced-mode serial communications standards that support multidrop applications
RSL	received signal level
RSSI	received signal-strength indicator
RTS	request-to-send
RZ	return-to-zero code

S

SACCH	slow-associated control channel
SAT	signal-audio tone
SATCOM	satellite communications
SAW	surface acoustic wave
SCADA	supervisory control and data-acquisition systems
SCCP	signaling connection control part
SCH	synchronization
SCPI	Standard Commands for Programmable Instruments *or* small-computer programmable instrument
SCR	silicon-controlled rectifier
SCSA	Signal Computing Systems Architecture [industry-standard architecture for deploying computer telephone integration (CTI)]
SCSI	small computer system interface
SDCCH	standalone dedicated control channel
SDH	synchronous digital hierarchy
SDLC	synchronous data link control
SDMA	space-division multiple access
SDR	signal-to-distortion ratio *or* software-defined radio
SDSL	single-line digital subscriber line
SDTV	standard definition television
SER	segment error rate
SFDR	spurious-free dynamic range
SG	signal ground
S/H	sample and hold
SHF	super-high frequency
SIM	subscriber identity module
SIMOX	separation by implantation of oxygen
SINAD	signal plus noise and distortion
SLA	sealed lead acid
SLIC	subscriber-line interface circuit
SMART	system monitoring and remote tuning
SMD	surface-mount device
SMP	surface-mount package
SMPTE	Society of Motion Picture and Television Engineers

SMR	specialized land-mobile radio
SMSR	side-mode suppression ratio
SMT	surface-mount technology or surface-mount toroidal
S/N	signal to noise
SNR	signal-to-noise ratio
SOE	stripline opposed emitter (package)
SOI	silicon-on-insulator
SOIC	small-outline integrated circuit
SONET	Synchronous Optical Network
SOS	silicon on sapphire
SPDT	single pole, double throw
SPICE	Simulation Program with Integrated Circuit Emphasis
SPST	single pole, single throw
SRAM	static random-access memory
SS	spread spectrum
SS7	Signaling System 7 (a signaling system used to administer the PSTN)
SSB	single sideband
SS/TDMA	satellite-switched TDMA
SSTV	slow-scan television
STM-1	synchronous transmission module, level one
STS	synchronous transport signal
SVC	switched virtual circuit
S-video	separate luminance and chrominance
SWR	standing wave ratio

T

TACS	Total Access Communication System (U.K. analog)
T&M	test and measurement
TBR	Technical Basis for Regulation (European TETRA Standards)
TC	temperature coefficient
TCAP	transactions capabilities application part
TCP/IP	Transmission Control Protocol/Internet Protocol
TCR	temperature coefficient of resistance
TCVCXO	temperature-compensated voltage-controlled crystal oscillator
TCXO	temperature-compensated crystal oscillator
TD	transmit data
TDD	time-division duplex
TDM	time-division multiplex
TDMA	time-division multiple access
TDR	time-domain reflectometer
TD-SCDMA	time-division, synchronized CDMA
TE	transverse electric
TEM	transverse electromagnetic
TETRA	trans-European trunked radio system (for public service applications)
THD	total harmonic distortion
3D	three-dimensional
3G	the third generation in wireless connectivity
TIA	Telecommunications Industry Association

TIMS	transmission-impairment measurement set (an interface for PCM)
TIR	total internal reflection
TQFP	thin-quad flat pack
TR	transmit/receive
TRF	tuned radio-frequency
TSS	tangential signal sensitivity
TSSOP	thin-shrink small-outline package
TT&C	telemetry, tracking, and control (or command)
TTC&M	telemetry, tracking, control, and monitoring
T-3	ATM industry standard—44.736 Mb/s
TTL	transistor-transistor logic
TVI	television interference
TVRO	television receive only
2D	two-dimensional
TWT	traveling wave tube
TWTA	traveling wave tube amplifier

U

UART	universal asynchronous receiver-transmitter
UDLT	universal digital-loop transceiver
UHF	ultra-high frequency
UMTS	Universal Mobile Telephone Service
U-NII	unlicensed national information infrastructure
USB	universal serial bus
UTOPIA	Universal Test and Operations Physical-Layer Interface for ATM
UTP	unshielded twisted-pair

V

VA	voltampere
VAR	volt-ampere reactive
VBT	variable-bandwidth tuning
VCC	virtual channel connection
VCI	virtual channel identifier
VCO	voltage-controlled oscillator
VCSEL	vertical cavity surface-emitting laser
VCXO	voltage-controlled crystal oscillator
VDSL	variable-bit-data-rate digital subscriber line
V/F	voltage-to-frequency
VGA	video graphics array
VHF	very-high frequency
V/I	voltage/in-current
VLB	vesa local bus
VLF	very-low frequency
VNA	vector network analyzer
VPC	virtual path connection
VPI	virtual path identifier
VPSK	variable phase-shift keying
VSAT	very small aperture terminal
VSB	vestigial sideband (modulation)
VSWR	voltage-standing-wave ratio
VVA	voltage-variable attenuator

W

WAN	wide-area network
WAP	wireless application protocol
WCDMA	wideband code division multiple access
WCPE	wireless customer premises equipment
WDM	wavelength-division multiplex(er)
WiMAX	Worldwide Interoperability for Microwave Access
WLA	wireless LAN adapter
WLAN	wireless local-area network
WLL	wireless local loop
WML	wireless markup language
WTA	wireless transaction protocol
WTLS	wireless transport layer security
WTP	wireless transaction protocol

X

X.25	a packet-switched protocol designed for data transmission over analog lines
xDSL	a generic type of digital subscriber line
xoR	exclusive-OR

Y

YAG	yttrium-aluminum garnet
YIG	yttrium-iron garnet

Z

ZC	ZigBee coordinator
ZED	ZigBee end device
ZR	ZigBee router

GLOSSARY

1/*f* filter a type of low-pass filter used to offset the additional frequency deviation created by the intelligence signal frequency in indirect FM (phase modulated) transmitters. Also known as a predistorter or frequency-correcting network

12-dB SINAD industry-standard measurement of the minimum usable sensitivity of a narrowband FM receiver in which the received signal is 12 dB higher than the sum of noise and distortion

absorption as applied to fiber optics, the loss of light energy caused, in part, by the interaction of light energy with hydroxyl ions dissolved in the glass core of the fiber-optic line

acquisition time amount of time it takes for the hold circuit to reach its final value

active attack the bad guy is transmitting an interfering signal disrupting the communications link

aliasing distortion the distortion that results if Nyquist criteria are not met in a digital communications system using sampling of the information signal; the resulting alias frequency equals the difference between the input intelligence frequency and the sampling frequency

aliasing errors that occur when the input frequency exceeds one-half the sample rate

alternate mark inversion (AMI) line coding format in which each logic-one level is represented by a pulse whose polarity is opposite to that of the pulse preceding it

American Standard Code for Information Interchange (ASCII) an industry standard representation of alphanumeric characters

amplitude companding process of volume compression before transmission and volume expansion after detection

amplitude modulation (AM) the process of impressing low-frequency intelligence onto a high-frequency carrier so that the instantaneous changes in the amplitude of the intelligence produce corresponding changes in the amplitude of the high-frequency carrier

antenna array group of antennas or antenna elements arranged to provide the desired directional characteristics

antenna gain a measure of how much more power in dB an antenna will radiate in a certain direction with respect to that which would be radiated by a reference antenna, i.e., an isotropic point source or dipole

aperture time the time that the S/H circuit must hold the sampled voltage

apogee farthest distance of a satellite's orbit to earth

Armstrong a type of LC oscillator in which transformer coupling is used to provide in-phase feedback to the input of a gain stage at the desired frequency of operation

asymmetric DSL a form of digital subscriber line data communication where the uplink and downlink data rates are unequal

asymmetric operation a term used to describe the modem connection when the data transfer rates to and from the service provider differ

asynchronous system the transmitter and receiver clocks free-run at approximately the same speed

asynchronous transfer mode (ATM) a cell relay network designed for voice, data, and video traffic

asynchronous a mode of operation implying that the transmit and receiver clocks are not locked together and the data must provide start and stop information to lock the systems together temporarily

atmospheric noise external noise caused by naturally occurring disturbances in the Earth's atmosphere

attenuation distortion in telephone lines, the difference in gain at some frequency with respect to gain at a reference tone of 1004 Hz

attenuation the loss of power as a signal propagates through a medium such as copper, fiber, and free space

attitude controls used for orbital corrections (station keeping) on the satellite

automatic frequency control (AFC) negative feedback control system in FM receivers used to achieve stability of the local oscillator

auxiliary AGC diode reduces receiver gain for very large signals

backhoe fading total loss of data flow because the cable was dug up by a backhoe

backscatter refers to the reflection of the radio waves striking the RFID tag and reflecting back to transmitter source

balanced line the same current flows in each of two wires but 180° out of phase

balanced mode neither wire in the wire pairs connects to ground

balanced modulator modulator stage that mixes intelligence with the carrier to produce both sidebands with the carrier eliminated

baluns circuits that convert between balanced and unbalanced operation

bandwidth term used to express a usable frequency range. (a) In the context of filters, bandwidth defines the range of frequencies above a predefined minimum power level (usually -3 dB) passed from input to output; (b) the range of frequencies occupied by modulated signal

Barkhausen criteria two requirements for oscillations: loop gain must be at least unity and loop phase shift must be zero degrees

base modulation a modulation system in which the intelligence is injected into the base of a transistor

base station generates the signal that enables communication between the mobile and the phone system

baseband the signal is transmitted at its base frequency with no modulation to another frequency range. Entire bandwidth of transmission medium is used to carry one signal

baud unit expressing the rate of modulation by multiple-bit symbols

Baudot code fairly obsolete coding scheme for alphanumeric symbols

bit energy the amount of power in a digital bit for a given amount of time

broadband transmission method in which multiple signals from multiple channels are transmitted simultaneously over a single medium; also, in radio-frequency communications, an expression of the frequency-spreading effect produced by modulation (cf. baseband)

broadband wireless access (BWA) method for providing "last-mile" high-data-rate connectivity by means of WiMAX

carrier sense multiple access with collision avoidance (CSMA/CA) access protocol used by local-area networks in which packet collisions are avoided by having network devices wait for an acknowledgment that a packet has arrived intact before initiating another transmission

category 6 (CAT6) specification for unshielded twisted-pair cable intended to support data rates up to 1000 Mbps for a maximum length of 100 m

channel path for signal transmission

character insertion addition of bits to a data signal for the purposes of synchronization and control character recognition (also known as character-stuffing or bit-stuffing)

Class A as applied to amplifiers, a designation that the active device conducts current for the full 360-degree conduction angle of the applied signal; the most linear but least efficient form of amplification

Class B conduction of current by the active amplifying device for 180 degrees of the input cycle

Class D an amplifier designation indicating that active devices are switched alternately between saturation and cutoff

composite in the context of M-ary systems, the combination of orthogonal, in-phase (I) and quadrature (Q) data streams in preparation for modulation

conduction angle portion of input cycle in which an active device conducts current; used to define classes of amplification

continuous wave a type of transmission where a continuous sinusoidal waveform is interrupted to convey information

core the portion of the fiber strand that carries the light

corona discharge luminous discharge of energy by an antenna caused by ionization of the air around the surface of the conductor

cosmic noise space noise originating from stars other than the sun

counter measures as applied to data security, cryptography to prevent eavesdropping

counterpoise reflecting surface of a monopole antenna if the actual Earth ground cannot be used; a flat structure of wire or screen placed a short distance above ground with at least a quarter-wavelength radius

conversion frequency another name for the carrier in a balanced modulator

critical angle the highest angle with respect to a vertical line at which a radio wave of a specified frequency can be propagated and still be returned to the earth from the ionosphere

critical frequency the highest frequency that will be returned to the earth when transmitted vertically under given ionospheric conditions

Crosby systems FM systems using direct FM modulation with AFC to control for carrier drift

cross-modulation distortion that results from undesired mixer outputs

crossover distortion distortion produced when both active devices in a push-pull stage are biased off

crosstalk unwanted coupling between adjacent conductors by overlapping electric and magnetic fields

cyclic prefix the end of a symbol is copied to the beginning of the data stream, thereby increasing its overall length and thus removing any gaps in the data transmission

cyclic redundancy check (CRC) method of error detection involving performing repetitive binary division on each block of data and checking the remainders

D4 framing the original data framing used in T1 circuits

damped the gradual reduction of a repetitive signal due to resistive losses

dark current the very little current that flows when a *pn* junction is reverse-biased and under dark conditions

data bandwidth compression in QPSK transmission, the compression of more data into the same available bandwidth as compared to BPSK

data communications equipment refers to peripheral computer equipment such as a modem, printer, mouse, etc.

Data Encryption Standard (DES) an encryption method employing a 56-bit encryption key

data terminal equipment (DTE) refers to a computer, terminal, personal computer, etc.

dBd antenna gain relative to a dipole antenna

dBi antenna gain relative to an isotropic radiator

dBm a method of rating power or voltage levels with respect to 1 mW of power

deemphasis process in an FM receiver that reduces the amplitudes of high-frequency audio signals down to their original values to counteract the effect of the preemphasis network in the transmitter

delay distortion when various frequency components of a signal are delayed different amounts during transmission

delay equalizer an *LC* filter that removes delay distortion from signals on phone lines by providing increased delay to those frequencies least delayed by the line, so that all frequencies arrive at nearly the same time

delay line a length of a transmission line designed to delay a signal from reaching a point by a specific amount of time

delay skew measure of the difference in time for the fastest to the slowest pair in a UTP cable

delayed AGC an AGC that does not provide gain reduction until an arbitrary signal level is attained

delta match an impedance matching device that spreads the transmission line as it approaches the antenna

demodulation process of removing intelligence from the high-frequency carrier in a receiver

demultiplexer (DMUX) a device that recovers the individual groups of data from the TDMA serial data stream

dense wave division multiplexing (DWDM) incorporation of the propagation of several wavelengths in the 1550-nm range of a single fiber

detection another term for demodulation

deviation in FM, the amount by which a carrier frequency increases or decreases from its center reference value in response to intelligence

deviation constant definition of how much the carrier frequency will deviate for a given modulating input voltage level

deviation ratio (DR) the maximum possible frequency deviation divided by the maximum input frequency

dibits data sent two bits at a time

dielectric waveguide a waveguide with just a dielectric (no conductors) used to guide electromagnetic waves

difference equation makes use of the present digital sample value of the input signal along with a number of previous input values and possibly previous output values to generate the output signal

differential GPS a technique where GPS satellite clocking corrections are transmitted so that the position error can be minimized

differential phase-shift keying (DPSK) a form of PSK in which any given logical state (bit value) is derived from the logical state immediately preceding it

diffraction the phenomenon whereby waves traveling in straight paths bend around an obstacle

digital subscriber line a wired telephone line capable of passing digital signals at high bit rates

digitized process of converting an analog (continuously varying) waveform to a series of discrete data values

dipole a type of antenna composed of two conductive elements

direct conversion a receiver design in which intelligence is recovered from the output of a single mixer stage rather than from an intermediate frequency mixing product

direct digital synthesis (DDS) frequency synthesizer design that has better repeatability and less drifting but limited maximum output frequencies, greater phase noise, and greater complexity and cost

direct FM a form of frequency modulation in which the modulating signal is applied directly to the frequency-determining stage of the transmitter (usually the oscillator)

directional concentrating antenna energy in certain directions at the expense of lower energy in other directions

director the parasitic element that effectively directs energy in the desired direction

direct sequence spread spectrum (DSSS) a form of wideband modulation in which a carrier is spread by a high-bit-rate spreading code

discrete multitone (DMT) an industry standard data-modulation technique used by ADSL that uses the multiple subchannel frequencies to carry the data

discriminator stage in an FM receiver that creates an output signal that varies as a function of its input frequency; recovers the intelligence signal

dispersion compensating fiber acts like an equalizer canceling dispersion effects and yielding close to zero dispersion in the 1550-nm region

dispersion the broadening of a light pulse as it propagates through a fiber strand

dissipation inverse of quality factor

distributed feedback laser (DFB) a more stable laser suitable for use in DWDM systems

diversity reception transmitting and/or receiving several signals and either adding them together at the receiver or selecting the best one at any given instant

Doppler effect phenomenon whereby the frequency of a reflected signal is shifted if there is a relative motion between the source and the reflecting object

double conversion superheterodyne receiver design with two separate mixers, local oscillators, and intermediate frequencies

double range echoes echoes produced when the reflected beam makes a second trip

double-sideband suppressed carrier output signal of a balanced modulator

double-stub tuner has fixed stub locations, but the position of the short circuits is adjustable to allow a match between line and load

downlink a satellite sending signals to earth

downward modulation the decrease in dc output current in an AM modulator usually caused by low excitation

driven array multi-element antenna in which all the elements are excited through a transmission line

dummy antenna resistive load used in place of an antenna to test a transmitter without radiating the output signal

duty cycle the ratio of pulse width to pulse repetition time

dynamic range (DR) in a PCM system, the ratio of the maximum input or output voltage level to the smallest voltage level that can be quantized and/or reproduced by the converters; for a receiver, the decibel difference between the largest tolerable receiver input level and its sensitivity (smallest useful input level)

earth station the satellite uplink and downlink

E_b/N_o the bit energy to noise ratio

echo signal part of the returning radar energy collected by the antenna and sent to the receiver

effective aperture as applied to a parabolic reflector, a measurement of the effective signal capture area as a function of diameter and reflector efficiency

electrical length the length of a line in wavelengths, not physical length

electromagnetic interference (EMI) unwanted signals from devices that produce excessive electromagnetic radiation

energy per bit, or bit energy amount of power in a digital bit for a given amount of time for that bit

envelope detector another name for diode detector

event a disturbance in the light propagating down a fiber span, which results in a disturbance on the OTDR trace

excess noise noise occurring at frequencies below 1 kHz, varying in amplitude and inversely proportional to frequency

exciter stages necessary in a transmitter to create the modulated signal before subsequent amplification

Extended Binary-Coded Decimal Interchange Code (EBCDIC) standardized coding scheme for alphanumeric symbols

extended superframe (ESF) sequence of 24 bits used to defne multiplexed T1 voice data

external attack an attack by someone who doesn't have access to the network

external noise noise in a received radio signal that has been introduced by the transmitting medium

eye patterns using the oscilloscope to display overlayed received data bits that provide information on noise, jitter, and linearity

fading variations in signal strength that may occur at the receiver over a period of time

far field region greater than $2D^2/\lambda$ from the antenna; effect of induction field is negligible

fast Fourier transform (FFT) a technique for converting time-varying information to its frequency component

feed line transmission line that transfers energy from the generator to the antenna

fiber Bragg grating a short strand of fiber that changes the index of refraction and minimizes intersymbol interference

fiber to the curb (FTTC) a fiber network architecture where optical fibers are terminated at points in the vicinity of subscriber homes but with final connections to the end user's location made with copper wire digital subscriber lines

fiber to the home (FTTH) a fiber network architecture where the optical fibers are terminated at the end user's location

figure of merit a way to compare different earth station receivers

FireWire A (IEEE 1394a) a high-speed serial connection that supports data transfers up to 400 Mbps

FireWire B (IEEE 1394b) a high-speed serial connection that supports data transfers up to 800 Mbps

first detector the mixer stage in a superheterodyne receiver that mixes the RF signal with a local oscillator signal to form the intermediate frequency signal

flash ADC a type of analog-to-digital converter in which a complete digital word is produced with each sample

flash OFDM a spread-spectrum version of OFDM

flat line condition of no reflection; VSWR is 1

flat-top sampling holding the sample signal voltage constant during samples, creating a staircase that tracks the changing input signal

flow control protocol used to monitor and control rates at which receiving devices can accept data

flywheel effect repetitive exchange of energy in an *LC* circuit between the inductor and the capacitor

foldover distortion another term for aliasing

footprint a map of the satellite's coverage area for the transmission back to earth

forward error-correcting error-checking techniques that permit correction at the receiver, rather than retransmitting the data

fractional T1 a term used to indicate that only a portion of the data bandwidth of a T1 line is being used

frame relay a packet switching network designed to carry data traffic over the public data network

framing separation of blocks of data into information and control sections

free space path loss this is a measure of the attenuation of the RF path loss as it propagates through space, the Earth's atmosphere to and from the satellite

frequency deviation amount of carrier frequency increase or decrease around its center reference value

frequency domain representation of signal power or amplitude as a function of frequency

frequency hopping spread spectrum (FHSS) form of wideband modulation in which a carrier is rapidly switched among a large number of possible frequencies

frequency multiplexing process of combining signals that are at slightly different frequencies to allow transmission over a single medium

frequency multipliers amplifiers designed so that the output signal's frequency is an integer multiple of the input frequency

frequency reuse in cellular phones, the process of using the same carrier frequency in different cells that are geographically separated

frequency shift keying a form of data transmission in which the modulating wave shifts the output between two predetermined frequencies

frequency-correcting network another term for 1/*f* filter

frequency-division multiplexing simultaneous transmission of two or more signals on one carrier, each on its own separate frequency range; also called frequency multiplexing

Friiss's formula method of determining the total noise produced by amplifier stages in cascade

front-to-back ratio the difference in antenna gain in dB from the forward to the reverse direction

fundamental the minimum frequency of a waveform, equal to the reciprocal of its period; the first frequency term in a Fourier series

fusion splicing a long-term splicing method where the two fibers are fused or welded together

Gaussian minimum shift keying (GMSK) form of frequency-shift keying employing Gaussian filters and used in the Global System for Mobile Communications cellular standard

generating polynomial defines the feedback paths to be used in the CRC-generating circuit

geosynchronous orbit another name for synchronous orbit

glass, fiber, light pipe common synonymous terms for fiber-optic strand

graded-index fiber the index of refraction is gradually varied with a parabolic profile; the highest index occurs at the fiber's center

Gray code numeric code for representing decimal values from 0 to 9

ground wave radio wave that travels along the Earth's surface

GSM the global system for mobile communications

guard bands 25-kHz bands at each end of a broadcast FM channel to help minimize interference with adjacent stations

guard times time added to the TDMA frame to allow for the variation in data arrival

half-wave an antenna that is electrically one-half wavelength

Hamming code a forward error-checking technique named for R. W. Hamming

Hamming distance the logical distance between defined states; also called minimum distance and D_{min}

handoff the process of changing channels to a new cell site

handshaking procedures allowing for orderly exchange of information between a central computer and remote sites

harmonics sinusoidal waves whose frequencies are a multiple of the fundamental frequency

Hartley's law information that can be transmitted is proportional to the product of the bandwidth times the time of transmission

high-level modulation in an AM transmitter, intelligence superimposed on the carrier at the last possible point before the antenna

hot-swappable a term used to describe that an external device can be plugged in or unplugged at any time

idle channel noise small-amplitude signal that exists due to the noise in the system

image antenna the simulated λ/4 antenna resulting from the Earth's conductivity with a monopole antenna

image frequency undesired input frequency in a superheterodyne receiver that produces the same intermediate frequency as the desired input signal

in-band on-channel (IBOC) original name for HD radio technology

in-band the same physical wires are used to multiplex both the voice traffic and the data traffic required to administer the system

independent sideband transmission another name for twin-sideband suppressed carrier transmission

index-matching gel a jellylike substance that has an index of refraction much closer to the glass than air

indirect FM form of frequency modulation in which modulating signal is applied to a stage subsequent to the frequency-determining oscillator stage

induction field radiation that surrounds an antenna and collapses its field back into the antenna

industrial, scientific, and medical (ISM) regulatory designation for specific frequency bands authorized for unlicensed, low-power operation

information theory the branch of learning concerned with optimization of transmitted information

infrared light extending from 680 nm up to the wavelengths of the microwaves

input intercept another name for third-order intercept point

inquiry procedure used by Bluetooth to discover other Bluetooth devices or to allow itself to be discovered

integrator a low-pass filter

intelligence low-frequency information that can be modulated onto a high-frequency carrier in a transmitter

interleaving generating color information around just the right frequency so that it becomes centered in clusters between the black-and-white signals

intermod intermodulation distortion

intermodulation distortion undesired mixing of two signals in a receiver resulting in an output frequency component equal to that of the desired signal

internal attack an attack by someone inside the network

internal noise noise in a radio signal that has been introduced by the receiver

Internet Assigned Numbers Authority (IANA) the agency that assigns the computer network IP address

intersymbol interference (ISI) the overlapping of data bits that can increase the bit error rate

ionosphere region of Earth's atmosphere consisting primarily of charged atomic particles

IP telephony (voice-over IP) the telephone system for computer networks

isolators in-line passive devices that allow power to flow in one direction only

isothermal region the stratosphere, considered to have a constant temperature

isotropic point source a point in space that radiates electromagnetic radiation equally in all directions

iterative algorithms that employ previous output values to generate the current output

Johnson noise another name for thermal noise, first studied by J. B. Johnson

key the secret code used in the encryption algorithm to both create and decode the message

latency the time delay from the request for information until a response is obtained

lattice modulator another name for balanced ring modulator

leakage loss of electrical energy between the plates of a capacitor

light pipe, glass, fiber common synonymous terms for a fiber-optic strand

limiter stage in an FM receiver that removes any amplitude variations of the received signal before it reaches the discriminator

limiting knee voltage another term for quieting voltage

line control procedure that decides which device has permission to transmit at a given time

linear device device whose behavior is characterized by a straight-line relationship between parameters, such as input versus output or current versus voltage

linear quantization level another name for uniform quantization level

loaded cable cable with added inductance every 6000, 4500, or 3000 feet

loading coil a series inductance used to tune out the capacitive appearance of an antenna or transmission line

lobes small amounts of RF radiation shown on a radiation pattern; generally undesirable

local area network network of users that share computers in a limited area

local loop another name for connection from the central office to the end user

local oscillator reradiation undesired radiation of the local oscillator signal through a receiver's antenna

long haul the intercity or interoffice class of system used by telephone companies and long-distance carriers

longitudinal redundancy check extending parity into two dimensions

look angle azimuth and elevation angles for the earth station antenna

loopback test configuration for a data link; the receiver takes the data and sends it back to the transmitter, where it is compared with the original data to indicate system performance; also, when data are routed back to the sender

low Earth orbit (LEO) satellite satellite orbiting the Earth at altitudes between 250 and 1000 miles

low excitation improper bias or low carrier signal power in an AM modulator

lower sideband band of frequencies produced in a modulator from the creation of difference frequencies between the carrier and information signals

low-level modulation in an AM transmitter, intelligence superimposed on the carrier; then the modulated waveform is amplified before reaching the antenna

low-noise resistor a resistor that exhibits low levels of thermal noise

macrobending loss due to the light escaping into the cladding

M-ary modulation term derived from "binary," expressing the number of possible symbol locations in I/Q space

matrix network adds and/or subtracts and/or inverts electrical signals

maximal length indicates that the PN code has a length of $2^n - 1$

maximum usable frequency the highest frequency that is returned to the Earth from the ionosphere between two specific points on Earth

maximum usable range the maximum distance before second return echoes start occurring in radar

m-derived filters type of filter in which a tuned circuit is used to provide nearly infinite attenuation at a specific frequency; m is the ratio of filter cutoff frequency to the frequency of near-infinite attenuation

mechanical splices splices joining two fibers together with an air gap, thereby requiring an index matching gel to provide a good splice

media access control (MAC) address a unique 6-byte address assigned by the vendor of the network interface card

metropolitan area network two or more LANs linked together in a limited geographical area

microbending loss caused by very small mechanical deflections and stress on the fiber

microbrowser analogous to a web browser that has been adapted for the wireless environment

microstrip transmission line used at microwave frequencies that has one or two conducting strips over a ground plane

microwave dish paraboloid antenna

millmeter (mm) waves microwave frequencies above 40 GHz; wavelength is often expressed in millimeters

minimum distance (D_{min}) the minimum distance between defined logical states

minimum ones density a pulse is intentionally sent in the data stream even if the data being transmitted is a series of 0s only

minimum shift keying (MSK) form of frequency-shift keying in which bandwidth is minimized by ensuring smooth phase transitions from mark to space frequencies or vice versa

mixing multiplication of two or more frequencies (as in a nonlinear device or linear multiplier stage) such that sum and difference frequencies are produced

mobile telephone switching office (MTSO) location where switching equipment for cellular telephone carriers, as opposed to wireline carriers, is present

modal dispersion the different paths taken by the various propagation modes in fiber-optic cable

mode field diameter the actual guided optical power distribution, which is typically about 1 μm or so larger than the core diameter; single-mode fiber specification sheets typically list the mode field diameter

modulated amplifier stage that generates the AM signal

modulation factor another name for modulation index

modulation index measure of the extent to which a carrier is varied by the intelligence

modulation impressing a low-frequency intelligence signal onto a higher-frequency carrier signal

monopole antenna usually a quarter-wave grounded antenna

multilevel binary codes that have more than two levels representing the data

multimode fibers fibers with cores of about 50 to 100 μm that support many modes; light takes many paths

multiple-input multiple-output (MIMO) a arrangement using multiple antennas at both the transmit and receive ends to set up a potentially very large number of wireless paths for data transfer

multiplex operation simultaneous transmission of two or more signals in a single medium

multipoint circuits systems with three or more devices

narrowband FM FM signals used for voice transmissions such as public service communication systems

natural sampling when the tops of the sampled waveform or analog input signal retain their natural shape

near field region less than $2D^2/\lambda$ from the antenna

near-end crosstalk (NEXT) a measure of the level of crosstalk or signal coupling within the cable; a high NEXT (dB) value is desirable

network interface card (NIC) the electronic hardware used to interface the computer to the network

(n, k) cyclic codes nomenclature used to identify cyclic codes in terms of their transmitted code length (n) and message length (k)

neutralizing capacitor a capacitor that cancels fed-back signals to suppress self-oscillation

noise undesired disturbances of a signal such that information content is obscured

noise figure a figure describing how noisy a device is in decibels

noise floor the baseline on a spectrum analyzer display, representing input or output noise of the system under test

noise ratio a figure describing how noisy a device is as a ratio having no units

nonlinear behavior characterized by a non-straight-line relationship between parameters

nonlinear device characterized by a nonlinear output versus input signal relationship

nonline-of-sight (NLOS) path in which a distant location cannot be seen directly because of obstructions or the curvature of the Earth

nonresonant line one of infinite length or that is terminated with a resistive load equal in ohmic value to its characteristic impedance

nonuniform coding another name for nonlinear coding

normalizing as applied to Smith charts, dividing impedances by the characteristic impedance

null modem type of cable in which transmit and receive paths are reversed between ends

null a direction in space with minimal signal level

numerical aperture a number less than 1 that indicates the range of angles of light that can be introduced to a fiber for transmission

Nyquist rate the sampling frequency must be at least twice the highest frequency of the intelligence signal or there will be distortion that cannot be corrected by the receiver

O-, E-, S-, C-, L-, and U-bands new optical band designations that have been proposed

OC-1 optical carrier level 1, which operates at 51.84 Mbps

octave range of frequency in which the upper frequency is double the lower frequency

offset feed type of microwave antenna in which antenna and amplifier are offset from the focal point of the reflector

offset QPSK (OQPSK) variation of QPSK modulation in which I and Q data streams are offset by one-half symbol to prevent transitions through the origin in I/Q space

omnidirectional a spherical radiation pattern

100BaseT 100-Mbps baseband data over twisted-pair cable

open systems interconnection reference model to allow different types of networks to be linked together

optical carrier (OC) designation for data rates in the SONET fiber-optic telecommunication hierarchy

optical spectrum light frequencies from the infrareds on up

optical time-domain ll reflectometer (OTDR) an instrument that sends a light pulse down the fiber and measures the reflected light, which provides some measure of performance for the fiber

optimum working frequency the frequency that provides for the most consistent communication path via sky waves

orthogonal frequency division multiplexing (OFDM) a technique used in digital communications to transmit the data on multiple carriers over a single communications channel

orthogonal two signals are orthogonal if the signals can be sent over the same medium without interference

oscillator circuit capable of converting electrical energy from dc to ac

out-of-band various increments of time are dedicated for signaling and are not available for voice traffic

overmodulation when an excessive intelligence signal overdrives an AM modulator producing percentage modulation exceeding 100 percent

packet switching packets are processed at switching centers and directed to the best network for delivery

packets segments of data

paging procedure used to establish and synchronize a connection between two Bluetooth devices

parasitic array when one or more of the elements in an antenna array is not electrically connected

parasitic oscillations undesired higher frequency self-oscillations in amplifiers

parity a common method of error detection, adding an extra bit to each code representation to give the word either an even or odd number of 1s

Part 15 section of FCC Rules pertaining to maximum permissible electromagnetic radiation emanating from electronic devices

passive attack the bad guy is just listening and picking up what information can be obtained

patch antenna square or round "island" of a conductor on a dielectric substrate backed by a conducting ground plane

payload another name for the data being transported

peak envelope power method used to rate the output power of an SSB transmitter

peak power the useful power of the transmitter contained in the radiated pulses

peak-to-valley ratio another name for ripple amplitude

percentage modulation measure of the extent to which a carrier voltage is varied by the intelligence for AM systems

perigee closest distance of a satellite's orbit to earth

personal-area network (PAN) network of wireless devices located within the immediate vicinity of the user, such as Bluetooth-enabled computer peripherals

phase modulation (PM) superimposing the intelligence signal on a high-frequency carrier so that the carrier's phase angle departs from its reference value by an amount proportional to the intelligence amplitude

phase noise spurious changes in the phase of a frequency synthesizer's output that produce frequencies other than the desired one

phase shift keying method of data transmission in which data causes the phase of the carrier to shift by a predefined amount

phased array combination of antennas in which there is control of the phase and power of the signal applied at each antenna resulting in a wide variety of possible radiation patterns

phasing capacitor cancels the effect of another capacitance by a 180° phase difference

piconet an ad hoc network of up to eight Bluetooth devices

piezoelectric effect the property exhibited by crystals that causes a voltage to be generated when they are subject to mechanical stress and, conversely, a mechanical stress to be produced when they are subjected to a voltage

pilot carrier a reference carrier signal

pi-over-four differential QPSK (π/4 DQPSK) a form of differential QPSK in which symbols are offset by π/4 radians in I/Q space to minimize transitions through the origin

PN sequence length the number of times a PN generating circuit must be clocked before repeating the output data sequence

point of presence the point where the user connects data to the communications carrier

polar pattern a circular graph that indicates the direction of antenna radiation

polarization dispersion broadening of the pulse due to the different propagation velocities of the X and Y polarization components

polarization the direction of the electric field of an electromagnetic wave

poles number of RC or LC sections in a filter

power-sum NEXT testing (PSNEXT) a measure of the total crosstalk of all cable pairs to ensure that the cable can carry data traffic on all four pairs at the same time with minimal interference

predistorter another term for 1/f filter

preemphasis process in an FM transmitter that amplifies high frequencies more than low-frequency audio signals to reduce the effect of noise

preselector the tuned circuits prior to the mixer in a superheterodyne receiver

prime focus feed a type of horn-fed satellite dish in which the antenna and amplifier are placed at the focal point of the reflector

product detector oscillator, mixer, and low-pass filter stage used to obtain the intelligence from an AM or SSB signal

protocols set of rules to allow devices sharing a channel to observe orderly communication procedures

pseudonoise (PN) codes digital codes with pseudorandom output data streams that appear to be noiselike

pseudorandom the number sequence appears random but actually repeats

public data network (PDN) the local telephone company or a communications carrier

public switched telephone network (PSTN) designation for the wireline telephone network used for dial-up telephone service by the public

pulse dispersion a stretching of received pulse width because of the multiple paths taken by the light

pulse modulation the process of using some characteristic of a pulse (amplitude, width, position) to carry an analog signal

pulse repetition frequency (PRF) the number of radar pulses (waveforms) transmitted per second, also known as pulse repetition rate (PRR)

pulse repetition time (PRT) the time from the beginning of one pulse to the beginning of the next

pulse-amplitude modulation sampling short pulses of the intelligence signal; the resulting pulse amplitude is directly proportional to the intelligence signal's amplitude

pulse-duration modulation another name for pulse-width modulation

pulse-length modulation another name for pulse-width modulation

pulse-position modulation sampling short pulses of the intelligence signal; the resulting position of the pulses is directly proportional to the intelligence signal's amplitude

pulse-time modulation modulation schemes that vary the timing (not amplitude) of pulses

pulse-width modulation sampling short pulses of the intelligence signal; the resulting pulse-width is directly proportional to the intelligence signal's amplitude

pump chain the electronic circuitry used to increase the operating frequency up to a specified level

pyramidal horn type of horn antenna with two flare angles

quadrature amplitude modulation (QAM) method of achieving high data rates in limited bandwidth channels, characterized by data signals that are 90° out of phase with each other

quadrature phase-shift keying (QPSK) a form of phase-shift keying that uses four vectors to represent binary data, resulting in reduced bandwidth requirements for the data transmission channel

quadrature signals at a 90° angle

quality of service expected quality of the service

quality ratio of energy stored to energy lost in an inductor or capacitor

quantile interval another name for quantile

quantile a quantization level step-size

quantization levels another name for quantile

quantization process of segmenting a sampled signal in a PCM system into different voltage levels, each level corresponding to a different binary number

quantizing error an error resulting from the quantization process

quantizing noise another name for quantizing error

quarter-wavelength matching transformer quarter-wavelength piece of transmission line of specified line impedance used to force a perfect match between a transmission line and its load resistance

quiet zone between the point where the ground wave is completely dissipated and the point where the first sky wave is received

quieting voltage the minimum FM receiver input signal that begins the limiting process

radar mile unit of measurement equal to 2000 yd (6000 ft)

radar using radio waves to detect and locate objects by determining the distance and direction from the radar equipment to the object

radiation field radiation that surrounds an antenna but does not collapse its field back into the antenna

radiation pattern diagram indicating the intensity of radiation from a transmitting antenna or the response of a receiving antenna as a function of direction

radiation resistance the portion of an antenna's input impedance that results in power radiated into space

radio frequency identification (RFID) a wireless, non-contact system that uses radio waves to identify objects

radio horizon a distance about 4/3 greater than line-of-sight; approximate limit for direct space wave propagation

radio-frequency interference (RFI) undesired radiation from a radio transmitter

radome a low-loss dielectric material used as a cover over a microwave antenna

raised-cosine a mathematical function describing a type of filter whose impulse response rings at the symbol rate, thus minimizing intersymbol interference

ranging a technique used by a cable modem to determine the time it takes for data to travel to the cable head end

Rayleigh fading rapid variation in signal strength received by mobile units in urban environments

received signal level (RSL) the input signal level receiver

receiver time in radar, rest time

reciprocity an antenna's ability to transfer energy from the atmosphere to its receiver with the same efficiency with which it transfers energy from the transmitter into the atmosphere

recursive or iterative algorithms that employ previous output values to generate the current output

reflector the parasitic element that effectively reflects energy from the driven element

refractive index ratio of the speed of light in free space to its speed in a given material

regeneration restoring a noise-corrupted signal to its original condition

rejection notch a narrow range of frequencies attenuated by a filter

relative harmonic distortion expression specifying the fundamental frequency component of a signal with respect to its largest harmonic, in dB

resonance balanced condition between the inductive and capacitive reactance of a circuit

resonant line a transmission line terminated with an impedance that is not equal to its characteristic impedance

rest time the time between pulses

return loss a measure of the ratio of power transmitted into a cable to the amount of power returned or reflected

ring modulator another name for balanced ring modulator

ring the nongrounded wire in two-wire phone service

ripple amplitude variation in attenuation of a sharp bandpass filter within its 6-dB bandwidths

RJ-45 the four-pair termination commonly used for terminating CAT6/5e cable

RS-232 a standard of voltage levels, timing, and connector pin assignments for serial data transmission

RS-422, RS-485 balanced-mode serial communications standards that support multidrop applications

satellite link budget used to verify the received C/N and signal level to the satellite receiver will be met

SC, ST currently the most popular full-size fiber connectors on the market

scattering caused by refractive index fluctuations and accounts for 85 percent of the attenuation loss

second return echoes echoes that arrive after the transmission of the next pulse

sectoral horn type of horn antenna with top and bottom walls at a 0° flare angle

selectivity the extent to which a receiver can differentiate between the desired signal and other signals

sensitivity the minimum input RF signal to a receiver required to produce a specified audio signal at the output

sequence control keeps message blocks from being lost or duplicated and ensures that they are received in the proper sequence

shadow zone an area following an obstacle that does not receive a wave by diffraction

Shannon–Hartley theorem expression for determining channel capacity as a function of bandwidth and signal-to-noise ratio

shape factor ratio of the 60-dB and 6-dB bandwidths of a high-Q bandpass filter

shot noise noise introduced by carriers in the *pn* junctions of semiconductors

sideband splatter distortion resulting in an overmodulated AM transmission creating excessive bandwidths

signaling systems a system used to administer calls on the telephone network

signal-to-noise ratio relative measure of desired signal power to noise power

single-mode type of fiber-optic line where primary mode of propagation is at a single wavelength (low-order mode); used in high-data-rate, long-distance systems

single-stub tuner the stub's distance from the load and the location of its short circuit are adjustable to allow a match between line and load

skin effect the tendency for high-frequency electric current to flow mostly near the surface of the conductive material

skip zone another name for quiet zone

skipping the alternate refracting and reflecting of a sky wave signal between the ionosphere and the Earth's surface

sky-wave propagation those radio waves radiated from the transmitting antenna in a direction toward the ionosphere

slotted aloha a network communications protocol technique similar to the Ethernet protocol

slotted line section of coaxial line with a lengthwise slot cut in the outer conductor to allow measurement of the standing wave pattern

small-form factor a family of connectors about half the size of ST and SC connectors

Smith chart impedance chart developed by P. H. Smith, useful for transmission line analysis

software-defined radio (SDR) communications receivers in which many functions of a traditional receiver are implemented as signal-processing algorithms in software

solar noise space noise originating from the sun

space noise external noise produced outside the earth's atmosphere

specialized mobile radio (SMR) regulatory designation for narrowband FM, two-way radio services used for public-safety communications or business operations, rather than entertainment or broadcasting to the public

spectral regrowth the production of wide modulation sidebands as the result of nonlinearities created by amplitude excursions in amplifiers, requiring excursions through the origin in I/Q space

spectrum analyzer instrument used to measure the harmonic content of a signal by displaying a plot of amplitude versus frequency

spread spectrum communication system in which the carrier is periodically shifted about at different nearby frequencies in a random-like manner determined by a hidden code; the receiver must decode the sequence so that it can follow the transmitter's frequency hops to the various values within the specified bandwidth

spread the RF signal is spread randomly over a range of frequencies in a noiselike manner

spurs undesired frequency components of a signal

squelch a circuit in a receiver that cuts off the background noise in the absence of a desired signal; often found in FM receivers

SS7 Signaling System 7, the suite of protocols and standards used for the management and administration of phone calls and for allowing interconnected telephone switching equipment to communicate with each other

ST designation for a type of fiber connector

standing wave ratio expression of voltage standing wave ratio in decibel form: SWR = 20 log VSWR

standing wave waveforms that apparently seem to remain in one position as the result of an impedance mismatch, varying only in amplitude

start bit, stop bit used to precede and follow each transmitted data word

statistical concentration processors at switching centers directing packets so that a network is used most efficiently

step-index referring to the abrupt change in refractive index from core to clad in a fiber

stray capacitance undesired capacitance between two points in a circuit or device

stripline transmission line used at microwave frequencies that has two ground planes sandwiching a conducting strip

sub-satellite point point on the Earth's surface where a line drawn from the satellite to the center of the Earth intersects the Earth's surface

subsidiary communication authorization (SCA) an additional channel of multiplexed information authorized by the FCC for stereo FM radio stations to feed services to selected customers

successive approximation a type of analog-to-digital converter in which a digital word is created sequentially, rather than all at once (cf. flash ADC)

surface wave another name for ground wave

surge impedance another name for characteristic impedance

switch MTSO, msc, and switch refer to the same equipment

switched virtual circuits (SVCs) a packet mode of network operation in which a circuit is established and disconnected on a per-call basis

symbol in digital communications, a grouping of two or more bits in preparation for modulation of a carrier

symbol substitution displaying an unused symbol for the character with a parity error

synchronous detector a complex method of detecting an AM signal that gives low distortion, fast response, and amplification

synchronous optical network (SONET) standard for long-haul transport for communications data

synchronous system the transmitter and receiver clocks run at exactly the same frequency

synchronous transport signals (STS) used for transporting data in fiber-optic transmission; has equivalence to OC-number specifications

synchronous a system in which the transmitter and receiver clocks run at exactly the same frequency because the receiver derives its clock signal from the received data stream

syndrome the value left in the CRC dividing circuit after all data have been shifted in

systematic codes the message and block-check character transmitted as separate parts within the same transmitted code

T3 a digital data rate of 44.736 Mbps

tank circuit parallel *LC* circuit

thermal noise internal noise caused by thermal interaction between free electrons and vibrating ions in a conductor

third-order intercept point receiver figure of merit describing how well it rejects intermodulation distortion from third-order products resulting at the mixer output

threshold voltage another term for quieting voltage

threshold in FM, the point where S/N in the output rapidly degrades as S/N of received signal is degrading

time domain reflectometry technique of sending short pulses of electrical energy down a transmission line to determine its characteristics by observing on an oscilloscope for resulting reflections

time slot a fixed location (relative in time to the start of a data frame) provided for each group of data

time-division multiple-access (TDMA) a technique used to transport data from multiple users over the same data channel

time-division multiplexing (TDM) two or more intelligence signals are sequentially sampled to modulate the carrier in a continuous, repeating fashion

tip the grounded wire in two-wire phone service

topology architecture of a network

total harmonic distortion a measure of distortion that takes all significant harmonics into account

transceiver transmitter and receiver sharing a single package and some circuits

transducer device that converts energy from one form to another

transit-time noise noise produced in semiconductors when the transit time of the carriers crossing a junction is close to the signal's period and some of the carriers diffuse back to the source or emitter of the semiconductor

transmission line the conductive connections between system elements that carry signal power

transparency control character recognition by using the character insertion process

transponder electronic system that performs reception, frequency translation, and retransmission of received radio signals

transverse when the oscillations of a wave are perpendicular to the direction of propagation

traveling waves voltage and current waves moving through a transmission line

trunk the circuit connecting one central office to another

tunable laser when a laser's fundamental wavelength can be shifted a few nanometers; ideal for traffic routing in DWDM systems

tuned radio frequency (TRF) the most elementary receiver design, consisting of RF amplifier stages, a detector, and audio amplifier stages

twin lead standard 300-Ω parallel wire transmission line

twin-sideband suppressed carrier the transmission of two independent sidebands, containing different intelligence, with the carrier suppressed to a desired level

Type A connector the USB upstream connection that connects to the computer

Type B connector the USB downstream connection to the peripheral

unbalanced line the electrical signal in a coaxial line is carried by the center conductor with respect to the grounded outer conductor

U-NII unlicensed national information infrastructure

universal serial bus (USB) a hot-swappable, high-speed serial communications interface

up-conversion mixing the received RF signal with an LO signal to produce an IF signal higher in frequency than the original RF signal

uplink sending signals to a satellite

upper sideband band of frequencies produced in a modulator from the creation of sum-frequencies between the carrier and information signals

V.44 (V.34) the standard for an all analog modem connection with a maximum data rate of up to 34 Kbps

V.92 (V.90) the standard for a combination analog and digital modem connection with a maximum data rate up to 56 Kbps

vane a thin card of resistive material used as a variable attenuator in a waveguide

varactor diode diode with a small internal capacitance that varies as a function of its reverse bias voltage

variable bandwidth tuning (VBT) technique to obtain variable selectivity to accommodate reception of variable bandwidth signals

varicap diodes another name for varactor diodes

velocity constant ratio of actual velocity to velocity in free space

velocity factor another name for velocity constant

velocity of propagation the speed at which an electrical or optical signal travels

vertical cavity surface emitting lasers (VCSELs) a light source exhibiting the simplicity of LEDs combined with the performance of lasers

virtual channel connection (VCC) carries the ATM cell from user to user

virtual circuit identifier (VCI) sixteen-bit word used to identify connection between two ATM stations

virtual path connection (VPC) used to connect the end users

virtual path identifier (VPI) eight-bit word used to identify the circuit established to deliver cells in an ATM network

voltage standing wave ratio ratio of the maximum voltage to minimum on a line

VVC diodes another name for varactor or varicap diodes

wavefront a plane joining all points of equal phase in a wave

waveguide a microwave transmission line consisting of a hollow metal tube or pipe that conducts electromagnetic waves through its interior

wavelength the distance traveled by a wave during a period of one cycle

white noise another name for thermal noise because its frequency content is uniform across the spectrum

wide area network two or more LANs linked together over a wide geographical area

wideband FM FM transmitter/receiver systems that are set up for high-fidelity information, such as music, high-speed data, stereo, etc.

wireless local area networks (WLANs) a local-area network that features wireless access

wireless markup language (WML) the hypertext language for the wireless environment

X.25 a packet-switched protocol designed for data transmission over analog lines

xDSL a generic representation of the various DSL technologies available

zero-dispersion wavelength the wavelength where material and waveguide dispersion cancel each other

zero-IF receiver another term for direct-conversion receiver

zoning a fabrication process that allows a dielectric to change a spherical wavefront into a plane wave

INDEX